First International Conference on

Optical Communications and Networks (ICOCN 2002)

First International Conference on

Optical Communications and Networks (ICOCN 2002)

Shangri-La Hotel, Singapore
11–14 November 2002

Editors

C. G. Omidyar
Institute for Communications Research

H. G. Shiraz
Nanyang Technological University, Singapore

W. D. Zhong
Nanyang Technological University, Singapore

Published by

World Scientific Publishing Co. Pte. Ltd.
5 Toh Tuck Link, Singapore 596224
USA office: Suite 202, 1060 Main Street, River Edge, NJ 07661
UK office: 57 Shelton Street, Covent Garden, London WC2H 9HE

British Library Cataloguing-in-Publication Data
A catalogue record for this book is available from the British Library.

First International Conference on
OPTICAL COMMUNICATIONS AND NETWORKS (ICOCN 2002)

ISBN 981-238-232-1

Printed by FuIsland Offset Printing (S) Pte Ltd, Singapore

Message from Conference Chairs

Welcome to the beautiful garden city, Singapore, a wonderful place to be in, while attending the First International Conference on Optical Communications and Networks ICOCN 2002. ICOCN has established itself as the first conference of the Institute for Communications Research in the field of optical communications and networks.

The conference has received a large number of submissions, thanks to the efforts of the International Advisory and Technical Committee. After careful review and selection of papers, the conference was organized around a two-track program which was complemented with keynote and invited papers. The conference has been arranged with one day for tutorials and three days for presentation.

The Optical Communications and Networks 2002 conference is a gathering together of scientists, the industry and developers to report on their findings, discuss their experiences and sharing of lessons to be learned from one another. The ICOCN 2002 provides a forum for tutorials, discussions and presentations of the new developments in optical communications research and industry. Optical communications networks have created a major breakthrough in new telecom services in both local and wide area networks.

ICOCN 2002 was organized by the Institute for Communications Research in co-operation with the Nanyang Technological University, National University of Singapore, Laboratories for Information Technology and IEEE Communications Society.

We hope we have done our best to bring you an outstanding ICOCN 2002 program and that you will enjoy it.

Cambyse Guy Omidyar
Conference and Steering Committee Chair

Cheng Tee Hiang
Conference Co-Chair

Lu Chao
Conference Vice-Chair

Message from the Technical Program Chairs

On behalf of the Technical Program Committee, we would like to welcome all of the conference participants to Singapore. ICOCN2002 is the inaugural conference on Optical Communications and Networks held in Singapore. A total of 121 papers were received from 18 countries. 107 papers are accepted and scheduled into 20 sessions together with one plenary session. All selected papers for oral presentation have gone through comprehensive reviewing process and hence are of high quality.

The technical papers have been arranged into 2 parallel sessions, covering the main technical areas of the conference i.e. optical networking, optical switching, optical transmission, passive WDM devices and active WDM devices and systems. The conference starts off with tutorials on Monday 11th of November, where 4 well known experts present some topics related to the conference theme.

We are delighted to have three keynote speakers addressing on very topical issues that are currently being researched. On Tuesday morning, three keynote speakers Prof R. S. Tucker, Dr. G. Keiser and Dr. G. Ishak address on "Towards an all-optical internet", "Perspectives on education and training in optical communications", and "Micro photonic integrated circuits, MicPIC", respectively.

The Technical Program Committee members as well as all our reviewers contributed substantially to bringing you a top quality technical program for this conference. We would like to express our sincere thanks and appreciation to each member of the committee and all the reviewers. We would also like to express our deep appreciation to our keynote and invited speakers for sharing their valuable time with us. Last but not the least, we thank all the authors of regular papers, as well as the session chairs for enabling us to present you with an exciting conference.

We hope this conference becomes a bridge for you so that not only meet your old friends, but also find a lot of new ones. Finally may we wish you having a pleasant time in Singapore and that ICOCN2002 be a fulfilling and memorable experience. Thank you all.

A/PROF. H. GHAFOURI-SHIRAZ AND A/PROF. W. D. ZHONG
Technical Program Chairs of ICOCN 2002

LIST OF COMMITTEE MEMBERS

Organizing Committee

General Chair:	Cambyse Guy Omidyar, gomidyar@icr.a-star.edu.sg Institute for Communications Research, Singapore
General co-chair:	Cheng Tee Hiang, ethcheng@ntu.edu.sg Institute for Communications Research, Singapore
General Vice Chair:	Lu Chao, eclu@ntu.edu.sg Institute for Communications Research, Singapore
Program Chairs:	H. G. Shiraz, ehgshiraz@ntu.edu.sg Nanyang Technological University, Singapore W. D. Zhong, ewdzhong@ntu.edu,sg Nanyang Technological University, Singapore
Tutorials Chairs:	P. Shum, epshum@ntu.edu.sg Nanyang Technological University, Singapore N. Q. Ngo, eqnngo@ntu.edu.sg Nanyang Technological University, Singapore
Secretariat:	Charlene Boh, bohcharlene@icr.a-star.edu.sg Institute for Communications Research, Singapore
Treasurer:	S. Kunaselvam, kuna@icr.a-star.edu.sg Institute for Communications Research, Singapore
Local arrangement:	Institute for Communications Research Nanyang Technological University National University of Singapore
Publication:	Charlene Boh, bohcharlene@icr.a-star.edu.sg Institute for Communications Research, Singapore
Publicity:	Charlene Boh, bohcharlene@icr.a-star.edu.sg Institute for Communications Research, Singapore

Technical Programme Committee

Chairs: H. G. Shiraz, Nanyang Technological University, Singapore
W. D. Zhong, Nanyang Technological University, Singapore

Members: J. Biswas, Laboratories for Information Technology, Singapore
T. H. Cheng, Institute for Communications Research, Singapore
K. Chiang, City University of Hong Kong, Hong Kong
S. J. Chua, National University of Singapore, Singapore
M. Gagnaire, Ecole Nationale Superieure des Telecommunications (ENST) Paris, France
W. Gu, Beijing University of Posts & Telecommunications, People's Republic of China
M. K. Iyer, Institute of Microelectronics, Singapore

R. Jain, Nayna Networks, USA
V. Li, University of Hong Kong, Hong Kong
A. Q. Liu , Nanyang Technological University, Singapore
C. Lu, Institute for Communications Research, Singapore
Y. F. Lu, National University of Singapore, Singapore
M. Ma, Nanyang Technological University, Singapore
M. Gurusamy, National University of Singapore, Singapore
Y. Nakano, University of Tokyo, Japan
B. Narayanan, Institute of Microelectronics, Singapore
F. Neri, Politecnico di Torino, Italy
N. Q. Ngo, Nanyang Technological University, Singapore
L. H. Ngoh, Laboratories for Information Technology, Singapore
T. Nirmalathas, University of Melbourne, Australia
G. S. Poo, Nanyang Technological University, Singapore
P. Shum, Nanyang Technological University, Singapore
J. Singh, Institute of Microelectronics, Singapore
D. Tang, Nanyang Technological University, Singapore
M. Tatipamula, Cisco Systems, Inc, USA
S. Z. Xie, Tsinghua University, People's Republic of China
A. Xu, Peking University, People's Republic of China
B. X. Xu, Data Storage Institute, Singapore
H. Zhang, Tsinghua University, People's Republic of China
L. Zhang, Nanyang Technological University, Singapore
L. Zhou, Laboratories for Information Technology, Singapore

Advisory Committee

Members: P. Chu, City University of Hong Kong, Hong Kong
A. Ghatak, Indian Institute of Technology Delhi, India
H. Kobayashi, Princeton University, USA
M. O'Mahony , University of Essex, UK
R. S. Tucker, University of Melbourne, Australia
H. van As, Vienna University of Technology, Austria
K. Yukimatsu, Akita University, Japan

Steering Committee Chair

Cambyse Guy Omidyar

Towards an All-Optical Internet

Rodney S. Tucker

ARC Centre for Ultra-Broadband Information Networks
Department of Electrical and Electronic Engineering
University of Melbourne, Vic. 3010, Australia

Photonics is moving towards a new era. Photonics can be compared to the field of electronics in the 1960's, when transistor device technologies were maturing, and single-chip integration of devices was becoming commercially viable. The electronic circuits revolution was about to begin. Today, photonic integration is a reality and photonic circuits are used throughout the global telecommunications system. The new era for photonics will be marked by the ultimate photonic circuit – the all-optical Internet – a seamless connection of interacting photonic devices, circuits, and systems spanning the globe.

The all-optical Internet (or perhaps the almost-all-optical Internet) will become a reality only if it can provide a cost-effective alternative to the present telecommunications network. Promises of vast increases in bandwidth are not enough. Also needed is a substantial reduction in the unit cost for bandwidth, together with flexibility and low operating costs. For many researchers in the field, it is easy to see how photonics and an all-optical Internet create a path to increased bandwidth. The challenge for all of us is to find ways to reduce the cost of the bandwidth. In this talk, I will give a personal perspective on some of the photonic technologies, circuits, sub-systems, architectures, protocols, and control strategies that will be needed to implement a cost-effective all-optical Internet.

Perspectives on Education and Training in Optical Communications

Gerd Keiser

Abstract--This talk will address the different aspects of what education and training programs, courses, and other materials are needed to prepare and help all types of people to work in the optical communications field. The people who are involved with this technology range from the researchers and engineers who pursue new ideas to the individuals who actually have to make a product work for a customer, that is, those people who install and maintain the fiber optic components and equipment. In between these categories is a wide range of professionals such as sales and marketing personnel, business managers in telecommunications, investment people, cable installation engineers, cable plant design engineers, test and measurement technicians, network operations people, and troubleshooting and repair personnel.

I. Extended Abstract

Many different sets of skills come to mind when we think of optical communications. A basic concern is how these individuals get the knowledge and skills they need to work effectively and productively in the optical communications field. This talk will address the techniques and resources used in three broad training categories, which may be called general background learning, random training, and detailed hands-on training. Included in the discussions of these three categories are descriptions of the various resources, an overview of what courses universities and professional hands-on training organizations offer in the US, considerations of who might benefit from the learning resources of a category, and what further training would be helpful for the people to be more effective in their work.

General background learning includes the contents of university courses, short courses at conferences, conference tutorials, and professional journal or trade magazine articles. This category requires a long-term commitment of either two years if the person is pursuing an associates degree or at least four years for a bachelors degree or higher. However, very few institutions in the US offer engineering degrees with a concentration in optical communications, so there is room for growth in this area. Since education and training is an ongoing task no matter what position one has, the short courses, tutorials, and journal articles are a necessity to keep up to date and to avoid some of the problems associated with the random-training category.

Random training encompasses applications in an everyday work environment. This includes learning on-the-job from knowledgeable peers, learning by trial-and-error when there is no help, extracting information from product brochures or user manuals, and watching vendor demonstrations. In contrast to taking formal structured courses or reading refereed journal articles, this category has little structure to it and also puts a lot of pressure on technical personnel. In the classroom the least one needs to do is get a passing grade. At work there is no solutions manual to the problem sets, and a person's job depends on making the correct decision. Unless an employer ensures that employees have the proper background to use equipment properly or to provide product performance parameters and application information correctly to potential clients, the product could be made, installed, or used incorrectly. In addition to taking formal higher-education courses, some of these problems can be mitigated with detailed hands-on training.

Detailed hands-on training can include 2-day to 4-day courses from professional training organizations, training courses from equipment vendors, special in-house training programs, and Web-based training courses. To be highly effective these courses should take several days to complete and should include an exam to verify that the hands-on problem-solving skills being presented actually were learned. Ideally in such a course the participants should work with actual products, test equipment, or networking gear and be tested on a realistic problem that they might encounter when working. In addition, it would be beneficial to have the participants solve the problem within a fixed time period in order to simulate the pressure they would encounter on the job.

If appropriate education and training programs are available, employees will have a higher morale and more confidence in what they are doing, products will function better with a higher reliability, and customers will sign purchase orders since they recognize that the higher knowledge and skills of the personnel produce better products or services.

PhotonicsComm Solutions, Inc.
gkeiser@photonicscomm.com.

Micro Photonic Integrated Circuits, MicPIC

Waguih S. Ishak
Director
Communications & Optics Research Laboratory
Agilent Technologies
3500 Deer Creek Road, Palo Alto, California 94304
waguih_ishak@agilent.com
(650) 485 - 4449

The current economic downturn in telecommunications will result in a shakeup both in network architectures and in optical and optoelectronic component designs. It is almost guaranteed that service providers will pressure network equipment manufacturers (NEMs) to lower their product costs and, in turn, the NEMs will pressure the component manufacturers to develop low-cost components; all in an attempt to lower the cost per managed bit in telecommunication networks.

The three most important factors in lowering the cost of optical components are: Miniaturization, Integration and Automated Assembly (MI & AA). Miniaturization will allow smaller devices and, hence, higher yields. Integration will result in more functionality and, therefore, better subassembly architectures. Automated assembly will be needed to produce volume photonic devices.

The talk will address the 3 areas (MI & AA) with examples of new devices for the next generation networks.

CONTENTS

Session 14B: Optical Signal Processing 283

Session 14C: Network Protection and Restoration 303

Session 12A

Optical Networking I

Concept and management of Ethernet and WDM based wide area access network

Takeshi Kinoshita, Yuuzou Miyagawa, Yasunao Suzuki, and Hitoshi Uematsu

NTT Network Innovation Laboratories

1-1Hikarinooka, Yokosuka, Kanagawa, 239-0847 Japan

Telephone: +81-468-59-2577, Facsimile: +81-468-55-1282, Email: kinosita@exa.onlab.ntt.co.jp

Abstract: 'Ethernet over WDM' is very promising as a next-generation network technology. After some prospective Ethernet-based services are introduced, a novel WDM-based wide-area access network, combined with a regional transport network, is introduced. This network enables the efficient deployment of such services. We show a detailed system configuration of the network and OAM information channels that are effective for it. We then present its management information model in reference to the architectural transport model. This model will be useful in building network management systems.

Key Words: Access Network, Ethernet, WDM, OAM

1. INTRODUCTION

Ethernet is being used extensively in local area networks (LANs) and is expanding its application area. It is quite natural that many carriers are adopting Ethernet as the user-to-network interface (UNI) for their network services. This can be seen not only in services targeting consumers but also those targeting large companies. One such service is the 'transparent LAN' service, where customers' traffic is carried and switched in the carrier's networks according to its layer-2 (L2) information.

Against this background, several standardization bodies and their support organizations are working to design and specify technologies related to Ethernet. For example, the IEEE 802.3ah Task Force is attempting to apply Ethernet to access networks in the framework of the 'Ethernet in the First Mile (EFM) [1].'

Ethernet-based network services are likely to become more common among all kinds of customers. Many services like the leased line, the virtual private network (VPN), contents delivery, and others, could be provided so as to make the best use of Ethernet-based (or related) technologies.

As to capacity, there is a high possibility that the demand for gigabit-class capacity will increase significantly. The wavelength division multiplexing (WDM) technique is an attractive way of accommodating such demand. Initial moves to apply the WDM technique to access networks are underway [2].

In this paper, we introduce a 'regional network' that is designed to provide various kinds of Ethernet-based services, as well as a novel WDM-based 'wide-area access network.' This access network enables the efficient deployment of services, especially high-end ones. After a detailed system configuration is shown, a novel OAM (operation, administration, and management) information channel that is effective in that system is described. We then present its management information model, as well as the architectural transport model.

2. ETHERNET AND WDM BASED REGIONAL NETWORK

2.1 Service classification

Table 1 classifies several Ethernet-based services. There are two major categories; those on 'dedicated links' and those on 'shared links.' We assume both to be prospective, especially among large companies and/or small office home office (SOHO) customers. The difference between the two categories is whether the optical channel (equivalent to wavelength) is occupied by a single link or shared by multiple links in a carrier's networks. In the shared link category, several links of different users are accommodated together in an optical channel. It means that a method is required to distinguish a closed user group (CUG), since the link is formed among user devices that belong to a specific CUG. An example of this method is the well-known virtual

Table 1 Classification of Ethernet-based services.

Category	Optical Channel	CUG	Link Type	Bandwidth	Example of Applications
Dedicated Link	Occupied by a single link	Not Required	Point-to-Point	Guaranteed	Leased Line (for critical use)
Shared Link	Shared by multiple links	Required	Point-to-Point	Partly Guranteed	Leased Line (for non-critical use)
			Mutipoint-to -Multipoint	Partly Guranteed Non-guranteed	VPN (Intranet/Extranet) VPN (Community Network)

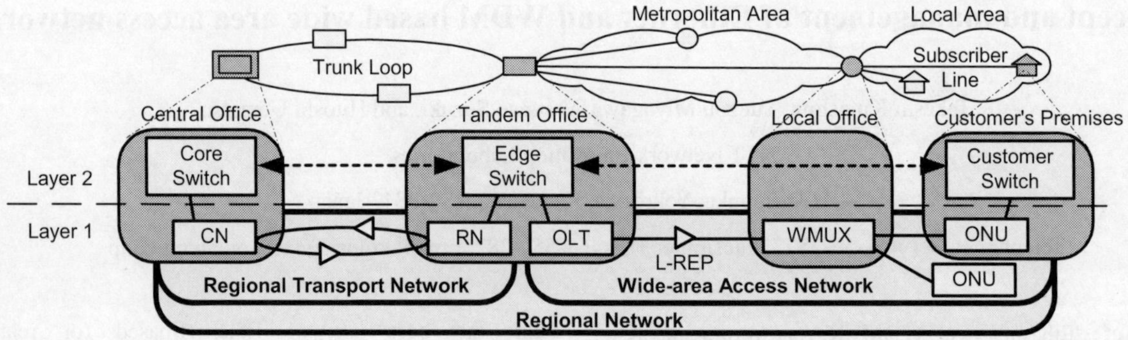

Figure 1 Overall configuration of the regional network.

LAN (VLAN) technology [3].

Table 1 presents an example of applications for the service categories.

2.2 Conceptual configuration of the regional network

Figure 1 shows the overall configuration of the 'regional network' introduced here. This network is designed to provide various kinds of Ethernet-based services. We focus here, however, on the services presented in Table 1. This establishes certain criteria for network configuration.

The regional network consists of two main parts, that is, the 'regional transport network' and the 'wide-area access network.' Traffic from these two constituent networks is bridged at a 'tandem office.' In the tandem office, the 'edge switch' processes Ethernet frames into adequate L2-formats and directs them to individual output ports. For example, this process may include adding and removing VLAN tags that are only effective within the regional transport network.

The regional transport network is a core network that is configured with the center node (CN) at the central office, the remote nodes (RNs) at the tandem offices, and the trunk loop. The 'core switch' is set at the central office. This switch will be equivalent to a root bridge if the spanning tree protocol [4] is supported.

We introduce a WDM-based ring network in which layer-1 (L1) traffic is carried between the CN and a specific RN (with unicasting), or between the CN and several RNs (with multicasting). It enables the efficient transport of Ethernet frames, because an Ethernet-based L2-switching network basically has a logical tree topology. The WDM technique is useful in this network not only for increasing capacity but also for adding network capabilities, like service class differentiation. Details of this network are not described in this paper. However, an example of such an L1-network can be seen in [5].

2.3 Concept of the wide-area access network

In the current network hierarchy, an access network is usually configured within a local area. Transmission equipment that terminates each subscriber's signal is installed at a local office. In a metropolitan area, a relay network is configured in order to transfer traffic between local offices and a tandem office.

The proposed wide-area access network is, however, configured to contain a metropolitan area and several local areas. The tandem office hosts an optical line terminal (OLT) to terminates each subscriber's optical signal. Each local office hosts a WMUX (wavelength multiplexer), which consists of some simple optical devices, to collect the multiple subscribers' optical signals and pass them to the tandem office while splitting the downstream signal. This process does not include optical/electric conversion. Each optical network unit (ONU) at a customer's premises is assigned with a different wavelength, or an equivalent optical channel. Each wavelength is routed passively through the WMUX. In other words, a virtual point-to-point connection is formed between an OLT and an ONU via the WMUX. If an ONU is to support one or more service classifications with several CUGs, the optical channel assigned to that ONU should be shared among these services and/or CUGs.

In order to enlarge the coverage of the access network, a linear optical repeater (L-REP) may be employed in metropolitan areas when the distance between the tandem office and the WMUX becomes excessive.

One of the advantages of the wide-area access network is that the passive optical devices employed in the WMUX and L-REP need less maintenance than other transmission equipment. Another advantage is that it can help promote service deployment, because any equipment for a new service need be installed only at the tandem office, not at a number of local offices. This advantage works well especially when the demand is unevenly distributed. In other words, the wide-area access network is a network that strongly supports high-end services.

3. MODELING THE WIDE-AREA ACCESS NETWORK

3.1 System configuration

An example of the system configuration of the proposed wide-area access network is shown in figure 2

Optical carriers that are used as individual optical channels are generated in the optical carrier supply module (OCSM) [6], and provided to the OLT. Two separate

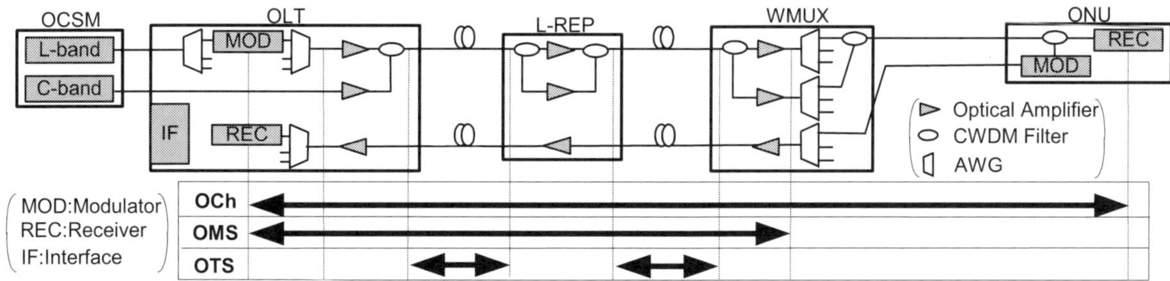

Figure 2 System configuration of the wide-area access network.

wavelength bands are used in this network. The C-band (around 1550nm) is allocated for upstream signals, and the L-band (around 1580nm) for downstream signals. The optical carrier used to form the upstream signal is supplied to the ONU from the OLT on the same optical fiber as the downstream signal. The upstream signal is carried on the other optical fiber. In this system configuration, the ONU employs a modulator instead of a light source.

The optical channels associated with individual ONUs are multiplexed or demultiplexed in the wavelength realm by arrayed waveguide gratings (AWGs) in the OLT or in the WMUX. The two wavelength bands also have to be multiplexed or demultiplexed to share the same optical fiber. This can be realized by coarse WDM (CWDM) filters.

We assume that each optical channel carries a Gigabit-Ethernet signal. More specifically, both downstream and upstream signals follow the 1000BASE-X format [7] (8B/10B coding), and have the transmission rate of 1.25Gbps.

More details are given in [2].

3.2 OAM information channel

We describe the OAM information channels for the system configuration described above. The overall network has a layered structure and each layer is applied to a specific section of the network. An OAM information channel supports one or more layers.

The terminology here is adopted from the description of network architecture of optical transport network (OTN) presented in ITU-T recommendation G.872 [8]. The terms 'OCh', 'OMS', and 'OTS' represent the optical channel, optical multiplex section, and optical transmission section layers, respectively. An OCh has a defined wavelength and realizes the transport of client information. An OMS has a defined aggregate optical bandwidth and bundles multiple OCh's. An OTS has a defined physical interface with optical parameters and supports an OMS. The section to which each layer is applied is shown in figure 2.

The OAM information channel for the OTS layer is carried on an optical supervisory channel (OSC) [9]. The OSC is a channel that is independent from other subscriber related channels. This channel can be used to carry other kinds of information, for example, an auxiliary channel that is operator specific. We consider utilizing this to manage some of the optical devices. The OAM information channel related to the OMS layer is also carried on the OSC. This includes maintenance information, for instance, fault notification from a WMUX.

The OCh layer corresponds to the Gigabit-Ethernet physical layer. An OCh accommodates a physical bit-stream of Gigabit-Ethernet. In order to handle this bit-stream directly, we introduce a sublayer within the physical layer and apply a unique technique to it. This sublayer is called the link signaling sublayer (LSS).

This new technique replaces a portion of a bit-stream with a set of fixed bit patterns that contain OAM information. This portion is in an interframe gap (IFG) on the layer-2, which is specified to exist between two media access control (MAC) frames. An IFG is converted into some ordered sets of fixed bit patterns in a bit-stream through 8B/10B coding. These ordered sets are basically idle patterns, although some of them are used for other particular purposes. Our technique utilizes some of these idle patterns. Since this process does not involve MAC frames, it does not interfere with the main information, or user bandwidth. This process is completely performed within the LSS. A similar idea for 10Gigabit-Ethernet can be seen in [10]. This can be introduced in such a way so as to retain compatibility with the IEEE standard. By the use of this technique, an OCh can carry OAM information equivalent to that defined in the synchronous digital hierarchy (SDH) overhead.

We do not describe OAM information channels in layer-2. Such information channels may include some standard ones (e.g. flow control), and others that are being discussed or are vendor specific. They could be effective and useful in our system, but we think that an OAM information channel inside layer-1 is even more essential.

3.3 Management information model

Figure 3 shows the architectural transport model of the wide-area access network. For simplicity, only the OLT part is depicted. This model conforms to the generic method described in ITU-T recommendation G.805 [11]. The model decomposes network elements into their architectural components. Information transfer functionalities are attributed to individual architectural components, and an OAM information channel is set within each of them.

The trapezoid notation and the triangle notation represent the adaptation function and the trail termination function, respectively. Adaptation corresponds to the process in which a client layer is accommodated in a server layer. Trail termination includes the process in which an OAM information channel is added and removed on a trail. A trail is formed between two ends that belong to the same layer. For example, an OCh trail is formed between an OLT and an ONU.

Figure 3 also describes the management information model. The suffixes TTP and CTP represent 'trail termination point' and 'connection termination point,' respectively. A CTP is associated with an adaptation function and supported by a TTP of a server layer. Each management entity defined in this model has a specific name with a suffix of either TTP or CTP.

MAC frames can be exchanged between the two sides, the Ethernet Switch side and the Transmission Line side. This is modeled with the Mac Bridge Matrix and the Mac Bridge Ports.

This model will provide a basis for designing network management systems.

4. CONCLUSION

We presented the concept of a novel WDM-based wide-area access network that efficiently provides Ethernet-based services. We also presented its management information model along with its architectural transport model. When a Gigabit-Ethernet signal is carried on an optical channel, it is very effective to introduce a sublayer (LSS) inside the physical layer and set an OAM information channel that is handled in that layer. This realizes OAM functions equivalent to SDH overhead.

ACKNOWLEDGEMENT

The authors would like to thank Nobuo Fujii of NTT Network Innovation Laboratories for his continuous encouragement, and Tatsuhiko Yoshida and Hidetaka Kato of NTT Advanced Technology Corporation for their useful advice.

REFERENCES

[1] IEEE 802.3ah Task Force; http://www.ieee802.org/3/efm/public/index.html

[2] J. Kani, M. Teshima, K. Akimoto, M. Ishii, N. Takachio, and K. Iwatsuki, "Super-Dense WDM Access Network for Wide-area Gigabit Access Services," Proc. 14th ISSLS Conf., Seoul, Korea, S9-2, 2002

[3] IEEE Standard 802.1Q, *Virtual Bridged Local Area Networks*: IEEE, 1998

[4] IEEE Standard 802.1D, *Media Access Control (MAC) Bridges*: IEEE, 1998

[5] N. Takachio, H. Suzuki, M. Fujiwara, J. Kani, M. Teshima, and K. Iwatsuki, "12.5-GHz-spaced super-dense WDM ring network handling 256 wavelengths with tapped-type OADMs," Proc. OFC2002 Conf., Anaheim, CA, WW2, 2002

[6] M. Teshima, M. Fujiwara, J. Kani, H. Suzuki, M. Amemiya, N. Takachio, and K. Iwatsuki, "Optical carrier supply module applicable to over 100 super-dense WDM systems of 1000 channels," Proc. 27th ECOC, Amsterdam, Netherlands, Mo.L.3.7, 2001

[7] IEEE Standard 802.3, *Carrier sense multiple access with collision detection (CSMA/CD) access method and physical layer specifications*: IEEE, 2000

[8] ITU-T Recommendation G.872, *Architecture of optical transport networks*: ITU-T, 2001

[9] ITU-T Recommendation G.709, *Interface for the optical transport network (OTN)*: ITU-T, 2001

[10] O. Ishida, "LSS for Link Status Mechanism," Proc. IEEE P802.3ae Plenary Meeting, Tampa, FL, 2000; http://grouper.ieee.org/groups/802/3/ae/public/nov00/ishida_1_1100.pdf

[11] ITU-T Recommendation G.805, *Generic functional architecture of transport networks*: ITU-T, 2000

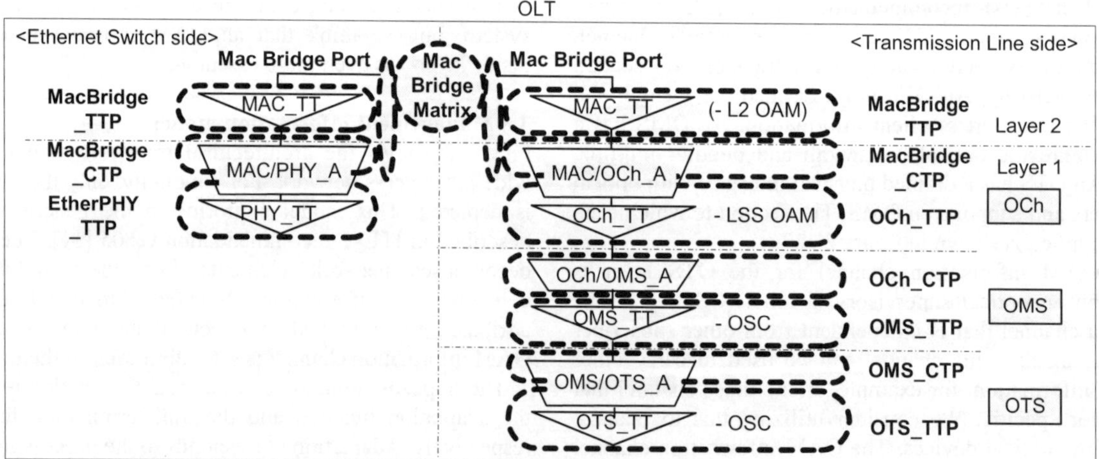

Figure 3 Management information model with reference to architectural tranport model.

Fairness Control Based on Link Scheduling Statistics in WDM Optical Burst Switching Networks

G. Mohan[1] Li Zi Li[2]
Electrical and Computer Engineering Department
National University of Singapore, Singapore 117576
Email: elegm@nus.edu.sg

***Abstract*— The recent explosive growth in the Internet technology has made it imperative to have a network which provides a very high bandwidth. Optical networks employing *wavelength division multiplexing* (WDM) technology are becoming a viable solution to meet this ever-increasing bandwidth demand. *Optical burst switching (OBS)* networks provide a good solution for a very high traffic influx which is generally bursty in nature. In this, a single control packet is sent for a group of packets which constitute a data burst with a time gap between them to ensure pure optical switching of bursts. Usually bursts that traverse longer paths are more likely to be dropped than those that traverse shorter paths resulting in *fairness* problem. Fairness means that for all ingress-egress node pairs in a network a burst must have equal likelihood to get through independent of the hop length involved. In this paper, we develop an efficient fairness method to achieve fairness among bursts with different hop counts. The key idea is to collect scheduling statistics by observing the bursts, and based on this information assign different offset time to bursts with different hop counts. Apart from ensuring fairness our method has several attractive features. By using the online statistics, our methods inherently captures the traffic loading patterns and network topological connectivity. It also ensures that the delay experienced by a burst is low and shorter-hop bursts are not over-penalized while improving the performance of longer-hop bursts. We demonstrate the effectiveness of our fairness method through extensive simulation experiments for a number of random networks which differ in the number of nodes and topological connectivity.**

Key Words: **Wavelength division multiplexing, optical burst switching, fairness control, scheduling.**

I. Introduction

Potential bottlenecks of electronic processing to carry IP packet traffic over WDM optical networks can be overcome by aggregating multiple data packets into a super packet, called burst. Bursts are a collection of data packets assembled at an ingress router, having the same network egress address and some common attributes, like QoS requirements [1], [2], [3]. A network which supports this kind of switching is called an *optical burst switching* (OBS) network. The basic difference between OBS and circuit switching is that in burst switching, bandwidth is reserved in a one-way process; that is, a burst can be sent out without waiting for the acknowledgement of a successful reservation, whereas in circuit switching, bandwidth is reserved in a two-way process. In other words, data can only be sent after a circuit has been successfully established. This results in a longer latency. As compared to optical packet switching, a burst cuts through the intermediate nodes without being buffered, whereas in packet switching, packets are optically stored and forwarded at each intermediate node, leading to increased processing complexity, buffering, and synchronization problems [4], [5], [6].

[1] Author for correspondence.
[2] This work was carried out when this author was a student at NUS.

A burst consists of a burst header and a burst payload which are referred to as *control packet* and *data burst*, respectively. In the literature several optical burst switching protocols have been proposed. Among these protocols, time-offset-based Just-Enough-Time (JET) is an attractive one [2], [4]. In an OBS network, a data burst and its control packet are transmitted by an ingress node separately with a time lag. This time lag is called the *offset time*. The minimum offset time is given by $h\delta$, where δ is the processing time of the control packet at each node along the path with h hops. This processing time is the difference between the time when the control header enters a node and the time when it leaves the node after scheduling all the resources such as wavelengths. By extending *multi-protocol label switching* (MPLS) capabilities to an OBS network, explicit routing can be used at the ingress nodes [7]. The data burst and the control packet are sent on different wavelengths/channels with the control packet slightly ahead in time, and are switched in optical and electronic domains, respectively, at each core router they traverse. The control packet contains all the necessary routing information such as burst duration and burst arrival time to be used by the *switch control unit* (SCU) at each hop to schedule the data burst and configure the optical switching matrix to switch the data burst optically [1], [2].

The separate transmission and switching of data bursts and their headers help to facilitate the electronic processing of headers before the corresponding data bursts arrive and to lower the opto-electronic processing capacity required at the routers. Further, it provides ingress to egress clear optical paths for transporting data bursts in optical domain without requiring opto-electronic-opto conversions at the intermediate routers. A data burst can be optically buffered by using *fiber delay lines* (FDLs). But, FDLs are very expensive and they can delay only for a short time which is on the order of micro seconds. In a multi-class environment, service differentiation can be achieved by using different offset time to different classes [2]. A large offset time could be assigned to bursts of a high priority class so that the control packet can reserve resources (wavelengths and FDLs) well in advance, resulting in lower dropping probability.

In the event of no wavelength or FDL being available at the time of arrival of a burst at a node, the burst is dropped. A scheduling algorithm is used at nodes to choose the best wavelength (and/or FDL) on the outgoing link for the data burst. In the literature, a few scheduling algorithms have been proposed [1], [3]. These algorithms differ in their burst dropping performance and computational complexity. Usually, bursts that traverse longer paths are more likely to be dropped than those that traverse shorter paths resulting in *fairness* problem. Fairness means that for all ingress-egress node pairs in a network a burst must have equal likelihood to get through independent of the hop length involved.

In [4], a variation of JET protocol namely JET-FA has been proposed to achieve fairness. The key idea used in this fairness method is to prioritize bursts by assigning extra offset time proportional to the hops. An extra offset time of $5hL$ or $10hL$, where L is the mean burst duration has been recommended to achieve fairness. This method has some drawbacks. The extra offset time is independent of traffic load and network topological connectivity. Further, this offset time is very large requiring longer queuing delay and large buffers at the ingress nodes. This will increase the delay of a burst and result in poor support for QoS. For example, a 3-hop burst for a mean duration of $100\mu s$ and control processing time of $10\mu s$ will require the data burst to be queued at the ingress node for 3.030 ms. Another undesirable feature of this method is that it tends to penalize shorter-hop bursts heavily by assigning a comparatively longer offset for longer-hop bursts.

In this paper, we develop an efficient fairness method called *link scheduling statistics- based fairness* (LSS-FA) method to achieve fairness among bursts with different hop length. The key idea is to collect scheduling statistics by observing the bursts, and based on this information assign different offset times to bursts with different hop counts. Apart from ensuring fairness our method has several attractive features. By using the online statistics, our method inherently captures the traffic loading patterns and network topological connectivity. It also ensures that the delay experienced by a burst is low and shorter-hop bursts are not over-penalized while improving the performance of longer-hop bursts. We demonstrate the effectiveness of our fairness method through extensive simulation experiments for a number of random networks which differ in the number of nodes and topological connectivity.

The rest of the paper is organized as follows. In Section II, the proposed fairness control method is discussed. Section III presents the results of performance study while Section IV makes concluding remarks.

II. Proposed Fairness Control Method

We consider a burst switching network with no FDL optical buffers. We use a time-offset based burst switching protocol to transmit control packet and data burst with an initial offset to be set at the ingress node. We recall that, in order to transmit a burst between an ingress-egress pair separated by h hops, the initial offset should be at least $h\delta$, where δ is the control packet processing time. We note that the offset time depends on the hop count and this offset time decreases by δ whenever it traverses a node. For instance, if the control packet and data burst are separated by an initial offset time T_{init} at the ingress node, then the offset time is $T_{init} - \delta$ at the first next node and it is $T_{init} - 2\delta$ at the second next node, and so on. This means that at any instance of time the offsets used by bursts at a node are different depending on the number of remaining hops to be traversed.

With the varying offset time used by different bursts at a node, it is more likely that longer is the offset time, better is the chance of finding a free wavelength on a link. In such a case, a longer-hop burst is more likely to find a free wavelength initially and as it traverses further it is less likely to find a free wavelength because of shorter offset time. Also, the probability of finding free wavelengths on every hop on a path is becoming lesser as the hop length increases. Our fairness method LSS-FA allows longer-hop bursts to use higher initial offset time than the minimum required in order to achieve fairness among bursts with different hop counts. The offset time values assigned by our LSS-FA method are calculated based on link scheduling statistics rather than a linear function of hops and mean burst duration as in the case of the earlier method JET-FA. It ensures that the initial offset time is not too long. It may be recalled that longer offset time increases queuing delay of a burst at the ingress node leading to longer end-to-end delay and poor quality of service. Also, long queuing delay requires large buffers at the ingress edge nodes.

An ingress node collects scheduling information of bursts on an outgoing link based on the bursts assembled or pass through it. Whenever a burst arrives, a node determines if it could find a free wavelength on the outgoing link at each of the time (from the current time) t = δ, t = $\delta + M$, t = $\delta + 2M$, t = $\delta + 3M \ldots \delta + kM$, where the values of M and k are predefined. This observation is made for a specific period. The length of this period and frequency of repeated observations can be chosen depending the level of accuracy required and the frequency of change of traffic load patterns. Let x be the total number of bursts processed at a node during the observation period. Let x_i be the number of bursts that could possibly be successful in finding a free wavelength on the outgoing link if they were scheduled with offset time $\delta + iM$. This measured scheduling statistics lead us to define $p_i = \frac{x_i}{x}, 0 \leq i \leq k$ as the probability that a burst is successfully scheduled on the link if an offset time of $\delta + iM$ is used. Since longer offset time means earlier reservation, it is likely that $p_0 \leq p_1 \leq p_2 \ldots \leq p_k$.

We now describe how we determine the initial offset time for the bursts at an ingress node. Let $t_h = \delta + zM$ be the initial offset time to be used by a burst which needs to traverse from an ingress node to an egress node separated by h hops. We need to determine the value of z from the p values calculated from the measured scheduling statistics. Let q_h be the success probability of a h-hop burst on the entire path. For single-hop bursts, we use an initial offset time of $t_1 = \delta$. From the scheduling statistics, the success probability q_1 associated with a one-hop burst is given by p_0. In order to achieve fairness among bursts with different hop counts we expect that $q_1 = q_2 = q_3 = q_h = \ldots = p_0$. We assume that the success probability p_i on a link at the ingress node is the same on the subsequent links and this probability on different links are independent. This assumption may lead to inaccuracies. However, this will not affect the results significantly (as evident from the experimental results presented in the next sec-

TABLE I

AN EXAMPLE OF OFFSET CALCULATIONS.

Offset	2	3	4	5	6	7
i	0	1	2	3	4	5
p_i	0.48	0.51	0.62	0.73	0.82	0.94

tion). Moreover, this assumption makes computations simpler which are desirable in a dynamic traffic scenario. The offset time used at nodes starting from the ingress node is $t_h, t_h - \delta, t_h - 2\delta, \ldots t_h - (h-1)\delta$, respectively. Let the associated probabilities be $p_z^0, p_z^1, p_z^2, \ldots p_z^{h-1}$ where $p_z^0 = p_z$. The success probability q_h on the entire path traversing h hops is then given by $q_h = p_z^0 \times p_z^1 \times p_z^2 \times \ldots \times p_z^{h-1}$. We choose z such that $t_h \geq h\delta$ and q_h is closest to $q_1 (= p_0)$.

The value chosen for k determines the maximum number of p values calculated. This value is based on the longest permissible offset time to be used by any burst. We note that the computational complexity of the above method is a linear function k and h. The value of M could be chosen to be δ or less to aid the above calculation and to get better results.

We illustrate the offset calculations using a simple example. Let $\delta = 2$, $M = 1$, and $k = 5$. The offset values and the probabilities based on the statistics collected are tabulated in Table I. We now determine the best offset value for a 2-hop burst, i.e. the best value for t_2.

From the table, $q_1 = p_0 = 0.48$. We express offset t_2 as $\delta + zM = 2 + z$. For several possible values of z, we calculate q_2 the probability that the burst is successful on a 2-hop path.

For $z = 2$, offset = 4 and $q_2 = p_2 \times p_0 = 0.2976$.
For $z = 3$, offset = 5 and $q_2 = p_3 \times p_1 = 0.3723$.
For $z = 4$, offset = 6 and $q_2 = p_4 \times p_2 = 0.5084$.
For $z = 5$, offset = 7 and $q_2 = p_5 \times p_3 = 0.6862$.

Among the four possibilities given above, the offset of 6 is chosen as the corresponding q value is closest to $p_0 = 0.48$.

III. PERFORMANCE STUDY

We study the performance of the proposed fairness method LSS-FA through extensive simulation experiments. We use three performance metrics, *burst dropping probability, burst offset time,* and *standard deviation in dropping probability*. The burst dropping probability and offset time are measured for different classes of bursts that correspond to different hops, i.e. the dropping probability of bursts with the same hop count are considered. The standard deviation in dropping probabilities encountered by bursts with different hop counts shows how good the fairness is. We use six random networks that differ in the number of nodes and topological connectivity. We compare the performance of our fairness method LSS-FA with the earlier method JET-FA that assigns an extra offset time of $10hL$ to a h-hop burst, where L is the mean burst duration.

The bursts are assumed to arrive as a Poisson process with exponentially distributed mean duration of $100\mu s$. The destination nodes are generated using uniform distribution. The control packet processing time is assumed to be $10\mu s$. A link is assumed to carry 4 wavelengths. We use the Poisson traffic as it has been used widely by the research community. Although models such as Gaussian noise self-similar traffic exist for IP traffic, for burst traffic the research is still going on to develop an accurate model as it depends on several factors including burst assembly time.

The burst dropping probability for different hop counts is plotted in Fig. 1 for a random network G36-5. We label a random network as $G\ n - p$, where n is the number of nodes and p is the density of topological connectivity defined as the probability with which an edge exists between two nodes. We choose burst arrival rates so that the dropping probability is reasonably low. When no fairness method is employed, the dropping probability increases with the the increasing hop count. This justifies the need for a fairness method. The fairness achieved by our algorithm LSS-FA is significantly better than the earlier method JET-FA. The figure also shows that JET-FA penalizes shorter-hop bursts heavily whereas our algorithm LSS-FA is reasonably fair to all hop counts. The fairness achieved by our method LSS-FA improves when the network becomes denser as evident from Fig. 2 which depicts the performance for a random network G36-10. This is because, many ingress-egress node pairs become closer in terms of hop count and the average hop count decreases when network becomes denser.

The offset time (in μs) used by bursts initially at the ingress edge nodes for different hop counts is plotted in Fig. 3 and Fig. 4 for random networks G36-5 and G36-10, respectively. The curves show that our method LSS-FA requires much lower offset time (and hence lower queueing and end-to-end delay) when compared to the earlier JET-FA. While LSS-FA uses offset time on the order of hundreds of microseconds, JET-FA uses offset time on the order of thousands of microseconds. The offset time used when no fairness method is employed is low but it is at the cost of poor fairness.

The performance in terms of burst dropping probability and offset time are shown only for 2 random networks due to space limitation. However, we give the standard deviation in dropping probability for different hop counts in Table II for six random networks. It can be observed that the standard deviation is the lowest in the case of the proposed LSS-FA method among all. Also, for a given number of nodes when the degree of connectivity increases from 5% to 10%, the deviation decreases and hence the fairness improves due to the reasons discussed above.

IV. CONCLUSIONS

In this paper we addressed the problem of fairness control in WDM optical burst switching networks. We developed an efficient algorithm to achieve fairness among bursts transmitted between ingress-egress node pairs with different hop distances. By collecting scheduling statistics online, the proposed fairness method is able to assign reasonably low but different offset times to bursts with varying hop counts. This leads to improved fairness, low burst delay while avoiding over-penalizing shorter-hop bursts. The need for fair-

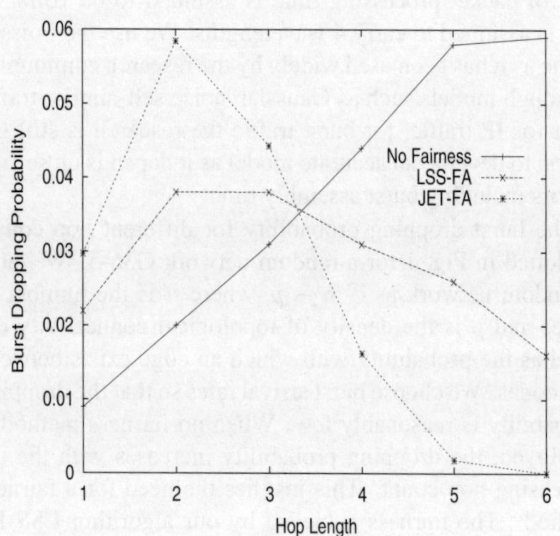

Fig. 1. Dropping performance of bursts with different hop counts in a random network G36-5.

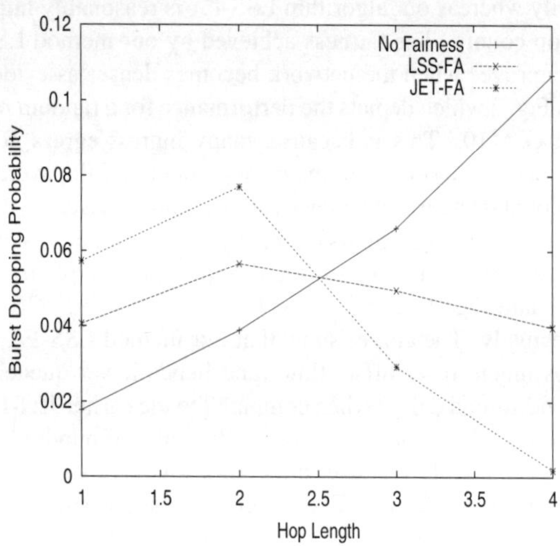

Fig. 2. Dropping performance of bursts with different hop counts in a random network G36-10.

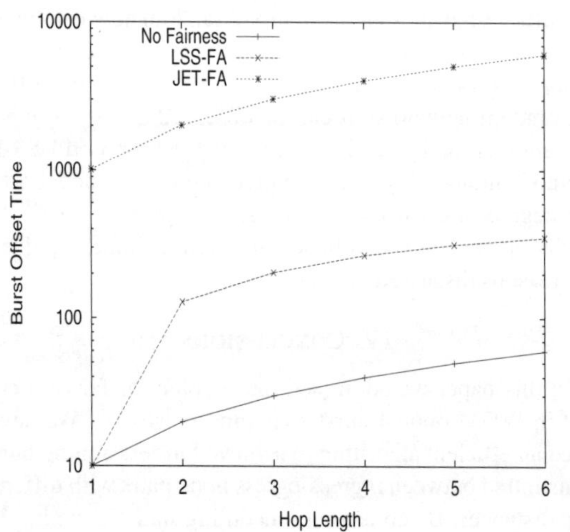

Fig. 3. Offset time used by bursts with different hop counts in a random network G36-5.

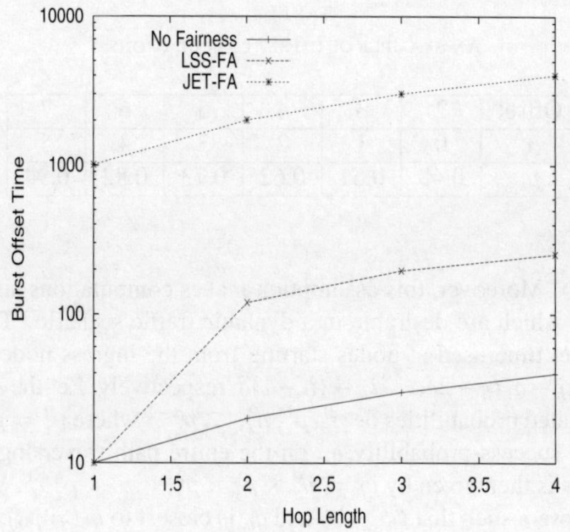

Fig. 4. Offset time used by bursts with different hop counts in a random network G36-10.

TABLE II

STANDARD DEVIATION IN BURST DROPPING PROBABILITIES OF BURSTS WITH DIFFERENT HOP COUNTS.

	No Fairness	LSS-FA	JET-FA
G32-5	0.053968	0.025204	0.054271
G32-10	0.024308	0.014272	0.027227
G36-5	0.018632	0.008078	0.021421
G36-10	0.033258	0.006939	0.02844
G40-5	0.036262	0.019717	0.043408
G40-10	0.051949	0.015343	0.04725

ness and the effectiveness of our fairness method have been demonstrated through extensive simulation experiments on a number of random networks. The results show that our method achieves significant level of fairness with low burst delay. Further, the fairness improves when the networks become denser.

REFERENCES

[1] Y. Xiong, M. Vandenhoute, and H. C. Cankaya, "Control Architecture in Optical Burst-Switched WDM networks," *IEEE Journal on Selected Areas in Communications*, pp. 1838-1851, October 2000.

[2] M. Yoo and C. Qiao, "QoS Performance in IP over WDM Networks," *IEEE Journal on Selected Areas in Communications*, pp. 2062-2071, October 2000.

[3] J. S. Turner, "Terabit Burst Switching," *Journal of High Speed Networks*, 1999.

[4] C. Qiao and M. Yoo, "Optical Burst Switching - A New Paradigm for an Optical Internet," *Journal of High Speed Networks*, pp. 69-84, 1999.

[5] L. Xu, H. G. Perros, and G. Rouskas, "Techniques for Optical Packet Switching and Optical Burst Switching," *IEEE Communication Magazine*, pp. 136-142, January 2001.

[6] C. Siva Ram Murthy and Mohan Gurusamy, "WDM Optical Networks: Concepts, Design, and Algorithms," *Prentice Hall PTR, NJ, USA*, November 2001.

[7] C. Qiao, "Labeled Optical Burst Switching for IP-over-WDM Integration," *IEEE Communication Magazine*, pp. 104-114, September 2000.

Packet Aggregation in All Optical Networks

Gérard Hébuterne, Hind Castel
Institut National des Télécommunications
9, rue Charles Fourier, 91011 Evry, France
phone: 0160764583, fax: 0160764291
email:{Gerard.Hebuterne}@int-evry.fr; {Hind.Castel}@int-evry.fr

Abstract

This study focuses on the edge router of an optical network. This device works as an interface between the electronic and optical domains : it takes packets coming from client layers and converts them into optical packets to be sent into the optical network.

Arriving variable length client packets fill the optical packet until the latter is full or cannot accommodate the arriving block. The aim of this work is to measure the efficiency with which electronic packets can be carried into optical packets.

Performance measures such as the packetisation efficiency and the mean time of packetisation give good results.

Keywords: Optical networks, Packet aggregation, Edge node, Performance evaluation.

1 Introduction

In any packet switching network, information is carried and processed in blocks, incorporating the useful *payload* and the *header* needed to process and forward the packet through the network. As the header has no usefulness from the end-to-end viewpoint, it corresponds to a portion of the total bandwidth which may be seen as wasted – and thus its relative importance has to be minimized, which calls for rather long packets.

Now, for real-time services, lengthening packets amounts to increasing the time to fill it, and so the end-to-end delay, so that short packets would be preferable.

On the other hand, packet length depends also on the network technologies and protocols in use. IP allows variable length frames while using ATM imposes to segment data in fixed length cells. Of prime importance here are the questions of *packet format* (fixed or variable length) and – in the latter case – of maximum length.

1.1 Fixed vs variable length packets

Fixed length packets have been extensively considered with the ATM technology. They result in the need to implement the Segmentation and Reassembly (SAR) layer, as most data blocks (coming from the IP layer, typically) have much higher length than the ATM cell. Another more interesting configuration is provided by the optical packet technologies, as in this case the optical packet is likely to be much larger than the maximum data block, which introduces the inverse possibility of aggregation at the network edges.

The optical network will directly provide a global transport infrastructure for legacy and new IP services. In [1] an optical edge router is proposed which provide an efficient traffic aggregator. Packets with the same destination and quality of service (QoS) are grouped together forming an optical packet.

In this study, we analyse the performance of an optical packet switching network where aggregation is performed into fixed length packets. The issues to be addressed are the efficiency of the aggregation (related with wasted bandwidth) and the impact on the delay (as packets of the client layer have to wait until the optical packet is completed).

1.2 The optical packet

In the present technology of optical networks, the information travels in bursts of light, without bit synchronisation. This allows more simple switch architectures, as it eases to deal with various physical constraints such as chromatic dispersion or thermal drift. This imposes however to insert *guard times* before each information block. Also, a synchronisation field is to be provided, so that the optical packets is built as in Figure 1. Current estimations (see e.g. [2]) show that the order of magnitude of the guard time should be around 50 ns.

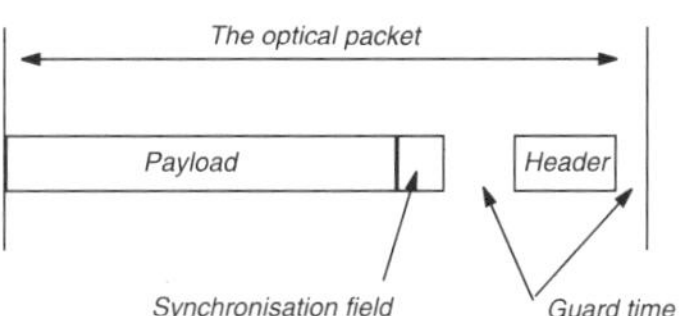

Figure 1: The optical packet format

2 Problem formulation

The study considers the aggregation of blocks of data in a single packet. The typical application is the building of optical packets in the edge router of optical backbone: IP packets, of variable length, are put together in fixed-size packets before being sent in the network. Most often an Ethernet link is used so that IP packets are segmented according to the maximum size of Ethernet frames. For convenience, the entering packets from the client layer are named "blocks", while the term of "packet" is reserved to the constant-size optical packet.The operating mode is as follows:

- Individual blocks, with variable length, arrive in the aggregation unit. The block is tentatively inserted in the packet under building.

- If there is enough room for the block, the packet remains in the unit until next block. If the block cannot be inserted because of its size, then the current packet is considered as being completed and is sent in the network. The arriving block is denoted as a *trigger block*, as it provokes the packet sending . The part of the packet which is still not filled is wasted. At the same time a new packet is created, and the trigger block is inserted in it.

- When a new packet is created, a time-out is initialized. If the time-out fires before the packet is completed, the packet is immediately sent, whichever its current filling.

There are two problems related with this process, namely the choice of the packet size and the choice of the time-out value. The point is that long packets yield an optimal use of the available bandwidth, at the price of an increasing end-to-end delay, while too short packets increase the burden of headers and wasted bandwidth resource.

3 Assumptions and notations

Let the individual blocks have their lengths distributed according to a common probability distribution function F, with density f. The successive block sizes are independent and identically distributed (i.i.d.). Let m be the average block length. The constant size of the optical packet is denoted as K. The units for measuring K or the block lengths may be either bits, or bytes, or larger units.

For the analysis of delay performance, one has to specify the arrival process of the client blocks. When needed, one assumes they arrive according to a Poisson process, with rate λ.

4 Mathematical analysis

4.1 Preliminary relations

In a time interval T, λT blocks arrive, which represent a total amount of λTm bits. They will be sent in $N(T)$ packets, carrying both the useful data, the header information and the padding bits. Let E denote the average pad length.

The average number of blocks per packet, p, gives the average useful payload pm (using Wald's relation proves this result), which is related with the average waste, by the relation : $E + pm = K$.

The packetisation efficiency ϵ is defined as the ratio of the average filled payload to the total payload length : $\epsilon = \frac{pm}{K}$. $N(T)$ is given by the simple relation : $N(T) = \frac{\lambda T}{p}$.

The average packetization delay θ is related with p: The packet begins with block 1 and is sent when block $p+1$ arrives, representing thus p interarrival periods. So the average delay θ is equal to $\frac{p}{\lambda}$. If D represents the bandwith (in bit/second), then $D = \lambda m$, and we obtain that: $\theta = \frac{pm}{D}$.

4.2 A model for packet length

The study considers the process (X_n) of the cumulated packet size after the arrival of the n-th block (whether the optical packet is sent or not). Let $H_n(x)$ denote the probability distribution function of the (X_n). The process (X_n) is a discrete-time Markov process, which obeys the following recurrence relation (f_n stands for the size of the n-th arriving block):

$$X_n = \begin{cases} X_{n-1} + f_n & \text{if } X_{n-1} + f_n \leq K \\ f_n & otherwise \end{cases} \quad (1)$$

The above recurrence yields the following relation between pdf's :

$$P\{X_n \leq x\} = \int_{y=0}^{x} f_n(y) P\{X_{n-1} > K - y\} \mathrm{d}y$$

$$+ \int_{y=0}^{x} f_n(y) P\{X_{n-1} \leq x - y\} \mathrm{d}y$$

Now, we assume that a stationary limit exists. This gives the fundamental relation: $H(x) = \int_{y=0}^{x} f(y)[1 + H(x-y) - H(K-y)]\mathrm{d}y$.

From the solution of the precedent equation , the major performance figures are derived. Especially, the distribution of the packet size is of prime importance. Let $Q(x)$ be the probability that a packet is sent with size lower or equal to x.

First, the probability Π that a packet is sent at epoch n is equal to the probability that the n-th arriving block is a "'trigger block". So $\Pi = \int_{y=0}^{K} f(y)[1 - H(K-y)]\mathrm{d}y$.

Now, Q is derived from H: this is the probability that the X_n is lower than x, given that a packet is actually sent: $Q(x) = P(X_{n-1} \leq x \text{ and } f_n \text{ trigger block}|\text{a packet is sent at } n-1)$.

So $Q(x)$ is : $\frac{1}{\Pi}\int_{u=0}^{x}\int_{y=K-u}^{K} \mathrm{d}H(u)f(y)\mathrm{d}y$ which is equal to : $\frac{1}{\Pi}\int_{y=K-x}^{K} f(y)\left[1-H(K-y)\right]\mathrm{d}y$

4.3 The case where block sizes have a discrete distribution

When the f_n take only discrete values, a simpler set of equations may be used. First, one can assume that the distribution of the arriving block is an integer multiple of a basic block (a byte, or more – e.g. 64 bytes). In this case, all the problem is reformulated using the basic block as unit. The packet size K is then expressed as an integer number of basic blocs. The recurrence relation (1) is still valid, but it allows a more tractable one-dimension Markov chain formulation. Especially, $H(x)$ is given by:

$$H(x) = \sum_{y=0}^{x} P\{f = y\}[1 + H(x-y) - H(K-y)]$$

As (X_n) can take only discrete values, the following relation holds: $P\{X_n = x\} = H(x) - H(x-1)$, which is equal to :

$$\sum_{y=0}^{x} P\{f_n = y\}\ P\{X_n = x-y\}$$

$$+P\{f_n = x\}\,P\{X_n > K-x\}$$

This relation corresponds to the local balance equation on state $\{X_n = x\}$. In other words, in the case of discrete values of ingoing blocks, the relations given for $H(x)$ generate the resolution of the discrete Markov chain which governs the evolution of (X_n) .

Let us explain now how to compute performance measures, such as the mean size of the payload of optical packets, and the average packetisation delay. We must give $Q(x)$ in the case of discrete size of packets : it will represent the probability that a packet of size $x \in N$ will be sent. The relations given in the continuous case still hold. First let us write Π, the packetization probability, which becomes in the discrete case : $\Pi = \sum_{y=0}^{K} P\{f = y\}\left[1 - H(K-y)\right]$.

So $Q(x)$ is given by the following relation :

$$Q(x) = \frac{\sum_{y>K-x} P\{f = y\}\,P\{X = x\}}{\Pi}$$

Let N be the mean size of the effective payload in optical packets :$N = \sum_{x=0}^{K} x\ Q(x)$.

The packetisation efficiency ϵ is : $\epsilon = \frac{N}{K}$. The average packetization delay θ is $\frac{N}{m\lambda}$, which is equivalent to : $\frac{N}{D}$.

5 Results

In this section, we give some numerical results in order to see the performance of packet aggregation. We have supposed that the block sizes have a discret distribution. We have represented two important performance measures : the packetisation efficiency and the mean time of packetisation. The steady state distribution of (X_n) is computed using the GTH algorithm [3]. Performance measures are obtained from equations given in section 4.3.

In Figure 2, we have plotted several curves representing the packetisation efficiency versus the optical packet size. For all the experiments, we have supposed that the arrival rate λ of clients blocks is 10/μs. The unit for measuring the block lengths and K is 64 bytes. Performance measures are computed for optical packet sizes which vary from 1600 bytes to 6400 bytes (25 packet units to 100 packet units in figures).

Each curve represents typical values for the probability distribution of the block length. The curve "C_i" for $i \in [1\ldots4]$ corresponds to the C_i configuration for the block length distribution. We have chosen block lengths equal to : 64 bytes, 576 bytes, and 1600 bytes. C_1 : 0.25 (64 bytes), 0.75 (576 bytes), 0 (1600 bytes), C_2 : 0.25 (64 bytes), 0.5 (576 bytes), 0.25 (1600 bytes), C_3 : 0.25 (64 bytes), 0.25 (576 bytes), 0.5 (1600 bytes), C_4 : 0.25 (64 bytes), 0 (576 bytes), 0.75 (1600 bytes).

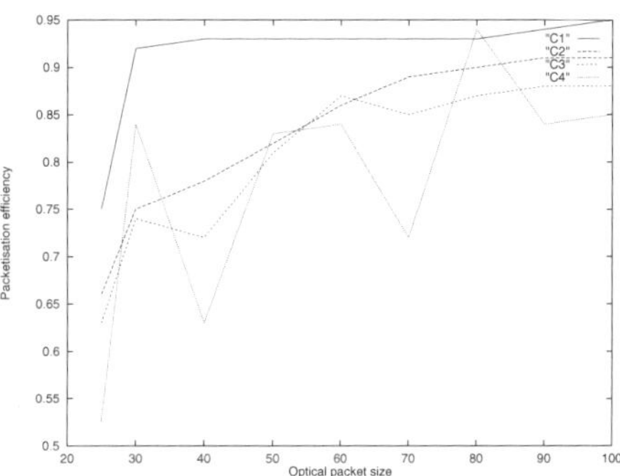

Figure 2: Packetisation efficiency versus optical packet size

We notice that when the ratio of blocks whith a high length increases, then the packet efficiency decreases. For the curve "C4", we can also see an oscillation phenomenon. For a given value of the optical packet size, as we have often arrivals of large length block then we have two cases : the packet is either well filled, or not (he has an important free places but not enough to be filled with the block). Except for this case, we can sec that the block length distribution has not an important impact on the packetisation efficiency. As it seems that current IP flows exhibit a rather high proportion of short IP packets (acknowledgements typically), configurations C1 and C2 are the most likely to be observed.

However, the partial filling effect is to be mixed with

the bandwidth waste related to guard bands and headers. Assume for instance that they result in a 200 ns (100 ns for guard times and 100 ns for the header), and that the link speed is 10 Gbits/s. In Table 1 we give the following global performance figures in the case of the (probably most typical) configuration C2.

Table 1: Global performance figures

Payload length	Packet duration	Global efficiency
1600 bytes	1.48 μs	0.57
3200 bytes	2.76 μs	0.77
6400 bytes	5.32 μs	0.87

Figure 3 shows the mean packetisation delay (in μ s) versus the optical size. We have plotted curves "C_i" for $i \in [1 \ldots 4]$ corresponding to the block length distributions given above. We can easily see that when the ratio of blocks whith a high length increases, then the mean time of packetisation decreases.

As can be expected, the longer the packet, the longer the aggregation delay. Also, as the aggregation efficiency increases, so does the delay-since more data is needed to fill the packet. Therefore, efficiency is obtained at the price of additional delay. We can see in this figure, that for an optical packet size of 1600 bytes, the packetisation delay varies from 0.26 μs (curve C1) to 0.069 μs (curve C4). And for 6400 bytes it varies from 1.358 μs (curve C1) to 0.45 μs (curve C4). So we deduce that the delay is quite negligible for λ=10 blocks/μs, but it increases if λ decreases (for λ=10 blocks/s, the delay will have the same values but given in secondes).

Suppose now that the flow to carried through the aggregation process is 1Gbits/s (so D=1Gbits/s), we obtain the following values of the mean packetisation delay given in Table 2.

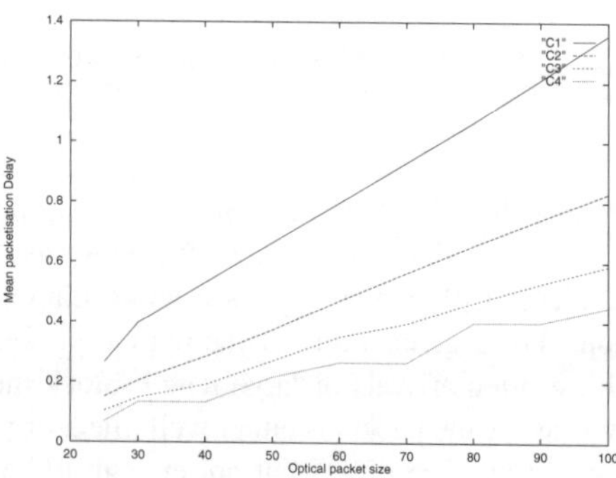

Figure 3: Average packetisation delay versus optical packet size

We can see that for D=1 Gbit/s, the delay is quite negligible, but as for λ, if D decreases, the delay increases.

6 Conclusion

The goal of this paper is to study the performance of the packet aggregation on an edge router in an optical network. We have supposed the following distributions : the arrivals of blocks are Poisson , and the block length have discret values. With these assumptions, the evolution of the number of packet units in the optical packet is represented by a finite Markov chain. We have computed two performance measures on the stationnary distribution : the packetisation efficiency and the mean time of packetisation. Values have been obtained for several probability distributions of the block length. The analysis allows to estimate the efficiency of the process, in terms of bandwidth utilisation and to evaluate the additional delay due to the aggregation task.

The performance values that we have obtained prove that the packet aggregation is efficient. Nevertheless, the assumptions on the block arrivals are too simple. As we focus on the edge node, it will be interesting to take more variable traffics, and then to consider a time-out parameter for the packetisation. It is a time counter which is initiated as a new optical packet begins to be filled, and if the time needed to fill the packet happens to exceed some limit, the packet is sent before being entirely filled in order to avoid too large aggregation delay, especially in case of highly bursty flows. So in this case, the aggregation efficiency will decrease.

References

[1] M.J.O'Mahony D.Simeonidou D.K.Hunter A.Tzanakaki, *The Application of Optical Packet Switching in Future Communication Networks*, IEEE Communication Magazine, 2001.

[2] P. Gravey an al, *Multiservice Optical Network : main concepts and first achievements of the ROM program*, vol. 19, Lightwave Technology, 2001.

[3] D.P.Heymann, *Further comparisons of direct methods for computing stationnary distributions of Markov chains*, vol. 8, SIAM J. Alg. Disc. Meth., 1987.

Table 2: Packetisation delay in μs for 1 Gbit/s flow

Payload length	C1	C2	C3	C4
1600 bytes	1.19	1.06	1.01	0.84
3200 bytes	2.98	2.64	2.60	2.64
6400 bytes	6.09	5.84	5.67	5.48

A New Packet Fair Queueing Algorithm using Virtual Start Time for High-Speed Packet Networks

Nam-Seok Ko[†], Dong-Yong Kwak[†] and Hong-Shik Park[††]
Network Technology Lab, Electronics and Telecommunications Research Institute
161 Gajeong-dong Yusong-gu, Daejeon, 305-350, Korea
e-mail: nsko@etri.re.kr
[††] School of Engineering, Information and Communications University
Yusong P.O. Box 77, Daejon, 305-600, Korea
e-mail : hspark@icu.ac.kr

Abstract: In this paper, we propose an efficient and simple fair queueing algorithm, called Minimum Possible Virtual Start Time Fair Queueing (MPSFQ), which has O(1) complexity for the virtual time computation while it has good delay and fairness properties. The key idea of MPSFQ is that it has an easy system virtual time recalibration method while it follows a rate-proportional property. MPSFQ algorithm recalibrates the system virtual time to the minimum virtual start time among all possible virtual start times of HOL packets in backlogged sessions. We will show our algorithm has good delay and fairness properties by analysis.

Key Words: Packet Scheduling, Fair Queueing

1. INTRODUCTION

Traffic scheduling algorithm is very important within individual switches or routers in the high-speed integrated services packet networks for providing a wide range of Quality-of-Service (QoS) guarantees. A function of a scheduling algorithm is to select, for each outgoing link of the switch, the packet to be transmitted in the next cycle from the available packets belonging to the flows sharing the same output link.

Until now many packet scheduling algorithms have been presented in the literature [1]-[5]. Among them, Weighted Fair Queueing (WFQ) [2] is known as an ideal traffic scheduling algorithm in terms of its delay and fairness properties. However, timestamp computation in the WFQ scheduler serving N sessions has a complexity of O(N) per packet transmission time and this makes its implementation difficult. Recently, a lot of algorithms have been proposed to simplify the implementation of WFQ, such as Self-Clocked Fair Queueing (SCFQ)[3], Frame-based Fair Queueing (FFQ)[4], Starting Potential Fair Queueing (SPFQ)[4], and Minimum Delay Self-Clocked Fair Queueing (MD-SCFQ)[5]. However, most of the efforts to reduce the complexity of the virtual time computation deteriorate the system performance characteristics such as delay and fairness properties, and most of the efforts to get a good performance characteristics require additional complexities.

In this paper, we propose an efficient and simple fair queueing algorithm, called Minimum Possible Virtual Start Time Fair Queueing (MPSFQ), which has O(1) complexity for the virtual time computation without any additional computation complexity while it has comparable delay and fairness properties to those of WFQ.

This paper is organized as follows. In Section II, we look over the existing fair queueing algorithms and their problems. In Section III, the concept and operational principle of the proposed algorithm are presented and the performance analysis of the algorithm is given. Finally, conclusions are presented in Section IV.

2. FAIR QUEUEING ALGORITHMS

A. Backgrounds

We study about the scheduling algorithms that depend on the timestamp allocated per packet on its arrival to decide the transmission order for packets waiting in queues. In those scheduling algorithms, a packet is provided with the timestamp before it is located at the queue for service and the server chooses a packet whose timestamp is minimum among all HOL packets from backlogged sessions. As a reference value for the calculation of timestamp, each algorithm maintains its own system virtual time (or system potential), which tracks the amount of work that is done by the server. The system virtual time is updated at every event such as a packet arrival or a departure during the server busy period.

Generalized Packet Sharing (GPS) is an idealized scheduling algorithm where packets are considered to be infinitely divisible and all backlogged session are serviced at each instant of time proportional to their reserved rates. Therefore, this algorithm provides the perfect isolation, fairness and low end-to-end delays.

WFQ or Packet-by-packet GPS (PGPS) scheduling algorithm is the packet-by-packet equivalent of GPS, that is, it derives the system virtual time from the background simulation of a GPS server [2]. The system virtual time v(t) of WFQ evolves as that of the corresponding GPS system, whose derivative is as follows.

$$\frac{dv(t)}{dt} = r \Big/ \sum_{i \in B(t)} r_i , \qquad (1)$$

where $B(t)$ is the set of sessions that are backlogged in the GPS server at time t and r is the server rate. WFQ, therefore, achieves the delay bound and fairness properties very close to those of GPS. However, since all N sessions can join or leave the set of backlogged sessions during a packet transmission time, the worst-case complexity of maintaining the system virtual time is *O(N)*, which makes the algorithm not suitable for practical deployment in high-speed packet networks.

There have been a lot of efforts to reduce the complexity while maintaining the performance characteristics close to WFQ, such as SCFQ, FFQ, SPFQ, and MD-SCFQ etc. SCFQ uses the virtual finish time of the packet that is currently being transmitted as the system virtual time. As a result, the complexity of computing the system virtual time of SCFQ is *O(1)* and it becomes more feasible for the high-speed network than WFQ. It also has the optimal fairness properties. However, the price paid to the easy system virtual time computation is the reduced level of isolation among the sessions, which causes the end-to-end delay bounds to grow linearly with the number of sessions that share the outgoing link.

FFQ, SPFQ, and MD-SCFQ are all included in RPS class [4]. If a scheduling algorithm is included in RPS class, the algorithm has the same delay properties to those of WFQ, and basically unfairness is bounded. Therefore, the latencies of all these algorithms are the same to that of WFQ. The fairness properties of each of them depend on how the system virtual time is recalibrated at any events. FFQ recalibrates the system virtual time when the session virtual times of all the backlogged sessions exceed the fixed value of fame time T. Therefore, the fairness bound is affected by the frame time T, while the complexity of the system virtual time computation is *O(1)*. SPFQ derives the system virtual time from the minimum virtual start time of all backlogged sessions. It has comparable fairness characteristics to WFQ. However, it needs another sorting structure to maintain the minimum virtual start time of all backlogged sessions. The resulting complexity of system virtual time computation is of *O(log N)*. MD-SCFQ recalibrates the system virtual time based on the weighted average virtual start time of all backlogged sessions. It also has the comparable fairness characteristics to WFQ. However, the computation of weighted average of virtual start time of all backlogged sessions need some complexities.

As explained above, most of the efforts to reduce the complexity of the virtual time computation deteriorate the system performance characteristics such as delay and fairness properties, and most of the efforts to get a good performance characteristics require additional complexities.

The purpose of this paper is to propose a new scheduling algorithm, which has a low virtual time computation complexity while having good performance characteristics.

3. A NEW PACKET FAIR QUEUEING ALGORITHMS

The key idea of MPSFQ is that recalibrates the system virtual time to the minimum virtual start time among all possible virtual start times of HOL packets in backlogged sessions while maintaining the PRPS properties.

A. The Service Discipline of MPSFQ

As in all GPS-related schedulers, in MPSFQ, the timestamps are assigned to the arriving packets according to the following equation.

$$F_i^k = \max\left(F_i^{k-1}, v(a_i^k)\right) + \frac{l_i^k}{r_i}, \qquad (2)$$

where l_i^k is the length of the k-th packet from session *i*, $v(a_i^k)$ is the system virtual time at arrival time of k-th packet from session *i* and r_i is the reserved rate of session *i*.

The system virtual time is maintained according to the same rule as other packet-by-packet Rate Proportional Servers (PRPS) while the system virtual time at each recalibration time, τ_i, is calculated easily as will be shown below.

The detailed procedures of MPSFQ at the packet arrival time and at the end of the packet transmission are given below.

[Procedures at the packet arrival time]

① When a packet arrives at *t*, update the system virtual time using the recent system virtual time calculated at τ_{j-1}.

- $v(t) = v(\tau_{j-1}) + (t - \tau_{j-1})$

where $\tau_{j-1} \le t < \tau_j$ and $v(\tau_0) = 0$.

② Calculate the virtual finish time (i.e., timestamp) of the packet.

- $F_i^k = S_i^k + l_i^k / r_i$, where $S_i^k = \max\left(F_i^{k-1}, v(t)\right)$

③ Place the packet with the timestamp to the related queue.

[Procedures at the end of each packet transmission]

① Increase the system virtual time by the transmission time of the packet just completed

- $v(\tau_j^-) = v(\tau_{j-1}) + (\tau_j - \tau_{j-1})$

② Retrieve a packet with a minimum virtual finish time (F_{cur}) from the HOL packets and transmit.

③ Recalibrate the system virtual time

- $v(\tau_j) = \max(v(\tau_j^-), F_{cur} - \max_{i \in B(\tau_j)} \frac{L_i}{r_i})$

where L_i is the maximum packet length of session *i* and the right term of max( ) is for the minimum virtual start time among all possible virtual start times of HOL packets in backlogged sessions.

As we can see, the complexity of computing the system virtual time in MPSFQ is *O(1)* and the only information maintained is the maximum packet length of each backlogged session, which is easily obtained at each flow add or delete time. Thus, MPSFQ requires a much smaller amount of information and less computation complexity than WFQ and SPFQ. In the following chapter, we will show our algorithm have good performance characteristics while having this simplicity.

B. Delay Property of MPSFQ

Even though we should prove the MPSFQ algorithm follows the RPS properties, we would not prove here due to the space limitation. However, since MPSFQ algorithm recalibrate the system virtual time to the value not less than the minimum virtual start time of the backlogged sessions, it is easy to show that this algorithm follows the RPS properties. Therefore, MPSFQ algorithm has all the delay properties PRPS, which is the same to those of WFQ.

C. Fairness Property of MPSFQ

Since the recalibrations make the system virtual time not to lag indefinitely behind the virtual times of the sessions that are currently backlogged, only with this fact that MPSFQ algorithm is a PRPS, we can say that MPSFQ is a fair scheduler. However, we need to define the formal fairness index. We adopt the Golestani's definition of the fairness index [3]. The fairness index $F_{i,j}$ is defined as

$$\left|\frac{W_i(t_1,t_2)}{r_i} - \frac{W_j(t_1,t_2)}{r_j}\right| \le F_{i,j}, \tag{3}$$

where $W_i(t_1,t_2)/r_i$ is the normalized service received by session *i* during the continuously backlogged time interval (t_1, t_2).

Lemma 1 In a PRPS system, at any time *t*, the difference between the system virtual time $v(t)$ and the timestamp F_i^k of any packet p_i^k of session *i* that is currently in the system is bounded as follows;

$$v(t) - F_i^k \le \frac{L_{\max}}{r_i} \quad \forall i \in B(t),$$

where r_i is the reserved rate of session *i* and $L_{\max}$ is the maximum packet length of the system.

Proof: The result of this Lemma is the same to the Lemma 1 in [5]. However, the proof is somewhat different. Actually this result could be applied to the any RPS system.

For any PRPS and corresponding RPS, the following holds [4];

$$t_i^k \le \hat{t}_i^k + \frac{L_{\max}}{r} \tag{4}$$

where t_i^k is the time when packet p_i^k completes transmission in the considered PRPS, and $\hat{t}_i^k$ is the time when the same packet completes transmission in the corresponding RPS.

Assume n-th recalibration of the system virtual time, τ_n, occurs at time t_i^k. The session virtual time $v_i(t_i^k)$ is as follows.

$$v_i(t_i^k) = v_i(\hat{t}_i^k) + \frac{r}{r_{M(t)}}(t_i^k - \hat{t}_i^k) \tag{5}$$

where $M(t) = \{j \mid v_j(t) = \min_{k\in B(t)} v_k(t)\}$

In an RPS system, when packet p_i^k completes transmission in the fluid system, its timestamp F_i^k is greater than or equal to the system virtual time:

$$v(\hat{t}_i^k) \le F_i^k \tag{6}$$

Therefore, according to (5) and (6) with the second condition of RPS system, $v(t) \le \min_{i\in B(t)} v_i(t)$, the system virtual time at time t_i^k is bounded as follows.

$$\begin{aligned} v(t_i^k) &\le v(\hat{t}_i^k) + \frac{r}{r_i}(t_i^k - \hat{t}_i^k) \\ &\le F_i^k + \frac{L_{\max}}{r_i} \end{aligned} \tag{7}$$

Since the system virtual time between the two recalibration times increases by the real time elapsed, the following corollary comes directly.

Corollary 1 In a PRPS system, at time s_i^k, when packet p_i^k starts transmission, the difference between the system virtual time and its timestamp F_i^k is as follows.

$$v(s_i^k) - F_i^k \le \frac{L_{\max}}{r_i} - \frac{l_i^k}{r} \tag{8}$$

We can evaluate the fairness index of MPSFQ following the procedures presented in [5].

Theorem 1 The fairness index of MPSFQ is

$$FI_{(MPSFQ)} = \max(f_{i,j}, f_{j,i}),$$

where $f_{i,j} = \frac{L_i}{r_i} - \frac{L_i}{r} + \max\left(\frac{L_j + L_{\max}}{r_j}, \max_{1\le n\le N}\frac{L_n}{r_n}\right)$

,which means the maximum amount by which the normalized service received by session *j* exceeds the normalized service received by session *i*.

Proof: We can derive $FI_{(MPSFQ)}$ from the difference of normalized service received by session *i* and *j* at any time interval, (t_1, t_2) during when the two sessions are continuously backlogged. All the possible cases which we can consider are as follows.

1. One session becomes backlogged at time t_1, while the other session is already backlogged before time t_1.
2. Both sessions becomes backlogged at time t_1.

Even though we can consider one more case, it is included in the case 1 at any time $t' < t_1$.

Case 1: Session *j* becomes backlogged at time t_1, whereas session *i* is already backlogged at time t_1, and F_i^k is the timestamp of its HOL queue packet p_i^k. We must consider two subcases.

Subcase a) Packet p_j^m of session *j* receives a timestamp $F_j^m > F_i^k$.

In order to maximize the amount of normalized service provided to session *i* before session *j* receives its first service, packet p_i^k must reach the system at the exact time when the packet p_j^m is picked for transmission. According to Lemma 1 and Corollary 1, at the exact time satisfying the condition, virtual finishing time of session j, F_j^m is as follows.

$$F_j^m \le F_i^k + \frac{L_{\max}}{r_i} - \frac{l_i^k}{r} + \frac{l_j^m}{r_j} \qquad (9)$$

where l_i^k and l_j^m are the length of packet p_i^k and p_j^m, respectively.

The service provided to packet p_i^k (equal to l_i^k / r_i) contributes to the difference of normalized service, and all the following packets of session i having timestamp not greater than F_j^m are transmitted before p_j^m.

From eq. (9)

$$F_j^m - S_i^k = F_j^m - (F_i^k - \frac{l_i^k}{r_i}) \le \frac{L_{\max}}{r_i} + \frac{l_j^m}{r_j}$$

where S_i^k is the virtual start time of packet p_i^k.

The resulting difference of received normalized service before packet p_j^m is picked for transmission is, therefore, bounded by

$$f_{j,i}^{(1a)} = \frac{L_i + L_{\max}}{r_i} - \frac{L_i}{r} + \frac{L_j}{r_j} \qquad (10)$$

Subcase b) Packet p_j^m of session j receives a timestamp $F_j^m < F_i^k$.

In order to maximize the amount of normalized service provided to session j before session i receives service again, packet p_j^m must reach the system just before packet p_i^k completes transmission and the system virtual time is consequently recalibrated. Let l_i^k be the length of the packet p_i^k, and l_i^{k+1} be the length of p_i^{k+1}. Packet p_i^k started transmission at time $t_s = t_1 - l_i^k / r$; at that time, according to the definition of MPSFQ, the minimum possible value of the system virtual time is

$$v(t_s) = F_i^k - \max_{1 \le n \le N} \frac{L_n}{r_n}$$

Packet p_i^{k+1} receives a timestamp $F_i^{k+1} = F_i^k + l_i^{k+1} / r_i$. If p_j^m reaches the system at time t_1 just before p_i^k completes transmission, the minimum possible value of the system virtual time is

$$v(t_1) = v(t_s) + \frac{l_i^k}{r} \qquad (11)$$

so that the resulting difference of normalized service between session j and session i can be expressed as

$$F_i^{k+1} - v(t_1) = \frac{l_i^{k+1}}{r_i} + \max_{1 \le n \le N} \frac{L_n}{r_n} - \frac{l_i^k}{r} \qquad (12)$$

The whole difference of normalized service in above eq. (12) increases with l_i^k and l_i^{k+1} so that

$$f_{i,j}^{(1b)} = \frac{L_i}{r_i} + \max_{1 \le n \le N} \frac{L_n}{r_n} - \frac{L_i}{r} \qquad (13)$$

Case 2: Both session i and j become backlogged simultaneously at time t_1.

In this case, both sessions use the same system virtual time to calculate the virtual finish times. Therefore, the maximum difference of the received normalized service between them is decided by their virtual finish times. Without loss of generality, assuming that $l_i^k / r_i > l_j^k / r_j$, session i begins service before session j. Consequently, the maximum difference of received normalized service between them at any time t_2, when p_i^k ends the transmission, is

$$f_{i,j}^{(2)} \le \frac{L_i}{r_i} \qquad (14)$$

Therefore, with eq. (10), (13) and (14), we conclude the proof of theorem 1.

4. CONCLUSIONS

In this paper, we proposed the new fair queueing algorithm, called Minimum Possible Virtual Start Time Fair Queueing (*MPSFQ*). The *MPSFQ* algorithm is a simple scheduling algorithm which follows the basic properties of PRPS while the system virtual time is recalibrated to the minimum virtual start time among all possible virtual start times of HOL packets in backlogged sessions. It does not need any additional sorting structure differently from SPFQ. The only information needed is the maximum packet length of each backlogged session, which is easily obtained at each flow add or delete time. Since *MPSFQ* belongs to the PRPS, we obtained the delay bounds of it, which are the same as those of WFQ. We also provided the fairness bound, which means that the difference in normalized service offered to any two sessions that are continuously backlogged is always bounded and this bound is comparable to that of WFQ.

REFERENCES

[1] H. Zhang, "Service disciplines for guaranteed performance service in packet switching networks", Proceedings of the IEEE, vol.83, no.10, pp. 1374-1396, October 1995.

[2] A. K. Parekh and R. G. Gallager, "A generalized processor sharing approach to flow control in integrated services networks: The single node case," Proceedings of IEEE INFOCOM' 92, vol. 2, pp. 915-924, May 1992.

[3] S. J. Golestani, "A self-clocked fair queueing scheme for broadband applications," Proceedings of IEEE INFOCOM ' 94, pp. 636-646, April 1994.

[4] D. Stiliadis and A. Varma, "Efficient fair queueing algorithms for packet-switched networks," IEEE/ACM Trans. on Networking, vol.6, no.2, pp. 175-185, April 1998.

[5] F. M. Chiussi and A. Francini, "Minimum-delay self-clocked fair queueing algorithm for packet-switched networks," Proceedings of IEEE INFOCOM '98, vol. 3, pp. 1112-1121, March 1998.

A Real Time Transmission Scheduling Scheme for Multiple Services in WDM Passive Star Networks

NamUk Kim,YeongSuk Choe, Minho Kang

Optical Internet Research Center, Information and Communication University, P.O.Box 77,Yusong,Daejon, (305-732), Korea.Tel.+82-42-866-6208,Fax.+82-42-866-6230,E-mail:niceguy@icu.ac.kr

Abstract: In WDM passive star networks(PSN), Two critical collision problems are a data channel collision during packet transmission and a receiver collision during receiving at a receiver node. We propose a new transmission scheduling scheme that avoids two collision problems and provides the differentiated services scheme with guaranteed reservation time. The proposed protocol also introduces the self collision prediction mechanism with a fixed roundtrip delay by using the FDL and uses the multiple control slots mechanism for real time transmission of high priority traffic.

Key Words–WDM-PSN, Self Collision Prediction, Multiple Services, Delay Equalizer

1. INTRODUCTION

The huge amount of Internet traffic from every parts of current business and home areas ceaselessly demands for an efficient internet protocol and network architecture which supports the wide bandwidth and high speed in whole network domain. Due to the large bandwidth and enormous repeats spacing, the optical fiber communication technology outperforms over every other transmission medium. Based on this, many new network architectures and protocols have been studied for an efficient use of optical fiber's wide bandwidth. First of all, the Wavelength Division Multiplexing technology (WDM) that can furnish multiple data channels and high transmission speed let these challenges to be more practical [1, 2, 3]. But, in the past, the use of WDM technology has been limited to core networks since the high bandwidth needs not to be used in access network. But, as the technical and economic problems are solved by successive evolutions of WDM device technology, many protocols that are based on the WDM access networks have been proposed.

In the backbone network, the optical burst switching and GMPLS are promising for high performance and low end-to-end delay [4-5]. The routing algorithm, wavelength assignment, and the real time signaling are the major topics since the backbone network configures originally multiple hops [4]. Oppositely, the single-hop local access and metropolitan network have other protocol requirements. The most important role of protocol of access network is an efficient media access mechanism that can guarantee a low delay time and low blocking probability while supporting impartial access chances to every access nodes [6-9]. In addition, the differentiated services must be supported in future access network to meet various demands of end users while using the network resources efficiently. The importance of a reasonable transmission scheduling for differentiated services is more emphasized on Ethernet PON or future WDM PON in these days. We propose a new transmission scheduling mechanism for differentiated service based on WDM passive star network (PSN) in Section 2. Finally we show the performance of the proposed protocol in Section 3.

To resolve collision problems, we introduced the collision prevention mechanism mainly based on [1], [2] and we adopt those schemes to multiple control mini-slots system for real time and differentiated services.

2.MULTIPLE SERVICES WITH THE SELF COLLISION PREDICTION MECHANISM

All access nodes of the WDM PSN are synchronized by a global clock. For easy configuration, each channel is assumed to be associated with WDM 's specific wavelength.

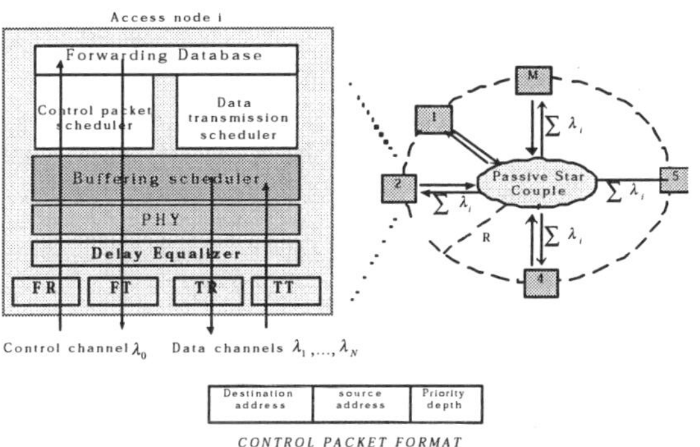

Fig. 1. Network topology and configuration of access node

The whole network topology and access node's layered function blocks are shown in Fig.1. The wavelength field does not exist in control packet since the sequence of control mini-slot is mapped to the data channel number. There are some operational blocks in the access node's configuration.

Data Transmission Scheduler: It determines the start time of the data frame transmission via the data channel, which is based on the information of Forwarding Database. It also decides the data frame configuration and length by controlling the operation rule of Buffering scheduler. It uses the optical tunable transmitter, specifically a tunable laser, and optical tunable receiver since it changes the channel dynamically.

Control Packet Scheduler: It continuously sends and received the control packet on the fixed control channel λ_0 with optical devices, FT (Fixed tuned Transmitter) and FR(Fixed tuned Receiver).

Buffering scheduler: It controls the possible buffer size per class based on the total buffered traffic's situation. For differentiated services, it can use weighted fair queuing scheme based on the current information of the Forwarding database.

Forwarding Database: It manages all information about the current data channel's usage condition and the future transmission schedules per each data channel by checking all incoming control packets and updates its information to the database. So, when there is a data frame to send, the control packet scheduler and the data transmission scheduler must check the current information of database first.

Delay Equalizer: It makes the propagation time between each access node and optical splitter to be equal to the network's reference delay time by setting the length of the FDL(Fiber Delay Line) in initialization process. By this, we can assume that all access nodes are located at the same distance, R, from the passive star coupler whose role is an optical hub. Consequently, the end-to-end propagation time among all access nodes is same to t_{2R} regardless of the real geographical distance.

As shown in Fig.2, There are two types of time slots, control time slot and data transmission time slot in one cycle. Between them, a tuning time exists to supplement for TR's tuning time fluctuation. These three time slots consist of one transmission cycle. A control time slot is again divided into the N mini-slots where N is an initial of data channel. For example, the i_{th} mini-slot of control time slot is allocated to the data transmission channel $?_i$ .By this fixed dedicated channel method between mini-slot and data channel, we can simplify the intricate data channel allocation problem.

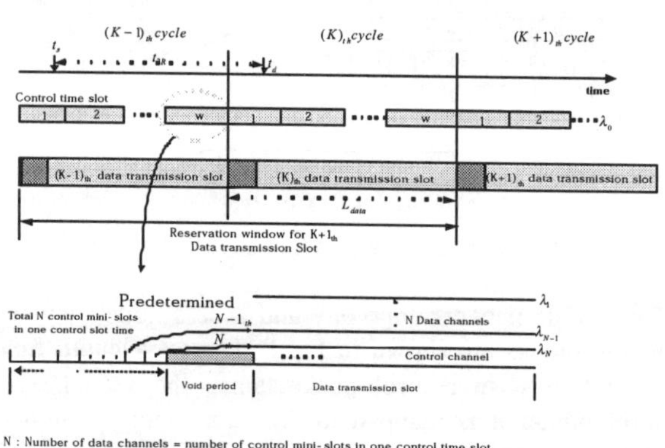

Fig. 2. Transmission cycle configuration

Here,

t_{2R}: propagation time from source to destination

t_s: control packet sending time

t_d: control packet receiving time at destination.

Basically, all t_{2R} is set to the equal value, $t_d - t_s$ by the Delay Equalizer. A packet transmission is performed through 3 transmissions time cycles. Access nodes contend for the control mini-slot during $(K\text{-}1)_{th}$ *cycle* and $(K)_{th}$ *cycle*, and the resulting data transmissions which will practically be done in $(K+1)_{th}$ cycle via allocated data channels are determined at the end of *(K) cycle*.

The self collision prediction mechanism consists of the following two steps for the differentiated services.

A. pre-estimation step

The access node's control packet scheduler first calculates a current usable data channel and a control packet transmission timing based on the current information of the Forwarding Database and buffer condition. The operational configuration between buffering discipline and scheduling method is shown in below Fig.3.

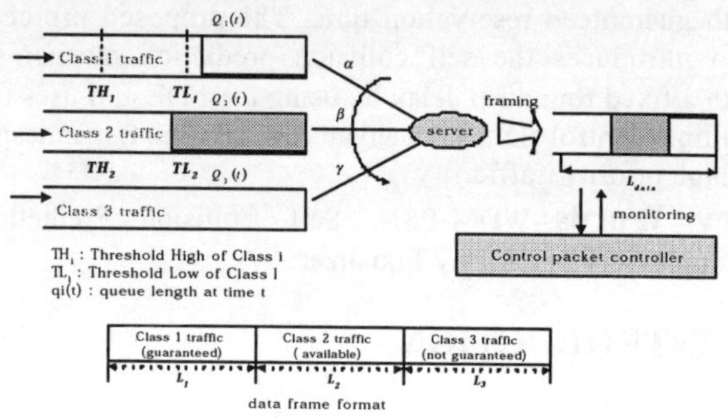

Fig. 3. Configuration of data frame

The scheduler must always guarantee the relatively low reservation delay time for Class 1 traffic with high transmission success probability. It also must dynamically control the service rate between Class 2 and Class 3 traffic using weighted round-robin method.

The pseudo code of control packet transmission operation for multiple services in Fig.4

If (total sum of buffered length = L_{data}*)*
Switch (class)
Case 1: available transmission
If $(q_1(t) > TL_1)$
Set priority depth = 2;
Send control packet at nearest control slot time first. Send it again at $t_{2R/2}$ *based on changed forwarding database' information at that time. It relatively reduces the contention between control packets.*
Case 2: urgent transmission
If $(q_1(t) > TH_1)$
Set priority depth =1;
Send control packet at all incoming control slot. By this, it can inform other nodes of its buffer situation and take priority to other transmission.
Case 3: default transmission
Send control packet at nearest control slot time.
If ($q2(t) > TL_2$*) set priority depth = 4;*
If ($q2(t) > TH_2$*) set priority depth = 3; It informs other nodes of its buffer situation. So, other nodes*

can avoid choosing data transmission schedule that has a high probability to contend with high priority traffic node in future transmission.

Fig. 4. Pseudo code of control scheduler operation

By this differentiated control packet transmission mechanism per each class, the access node that has the high priority real time traffic can take more transmission chances than others, which results in guaranteeing high probability of reservation success probability in relatively short time.

B. Confirmation and waiting step

During t_{2R}, the control packet scheduler inspects whether there is any incoming control packet that was sent before its own control packet transmission time, since those have not been recorded at that time. If two types of control packets which are same destination or data channel are found, the access node cancels an existing transmission schedule and prepares another transmission schedule based on current changed forwarding database information. The incoming control packet's priority depth also informs the forwarding database of other access node's buffer situation. Based on this, the control packet scheduler can avoid possible high probability channel contention. Fig. 5 shows a specific example. In this figure, the access node p sends its control packet whose destination is q at $t_{s,p}$. If a control packet of which destination or channel is same to that of p arrives at $t_{d,other}$, access node p recognizes its failure at time $t_{d,other}$ and all the other access nodes also recognizes it well.

Specifically, if there is another access node j whose destination is q or data channel is $?_i$, the node j also knows its failure at $t_{d,other}$. It means that the node j as well as the node p can change its own transmission scheduling before its own delay t_{2R}.

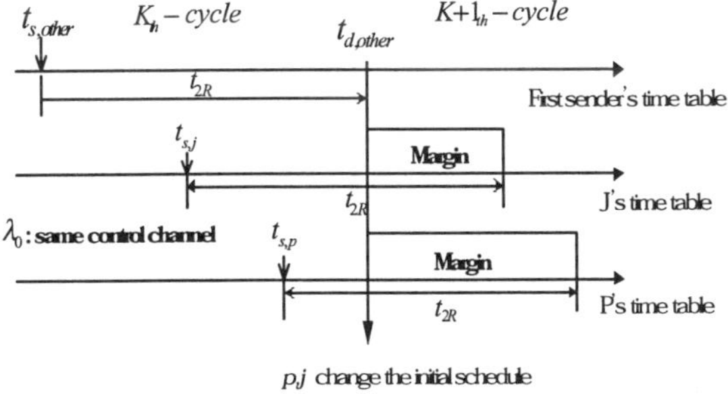

Fig. 5. Margin gain mechanism

The Margin means that there are some available control slot times by which access nodes try a new transmission scheduling procedure with ceaselessly coming multiple mini-slots in one cycle within t_{2R}.

3. PERFORMANCE EVALUATION

We use the following notations for analysis.

P_c = success probability of a control packet
P_{cr}= receiver collision free probability of control packet
P_{cd}= data channel collision free probability of control packet
P_d = success probability of a data frame transmission when corresponding control packet is successful
P_s = total transmission success probability in a data transmission cycle.
T_{reser} = the average time for success of reservation procedure

We assume that M is the number of access nodes and N is a number of data channels. The control packets arrive with the general Poisson arrival process and a is the mean value. Each access node chooses the destination node and data channel uniformly. In this analysis, we calculate the average reservation delay time per class and the Margin gain that represents the utilization of data channels.

Analysis: In our protocol, all access nodes must wait for at least t_{2R} (NA-1 mini-slot time) to verifying reservation result. To calculate P_d, let's assume the data channel $?_i$ to transmit data frame in *K+2 cycle* and its destination is currently available q.

$P_{cr}=P_r$ (no control packet whose destination is q arrive in next $NA-1$ mini-slot time) = $[P_r$ (no control packet whose destination is q arrive in one mini-slot time)$]^{NA-1}$
P_r (no same destination packet / k arrivals in one minislot) $= (1-(1/M))$, P_r (k arrivals) $= a^k e^{-a}/k!$
Therefore, the total reservation success probability in t_{2R} of first control packet transmission is

$$P_{cr}=\{\sum_{k=0}^{\infty}(1-1/M)^k a^k e^{-a}/k!\}^{(NA-1)} = \exp[-a(NA-1)/M]$$

$P_{cd} = P_r$ (no control packet whose data channel is $?_i$ arrive during next $(A-1)$ control-slot) $= exp[-a]^{A-1}$
$P_c=P_{cr}P_{cd}$ / based on the conditional probability $= exp\{-a\,([(N-1)/M]+1)(A-1)\}$
$P_s=P_cP_d = Pc$ (by dedicated channel allocation) = $exp\{-a\,([(N-1)/M]+1)(A-1)\}$

T_{reser} of class 1 traffic

$$= t_{2R}+(t_{2R}+t_{2R}/A)P_1+(t_{2R}+2t_{2R}/A)(1-P_1)P_2+(t_{2R}+3t_{2R}/A)(1-P_1)(1-P_2)P_3...+(t_{2R}+(A-1)t_{2R}/A)(1-P_1)(1-P_2)..(P_{A-1})$$
$$= t_{2R}+\sum_{k=1}^{A}\prod_{i=1}^{k}(1-P_{i-1})P_k(t_{2R}+kt_{2R}/A)$$

when, $P_k= exp\{-a\,([(N-1)/M]+1)(A-1)(1-k/A)\}$

T_{reser} of class 2 traffic

$$=t_{2R}+(t_{2R}+1t_{2R}/A)P_S+(t_{2R}+3t_{2R}/A)(1-Ps)P_1+..+(t_{2R}+(A-1)t_{2R}/A)(1-P_1)(1-P_2)..(P_{(A-1)/2})$$

T_{reser} of class 3 traffic

$$= P_s\sum_{k=0}^{\infty}(1-P_s)^k t_{2R}$$

when, $Ps = exp\{-a[(NA-1)/M+(A-1)/N]\}$

$$Margin=t_{2R}-\{P_k+[(1-P_k)P_k+1]+[(1-P_k)^2P_k+2]..\}t_{2R}/NA = t_{2R}-\{P_k+[\sum_{k=1}(1-P_k)^kP_k+k]\}$$

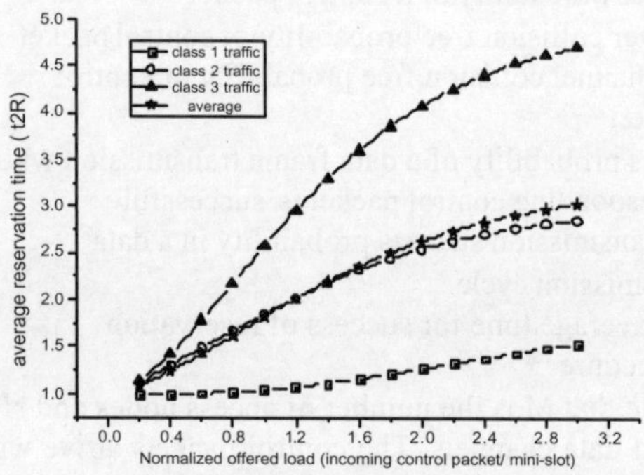

Fig. 6. Average reservation time per class (M=40, average value of N=4,8,12)

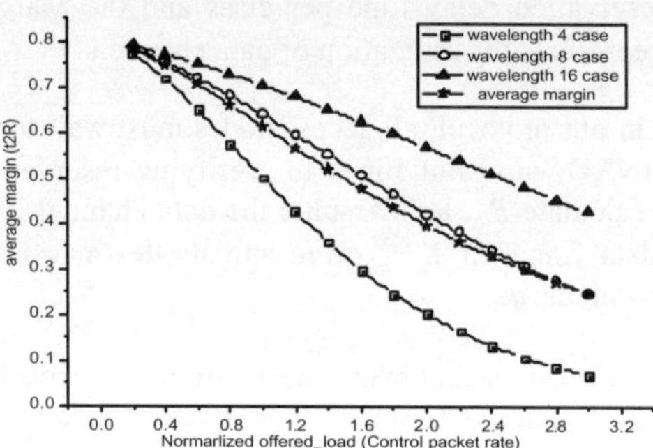

Fig.7. Average margin per reservation procedure

As shown in Fig.6, our control scheduler can guarantee the different reservation time for each traffic class by the differentiated control packet transmission scheme. The average reservation time of class 1 traffic is always less than $2t_{2R}$. This means that class 1 traffic will be sent without any reservation delay time since the reservation always finished before target data transmission cycle. Although class 2 and class 3 traffic's average reservation time is somewhat longer than that of class 1 traffic, those are relatively short than the average waiting time of the classless reservation fixed type Aloha mechanism in which we can not control and support the multiple service. In our scheme, the reservation procedure starts in two cycles before real data frame transmission cycle, the waiting time is not long in all classes and the packet loss does not easily happen in class 1 traffic's buffer.

Fig.7 shows the average margin of data channel per one reservation procedure. As we can expect, the margin decreases as the offered load increases since there are relatively many access nodes that attempt another scheduling based on multiple channels. So, the resulting utilization increases and the margin decreases. If the number of data channels increases, the contention probability for the same data channel will be low and utilization per each channel decreases. By this reason, the margin of data channel that means the new chances for another schedules increases.

Based on these two performance enhancements, our scheme can support multiple services within the guaranteed delay time by which it is possible to manage buffers efficiently for avoiding buffer overflow, and it also improves the utilization of transmission channels.

4. CONCLUSION

We proposed an efficient transmission scheduling scheme that supports multiple services based on the guaranteed data channel reservation time per each traffic class. The collision problems are solved out by adopting the self collision prediction mechanism to the reservation slotted Aloha not by general receiver initiated ACK mechanism of the general reservation Aloha. By using this scheme, the access node does not wait for a long time for ACK as much as that of the existing protocols since access nodes can predict the collision of its control packet in advance. The 3 cycle pre-reservation method and multiple control slots also enhance the channel utilization and reduce the each access node's waiting time which results in decrease of packet loss probability of real time traffic also.

ACKNOWLEDGEMENT

This work was supported in part by KOSEF-OIRC, ETRI, and KT.

REFERENCES

Journal papers:

[1] Mon-song Chen and Nicolas R.dono and Rajiv Ramaswami "A Media-Access Protocol for Packet-Switched Wavelength Division Multiaccess Metropolitan Area Networks" IEEE Journal of Selected Areas in Communications. Vol.8, NO.6, AUGUST 1990

[3] Ioannis E Pountourakis "Multiwavelength control protocols for WDM networks with restricted data channel collision" International Journal of Communication Systems Vol.13 (2000)

Proceeding Papers:

[2] Aura Ganz and Gao, Y "A time-wavelength assignment algorithm for a WDM star network " IEEE INFOCOM 1992

[4] Sungchang Kim, NamUk Kim, Minho Kang "Contention Resolution for Optical Burst Switching Networks Using Alternative Routing" IEEE ICC, 2002

[5] Se-yoon Oh, Minho Kang "A Burst Assembly Algorithm in Optical Burst Switching Networks" OFC 2002

[6] Biswanath Mukherjee " A WDM-Based Local Lightwave Networks Part I: Single-Hop Systems" IEEE Network Magazine, May 1992

[7] Jonathan C.Lu and Leonard Kleinrock "On The Performance of Wavelength Division Multiple Access Networks" IEEE ICC 1992

[8] David A.Levine and Ian F.Akyildiz "PROTON: A Media Access Control Protocol for Optical Networks with Star Topology" IEEE/ACM Transactions on Networking, Vol.3, NO.2, APRIL 1995

[9] Kemal Bengi " Medium access control protocols for passive star WDM single hop networks efficiently integrating real time and data service" Optical Networks Magazine July/August 2001

Session 12B

Chromatic Dispersion

GVD compensation schemes with considering PMD

Aiying Yang, Deming Wu, Anshi Xu
(National Laboratory on Local Optic-Fiber Communication Networks & Advanced Optical Communication Systems, Department of Electronics, Peking University, Beijing, China, 100871,
Tel: +86-10-62763334 Fax: +86-10- 62751763 E-mail: qiaoyay@water.pku.edu.cn)

Abstract: In this paper, we discuss the three GVD compensation schemes, including the post-compensation, pre-compensation and hybrid-compensation schemes with considering PMD. Our analysis is based on the 10Gbit/s and 40Gbit/s NRZ OOK systems, in which we only consider three physical factors, Kerr effect, GVD and PMD. By comparison, we find that, taking the impact of PMD into account, the GVD pre-compensation scheme performs best when the bit rate is 10Gbit/s, but the GVD post-compensation scheme performs best for the case of 40Gbit/s. The results are different from those already reported under the assumption that PMD is neglected. Therefore our researches provide a different insight into the system optimization when the impact of PMD, Kerr effect and GVD are considered altogether.

Key words: GVD, PMD, Kerr Effect, EOP

1. Introduction

Researches on group velocity dispersion (GVD) and Kerr effect have been extensively carried before the impact of polarization dispersion (PMD) draws extensive attention. In particular, many results have revealed that the GVD hybrid-compensation scheme has the best performance compared with the GVD post-compensation and pre-compensation schemes [1,2]. Recent studies indicate that, as the bit rate in a single channel increases to 10Gbit/s or beyond, the impairment of PMD will form an ultimate limitation to the design of an ultra high-speed long haul light-wave transmission systems. Many theoretical and experimental researches on the impact of PMD have been reported in the past few years [2]. However, it is firstly noted by the author that, when the limitation induced by PMD is considered, the GVD compensation schemes have the different character from that has been reported in [1,3]. By numerical simulating, we find that in the system of 10Gbit/s, the GVD pre-compensation scheme performs best. However in the 40Gbit/s system, the GVD post-compensation scheme performs best. It is believed indispensable to consider the interactions of the PMD, GVD and Kerr effect when the optimization of the systems with bit rate of 10Gbit/s or beyond is studied. Therefore this paper provides a valuable instruction of choosing a suitable GVD management scheme for different bit rate systems when PMD has ill-negligible impact. [*]

2. Three GVD compensation schemes with considering PMD

Starting from the nonlinear coupling equations [4], we use as fiber model a cascade of short wave-plates with well-determined birefringence $\Delta\beta = \beta_x - \beta_y$ with randomly changed birefringence axes between the adjacent sections. To characterize the performance of the different GVD compensation scheme, we first define a parameter eye-opening penalty (EOP).

$$EOP = -20 \log[B_L / B_0] \quad (1)$$

Where B_0 is the eye opening of the input signal and B_L is the eye opening after the transmission distance L. We make use of the split-step method and assume that the launching signal is a $2^6 - 1$ PRBS. The polarization state of the input signal is random but fixed all though the simulations, and in this paper the input polarization state is a 50:50 combination of the two input principal states of polarization. The transmission span is made up of 100km SMF and 20km DCF, and the parameters of the fibers are listed in table 1. In this paper, we use the EDFA as the pure power amplification and exclude the impact of all the noise. The cutoff frequency of the electrical filter at the receiver end is 0.75bit rate. Considering the statistics character of the PMD, we obtain the results averaged on 1000 times simulation. Fig.1 and Fig.2 show the relation of average EOP and input power in the three GVD compensation systems of 10Gbit/s and 40Gbit/s respectively.

Table.1

Parameters@ $\lambda = 1550nm$	SMF	DCF
Fiber loss (dB/km)	0.2	0.5
Dispersion (ps/km/nm)	16.0	-80.0
Dispersion slope ($ps/nm^2/km$)	0.057	-0.216
Nonlinear index	2.9×10^{-20}	4.3×10^{-20}
Effective core area (um^2)	80.0	14.3

Then we define the cumulative probability P_X as $P_X = P(EOP \geq X)$. By linearly fitting the curve of P_X versus EOP, we obtain the value of EOP corresponding to the cumulative probability of 1e-5, which reflects the trailing of the DGD's Maxswell distribution. Fig.3 and Fig.4 show the EOP at the cumulative probability of 1e-5 in the three GVD compensation schemes.

3. Analysis of the simulated results

We first analyze the results for the 10Gbit/s systems. Fig.1 and Fig.3 show that the EOP is the least in the GVD pre-compensation system of 10Gbit/s, which means that the GVD pre-compensation scheme is best when PMD, Kerr

[*] Supported by National High technology Development Program of China (No.2001 AA122012, 2001AA120205)

effect and GVD are take into account. This result is different from the conclusion, as reported in many papers, that GVD hybrid-compensation scheme is best when PMD is not considered. This can be explained that in the 10Gbit/s system, if the value of PMD is not too large, the signal is more vulnerable to the distortion caused by the GVD and Kerr effect than that induced by the PMD. Therefore optimum design of the GVD management system should make the net frequency chirp least, including those caused by the chromatic dispersion and nonlinear effect. As is known in [6] that, the signal spectrum is modulated by the effect of PMD, which results in the reduced frequency chirp caused by the normal chromatic dispersion and nonlinear effect. In the earlier stage of the GVD pre-compensation scheme, the frequency chirp induced by the normal GVD and Kerr effect is effectively reduced by the modulation of PMD, so is the total chirp of the whole fiber span. This leads to the signal distortion induced by GVD and Kerr effect greatly reduced. The point can be verified by the fact that the EOP difference between the pre-compensation scheme and other two compensation schemes becomes uniformly greater with the increased input power.

We also observe from Fig.1 that in the GVD pre-compensation system of 10Gbit/s, the EOP first decreases then increases with the increased input power. This can be explained that PMD can be suppressed to an extent by increasing the Kerr effect; but if the input power is too large, the Kerr effect will be dominant and result in the worse performance. For example, Fig.1 shows that when total PMD value equals 24ps or 32ps, input power can be increased up to 10dBm to decrease the value of EOP, i.e., to suppress the impact of PMD. However when the input power is larger than 12dBm, EOP begins to increase, which means the worse performance.

Now we analyze the results for the 40Gbit/s system. Fig.2 and Fig.4 indicate that GVD post-compensation has the smallest EOP. Because of the shorter bit duration, the signal in 40Gbit/s system is more sensitive to the pulse broadening induced by the walk-off of the two polarized components than that in 10Gbit/s system. It is shown in [5,6,7] that nonlinear effect can effectively inhibit the broadening induced by PMD in presence of abnormal GVD, which is dominant in the earlier stage of GVD post-compensation scheme before the signal power becomes too small because of the fiber loss. Therefore, the post-compensation and hybrid- compensation performs better than the pre-compensation. However as the input power increased to a certain value that is related to the value of PMD, the nonlinear effect becomes dominant, and the pre-compensation performs better, as is similar to the case in 10Gbit/s system.

We also notice that in both 10Gbit/s and 40Gbit/s, the difference among the three compensation schemes is small when the input power is less than 0dBm.This is because the nonlinear effect is negligible when the input power is small and makes little difference among different compensation schemes.

4. Conclusion

Based on the above discussions, we can conclude that, in 10Gbit/s system, the GVD pre-compensation scheme has the best performance, and the impairment of PMD can be effectively suppressed by increasing the input power. However in the 40Gbit/s system, GVD post-compensation scheme performs best within a certain power range that is related to the value of PMD; and GVD pre-compensation will be better if the input power increased beyond the range.

Reference:

[1] A.Nake and S. Saito, J.L.T., Vol.13 (5), pp862~867, 1995

[2] Nuyts, R.J.; Yong Kwan Park; Gallion, P. J.L.T., Vol.15 (1), pp31~42, 1997

[3] Dal Forno, A.O.; Paradisi, A.; Passy, R.; von der Weid,J.P.; P.T.L.,Vol.12,No.3, 2000

[4] Govind P. Agrawal, *Nonlinear Fiber Optics*, San Diego, Calif: Academic 1989, pp265

[5] Aiying Yang, Xiaoxu Li Deming Wu, 9th IEEE ICT2002 pp1238~1242

[6] F. Matera and M. settembre, Optics letters Vol.20, N0.1, Jan.1995 pp28~30

[7] W. Weriershausen, R. Leppla, F. Kuppers, and H. Scholl, ECOC'1999 P130~133

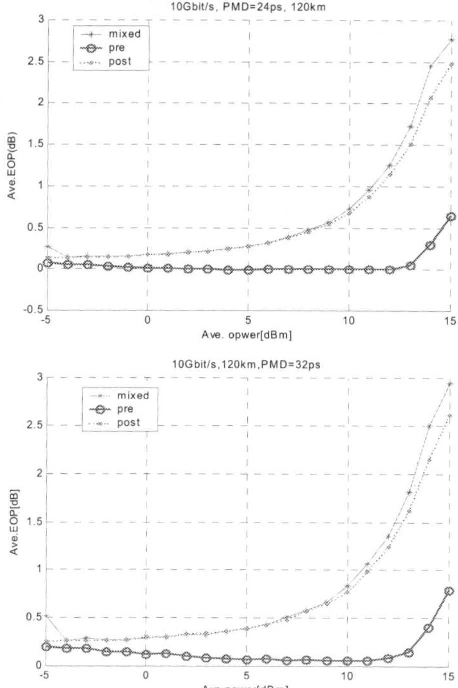

Fig.1 Mean EOP in three GVD Comp. schemes after 120km fiber span; the bit rate is 10Gbit/s.

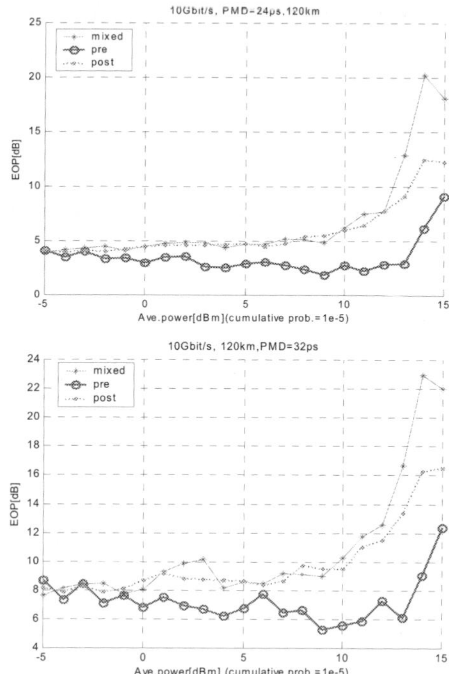

Fig.3 EOP @1e-5 cumulative prob. Total PMD of the fiber span is 24ps and 32ps respectively in 10Gbit/s system.

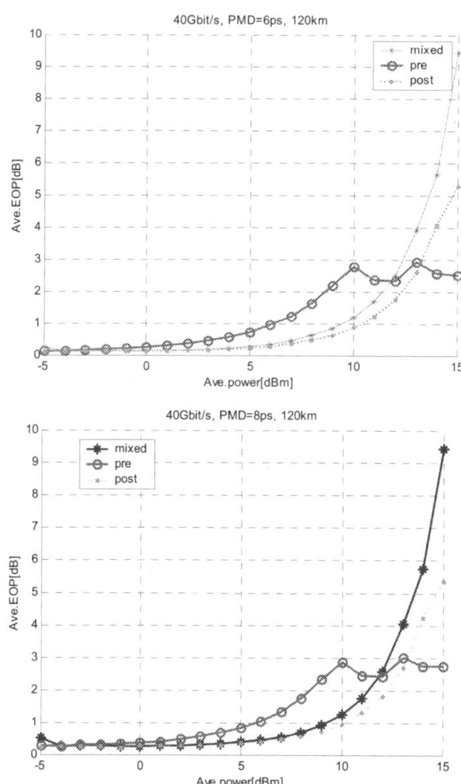

Fig.2 Mean EOP in three GVD Comp. schemes after 120km fiber span; the bit rate is 40Gbit/s.

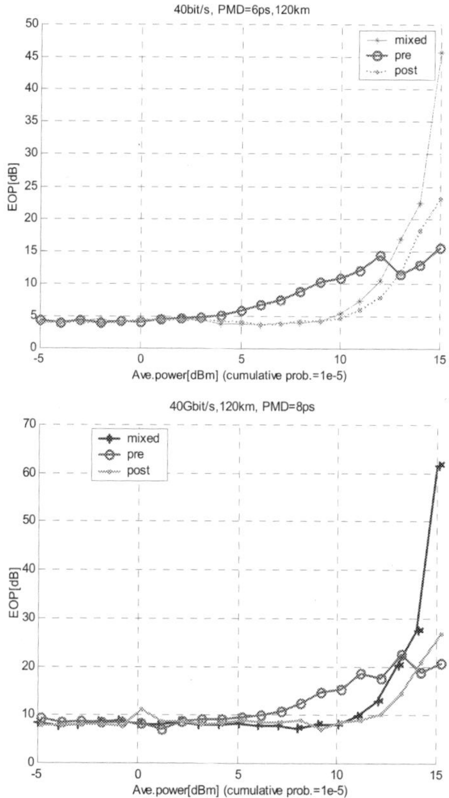

Fig.4 EOP @1e-5 cumulative prob. Total PMD of the fiber span is 6ps and 8ps respectively in 40Gbit/s system.

A Simple Method for Chromatic Dispersion Measurement Employing Photonic Microwave Technology

Xiaoke Yi, Chao Lu, Wen De Zhong, Wei Fang, Yixin Wang
Network Technology Research Center (S2-B3c-23), Nanyang Technological University, Nanyang Avenue 639798, Singapore
Tel.: (65) 7904765, Fax: (65) 7926894, e-mail: p148990070@ntu.edu.sg or eclu@ntu.edu.sg

Abstract:

A simple chromatic dispersion measurement method based on photonic microwave technique is presented in the paper. The dispersion is directly determined from the measured frequency response at a specified modulation frequency. The dispersion measurement range and sensitivity can be flexibly adjusted, without reconfiguring the measurement system. The validity of our method is demonstrated experimentally, and the results show that the dispersion measured by the proposed method agrees well with that measured by the conventional measurement technique.

Keywords: Chromatic dispersion, optical fiber communication, optical fiber measurement

1. Introduction

With the increased transmission speed of wavelength division multiplexing (WDM) systems, accurate measurements of group delay and dispersion characteristics of WDM devices become very important [1][2]. Such measurements are commonly carried out using a modulated phase-shift technique. In this scheme, to provide high phase accuracy, a vector network analyzer is normally used to generate a modulation signal and to detect its phase shift [3][4], which significantly increases the system cost. Furthermore, curve fitting and numerical differentiation are required to obtain the dispersion from the measured group delay, resulting in additional inaccuracy.

The baseband-AM-response method has been reported in [5] to directly measure the dispersion of optical fibers using a 20GHz scalar network analyzer. This method is limited to the measurement of large amounts of accumulated dispersion. The limitation arises from the fact that for a given amount of dispersion, the phase mismatch between the sideband required for cancellation of the photocurrent is only achieved when the sideband separation is large enough. The minimum measurable dispersion of this method is constrained by the bandwidth of the equipment used. To enable the determination of small dispersion values in optical devices, it requires high-speed modulator, wideband photodetector and high-frequency scalar network analyzer, and a constant dispersion offset should be added [6]. These significantly increase the system cost and complexity.

In this paper, we demonstrate a simple method for chromatic dispersion measurement employing photonic microwave technique. The dispersion can be directly determined from the measured frequency response value at a specified modulation frequency (f_{sm}), without the restriction of the carrier cancellation position in AM response method [5]. Furthermore, by changing f_{sm}, the measurable range and sensitivity can be adjusted flexibly, without reconfiguring the measurement system. Its performance is verified experimentally using optical fiber as the tested device, and the experimental results show a good agreement with that measured by the conventional measurement technique.

2. Theoretical analysis of the measurement technique

Fig.1 shows the schematic diagram of our proposed system. A single-mode fiber (SMF) of length L is used as the device under test (DUT) to verify the system performance. The RF output of a 10GHz scalar network analyzer drives a zero-chirp electro-optic modulator (EOM) that amplitude modulates the lightwave signal from a tunable laser. After propagating through the DUT, the modulated signal is detected by the photodetector.

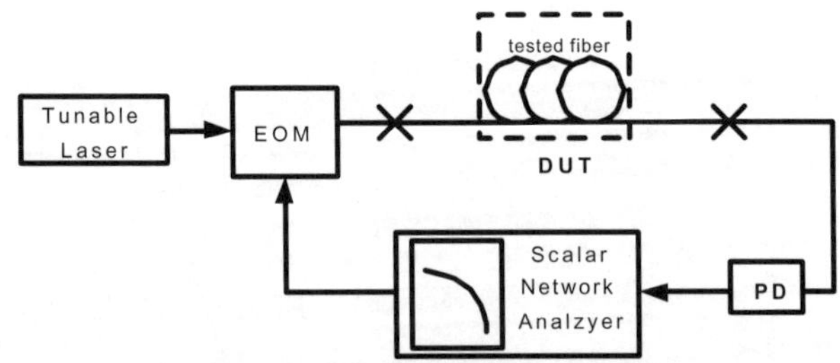

Fig.1 Experimental setup
PD: photodetector, EOM: eletro-optic modulator

The electric field of the modulated optical signal can be expressed as

$$\vec{E}(z,t) =$$
$$\vec{E}_0\{\cos(\omega_0 t - \beta_0 z) + \frac{m}{2}[\cos((\omega_0 + \omega_m)t - \beta_+ z) +$$
$$\cos((\omega_0 - \omega_m)t - \beta_- z)]\} \quad (1)$$

where m is the modulation index of the amplitude modulation, $\omega_m = 2\pi f_m$, f_m is the modulation frequency, β_0 is the propagation constant for the carrier signal (ω_0), and β_+, β_- represent the propagation constants for the upper and lower sidebands respectively.

The second order expansion of the propagation constant around the optical carrier frequency is given by

$$\beta(\omega) = \beta_0 + \beta_1(\omega - \omega_0) + \frac{\beta_2}{2}(\omega - \omega_0)^2 \quad (2)$$

where β_1 and β_2 are the first and second derivation of the propagation constant with respect to the frequency. The optical signal propagates in the fiber with the group velocity $1/\beta_1$. The total dispersion value D is related to β_2 through $D = -2\pi c\beta_2 L/\lambda^2$, where λ is the wavelength in vacuum.

After passing through the fiber under test ($z = L$), the detected photocurrent at the frequency ω_m can be expressed as:

$$i(t) = I_0 m\cos(\frac{\omega_m^2\lambda^2 D}{4\pi c})\cos(\omega_m t - \beta_1\omega_m L) \quad (3)$$

Thus the normalized frequency response measured by our setup described above is given by:

$$|H(f_m)| = |\cos(\frac{\pi f_m^2\lambda^2 D}{c})| \quad (4)$$

Eq.(4) shows that the dispersion can be obtained from the frequency response value measured by the scalar network analyzer, at the specified modulation frequency (f_{sm}). For the unambiguous measurement requirement, $\frac{\pi f_{sm}^2\lambda^2 D}{c} \le \frac{\pi}{2}$ has to be satisfied. Therefore, the measurement range can be derived as

$$D \le \frac{c}{2 f_{sm}^2\lambda^2} \quad (5)$$

and the dispersion is calculated by

$$D = \frac{\cos^{-1}(H(f_{sm}))c}{\pi f_{sm}^2\lambda^2} \quad (6)$$

Compared with the baseband-AM-response method reported in [5], our method does not need to find the carrier canceling position. It means that it has no restriction on the minimum measurable dispersion value, and it requires smaller modulation frequency range. Accordingly, it requires lower frequency equipment (scalar network analyzer, modulator and photodetector). Another advantage is that through adjusting f_{sm}, our technique can flexibly select the measurement range and sensitivity (to be described in Section 3).

3. Experimental results

In our experiments, in order to cancel the roll-off effect of the modulator and photodetector [7], the network analyzer should be calibrated before measurement [8]. The measured and theoretical frequency responses of the scalar network analyzer at 1530nm input wavelength are shown in Fig.2, where the dispersions are 613.61ps/nm and 915.52ps/nm respectively. It is worth to note that, for the frequency response at 613.61ps/nm dispersion, the baseband-AM-response method [5] can not be used, since the carrier canceling position is beyond the measurement frequency range of the scalar network analyzer used.

At the specified modulation frequency f_{sm}=7.162GHz, when the dispersion changes from 613.61ps/nm to 915.52ps/nm (increasing 301.91ps/nm), the frequency response varies 4.9dB as illustrated in the Fig.2. This means that a 61.63ps/nm increase in the dispersion gives 1dB decreasing in frequency response. Therefore, the resolution of dispersion to the RF frequency response at f_{sm}=7.162GHz is 61.63ps/nm/dB. The same consideration for the f_{sm}=8.1GHz leads to a resolution of 20.142ps/nm/dB. This implies that the measurement sensitivity of the system can be improved by enhancing f_{sm}, however it is at the expense of reducing the measurement range. The relationship between modulation frequency and measurable dispersion range is shown in Fig.2 as the circle line. As can be seen, smaller f_{sm} provides a wider measurement range: the measurement range for f_{sm}=7.162GHz and 8.1Ghz are respectively 1249.9ps/nm and 976.6ps/nm (as reduction of 273.3ps/nm). Consequently, by changing f_{sm}, our proposed system allows flexible selection of the measurement range and sensitivity according to practical requirements, without modifying the measurement setup.

In the experiment, we transmit optical signal with different wavelengths through the SMF fiber under test with 40.15756Km length, and measure the $H(f_{sm})$ for each wavelengths. f_{sm} is chosen at 8.76GHz, which provide a measurement range of 798.1ps/nm at the 1565nm input wavelength. Due to the inverse proportional relationship between the input wavelength and the dispersion measurement range (as stated in Eq(5)), we only give the dispersion measurement range corresponding to the maximum input wavelength (1565nm in this experiment), which provide a minimum measurable dispersion range at the same f_{sm} for the different input wavelengths. The dispersion of the DUT should be within this range to avoid ambiguous measurement. This issue can be neglected for the narrowband device testing, however it should be considered for the measurement of broadband devices, such as optical fibers.

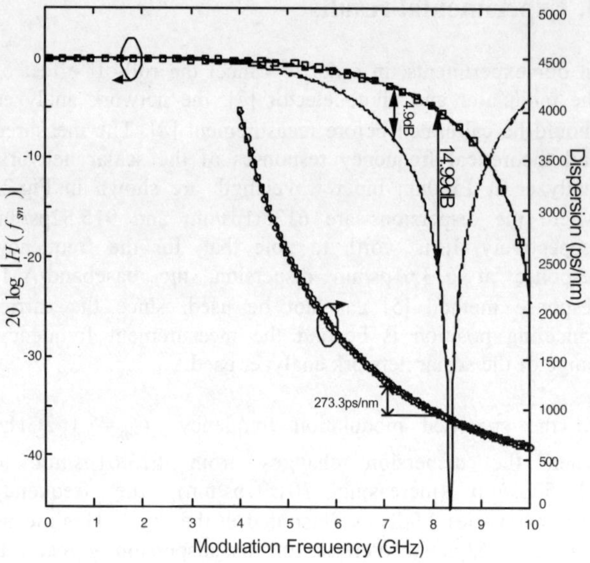

Fig.2 Frequency response and measurement range versus modulation frequency at 1530nm input wavelength
Frequency response at 613.61ps/nm dispersion: Square line (theoretical results), dashed line (Measured results)
Frequency response at 915.52 ps/nm dispersion: Solid line (theoretical results), Dotted line (Measured results)
Theory prediction of the measurement range: Circle line

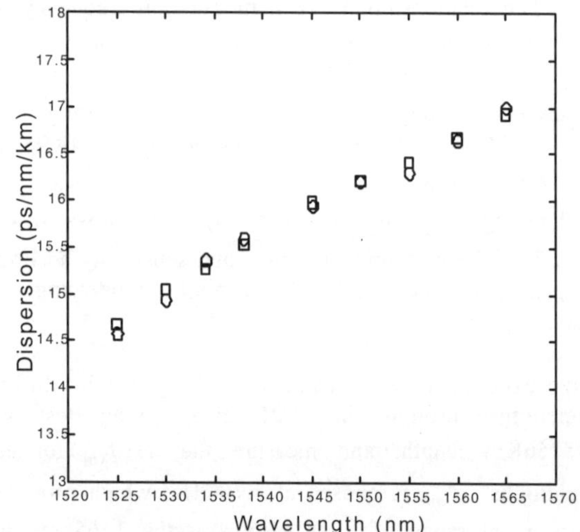

Fig.3 Dispersion of the tested fiber
Squares: baseband-AM-response method
Circles: the proposed method

Fig.3 shows the dispersion measurement results for the DUT as a function of wavelengths. The circles in the figure are the data obtained with our proposed method. For comparison, the results measured by the baseband-AM-response method [5] are also shown as squares in Fig.3. The data obtained by two different techniques agree quite well with each other, which confirms the validity of the proposed method. The discrepancy is within a $\pm 2\%$. These discrepancies are mainly resulted from the error of the wavelength measurement, and the incoherent noise added by the receiver.

4. Conclusions

We have demonstrated a simple method for chromatic dispersion measurement of optical devices, based on photonic microwave technology. This method can directly determine the dispersion from the measured frequency response of the scalar network analyzer, without the need for the phase information, numerical difference of group delay, and the minimum measurable dispersion restriction. The dispersion of the tested SMF has been successfully measured using this technique, and the accurate measurement results confirm its validity. We expect this method will be an effective tool to evaluate the chromatic dispersion of optical devices in practical optical communication systems.

5. References

[1] Y. Horiuchi, Y. Namihira, and H. Wakabayashi, "Chromatic dispersion measurement in 1.55 um narrow-band region using a tunable external-cavity laser ", IEEE Photon.Tech. Lett. Vol.1, No.12, pp458-460, 1989.

[2] Xiaoke Yi, Fang Wei, Yixin Wang, Chao Lu, Wen De Zhong, "Chromatic dispersion measurement using photonic microwave technique", Proc. OFC, WE3, pp194-196, 2002.

[3] Agilent 86037C chromatic dispersion test solution, Product Overview, Agilent Technology.

[4]T. Yamashita, "High-speed, high precision measurement technique for chromatic dispersion", Technical Report, Advantest Corporat ion.

[5] B. Christensen, J. Mark, P.D. Trinh and B. Jalali, " Simple dispersion measurement technique with high resolution", Electronic Lett., Vol.29, pp.132-133, 1993.

[6] C. Peucheret, F. Liu and R.J.S. Pedersen, "Measurement of small dispersion values in optical components", Electronic Lett., Vol.35, pp.409-411, 1999.

[7] D. Pastor, J. Capmany, and B. Ortega, "Experimental demonstration of parallel fiber-optic-based RF filtering using WDM techniques", IEEE Photon. Tech. Lett., Vol.6, pp.831-832, 1994.

[8] E-O Measurement Application Notes. Microwave Measurements Division, Anritsu Company.

Pulse-amplitude-equalization in a rational harmonic mode-locked σ-cavity fiber laser by using modulator as both mode-locker and equalizer

Yang Shiquan, Li Zhaohui, Yuan Shuzhong, Dong Xiaoyi, Kai Guiyun and Zhao Qida

Institute of Modern Optics, Nankai University, Tianjin, 300071, P. R. China

(Tel: 86-22-23509849; Fax: 86-22-23508770; email: ks0911@sina.com)

Abstract—A simple method, by using an electrooptic intensity modulator as both a mode-locker and a pulse-amplitude-equalizer, for the amplitude equalization of high repetition rate pulses generated from a rational harmonic mode-locked σ-cavity fiber laser was proposed and experimentally demonstrated in this letter.

I. INTRODUCTION

There has been considerable interest in developing actively mode-locked fiber lasers for high-speed optical time division multiplexing (OTDM) communication systems. To improve the pulse repetition rate that is usually limited by the bandwidth of the modulator used in the fiber lasers one can use a method called rational harmonic mode locking [1]. By applying *RF* drive frequency slightly deviated from the multiple of the fundamental cavity frequency, we can achieve very high rate pulses. Recently, Wu *et al.* have obtained the 22nd rational harmonic mode locking [2] while Yoshida *et al.* have obtained rational harmonic mode locking pulses up to 80-200GHz [3]. However, the pulse amplitude becomes severely uneven when the repetition rate is greater than twice that of the *RF* frequency. Uneven amplitudes create difficulties in the application of these pulses. So some methods, such as a configuration based on semiconductor laser amplifier in a loop mirror [4], nonlinear optical loop mirror [5,6] and a technique based on nonlinear polarization rotation [7], were proposed to solve the problem. In this letter, we demonstrate a simple method to equalize the pulse amplitudes in a rational harmonic mode-locked σ-cavity fiber laser when the rational harmonic order was less than 5, which let the pulses pass through the modulator two times in one round trip so that the modulator carried out the functions of mode-locker and pulse amplitude equalizer simultaneously.

II. EXPERIMENTS

The experimental setup is shown in Fig.1. The gain of the fiber laser was provided by a 22m long Er-doped fiber that was pumped by a 980nm pump laser diode. The Mach-Zehnder electrooptic intensity modulator was DC biased such that it operated on the linear part of its characteristic curve and a sinusoidal signal derived from a synthesizer was added on its *RF* port. Because of the polarization sensitivity of the $LiNbO_3$ modulator, two polarization controllers (PC1 and PC2) were inserted at both ports of the modulator to obtain required modulation. A fiber Bragg grating (FBG) whose reflectivity was 99.7% at 1556.8nm was used as both wavelength selector and reflector. An electrically controlled delay line was inserted into the cavity to adjust the optical distance between the modulator and the FBG. The output laser was taken via a 9:1 coupler and simultaneously measured by an optical spectrum analyzer and a sampling oscilloscope with optical input port. Compared with the conventional σ-cavity mode-locked fiber laser [8,9], the most distinct part of this scheme was that the pulses passed through the modulator two times, along positive and negative directions respectively, in one round trip.

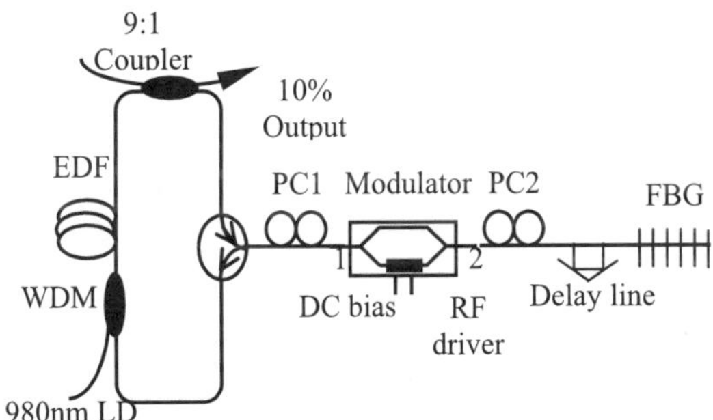

Fig.1 Experimental setup of the mode-locked σ-cavity fiber laser.

Fig.2 illustrates the working principle of the pulse-amplitude-equalization structure. Although only one *RF* source was used to drive the modulator, the transmission characteristics of the modulator from port 1 to port 2 may be different with that from port 2 to port 1 because the optical polarization was different along these two directions. So, by adjusting the status of PC1 or PC2, we could control the insertion losses and the modulation depths of the modulator along these two directions respectively. So the modulator can be thought as two separate modulators. Firstly, we consider the pulse train in the cavity for many round trips. For an nth order rational harmonic mode locking, the number of pulses in the cavity was np+1 (n, p are integers), the pulses experienced the same transfer function after n round trips although they passed the modulator two times in one round trip. So the rational harmonic mode locking can be achieved in this structure. Next, we consider pulse train in the cavity in only one round trip (pulse A, B and C shown in Fig.2). In a general way, the PC1 was adjusted to ensure optimum modulation from port 1 to port 2, whose transmission function was given by $T_{12}=M_{12}[1-\cos(\omega_m t)]$, where M_{12} is the modulation depth and ω_m is the RF circular frequency. Because the repetition rate of the pulses in rational harmonic mode locking was several times of the modulation frequency, some pulses may got small losses (pulse A), some may got medium losses (pulse B) and others may got large losses (pulse C). While the pulses reflected by the FBG and re-enter the modulator from port2, they were modulated again, but the transmission characteristic of the modulator was changed to $T_{21}=M_{21}[1-\cos(\omega_m t+\phi)]$, where $\phi=2n_{eff}l\,\omega_m/c$ is the phase of the modulation when the pulse re-enter into the modulator and c, n_{eff} and l are the light velocity in vacuum, the refractive index of the fiber and the fiber length between the modulator and the FBG, respectively. Thus, by adjusting the delay line to change l and finding an appropriate value of ϕ, we can let pulse A got larger loss this time, pulse B got medium loss again while pulse C got smaller loss. So the pulse amplitudes tended to became unique after they came out from port 2 of the modulator. In order not to counteract the effect of mode locking when pulses passed the modulator from port 1 to port 2, the modulation depth M_{21} should be smaller than M_{12}. By adjusting the status of PC2 and the delay line, we can get appropriate values of M_{21} and ϕ so that the pulse amplitudes were equalized in the cavity when the laser operated stably.

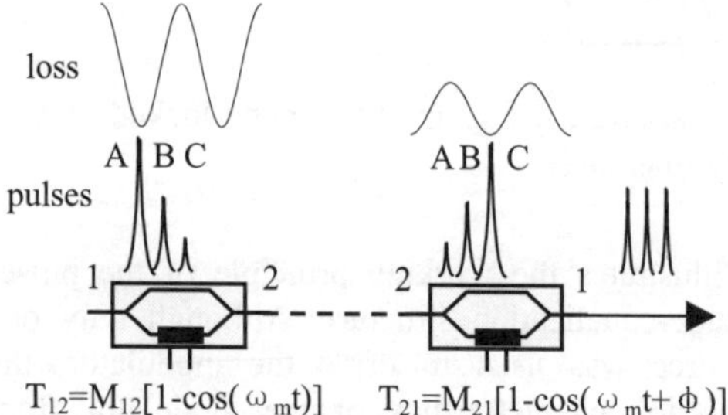

Fig.2 The working principle of the pulse-amplitude-equalization by using the modulator as both a modelocker and amplitude equalizer.

In the experiments, the fundamental cavity frequency of the mode-locked fiber laser was measured to be f_{cav}=2.9766MHz, and the corresponding total laser cavity length was thus determined to be 69.2m. Fig.3 shows the waveforms of pulses when the modulation frequency were 2488.44MHz ($836f_{cav}$), 2489.93MHz ($(836+1/2)f_{cav}$), 2001.44MHz ($(672+1/3)f_{cav}$) and 2002.62MHz ($(673-1/4)f_{cav}$). These figures show that the structure not only could maintain the good effect of the harmonic and the 2nd rational harmonic mode locking as conventional σ-cavity fiber laser, but also could equalize the pulse amplitudes of 3rd and 4th rational harmonic mode locking by appropriately adjusting the PC2 and the delay line. It should be mentioned that the change of the delay line might influence the cavity length of the fiber laser and thus influence the fundamental frequency. But the influences were little and could be easily compensated by slightly tuning the frequency of the *RF* driver.

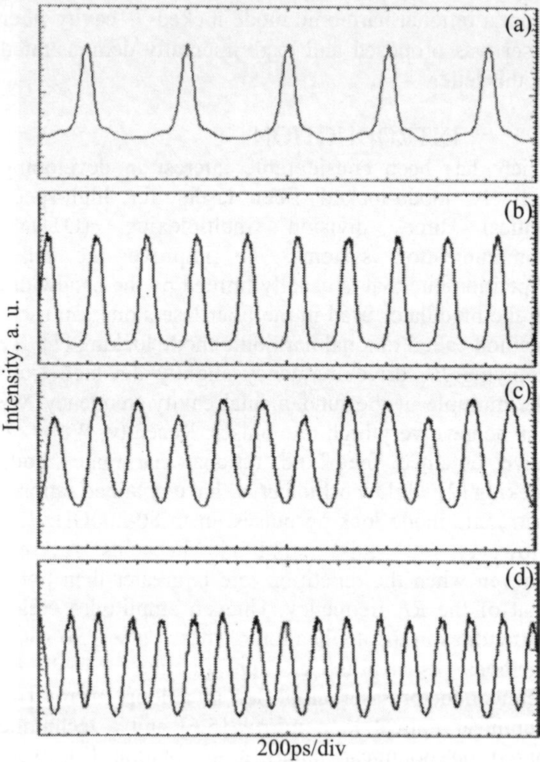

Fig.3 Waveforms of pulses at various RF drive frequencies. (a) f_{RF}=2488.44MHz, harmonic mode locking; (b) f_{RF}=2489.93MHz, 2nd rational harmonic mode locking; (c) f_{RF}=2001.44MHz, 3rd rational harmonic mode locking; (d) f_{RF}=2002.62MHz, 4th rational harmonic mode locking;

III. CONCLUSION

We have successfully demonstrated a simple scheme to utilize the modulator as both mode locker and equalizer. in a rational harmonic mode-locked σ-cavity fiber laser. Thus it was possible to realize pulse-amplitude-equalization without any other additional amplitude-equalization components when the rational harmonic order was less than 5. This technology may be useful for high-bit-rate optical communication systems.

REFERENCES

1 Z. Ahmed and N. Onodera.: 'High repetition rate optical pulse generation by frequency multiplication in actively modelocked fibre ring lasers', *Electric. Lett.* 1996, **32**, (5), pp. 455-457

2 C. Wu and N. K. Dutta.: 'High-repetition-rate optical pulse generation using a rational harmonic mode-locked fiber laser', *IEEE J. Quantum Electron.* 2000, **36**, (2), pp. 145-150

3 E. Yoshida and M. Nakazawa.: '80-200GHz erbium doped fibre laser using a rational harmonic mode-locking technique', *Electric. Lett.* 1996, **32**, (18), pp. 1370-1372

4 H. J. Lee, K. Kim and H. G. Kim.: 'Pulse-amplitude equalization of rational harmonic mode-locked fiber laser using a semiconductor optical amplifier loop mirror', *Opt. Commun.* 1999, **160**, pp. 51-56

5 M. Y. Jeon, H. K. Lee, J. T. Ahn, D. S. Lim, H. Y. Kim, K. H. Kim and E. H. Lee.: 'External fibre laser based pulse amplitude equalization scheme for rational harmonic modelocking in a ring-type fibre laser', *Electric. Lett.* 1998, **34**, (2), pp. 182-184

6 M. Y. Jeon, H. K. Lee, J. T. Ahn, K. H. Kim, D. S. Lim and E. H. Lee.: 'Pulse-amplitude-equalized output from a rational harmonic mode-locked fiber laser', *Opt. Lett.* 1998, **23**, (11), pp. 855-857

7 Z. Li, C. Lou, K. T. Chan, Y. Li and Y. Gao.: 'Theroetical nd experimental study of pulse-amplitude-equalization in a rational harmonic mode-locked fiber ring laser', *IEEE J. Quantum Electron.* 2001, **37**, (1), pp. 33-37

8 D. Foursa, P. Emplit, R. Leners and I. Meuleman.: '18GHz from a σ -cavity Er-fibre laser with dispersion management and rational harmonic active node-locking', *Electric. Lett.* 1997, **33**, (6), pp. 486-487

9 Y. Zhao and C. Shu.: 'A fiber laser for effective generation of tunable single- and dual-wavelength mode-locked optical pulses', *Appl. Phys. Lett.* 1998, **72**, (13), pp. 1556-1558

Field Measurement of Chromatic Dispersion with a Novel Phase-Shift Measurement Technique

Robert Baribault, François Babin, André Girard
EXFO Electro-Optic Engineering, 465 Godin Avenue, Vanier (Quebec), G1M 3G7 CANADA

ABSTRACT

The well-known modulation phase-shift technique is adapted for field use. This modification uses measurements of rf optical power-modulation amplitudes, from which relative group delays (RGD) are computed. It doesn't require any control or calibration of relative phases in the electrical domain. On account of this simple implementation, one can determine precise values of RGD over a wide range of link losses and in field conditions. A broadband emitting LED transmitter - tunable filter receiver combination is used. With this simple set-up, accurate RGD measurements are possible over un-amplified spans of more than 180 km of non-zero-dispersion-shifted fiber from 1530 to 1560 nm. Measurements through optical amplifiers are also presented.

Keywords: Chromatic dispersion, phase-shift, field tests.

1. INTRODUCTION

Chromatic Dispersion (CD) management in high-bit rate and/or long or ultra long-haul WDM systems is an essential element of a reliable network design, construction and operation. This calls for field-portable CD test sets to perform fast, well-documented measurements. CD field measurements must be applicable to a wide variety of link configurations, where different fiber types, concatenations of fiber types, in-line compensators, in-line and unidirectional EDFAs, and bandpass filtering may be used.

The CD of an optical fiber or link is usually done by measuring a relative propagation delay of a modulated signal determined as a function of wavelength and then, from this data, computing the CD. Different techniques have been developed over the past several years. The more commonly used and standardized technique (TIA-455-169[1], TIA-455-175A[2], IEC 60793-1-42[3], ITU-T Recommendation G.650[4] and references [5-11] and references therein) is the phase-shift technique in which the optical power is modulated (with one major Fourier component, including pulsed). In this case, phase variations (in the optical power modulation) with respect to wavelength are measured at some high frequency (typically 10 to 100 MHz, but frequencies can go up to many GHz). Since phase-shift measurements can be made to very high accuracy, this technique is more accurate than the pulsed (time-of-flight) approach.

Up to now, implementations of the phase-shift technique have used phase-difference measurements in the electrical domain. An optical receiver (or receivers if the reference comes from an optical signal) detects the modulated light and converts it into an electrical signal. The reference signal for electrical phase measurements and the phase varying signals (from the wavelength varying optical signal) generally do not follow the same electrical paths (with filters, amplifiers...). Electrical delays are generated that are difficult to control and calibrate, especially at high modulation frequencies. These extra delays are sensitive to external perturbations such as varying temperatures and may lead to measurement errors.

The CD measurement results presented here are based on an innovative modification of the modulation phase-shift method. With this new adaptation, phase measurements start by detecting various combinations of optical power-modulated signals on a single photo-detector. The phase is computed using the various measured RF amplitudes. This new technique for measuring phase differences between optical power-modulated signals can be coupled with a broadband transmitter and optical filtering at the receiver end to enable measurements to be taken over long links without end-to-end communication, unlike most other phase-shift method implementations. Measurements through optical amplifiers greatly benefit from optical filtering of the amplified spontaneous emission at the receiver end. These represent key advantages for cost-effective field CD measurements, for instance along segments of a long-haul link having amplifiers and other unidirectional in-line components and subsystems.

2. AMPLITUDE TO PHASE CONVERSION

When combining two optical-power-modulated light signals, the total resulting optical power modulation depends on the phase difference of the power modulations of the two original signals. If (P_i, ϕ_i), (P_j, ϕ_j) and (P_{ij}, ϕ_{ij}) are the amplitude and phase of first, second and total signals, respectively, then we can write;

$$P_{ij}\cos(2\pi f_{mod}t+\phi_{ij}) = P_i\cos(2\pi f_{mod}t+\phi_i) + P_j\cos(2\pi f_{mod}t+\phi_j)$$

$$\cos(\phi_i-\phi_j) = \frac{P_{ij}^2 - P_i^2 - P_j^2}{2\cdot\sqrt{P_i^2\cdot P_j^2}}$$

where f_{mod} is the optical-power-modulation frequency. These equations show that by measuring the amplitude of the optical power modulations, separately and in combination, we can compute the phase difference between two optical signals. For added accuracy, a third signal, close to 90 degrees out of phase with one of the two other signals is used. If i≠j≠k, then with three signals we get;

$$\sin(\phi_i - \phi_j) = \frac{\cos(\phi_i - \phi_k) - \cos(\phi_i - \phi_j)\cdot\cos(\phi_k - \phi_j)}{\sqrt{1-\cos^2(\phi_k - \phi_j)}}$$

By taking the argument of $\tan(\phi_i-\phi_j)$, it is always possible to determine with high accuracy the phase difference between two optical-power-modulated signals. Detecting the different light signals, separately and in combination, with a single photo-detector and then measuring modulation amplitudes enhances accuracy. This minimizes the electrical delays that influence time measurement accuracies. This novel adaptation makes use of two optical signals simultaneously. Both signals travel down the fiber and are detected on the same photo-detector simutaneously. Length variations are thus perfectly compensated. A limiting factor would be variations in optical-power modulation, since the different signals are measured separately and in combination in a sequential (albeit rapid) manner. The linearity of detection will ultimately determine the accuracy of the RGD. The optical signals could be of different polarizations or have traveled paths of different lengths, rendering possible PMD or elongation-length measurements. However, in this paper, we concentrate on CD determination via RGD measurements at different wavelengths.

3. MEASUREMENT SET-UP

Another design consideration for this CD measuring apparatus was the need to have no communication between the emitter and the receiver. In a typical CD field measurement using the phase-shift technique, the source would be at one end of the link and the detector at the other end, especially if there are unidirectional devices in the link (as with most amplifiers). The use of a second fiber for transmitting a reference signal or wavelength data is detrimental in most cases, adding noise or cost to the measurement. In order to avoid the need for communication and a second fiber, the broadband optical source is always actively modulated and transmitting over its full wavelength range, and the optical filtering is performed at the receiver end. A drop filter at the receiver end generates the reference signal.

In our systems, the source is a broadband edge SuperLEDs with a total power of more than 5 mW. In some cases, two of these directly modulated SuperLEDss are combined with a C+L band combiner, one for the C band, the other for the L band, allowing CD measurements to be carried out over a very wide spectral range. The modulated broadband optical signal is sent through the FUT or DUT. The rf modulation frequency is at 100 MHz. It then enters the compact and rugged receiver (Figure 1), built for field use. It has two inputs, the two input branches of a 90/10 fiber optic coupler. The 10% input is for larger optical signals that might otherwise saturate the detection circuitry, hence enhancing the dynamic range of the electronics. After the input coupler, the light meets a first drop filter at a fixed reference wavelength. This first filtered light is again split in two with a 50/50 coupler in the case of a C+ L measurement. The broadband output signal from the drop filter passes through a second drop filter (not shown). This second reference signal goes to the same splitter as the first reference signal. These two reference signals are recombined with another 50/50 coupler. One of the reference light paths has a length that is an odd integer times a value of close to 90° at the high frequency modulation (depending on the exact value of the phase difference between the two paths, some signs will change in the equations). Along the third path, the broadband signal goes to a double-pass bulk diffraction grating tunable filter having a 1.2 nm optical bandpass. This filter exhibits very low polarization dependent loss. The different light paths pass through a slotted wheel. These signals are recombined in different ways and detected on a single fiber pigtailed InGaAs photo-detector. The electrical signal is filtered at the rf modulation frequency and the *electrical* power (a signal proportional to the square of the current generated at the photo-detector) is measured. From these electrical power measurements and the known modulation frequency, a processor computes the phase differences and then the relative delays.

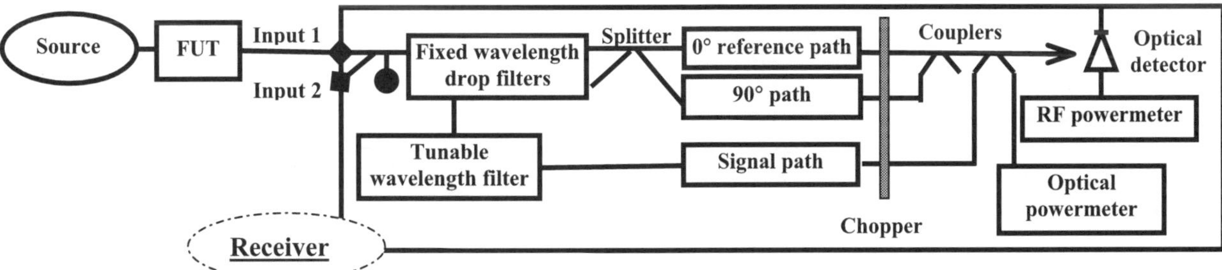

Figure 1. Schematic of measurement set-up.

The slotted wheel has a special pattern of holes. When the wheel turns, seven different combination of optical signals are recombined; 1) First reference signal, 2) second reference signal, 3) tunable signal, 4) first reference and tunable signal together, 5) second reference and tunable signal together, 6) first and second reference signal together and finally, 7) no signal at all (this last one to measure the background noise level and compensate for it). This slotted wheel turns at a specified speed so that approximately 100 cycles are measured per second. This low frequency switching has the advantage of reducing the 1/f noise, since all the signals are modulated at this low frequency. The user can set the appropriate level of averaging in order to improve the accuracy of the system without worrying about electronic 1/f noise, thus enhancing the measurement range. The total measurement time can be set between one cycle and tens of seconds. There is also an independent optical power-meter to measure the optical power on the detector.

The RGD value is stored along with the center wavelength values of the tunable filter. Varying the wavelength of the filtered light and computing a series of wavelength dependent phase differences enables us to compute CD in the standard ways. The tunable optical filter is calibrated between 1530 and 1625 nm.

4. RESULTS

Before making a CD measurement, the baseline of the instrument must be measured. Short lengths of fiber, the components (optical combiners, optical filters, couplers...), and the emitter all contribute to the baseline. It was shown [17] that direct modulation of these high power superLEDs introduces significant intrinsic phase variations of the modulation with wavelength. The baseline is subtracted from the measurements taken with FUT in place.

Figures 2 presents typical results of measurements of RGD. There are four curves. The top curve is 160 km of standard fiber plus 20 km of standard fiber, previously calibrated by the Swiss Metrology Bureau - METAS. The slightly lower curve is the relative delay curve of the 160 km of standard fiber alone. The two bottom curves are 1) the difference between the two top curves, thus the contribution of the 20 km of calibrated fiber (triangles) and 2) the calibrated curve (crosses). The difference between the measured curves is very close to the calibrated curve, the worst-case deviation being no more than 3.6% between 1535 and 1560 nm. Between 1530 and 1535 nm the signal from the tunable path was very low. Measurements were taken at every 0.5 nm and each of the displayed points represents an average of ten measurements (with 4 seconds per point) for the 160 km measurement and 14 measurements (16 seconds per point) for the 180 km. Standard deviations for points of the top curve (160 km + 20 km) are always less than few % of the total RGD between 1530 and 1560 nm (100000 ps). Repeatability is thus excellent, even for these long lengths of fiber. These curves show that there is no significant loss of measurement accuracy for CD after 180 km of standard fiber.

Figure 3 shows the same type of measurement, but this time

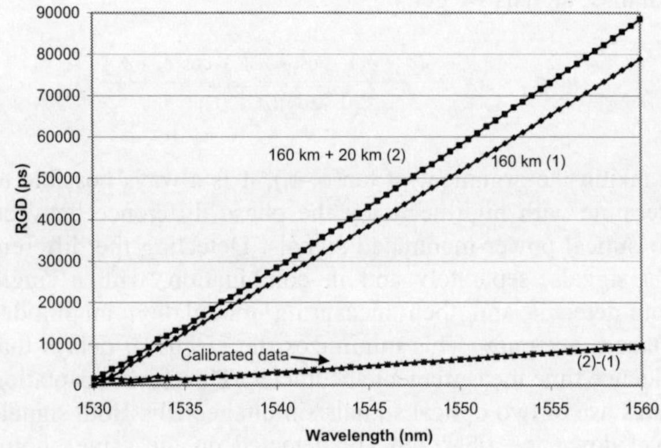

Figure 2. RGD data for long links.

with an amplifier in the link. This was a non-flattened C band Erbium-doped fiber amplifier. The differences in relative delay between the 2 X 80 km and 80 km + amplifier + 80 km are negligible. The measurement for 200 km of fiber with an amplifier at mid-span is excellent. Standard deviations of less than 0.01% of total RGD between 1530 and 1560 nm are easily obtained with 80 km of standard non-dispersion shifted fiber. Nonetheless, since the tunable (and reference) filter has a FWHM (3-dB bandpass) of only 1.2 nm, the apparatus can measure long lengths of highly dispersive fiber without adding significant loss of modulation due to dispersion. At the superLED's peak wavelength (1550 nm), over 200 km of 17 ps/nm.km fiber could probably be measured, for a loss measurement range of more than 40 dB with a filtered superLED. Results have also been taken in the L band and could be extended to the entire 1250-1650 nm range, superLEDs being available all across this range.

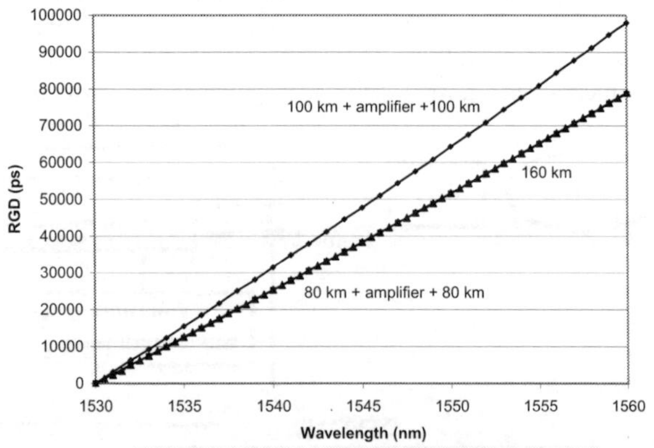

Figure 3. RGD data with amplifiers.

5. DISCUSSION AND CONCLUSION

The measurements presented here show that delays (and thus CD) can be measured precisely using only electrical power measurements. The detection electronics are hence greatly simplified, not having to control or calibrate the phases of electronic signals. In addition to detection linearity, other key factors for accurate CD determination are absolute wavelength accuracy and relative wavelength accuracy. To obtain correct values of CD, the optical power modulation should be constant over the tunable filter bandpass, which is not the case, as the broadband source has a spectral power distribution. Upon passage through the DUT, the power modulation distribution has changed. This distribution should be measured and used to compute the exact center wavelength of the filtered light to compensate for the power modulation lineshape. This technique lends itself naturally to this type of compensation. One of the signals used to compute delays is the filtered tunable optical power modulation level. The power modulation attenuation curve can be measured at the same time as the delays. From this, compensating for the power modulation distribution is possible.

In this paper, no CD results are presented. CD is computed from the delays and from the wavelengths. It is straightforward to compute CD from the data shown in the figures. There are two basic ways to compute CD. The first is to fit a curve on the delay versus wavelength data and find the derivative of the fit function. This works well for a small number of known fit functions, especially for fiber. If there is a mix of fibers or if components contribute significantly to the CD, then it is usually preferable to compute local CD, using local derivatives (on a restricted number of delay measurements over a short wavelength span). The apparatus described here can be used in both modes. The tunable filter is calibrated for steps of 0.2 nm (although steps can be much smaller). If using only two closely spaced wavelengths, relative wavelength accuracy will directly impact CD accuracy. On the other hand, it is highly recommended to measure with very small steps if the dispersion is high, so as to avoid missing 360-degree phase jumps.

CD in DWDM links could be measured by reducing the tunable filter bandpass. Going to a 100-pm bandpass would reduce the total optical power by 10 dB but sensitivity. The attenuation measurement range could be enhanced by using an APD (Avalanche Photo-Diode) instead of a PIN detector. CD measurements would then be limited by wavelength noise. It is to be noticed that it is straightforward to go to higher frequencies with this set-up (and thus to higher delay accuracies), as high as the emitter can respond efficiently to the modulation. All that needs to be changed is the input trans-impedance of the detection circuit and the frequency of the local oscillators. Work is underway to use this phase measurement technique applied to PMD and length measurements.

This paper demonstrates a new implementation of the phase-shift method for measuring relative propagation delays of optical modulations in fiber-optic links. The phase information is gathered through the measurement of the amplitudes of the power modulations of different combinations of signal. The set-up is simple, accurate and there is no need for communication via another fiber between the emitter and the receiver. Up to 180 km long unamplified spans were measured between 1530 and 1560 nm with less a few percent of standard deviation per point with respect to total relative group delay. Filtering at the output renders measuring through many amplifiers possible. Although measurements through only one amplifier are presented here, test were successfully carried out with up to eight amplifiers.

REFERENCES

1. EIA/TIA-455-169A, *FOTP-169*, Electronic Industries Association, Washington, D.C., October 1992.
2. TIA/EIA-455-175A, *FOTP-175*, Telecommunications Industry Association, Washington, D.C., November 1992.
3. IEC 60793-1-42, Ed.1: Optical Fibres 1-42: *Measurement methods and test procedures - Chromatic dispersion*, 2001.
4. ITU-T Recommendation G.650, *Definition and test methods for the relevant parameters of single-mode fibres*, 1997.
5. L.G. Cohen, "Comparison of single-mode fiber dispersion measurement techniques", *J. of Lightwave Technology*, LT-3(5), 958-966, (1985).
6. B. Costa, D. Mazzoni, M. Puleo, E. Vezzoni, "Phase shift technique for the measurement of chromatic dispersion in optical fibers using LEDs", *IEEE J. Quantum Electronics*, QE-18, 1509-1514 (1982).
7. B. Costa, M. Puleo, E. Vezzoni, "Phase shift technique for the measurement of chromatic dispersion in single-mode optical fibers using LEDs", *Electronics Letters*, 19(25/26), 1074-1076 (1983).
8. S. Tanaka, Y. Kitayama, "Measurement accuracy of chromatic dispersion by the modulation phase technique", J. of Lightwave Technology, LT-2(6), 1040-1044 (1984).
9. A.J. Barlow, R.S. Jones, K.W. Forsyth, "Technique for direct measurement of single-mode fiber chromatic dispersion", *J. of Lightwave Technology*, LT-5(9), 1207-1213 (1987).
10. M. Fujise, M. Kuwazuru, M. Nunokawa, Y. Iwamoto, "Chromatic dispersion measurement over a 100 km dispersion-shifted fibre by a new phase-shift technique", *Electronic Letters*, 22(11), 570-572 (1986).
11. Y. Horiuchi, Y. Namihira, H. Wakabayashi, "Chromatic dispersion measurement in 1.55 μm narrow-band region using a tunable external-cavity laser", *IEEE Photonics Technology Letters*, 1(12), 458-460 (1989).
12. N.Cyr, F.Babin, R.Baribault, F.Audet, A.Girard, "An Innovative Implementation of the Phase-Shift Method for Field Measurement of Chromatic Dispersion", to be published in the proceeding of APOC 2002.

Electrically Tunable Dispersion Compensator based on Fiber Bragg Grating without Central Wavelength Shift

S. Y. Li, N. Q. Ngo, R. T. Zheng, S. R. Natarajan, S. C. Tjin and P. Shum
Tel: 65-67904242, Fax: 65-67904161, E-mail: songyang_li@pmail.ntu.edu.sg
Photonics Research Lab (S1-B3a-08), School of Electrical & Electronic Engineering, Nanyang Technological University, Nanyang Avenue, Block, Singapore 639798

Abstract: We demonstrate a new electrically tunable dispersion compensator by electrically tuning the chirp of a uniform fiber Bragg grating (FBG) without center wavelength shift. A distributed on-fiber thin film heater was deposited on the surface of a taper FBG by means of controllable electrical plating to adjust the chirp and hence dispersion of the FBG. A negative thermal-expansion coefficient (NTEC) ceramic was used to fix the central wavelength in the tuning. With an applied electrical power smaller than 0.68 W, the dispersion tuning range from −178 ps/nm to −302 ps/nm with a dispersion ripple of less than 30 ps was achieved. The corresponding central wavelength shift was as small as 0.173 nm.

Key Words: Tunable filter, fiber Bragg grating, dispersion compensation

1. INTRODUCTION

Dispersion management is becoming paramount in wavelength-division-multiplexed (WDM) systems, operating at a per-channel rate of 40 Gb/s or higher. Dispersion compensating devices that can provide dynamically adjustable dispersion and that can be incorporated directly into the fiber link are highly desirable in high-capacity WDM systems. Chirped FBGs are attractive as a dispersion compensator because of its compact size, low insertion loss and ease of fabrication. Dynamically tunable dispersion compensation can be achieved through variation of the grating chirp, typically by applying a strain gradient [1]-[3], or a temperature gradient [4]-[6], along the length of a uniform grating.

Electrically tunable-chirped fiber gratings have been proposed recently [7]. However, there is a serious problem in that the reflection wavelength of the FBG shifts to the longer wavelength side when the dispersion is adjusted by the methods mentioned above. To solve this problem, a tunable fiber grating that allows the chirp and the central wavelength to be adjusted independently has been demonstrated by using two electrically insulated thin-film heaters deposited in a multi-layer geometry on the surface of the fiber [8]. This device used an additional power source to compensate for the central wavelength shift, which makes the control mechanism complicated and introduces more power consumption.

Here, we demonstrate a novel technique to eliminate the central wavelength shift when electrically tuning the dispersion of an FBG with taper thin-film on-fiber metal coating. The taper uniform FBG with the taper thin-film metal coating on its surface was mounted under tension in the NTEC ceramic. With the applied voltage increase, the chirp of the FBG can be adjusted because of the strain gradient and temperature gradient changes. At the same time, the shrinkage of the NTEC ceramic will compress the FBG, which will compensate for the central wavelength shift because of temperature rise.

2. PRINCIPLE OF DESIGN AND EXPERIMENT

Fig. 1 shows the schematic diagram of our novel dispersion compensating FBG.

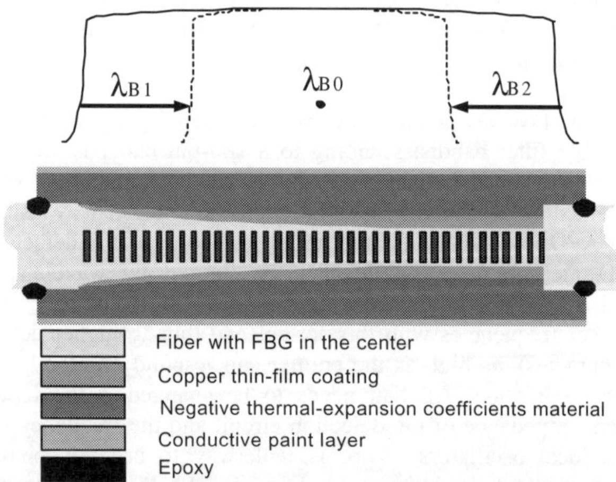

Fig. 1 The diagrammatic sketch of the dispersion compensating FBG with fixed central wavelength. The diameter profile of the fiber with FBG and the thickness profile o f copper thin -film coating are predetermined to ensure the linear temperature gradient and strain gradient. When the current flow through the thin -film coating and conductive paint layer, the reflective spectrum shift from the straight line to the dashed line.

A fiber with a uniform Bragg grating (2.5 cm long, with a reflection peak near 1546 nm) is merged into ~49% concentration hydrofluoric acid with a prescribed speed for etching it to a predetermined diameter profile. The fiber diameter in the FBG region varies from 90 to 125 μm. To metallize the taper fiber, titanium (~200 Å as adhesion promoter) and gold (~2500 Å) were evaporated onto the outer surface of the fiber in an electron beam evaporator. Subsequently, deposit a copper coating with prescribed thickness profile on the fiber with controllable electrical plating technique by pulling the taper fiber from a plating path in a predetermined speed, while supplying a constant current. Then the FBG with on-fiber thin-film thickness variation coating was mounted in a NTEC ceramic (3×3×50 mm, with 0.3×0.3 mm groove) under tension [9]. The thermal-expansion coefficient of the NTEC material is -7.6×10^{-6}/°C.

To heat the NTEC ceramic efficiently, a conductive paint layer was deposited on the surface of the NTEC ceramic. After the packaging, the Bragg wavelength shifted 3.6 nm to the longer wavelength side, and bandwidth was broadened to about 1 nm because of the taper shape of the fiber. The resistance of the on-fiber thin-film coating is 4.0 Ω, and the resistance of conductive paint layer deposited on the NTEC ceramic is 4.1 Ω.

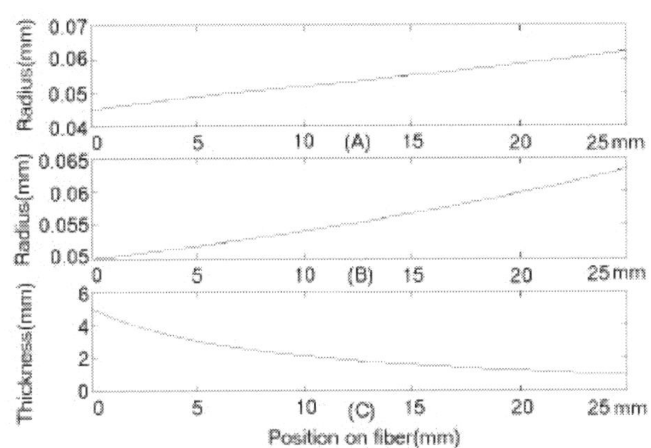

Fig. 2 (A) Profile of radius of a 25 mm length FBG, prescribed radius at 0 and 25 mm position are 45 um and 62.5 um respectively; (B) The profile of total radius with thin-film coating; (C) Profile of copper thin-film thickness, prescribed thickness at 0 and 25 mm position are 5 μm and 1 μm respectively.

When an electrical current flows through the thin-film coating and the conductive paint layer, the temperature rise will shift the Bragg wavelength to the longer wavelength side. At the same time, the shrinkage of the NTEC ceramic will compress the FBG, which will cause the Bragg wavelength shift to shorter wavelength side. In Fig. 1, at the thinner side of the thin-film coating, the thermal effect is larger than the strain effect, so the Bragg wavelength at this point, λ_{B1}, will shift to the longer wavelength side. At the thicker side of the thin-film coating, the strain effect is larger than the thermal effect, so the Bragg wavelength at this point, λ_{B2}, will shift to the shorter wavelength side. And at the middle of the FBG, the strain effect and the thermal effect offset each other, so the Bragg wavelength at this point, λ_{B2}, will be kept fixed.

The diameter profile of the fiber with FBG $D(z)$ and the thickness profile of the thin-film coating $T(z)$ are determined by the Eqs. (1) and (2)

$$D(z)=2\cdot\sqrt{\frac{1}{\pi\cdot E_f(m\cdot z+n)}-\frac{E_c}{\pi\cdot E_f}\cdot A_c(z)} \tag{1}$$

$$T(z)=\frac{A_c(z)}{\pi\cdot D(z)} \tag{2}$$

where

$A_c(z)$ the cross-section area of the thin-film coating
z the position along the fiber
E_c the Young's modulus of the thin-film coating
E_f the Young's modulus of the fiber with FBG

$A_c(z)$, m and n are defined by Eqs. (3)-(5)

$$A_c(z)=\frac{l\cdot A_c(l)\cdot A_c(0)}{[A_c(0)-A_c(l)]\cdot z+l\cdot A_c(l)} \tag{3}$$

$$m=\frac{1}{E_cA_c(l)+E_fA_f(l)}\cdot\frac{1}{l}-\frac{1}{E_cA_c(0)+E_fA_f(0)}\cdot\frac{1}{l} \tag{4}$$

$$n=\frac{1}{E_cA_c(0)+E_fA_f(0)} \tag{5}$$

where $A_c(l)$ and $A_c(0)$ are the cross-section areas of the thin-film coating at $z=l$ and $z=0$ positions respectively, $A_f(l)$ and $A_f(0)$ are the cross-section areas of the fiber with FBG at $z=l$ and $z=0$ position respectively. The ideal fiber diameter profile and thin-film thickness profile are demonstrated in Fig. 2.

Fig. 3 demonstrates the change in the reflection spectrum of this tunable dispersion compensating FBG under different DC voltages. Fig. 4 shows the central wavelength shift and bandwidth change when raising and reducing applied voltage. When the applied voltage was raised from 0 V to 0.8 V, the central wavelength shifted by as small as 0.079 nm and the bandwidth broadened by up to 0.309 nm. When the applied voltage was reduced from 0.8 V to 0 V, the central wavelength shifted by 0.066 nm and the bandwidth changed by 0.303 nm. The central wavelength shift is negligibly small in the tuning. And there is no significant difference between the curves of ascending applied voltage and descending applied voltage, which shows repeatability of the device. When adjusting the applied voltage from 0 V to 1.2 V, the dispersion of this device varies from −178 ps/nm to −302 ps/nm (Fig. 5).

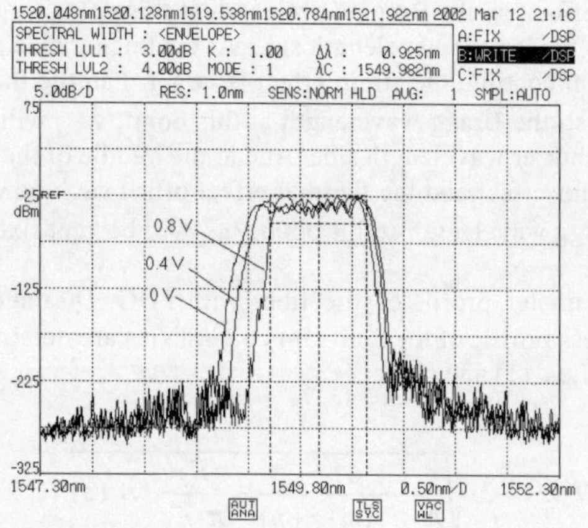

Fig. 3 Reflection spectrum of dispersion compensating FBG with applied voltage of 0 V, 0.4 V and 0.8 V.

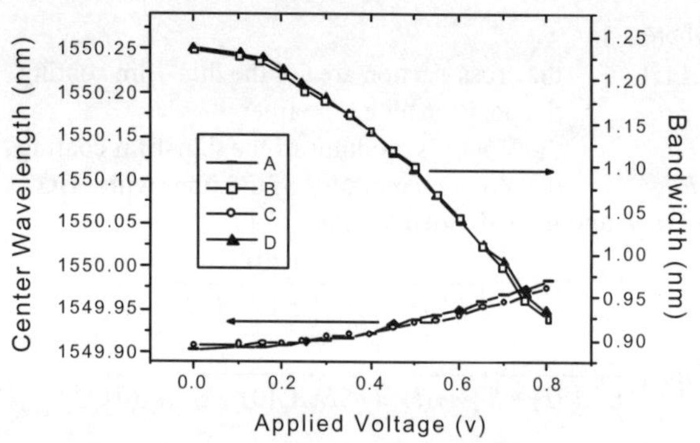

Fig. 4 Central wavelength and bandwidth changes with applied voltage. A: central wavelength shift when raising the voltage form 0 V to 0.8 V; B: Bandwidth change when raising the voltage from 0 V to 0.8 V; C: Central wavelength shift when reducing the voltage from 0.8 V to 0 V; D: Bandwidth change when reducing the voltage from 0.8 V to 0 V.

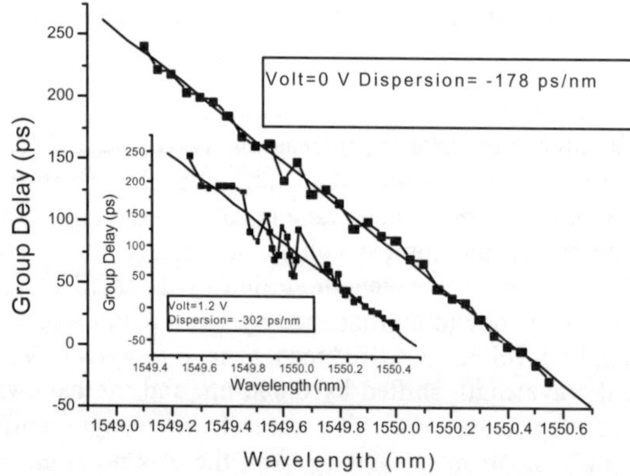

Fig. 5 The group delay lines of this device with applied voltages of 0 V and 1.2 V.

3. CONCLUSIONS AND DISCUSSIONS

A novel electrically tunable dispersion compensating FBG with a fixed central wavelength has been developed. With distributed on-fiber resistive heaters and NTEC ceramic, this device allows the tuning of the dispersion electrically with central wavelength fixed. This device is compact, electrically tunable, highly repeatable, and has central wavelength almost fixed and has low insertion loss. The tuning range of dispersion can be improved by using longer FBG and employing high temperature stable epoxy to mount the FBG on the NTEC ceramic. And with different thickness profile of the thin-film coating and different diameter profile of the taper FBG, high-order dispersion, which is important in high-speed long-haul communication systems, can also be compensated for by this technique.

REFERENCES

[1] P. C. Hill and B. J. Eggleton, "Strain gradient chirp of fiber Bragg grating," *Electron. Lett.,* vol. 30, pp. 1172-1174, 1994.

[2] M. Pacheco, A. Medez, L. A. Zenteni, and F. Mendoz-Santoyo, "Chirping Optical fiber Bragg gratings using tapered-thickness piezoelectric ceramic," *Electron. Lett.,* vol. 34, pp. 2348-2350, 1998.

[3] Yinian Zhu, Pieter L. Swart, and Beatrys M. Lacquet, "Chirp tuning of a fiber Bragg grating by using different tapered transducers and loading procedures: an application in the accelerometer," *Opt. Eng.*, vol. 40, pp. 2092-2096, 2001.

[4] J. Lauzon, S. Thibault, J. Martin, and F. Ouellette, "Implementation and characterization of fiber Bragg gratings linearly chirped by a temperature-gradient," *Opt. Lett.,* vol. 19 pp. 2027-2029, 1994.

[5] S. Barcelos, M. N. Zervas, R. I. Laming, D. N. Payne, L. Reekie, J. A. Tucknott, R. Kashyap, P. F. McKee, F. Sladen, and B. Wojciechowicz, "High-accuracy dispersion measurements of chirped fiber gratings," *Electron. Lett.* Vol. 31, p. 1280, 1995.

[6] Z. H. Chen, C. Lu, J. H. Ng, X. F. Yang, T. H. Cheng, "Tunable chirped fiber Bragg gratings by a distributed heater," *Opt. Eng. Lett.,* Vol. 40, pp. 1156-1157, 2001.

[7] John A. Rogers, Benjamin J. Eggleton, Janet R. Pedrazzani, and Thomas A. Strasser, "Distributed on-fiber thin film heaters for Bragg grating with adjustable chirp," *Appl. Phys. Lett.,* Vol. 74, pp. 3131-3133, 1999.

[8] Eggleton,Rebecca J. Jackman,Glen R. Kowach and Thomas A. Strasser, 'Dual on-fiberthin-film heaters for fiber gratings with independently adjustable chirp and wavelength," *Opt. Lett.,* Vol. 24, pp. 1328-1330, 1999.

[9] Zhanxione Wei, Yongsen Yu, Hua Xing, Zhongchuang Zhuo, Yuanda Wu, Liang Zhang, Wei Zheng and Yushu Zhang, "Fabrication of chirped fiber grating with adjustable chirp and fixed central wavelength," IEEE Photon. Technol. Lett., Vol. 13, pp. 821-823, 2001.

Effect of second-order dispersion on transmission performance of dispersion managed 40Gb/s prechirped return-to-zero (CRZ) format with very low duty-cycle

Zhihong Li (1), Caiyun Lou (2), Yizhi Gao (2), Lu Chao (1), Cheng Tee Hiang (1)
1 : Institute. of Communication. Research, BLK 2, 18 Nanyang drive, Unit 230, Innovation Centre, Singapore 637732, ezhli@ntu.edu.sg
2 : Dept.of Electronic Engineering, Tsinghua University, loucy@ee.tsinghua.edu.cn

Abstract: Transmission performance of 40Gb/s CRZ format with low-duty-cycle was studied numerically. It was found that, pulse prechirping will enhance the effect of second-order dispersion in 40Gb/s longhaul transmission. So dispersion slope compensation is very important even for single channel ultra-high-speed transmission.

Key words: transmission, CRZ format, dispersion management, second-order dispersion, simulation

1. Introduction

The demand for high capacity transmission is growing rapidly. Currently, there's much interest in increasing the single channel bit rate to 40Gb/s. A promising technique for optical data transmission at this bit rate is to use chirped RZ (CRZ) format together with dispersion management. Till now, most research are focused on conventional CRZ format[1], which possess a duty cycle of about 0.5, and prechirping is demonstrated to be helpful both in reducing pulse distortion and nonlinear interaction among pulses. While in this paper, the propagation of 40Gb/s single channel CRZ signal with much lower duty cycle (<0.25) is studied in detail. We demonstrated that, when the pulse is very short, prechirp will greatly enhance the effect of second-order dispersion and severely deteriorate the transmission performance. Under such situation, second-order dispersion compensation appears to be very important even for single channel transmission system. Only when dispersion slope is well compensated, can the transmission performance be improved by using CRZ format.

2. Theory

In this article, two dispersion allocation maps -- F1+F2 and F1+F3, were investigated. The parameters for the fiber span were listed in table 1. where D is the first-order chromatic dispersion and S is dispersion slope corresponding to the second order dispersion. The length of the fiber are carefully choosen to achieve zero average dispersion. The amplifier spacing, also the dispersion management span, is 80km.

Table1. Parameters for F1, F2 and F3.

Fiber	F1	F2	F3
D (ps/nm/km)	1.75	-2	-2
S (ps/nm^2/km)	0.07	0.07	-0.07
Loss (dB/km)	0.22	0.22	0.22
A_{eff} (μm^2)	50	50	50
K (nm)	25	-28.6	28.6

K is defined as the ratio between the dispersion and dispersion slope. If the two fibers have the same value of K, it means that the dispersion and slope can be compensated simultaneously. From the table we can find that K values for F1 and F2 have opposite sign, while for F1 and F3, they are nearly the same. This' means that the dispersion slope is compensated to a good extent only for dispersion map2 (F1+F3). Although these alternating dispersion values are not currently realistic, we have found that the simulation results are also applicable in conventional dispersion management regime.

Propagation of optical pulses in fiber is determined by the Nonlinear Schordinger's Equation (NLSE)[2] which can be solved numerically using split-step fast Fourier transform algorithm. Simulations were done for a 40Gb/s PRBS of length 2^5-1, containing 16 ones.

The Gaussian input pulses can be described as:

$$U(0,T)=\exp\left[-\frac{(1+iC)}{2}\frac{T^2}{T_0^2}\right]$$

Where C is the linear chirp parameter. And T_0 , the half width at 1/e amplitute is 3ps in the calculation.

The amplifier noise (ASE) was modeled as white noise created by a Gaussian random generators and the maximum transmission distance of the system can be determined by Q value[3] (Q>9).

3. Result and discussion

Neglecting nonlinear effect, for a Gaussian pulse with prechirp C, the relationship between rms pulse versus transmission distance can be described as follow[3]:

$$\frac{\sigma}{\sigma_0}=\left[\left(1+\frac{C\beta_2 z}{T_0^2}\right)^2+\left(\frac{\beta_2 z}{T_0^2}\right)^2+(1+C^2)\left(\frac{\beta_3 z}{2T_0^3}\right)^2\right]^{1/2}$$

where σ_0 is the rms pulse-width of the initial chirpped pulse $\sigma_0=T_0/\sqrt{2}$. The third term in the right-hand side of the equations shows that if the dispersion slope was not compensated perfectly and the input pulse is prechirpped, then the effect of second-order dispersion will be enhanced by the chirp, which will greatly influence not only the width but also the waveform of the pulse.

Fig1 illustrate the pulse waveform evolution with and without second-order dispersion compensation under the condition of C=3. In Fig1a, where the slope is not compensated, the accumulated second-order dispersion asymmetrically broadened the pulse and result in a severely oscillating tail. However, in Fig1b when dispersion slope is compensated, oscillating tail is greatly depressed and the pulse broadening is also decreased. No obvious difference, however, can be found between these two cases if C=0. So we can say that it is the existence of the prechirping that enhance the effect of second-order dispersion.

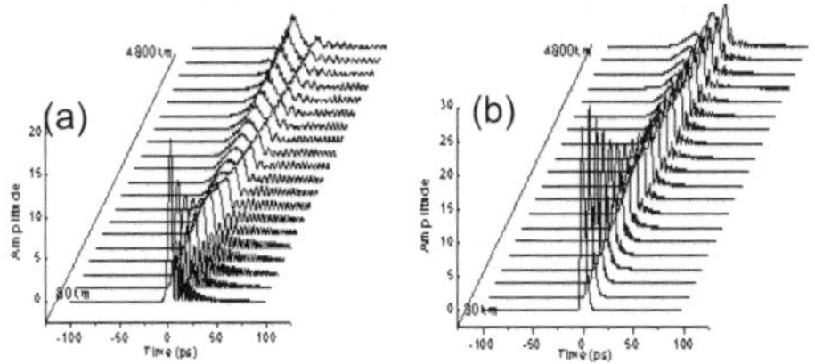

Figure 1 Wave form evolution after transmission of 4800km with prechirping of C=3, (a) Without slope compensation (F1+F2), (b) With slope compensation (F1+F3).

We plot in Fig2 the maximum transmission distance versus average signal input power after slope compensation by numerical simulation. It can be found that, within a certain range of input power (here, 0-7dBm), the transmission performance can be increased to a certain extend by using the prechirped input pulse.

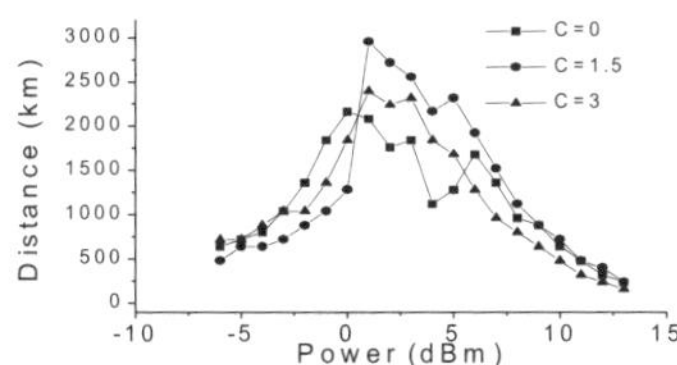

Figure 2 Transmission distance versus input power under dispersion slope compensation.

With the increase of power, nonlinearity takes the place of ASE to become the main factor to hamper the transmission performance. If properly chirpped, pulse will be broadened more rapidly when propagate through the fiber link and experience little nonlinearity and thus enables the transmission of signal over longer distance. When the power is further increased to exceed 7dBm, the interaction of the overlapped pulse during transmission become evident and limits the transmission distance.

Fig3(a-c) show the eye-diagrams of pulse transmitted after 1440km at average power of 2dBm with prechirp of 0,1.5 and 3.0 respectively without second-order dispersion compensation. Apparently, Fig3c is the worst among these three figures, from which we can observe the asymmetrical pulse waveform caused by prechirpping enhanced accumulated second-order dispersion. Further more, the pulse oscillating tails enter to adjacent bit and even further bits and make sever fluctuation of the pulse amplitude.

We also investigate cases under the condition of slope compensation. In Fig3(d-f), we depict the eye-diagram for the pulse after 2000km transmission. In Fig3d (where C=0), the eyes are not well opened since the pulses were significantly broadened by accumulated nonlinearities. This accumulation for the

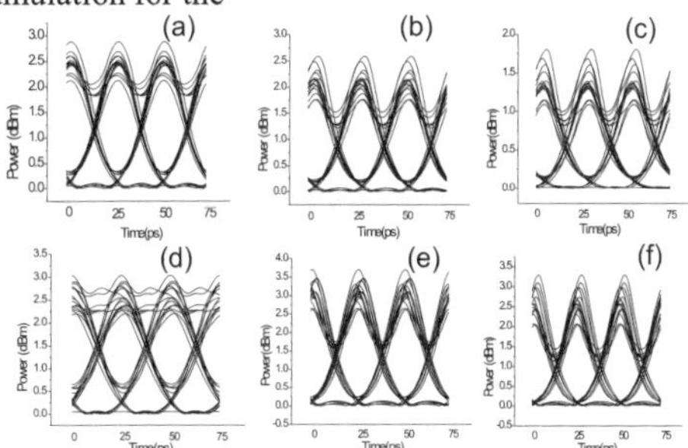

Figure 3 (a-c) Eye-diagram after transmission of 1400km without dispersion slope compensation, C=0, 1.5 and 3 respectively. (d-f). Eye-diagram after transmission of 2000km with dispersion slope compensation, C=0, 1.5 and 3 respectively.

prechirped pulse, however, is relatively low. Therefore, observed from Fig3(e,f), the pulse-width was not increased notably after long distance of transmission. At the same time, the pulse waveform is more symmetric than that in Fig6c. So it can be state that, after dispersion slope compensation, the transmission performance of properly chirpped low-duty-cycle RZ format is better than that without prechirping.

The prechirp should be big enough to reduce the nonlinear effect. But there exist an optimum value. If the prechirp is bigger than the proper value, pulse will be broadened severely and the nonlinear interaction among pulse will be the main source of signal degradation and affect the transmission performance.

4. Conclusions

For the first time to the best of our knowledge, numerical simulation of the transmission performance has been made to 40Gb/s low-duty-cycle CRZ format. We found that pulse prechirping will enhance the effect of second-order dispersion in longhaul transmission and the dispersion slope compensation is very important even for single channel ultra-high-speed transmission system. After compensation, prechirping of pulses is found to be helpful both in reducing pulse distortion and nonlinear interaction, and the transmission characteristic of the properly prechirped signals with certain power level is better than that without prechirping.

References

[1] R M Mu, Yu, T.; Grigoryan, V.S.; Menyuk, C.R, "Dynamics of the chirped return-to-zero modulation format " IEEE Journal of Lightwave Technology, Vol 20, No. 1, 2002 P 47-57.

[2] Govind P.Agrawal, Nonlinear fiber optics, Academic Press, third edition. 2001

[3] N.S.Bergano, Kerfoot, F.W.; Davidsion, C.R." Margin measurements in optical amplifier system" IEEE Photon. Techno. Lett, Vol.5, No.3, 1993, P304-306

Session 12C

Optical Networking II

Light Path Protection for IP/DWDM Networks

Luying Zhou, Yu Bei* and Yu Jia*
Laboratories for Information Technology, 21 Heng Mui Keng Terrace, Singapore 119613
Department of Computer Science*, National University of Singapore, Singapore 117543

Abstract -- This paper presents a new algorithm that dynamically allocate restorable bandwidth guaranteed paths in IP over DWDM network. This paper tackles the bandwidth efficiency by considering sharing of the bandwidth on the backup lightpaths. The amount of sharing that can be achieved is related to the information available to the algorithm. Three information scenarios are considered here: complete information scenario, partial information scenario and no information scenario. Simulation results show that the developed algorithm with a partial information scenario which uses only aggregated and not per-path information performs very well compared with the complete information scenario.

I. INTRODUCTION

Currently, most IP transport architectures are based on SONET/SDH, encapsulating IP packets(or ATM cells carrying IP packets) in SONET/SDH frames. However, this "full stack" approach reduces efficiency and poses increased management/operation costs. Therefore, designing a single, ubiquitous WDM access layer with tight IP interworking can significantly reduce the intermediate layering requirements. In order to provision a full range of network functionality, the optical layer must subsume some key functions which currently reside in different network layers. These include channel routing, channel monitoring, and fault detection and recovery capabilities. [1]. Recent years, a lot of work has been done on extending Multiprotocol Label Switching (MPLS) as a control plane that can be used not mealy with routers, but also with SONET equipment and optical devices like OXCs [2,3]. The control planes in the IP and optical networks can be loosely or tightly coupled. At one hand, the optical network primarily offers high bandwidth connectivity in the form of lightpaths. Standard signaling across the UNI is used to invoke the services. IP routers at the edge of the optical networks must necessarily establish such paths before communication at the IP layer can begin. On the other hand, the IP and optical networks are treated together as a single integrated network that is managed and traffic engineered in a unified manner. There is no distinction between UNI, NNI and any other router-to-router interface. It is assumed this control plane is MPLS-based, IP routers and OXCs can have a peer relation on the control plane.

A. *IP/WDM Network Survivability*

The network survivability issue has long been of primary concern in transport networks. SONET/SDH networks mainly provide dedicated-resource protection, such as self-healing ring and automatic protection switching (APS), which relies on fixed (redundant) backup resources and a rigid electronic framing/synchronization format, and provides mainly point-to-point or ring based protection.

Similar protection and restoration mechanisms can be carried out on IP/WDM networks to provide survivability [4]. In the path protection scheme, the source and destination nodes of each connection statically reserve backup paths on an end-to-end basis during call setup. There is dedicated path protection, in which payload data is transmitted simultaneously over two disjoint paths, and a selector is used at the receiving node to choose the best signal. And there is also shared-path protection, at the time of call setup for a primary path, a link-disjoint backup path and wavelength are also reserved. However, the backup wavelength reserved on the links of the backup path may be shared with other backup paths.

In the path restoration scheme, the source and destination nodes of each connection traversing the failed link participate in a distributed algorithm to dynamically discover a backup route and wavelength on an end-to-end basis. If no new route is discovered for a broken connection, that connection is blocked.

The distinction between protection and restoration is centered on the different time scales in which they operate. Protection requires preallocated resources, while restoration relies on dynamic resource establishment.

WDM layer survivability cannot protect against failures at higher layers. It cannot handle all types of faults, can not be able to detect all types of faults, and only protect traffic in units of wavelength. So, multilayer survivability incorporates survivability mechanisms at multiple layers, and two escalation or interworking strategies can be used. One is bottom-up strategy, where recovery starts at the layer closest to the failure, and escalates upward upon expiration of a holdoff timer. This timer allows the lower layers time to recover form a fault before triggering recovery mechanisms at a higher layer, it ensuresquick activation of the recovery process. Another is top-down

This work was partially supported by A-Star under ONFIG GMPLS Project

strategy, where recovery always starts at the uppermost layer and escalates downward. Holdoff timers are not necessary in this strategy, but a disadvantage is the potentially large number of traffic streams that must be restored at the higher layers.

B. *Related Work*

Bandwidth sharing routing algorithm for providing protection in the traditional MPLS network has been developed in [5] and [7]. Bandwidth efficiency and the amount of information dissemination are two main concerns of the bandwidth sharing routing algorithm. A suitably developed algorithm in [5] can achieve good bandwidth utilization according to sharing backup bandwidth with only aggregated and not per-path information.

There exist some differences about sharing backup bandwidth between traditional MPLS network and IP/DWDM network. Firstly, routing in the traditional MPLS network only involves one layer of network, ie. IP layer, whereas routing in the IP/DWDM network considers two layers of network, the DWDM layer and the IP layer [9]. Secondly, the optical network has special properties that are different from the traditional MPLS network. Path selecting in the optical network is the problem of *routing and wavelength assignment* (RWA) [8], not simply shortest path problem [6]. Lightpath is the unit of end to end transmission in the optical network, and it forms the virtual connection between edge routers in the IP network. So the backup bandwidth that can be shared is reserved on lightpath instead of on each link as stated in [5].

So the new architecture of IP/DWDM network motivates us to extend the algorithm in [5] to provide protection against single fiber cut in the IP/DWDM network and optimize using network resources with feasible information dissemination as well.

II. SHARED PROTECTION PATH

Consider an IP/DWDM network, the optical layer is a mesh of *n* nodes (OXC), which do not perform opto-electronic-optic conversion, and *m* fiber links, and each fiber link contains *W* wavelengths. At the IP layer, ie., virtual network layer, there are *n'* nodes. The nodes at IP layer are edge routers, and each of them is connected to one OXC in the optical layer.

We assume the IP requests with bandwidth requirement arrive one by one dynamically at the ingress edge router. The goal of the algorithm is allocating a bandwidth guaranteed primary connection (ie. Lightpath) and backup connection for each request. The backup connection reserves the amount of bandwidth necessary to protect the traffic of the request on the primary connection against single fiber cut in the optical network. At the optical layer, the lightpaths of the primary virtual connection and backup virtual connection should be link-disjoint. When trying to process an IP request with protection requirement at ingress edge router, if either of the primary or backup connection cannot be allocated, the request is rejected.

Suppose current request arriving at the edge router requires *b* units of bandwidth, the primary connection for the request has to reserve *b* units of bandwidth for transmitting the packets. However, the backup virtual connection does not necessarily have to reserve *b* units of bandwidth. This is because the potential sharing of the bandwidth of the backup lightpath.

The information of each existing connection at the IP layer can be defined as follows. The capacity of the connection is the capacity of the corresponding lightpath. For each existing connection p, let G_p denote the amount of bandwidth reserved for backup purpose, R_p be the amount of residual bandwidth, which is the difference between the capacity of the connection and bandwidth already consumed, whether for primary transmission or backup.

A. *No Information Model*

In no information scenario, we only know residual bandwidth R_p for each existing connection *p*. So we cannot share the backup bandwidth in this information model. When a request (*s, d, b*) arrives, where s and d are ingress and egress routers, and b is bandwidth requirement. We first find out if there exists primary connection between edge router *s* and *d*. If there is, we reserve *b* units of bandwidth on the connection for transmitting the packets. If there is not, we dynamically establish a lightpath between the ingress edge router and egress edge router as the primary connection. Then we find out if there exists backup connection. If there is, we reserve *b* units of bandwidth on the connection to backup the traffic on the primary connection. Otherwise, we have to dynamically establish a link-disjoint lightpath as backup connection.

B. *Complete Information Model*

We have the complete connection information for all the requests currently in progress in complete information model. With the information, we can fully utilize the network resources by sharing the bandwidth of backup lightpaths. Then the problem is how to get the extra bandwidth needed to be reserved on backup connection.

Let A_{ij} denote the set of IP requests whose primary connections use fiber link (i, j). Let B_p represent the set of IP requests that use existing connection *p* as backup connection. Let ϕ_{ij}^p represent the set of IP requests that use existing connection *p* as backup connection, and their primary connection use fiber link (i, j). We have

$$\phi_{ij}^p = A_{ij} \cap B_p .$$

Let b_k denote the k-th request's bandwidth demand in set ϕ_{ij}^{p}, then the sum of all the bandwidths demands in set ϕ_{ij}^{p} is $\delta_{ij}^{p} = \sum_{k \in \phi_{ij}^{p}} b_k$. So, if the lightpath of the primary connection for the current request that has bandwidth demand b uses link (i, j) in the route, the bandwidth necessary to be reserved on backup connection p for protecting the traffic on link (i, j) is,

$$\theta_{ij}^{p} = \begin{cases} 0, & \delta_{ij}^{p} + b \le G_p \\ \delta_{ij}^{p} + b - G_p, & \delta_{ij}^{p} + b > G_p \ \& \ R_p \ge \delta_{ij}^{p} + b - G_p \end{cases}$$

If we use a link-disjoint primary connection which is denoted as PRM for current request, the bandwidth necessary to be reserved on backup connection p is the maximum θ_{ij}^{p} for each link (i, j) of the route of PRM, that is $\max_{ij \in PRM} \{\theta_{ij}^{p}\}$.

C. *Partial Information Model*

Partial information model is the primary focus of this paper. In this model, the available information for each existing connection p is the residual bandwidth R_p, and backup bandwidth G_p.

We have following notations to define the fiber link state information in the optical network. Each fiber in the optical network has W wavelengths, which are numbered as 0, 1, ..., W-1. Let A_{ij}^{w} denote the set of IP requests whose primary connections use fiber (i, j) and wavelength w. Then the total amount of bandwidth reserved on fiber link (i, j) and wavelength w for primary purpose is $f_{ij}^{w} = \sum_{k \in A_{ij}^{w}} b_k$. For each fiber link (i, j), the available information to the algorithm in partial information model is the aggregate bandwidth f_{ij}^{w} (w = 0, 1,..., W-1). The main difference of the routing algorithm in partial information model from the one in complete information model is how to get the extra bandwidth needed to be reserved on backup connection p for the current request with limited information.

In order to achieve bandwidth sharing of backup lightpath with only partial information, we add a constraint when allocating connections for IP request. The constraint is that each virtual connection p can only backup the traffic on at most one wavelength for each link fiber (i, j). If we use $\phi_{ij}^{w,p}$ to denote the set of requests that use p as backup connection and their primary connection traverse link (i, j) and occupy wavelength w on link (i, j), the constraint is expressed as,

$$\phi_{ij}^{w_1,p} \cap \phi_{ij}^{w_2,p} = \phi \qquad (w_1 \ne w_2).$$

With this constraint, if we use primary connection PRM which occupies wavelength w and backup connection BAK for current request with b units of bandwidth demand, BAK can only protect the traffic on wavelength w for each fiber of the route of PRM. That is, for each fiber (i, j) of the route of PRM, we have $\phi_{ij}^{BAK} = \phi_{ij}^{w,BAK}$ and $\delta_{ij}^{BAK} = \delta_{ij}^{w,BAK}$, where $\delta_{ij}^{w,BAK} = \sum_{k \in \phi_{ij}^{w,BAK}} b_k$.

Note that $f_{ij}^{w} \ge \delta_{ij}^{w,p}$, because the traffic of wavelength w on fiber link (i, j) may be protected by the connections other than p. So we have $f_{ij}^{w} \ge \delta_{ij}^{p}$, if w is wavelength of the primary connection and p is the backup connection. Thus we can use f_{ij}^{w} to get the extra bandwidth $\theta_{ij}^{p\prime}$ needed to be reserved on backup connection p to protect the traffic on fiber (i, j), which is expressed as follows:

$$\theta_{ij}^{p\prime} = \begin{cases} 0, & f_{ij}^{w} + b \le G_p \\ f_{ij}^{w} + b - G_p, & f_{ij}^{w} + b > G_p \ \& \ R_p \ge f_{ij}^{w} + b - G_p \end{cases}$$

So we can reserve $\max_{ij \in PRM} \{\theta_{ij}^{p\prime}\}$ bandwidth on backup connection p if we use PRM as the primary connection of the current request.

The algorithm for allocating connections for each request is as follows. Each edge router stores all the existing connections starting from it in LSP table. Each primary connection is linked with a backup connection. When requests come in, the edge router tries to find a pair of existing primary connection and backup connection at first. If edge router cannot find a pair of connections, it tries to find a primary connection p, then scans through the LSP table to find a suitable backup connection. If no existing connections can satisfy the current request, the edge router will launch the RWA process to find a new primary connection, and scan through LSP table to find a suitable backup connection. If there is no available backup connection, the edge router will call RWA again to find a new backup connection. The suitable backup connection means the extra bandwidth necessary to be reserved on it is minimum. The routing algorithm in this information model can be executed in polynomial time, since it only involves searching in table and finding shortest path in the network.

III. SIMULATION RESULTS

This section compares the performance of the algorithms in the three different information scenarios. We test the performance in the mesh-based network topology shown in Figure 1. The optical network consists of 15 optical switches and 21 physical links. Each adjacent node pair is connected by a physical link that consists of bi-directional working fibers. We assume there is only one fiber in one direction, and each fiber contains 4 wavelengths. In the IP layer, there are ten edge routers, which are connected with OXCs. In our simulation, we set the capacity of the lightpath as 100. We assume that the bandwidth requirement of the requests is evenly generated between 1 and 10.

We set 70% of the requests go from edge router 1 to edge router 2, and 30% are evenly distributed among all pairs of edge routers in the IP network. There are totally 150 requests loaded to the network. Figure 2 shows the number of dropped requests for 10 random experiments. From the figure, we can see that partial and complete information models dropped fewer requests than the algorithm in no information model. This is because there are plenty of disjoint paths between edge router 1 and 2, and the backup lightpath can be shared by many requests. And the performance of the algorithm in partial information model is almost as good as that in complete information.

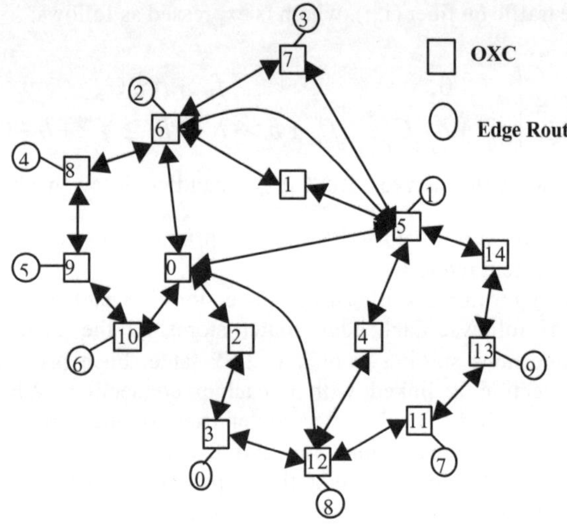

Figure 1 Testing IP/DWDM

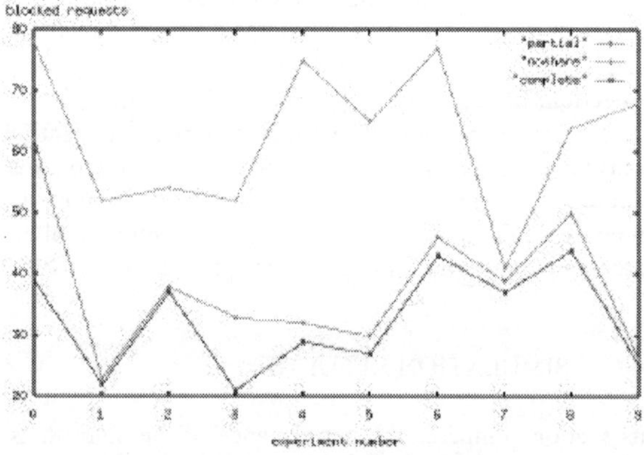

Figure 2 Number of rejected requests for 10 random experiments in 15-node DWDM network (70% traffic is from edge router 1 to edge router 2, 30% random)

IV. CONCLUSION

This paper has introduced the survivability problem at IP/DWDM network. We consider providing protection to the connections in IP/DWDM network, and meanwhile, we attempt to optimize the usage of network resources through sharing backup bandwidth. Three information models are defined, and we presented a simple bandwidth sharing algorithm for sharing the bandwidth of backup lightpaths at IP/DWDM network under partial information model to achieve good performance as well as little information dissemination in the network. We implemented the algorithms of the three scenarios and compared their performance. The simulation results show that when the network has plenty of disjoint paths between edge routers our algorithm has almost the same good performance as the algorithm in complete information model.

REFERENCES

[1] Ayan Banerjee, Kireeti Kompella, Yakov Rekhter, etc, "*Generalized Multiprotocol Label Switching: An Overview of Routing and Management Enhancements*", IEEE Communications, January 2001, pp144-150

[2] Nasir Ghani, et al, "*On IP-over WDM Integration*", IEEE Communications Magazine, March 2000, pp72-84

[3] Eric Mannie (Editor), "*Generalized Multi-Protocol Label Switching (GMPLS) Architecture*", draft-ietf-ccamp-gmpls-architecture-02.txt, March 2002

[4] Didier Colle, et al, "*Data-Centric Optical Networks and Their Survivability*", IEEE Journal SAC, Vol. 20, No. 1, January 2002, pp6-20.

[5] Murali Kodialam, T.V.Lakshman, *Dynamic Routing of Bandwidth Guaranteed Tunnels with Restoration*, IEEE INFOCOM 2000.

[6] Thomas H. Cormen, Charles E. Leiserson, Ronald L. Rivest, Clifford Stein, *Introduction to Algorithms,* McGraw-Hill, ISBN 0-07-013151-1.

[7] Murali Kodialam, T.V.Lakshman, *Dynamic Routing of Locally Restorable Bandwidth Guaranteed Tunnels using Aggregated Link Usage Information*, IEEE INFOCOM 2000.

[8] Hui Zang, Jason P. Jue, Biswanath Mukherjee, A Review of Routing and Wavelength Assignment Approaches for Wavelength-Routed Optical WDM Networks, SPIE Opt. Net. Mag., vol. 1, Jan. 2000.

[9] Yinghua Ye, *A Simple Dynamic Integrated Provisioning/Protection Scheme in IP over WDM Networks*, IEEE Communications Magazine, November 2001.

The Advantage of the Burst Segmentation Option in Optical Burst Switching

Marcel Neuts, Zvi Rosberg, Hai Le Vu, Jolyon White and Moshe Zukerman
ARC Special Research Centre for Ultra-Broadband Information Networks
Deptartment of Electrical and Electronic Engineering
The University of Melbourne, Victoria 3010, Australia
Tel: +61 3 8344 3819, Fax: +61 3 8344 3821
E-mail:{m.neuts,z.rosberg,h.vu,j.white,m.zukerman}@ee.mu.oz.au

***Abstract*— This paper provides an analytical basis for comparison of reservation policies in Optical Burst Switching (OBS) networks. In particular, the Just-Enough-Time (JET) and Burst Segmentation policies are analysed. We first compare these two policies using single link models and then we extend these models to a network scenario using a reduced load fixed point approximation model to evaluate blocking probabilities. We use this analysis framework to provide insights into the significant advantages provided by the Burst Segmentation policy for OBS networks, taking into account network related effects which cannot be studied by single link models alone.**

***Index Terms*—Optical Burst Switching, Erlang Fixed Point Approximation, Reduced Load, Just Enough Time (JET), Burst Segmentation.**

I. Introduction

OPTICAL Burst Switching (OBS) (see [1], [2], [7], [8] and [9]) has been proposed as an efficient switching technique to exploit the terabit bandwidth of Wavelength Division Multiplexing (WDM) transmission technology. In OBS, IP packets with a common destination arriving at the same ingress node are aggregated into large bursts, switched and routed as one unit. Since only one header is associated with each burst, the header processing is amortized over more data, and the switch fabric can be reconfigured on a longer timescale. The header (or control packet) precedes the burst payload and attempts to reserve switching and transmission resources at each switch and output port along the route.

A key feature in OBS is that the header precedes the payload by an offset time and is usually transmitted on a dedicated signalling wavelength. The payload follows the header without waiting for acknowledgment. At every switch, if the requested resources are available, the burst is transparently switched to its next hop; otherwise, the burst is blocked and some fraction (possibly all) of the data is lost. OBS sits somewhere between circuit switching and packet switching: the path is not acknowledged end-to-end, leading to resource usage that is ultimately wasted when a burst is blocked; and the current limited ability for buffering in the optical domain means that burst data cuts through the switch and uses route links in a time-synchronous manner, whereas packets use them asynchronously due to buffering.

Note that while optical micro-electro-mechanical switches (MEMS) can only provide fabric reconfiguration time in the order of a few milliseconds, switches based on Semiconductor Optical Amplifiers (SOAs) achieve reconfiguration time in the order of a few nanoseconds. More recent techniques [10] even achieve reconfiguration time in the order of a few hundred picoseconds. Moreover, Electro-Absorption Modulator based devices capable of reconfiguration time in the order of a few picoseconds [3], [4], are already being used in research laboratories. With support from these technologies, OBS is likely to become a feasible switching technique in the near future.

To this end it is important to maximise usage of the optical bandwidth under OBS and to be able to understand and model the reservation protocols involved. Several OBS reservation protocols have been proposed by researchers. In this paper we consider the merits of two major schemes, the Just-Enough Time (JET) procotol and the Burst Segmentation protocol and we provide analytical results which demonstrate the improved performance offered by the Burst Segmentation policy. Under JET [7] the control packet contains the burst length and requests link bandwidth from the predetermined time offset and for the duration of the burst transmission. In Burst Segmentation [9] and [1] on the other hand, the control packet contains the burst length and reserves capacity from the first instant when a wavelength becomes available. The initial portion of the burst which is not served before a wavelength becomes free is discarded and lost, i.e. the burst is segmented. The remainder is transmitted successfully as a truncated burst.

In Section II we provide single node, single link analyses of JET and Burst Segmentation. Whilst these techniques provide crude upper bounds on link blocking probability they cannot capture network effects due to blocking along the network routes, and hence in Section III we use the analysis framework for OBS networks presented in [5] to make a comparison of JET and Burst Segmentation in a network scenario. Finally, the conclusions are presented in Section IV.

II. Single Node Analysis

Since OBS is a bufferless system, the $M/M/k/k$ queueing model is a standard analysis approach [7], [8]. This model assumes that the burst arrival process at a given output port of an optical burst switch is a Poisson process with rate λ. If the burst duration is exponentially distributed with mean $1/\mu$ and the number of wavelengths on the output fibre is k then the Erlang B formula gives the probability of burst blocking:

$$P_B(k, A) = \frac{A^k/k!}{\sum_{m=0}^{k} A^m/m!}, \tag{1}$$

where $A = \frac{\lambda}{\mu}$ is the offered traffic load. Due to the insensitivity of the Erlang B formula to the service time distribution, the burst duration distribution can be relaxed to be a general distribution. This model is suitable for the JET and JIT reservation policies in which a burst is either accepted or rejected in its entirety. In this case, the packet loss probability is the same as the burst loss probability [9].

The Burst Segmentation policy cannot be modelled by (1) since blocked bursts are segmented and may be partially rejected under this regime [9]. However, this system can be modelled simply using the $M/G/\infty$ queueing model. In this model, the first k servers of the $M/G/\infty$ system are real servers representing actual wavelength channels. The remaining servers are pseudo-servers which are used simply to record the arrival of a burst to the full system. When a burst is "served" by a pseudo server, its payload is being lost and the burst itself is segmented. If the system is full and there are occupied pseudo-servers, when one of the real servers becomes free, the remainder of the first blocked burst is moved to the free real server so that its truncated portion is given service and is not lost. In this way the $M/G/\infty$ system is able to model the segmentation of bursts which are blocked, throwing away the blocked portion and retaining the useable part.

In the $M/G/\infty$ model, if k or less servers are busy then all bursts in the system are being successfully transmitted on a wavelength channel and no loss is occurring. In the case of $k+j$ bursts in the system, $j \geq 1$, then j of the bursts are losing packets, waiting for one of the k wavelengths to become free. Thus j out of every $k+j$ packets are lost in this case. The packet loss probability is

$$P_{BS}(k, A) = \frac{E[L]}{A}. \tag{2}$$

In this equation $E[L]$ is the mean loss rate given by

$$E[L] = \sum_{j=1}^{\infty} j \cdot P(k+j), \tag{3}$$

where $P(k+j)$ is the probability that $(k+j)$ servers are busy in an $M/G/\infty$ model, which is well known [13] to be

$$P(k+j) = A^{k+j} \frac{e^{-A}}{(k+j)!}, \quad j = 1, 2, \ldots \tag{4}$$

Thus we have a simple formulation for the packet loss probability in Burst Segmentation for a Poisson arrival process. Figure 1 compares the performance of JET and Burst Segmentation for $k = 120$ wavelengths. For such a large system we see that loss rates of 10^{-9} can be achieved at quite high utilisations, around 0.55 for JET and 0.59 for Burst Segmentation. Furthermore, segmentation allows a 7% improvement in utilisation for the same low loss rate.

It is interesting to note that Detti et. al. [9] have analysed Burst Segmentation by supposing that a number of ON/OFF sources contribute to the input process at a switch output, which models the fact that the input traffic comes from individual wavelengths on a number of switch input ports. This analysis arrives at the Engset formula, with the exact value of traffic

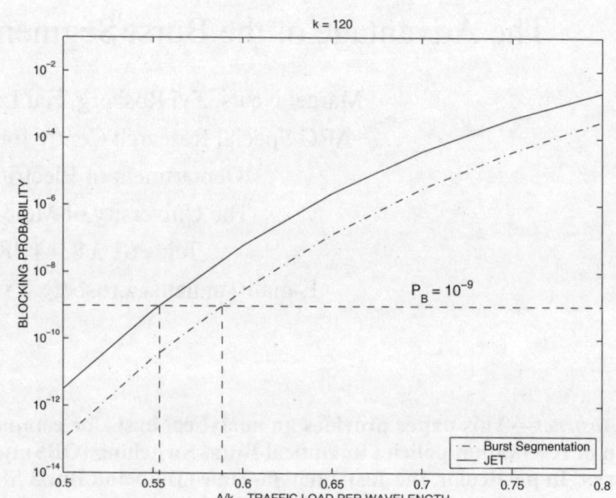

Fig. 1. Comparison of JET vs. Burst Segmentation for 120 wavelengths.

lost depending on the number of contributing sources. Since the superposition of a large number of ON/OFF sources tends to a Poisson process in the limit, we can regard the $M/G/\infty$ model as a simplified limiting case of many sources which can provide a more convenient basis for network modelling, as in Section III.

Having established the burst and packet loss rates for both JET and Burst Segmentation, we are interested in the relative benefit of Burst Segmentation over JET. A very simple expression for the ratio of blocking probabilities can be obtained under very light load conditions, when $k \gg A$. We use (1) and (2) to calculate the ratio

$$\frac{P_B(k, A)}{P_{BS}(k, A)} = \frac{P_B(k, A)}{E[L]/A} \tag{5}$$

We notice from (4) that $P(k+j) = [A/(k+j)] \cdot P(k+j-1)$, $j \geq 1$, so that if $k \gg A$ then $P(k+1) \gg P(k+2) \gg P(k+3) \gg \ldots$, so that we can write

$$E[L] \approx P(k+1). \tag{6}$$

$P_B(k, A)$ is the proportion of work lost in a system without burst segmentation. $P(k+1)/A$ is the proportion of work lost when burst segmentation is used assuming $P(i)$ is negligible for $i > k+1$, namely, the overall proportion of work lost is very small. This will happen if $k \gg A$. The ratio between the two is

$$\frac{P_B(k, A)}{P(k+1)/A} = \frac{(k+1)}{e^{-A} \sum_{m=0}^{k} \frac{A^m}{m!}}. \tag{7}$$

Hence

$$\lim_{k \to \infty} \frac{P_B(k, A)}{(k+1)P(k+1)/A} = 1. \tag{8}$$

Note that the simple result of (8) applies only to cases where the load is unrealistically low.

III. Network Modelling and Numerical Results

Whilst the single node models of Section II give much insight into the performance of OBS, they cannot capture network-related effects such as increase in load due to unsuccessful bursts which still consume resources along the route until they are blocked, and reduced load due to blocking of bursts earlier in their routes. A reduced load, fixed point approximation technique has been introduced in [5] for analysis of OBS networks, taking account of network-related issues.

In this framework we consider a network with J directional links. Each link $j \in \{1, \ldots, J\}$ is capable to transmit data only in one fixed direction and serves up to N_j concurrent logical channels using WDM.

A route $\boldsymbol{r} = (r_1, r_2, \ldots, r_n)$ having n hops is an ordered set of links that connect the source node to the destination node. Let $\mathcal{R}$ be the set of all possible routes. We assume that bursts offered to route $\boldsymbol{r}$ arrive according to a Poisson process with rate $\lambda_{\boldsymbol{r}}$ and all arrival processes are independent. A burst offered to route $\boldsymbol{r}$ uses a single wavelength from each link along the route until the first link where it is being blocked or until it exits the network. That is, if the burst is first blocked by link r_j it uses a single wavelength channel from links $r_1, r_2, \ldots, r_{j-1}$ in succession. We further assume that burst transmission times are independent and exponentially distributed and we denote by μ_j the transmission rate of link j.

We denote by $\boldsymbol{B} = (B_1, B_2, \ldots, B_J)$ the vector of stationary link blocking probabilities. Given the difficulty in obtaining exact mathematical results for this problem, the model assumes as in [6], [11], [12] and [14] that each blocking event occurs independently from link to link along any route. It was shown in [5] that under stationary conditions the blocking probability of an arbitrary burst offered to route $\boldsymbol{r}$ can be calculated by computing the reduced load ρ_j offered to link j:

$$\rho_j = {\mu_j}^{-1} \sum_{\boldsymbol{r} \in \mathcal{R}} \lambda_{\boldsymbol{r}} \prod_{i=1}^{J} I(i,j,\boldsymbol{r}) \cdot (1 - B_i) \,, \tag{9}$$

where $I(i, j, \boldsymbol{r})$ equals 1 or 0 depending whether or not $i, j \in \boldsymbol{r}$ and link i strictly precedes link j along route $\boldsymbol{r}$, respectively.

Furthermore, the independence assumption implies that the offered load to each link j is a Poisson process with rate ρ_j. Thus, the blocking probability is given by (1), i.e.

$$B_j \stackrel{\text{def}}{=} P_B(N_j, \rho_j). \tag{10}$$

Combining equations (9) and (10) yields the following Erlang Fixed Point (EFP) equations satisfied by the approximate link blocking probabilities,

$$B_j = P_B\Big(N_j,\, {\mu_j}^{-1} \sum_{\boldsymbol{r} \in \mathcal{R}} \lambda_{\boldsymbol{r}} \prod_{i=1}^{J} I(i,j,\boldsymbol{r}) \cdot (1 - B_i)\Big) \,. \tag{11}$$

Resolving the vector $\boldsymbol{B}$ from the EFP equations (11) and invoking the independence assumption again, the approximate blocking probability of bursts offered to route $\boldsymbol{r}$, $B(\boldsymbol{r})$, satisfies

$$B(\boldsymbol{r}) = 1 - \prod_{i \in \boldsymbol{r}} (1 - B_i) \,,$$

and the blocking probability of an arbitrary burst B, satisfies

$$B = \frac{1}{\Lambda} \sum_{\boldsymbol{r} \in \mathcal{R}} \lambda_{\boldsymbol{r}} \cdot B(\boldsymbol{r}) \,,$$

where $\Lambda = \sum_{\boldsymbol{r} \in \mathcal{R}} \lambda_{\boldsymbol{r}}$.

Notice that the OBS reduced load offered to link j as given in equation (9) is larger than ${\mu_j}^{-1} \sum_{\boldsymbol{r} \in \mathcal{R}} \lambda_{\boldsymbol{r}} \prod_{i \in \boldsymbol{r} \setminus \{j\}} (1 - B_i)$, the reduced load offered to link j in a conventional circuit switching network.

It was demonstrated in [5] that the fixed point equations (11) can be solved for the blocking probabilities $\boldsymbol{B}$ by a successive substitution procedure, and although uniqueness of the solutions has not been established, the existence of a fixed point has been shown and in numerical studies the procedure is always observed to converge to a unique fixed point. By Theorem 1 of [5], it was shown that an upper and lower bound on the vector of blocking probabilities can always be established.

The definition of the link blocking probabilities (10) can be replaced with other expressions for systems wherein the $M/M/k/k$ model is not appropriate. To analyse Burst Segmentation networks in this framework we replace Equation (10) with (2) above to define $B_j = P_{BS}(N_j, \rho_j)$ and form the fixed-point equations from this expression. Convergence to the upper and lower bounds is also guaranteed so long as the transformation defined by the fixed point equations is decreasing.

We used the *NSFNET backbone network* topology depicted in Fig. 2 to compare the results of the reduced load fixed point approximation for both JET and Burst Segmentation. The network topology comprises 13 OBS switches and 32 unidirectional fibre links, each comprised of 8 wavelength channels. We evaluated the blocking probabilities of the 12 routes defined in Table I. The selected routes represent a variety of path lengths, link sharing degrees and mixtures of external and on-route internal traffic processes. All routes are shortest paths, except for R_3 and R_7 that are selected to obtain better route diversity.

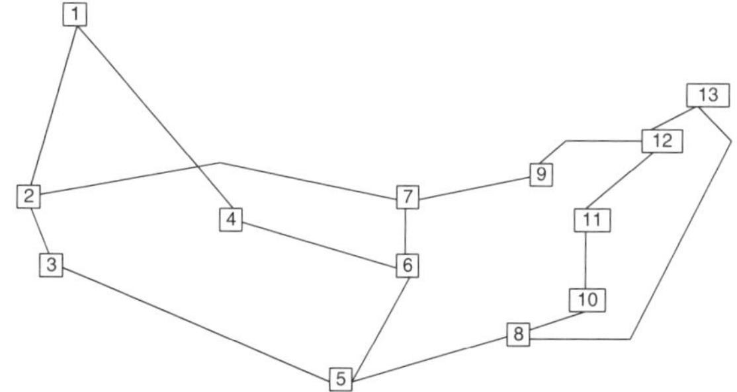

Fig. 2. NSFNET backbone network (T3, 45 Mb/s, April 1995).

Table II gives the overall blocking probability for the network as calculated by the reduced load fixed point approximation, alongside simulation results for the same network and routes. We considered symmetric traffic scenarios where all routes are offered the same burst external arrival rate, i.e. $\lambda_{\boldsymbol{r}} = \lambda$ for all $\boldsymbol{r}$. Two values of arrival rate were investigated in this table, $\lambda = 30$ and $\lambda = 50$ bursts per second. This corresponds to a moderate load, with the channel capacity and burst length set to yield a burst service rate of $\mu = 25$ bursts per second. It can be seen that the approximation provides accurate predictions of overall blocking probabilities and that burst segmentation provides significant benefit in overall blocking. Table III shows the blocking probability for each route under low load conditions of $\lambda = 5$ and $\lambda = 10$ bursts per second, as calculated by

Route Name	Route hops
R_1	1→4→6→5→8→10
R_2	2→3→5→6→7
R_3	2→7→9→12→13
R_4	3→5→8→13
R_5	5→6→7→9→12
R_6	8→10→11→12→13
R_7	10→8→5→6→4→1
R_8	7→6→5→3→2
R_9	13→12→9→7→2
R_{10}	13→8→5→3
R_{11}	12→9→7→6→5
R_{12}	13→12→11→10→8

TABLE I

NSFNET ROUTES

Policy	Approx.	Sim.
	$\lambda = 30$ /sec	
JET	0.0151	0.0147 ±0 .00033
Seg.	0.0035	0.0034 ± 0.00013
	$\lambda = 50$ /sec	
JET	0.1108	0.1060±0 .00094
Seg.	0.0427	0.0418 ± 0.00071

TABLE II

OVERALL BLOCKING PROBABILITIES WITH JET AND BURST SEGMENTATION (APPROXIMATION AND SIMULATION)

Routes\\λ	$\lambda = 5$ /sec		$\lambda = 10$ /sec	
	JET $\times 10^{-9}$	Seg. $\times 10^{-9}$	JET $\times 10^{-6}$	Seg. $\times 10^{-6}$
R_1	251	27.8	35.9	4.98
R_2	251	27.8	35.9	4.98
R_3	32.7	3.64	5.62	0.624
R_4	21.8	2.43	3.75	0.417
R_5	261	29.0	37.73	5.19
R_6	21.9	2.43	3.76	0.418
R_7	251	27.8	35.9	4.98
R_8	251	27.8	35.9	4.98
R_9	32.7	3.64	5.62	0.624
R_{10}	21.8	2.43	3.75	0.417
R_{11}	261	29.0	37.73	5.19
R_{12}	21.9	2.43	3.76	0.418
Overall	140	15.5	20.4	2.77

TABLE III

ROUTE BLOCKING PROBABILITIES WITH JET AND SEGMENTATION – APPROXIMATION

the approximation. Again, segmentation provides large benefits on all routes, and the overall improvement is almost an order of magnitude.

IV. CONCLUSION

We have provided an analytical comparison of JET and Burst Segmentation for the single node, single link scenario, and we have shown that in the limit of low load, $k \gg A$, Burst Segmentation offers a factor of $k + 1$ improvement, which can be more than two orders of magnitude for foreseeable dense WDM systems. We have extended these two models using the reduced load fixed point approximation for network analysis and shown that segmentation again provides significant benefits on a per-route and an overall network blocking probability basis.

REFERENCES

Proceedings Papers:

[1] V. M. Vokkarane, J. P. Jue and S. Sitaraman, "Burst segmentation: an approach for reducing packet loss in optical burst switched networks", *in Proc. of ICC'02*, Apr. 2002.

[2] J. Y. Wei, J. L. Pastor, R. S Ramamurthy and Y. Tsai, "Just-in-time optical burst switching for multi-wavelength networks", *Proceedings of the 5th Int. Conf. on Broadband Commun. (BC'99)*, pp. 339–352, 1999.

[3] L. Rau et al., "Two-hop all-optical label swapping with varaible length 80 Gb/s packets and 10 Gb/s labels using nonlinear fiber wavelength converters, unicast/multicast output and a single EAM for 80- to 10Gb/s packet multiplexing", *Proceedings of OFC 2002*, Anaheim, Cal., pp. DF2 1–3, March, 2002.

[4] J. Yu, K. Kojima, N. Chand, A. Ougazzaden, C. W. Lentz, J.M. Freund, and B. Mason, "Simultaneous Demultiplexing and Clock Recovery of 80 Gb/s OTDM Signals Using a Tandem Electro-Absorption Modulator", *14th Annual Meeting of the IEEE Lasers and Electro-Optics Society.*, San Diego, USA, November 2001.

[5] Z. Rosberg, H. L. Vu, M. Zukerman and J. White, "Reduced load fixed point approximation for optical burst switching networks", *Proceedings of COIN-PS 2002*, Cheju Island, Korea, July 2002.

[6] R. A. Barry and P. A. Humblet, "Models of blocking probability in all-optical networks with and without wavelength changers", *Proc. INFOCOM'95*, pp. 4a.1.1–4a.1.11, 1995.

Journal Papers:

[7] C. Qiao, "Labeled Optical Burst Switching for IP-over-WDM Integration", *IEEE Communications Magazine*, pp. 104–114, Sept. 2000.

[8] K. Dolzer, C. Gauger, J. Spath and S. Bodamer, "Evaluation of reservation mechanisms for optical burst switching", *AEU Int. J. of Electron. and Commun.*, vol. 55, no. 1, 2001.

[9] A. Detti, V. Eramo and M. Listanti, "Performance evaluation of a new technique for IP support in a WDM optical network: optical composite burst switching (OCBS)", *IEEE Journal of Lightwave Technology*, vol. 20, issue 2, pp. 154–165, Feb. 2002.

[10] C. M. Gallep and E. Conforti, "Reduction of Semiconductor Optical Amplifier Switching Times by Preimpulse Step-Injected Current Technique", *IEEE Photon. Technol. Lett.*, vol. 14 pp902–904, July 2002.

[11] F. P. Kelly, "Blocking probabilities in large circuit-switched networks", *Adv. in Appl. Probab.*, vol. 18, pp. 473–505, 1986.

[12] W. Whitt, "Blocking when service is required from several facilities simultaneously", *AT&T Technical Journal*, vol. 64, no. 8, pp. 1807–1856, Oct. 1985.

Books:

[13] D. Bertsekas and R. Gallager, *Data Networks*, 2nd Edition, Prentice Hall, 1992.

[14] K. W. Ross, *Multiservice loss models for broadband telecommunication networks*, Springer-Verlag London Limited, 1995.

Blocking Performance Analysis of Distributed Wavelength Reservation in Multi-fiber WDM Networks

Feifei Feng, Xiaoping Zheng, Hanyi Zhang
Department of Electronic Engineering, Tsinghua University, 100084, Beijing, China
Tel. +86-10-62772670, Fax +86-10-62770317 E-mail: fff99@mails.tsinghua.edu.cn

Abstract: Distributed wavelength reservation protocols are employed to dynamically set up and tear down lightpaths in WDM networks. In this paper, we develop an analytical model, which can analyze the blocking performance of backward reservation protocols (BRP) under both single and multi-fiber networks. The effects of different wavelength selection polices are also investigated.

Key Words: Wavelength Reservation; Blocking Probability; Multi-fiber networks

I. INTRODUCTION

Wavelength Division Multiplexing (WDM) networks have emerged as promising solutions for future transport networks. To support the bursty nature of the Internet traffic efficiently, WDM lightpaths should be dynamically set up and torn down on-demand. Therefore, an on-line signaling protocol is required to reserve or release wavelength resources along the optical path. Considering the weak scalability and poor survivability of centralized control mechanisms, distributed control protocols are essential in large-scale optical networks [1].

In [2-4], various distributed wavelength reservation protocols are proposed along with their simulation results. These protocols used to be classified into two categories: forward reservation protocols (FRP) and backward reservation protocols (BRP). Simulation results in these studies [2-4] show that in single-fiber networks the performance of backward scheme excels the forward scheme in terms of network blocking probability.

Analytical models for the centralized control in WDM networks have been well studied. However, the analyses will be more complex for distributed control schemes. In distributed circumstances, the influence of signaling transmission delay and processing time must be considered. In [5,6], theoretical models are proposed to analyze the performance of distributed protocols. However, all these work focus on the single fiber networks and consider only random wavelength assignment.

In practical networks it is likely to be multiple fibers in each link for cost-effectiveness, scalability and survivability [7]. In this paper, we extend the analysis model in [5,6] to study the performance of distributed wavelength reservation protocols in both single and multi-fiber networks.

II. DISTRIBUTED WAVELENGTH RESERVATION

Distributed wavelength reservation protocols can be classified into FRPs and BRPs [2,3]. In FRP, a reservation signaling packet (*Resv*) is sent from the source node to the destination node hop-by-hop. This *Resv* packet carries an *AvailSet* structure that indicates which wavelengths are currently available. Each node, after receiving the *Resv* packet, updates the *AvailSet* by removing currently unavailable wavelengths. The node also reserves these wavelengths remaining in *AvailSet*; meanwhile, the node will originate a releasing-signaling packet (*Rls*) to release those removed wavelengths for upstream nodes. If the *AvailSet* becomes empty before the *Resv* reaches the destination node, the connection blocking occurs and a negative acknowledgement-signaling packet (*NACK*) will be sent towards the source node. If the *Resv* packet arrives at the destination node and the final *AvailSet* is not empty, the destination node will choose one wavelength from the remaining *AvailSet*. Then the destination will send a confirmation-signaling packet (*Conf*) towards the source node to confirm the selected wavelength and release these unnecessary wavelengths. The source node starts data transfer once it receives the *Conf* packet.

In contrast to FRP, the BRP scheme doesn't reserve wavelengths along the forward path; it just collects the usage information of wavelengths in this course. The reservations are done along the backward path as shown in Figure 1(b). The source node sends out a probe-signaling packet (*Prob*) with an initial *AvailSet*, each node removes the wavelengths currently unavailable from the *AvailSet* in the incoming *Prob* packet. However, it doesn't do any reservation this moment.

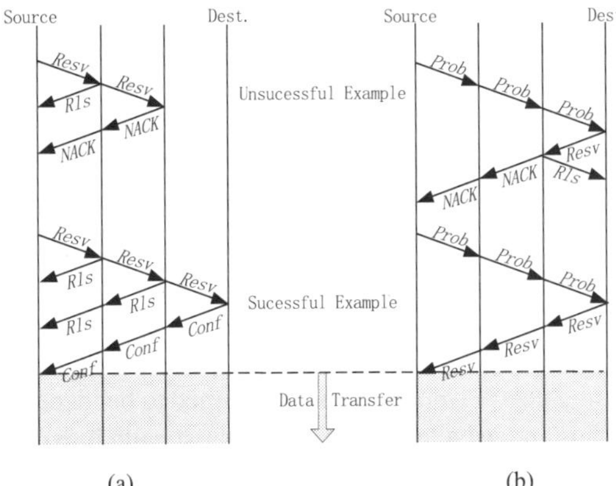

Fig. 1. Illustration of Forward Reservation (a) and Backward Reservation (b)

Once the *Prob* packet arrives at the destination node and the final *AvailSet* is held still not empty, the destination node will choose someone wavelength out of the *AvailSet*. Then the destination node will send a reservation-signaling packet (*Resv*) towards the source to reserve the selected wavelength. During the configuration process, it's possible that the selected wavelengths, which were available during the probing course, are now being occupied by some other connections. In this situation a releasing-signaling (*Rls*) packet will be sent towards the destination node to release them. And a *NACK* packet is also sent out to notify the source node that this call has been blocked. If the *Resv* packet arrives at the source node successfully, data transfer can be started. Previous studies [2-4] indicated that the BRP schemes excel the FRP schemes in network blocking probability. This is due to the over-reservation behavior of FRP schemes.

The policy of wavelength selection is an important parameter to the blocking probability performance of distributed wavelength reservation protocols. There are two proposed wavelength selection policies named First-Fit (FF) and Random (Rd) respectively. The First-Fit policy searches wavelengths according to a predefined rule and the first available wavelength will be selected. Whereas Random policy is to select randomly a wavelength among all available wavelengths.

III. Performance Analysis of BRPs under Multi-fiber situation

As far as a practical network is concerned, there is often more than one fiber installed between two adjacent nodes. It's a cost-effective way to install bundles of fibers for future network growth and fault tolerance. In the literature [5,6], however, performance analyses of distributed wavelength reservation protocols are dedicated to single-fiber networks. In this paper, we develop an analytical model for evaluating the performance of BRPs in multi-fiber network. Our model can accommodate both First-Fit and Random policy.

The following notations and assumptions are used in our analytical model: (1) The number of fibers per hop, F, and the number of wavelengths per fiber, W, is the same on all links. The words "hop" and "link" will both stand for all fibers between two neighbor nodes; (2) Poisson arrival with rate λ of connection requests; Call holding time is exponentially distributed with mean $1/\mu$. (3) Uniform traffic distribution across all source-destination pairs. (4) The fixed route for every node pair ***a***, is chosen according to the shortest path. Denote the set of all routes by ***R***; (5) The propagation and processing delay τ on each hop is assumed to be identical. (6) Link loads and wavelength loads are statistically independent. We further suppose that the arrival of *Resv* packet at a certain wavelength ***w*** on given link ***l*** is a Poisson process, whose arrival rate is notated as λ_w^l. The holding time of each reservation is assumed to be generalized distribution with mean of μ_w^l. Then each wavelength on certain link can be modeled as an M/G/C/C queueing system. The steady state probability that *m* channels are used on wavelength ***w*** of link ***l*** is given by:

$$P_w^l(m) = \frac{(\lambda_w^l/\mu_w^l)^m/m!}{\sum_{i=0}^{F}\left[(\lambda_w^l/\mu_w^l)^i/i!\right]} \tag{1}$$

λ_w^l and μ_w^l can be further determined according to the following steps.

The probability that wavelength ***w*** is available on route ***a*** can be represented as:

$$A_w^a = \prod_{i=1}^{h_a}\left(1 - P_w^{a_i}(F)\right)$$

where $\boldsymbol{h_a}$ is the hop count of route ***a***, and $\boldsymbol{a_i}$ denote the link which is the ***i***-th hop of route ***a***.

Obviously with Random wavelength selection policy, we have:

$$A_1^a = A_2^a = \cdots = A_W^a = A^a$$

Then the probability that a given wavelength ***w*** is selected for the backward reservation process of route ***a*** can be calculated as:

with First-Fit policy,

$$S_w^a = A_w^a \times \prod_{i=1}^{w-1}\left(1 - A_i^a\right)$$

with Random policy,

$$S_w^a = A^a \times \sum_{i=0}^{F-1}\left[\frac{C_{F-1}^i}{i+1} \times \left(A^a\right)^i \times \left(1 - A^a\right)^{F-i-1}\right]$$

During the reservation process, it's possible that the selected wavelengths, which were available on the path during the probing course, are now being occupied by some other connections. To determine the blocking behavior during the backward reservation process, we define ***H***:

$$H_{h_j}^{a,w} = e^{Q_w^l \times 2 \times (h_a - h_j + 1) \times \tau}$$

where Q_w^l is the rate transition matrix:

$$Q_w^l = \begin{bmatrix} -\lambda_w^l & \lambda_w^l & 0 & \cdots & 0 \\ \mu_w^l & -(\lambda_w^l + \mu_w^l) & \lambda_w^l & \ddots & \vdots \\ 0 & 2\mu_w^l & -(\lambda_w^l + 2\times\mu_w^l) & \ddots & 0 \\ \vdots & \ddots & \ddots & \ddots & \lambda_w^l \\ 0 & \cdots & 0 & F\times\mu_w^l & -F\times\mu_w^l \end{bmatrix}$$

Then the probability that a certain wavelength ***w*** is held still free at the h_j-th hop of route a during the backward reservation process is approximated as:

$$K_{h_j}^{a,w} = 1 - \sum_{i=0}^{F-1} \left[P_w^{a_{h_j}} \times H_{h_j}^{a,w}(i+1, F+1) \right]$$

Therefore, the arrival rate of *Resv* signaling packets for wavelength w on the h_j-th hop of route a can be determined as:

$$\Lambda_w^{a_{h_j}} = \lambda \times S_w^a \times \left[1 - P_w^{a_{h_j}}(F)\right]^{-1} \times \prod_{i=h_j}^{h_a} K_i^{a,w}$$

These connection requests whose *Resv* packets arrive at a wavelength can be divided into two categories. One category includes these requests that will be finally setup successfully. The other category includes these requests that will be blocked on some upstream links. The arrival rate of each category can be represented as:

$$\Lambda 1_w^{a_{h_j}} = \lambda \times S_w^a \times \left[1 - P_w^{a_{h_j}}(F)\right]^{-1} \times \prod_{i=h_1}^{h_a} K_i^{a,w}$$

$$\Lambda 2_w^{a_{h_j}} = \Lambda_w^{a_{h_j}} - \Lambda 1_w^{a_{h_j}}$$

And the mean holding time of each category is given by:

$$T1_w^{a_{h_j}} = 2h_j \times \tau + \frac{1}{\mu}$$

$$T2_w^{a_{h_j}} = 2\tau \times \sum_{i=1}^{h_j - 1} \left[i \times \left(\prod_{m=1}^{i-1} K_{h_j - m}^{a,w}\right) \times (1 - K_{h_j - i}^{a,w}) \right]$$

Then the arrival rate and mean service time of each link l in network can be calculated as:

$$\lambda_w^l = \sum_{a_i = l, a \in R} \Lambda_w^{a_i} \tag{2}$$

$$\frac{1}{\mu_w^l} = \frac{\sum_{a_i = l, a \in R} (\Lambda 1_w^{a_i} \times T1_w^{a_i} + \Lambda 2_w^{a_i} \times T2_w^{a_i})}{\sum_{a_i = l, a \in R} \Lambda_w^{a_i}} \tag{3}$$

The above equations leads to an iterative algorithm to calculate the steady state distribution of each link:

1) Choose initial $P_w^l(m)$, λ_w^l and μ_w^l arbitrarily.

2) Calculate new λ_w^l and μ_w^l by (2) and (3).

3) Calculate new $P_w^l(m)$ by (1).

4) Repeat from Step2 until each $P_w^l(m)$ has converged.

Once we get the steady $P_w^l(m)$, λ_w^l and μ_w^l, we can get the blocking probability for each route a:

$$B^a = 1 - \sum_{w=1}^{W} \left(S_w^a \times \prod_{i=h_1}^{h_a} K_i^{a,w} \right)$$

and the blocking probability of the whole network is the mean value of B^a for all $a \in R$.

IV. NUMERICAL RESULTS

In this section, we evaluate the accuracy of our analytical models by comparing the analytical results with simulation results. Here we present our results associated with NSFNET network as shown in Fig. 2, which has 14 nodes and 21 bi-directional links. The number of wavelengths on each fiber is set to 8.

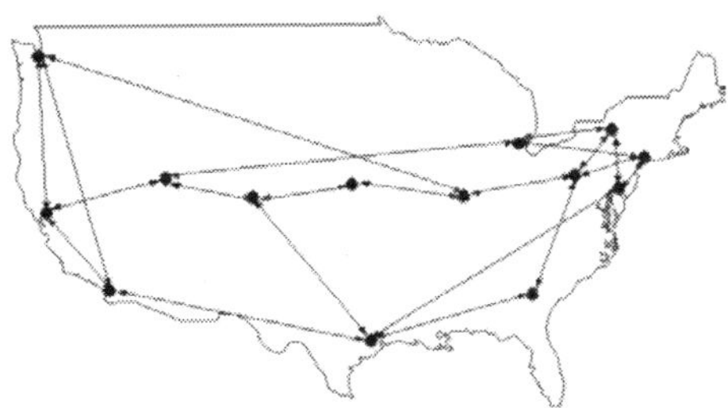

Fig. 2. NSFNET topology, 14 nodes and 21 links

We set $\mu = 1$, $\tau = 0.01$ and change the number of fibers each time. The results in Fig.3 show how the blocking probability performance changes with the offered load per source-destination pair in different scenarios. It shows that our analytical results are highly accurate.

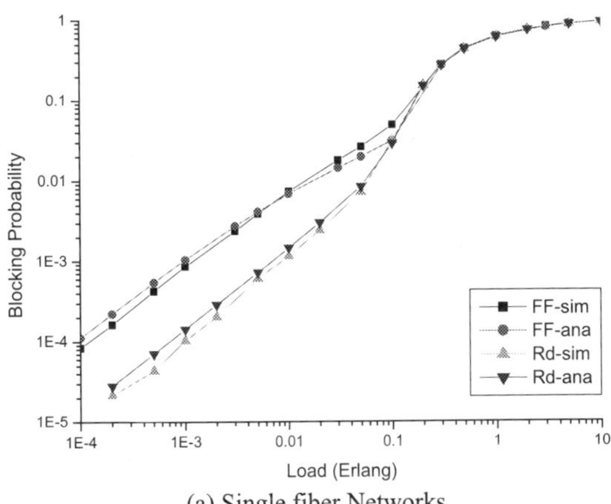

(a) Single fiber Networks

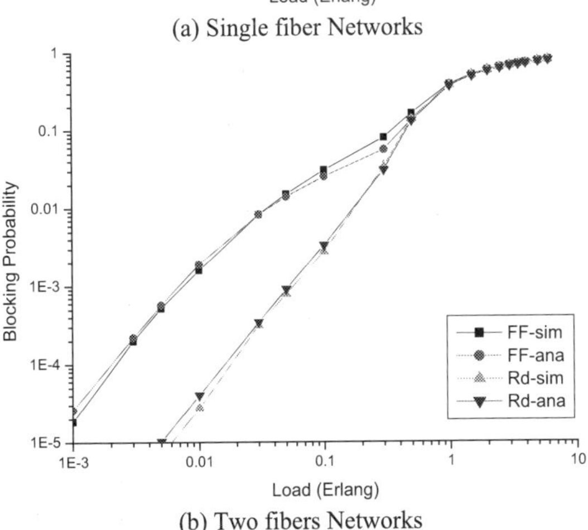

(b) Two fibers Networks

Fig. 3. Blocking Probability Performance of Distributed Wavelength Reservation Protocols

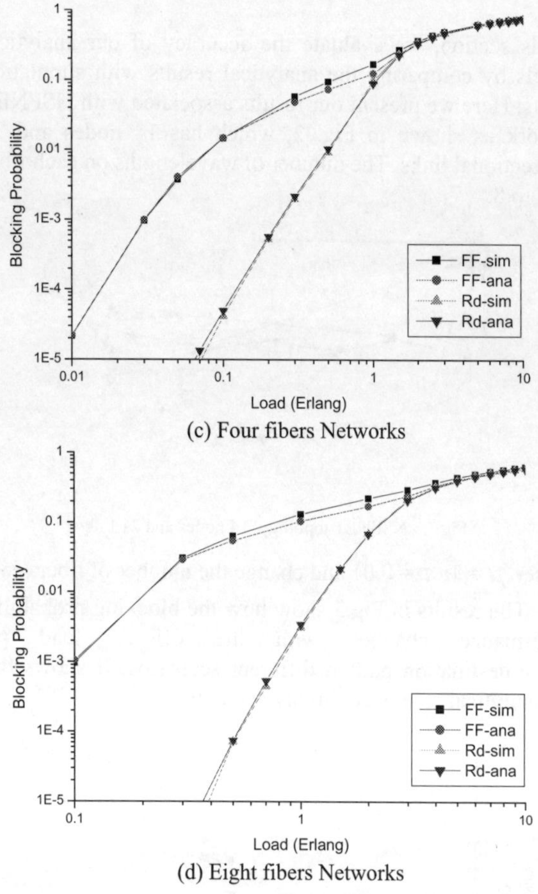

(c) Four fibers Networks

(d) Eight fibers Networks

Fig. 3 (*Continued*)

Fig.3 shows that the multi-fiber network can greatly improve the blocking performance. For example, with a given blocking probability such as 1e-4, the traffic load that a two-fiber network can accommodate is about 20 times larger than a single-fiber network.

The influence of wavelength selection policies on blocking probability is very notable. Random policy always performs much better than First-Fit policy in any network scenario. To

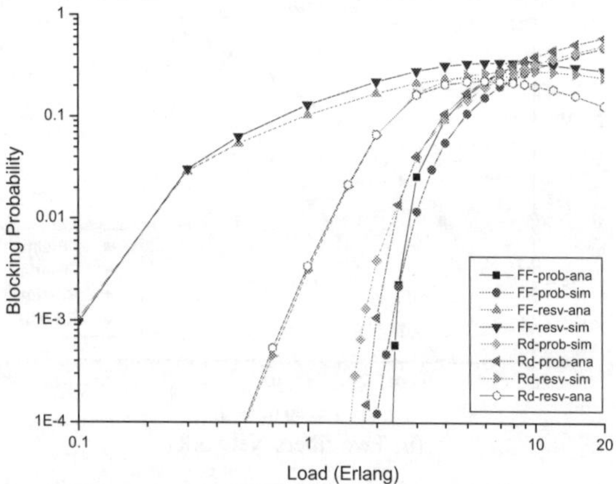

Fig.4. Blocking Probability Categories (F=8)

make this result clear, we classify the blocked calls into two categories according to the occurring time of blocking. One group includes the calls blocked during the forward probing course and the other includes the calls blocked during the backward reservation period. We can get the blocking probability for each group as following:

$$B^a_{prob} = \prod_{w=1}^{W}\left(1 - A^a_w\right),\ B^a_{resv} = B^a - B^a_{prob}$$

In Fig.4 we compare the blocking probabilities of these two groups for First-Fit and Random policy. It reveals that the blocking probability of forward probing group for both protocols is negligible in low load region. The difference of blocking performance between these policies is mostly due to the contribution of backward reservation group, especially in the low load region. The backward blockings result from the situation that the selected wavelength, which was available in the probing course, is now consumed by other simultaneous connections. First-Fit policy creates a higher likelihood that two simultaneous connection requests will choose the same wavelength, then it tends to increase the probability of resource contentions between simultaneous connection requests during the backward reservation period.

V. CONCLUSION

In this paper, we develop an analytical model to analyze the blocking performance of distributed wavelength reservation protocols. Our model can be used for both single- and multi-fiber networks and can accommodate both First-Fit and Random wavelength selection policy. Numerical results show that our model gives high accuracy. It's shown that multi-fiber network can greatly improve the blocking performance and the Random policy outperforms the First-Fit policy.

REFERENCES

[1]H. Zang, J.P. Jue, L. Sahasrabuddhe, R. Ramamurthy, B. Mukherjee, "Dynamic Lightpath Establishment in Wavelength Routed WDM Networks", IEEE Communications Magazine, vol. 39, no. 9, September 2001, pp. 100-108

[2]X. Yuan, R. Melhem and R. Gupta, "Distributed Path Reservation for Multiplexed All-Optical Networks", IEEE Transactions on Computers, Volume 48, No. 12, pages 1355-1363, Dec. 1999.

[3]Y. Mei and C. Qiao, "Efficient distributed control protocols for WDM all-optical networks," in Proceedings of Sixth ICCCN, pp.150-153.

[4]D. Saha," A Comparative Study of Distributed Protocols for Wavelength Reservation in WDM optical networks", Optical Networks Magazine, vol. 3, no. 1, pp. 45-52, Jan 2002

[5]S. Arakawa, K. Miyamoto, M. Murata and H. Miyahara, "Performance analyses of wavelength reservation methods for high-speed data transfer in photonic networks," in Proceedings of ITC-CSCC, (Sado, July 1999), pp.828-831.

[6]J.P. Jue and G. Xiao, "Analysis of Blocking Probability for Connection Management Schemes in Optical Networks," Proceedings, IEEE Globecom 2001, San Antonio, TX, November 2001.

[7]Y.W. Leung, G. Xiao, and K.W. Hung, "Design of Node Configuration for All-Optical Multi-Fiber Networks", IEEE Transactions on Communications, vol. 50, no. 1, pp. 135-145, January 2002

Design and Implementation of a Reliable and Integrated GMPLS-based Control Plane

Victor Foo, Prashant Agrawal, Luying Zhou and Jit Biswas
Laboratories for Information Technology
21 Heng Mui Keng Terrace, Singapore 119613
{sffoo, prashant, lzhou, biswas}@lit.a-star.edu.sg

Abstract: This paper describes the subtle design and implementation issues of a reliable and integrated GMPLS-based control plane. Although there are currently standard-track drafts that specify the protocols for controlling IP/optical network, there is limited work taking into account the applicability of these protocols to seamlessly and reliably control such a network. This paper is to study and provide a seamless and robust architecture for the different protocols to inter-operate, and to provide a tightly coupled, integrated and reliable control in IP/optical networks.

Key Words: IP/optical network, GMPLS-based control plane, OXCs, Resident module

1. INTRODUCTION

Future data and transmission networks will consist of network elements such as routers, DWDM systems, optical add-drop multiplexers (OADMs) and optical cross-connects (OXCs) to support the growing Internet IP traffic. Currently, there is tremendous interest within the research community in optical networks of OXCs and to integrate OXC backbones within the overall Internet infrastructure using an IP-centric control plane. The Internet Engineering Task Force (IETF) has proposed the use of Generalized Multi-Protocol Label Switching (GMPLS) [1] to support the optical transport networks. The Optical Domain Service Interconnection (ODSI) initiative and the Optical Internetworking Forum (OIF) have also provided standards to interface IP and optical domain control planes as well as those between the OXCs. International Telecommunications Union (ITU) also has an Optical Transport Network (OTN) project which deals with the studies and recommendations on the optical network. The use of an IP-centric control plane can also be applied to traditional IP networks. There are recent efforts [2] by the IETF to define a framework for standardizing the exchange of information between the logically separate functionality of the control plane and the forwarding plane.

The focus of this paper, however, will be mainly on the design and implementation of a reliable and integrated GMPLS-based control plane for IP/optical networks instead of the conceptual and functional requirement works by IETF. Similar research works [3,4] have been done in the past. However, our design and implementation of the GMPLS-based control plane aims to be as generic as possible to support all transport networks which consist of O-O, E-O, O-E and E-E conversion devices such as routers and OXCs because GMPLS control and forwarding paradigms are independent of the underlying data transmission. It will support effectively different protocols, such as signaling protocol RSVP-TE [5] or CR-LDP [6] extensions, routing protocol OSPF [7] or IS-IS [8] extensions, link management protocol LMP [9] and SNMP-based NMS supporting the defined MIBS [10]. The design will also support concepts of unnumbered links, link bundling, TE links, LSPs hierarchy and forwarding adjacency for scalability. Also, it must be extensible to support variations of protection and restoration methods, routing and wavelength assignment (RWA) methods and QOS mechanisms, and addresses the network management design issues to provide a reliable control plane using LMP or SNMP-based network management system (NMS).

The paper is organized as follows. In Section 2, we give an overview of our control plane architecture that consists of five main modules – resident module, signaling module, routing module, LMP module and NMS module. In Section 3, we go into detail the design and implementation of resident module. We then conclude with some of our proposed works for further research in Section 4.

2. CONTROL PLANE MODULES

Figure 1 shows an overview of the software modules for the control plane within a node. It consists of a Resident Module, Signaling Module (RSVP-TE/CR-LDP extensions), Routing Module (OSPF-TE/IS-IS extensions), Link Management Protocol (LMP) module and Network Management System (NMS) module. In actual implementation, all modules except the NMS module are daemons in the user space written in C/C++ or Java. The NMS module consists of an agent module in the user space and a separate SNMP-based manager. The implementation concerns for each of the modules are briefly described here. For Resident Module, we will discuss it in more detail in next Section.
The independent software module within the node communicates with each other through the Resident Module

This work was partially supported by A-Star under ONFIG GMPLS Project

by using sockets. For inter-node communications, the module in each node communicates with its respective module in another node by IP messages as stated in the IETF respective drafts. We will not go into the detail of that and the readers are encouraged to read the drafts.

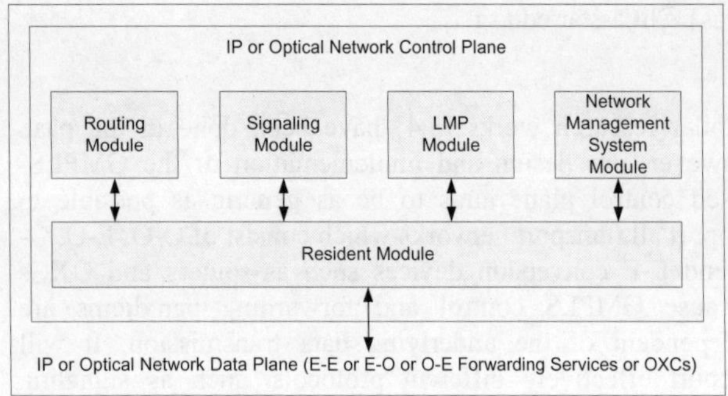

Figure 1 Control Plane Modules within a Node

A. *Routing Module*

The public version of GateD daemon software which supports both OSPF and IS-IS routing protocols was extended to advertise the additional information needed for the computation of contraint-based routes for optical network. However, the semantics of such information, as well as details of path selection are not specified though Dijkstra Shortest Path First algorithm is commonly adapted to path computation [11,12]. In our implementation, we use our design algorithm for integrated routing over the IP/optical networks for better network usage. The basic principle of the algorithm is to consider the protection and restoration issues in the IP/optical network when selecting the routes. Although reference [5] proposed to classify the link state information into quasi-static and dynamic information so that only dynamic information will be advertise frequently in order to reduce the large information overheads being advertised, we follow the routing drafts for our implementation as we found that the overhead is not substantial.

B. *Signaling Module*

For fast implementation, the public version of LDP source code in sourgeforge.net and RSVP source code by Fujitsu Labs are extended to provide explicit routing in an GMPLS-enabled network. The signaling protocol extensions includes the mechanisms to support the new classes of interfaces and switching, set up bi-directional LSP, support the concept of new suggested labels and label sets, and provide the specific procedure to support LSP hierarchy and link bundling. All these requires additional processing of the requests for LSP formation and additional functionality for label mapping.

We also pay a lot of extra attention on the communication between the signaling module and resident module as the signaling protocol is heavily involved in the connection management. The suggested labels and labels set are provided by the resident module which has information of the data forwarding device.

C. *LMP Module*

LMP module is used to perform four basic operations – control channel management, link connectivity verification (optional), link property correlation and fault management. For our design and implementation, we have adopted the approach that the control channels exist independently of data-bearing links (out-of-band). The resident module provides a list of all data links (ports or component links) that comprise the TE links and various link properties to the LMP module for link property correlation procedure. LMP module will confirm that two adjacent LMP nodes have consistent interface ID mappings and link properties after combining the results of automatic link rectification and the configuration of each node. For link connectivity verification procedure which is optional, the free data-bearing links must be opaque (i.e., able to be terminated). Thus, it requires the OXC to terminate the data link. We proposed an end-to-end way of verifying the link connectivity. An ingress node in an IP network can send a test message to another IP network egress node via the links in the optical network. If the message is transmitted correctly, this means that all the links within the route are functional. If not, this means that there is a failure in some links and the resident module is assumed to be able to provide these failure information.

For fault management procedure, the mechanism used for fault notification relies on the fact that a failure is detected by detecting loss of light (LOL). If one or more data links fail between two nodes, the resident module will notify LMP module so that appropriate localization of fault procedure by LMP module will be activated.

D. *Network Management System (NMS) Module*

In order to provide reliable control of the IP/optical network, we have adopted a dual mode control for the IP/optical network – GMPLS-based control using resident module or SNMP-based control using a SNMP manager. GMPLS-based control provides distributed control of the nodes whereas SNMP-based control provides centralized control of the nodes. Both types of control are automated but at one time, only one mode of control is activated. By default, the control mode is GMPLS-based. If the resident module fails, the SNMP manager will take over control from it. Under the worst case, if both mode of control fails, using the innovative design of our resident module which will be described in the next section, we provide a Data Plane Control sub-Module client to manual control the node using command line interface (CLI).

The NMS module consists of an agent which will maintain the Managed Objects (MO) and get/set the Physical Objects of resident module and its constituent modules, and a manager. The NMS agent module can interface with other

modules directly. The other modules save the data relevant to NMS agent module in a group of files. It is the function of NMS agent to read the files and retrieve the data it requires. In cases of link failures, module failures or other events, resident module informs NMS agent/manager module by sending an appropriate message.
In order to retrieve the data which is frequently updated and for which the resident module does not provide any implicit support, the NMS agent module provides its own back-end processing functions to collect and maintain the data in each of the node. Such operations, however, will not interfere with the existing GMPLS mechanisms

3. RESIDENT MODULE

The Resident Module consists of a Main Control sub-Module (MCM) and Data Plane Control sub-Module (DPCM). Its main function is to initialize the control and data plane and listen to messages from other modules in the control and data plane and co-ordinate their activities by taking appropriate actions. It provides protection and restoration support, TE mechanisms to support QOS and RWA. It also manages the information shared by two or more modules from the control plane and the supported data plane information. In particular, MCM maintains the generic device independent information by obtaining node specific information from DPCM to fill the node properties table. The DPCM maintains device dependent state and interface information of the device used for the data plane as well as all other related device dependent information of the node and provides a library of generic function calls for MCM to access the information maintained by it.

A. Main Control Sub-module (MCM)

The MCM consists of a web-based configurator sub-module, message handler sub-module, data handler sub-module and resource handler sub-module. The web-based configurator sub-module provides a web interface for the user to initialize and set manual configurations such as the TE links and OXC parameters in the node. The message handler sub-module handles the communication between the different control plane modules such as that between the signaling and routing module, LMP and resident module and etc.
The data handler sub-module will manage the information shared by two or more modules from the control plane and the supported data plane information. In particular, it maintains the generic device independent information by obtaining node specific information from DPCM to fill the node properties table. Two important types of table maintained by the data handler sub-module are the interface tables and TE-link tables. To support unnumbered link, we use a 32-bits identifier to identify the link. For link bundling, there is an interface stack table within the interface tables which maintains the link bundling information. The information in the TE-link tables will support the LSP hierarchy, forwarding adjacency and TE mechanisms.
The resource handler module provides protection and restoration support, TE mechanisms to support QOS and routing and wavelength assignment (RWA). When there is any detection of data plane failure, the resource handler module will be invoked and the respective protection and restoration algorithm is called. When an ingress node intends to set an explicit route using signaling, the resource handler module will look into the QOS traffic class and TE links database information and invoke the respective integrated routing and wavelength assignment method.

B. Data Plane Control sub-Module (DPCM)

DPCM contains a data plane interface sub-module, a data handler sub-module and provides a library of generic function calls for MCM to access the device dependent information maintained by it as well as controlling the data forwarding devices.

The data plane interface sub-module consists of a wrapper function to control Electrical-Electrical (E-E), Electrical-Optical (E-O) and Optical-Electrical (O-E) and Optical-Optical (O-O) conversion data forwarding devices. For E-E, E-O and O-E conversion data forwarding devices such as a router, we adopt an IP forwarding service model based on a separate control and data plane using netlink layer in Linux as shown in Figure 2. A library of functions using netlink layer to control the router will be wrapped inside the data plane interface sub-module.

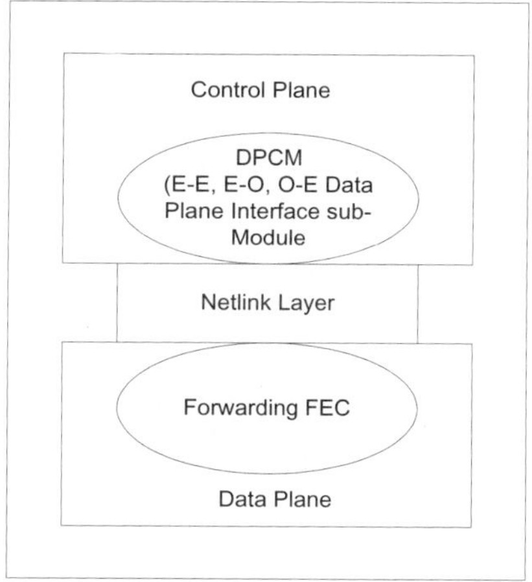

Figure 2 Data Plane Interface sub-Module (E-E, E-O, O-E)

For O-O conversion devices such as an OXC, we provide a device specific device module which is wrapped inside the data plane interface sub-module to control the OXC. The device module allows the data plane interface sub-module to initiate the pre-defined control actions on the OXC/OADM, establish and release connections across the OXC, manage OXC ports, request information and statistics, and to handle any exception message from the device such as a port going down. Hence, four classes of functions are defined – connection management, state and statistics management, configuration management and event management. The device module writes into any type of interface port such as RS-232 and USB port that are commonly used to control the OXC. The General Switch Management Protocol (GSMP) V3 [13] has provided plenty of insight for the message format and grouping functions though we implemented our lightweight version of device module control. An example of the device module wrapped in the data plane interface sub-module is shown in Figure 3.

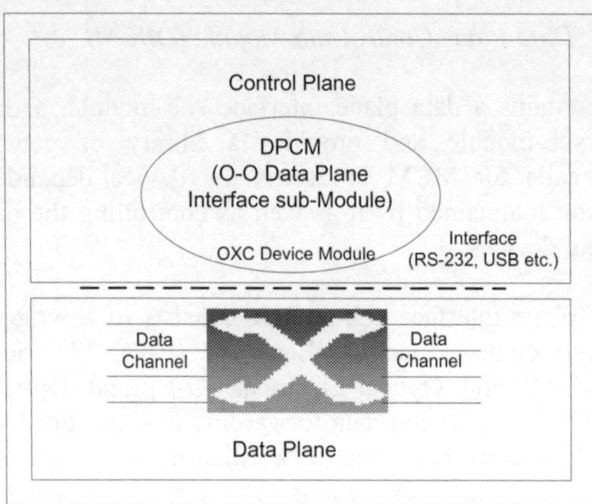

Figure 3 Data Plane Interface sub-Module (O-O)

The data handler sub-module maintains the device dependent information for the device and all the interfaces of the node. MCM can always call the functions provided by the DPCM to access the device information maintained by it as well as controlling the data forwarding devices.

C. *IP or Optical Network Data Plane (E-E, E-O, O-E forwarding services or OXC)*

The E-E, E-O, O-E forwarding services are provided by the forwarding module which resides inside the kernel of the operating system as shown in Figure 2. It provides forwarding and mapping functionality which is very similar to MPLS with additions of the notion of generalized label and FECs other than destination prefix. The public version of MPLS source code in sourgeforge.net was extended to support the above functionality. The OXC provides the forwarding services by means of the physical switching fabric.

4. CONCLUSION

We have discussed the subtle design and implementation issues of a robust and integrated control plane for IP/optical networks. The software architecture which consists of a Resident Module, Routing Module, Signaling Module, LMP module and NMS module, is generic and extensible enough to apply not only for network of OXCs, but also for the packet-based routers in a tightly coupled, integrated and reliable manner. It also supports the new concepts of link bundling, TE links, suggested labels, label sets etc. For future research, we will be working on the traffic engineering and protection and restoration issues by using the developed GMPLS control plane.

REFERENCES

[1] E. Mannie et al., "Generalized Multi-Protocol Label Switching Architecture", Internet Draft, draft-ietf-ccamp-gmpls-architecture-02.txt, September 2002, work in progress.

[2] T. Anderson et al., "Requirements for Separation of IP Control and Forwarding", Internet Draft, draft-ietf-forces-requirements-03.txt, October 2002, work in progress.

[3] G. Li, J. Yates, D. Wang and K. Charles, "Control Plane Design for Reliable Optical Networks", *IEEE Communications Magazine*, February 2002, pp. 90-96.

[4] C. Xin et al., "On an IP-Centric Optical Control Plane", *IEEE Communications Magazine*, September 2001, pp. 88-93.

[5] L. Berger et al., "Generalized MPLS Signaling - RSVP-TE Extensions", Internet Draft, draft-ietf-mpls-generalized-rsvp-te-07.txt, April 2002, work in progress.

[6] L. Berger et al., "Generalized MPLS Signaling - CR-LDP Extensions", Internet Draft, draft-ietf-mpls-generalized-cr-ldp-06.txt, April 2002, work in progress.

[7] K. Kompella et al., "OSPF Extensions in Support of Generalized MPLS", Internet Draft, draft-ietf-ccamp-ospf-gmpls-extensions-07.txt, November 2002, work in progress.

[8] K. Kompella et al., "IS-IS Extensions in Support of Generalized MPLS", Internet Draft, draft-ietf-isis-gmpls-extensions-12.txt, November 2002, work in progress.

[9] J.P. Lang et al., "Link Management Protocol (LMP)", Internet Draft, draft-ietf-ccamp-lmp-03.txt, March 2002, work in progress.

[10]D. Martin et al., "Link Management Protocol Management Information Base" Internet Draft, draft-ietf-ccamp-lmp-mib-02.txt, November 2002, work in progress.

[11] R. Ramaswami, K. Sivarajan, *Optical Networks: A Practical Perspective*, Morgan Kaufmann Publishers, 1998.

[12]Y. Ye, C. Assi, S. Dixit, M. Ali, "A simple Dynamic Integrated Provisioning/Protection Scheme in IP over WDM networks", IEEE *Commun. Mag.,Vol 39, No. 11,* November 2001, pp174-181

[13]A. Doria et al., RFC 3292 "General Switch Management Protocol V3", IETF.

PERFORMANCE EVALUATION OF LINE CODES IN OPTICAL ETHERNET SYSTEMS

Dae Ub Kim, Sung Soo Kang, Hae Won Jung, and Hyeong Ho Lee
artkdu(sskang, hw-jung, holee)@etri.re.kr
Network Technology Laboratory
Electronics and Telecommunications Research Institute (ETRI)
161 Gajeong-dong, Yuseong-gu, Daejeon, 305-350, KOREA
Tel. +82-42-860-1772/Fax. +82-42-860-5213

Abstract: This investigation evaluated the MB810 and the 8B/10B line codes used in optical Ethernet systems. Compared with the 8B/10B, the MB810 has the advantage of a sharp shape of the optical carrier frequency component and Nyquist bandwidth. These advantages make it possible to compensate for optical transmission impairments such as dispersion and distortion, and to improve the signal-detection capability at the optical receiver. We compared the performance of the MB810 and 8B/10B in an optical transmission system based on a application of 10Gigabit Ethernet. The experimental result showed that the power budget of the MB810 1.1dB better than that of the 8B/10B.

Key Words: Line code, 10Gigabit Ethernet, 8B/10B, MB810.

1. INTRODUCTION

The recent tremendous increase in data traffic requires a high capacity Wide Area Network (WAN) and 10Gb/s clock and data recovery (CDR) technology is already implemented. [7] The 10Gigabit Ethernet (10GbE) system is one of the major families of LAN technology with a high demand in the market. The 10GbE standardization project, namely the IEEE 802.3ae, will be completed soon and the line code is one of the important issues in 10GbE technologies.

Figure 1 shows a diagram of the 10GBASE-LX4 in the 10GbE system. The physical layer of the 10GBASE-LX4 is composed of a Physical Coding Sub-layer (PCS), Physical Medium Attachment (PMA), and Physical Medium Dependent (PMD). The PCS executes encoding or decoding of four parallel lanes to convey code-groups for transmission. The 10GbE standardization project proposes the 8B/10B as the PCS code. The PMA provides a medium-independent means that allows PCS to support the use of a range of serial-bit-oriented physical media. The PMD performs transmission or reception of a quad serial bit stream on the underlying medium and executes Wide Wavelength Division Multiplexing (WWDM). There are also interfaces such as 10 Gigabit Media Independent Interface (XGMII), 10 Gigabit Attachment Unit Interface (XAUI), and Medium Dependent Interface (MDI) among the sub-layers on the physical layer [1].

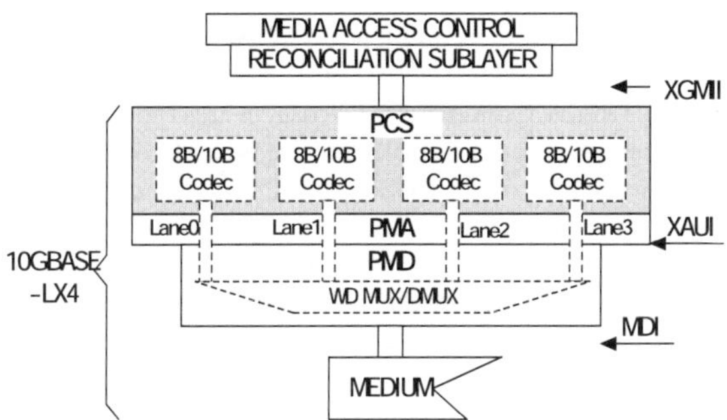

Figure 1. Diagram of 10GBASE-LX4 in the 10GbE System

To verify the advantages of MB810 in a real optical transmission channel, this paper evaluates two line codes, the 8B/10B and MB810, for 10GbE by comparing their Bit Error Rate (BER) performance. The organization of this paper is as follows: Section 2 summarizes the characteristics of two line codes and shows the difference between their power spectrum. The performance comparison between MB810 and 8B/10B is described in section 3. Finally, section 4 concludes and summarizes this paper.

This paper evaluates two line codes, 8B/10B and MB810, to be used in the 10GbE by way of comparing Bit Error Rate (BER) performance in order to verify advantages of MB810 in real optical transmission channel. The organization of this paper is as follows: Section 2 summarizes the characteristics of two line codes and shows the difference of optical power spectrum. The performance comparison between MB810 and 8B/10B is described in section 3. Finally, section 4 concludes and summarizes this paper.

2. Line Code: MB810 and 8B/10B

In transmission systems, line coding refers to the class of signal transformations designed to improve communications performance by enabling the transmitted signals to withstand better the effects of various channel impairments [11]. The required features of line codes are defined in [8]. In any practical digital optical fibre data link, line coding is important because the decision circuitry in the receiver must be able to extract precise timing information from the incoming optical signal. The three main purposes of timing are as follows: (1) Allow the signal to be sampled by the receiver at the time the signal-to-noise ratio is maximum, (2) Maintain the proper pulse spacing (3) Indicate the start and end of each timing interval. In addition, since errors resulting from channel noise and distortion mechanisms can occur in the signal-detection process, it may be desirable for the optical signal to have an inherent error detecting capability [12].

Two line codes are recommended in the IEEE 802.3ae. One is the 64B/66B code, which only performs scrambling in order to reduce redundancy, for use in the type 10GBASE-R/W. The other is the 8B/10B for the type 10GBASE-X. The usage of 8B/10B in the PCS is similar to that of the Gigabit Ethernet. However the PCS used in the 10GBASE-X divides the 10gigabit data for the purpose of WWDM transmission. The 8B/10B is a byte-oriented binary transmission code. It has the characteristics of dc-free and finite run-length but it requires a 25% redundant transmission capacity, which is excessive for transmission efficiency [9]. In spite of its extra transmission bandwidth, the 8B/10B is proposed for the 10GbE because it is suited for high-speed data links, where the information format consists of packets and it has been used in the Gigabit Ethernet.

Table 1. Comparison between MB810 and 8B/10B

	MB810	8B/10B
Input bits	8	8
Coded output bits	10	10
Maximum run-length	7	5
RAS (Running Alternate Sum)	2.5	Not limited
RDS (Running digital Sum)	3	3
ASV (Alternate Sum Variation)	5	Not limited
DSV (Digital Sum Variation)	6	6
Spectrum Bandwidth	1/(2T)	1/T (T: the pulse duration)

The key contribution of this paper is to present the properties of the MB810 that make it better than the 8B/10B in optical channels. The MB810 can be well compliant with the IEEE 802.3ae because it has the same structure, which translates an input of 8 bits into an output of 10 bits. The MB810 also has half of the resultant line bandwidth, which makes it similar to an optical duo-binary code [10] and it has almost all the advantages of the 8B/10B, such as dc-free and finite run-length. Table 1 gives the characteristics of two line-codes shown. The MB810 line code is dc-free and of minimum bandwidth because Alternate Sum Variation (ASV) and Digital Sum Variation (DSV) as system parameters of the code are finite [2~6].

Figure 2 shows the power spectra of the MB810 and 8B/10B at the rate of 3.125Gbps measured by an optical transmission simulation tool. A continuous wave (CW) laser and an optical external modulator are used the transmitter source in this simulation. We found that the bandwidth of the MB810 is half the bandwidth of the 8B/10B at a signal frequency component, that is, this component of the MB810 has first the value of null at nearby 8.9nm. The value of the spectral width is also achieved through equation (1) described below. The conversion of the spectral width in wavelength $\Delta\lambda$ to the bandwidth in frequency Δf (= 3.125/2GHz =1.5625GHz: half bandwidth) is given as follows [13]:

$$\Delta\lambda = \frac{\lambda}{f}\Delta f = \frac{\lambda^2}{v}\Delta f \quad (1)$$

where f is the centre frequency, λ the centre wavelength, and v the speed of the optical carrier.

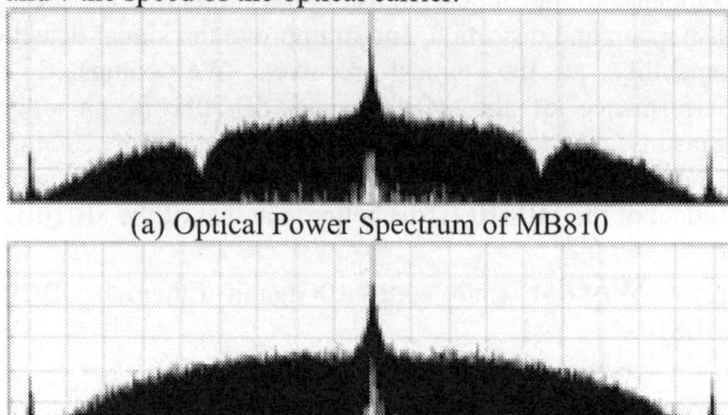

(a) Optical Power Spectrum of MB810

(b) Optical Power Spectrum of 8B/10B

Figure 2. Power Spectra of MB810 and 8B/10B

The optical carrier frequency component of the MB810 spectrum has a sharp slope in contrast with that of the 8B/10B. The peculiar properties of the MB810 make it possible for the system to be more profitable in optical transmission channels and optical receivers, since the receiver extracts timing information from the incoming optical signal suppressing various fibre channel impairments and improving the detection capacity in the signal-detection process.

3. Performance Evaluations and Discussion

The schematic diagram of the MB810 and 8B/10B coded transmission system is shown in Figure 3. The PCS of the 10GbE system splits the 10Gigabit data into 4 lanes of 2.5Gigabit data for the type 10GBASE-LX4. The 10Gbps MAC data stream is converted into four lanes at the XGMII. The byte stream of each lane is encoded by the XGMII Extender Sub-layer (XGXS) for transmission across the XAUI at a nominal rate of 3.125 Gbaud [1]. Therefore, the 10GbE system of type 10GBASE-LX4 transfers data-streams with a capacity of 3.125GHz per lane in an optical fibre.

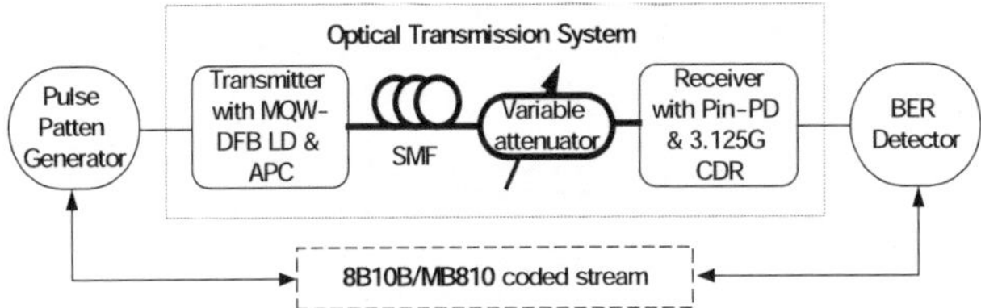

Figure 3. Experimental Set-up Diagram of MB810 or 8B/10B Coded Transmission System

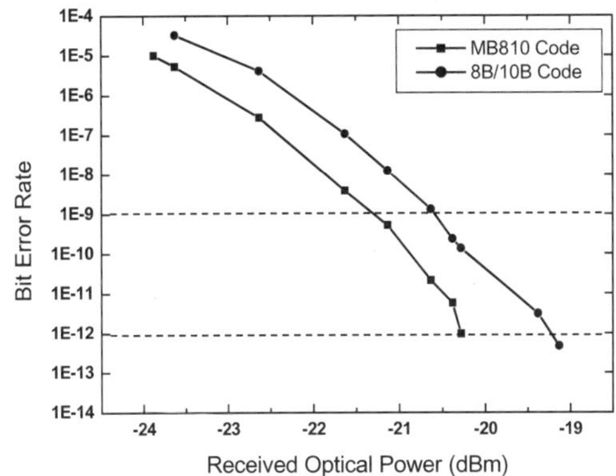

Figure 4. Bit Error Rate Performance for MB810 and 8B/10B over Standard Single Mode Fibre

To evaluate the performance of the MB810 and 8B/10B, we measure the BER by turns in the same optical transmission system. Two coded data-streams with a mark ratio of 0.5 and a sequence period of 100K are generated, and then the coded data-streams are set on the Pulse Patten Generator and BER Detector. The optical transmitter consists of an un-cooled Multiple Quantum-Well Distributed Feedback (MQW-DFB) Laser and automatic optical output power control (APC). The average output power of a laser diode is -1.45dBm and the centre wavelength is 1309.76nm. A standard Single Mode Fibre (SMF) with a length of 29.118Km and an attenuator are used as the transmission medium. The total insertion loss is about –16dB in the optical channel. The insertion loss includes the attenuation of the fibre cable, connection losses, and the inherent attenuator's loss of 4dB. In order to measure accurate performance between codes, a variable attenuator is used. The optical receiver is composed of a pin-photodiode and Clock & Data Recovery (CDR). The average sensitivity of photodiode is –17.9dBm in the NRZ coded system and its bit rate is 3.125Gbps.

Figure 4 shows the BER performance of the MB810 and 8B/10B over the standard SMF at a rate of 3.125Gbps. The receiver's sensitivity of the MB810 coded transmission system is -20.3dBm at a BER of 10-12. The receiver's sensitivity of the 8B/10B coded transmission system is -19.2dBm at a BER of 10-12. At a BER of 10-9, the sensitivity of the MB810 is –21.3dBm and that of the 8B/10B is –20.6dBm. We found that MB810 coded system performs better by 1.1dB per lane than the 8B/10B. This value corresponds to a transmission length of about 3Km. The results indicate that using the MB810 in all lanes gains about 4dB in the power budget in comparison with the 8B/10B in a 10GbE system of type 10GBASE-LX4.

4. Summaries and Conclusion

The 10GbE system has been suggested as an attractive, economical solution in service provider data centers and enterprise LANs. Ways to achieve cheaper systems are continually being studied.

This paper presented our analysis of the MB810 and 8B/10B in an optical transmission system. The MB810 spectrum had a sharp shape of the optical carrier frequency component as well as a Nyquist bandwidth of the signal frequency component in the optical power spectrum compared with the 8B/10B. The experimental result showed that the power budget of the MB810 code was 1.1dB better than that of the 8B/10B code per lane. We believe that using an MB810 coded Ethernet will produce cheaper optical transmission modules.

REFERENCES

[1] IEEE Draft P802.3ae/D5.0: "Media Access Control (MAC) Parameters, Physical Layer, and Management Parameters for 10 Gb/s Operation", Feb. 4, 2002.

[2] http://grouper.ieee.org/groups/802/3/10G_study/public

Proceedings Papers:

[3] Chan Goo Lee, DaeYoung Kim, Byungjun Ahn, Hae Won Jung, and Hyeong Ho Lee: "A Line code for high capacity optical transmission system", proc. of WTC/ISS2000, Birmingham UK, May 9-12, 2000.

[4] Chan Goo Lee, Hyeong Ho Lee, Dae Young Kim, Jung Whan Kim, and Hae Won Jung: "A New Line Code for 10-Gigabit Ethernet: MB810", proc. of ICC2000, vol. 3, pp.1774-1777, Jun. 2000.

[5] Chan Goo Lee, Dae Young Kim, Chun Sik Shin, Hae Won Jung, and Hyeong Ho Lee: "White paper on the MB810 Line Code for 10GbE", IEEE 802.3 HSSG Interim meeting, York, UK, Sept. 27-29, 1999.

[6] Chan Goo Lee, and Dae Young Kim: "Trend of 10Gigabit Ethernet Technologies", proc. of Korea Institute of Communication Sciences, vol. 16, no. 12, pp. 59-70, Dec. 1999.

Journal Papers:

[7] Jae Ho Songa, Tae Whan Yoo, Jeong Hoon Ko, Chang Soo Park, and Jae Keun Kim, Design and Characterization of a 10 Gb/s Clock and Data Recovery Circuit Implemented with Phase-Locked Loop, ETRI Journal, Vol. 21, No. 3, September 1999

[8] R. M. Brooks and A. Jessop, Line coding for optical fibre systems, International Journal of Electronics, vol. 55, pp. 81-120, 1983.

[9] A. X. Widmer and P. A. Franaszek, A dc-balanced, Partitioned-Block, 8B/10B Transmission code, IBM

Journal of Research and Development, vol. 27, no. 5, pp. 440-451, Sept. 1983.

[10] K. Yonegawa, and S. Kuwano, Dispersion-Tolerant Optical Transmission Sysetm Using Duobinary Transmitter and Binary Receiver, J. of Lightwave Tech., vol. 16, no. 8, pp.1530-1537, Aug. 1997.

Books:

[11] Bernard Sklar, *Digital Communications: Fundamentals and Applications*, Chapter 6, pp. 305-314, Prentice Hall, Second edition 2000.

[12] Gerd Keiser, *Optical Fiber Communications*, Chapter 8.2, pp. 335~339, McGraw-Hill, Third edition 2000.

[13] Jeseph C. Palais, *Fiber Optic Communications*, Chapter 3.2, pp. 54, Prentice Hall, Fourth edition 1998.

Session 12D

WDM Devices I

Superstructure fiber Bragg gratings (SSFBG) for DWDM systems

Shinji Yamashita[1,2]

[1]Department of Frontier Informatics, Graduate School of Frontier Sciences, The University of Tokyo
7-3-1, Hongo, Bunkyo-ku, Tokyo 113-8656, Japan
(Tel.): +81-3-5841-6659, (Fax): +81-3-5841-6025, e-mail:syama@ee.t.u-tokyo.ac.jp
[2]CREST,JST (Japan Science and Technology)

Abstract

Superstructure fiber Bragg gratings (SSFBG), in which the amplitude and phase in grating corrugation are controlled, can realize versatile functions for DWDM systems. We review our technique to fabricate densely-spaced SSFBG, multiple phase-shift (MPS) technique.

1. Introduction

Irradiation of UV light onto optical fiber from outside can induce the permanent change of the refractive index, which is called "photosensitivity." Fiber Bragg grating (FBG) is a fiber-type grating in which UV fringe forms a periodic refractive index change in the fiber core through the photosensitivity. FBG has a sharp wavelength-selective reflection spectrum, and is versatile for WDM optical fiber networks because of many advantages [1][2]. Superstructure fiber Bragg gratings (SSFBG), in which the amplitude and phase in grating corrugation are controlled, are especially attractive for dense WDM (DWDM) systems, because they can realize versatile functions, such as comb filter, interleaver, and dispersion compensators for WDM systems [3][4].

We recently proposed and demonstrated a novel multiple-phase-shift (MPS) technique, which can densify the channel spacing of SSFBGs without changing the FBG interval and the total length [5]. We also demonstrated that MPS technique is applicable to various kinds of complex SSFBGs, such as apodized, sinc-, and chirped SSFBGs [6]. Here we review the MPS technique.

2. Principle of MPS tecnbique

The comb-like filter response (Fig.1(1-a)) of SSFBGs having the total length, L_{total}, originates in multiple FBGs in their structures (Fig.1(1-b)). The channel spacing Δf of reflection spectrum is inversely proportional to the period of respective FBGs, L_g, expressed as

$$\Delta f = \frac{c}{2nL_g} \tag{1}$$

where c is the velocity of light and n is the effective refractive index. It can be viewed that the amplitude modulation (AM) with the period of L_g onto the grating corrugation generates many sidebands spaced by Δf. Therefore, to halve the channel spacing to $\Delta f/2$

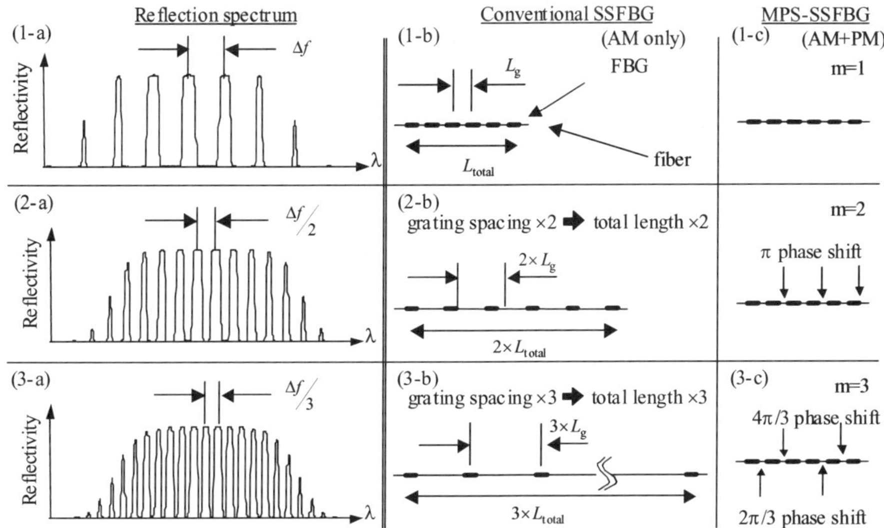

Fig.1: Comparison between conventional SSFBGs and MPS-SSFBGs.

(Fig.1(1-a)→ (2-a)), the FBG period have to be doubled to be $2\times L_g$ (Fig.1(1-b) → (2-b)). Thus, the total SSFBG length is doubled ($2\times L_{total}$) to keep the total SSFBG bandwidth. In the same way, the total length becomes $3\times L_{total}$ to realize $\Delta f/3$ (Fig.1(3-a,b)). The sections between each FBGs in the SSFBG are left unused, therefore the fiber and the phase mask are not used efficiently in the conventional SSFBGs.

By contrast, the MPS technique uses phase modulation (PM) along with AM onto the grating corrugation. For example, phase shift $0, \pi, 2\pi, 3\pi, \ldots$ (or $0, \pi, 0, \pi, \ldots$) between each FBGs enable to double the channel density without changing the FBG period. Therefore, the total length of MPS-SSFBGs can be kept constant (Fig.1(2-c)). In the same way, MPS-SSFBGs having triple channel density can be realized by giving phase shifts $0, 2\pi/3, 4\pi/3, 0, 2\pi/3, \ldots$ (Fig.1(3-c)). To reduce the channel spacing to $1/m$ with keeping the FBG period, the phase shifts (ϕ_k) between k-th and (k+1)-th FBG are,

$$\phi_k = \frac{2\pi}{m}(k-1) \qquad [0 \le k \le N_{\text{FBG}}] \qquad (2)$$

where N_{FBG} is the total number of FBGs. Note that phase shifts can be folded at 2π. Then, with keeping L_g, the channel spacing can be densified to $\Delta f/m$. It can be viewed that PM with the period of mL_g generates sidebands spaced by $\Delta f/m$. Note that m has to be smaller than N_{FBG}. Thus, MPS-SSFBGs can realize the dense channel spacing with short length and can use the phase mask and the fiber efficiently.

It is easy to calculate the reflection and dispersion characteristics of the MPS-SSFBGs by using the standard transfer-matrix method [4]. We found that reflection spectra are almost identical for strong SSFBGs having the reflectivity R~100%. For weak SSFBG, the reflectivity is reduced to $1/m$ times and the channel bandwidth is enhanced in the MPS-SSFBG compared with the conventional SSFBG having the same channel spacing. Dispersion characteristics are almost identical for both cases.

3. Experimental demonstration

We fabricated MPS-SSFBGs using a KrF excimer laser (λ= 248nm) as a UV light source. First, we fabricated a conventional (m=1) 30GHz-channel spaced SSFBG by scanning and irradiating the UV light onto

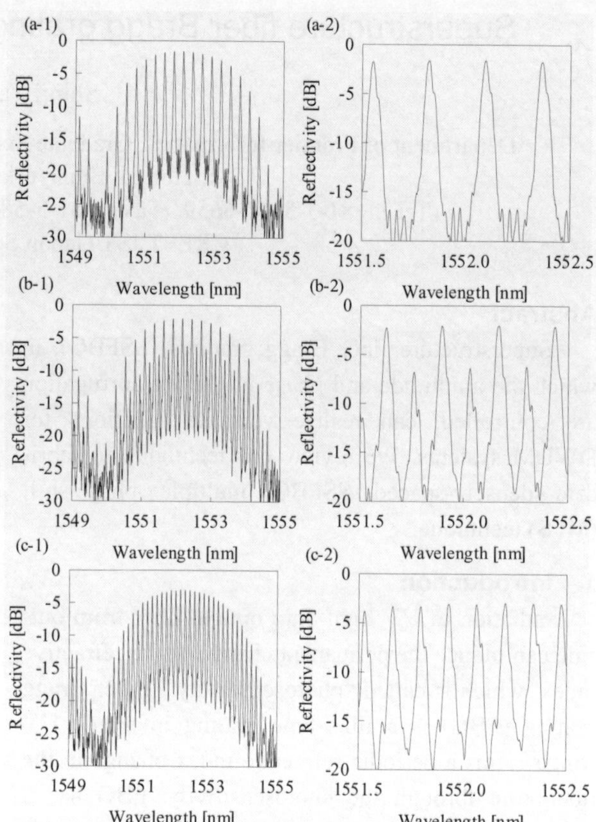

Fig.2: Reflection spectrum with m=2.
(a) SSFBG w/o MPS, (b) SSFBG with imperfect MPS, (c) SSFBG with perfect MPS.

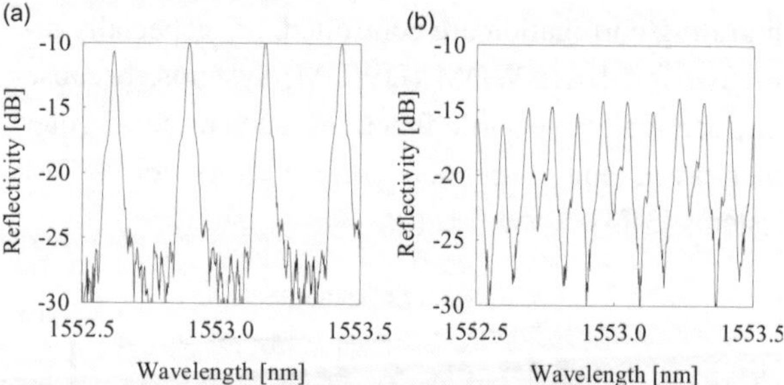

Fig.3: Reflection spectrum with m=3.
(a) SSFBG w/o MPS, (b) SSFBG with MPS.

the phase mask only at the FBG sections. The FBG length is 0.3mm, the superstructure period is 3mm and number of FBG is 15. The pitch of the phase mask is 1074nm. The reflection spectrum is shown in Figs.2(a-1,2). At this point, channel spacing is 30GHz. To halve to the channel spacing (m = 2), phase shifts ($\phi_{1,2,3,\ldots} = 0, \pi, 0, \pi, \ldots$) were introduced as a post-process by trimming the refractive index in the sections between FBGs with irradiation of uniform UV light. Magnitude of the phase shifts are adjusted by the number of irradiated UV pulses. Reflection

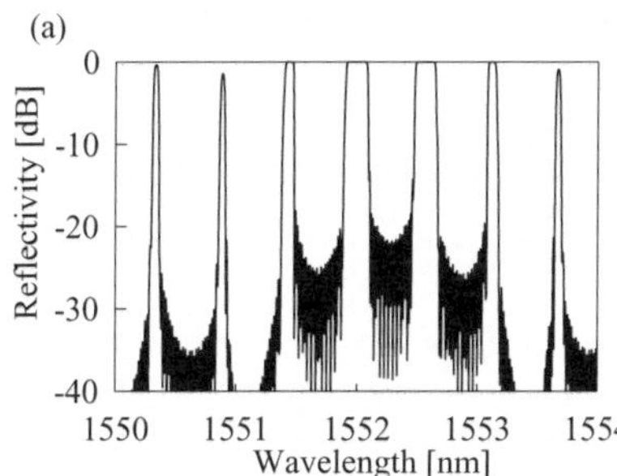

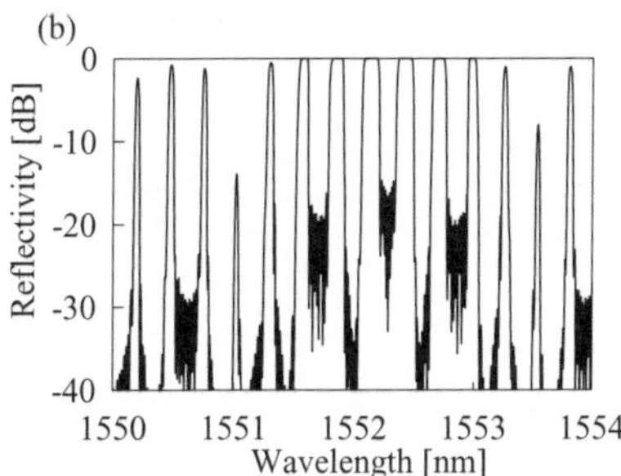

Fig.4: Calculated reflection spectrum of a gaussian apodized SSFBG. (a) w/o MPS, (b) with MPS (m = 2).

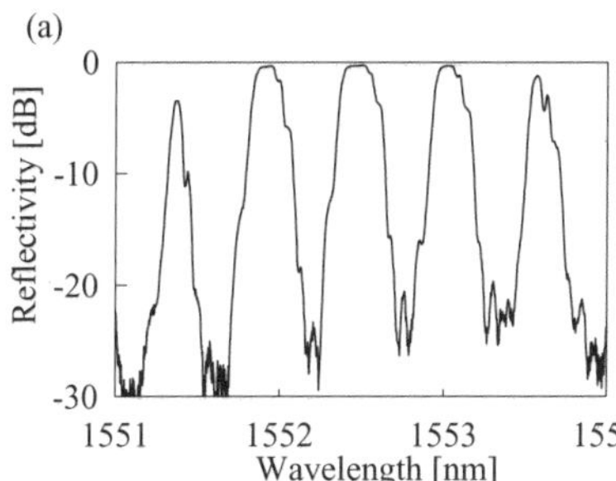

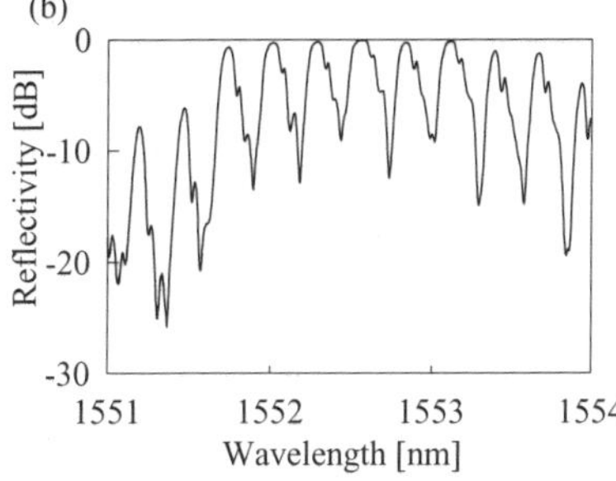

Fig.5: Experimental result of a gaussian apodized SSFBG. (a) w/o MPS, (b) with MPS (m = 2).

spectrum is monitored during the post-process, and the process is stopped when the spectrum reaches optimum. The reflection spectrum with imperfect phase shifts ($\phi_{1,2,3,4,...} \sim 0,\pi/2,0,\pi/2,...$) is shown in Figs.2(b-1,2). Growth of new channels is seen between original channels. Finally, we got the 15GHz-channel-spaced MPS-SSFBG as we designed (Figs.2(c-1,2)).

We also fabricated MPS-SSFBG with m = 3 by introducing phase shifts ($0, 2\pi/3, 4\pi/3, 0, 2\pi/3, ..$). Fig.3(b) shows the reflection spectrum whose channel spacing becomes 10GHz, one third of the conventional SSFBG.

4. Complex SSFBGs by MPS technique

Here we show and demonstrate that the MPS technique is applicable not only to simple sampled FBGs, but also to various kinds of complex SSFBGs, such as apodized, sinc-, and chirped SSFBGs.

Apodzation of SSFBGs can suppress the unwanted sidelobes. We confirmed that the MPS technique is applicable to apodized-SSFBGs both in a simulation (Fig.4) and an experiment (Fig.5). In the experiment, first we fabricated a conventional apodized-SSFBG (Fig.5(a)). Then we introduced phase shifts (m = 2: f = $0, \pi, 0, \pi, ...$) as a post-process. Irradiation of UV light was stopped when the reflection spectrum reached optimum. We could successfully halve the channel spacing, although the side-mode suppression becomes worse than the simulation (Fig3.(b)), which may be due to nonideal phase shifts.

Sinc-SSFBG has sinc-shaped modulation of the refractive index in order to suppress sidelobes and to equalize the each channel bandwidth. We also applied the MPS technique to sinc-SSFBGs. Fig.6(a) shows the reflection spectrum of 3-channel sinc-SSFBG having the total length of 100mm. When the phase shifts (m = 4) are introduced, the channel spacing is reduced to 1/4 in simulation (Fig.6(b)). To increase the number of channels in the conventional sinc-SSFBGs, more complex modulation in the refractive index is needed. In contrast, the MPS technique can increase the number of channels without complex modulation, at the cost of enhanced sidelobes.

Chirped-SSFBGs whose grating period changes along the fiber can be used for multichannel dispersion compensator [4]. We also applied the MPS technique to chirped-SSFBGs. Calculated reflection spectra and delay characteristics of chirped-SSFBG with and without MPS are shown in Fig.7. Chirp rate is assumed to be 0.01nm/cm, amplitude of refractive index modulation 10^{-4}, total length 200mm and FBG period 0.5mm. We found that the MPS technique can densify the channel spacing with keeping delay characteristics.

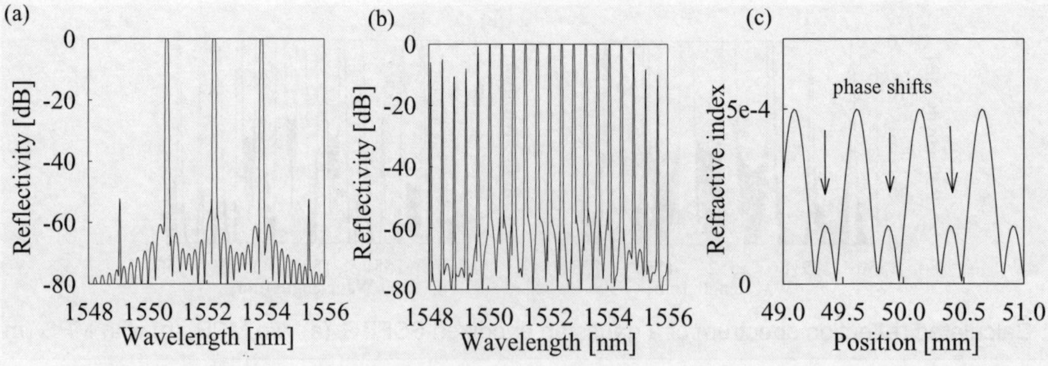

Fig.6: Sinc-SSFBG. (a) w/o MPS, (b) with MPS (m = 4), (c) Refractive index distribution.

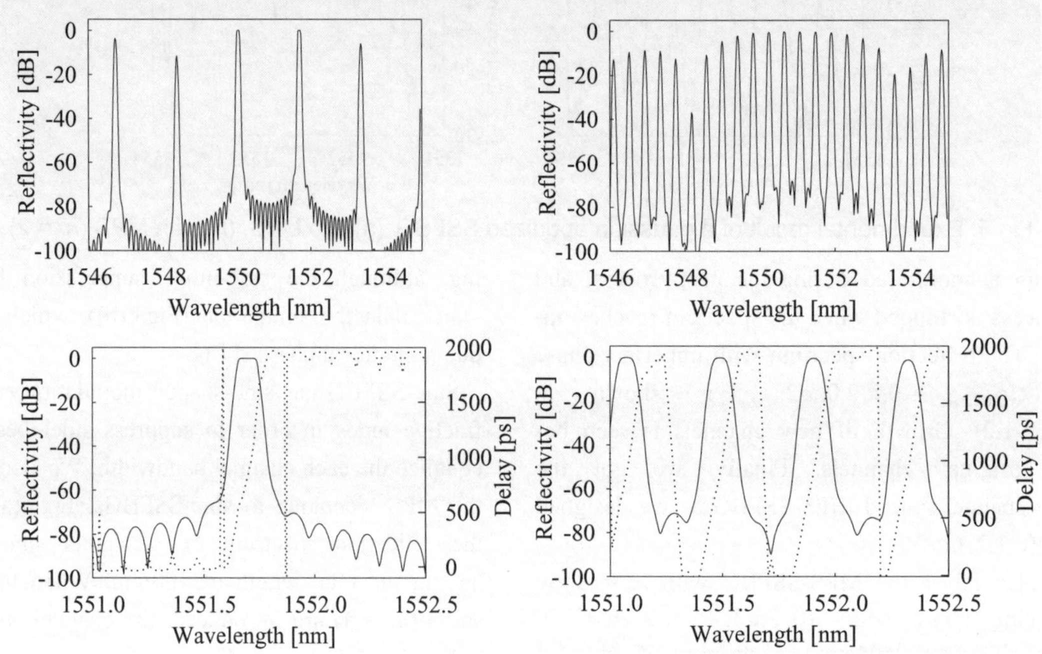

Fig.7: Chirped-SSFBG. (a) w/o MPS, (b) with MPS.

5. Conclusion

We reviewed our MPS technique to densify the channel spacing of SSFBGs, and demonstrated that it is applicable to various kinds of complex SSFBGs.

References

[1] R. Kashyap, *Fiber Bragg Gratings*, Academic Press, 1999.

[2] A. Othonos, K. Kalli, *Fiber Bragg Gratings*, Artech House, 1999.

[3] M. Ibsen, et al., "Sinc-sampled fiber Bragg gratings for identical multiple wavelength operation," *Photon. Tech. Lett.*, vol.10, no.6 , pp.842-844, 1998.

[4] W. H. Loh, et al., "Sampled Fiber Grating Based-Dispersion Slope Compensator," *Photon. Tech. Lett.*, vol. 11, no.10 , pp.1280-1282, 1999.

[5] Y. Nasu and S. Yamashita, "Multiple-phase-shift superstructure fiber Bragg grating for DWDM systems," *Electron. Lett.*, vol.37, no.24, pp.1471-1472, Nov. 2001.

[6] Y. Nasu and S. Yamashita, "Realization of various superstructure fiber Bragg gratings for DWDM systems using multiple-phase-shift technique," *OFC2002*, no.TuQ3, Mar. 2002.

Gain enhanced EDFA for WDM applications

K. Thyagarajan[1]
Department of Physics, Indian Institute of Technology, New Delhi 110016, India,
Fax: +91-11-686 5039, ktrajan@physics.iitd.ac.in

Jagneet Kaur Anand
Keshav Mahavidyalaya, University of Delhi 110035, India,
jagneetkaur@hotmail.com.

Abstract: We propose a new design of a gain enhanced EDFA for WDM applications by inserting a wavelength division multiplexing (WDM) filter (with periodic multiple pass band wavelengths coinciding with the ITU channel grid) and an isolator within the EDFA. It is shown that by placing the filter and isolator combination at a distance from the input end of about 30 % of the total length of the EDF, it is possible to achieve about 10 dB improvement in gain and a reduction of about 2 dB in noise figure compared to the case without the filter and isolator. Such an improved EDFA design should be of interest in fiber optic communication systems employing wavelength division multiplexing.

....

Key Words: EDFA, WDM, Filter, Isolator

1. INTRODUCTION

The development of erbium doped fiber amplifiers (EDFA) has truly revolutionized fiber optic communications with greatly increased capacities achievable by using wavelength division multiplexing (WDM). The different channel wavelengths in a WDM system are fixed by ITU and typical channel frequency spacing used today are 200 GHz and 100 GHz which correspond to wavelength spacing of approximately 1.6 and 0.8 nm respectively at 1550 nm. The performance of a WDM optical fiber communication system depends critically on the characteristics of EDFAs and there is constant effort to improve the performance of EDFAs in terms of gain, gain flatness [1-6] noise figure etc. Modification of the gain characteristics of EDFAs is usually achieved by using optical devices such as wavelength filters, isolators [7-9] etc. within and/or outside of the amplifier leading to improved performance

Since the channel wavelengths of a WDM system are fixed, in this work we show that it is possible to improve the gain and noise characteristics of an EDFA by using within the EDFA, a WDM filter with periodic, multiple pass bands (coinciding with the ITU wavelengths) with an isolator. The proposed design is shown to work as an efficient wavelength division multiplexed (WDM) gain enhanced EDFA with simulated results showing an increase in gain of about 10 dB and significantly improved noise figure.

2. THEORY

In order to show the improvement in the amplifier performance with the inclusion of the WDM filter and the isolator unit, we have considered an amplifier fabricated from a typical erbium doped fiber whose specifications are listed in Table 1. The amplifier is pumped in the forward direction by a pump at 980 nm through a WDM splitter and within which a combination of a WDM filter with pass bands separated by a frequency spacing $\Delta\nu_f$ and an isolator are placed at a distance x from the input end of the EDF. The extinction ratio of the isolator (in the backward direction) and that of the filter (in the stop band) is also mentioned in the same table.

We have employed the standard three-level rate equation model as described in Pedersen [10]. Assuming a cylindrical co-ordinate system with z-axis as the fiber axis and considering the fundamental mode LP_{01} for the signal, pump, and the ASE wavelengths, the steady state population concentrations N_1(r, z) and N_2(r, z) of Er^{3+} ions per unit volume in the ground and excited states, respectively are described by following equations:

$$N_1 = \rho_{Er} \frac{W_{a,s}(r,z)}{W_a(r,z)+W_e(r,z)} \quad (1)$$

$$N_2 = \rho_{Er} - N_1 \quad (2)$$

where ρ_{Er} is the total erbium ion concentration and W_e and W_a , which represent the total emission and absorption rates, respectively are given by

[1] Corresponding author

$$W_a(r,z) = \left[\frac{\sigma_a(\nu_s)}{h\nu_s}P_s(z) + \int_0^{\infty}\frac{\sigma_a(\nu)}{h\nu}S_{ase}(\nu,z)d\nu\right]I_s^{01}(r)$$
$$+\left[\frac{\sigma_a(\nu_p)}{h\nu_p}P_p(z)\right]I_p^{01}(r) \quad (3)$$

$$W_e(r,z) = \left[\frac{\sigma_e(\nu_s)}{h\nu_s}P_s(z) + \int_0^{\infty}\frac{\sigma_e(\nu)}{h\nu}S_{ase}(\nu,z)d\nu\right]I_s^{01}(r)$$
$$+ A_e \quad (4)$$

where A_e is the spontaneous emission power spectral density at frequency ν and position z, $I_{s,p}^{01}$ is the normalized intensity distribution at the LP_{01} signal/pump wavelength. The last term in $W_a(r,z)$ expression is the pump absorption rate R_{pa}. $P_{p,s}(z)$ is the pump/signal power at a position z along the fiber length, $\nu_{p,s}$ is the pump/signal frequency, and $\sigma_{a,e}$ are the wavelength dependent absorption and emission cross-sections. The spontaneous emission generated at any point is amplified in both forward and backward directions of the fiber axis. Therefore the total ASE spectral density at any value of z, $[S_{ase}(\nu,z)]$ can be written as a sum of its two components, namely forward (S_{ase}^{+}) as well as backward (S_{ase}^{-}) propagating ASEs. The equations governing the propagation of ASE, pump (in forward direction) and signal are given by Pedersen [10]

$$\frac{dS_{ase}^{\pm}(\nu,z)}{dz} = \pm 2h\nu\gamma_e(\nu,z) \pm [\gamma_e(\nu,z) - \gamma_a(\nu,z)]S_{ase}^{\pm}(\nu,z)$$
$$\frac{dP_k(z)}{dz} = [\gamma_e(\nu_k,z) - \gamma_a(\nu_k,z)]P_k(z) \quad (5)$$

where emission $\gamma_e(\nu,z)$ and absorption $\gamma_a(\nu,z)$ factors are determined from emission and absorption cross-sections as well as the overlap integrals between the signal mode and the population densities in the excited and ground states, respectively.

$$\gamma_e(\nu,z) = \sigma_e(\nu)\, 2\pi \int_a^b \int_0^{2\pi} N_2(r,z)I(r,\nu)\, r\, dr\, d\phi$$

$$\gamma_a(\nu,z) = \sigma_a(\nu)\, 2\pi \int_a^b \int_0^{2\pi} N_1(r,z)I(r,\nu)\, r\, dr\, d\phi \quad (6)$$

where a and b correspond to the inner and outer radii of the doped core of the fiber. $I(\nu,r)$ is the normalized intensity distribution at frequency ν.

In order to obtain the gain spectrum, the noise figure spectrum, and the forward ASE spectrum at the output of the EDF, we have solved Eqs, (5) over the whole emission bandwidth of ≈ 100 nm extending from 1500 nm to 1600 nm. For the case when a filter and an isolator are inserted at say position x from the input end of the EDF, these equations are first solved for EDF length 0 to x, then the effect of filter and isolator unit is introduced which introduces an insertion loss of 25 dB in the stop bands of the filter (in forward as well as in the backward direction) and for the isolator, an insertion loss of 25 dB only in the backward direction; the equations are then solved for the EDF length x to L, where L is the total length of EDF considered. To simplify the numerical calculations, the complete 100 nm bandwidth was divided into 125 slots of 0.8 nm width (which corresponds to Δν = 100GHz). These equations have been solved as a boundary value problem using fourth order Runge Kutta method. The accuracy of our procedure was checked by comparison with standard results for a conventional EDFA.

3. RESULTS

Table 1 lists the characteristics of EDF, the WDM filter and the isolator used in the simulations. The WDM filter is assumed to have a channel spacing of 200 GHz with 25 dB extinction ratio of the stopped channels. A typical filter with such characteristics can be designed using a pair of long period gratings [11]. Calculations of gain and noise figure were performed with the wavelength filter and the isolator placed within the amplifying fiber. Our calculations show that both gain and noise figure depend on the position of the filter and isolator within the amplifier. The optimum position for achieving maximum improvement in gain and noise figure has been found to be at about 30 % of the total length of the doped fiber from the input side. These are consistent with earlier analysis for optimum position of isolator in an EDFA or in designing of an ASE source or an amplifier [7,8]. As expected, the presence of the filter and the isolator modifies the amplified spontaneous emission (ASE) spectrum as well as the gain spectrum of the amplifier. Wavelengths corresponding to the stop band of the filter get filtered out from the amplifier and the isolator in turn allows light only in the forward direction. Thus forward propagating signals corresponding to the pass bands of the filter (and the corresponding ASE) have a larger fraction of population inversion available for their gain. Moreover, the isolator couples the backward ASE power into the forward direction, thus increasing the forward conversion efficiency of the amplifier.

Figure 1 shows the spectral gain of the amplifier with and without the filter-isolator combination. The continuous curve gives the gain variation without the filter and the points correspond to the gain at the pass wavelengths of the filter. One can clearly see that the presence of the filter-isolator combination has resulted in an increase in gain of about 10 dB at wavelengths corresponding to the pass wavelengths of the filter. Figure 2 shows the variation of noise figure in the range 1540 to 1560 nm with and without the filter and isolator. The noise figure at the wavelengths corresponding to the pass band also improves by about 2 dB compared to the case without the filter and the isolator. This improvement in noise figure is primarily due to the presence of the isolator within the amplifier. The noise figure in the stop bands is calculated by considering a signal of same strength at wavelengths corresponding to stop bands of the filter; these have been shown as peaks in the noise figure spectrum with the filter and isolator combination.

Although the calculations reported here correspond to a channel spacing of 200 GHz, calculations are underway for channel spacing of 100 GHz and the general conclusions drawn here are expected to remain valid. These and other calculations of evolution of ASE within the amplifier and saturation behavior will be reported at the conference.

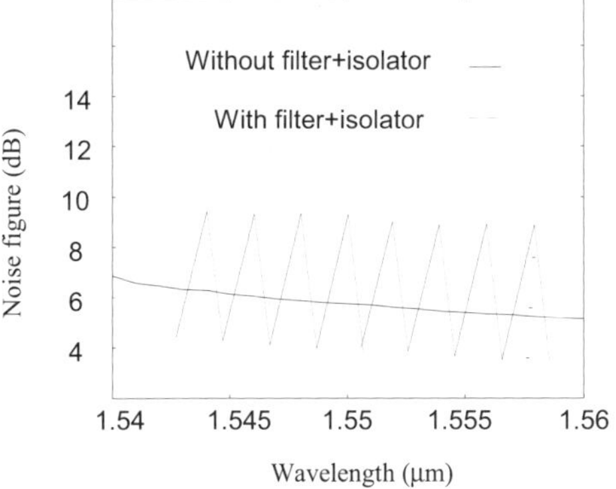

Figure 1. The gain spectrum with and without the isolator and filter unit. The dotted curve shows the gain only in the pass bands of the filter. The gain in the stop bands reduces significantly and is of no use to us.

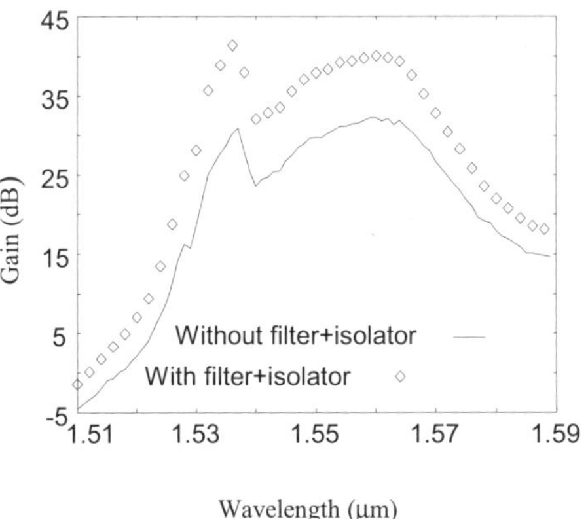

Figure 2. Spectral variation of noise figure with and without the filter and isolator. The figure shows that the noise figure reduces by about 2 dB in the pass bands of the filter. The peaks in the above dashed curve correspond to the stop bands and the dips to the pass bands respectively of the filter.

Fiber parameter	Value
Core radius (μm)	2.5
Numerical aperture	0.2
Doping concentration (m^{-3})	5.1×10^{25}
Pump Power (mW)	50
Signal power (nW)	100
Isolator extinction ratio (dB)	25
Filter extinction ratio (dB)	25
Length of the doped fiber (m)	4
Position of filter+isolator unit (m)	1.2

Table 1: Various fiber parameters and input power levels considered for simulation

4. CONCLUSIONS

We have proposed a new design of a gain enhanced EDFA for WDM applications by inserting a WDM filter (with periodic multiple pass wavelengths coinciding with the ITU channel grid) and an isolator within the EDFA. It is shown that by placing the filter and isolator combination at a distance from the input end of about 30 % of the total length of the EDF, it is possible to achieve about 10 dB improvement in gain and a reduction of about 2 dB in noise figure compared to the case without the filter and isolator. Such an improved EDFA design should be of interest in fiber optic communication systems employing wavelength division multiplexing.

REFERENCES

[1] Aisawa, S., Sakamoto, T., Fukui, M., Kani, J., Jinno, M., and Oguchi, K., Ultra-wideband, long distance WDM demonstration of 11 Tbit/s (50 × 20 Gbit/s), 600 km transmission using 1550 and 1580 nm wavelength bands, Electronics Letters, vol. 34, page 1127, 1998.

[2] Buxens, A., Poulsen, H. N., Clausen, A. T., and Jeppesen, P., Gain flattened L-band EDFA based on upgraded C-band EDFA using forward ASE pumping in an EDF section, Electronics Letters, vol. 36, page 822, 2000.

[3] Hodgson, C. W., and Vengsarkar, A. M., "Spectrally shaped high power amplified spontaneous emission sources incorporating long-period gratings", *Optical Fiber Communication Conf.,* 2, OSA Tech. Dig. Series (Optical Society of America, Washington, DC, 1996), page 29, 1996.

[4] K. Thyagarajan and Jagneet Kaur, A novel design of an intrinsically gain flattened erbium doped fiber, Optics Communications, vol. 183, page 407, 2000.

[5] Kashyap, R., Wyatt, R., and Campbell, R. J., Wideband gain flattened erbium fibre amplifier using a photosensitive fibre blazed grating, Electron Letters, vol. 29, no. 2, page 154, 1993.

[6] Tachibana, M., Laming, R. I., Morkel, P. R., and Payne, D. N., Erbium-doped fiber amplifier with flattened gain spectrum, IEEE Photonics Technology Letters, vol. 3, page 118, 1991.

[7] Jagneet Kaur, Suparna Sinha Roy, K. Thyagarajan, and B. P. Pal, Role of an isolator in optimization of forward conversion efficiency in an Er-doped SFS source, Optical Fiber Technology., vol. 5, page 390, 1999.

[8] Zervas, M. N., Efficient Er-doped fiber amplifier incorporating an optical isolator, IEEE J. Quantum Electronics, vol. 31, page 472, 1995.

[9] Digonnet, M. J. F.,Theory of Superfluorescent Fiber Lasers, IEEE J. Lightwave Technology, vol. LT-4, page 1631, 1986.

[10] Pedersen, B., Small signal erbium-doped fiber amplifiers pumped at 980 nm: a design study, Optical and Quantum Electronics, vol. 3, page S273, 1994.

[11] Mini Das and K. Thyagarajan, Wavelength-division multiplexing isolation filter using concatenated chirped long period gratings, Optics Communications, vol. 197, page 67, 2001.

Gain Enhancement in Double-pass Erbium-doped Fiber Amplifier

K. F. Chang, P. Poopalan, T. C. Teyo, T. Subramaniam, and H. Ahmad
Photonics Laboratory, Physics Department
University of Malaysia, 50603 Kuala Lumpur, Malaysia
Tel: +603-79676770
E-mail: kenchangkfui@hotmail.com

Abstract: A double-pass erbium-doped fiber amplifier (DP-EDFA) configuration for 980 nm pump is demonstrated. This amplifier provided gain up to 40 dB in the conventional band C-band. In this configuration the signals are looped back into the EDF with the aid of optical circulators and amplified twice by the EDF. Double-passing the signal through the EDF produces a net enhancement. The DP-EDFA gain is 6.5 dB higher than the standard single-pass fiber amplifier gain but at the cost of noise figure (NF) increasing by 1.4 dB.

Key words: *EDFA; C-band; Double-pass amplifier; optical circulator*

I. INTRODUCTION

The Erbium-Doped Fiber Amplifier (EDFA) in optical fiber communication is important because of its excellent performance parameters such as high signal gain, low noise and high saturation output [1]. However, conventional fiber amplifiers only utilize one-way signal amplification in the active fiber. The pump energy is wasted [2] and the performance is limited in this configuration. To improve the performance, a double-pass fiber amplifier (DP-EDFA) configuration has been developed.

In this paper, a DP-EDFA using optical circulators is proposed to achieve better utilization of wasted pump energy and effect an improvement in gain. The signals are looped back into the EDF to be amplified twice. The performance of EDFA is compared to a standard EDFA (SEDFA) of the same length.

II. EXPERIMENT SETUP

9 meters of alumino-germanosilicate EDF with an erbium ion concentration of 440 ppm and NA of 0.24 is used in this experiment. The EDF is pumped by a 980nm laser diode hence creating a three level laser system. Fig. 1(a) and 1(b) show the DP-EDFA configuration and SEDFA configuration, respectively. In this DP-EDFA the signals are amplified twice through the EDF, which is the active element of the amplifier. The signals are looped back into the EDF via circulator 2 (C2) and the signal output is taken from circulator 1 (C1). The loop of C2 is formed by connecting port 3 and port 1 together. The performance of both configurations are then measured. In order to compare both configurations and also to avoid uncertainty in the input and the output losses, the circulators are not removed for the SEDFA configuration. The only additional loss arises from port 1 to port 2 of C2 and from port 2 to port 3 of C1 in the DP-EDFA configuration.

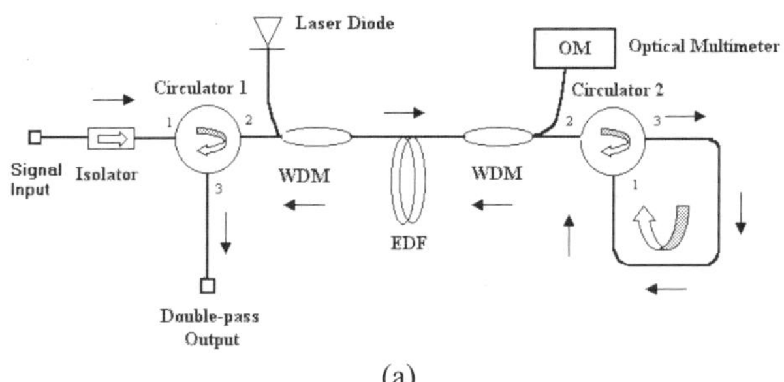

(a)

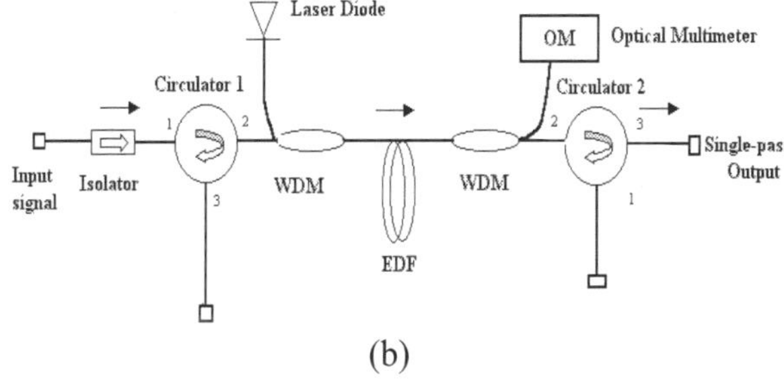

(b)

Fig. 1. EDFA configurations
(a) DP-EDFA configuration using Optical Circulator as a reflector
(b) SEDFA configuration
WDM= Wavelength Division Multiplexer
EDF = Erbium-Doped Fiber

III. RESULTS AND DISCUSSION

The ASE spectrum when pumped at 100 mW for both configurations is shown in the Fig. 2. In the SEDFA, only the forward ASE (ASE^+) is taken in the measurement but in DP-EDFA both backward ASE (ASE^-) and ASE^+ are integral in the measurement. The total ASE level in the DP-EDFA is

$$\text{ASE level} = ASE^- + \text{amplified } ASE^+$$

Therefore the total DP-EDFA ASE power is always greater than the SEDFA ASE [3]. Because of this, the ASE level for the DP-EDFA is 9 dB and 4 dB higher than the single-pass configuration at 1550 nm and 1530 nm, respectively.

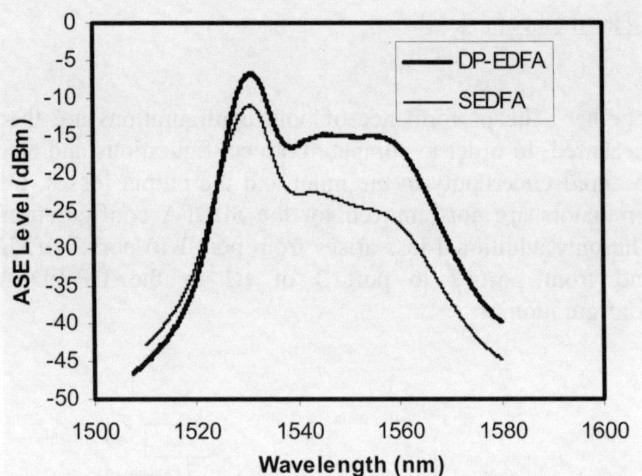

Fig. 2 Emission Spectrum of the both configurations pumped at 100mW

Fig. 3 shows the graph of gain as a function of pump power for an input signal power of –30 dBm at a signal wavelength of 1550 nm for SEDFA and DP-EDFA, respectively. The graph shows that the gain coefficient for both configurations is 2.1 dB/mW and 3.0 dB/mW, respectively. The graph also shows that gain starts to saturate above 30 mW of pump power. From the result, pump power is more efficient with the DP-EDFA. Fig. 3 also shows that DP-EDFA has enhanced gain by 6.5 dB and achieves a high gain of 40 dB in the DP-EDFA with noise figure (NF) also increasing but only by 1.4 dB. The DP-EDFA gain is given by [4]

$$G_{DP\text{-}EDFA}(\lambda) = L'(\lambda)\, G^2(\lambda)$$

where $G_{DP\text{-}EDFA}(\lambda)$ is the DP-EDFA gain, $L'(\lambda)$ is the total cavity loss of the feedback loop; $G(\lambda)$ is the SEDFA gain.

$$G(\lambda) = Exp\,\{\Gamma_s\,(\sigma_e\,[N_2] - \sigma_a\,[N_1])\,L\}$$

Where G is signal gain, N_1, N_2 is ground state and metastable population respectively, Γ_s is confinement factor, σ_e is stimulated emission cross-section, σ_a is absorption cross-section and L is fiber length. Therefore gain depends on the average inversion level of the erbium ion population. When pump power is increased, more erbium ions are excited to pump level. The excited ions will decay very fast to metastable level (level 2), and N_2 is increased. Therefore gain increases when pump power increases. However, the signal emission rate, which is proportional to its signal power, causes the gain to saturate since there are insufficient Er^{3+} ions. The gain of DP-EDFA can be improved if a low insertion loss optical circulator is used in the feedback loop. In this case, the cavity loss is 2.5dB, which allows only 44%

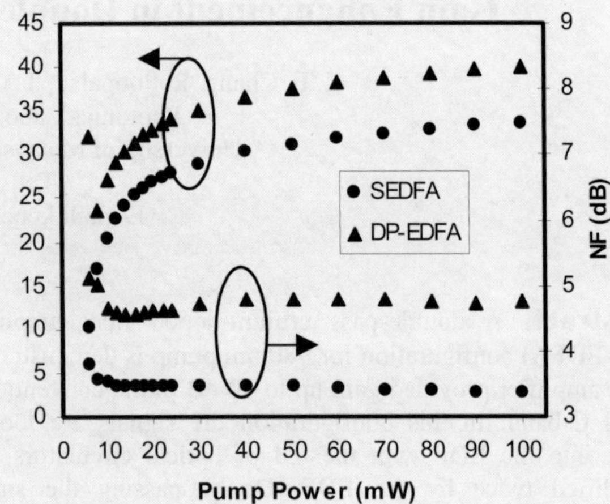

Fig. 3 Gain and NF versus pump power for both configurations at signal wavelength of 1550 nm with input signal power of –30 dBm

of the power to be reflected into the EDF. The NF of 4.7dB in the DP-EDFA when pumped at 100mW can be reduced near to the quantum limit (3dB) if a 2 stage reflective EDFA is used [5].

Fig. 4 shows that the gain and NF as a function of input signal power at a signal wavelength of 1550nm and pump power of 100mW comparison for both configurations. From Fig. 4, the gain is degraded for DP-EDFA due to amplifier saturation at large incident signals. On the other hand it is clearly seen, the small signal level gain is increased by 7 dB from 34 dB in the SEDFA to 41 dB in the DP-EDFA with the cost of NF also increasing but by only 1.3 dB. The spontaneous emission factor n_{sp} is given by [6]

$$n_{sp} = \frac{\sigma_e N_2}{\sigma_e N_2 - \sigma_a N_1}$$

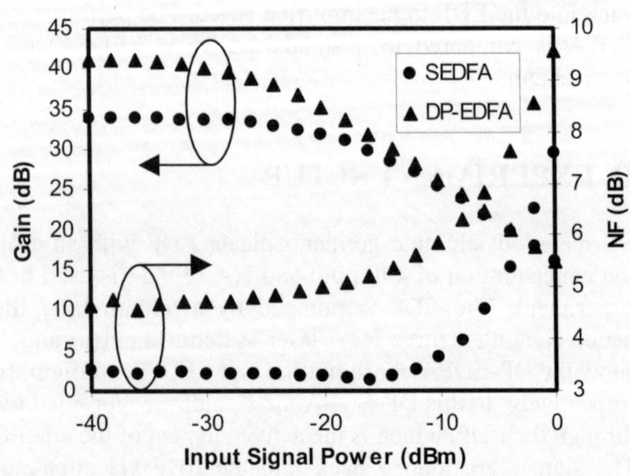

Fig. 4 Comparison gain and NF versus incident signal power for both configurations at 1550nm with pump power of 100mW.

where σ_e is stimulated emission cross section, σ_a is absorption cross-section, N_1 and N_2 are ground state population density and metastable state population density respectively. n_{sp} is a measure of quality of inversion of the EDFA and becomes unity (minimum) when complete inversion ($N_1 \sim 0$) occurs. When the input signal power becomes comparable to pump power, N_2 will return to ground state and n_{sp} increases. From Fig. 4, DP-EDFA gets saturated earlier than SEDFA where the dynamic range of DP-EDFA is –22dBm and SEDFA is –19dBm. So that NF in DP-EDFA is much higher than the SEDFA (3dB) with large signal powers.

Since the amplifier gain is proportional to N_2, a high incident signal power causes the signal emission rate for DP-EDFA configuration to be much higher than the signal emission rate for SEDFA. Because of this, N_2 is depleted very fast and therefore amplifier gain in the double-pass drops in respect to the incident signal power.

IV. CONCLUSION

A Double-pass EDFA configuration has been investigated and compared with standard EDFA configuration. A small signal gain of 40 dB has been achieved in the DP-EDFA configuration and the increment of signal gain is 6.5 dB with NF increases by 1.4 dB at signal wavelength of 1550nm compared to conventional configuration EDFA.

REFERENCES

[1] Bo Pedersen, Anders Bjarklev, Jorn Hedegaard Povlsen, Kristen Dybdal, and Carl Christian Larsen, "The Design of Erbium-Doped Fiber Amplifier", Journal of Lightwave Tech., Vol. 9, No. 9, pp. 1105-1112, 1991.

[2] V. Lauridsen, R. Tadayoni, A. Bjarklev, J. H. Povlsen, B. Pedersen, "Gain and Noise Performance of Fibre Amplifier Operating in New Pump Configuration", Electron Lett, Vol. 27, No. 4, pp. 327-329, 1991.

[3] E. Desurvire, "Erbium-Doped Fiber Amplifier", John Wiley & Sons, Inc, 1994.

[4] Qinghe Mao, John W. Y. Lit, " Amplification enhancement of L-band erbium-doped fiber amplifier by reflection scheme", Opt. Commun, Vol. 201, pp. 61-69, 2002.

[5] Joon Tae Ahn, Min-Yong Jeon, Kyong Hon Kim, "Two-stage reflective-type erbium-doped fiber amplifier with enhanced noise figure characteristics", Opt. Commun, Vol 197, pp.121-125, 2001.

[6] Dennis Derickson, "Fiber Optics Test and Measurement", Hewlett-Packard company 1998

Amplification bandwidth of below threshold regenerative erbium-doped fibre amplifiers

T. C. Teyo*, N. S. Mohd. Shah and H. Ahmad
Photonics Laboratory, Physics Department,
University of Malaya, 50603 Kuala Lumpur,
Malaysia.
Tel: +603-79676770
Fax: +603-79674146
*E-mail: teyo@time.net.my

Abstract: Amplification characteristics of regenerative erbium-doped fibre amplifier operating just below the oscillation threshold are presented. Resonant amplification is preserved only for small drive powers. At the higher drive powers, resonant amplification characteristics vanish and the amplification bandwidth increases.

....

Key Words: regenerative amplifiers, amplification bandwidth, below threshold

1. INTRODUCTION

Optical feedback has been introduced in the erbium-doped fiber amplifier systems to form the gain-clamped amplifiers [1-2] and regenerative amplifiers [3-4]. The regenerative amplifier systems differ from the gain-clamped amplifier systems in the sense that the drive signal experiences regenerative-feedback and thus regenerative amplification through the circulation in the cavity. To describe the behavior of the regenerative amplifier, it is necessary to include implicitly the saturation characteristics of the active medium. In this letter, we present the saturation-induced variation in the amplification bandwidth.

2. EXPERIMENT

Fig. 1 shows a schematic diagram of the regenerative erbium-doped fibre amplifier. The system consists of two wavelength division multiplexers (WDMs), three couplers: C_1 and C_2, with an output coupling ratio of 95%, and C_3 for monitoring the output spectrum in the anti-clockwise direction from the 1% port, and two optical isolators at the input and output ends. Another isolator, ISO 1, was placed in the cavity to ensure a unidirectional oscillation of the laser. A 15m long erbium-doped fibre with a cut-off wavelength of 950nm, a refractive index of 1.473 and an Er^{3+} concentration of +440ppm was used as the active medium. Pump power was provided by a laser diode from the 980nm port of WDM I. The drive signal used was from a tuneable laser source (ANDO AQ4321D) and output signal was monitored by an optical spectrum analyser (ANDO AQ6317B).

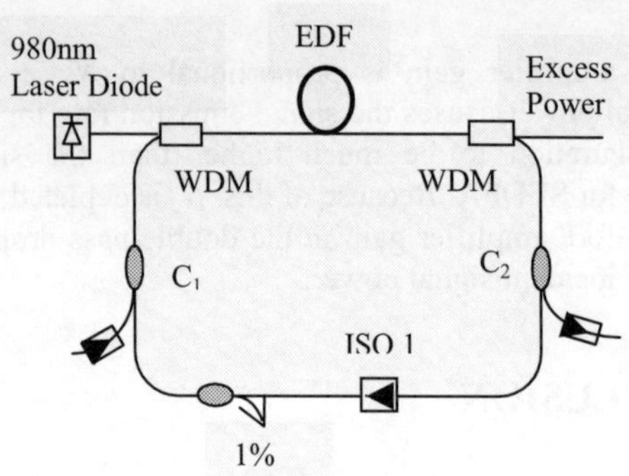

Fig.1 Experimental set-up

3. RESULTS AND DISCUSSION

The output spectrum without the drive power is shown with thin line in Fig. 2. Since the system is pumped at 29.4mW, just below the oscillation threshold, there is a potential laser mode at the wavelength of 1557.7nm. By injecting an external signal (thick line) with a power of –30dBm at the resonant wavelength (zero detuning), a significant of suppression of amplified spontaneous emission (ASE) power was observed. This is balanced by the energy needed to provide amplification of the drive power.

Fig. 3 shows the amplified output powers as a function of signal detuning for the drive powers of –10, -30 and –40dBm. With the small signal of –40dBm, resonant amplification is preserved where the signals with small detuning experience strong amplification. The amplification bandwidth is narrow in this case. With a higher drive power, the amplification bandwidth becomes broader. A flat amplified output power spectrum is obtained with a strong drive power of –10dBm. With such a high drive power, regenerative amplifier system behaves like a conventional single-pass amplifier system where the ASE-like shape is not longer preserved. The resonant amplification characteristic of the regenerative amplifier is modified as a result of the saturation of the gain medium by the strong drive power.

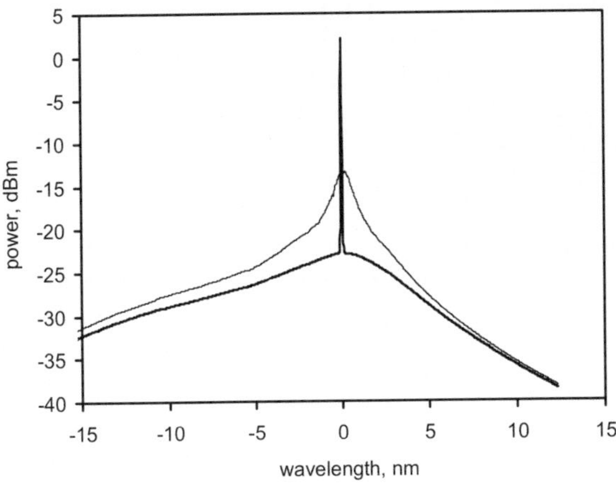

Fig. 2 Output spectrum at the pump power of 29.4mW. Thin line: without the drive signal; Thick line: with the drive power of –30dBm at the resonance

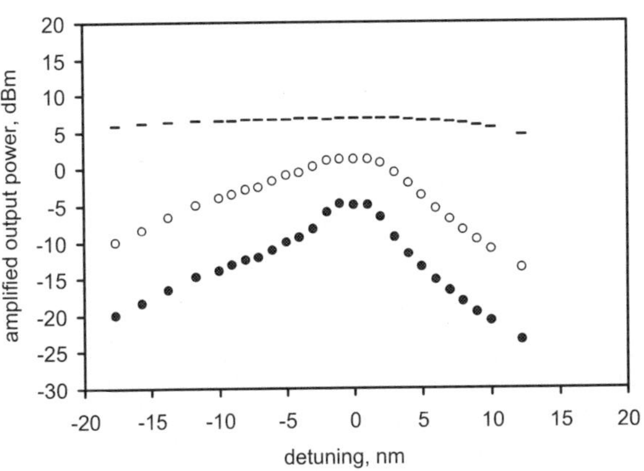

Fig. 3 Amplified output powers as a function of signal detuning for different drive powers at the pump power of 29.4mW (-: -10dBm; o: -30dBm; •: -40dBm)

Fig. 4 shows the signal gain and amplification bandwidth with 3-dB compression (or half-power bandwidth) as a function of the drive power. At the small signal gain regime, the amplifier bandwidth is only about 6nm. At the low drive powers, saturation can be neglected in the determination of the bandwidth. Above the drive power of –35dBm, the amplification bandwidth increases significantly. Regenerative-feedback allows drive signal inside the cavity to recirculate inside the cavity many times, extracting energy from the medium on each cycle. As a result, the recirculting signal builds up to very large amplitude inside the cavity. In the region of moderate saturation, the amplification bandwidth increases linearly. For these drive powers, the circulating signal power within the cavity significantly perturbs the small signal gain of the active medium.

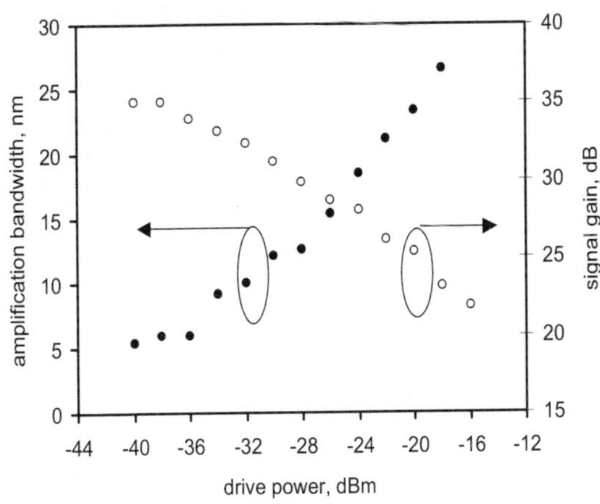

Fig. 4 Signal gain (empty circles) and amplification bandwidth (filled circles) with 3-dB compression as a function of drive power at the pump power of 29.4mW

4. CONCLUSIONS

Amplification characteristics of regenerative erbium-doped fibre amplifier operating just below threshold have been demonstrated. Although high resonant amplification can be achieved with small drive power, the system suffers from narrow amplification bandwidth. With the high drive powers, a flat amplification spectrum can be obtained. At small signal regime, the saturation effect is negligible. However, it is significant in the regime of moderate saturation where the amplification bandwidth increases markedly. In this stage, the regenerative amplifier system behaves like a conventional single-pass amplifier.

REFERENCES

1. M. T. Fatehi, Erbium-doped fibre amplifiers with wavelength-selective optical feedback, IEEE Photonics Technology Letters, vol. 8, page 1012-1014, 1996.
2. K. Inoue, Gain-clamped fiber amplifier with a short length of preamplification fiber, IEEE Photonics Technology Letters, vol. 11, page 1108-1110, 1999.
3. T. C. Teyo, P. Poopalan and H. Ahmad, Regenerative erbium-doped fiber ring laser-amplifier,' Electronics Letters, vol. 35, page 1471-1472, 1999.
4. B. P. Singh, Characteristics of regenerative erbium-doped fibre ring amplifier, Electronics Letters, vol. 36, page 1013-1015, 2000.

Double Pass L-Band EDFA with Unpumped EDF

S. W. Harun, N. Tamchek, P. Poopalan and H. Ahmad
Department of Physics, University of Malaya,
50603 Kuala Lumpur, Malaysia.
E-mail address: wadi72@yahoo.com

Abstract: A new gain improvement technique for long wavelength band erbium-doped fiber amplifier (L-band EDFA) is presented. It uses an unpumped erbium-doped fiber (EDF) in a double pass configuration in order to effectively suppress and completely exploit the backward amplified spontaneous emission (ASE). Compared to a conventional single pass amplifier, the gain coefficient can be improved from 1.5dB/mW to 3.1dB/mW without much noise figure penalties. The L-band EDFA can also reduce the pump power necessary to achieved a specified gain by about half however at the cost of noise figure penalty of 1.6dB. The optimum length for an unpumped EDF is experimentally measured to be 3m for both single pass and double pass system. By adding 3m of unpumped EDF, the maximum gain enhancements of 2.7dB and 5.7dB were obtained for single pass and double pass system at wavelengths of 1576nm and 1600nm, respectively.

Key Words: erbium-doped fiber, optical amplifier, L-band EDFA, double pass EDFA.

1. INTRODUCTION

Recently, extending the amplification bandwidth of an optical fiber amplifier is strongly desired to increase the transmission capacity in dense wavelength division multiplexing (DWDM) transmission systems. Long wavelength band erbium-doped fiber amplifiers (L-band EDFA) have recently gained interest as a potential candidate for increasing transmission capacity beyond the conventional band (C-band) in the DWDM systems. By integrating L-band in parallel with C-band, a gain bandwidth of about 80nm can be achieved [1]. Unfortunately, the L-band EDFAs are effectively inefficient due to the operating wavelengths that are far from the peak absorption of silica-based erbium-doped fiber (EDF) at 1531nm. To date, several designs have been implemented to increase the amplifier gain such as using an end reflector to feedback a fraction of the backward C-band ASE into the EDF as an ASE seed to suppress the backward C-band amplified spontaneous emission (ASE) [2].

More recently, we reported an efficient L-band EDFA design based on a double pass technique using a circulator [3]. In this design, good gain enhancement was achieved by double passing a forward ASE and signal in the EDF. In the present paper, an improved design for further gain improvement is proposed. It uses the wasted C-band ASE as the secondary pump source for the unpumped EDF located before a forward pumped EDF. The amplifier performance for different length of unpumped EDF is measured and compared to the single pass configuration without circulator.

2. EXPERIMENTAL PROCEDURE

Fig. 1 shows a configuration of the proposed double pass L-band EDFA with the unpumped EDF (EDF II), in which optical circulator C2 is located at the output end of the forward pumped EDF I. EDF with a peak absorption of 6.6dB/m at 1531nm is used in this experiment as EDF I and II. EDF I is fixed at 50m and EDF II is varied from 1 ~ 9m. EDF I is pumped at 980nm using a forward pumping scheme. The optical signal is input into EDF II, from port 1 to 2 of optical circulator C1. After having being amplified in the forward direction, the signal is reflected back into both EDF sections by a feedback loop which is formed by joining ports 3 and 1 of the optical circulator C2. Then the signal is amplified again in the backward direction, and finally exits from port 3 of C1. The feedback also reflects the forward ASE into the EDF. Insertion loss of the both optical circulators is about 1dB at the tested signal wavelengths. The gain and noise figure are not only measured for the double-pass system but also for single-pass system, which was modified by removing circulator C2, and the measurements taken from the output end of EDF I.

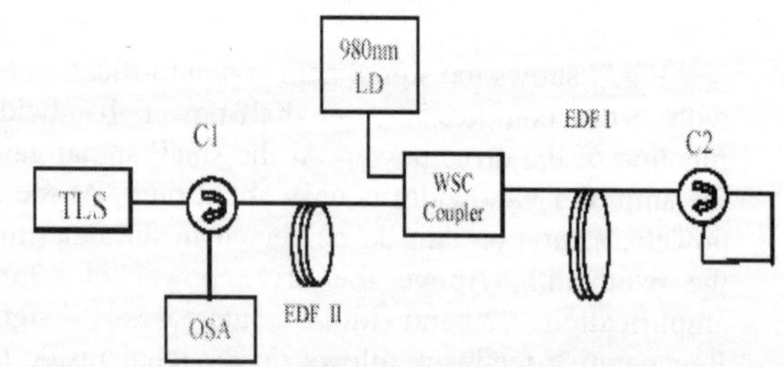

Fig. 1: Configuration of the proposed L-band EDFA

3. RESULT AND DISCUSSION

Fig. 2 shows the comparison of small-signal gain of single pass and double pass configurations, for various lengths of EDF II. The pump and input signal powers are fixed at 103mW and –30dBm, respectively. Both systems with an un-pumped EDF, show a considerable increase in small signal gain with strong dependence on the length of EDF II, when compared to the case of without EDF II. As is seen, the adding of unpumped EDF increases the gain by about 2.5dB and 5.0dB for single-pass and double-pass system, respectively. In addition, the proposed double-pass amplifier also achieved a gain enhancement of more than 10dB compared to the single pass amplifier. On the other hand, the noise figure increases with the length of unpumped EDF for both single-pass and double-pass system. From these results, it is found that the optimum length of unpumped EDF for both single-pass and double-pass system is about 3 m. At EDFII = 3 m, the noise figure penalties about 1.0dB and 0.5dB are obtained for single-pass and double-pass system, respectively. The noise figure in the double pass system, in which the signal is amplified twice in different directions, the noise figure penalty in the range of 0.3~1.2dB is obtained compared to the single-pass system, with strong dependence on the length of EDF II.

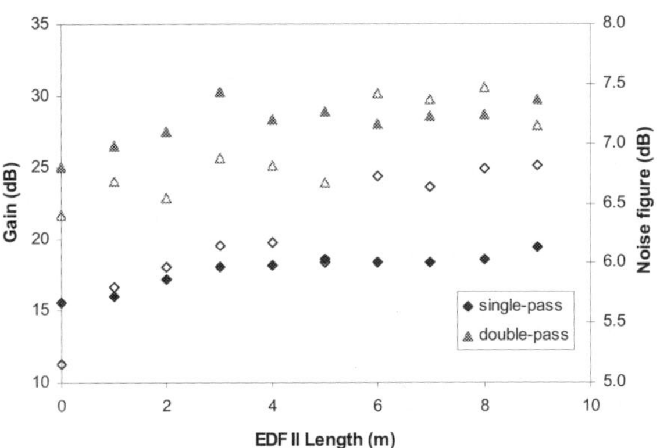

Fig. 2: Gain (shaded) and noise figure (clear) as a function of EDF II length for single-pass and double-pass system.

From the above results, it is evident that a very efficient use of pump power can be achieved by employing an un-pumped EDF before the forward pumped EDF in double pass system. The re-use of wasted backward ASE as a 1550nm pumping source for un-pumped EDF, generating photons in the 1600nm band, has improved the efficiency. The efficiency is further improved in the double-pass system, in which the signal is amplified twice in different directions, and the total gain is almost doubled due to the increase of the effective EDF length. This is also the result of the reuse of the backward ASE facilitated by the feedback loop. However, the use of unpumped EDF reduces population inversion at the input of an amplifier and therefore a higher noise figure is obtained. The total noise figure of the double-pass amplifier is made up of the noise of the single-pass amplifier, the insertion loss of the feedback loop and the accumulation of the cascaded amplifiers [4]. Here, optical circulator C2 is the main reason for the noise figure increase in the double-pass system, because it contributes about 1.0dB loss to the feedback signal.

In Fig. 3, the gain and noise figure are shown as a function of the pump power for the different L-band systems. The input signal wavelength and power are fixed at 1580nm and -30dBm, respectively. The conventional single-pass system (without EDF II) shows the worst gain coefficient of 1.5dB/mW. The gain coefficients are improved to 1.8dB/mW and 2.4dB/mW by adding a 3m of unpumped EDF and double propagation of the signal through the forward pumped EDF, respectively. By combining these two techniques as in the proposed system, the highest gain coefficient of 3.1dB/mW is obtained. Comparing the gain values in the Fig. 3, it is found that the proposed system can give a gain value equal to that for conventional single pass-system, using only about 50% of the pump power. On the other hand, the noise figure is increased about 0.8dB by adding a 3m of unpumped EDF in the single-pass system and 1.0dB in a conventional double-pass system. The worst noise figure penalty of 1.6dB was obtained for the proposed system when compared to the conventional single-pass system.

Fig. 4 shows the gain and noise figure enhancements, defined as the difference between the gain and noise figure with and without 3m length of unpumped EDF, as a function of signal wavelengths. During the measurement, the pump and signal powers were kept at 103mW and -30dBm, respectively. It can be seen from Fig. 4 that the enhancement is varied and depends on the signal wavelength. The maximum gain enhancements of 2.7dB and 5.7dB were obtained for single pass and double pass at wavelengths of 1576nm and 1600nm, respectively. On the other hand, the noise figure penalty is obtained with adding of unpumped EDF especially for single pass system.

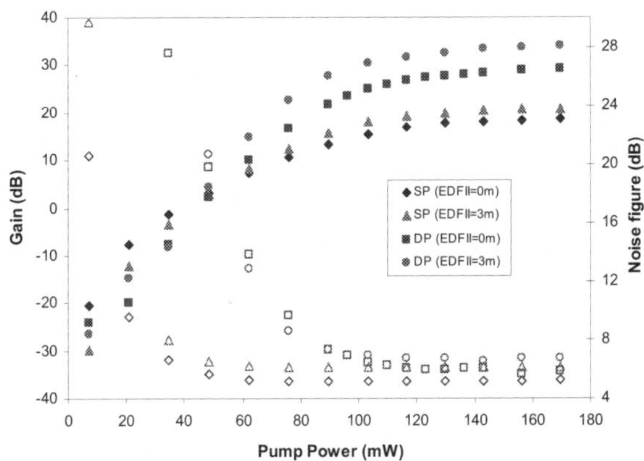

Fig.3: Gain (shaded) and noise figure (clear) as a function of pump power.

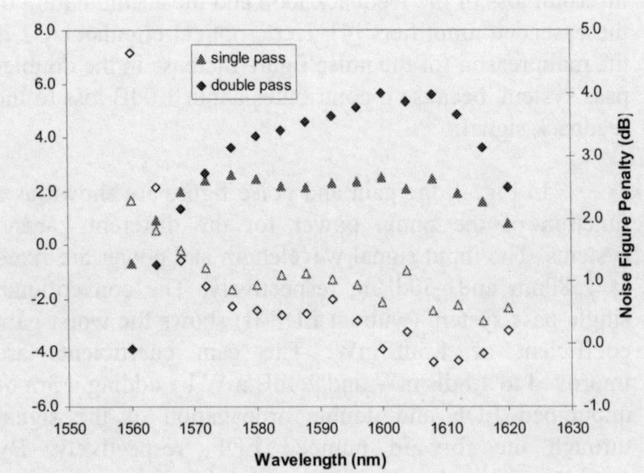

Fig.4: Gain enhancement and noise figure penalty as a function of signal wavelength.

4. CONCLUSION

The L-band EDFA with improved gain characteristic, which utilizes an unpumped EDF in double pass configuration, has been proposed. The proposed system has improved the gain coefficient from 1.5dB/mW to 3.1dB/mW, compared to a conventional single pass EDFA. The proposed amplifier reduces the pump power necessary to achieve a specified gain to about half at cost of noise figure increase of 1.6dB. The optimum length for an unpumped EDF is experimentally measured to be 3m for both single pass and double pass system. By adding 3m of unpumped EDF, the maximum gain enhancements of 2.7dB and 5.7dB were obtained for single pass and double pass at wavelengths of 1576nm and 1600nm, respectively. These results show the proposed amplifier will play an important role in the development of practical L-band EDFA from the perspective of economical usage of pump power.

REFERENCES

[1] M. Yamada, H. Ono, T. Kanamori, S. Sudo and Y. Ohishi, Broadband and gain-flattened amplifier composed of a 1.55 μm-band and a 1.58 μm-band Er3+- doped fiber amplifier in a parallel configuration, Electron. Lett., vol. 33, page 710-711, 1997.

[2] J. Nilsson, S. Y. Yun, S. T. Hwang, J. M. Kim and S. J. Kim, Long-wavelength erbium-doped fiber amplifier gain enhanced by ASE end-reflectors, IEEE Photon. Technol. Lett. vol. 10, page 1551-1533, 1998.

[3] S. W. Harun, S. K. Low, P. Poopalan and H. Ahmad, Gain enhancement in L-band EDFA through a double-pass technique, IEEE Photon. Technol. Lett. vol. 14, page 296-297, 2002.

[4] E. Desurvire, *Erbium-Doped Fiber Amplifiers, Principles and Applications*, New York: Wiley, 1993.

Session 13A

Network Architecture

Photonic MPLS Network Architecture Based on Hikari-Router [Invited]

Naoaki Yamanaka
NTT Network Innovation Laboratories
3-9-11 Midori-cho, Musashino-shi, Tokyo, 180-8585
Japan
Tel: +81-422-59-2047, Fax: +81-422-59-6387, yamanaka.naoaki@lab.ntt.co.jp

Abstract: This paper describes multi-layer traffic engineering and signaling technologies in the photonic-GMPLS-router (Hikari-Router) network. Multi-layer traffic engineering, which yields the dynamic cooperation of IP and photonic layers, is described with the goal of providing IP services cost-effectively. The establishment of photonic cut-through paths is triggered when the Layer 3 traffic loads become excessive which better uses the optical layer (L1) resources. To realize multilayer traffic engineering, we propose an OSPF extension that advertises both the number of total wavelengths and the number of unreserved wavelengths, and an RSVP-TE extension that minimizes the number of wavelength conversions needed. Networks are reconfigured by the proposed scheme so as to utilize the network resources cost-effectively.

Key Words: MPLS, MPλS, Photonic, GMPLS

1. INTRODUCTION

The explosion of Internet traffic has led to a greater need for high-speed backbone networks. The growth in Internet-protocol (IP) traffic exceeds that of IP packet processing capability. Therefore, the next-generation backbone networks should consist of IP routers with IP-packet switching capability and optical cross-connects (OXC) with wavelength-path switching capability to reduce the burden of heavy IP-packet switching loads. In addition, IP traffic fluctuates often over periods as short as hours.

These needs are satisfied by the photonic GMPLS router developed by NTT [1] that offers both IP packet switching and wavelength-path switching capabilities [2][3]. Wavelength paths, called optical label switch paths (OLSP) are set and released in a distributed manner based on the functions offered by the generalized multi-protocol label switch (GMPLS). Since the photonic MPLS router offers the functions of both switching and GMPLS, it enables us to create the optimum network configuration by considering IP and photonic network resources in a distributed manner.

An OSPF extension and a very sophisticated signaling mechanism that uses an extended form of RSVP-TE have been proposed and will be implemented. The OSPF extension is to advertise both the total number of wavelengths and the number of unreserved wavelengths for each link. This information is passed via OSPF packets to all ingress edge nodes so they are aware of the GMPLS link state. Based on the least-load rule, ingress edge nodes try to establish more efficient OLSPs by using the extended RSVP-TE signaling protocol. The signaling packet selects the wavelength that requires the least number of wavelength converters because wavelength conversion is a very expensive operation.

In addition, multi-layer traffic engineering, which ensures that IP and photonic layers dynamically cooperate, is required to provide IP services cost-effectively. This paper proposes a heuristics-based multi-layer topology design scheme that uses IP traffic measurements. The proposed scheme yields the optimum OLSP network topology, i.e. OLSP placement, so as to minimize network cost, in response to fluctuations in IP traffic demand. In other words, OLSP network topology is dynamically reconfigured to match IP traffic demand [4].

2. PHOTONIC GMPLS MULTI-LAYER ROUTER ARCHITECTURE (HIKARI-ROUTER)

The photonic GMPLS multi-layer router consists of an IP router, wavelength router, and photonic-GMPLS-router manager, as shown in Fig. 1 [1].

The router is based on a photonic universal platform with the addition of 3R (Retiming, Reshaping and Regeneration) functions, wavelength conversion, and layer-3 functions. These functions are used adaptively. In other words, if the signal is degraded by fiber loss as well as nonlinear effects such as PMD (polarization mode dispersion) or ASE (Amplified spontaneous emission), the 3R function is activated. In addition, wavelength conversion is also used when signaling is blocked by wavelength overbooking. Of course, layer 3 (L3) packet forwarding is adaptively used when L3 packet forwarding is performed.

Note that the photonic-GMPLS router can transfer three types of signals. Type 1 is transparent transfer when the functions of 3R and wavelength conversion are not needed. Type 2 is λ relay connections that need wavelength conversion. Type 2 connections are also supported by the adaptive use of the 3R function. Finally, type 3 involves layer 3 switching functions for operations such as MPLS path aggregation and L3 packet level forwarding. The photonic-GMPLS router can support all these transfer capabilities. In the photonic-MPLS-router manager, the GMPLS controller distributes own IP and photonic link states, and collects the link states of other photonic MPLS routers. IP traffic is always monitored, and the captured data is passed to the GMPLS controller through a low-pass filter [5]. A multi-layer topology design algorithm processes the

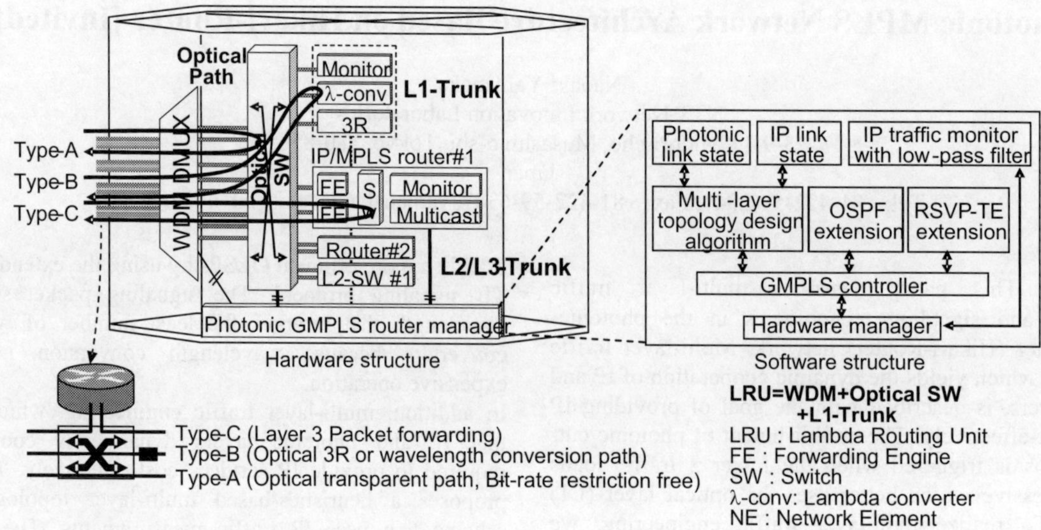

Fig. 1 Structure of photonic MPLS router with multi-layer traffic engineering based on IP traffic monitoring

collected IP and photonic link states and the collected traffic data.

3. GMPLS MULTI-LAYER NETWORK CONTROL AND A TOPOLOGY DESIGN SCHEME

3.1 FRAMEWORK

The bandwidth granularity of the photonic layer is coarse and equal to wavelength bandwidth, λ, i.e. 2.5Gbit/s or 10Gbit/s. On the other hand, the granularity of the IP layer is flexible and well engineered. Figure 2 shows that the center of the multi-layer network structure is a multi-layer photonic backbone network. This network is based on the multi-layer GMPLS router. At the edge node, all exchange traffic volumes between edge routers are monitored [5].

The volume of layer 3 traffic from edge #i to edge #j is exchanged by using the extended RSVP-TE protocol. The edge nodes use this information to establish the complete L3 traffic matrix.

There is a trade-off between the Layer 3 hop-by-hop network

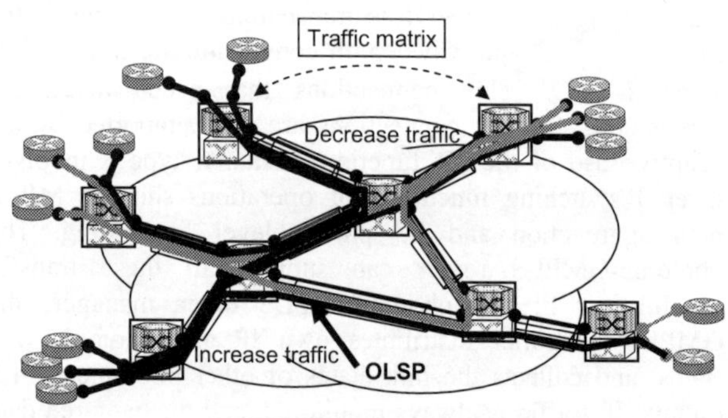

Fig. 2 Multi-layer network design

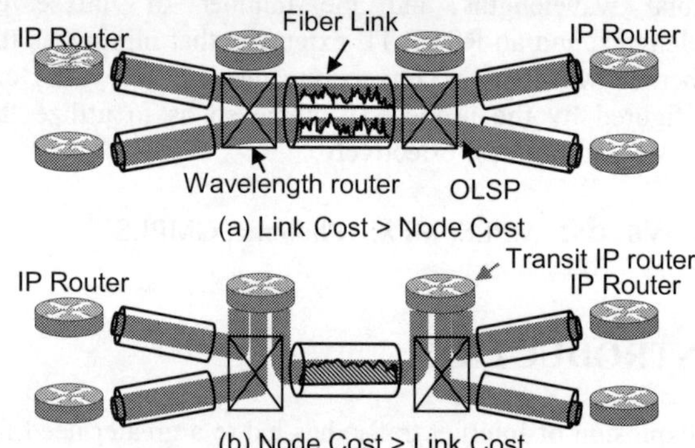

Fig. 3 Trade-off between Node Cost and Link Cost

and the photonic cut-through arrangement in terms of link cost and node cost as shown in Figure 3.

When the traffic demand between source and destination IP routers is much less than λ, the cut-through wavelength path between the source-destination IP routers is not fully utilized and so is not cost-effective. In this case, several IP traffic streams should be merged at some IP transit routers to better utilize the wavelength bandwidth at the cost of IP-packet processing at the transit nodes. On the other hand, when traffic demand between source and destination IP routers approaches or exceeds λ, such stream multiplexing is dropped.

Thus, the setting of cut-through paths between source-destination IP routers depends on IP traffic, and the OLSP topology of the photonic layer should change dynamically according to fluctuations in IP-traffic demand to optimize network resource utilization.

For this purpose, our objective is to minimize network cost Z, which is formulated as follows.

$$z = C_{node} + C_{link} = \alpha \sum_i r_{ip} + \beta \sum_i \sum_j \sum_p \sum_k l_{ijpk}, \qquad (1)$$

where r_{ip} is the cost of port p in router i, l_{ijpk} is the cost of wavelength path k at port p in fiber link ij. α/β is the Node/link ratio. α/β is set to more than one, because IP routers have IP-packet processing functions such as table look up and packet-based switching in addition to wavelength routing functions.

To minimize Z, we adopt the extended version of the BXCQ (branch exchange with quality-of-service constraints) scheme presented in [6], named EBXCQ. The BXCQ scheme was originally intended for multi-layer ATM network design. In BXCQ, the addition/elimination of links is iterated to solve a topological optimization problem with quality-of-service constraints, such as delay and blocking probability. In EBXCQ, the number of ports in both IP routers and wavelength routers, as well as the number of wavelengths per fiber are taken as additional constraints.

3.2 NUMERICAL RESULTS

We have confirmed the effectiveness of EBXBQ by using a LATA network model, see Fig. 4 [7]. Figure 4 also shows that the optimum OLSP topology changes with the IP traffic demand between source-destination IP routers. We assume that IP traffic demands between source-destination IP routers are evenly distributed and that each wavelength is converted at each photonic MPLS router. Wavelength bandwidth is set to 2.5Gbit/s. We used the metric of the average node degree, D, to characterize and evaluate OLSP network topologies. D is the average number of other IP routers to which individual IP routers are connected by OLSPs. As traffic demand increases, the optimum OLSP topology becomes a mesh because each OLSP bandwidth becomes more effectively utilized.

Changing the OLSP topology in response to traffic demand fluctuations dramatically reduces network cost. An estimation of the network cost reduction possible is shown in Fig. 5. Network cost is normalized by the fixed optimum topology T=300 [Mbit/s]. If the OLSP network topology is changed based on traffic measurements; the cost reduction can exceed 50%.

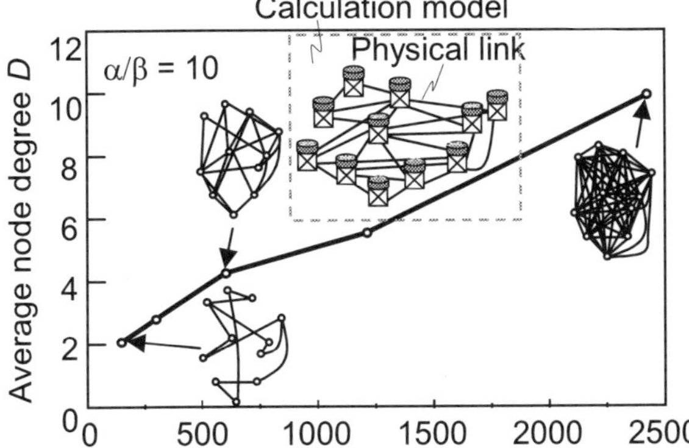

Fig. 4 Numerical model and optimum OLSP topology determined by IP traffic demand

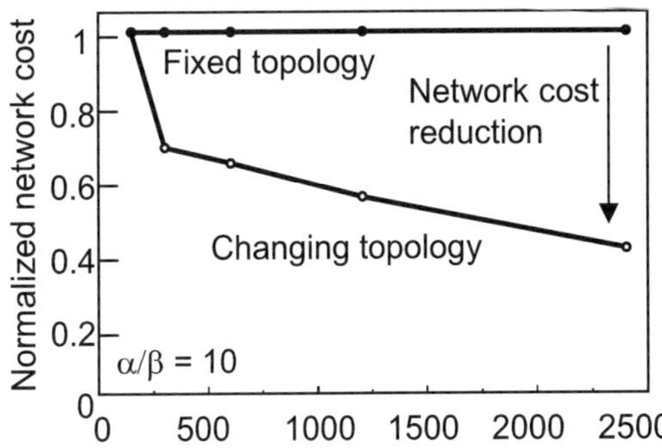

Fig. 5 Cost reduction effect of reconfigurable multi-layer network.

4. ROUTING AND SIGNALING FOR PROPOSED PHOTONIC GMPLS

Current available technologies make optical wavelength conversion very costly and a lot of hardware is needed. In other words, the transparent bit-rate restriction free network is very attractive from the viewpoints of low cost and high flexibility. Accordingly, our OSPF extension and RSVP-TE extension provide effective route selection as well as the least number of wavelength conversions.

First, route advertisement is achieved by the OSPF extension as shown in Fig. 6. Each photonic GMPLS router advertises its total number of used and unused wavelengths. According to this advertisement, edge nodes can discern GMPLS link state and can select the least expensive path. In addition, we proposed to extend the sub-TLV of IETF specification. The extensions include 3R resource information and statistical information such as utilization of each wavelength, and 3R and wavelength conversion resources. This information is used for source routing based on a combination of shortest path first and load information. For the signaling function, we offer the RSVP-TE extension. Link by link routing is used based on source routing. The first photonic GMPLS router sets the unused wavelength information in bit map format in the RSVP signals. Each transit photonic GMPLS router over-writes this information by placing "And" between arriving signaling unused wavelength bitmap and its own unused wavelengths. If there is no unused wavelength, wavelength conversion is used. The router, which offers wavelength conversion, creates a new unused wavelength bitmap and sends it to the next router [7]-[9].

This signaling and routing technique minimizes the frequency of wavelength conversion in the network and so can provide very cost effective photonic networks. The calculation results of the proposed method cost reduction is shown in Fig. 7. The results show that cost reductions of more than 70% can be expected.

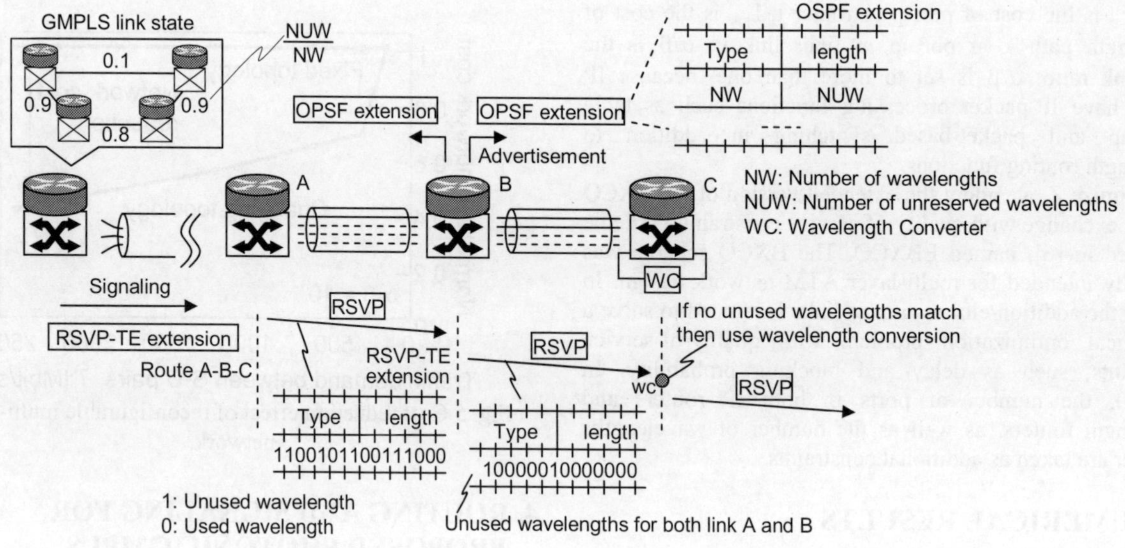

Fig. 6 Photonic GMPLS extensions for both routing and signaling based on OSPF and RSVP-TE

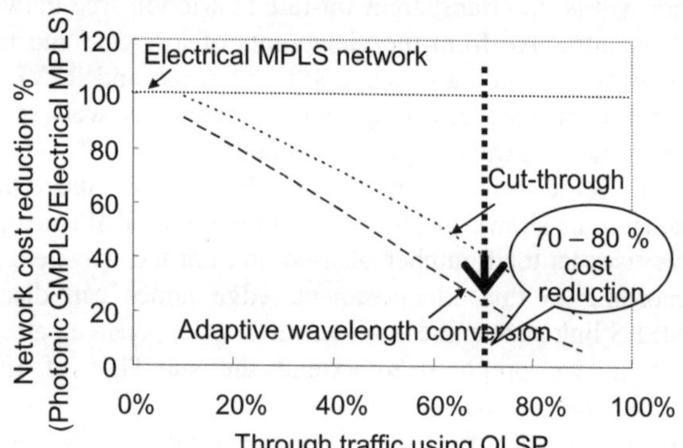

Fig. 7 Network cost reduction breakthrough

5. CONCLUSIONS

This paper presented multi-layer traffic engineering, signaling and routing technologies in the photonic-GMPLS-router (Hikari-router) networks. The photonic-GMPLS-router network architecture is based on OSPF and RSVP-TE extensions. The proposed routing and signaling technique yields the least use of wavelength conversion and so very cost effective photonic networks. This paper also described a heuristic-based multi-layer topology design scheme, called EBXCQ, for IP photonic networks. By monitoring IP traffic loads, photonic MPLS routers are controlled to dynamically change the network configuration; this allows network resources to be utilized efficiently in a distributed manner.

6. REFERENCES

[1] Y. Yamabayashi et al.,"Autonomously Controlled Multiprotocol Wavelength Switching Network for Internet Backbones," IEICE Trans. Commun., vol. E83-B, no. 10, pp. 2210-2215, 2000.

[2] K. Shimano et al., "MPLambdaS Demonstration Employing Photonic Routers (256X256 OLSPS) To Integrate Optical and IP Networks," National Fiber Optic Engineers Conf. 2001 Tech. Proc. p. 5.

[3] F. Kano et al., "Development of Photonic Router (256X256 OLSPS) Realize Next Generation Network with MPLambdaS Control," National Fiber Optic Engineers Conf.2001 Tech. Proc. p. 458.

[4] Naoaki Yamanaka et al., "Traffic Engineering and signaling technologies in photonic-MPLS-router Network," Proc. on HPSR2002, Kobe, Japan, Session 3A-4

[5] K. Shiomoto, "A Simple Bandwidth Management Strategy Based on Measurements of Instantaneous Virtual Path Utilization in ATM Networks," IEEE Trans. Networking, vol.6, no. 5, pp. 625-634, 1998.

[6] E. Oki and N. Yamanaka, "Impact of Multimedia Traffic Characteristic on ATM Network Configuration," J. Network Systems Management, vol. 6, no. 4, pp. 377-397, 1998.

[7] W. Imajuku et al., "Multi-layer routing using multiple-layer switch capable LSRs," IETF draft, draft-imajuku-ml-routing-00.txt, February 2002. (working in progress)

[8] E. Oki et al., "Requirements of optical link-state information for traffic engineering," IETF draft, draft-oki-ipo-optlink-req-00.txt, February 2002. (work in progress)

[9] N. Matsuura et al., "Requirements for using RSVP-TE in GMPLS," IETF draft, draft-matsuura-gmpls-rsvp-requirements-00.txt, February 2002. (work in progress)

THE MERGING OF A WAVELENGTH INTERLEAVED WDM FIBER-RADIO BACKBONE WITH A STANDARD WDM RING NETWORK

[1]Christina Lim, [1]Ampalavanapillai Nirmalathas, [2]Dalma Novak and [2]Rodney Waterhouse
[1]Australian Photonics Cooperative Research Centre, Photonics Research Laboratory,
Department of Electrical and Electronic Engineering, The University of Melbourne, VIC 3010, Australia
Tel: +61-3-8344-4486 Fax: +61-3-8344-6678 E-mail: c.lim@ee.mu.oz.au
[2]Dorsál Networks Inc., Columbia MD 21045, USA

Abstract: We investigate the issues related to the merging of a WDM ring fiber-radio backbone incorporating wavelength interleaving where each base station drives a sectorized antenna interface and a standard WDM infrastructure. The investigations show that by re-allocating the interleaved WDM channels to fit within a 100 GHz block, it enables the mm-wave fiber-radio system with sectorized antenna interface to integrate easily within standard WDM systems. In addition, such channel allocation also improves the overall network capacity.

Key Words: Fiber-radio technology, optical fiber communication, wavelength-division multiplexing

1. INTRODUCTION

Fixed wireless access technology operating at millimeter-wave (mm-wave) frequencies with optical fiber backhaul have been considered for the future provision of broadband interactive services with data rates in excess of 1 Gb/s [1]. The proposed architecture of a generic mm-wave fiber-radio access network consists of a central office (CO) that has the functions of frequency management, routing, switching, and control of the radio signals and an optical fiber network which interconnects a large number of functionally simple and compact remote antenna base-stations (BSs) for wireless distribution. Due to the higher propagation losses of radio signals at mm-wave frequencies, the implementation of pico- and micro- cellular network architectures is essential. The optical fiber network with an rf-over-fiber transport scheme that distributes the radio signals to and from the BSs as a mm-wave modulated optical signal has the advantage of a simplified BS design [2]. However such a transport scheme is susceptible to fiber chromatic dispersion that severely limits the performance of such links [3]. It has been demonstrated that such dispersion-induced power penalties can be overcome using optical single sideband with carrier (OSSB+C) modulation [4].

Due to line-of-sight constraints of radio signals at mm-wave frequencies, a sectorized antenna interface at the remote BS is essential to improve the radio coverage. Such a distribution scheme can be further simplified using wavelength-division multiplexing (WDM) in the optical distribution where each antenna sector in the BS can be driven by a different WDM channel. However implementation of WDM in mm-wave fiber-radio systems incorporating OSSB+C leads to poor optical spectral efficiency. This drawback can be overcome by a recently proposed wavelength-interleaving (WI) technique [5] and has the potential to improve the overall system capacity by threefold [6,7].

It is also essential that mm-wave fiber-radio systems are able to merge or integrate with current wired WDM infrastructure. For an optical WDM metropolitan area network (MAN) servicing a large metropolitan area, a number of mini switching centers are interconnected as shown in Fig. 1. Incorporating broadband pico- or micro-cellular network architectures with a large number of BSs into this network, requires the radio networks to be dimensioned as such a number of clusters of BSs are serviced by a switching node that assumes the role of a CO [8]. By doing so, WDM channels can be dropped off or added into the optical WDM MAN network via the switching center in terms of 100 GHz blocks or depending on the WDM channel spacing.

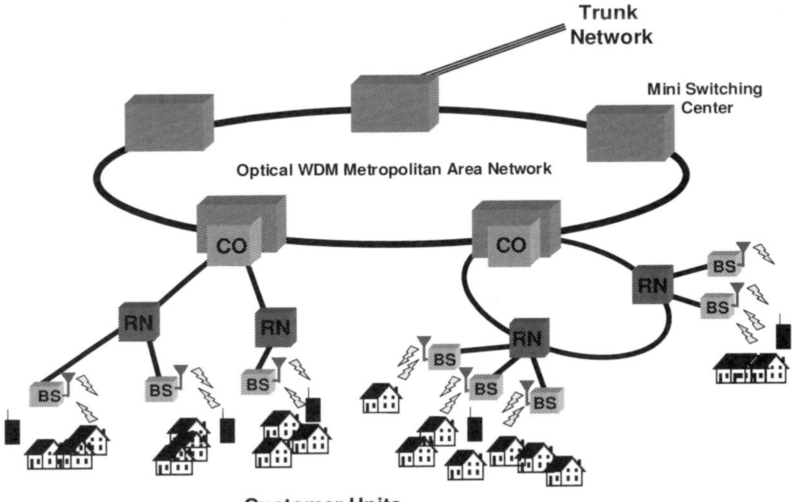

Fig. 1 Merging of optical WDM and broadband radio networks

In this paper, we present a mm-wave fiber-radio system incorporating WI and OSSB+C modulation with a sectorized antenna interface that is capable of being integrated with existing WDM infrastructure. By re-allocating the interleaved WDM channels such that these channels fall within the 100 GHz WDM channel spacing allocation, the mm-wave radio channels can be dropped off or added into the WDM optical MAN network with minimal interruption. We investigate the various channel allocation scenarios for optimum system performance.

2. OPTIMUM FREQUENCY ALLOCATION FOR WAVELENGTH INTERLEAVED WDM CHANNELS FOR A SECTORIZED ANTENNA INTERFACE

Fig. 2 shows a scenario of a WDM fiber-radio backbone with a primary ring architecture incorporating both WI and dispersion tolerant OSSB+C modulation connected to an optical WDM MAN network. The primary ring interconnects the mini switching center or the CO to the remote nodes (RNs) and a number of BSs are connected to each RN in a star-tree configuration. Assuming the optical WDM MAN network has a WDM channel spacing of 100 GHz, multiple blocks of 100 GHz containing the interleaved WDM channels are then dropped from the MAN network via the CO. These blocks are distributed to the RNs and a selection of wavelength blocks is dropped at each RN. At the RN, a wavelength band 100 GHz wide containing a few wavelength-interleaved channels will then be distributed to the designated BS to feed the sectorized antenna interface. Similarly, in the upstream direction, wavelengths originating from each antenna sector in the BS falling within the same 100 GHz bandwidth, are added to the network before being routed back to the trunk network. Such a mm-wave fiber-radio system enables the use of standard WDM components within the CO, RNs and BSs, thus reducing the complexity and the stringent requirements of the optical add-and-drop hardware.

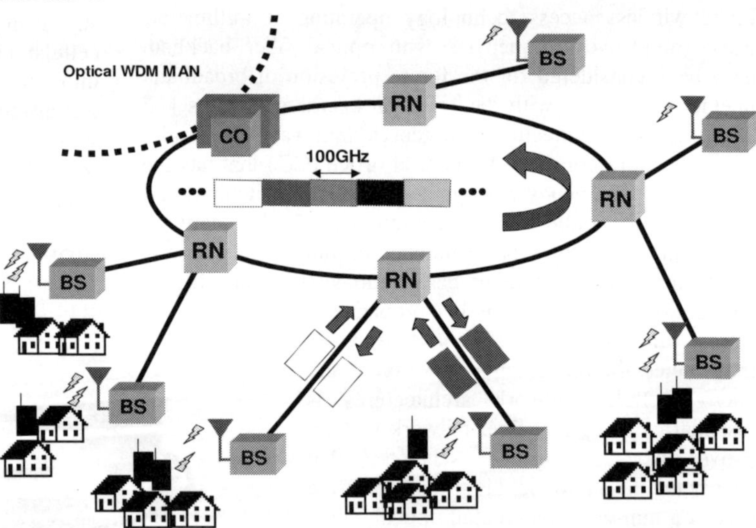

Fig. 2 Mm-wave fiber-radio backbone with ring and star-tree architecture with a 100 GHz wavelength block

Assuming a modulation frequency of 36 GHz and an erbium-doped fiber amplifier (EDFA) bandwidth of 30 nm, a total of 52 WDM channels with OSSB+C modulation can be accommodated in this network without WI. If these channels are interleaved at a channel spacing of 24 GHz, the total capacity can be theoretically tripled to 156 channels. However this does not take into account the limitations imposed by hardware based on 100 GHz WDM spacing. When this is taken into consideration, Fig. 3 shows that only three WI WDM channels are able to fit within a 100 GHz bandwidth with a guard band in excess of 15 GHz. This reduces the total capacity from 156 to 111 channels within the 30 nm optical amplifier bandwidth. In addition, such frequency allocation will allow only each 100 GHz block to serve a BS with a three-sector antenna interface and is not sufficient for a four-sector antenna interface.

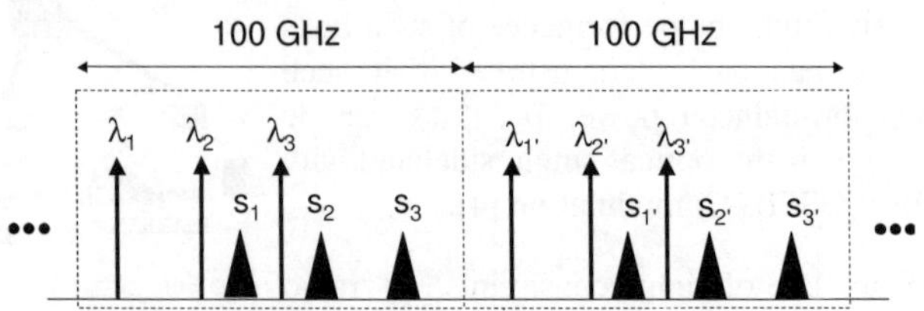

Fig. 3 Channel allocations with guard band incorporated for three WI channels within a 100 GHz bandwidth

To accommodate for a four-sector antenna interface with a modulation frequency of 36 GHz, modifications have to be made to the frequency allocation of the interleaved WDM channels. The interleaved channels can be allocated in six different configurations as depicted in Fig. 4, namely Configuration A (Fig. 4a), Configuration B (Fig. 4b), Configuration C (Fig. 4c), Configuration D (Fig. 4d), Configuration E (Fig. 4e) and Configuration F (Fig. 4f). In these frequency allocations, the distance between adjacent carriers or between adjacent sidebands is set to 9 GHz. Therefore, four interleaved channels are able to fit within a 100 GHz bandwidth and with a guard band in excess of 35 GHz. Now the total capacity of the network has increased to 148 channels compared to 111 channels.

3. SIMULATION RESULTS FOR OPTIMUM FREQUENCY ALLOCATION FOR A FOUR-SECTOR ANTENNA INTERFACE

To investigate the optimum channel allocation scheme with regards to those shown in Fig. 4 for a mm-wave fiber-radio backbone incorporating WI based on a 100 GHz WDM

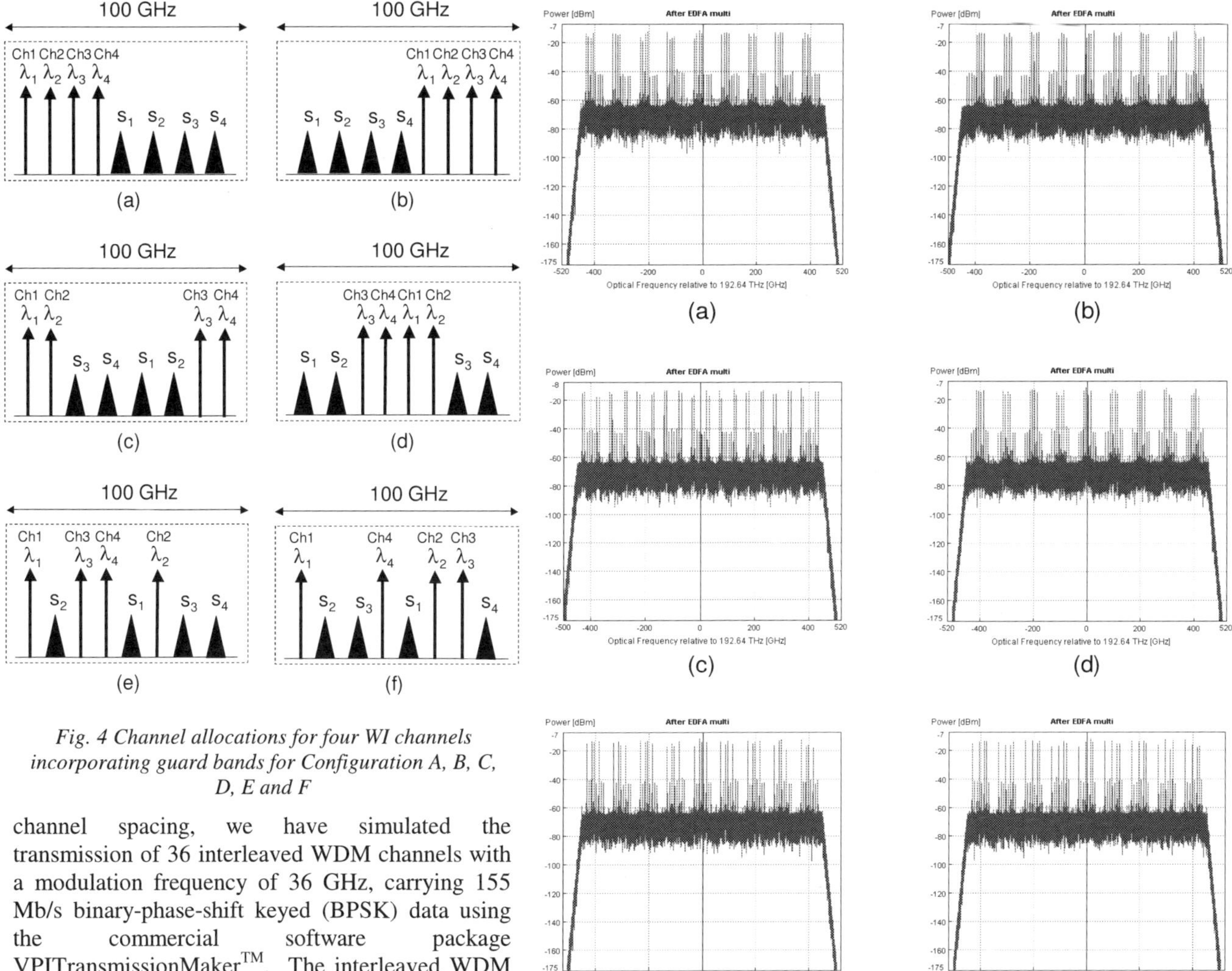

Fig. 4 Channel allocations for four WI channels incorporating guard bands for Configuration A, B, C, D, E and F

Fig. 5 Optical spectrum after transmission over 20 km of SMF and optical amplification for Configuration A, B, C, D, E and F

channel spacing, we have simulated the transmission of 36 interleaved WDM channels with a modulation frequency of 36 GHz, carrying 155 Mb/s binary-phase-shift keyed (BPSK) data using the commercial software package VPITransmissionMaker^{TM}. The interleaved WDM channels were transmitted over 20 km of single mode fiber (SMF) before one block (the middle block) of 4 interleaved channels within a 100 GHz bandwidth was dropped at the RN and subsequently distributed to the designated BS to feed a four-sector antenna interface. Shown in Fig. 5 are the optical spectra of the 36 channels after optical amplification and transmission over 20 km of SMF for the six configurations of Fig. 4. The center block containing four channels was then dropped at the RN and received at the BS. The bit-error-rate (BER) for each received channel was measured for the different frequency allocations. Channel 1 (Ch 1), Channel 2 (Ch 2), Channel 3 (Ch 3) and Channel 4 (Ch 4) are defined as the sets of optical carrier and the corresponding sideband or sideband and the corresponding optical carrier starting from the lowest frequency within the wavelength block as labelled in Fig. 4.

Figs. 6a-6d show the BERs for each received channel plotted as a function of average received optical power for the six different configurations. The best and worst performing configurations are different for each channel and similarly, the power penalty at a BER of 10^{-9} between these configurations also vary for each channel. From Fig. 6, it can be seen that the power penalties incurred by Ch 1, Ch 2, Ch 3, and Ch 4 are 1.8 dB, 0.7 dB, 1.5 dB, and 2.7 dB, respectively. These penalties could be attributed to higher order harmonics generated by the OSSB+C modulator that was not filtered out, falling within the channel bands and adjacent channel and block crosstalk. It was observed that the power penalty incurred by the end channels, i.e. Ch 1 and Ch 4 are larger compared to the middle channels, i.e. Ch 2 and Ch 3. This could be due to the fact that the optical carriers or sidebands of Ch 1 and Ch 4 are always located in the middle of the band where a large number of intermodulation products are located.

Table 1 shows the comparison of the power penalties between the best and worst performing received channels for each of the configuration, it can be seen that the power penalties incurred vary from 1 dB to 3 dB. Configuration A displays the lowest power penalty of 1 dB compared to Configuration C with the highest power penalty of 3 dB. These variations again show that the performance of the received channels is dependent on how the channels are allocated within the 100 GHz band and the incurred penalties could be attributed to adjacent channels or blocks interference, higher order harmonics and intermodulation products generated that fall within the band. From these results it can be concluded that Configuration A provides the best performance in terms of generating the lowest power penalty between the best and worst received channels.

4. CONCLUSIONS

We investigated the issues related to the optimization of system capacity and performance of a mm-wave fiber-radio backbone incorporating wavelength interleaving with a sectorized antenna interface when such a system is integrated to a standard WDM infrastructure. By changing the interleaved channel allocations to fit within a 100 GHz band, it enables the mm-wave fiber-radio system to be integrated easily with current WDM infrastructure. In addition, this scheme also increases the total capacity of the system with the inclusion of guard bands to separate the adjacent wavelength blocks. The investigations showed that the mm-wave fiber-radio system performance was dependent on how the channels were allocated within the band. Simulation investigations revealed that the incurred power penalty was larger for channels located at the end of band with the corresponding sidebands or optical carriers fall within the center of the band for a variety of channel allocations. There exists an optimum channel allocation scheme that minimizes the power penalty between the worst and best performing channels.

Configuration	A	B	C	D	E	F
Power penalty between best and worst channels (dB)	1.1	2.5	3	2.3	2	2.6

Table 1 Power penalty between best and worst performing channels for all configurations

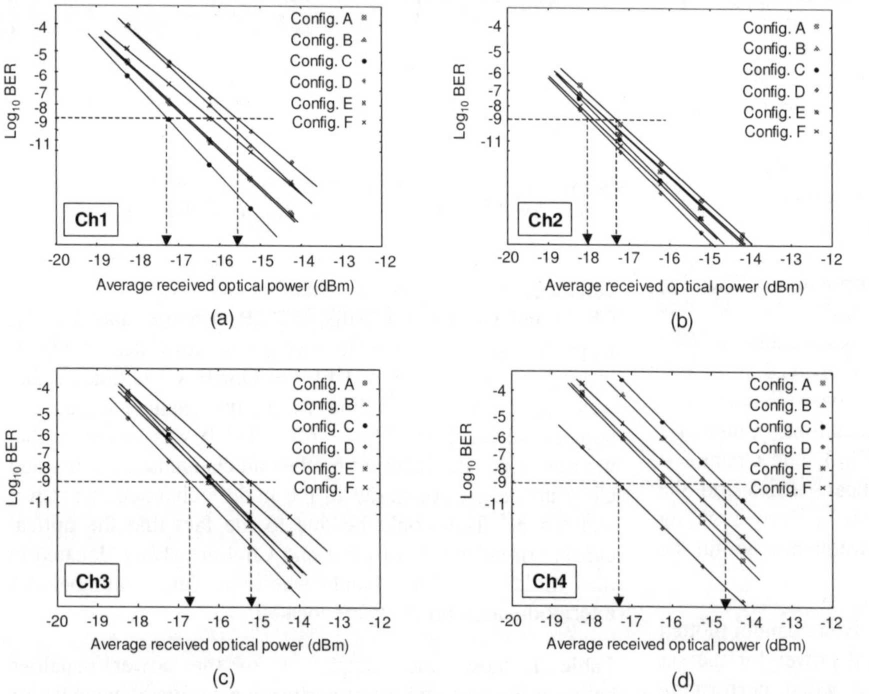

Fig. 6 BERs for (a) Channel 1 (b) Channel 2 (c) Channel 3 (d) Channel 4 for the six different channel allocations

5. REFERENCES

[1] R. Heidemann and G. Veith, "mm-wave photonic technologies for Gbit/s-wireless-local-loop", *Proc. OECC*, Chiba, Japan, pp. 310-311, Jun. 1998.

[2] G.H. Smith, C. Lim, and D. Novak, "A millimeter-wave full-duplex fiber-radio star-tree architecture incorporating WDM and SCM", *IEEE Photon. Technol. Lett.*, vol. 10, no. 11. pp. 1650-1652, Nov. 1998.

[3] H. Schmuck, "Comparison of optical millimetre-wave system concepts with regard to chromatic dispersion", *Electron. Lett.*, vol. 31, no. 21, pp. 1848-1849, Oct. 1995.

[4] G.H. Smith, D. Novak, and Z. Ahmed, "Technique for optical SSB generation to overcome dispersion penalties in fibre-radio systems", *Electron. Lett.*, vol. 33, no. 1, pp. 74-75, 1997.

[5] C. Lim, A. Nirmalathas, D. Novak, R.S. Tucker, and R.B. Waterhouse, "Technique for increasing optical spectral efficiency in millimetre-wave WDM fibre-radio", *Electron. Lett.*, vol. 37, no. 16, pp. 1043-1045, Aug. 2001.

[6] C. Lim, A. Nirmalathas, D. Novak, and R. Waterhouse, "Capacity analysis for a WDM fiber-radio backbone incorporating wavelength-interleaving", *Proc. OFC*, CA, USA, pp. 355-357, Mar. 2002.

[7] C. Lim, A. Nirmalathas, D. Novak, and R.B. Waterhouse, "Network performance and capacity analysis for WDM ring fiber-radio backbone incorporating wavelength-interleaving", to be presented at OECC 2002, Japan, Jul. 2002.

[8] A. Nirmalathas, C. Lim, D. Novak, R.B. Waterhouse, and D. Castleford, "The merging of photonic and radio technologies", *Proc. OECC*, Sydney, Australia, pp. 227-230, Jul. 2001.

Trial of an RPR-based Optical Metro Ring

R. Vasudha, H.S. Cheng and L.H. Ngoh

Laboratories for Information Technology
21, Heng Mui Keng Terrace
Singapore 119613
{vasu,hscheng,lhn}@lit.a-star.edu.sg

Abstract

This paper describes an experimental study of the Resilient Packet Ring (RPR) media access control (MAC) technology which is optimized for IP traffic in a Metropolitan Area Network (MAN). The study involved the deployment and trials of an RPR test-bed encompassing a public optical fiber infrastructure using Cisco Systems' Dynamic Packet Transport (DPT) Ring Technology – a pre-standard RPR implementation. The work described here focused on a number of important RPR protocol features that are vital to its future success as a MAN network technology. Related investigation works on RPR/DPT have so far been done through simulation studies only.

Key Words : RPR, DPT, Network Trial

1.0 INTRODUCTION

Resilient Packet Ring (RPR) is a new media access control (MAC) technology designed to optimize bandwidth utilization and facilitate services over a ring network [1]. Today, data traffic in the MAN is mainly transported over statically provisioned, circuit switched networks using the highly reliable Synchronous Optical Network (SONET) / Synchronous Digital Hierarchy (SDH) technology [2]. The allocation of fixed bandwidth to individual circuits for the transportation of data traffic is not an efficient approach since data traffic is bursty in nature. On the other hand, Ethernet (widely used in local area networks (LANs)) is optimized for data traffic, [2] but lacks the high level of reliability demanded by MANs. The combination of RPR's improved reliability mechanism and bandwidth efficiency therefore makes it a more ideal technology for MANs. Currently, RPR is still in the process of being standardized, and the standardization effort is expected to be completed by 2003.

While computer simulation studies are useful in our understanding of RPR's performance [3], field-trials and experimentations of RPR technologies would provide us with an insight into the practical aspects. Our testbed comprised of 3 router nodes running a pre-standard RPR implementation (DPT) from Cisco Systems, and connected over the public optical fibre infrastructure of Singapore Telecommunications in a ring topology spanning around 50 km. The motivation of the trial is to shed light on the suitability of RPR as a MAN technology by experimentally verifying the following key features: i)Spatial Reuse, ii) Dynamic Allocation of Bandwidth, iii) Simplified Service Provisioning, iv) Protection through Ring Wrapping or Steering [2].

2.0 KEY FEATURES OF RPR

In this section, we briefly describe the key RPR features that are under investigation in this trial network.

2.1 PROTECTION SWITCHING PROTOCOL

Any networking protocol designed for use in MANs must offer a high level of reliability against network failures through some form of protection scheme. For both SONET and RPR, the goal is to provide protection within 50ms in case of link or node failure. In SONET, protection is achieved by dedicating fifty percent of bandwidth as a backup resource. RPR does not require dedicated protection bandwidth [2]. The DPT technology evaluated by us supports *wrapping protection*[4].

2.2 SPATIAL REUSE

With spatial reuse, unicast packets traverse only the ring spans between the source and destination nodes instead of the whole ring [4,5,6]. At the destination node, unicast packets are stripped from the ring. Multiple source-receive pairs can simultaneously exploit the available raw speed of the network independently, and thus the overall aggregate bandwidth of the ring is effectively increased [4]. This feaure is important particularly in supporting environments where multi-source/destination peer-to-peer network traffic is predominant. The goal of evaluating this feature is to verify whether/how well the ring can support multiple concurrent transmit and receive operations in different segments of the network independently.

2.3 BANDWIDTH ALLOCATION

RPR allows dynamic allocation of bandwidth where all users share a common pool of bandwidth resources. The resulting statistical multiplexing effect will enable RPR to support more users than it could if each user is assigned a fixed amount of bandwidth. Equally, RPR can also support applications that have very bursty traffic. In RPR, this characteristic is also coupled with a fairness algorithm to ensure that all nodes on the ring can have their fair share of the raw bandwidth in the event where demand for bandwidth outstrips the maximum network capacity.

3.0 RPR TESTBED CONFIGURATION

Figure 1 shows the RPR testbed consisting of three nodes. A Cisco 7507 router node was positioned at the Centre for Internet Research (CIR), and a Cisco 7513 at the Singapore Telecommunications Office. A high-end Gigabit Switch Router (GSR 12008) was placed at the Laboratories for Information Technology (LIT). The ring circumference was approx. 50 km.

Each router was fitted with a DPT module having two bi-directional OC-12 ports with Multimode (MM) interfaces. To interconnect all the three nodes via the public fiber infrastructure based on Singlemode (SM) fibers, MM<->SM converters were used. The converters allow protocol-independent coupling of multimode and singlemode fibers.

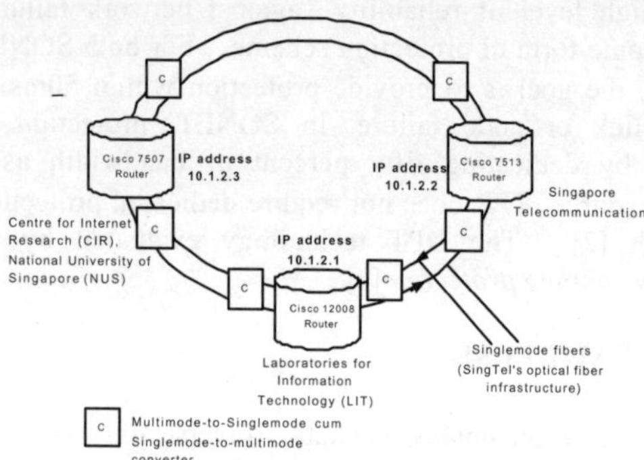

Figure 1. RPR Testbed (DPT Ring)

3.1 MEASURED POWER LOSS

Table 1.0 shows the power loss measured across various singlemode fiber segments of the ring.

Table 1.0 Measured Power Loss

Singlemode Fiber Segments	Measured Power Loss (dB)	Approx. Distance (km)
ST to LIT	17.25	20
LIT to ST	17.91	20
ST to CIR	14.26	15
CIR to ST	14.54	15
LIT to CIR	8.77	10
CIR to LIT	9.67	10

4.0 EXPERIMENTS

After the RPR testbed was deployed, a series of experiments was performed to evaluate its features and performance. The experiments were: i) Ring Wrap Test, ii) Pass-Through Test, iii) Fairness Test and iv) Spatial Reuse Test.

In the trial, a Smartbits Traffic Generator was used for throughput and packet-loss based measurements. This requires the packets generated at one port to return to another port on the same Smartbits chassis. To ensure that the IP packets are "forced" into the RPR ring, and not internally routed over the LIT router's backplane, a routing policy-map was installed at the LIT router's ATM interfaces that were connected to the Smartbits Traffic Generator. The policy states that the traffic originating at the Generator's subnet should be 'next-hopped' to one of the other RPR nodes. Static routes were also configured at the SingTel and CIR nodes.

4.1 RING WRAP TEST

Cisco's DPT protocol uses Ring Wrap for protection against failures caused by fiber cuts. In this test, we verified this feature by deliberately causing a disconnection between the SingTel and LIT nodes – Fig 2. The following observations were made:

i) Ring wraps occurred on Side B of SingTel router and Side A of LIT router.

ii) Router at CIR remained the same (i.e. Ring Wrap did not occur)

iii) Ping tests showed that there was IP connectivity between LIT and SingTel.

This test showed that the RPR's feature of protection through Ring wrapping was functioning correctly. There would not be IP connectivity between SingTel's and LIT's routers if the two router interfaces were not 'wrapped'.

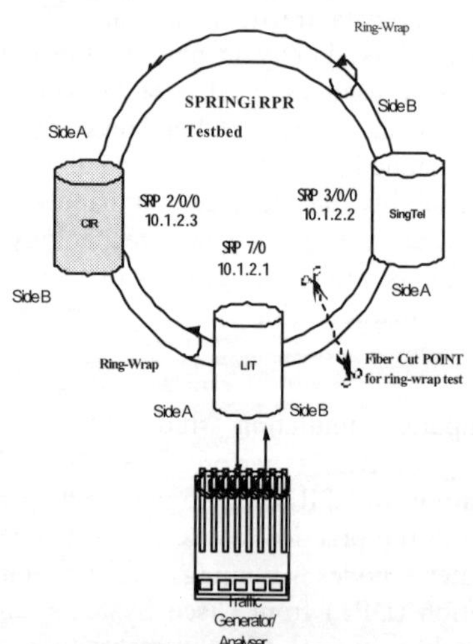

Figure 2. Ring Wrap Test

4.1.1 ESTIMATION OF RING WRAP TIME

Tests were done to get an estimate of the time it takes for the Intelligent Protection Switching (IPS) mechanism to detect a fibre-cut and perform a ring-wrap at the nodes adjoining the cut. The estimates in Table 2.0 and 2.1 below are based on packet loss measurements. Traffic Path before Ring-Wrap: Smartbits [Port 1] → LIT → CIR→ SingTel → LIT → Smartbits [Port 2]. The frame rate for the traffic was adjusted to ensure that all the router nodes were able to handle it without dropping any packets, and it was verified that no packet was dropped by the routers due to lack of buffers. This implies that any packets lost are due to the fibre-cut and kicking-in and completion of the ring-wrap. Hence we can get an upper-bound of the ring-wrap time from the number of lost frames and the frame rate, using the formula : RingWrapTime = Frames Lost/Frame Rate.

Table 2.0 Ring Wrap Time Estimation

Traffic Type: CBR, Frame Length: 50 bytes, Frame Rate: 100000 fps				
Test No.	No. of Frames Transmitted	No. of Frames Received	Frames Lost	Ring-wrap time (ms)
1	1389650	1389260	390	3.90
2	1253887	1253602	285	2.85
3	1448536	1447538	998	9.98
4	1647098	1645051	2047	20.47

Table 2.1 Ring Wrap Time Estimation

Traffic Type: CBR, Frame Length: 1518 bytes, Frame Rate: 8850 fps				
Test No.	No. of Frames Transmitted	No. of Frames Received	Frames Lost	Ring-wrap time (ms)
5	215013	214973	40	4.52
6	260613	260573	40	4.52
7	120471	120425	46	5.20
8	159306	159274	32	3.62

The ring-wrap time was found to be well within the specified 50 milliseconds.

4.2 SPATIAL REUSE TEST

With destination stripping, unicast packets traverse only the ring spans between the source and destination nodes instead of the whole ring [2, 3, 4]. The traffic generated by the Smartbits was destined for IP Address 10.1.2.2. where 10.1.2.2 is the IP address of the DPT interface of SingTel's node (Figure 3).

Verification was done by recording and comparing the values of the packet counters of all the routers' interfaces. No packets were received or transmitted by the CIR node at NUS. The receive counter at ST was equal to the transmit counter at LIT.

Table 3 Multiple Concurrent Streams' Throughput

Traffic Type : CBR, Frame Length : 1470 bytes		
Traffic Stream Description	Throughput of Stream A	Throughput of Stream B
1 Stream generated by Iperf [7] Source : CIR Destination : LIT	95.2 Mbps	0
1 Stream generated by Iperf Source : ST Destination : LIT	0 Mbps	95.6 Mbps
2 Streams generated by Iperf Stream A : CIR to LIT Stream B : ST to LIT	94.8 Mbps	95.8 Mbps

With destination stripping in place, and the absence of a shared token to control media access, it becomes possible for multiple streams of traffic to co-exist on the ring. This was verified by generating two concurrent traffic streams indicated Stream A and Stream B in Figure 3.

Each stream was able to attain throughput almost identical to a corresponding single traffic stream situation.

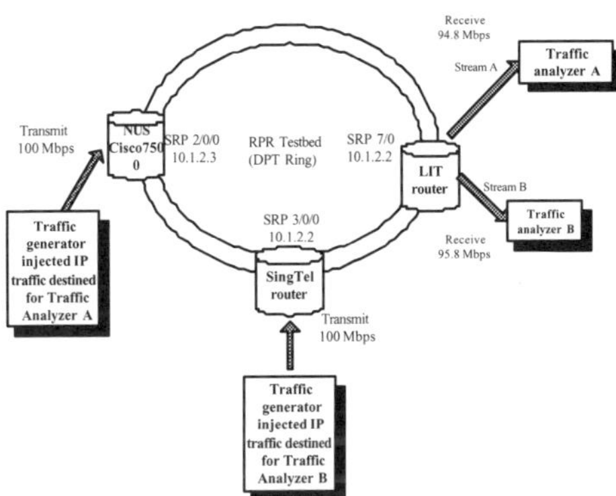

Figure 3. Spatial Reuse Test

4.3 BANDWIDTH ALLOCATION TEST

With many nodes sharing bandwidth on the RPR ring, it is necessary that each node get a fair share of the bandwidth without any complex and inefficient provisioning as in SONET. An RPR network can handle dynamically changing traffic patterns using a feedback mechanism [5].

To verify the effectiveness of the feedback mechanism in ensuring fairness among the various traffic flow, we performed the test shown in Figure 4. All the three nodes had transmitting as well as transiting traffic. All the three test streams are best effort in nature.

Although the transmitting rates of these three traffic streams were not the same, they were received at roughly the same rate (between 90.4 Mbps and 93.6 Mbps). This demonstrates that the DPT ring is able to allocate bandwidth fairly to all the nodes.

Table 4 Description of Test traffic streams

Path of traffic stream	Traffic Generator	Transmit Rate (Mbps)	Receive Rate (Mbps)
TG@CIR to TA@LIT	Iperf	100	92.7
TG@ST to TA@LIT	Iperf	100	90.4
TG@LIT->ST-> CIR-> TA@LIT	Smartbits	238.39	93.6

TG: Traffic Generator TA: Traffic Analyser

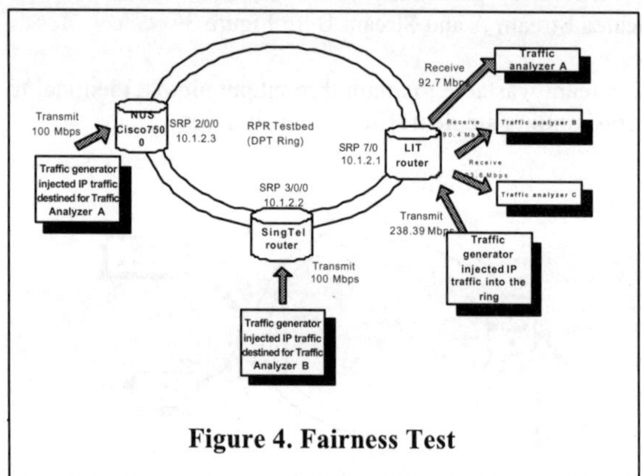

Figure 4. Fairness Test

5. CONCLUSIONS AND FUTURE WORK

Due to rapid growth of the Internet data traffic, it is necessary to develop and deploy networks that are optimized for IP data traffic. The Resilient Packet Ring (RPR) Metropolitan Area Network (MAN) media access control (MAC) standard provides service providers with an Internet-Protocol efficient technology with features comparable to SONET/SDH networks.

In this paper, some of RPR's key features highly desirable to a service provider were investigated using Cisco Systems' Dynamic Packet Transport (DPT) technology as the (pre-standard) RPR implementation. The experiments on these features verified that some of RPR's key features performed well over our trial network spanning 50 km.

Future work includes providing quality of service (QOS) guarantees to certain classes of traffic on the RPR Ring, and to further investigate the inter-working of IP-based solutions (e.g. Multi-Protocol Label Switching (MPLS)) and Traffic Engineering with RPR technology.

ACKNOWLEDGEMENTS

The authors would like to acknowledge the contributions of:
Centre for Internet Research, National University of Singapore: Mr. Lihua Yuan
Singapore Telecommunications: Mr. Chia Kok Soon, Ms. Chan Hwee Kheng, Mr. Lim Poon Wee, Mr. Visut Angkasuwan, Mr. Chia Ser Pan, and Dr. Koh Ping Chiek.
SingAREN/National University of Singapore: Mr. Yuan Zheng and Mr. Chen Gong.
Mr. Wong Chiang Yoon

Agency for Science, Technology and Research (A*STAR), Singapore, the funding agency.

REFERENCES

1. "Positioning RPR in the Technology Universe", A White Paper by the Resilient Packet Ring Alliance, 2001.http://www.rpralliance.com/articles/PositioningRPR.pdf
2. N. Cole, J. Hawkins, M. Green, R. Sharma, K. Vasani, "Resilient Packet Rings for Metro Networks", A White Paper by the Resilient Packet Ring Alliance, 2001, http://www.rpralliance.com/articles/RPR_AllianceGOC.pdf
3. D. Xie, B. J. Lee, "Cisco DPT/SRP Technology and Performance Evaluation" http://ww.cisco.com/warp/public/cc/techno/wnty/dpty/tech/dptm_wp.htm
4. J. Kao, ed., Proposed Draft Standard for Resilient Packet Ring Access Method and Physical Layer Specifications" Submitted to IEEE as the proposal Gandalf, Draft 0.4 P802.17/D0.4, November 8, 2001 http://grouper.ieee.org/groups/802/17/documents/drafts/gandalf_04Plus.pdf
5. An Introduction to Resilient Packet Ring Technology, A White Paper by the Resilient Packet Ring Alliance, 2001, http://www.rpralliance.com/articles/RPRWhitePaper10-01.pdf
6. Tsang, G. Suwala, "The Cisco SRP Mac Layer Protocol", www.faqs.org/rfcs/rfc2892.html.
7 Iperf: A TCP/UDP Bandwidth measurement tool. http://dast.nlanr.net/Projects/Iperf

Benefits of Optical Subnets in WDM Ring Networks

Ashwin Gumaste, Susumu Kinoshita and Gerardo Castanon
Photonics Networking Laboratory,
Fujitsu Network Communications, Inc.
2801 Telecom Pkwy, Richardson, TX. 75082, USA

Abstract: In this paper we propose the concept of optical subnets in WDM rings. By strategically placing few WDM optical add-drop multiplexers (OADMs), interleaved with many passive nodes, this hybrid ring can perform as well as a conventional OADM ring. The section of the ring between two given OADMs is called an optical subnet, while the OADM itself is called a wavelength gateway. We show analytically as well as through simulation that the capacity is still almost comparable to the case with only WDM OADMs, while the cost involved is greatly reduced. The architecture termed as optical subnets has salient features of low cost and good throughput.

I INTRODUCTION

WDM is a key technology in enhancing the network bandwidth at the optical layer. The topology in which WDM systems are used plays a key role in determining the extent to which the WDM network is utilized. Ring topologies are most common in today's networks. Wavelength reuse is one of the key benefits of WDM networking in rings. WDM add-drop units serve as network elements on the periphery of such optical rings. By using WDM add-drop equipment at each network element (node) the entire composite signal can be fully de-multiplexed into its constituent channels and switched (added/dropped or passed through). A generic WDM network element is built of multiplexers / de-multiplexers, switching matrix, set of transponder cards pre and post-line amplifiers in addition to some customized components. The number of transponder cards associated with each node is proportional to the net traffic added and dropped at that node. The cost of total number of transponders is thus dynamic. The main cost in the network is the multiplexer/ de-multiplexer and switching matrix. The theoretical upper bound on capacity of a WDM ring network is the product of number of wavelengths (assuming equal bandwidth per wavelength) and the number of nodes in the ring. However, this capacity can never be utilized because the average length of lightpaths [1] is generally greater than one hop. In other words, there are a sizable number of lightpaths for the establishment of which, WDM nodes need to configure their switches in pass-through mode. The practical capacity of a WDM ring is the product of number of wavelengths (bandwidth) and number of nodes divided by the average hop distance for all the lightpaths at a given time. This is a dynamic quantity with variation proportional to the traffic matrix. The system is under-utilized due to this offset in capacity. At a WDM network element, it is the switches (generally 2x2 per channel) which facilitate wavelength reuse by spatially separating two lightpaths on the same wavelength by positioning in the cross state. In [2] it was seen that there was a numerical limit to the number of lightpaths that could be set up on a single wavelength in a 2-fiber ring. For most practical networks where N is the numbers of nodes in the ring varying from 6-14, the number of lightpaths that could be set up on one wavelength is on average 2.43. This result proved the redundancy of multiplex and demultiplex sections as well as switches on each and every node in the ring. Due to the dynamic variations in traffic and the random distribution of length of lightpaths, it is difficult to predict the pure pass-through (transit) nodes and their placements. In this paper we attempt to address the issue of maximizing the system throughput with a new economical network topology. The network is divided into 'optical subnets', which are 'guarded' from each other using 'wavelength gateways'. This paper is divided as follows: section II describes the concept of optical subnet and wavelength gateways, section III describes the simulation model, section IV shows the analytical model to calculate the system degradation with respect to a conventional network, section V discusses the results of simulation.

II OPTICAL SUBNET: SYSTEM DESIGN

In [3] Slauze et. al. demonstrated a transparent ring network where peripheral nodes in the ring were constructed of 2x2 couplers and the cost of such nodes was low. In [4] Takeguchi et al demonstrated a unique combining and distributing amplifier inherently built of passive elements for dynamic OADM configuration. The passive node comprised of distributing and combining amplifier in addition to a set of acousto-optic tunable filters transverse to the signal flow. The passive node was used as a network element without wavelength reuse (as it lacked a switch to spatially separate two lightpaths). A single hubbed-ring network formed with any of the two proposed configurations proved to be a very low cost alternative, but was limited in capacity to λ_{max} the total number of wavelengths. The network could not deploy wavelength reuse on account of the passive node architecture. The spectrum of optical WDM ring networks thus has two extreme components, a low-cost passive network element with less system capacity and a high-cost conventional WDM OADM based network with excess system capacity. The prior needed more capacity while the excess capacity of the latter could never be utilized. We propose a subnet configuration, which essentially divides the ring logically into optical subnets. Each subnet consists of a number of passive nodes. A conventional WDM OADM demarks two subnets from one another. Fig. 1 shows a typical ring network with optical subnets internally comprising of passive low-cost nodes, and WDM OADMs between any two subnets. Inside a subnet, the fiber(s) acts as a shared medium. The OADM dividing two subnets breaks the spatial continuity between the two-shared mediums. Between any two OADMs are a number of passive nodes. The OADM can be viewed as a wavelength gateway on account of the spectral reuse available at the point of presence. For a given network the number of subnets needed depends on the maximum capacity of each node. Though network traffic is dynamic and hard to predict, the number of transponder cards needed at each node to provision lightpaths makes the upper bound on traffic an estimable quantity on account of the offline nature of demands. Let $(Tr_i)_{max}$ be the upper bound on traffic (in lightpaths) emanating from node 'i' and $(\text{Ó}Tr_i)_{max}$ be the cumulative maximum traffic in the ring. Further if the total number of wavelengths (assuming equal channel spacing) is λ_{max}, then the maximum number of subnets is given as $S_{max} = (\text{Ó}Tr_i)_{max}/\lambda_{max}+1$.

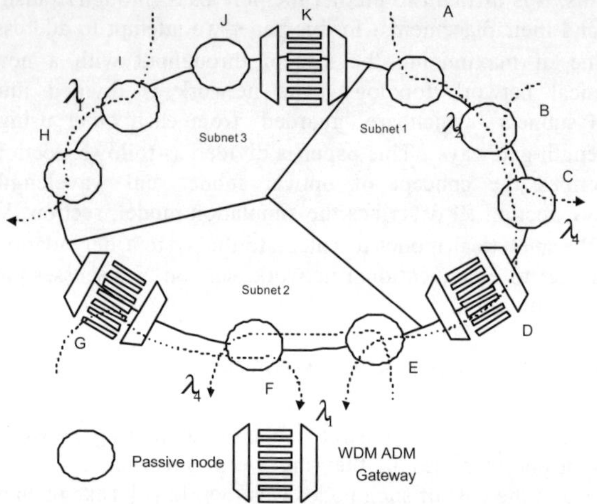

Fig. 1. Shows a 3 subnet ring configuration

```
Algorithm:
Initialize j← 1; t←1; sum←0
    for i=1:N
        sum=sum+cap(N_i)
        if sum>=λ_max
            subnet(j) = node(t....N_{i-1})
            sum←0;
            j=j+1
            t←N_i
        elseif i=N & j>1
            subnet(j) = node(t...N)
        end
    end
```

The algorithm above gives the nodes in each subnet as well as the total number of subnets. Consecutive nodes that have a cumulative bandwidth requirement approximately equal or slightly less than the total available bandwidth (in lightpaths) are grouped together into one subnet. The last node of each subnet is a WDM OADM (wavelength gateway). Moreover for an arbitrary network the last subnet may not be as heavily loaded as the other subnets (though this is a deficiency in the network, it serves as a good load balancing approach for hubbed traffic)

Two kinds of lightpath establishment deserve attention, intra-subnet lightpath establishment and inter-subnet lightpath establishment. The wavelength assignment algorithm [5] maximizes wavelength reuse. It also assigns wavelengths heuristically such that all intra-subnet (ingress and egress nodes in the same subnet) lightpaths are assigned on the lowest available wavelength. On the other hand inter-subnet lightpaths (those whose ingress and egress nodes are on different subnets or different rings for that matter) are assigned on the highest possible wavelengths. This way we have a static load balancing which also reduces the number of net transponder card type required in the ring [6]. We further assume that each ring network would have a two-fiber ring with half the channels in each fiber dedicated to shared path protection. For simplicity without loss of generality we assign even channels for work in the clock-wise fiber and odd channels for work in the counter-clockwise fiber. For lightpath establishment purpose we assume an optical service channel. As can be seen from Fig. 1. a WDM OADM guards the spectral contents of one subnet from the other. Hence to establish an inter-subnet lightpath (AE) the optical switch on that particular wavelength at the intermediate WDM OADM (node D) needs to be configured in the pass-through mode. The conventional service channel accomplishes this signaling procedure. So for lightpath AE, subnets 1 and 2 share the wavelength using the wavelength gateway node D. While establishing an intra-subnet lightpath IH, on say wavelength λ_l the optical switch corresponding to λ_l needs to be configured in the cross state. This ensures that λ_l can be reused by subnets 1 and 2 internally (lightpaths GF, IH). In this way wavelength reuse is permissible. When a fiber cut occurs, protection is achieved by loop-back function of the ring.

III SIMULATION MODEL

We developed a simulation program to estimate the performance of both subnet based network as well as conventional WDM OADM network for similar traffic requirements. For a given number of nodes, based on the maximum traffic constraints the network was divided into subnets or sectors as per the algorithm. All intra-subnet traffic and inter-subnet traffic was assigned on wavelengths in ascending and descending order respectively. The routing and wavelength assignment algorithm used only shortest paths for both cases, while maximizing wavelength reuse. The reason we assume lightpaths on only shortest paths is to reduce Q-penalty including OSNR degradation. For inter-subnet lightpaths wavelength continuity is an important constraint. Future versions would include wavelength conversion at the wavelength gateway devices resulting in slightly better performance in blocking. We calculated the throughput, blocking probability and wavelength reuse factor for both the networks. A cost function assigned normalized cost to both the networks in terms of the number of nodes and the components present at the nodes. From neutral vendor based calculations as well as the cost model provided in [3] a passive network element as demonstrated in [3] without transponders was roughly one-fourth the cost of a conventional WDM-OADM node. This cost model considered a conventional OADM network element to have full multiplex and de-multiplex sections in addition to amplifiers and switching fabric. The passive network element's cost analysis was done taking into account the passive devices and amplifiers mentioned in [4]. Both the network architectures required a pair of transponders for each source-destination pair to support lightpaths.

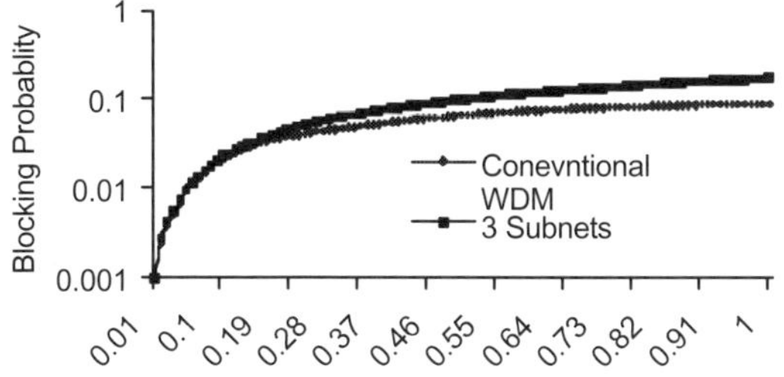

Fig. 2 Shows the blocking probability of a conventional WDM network and that of a subnet based network.

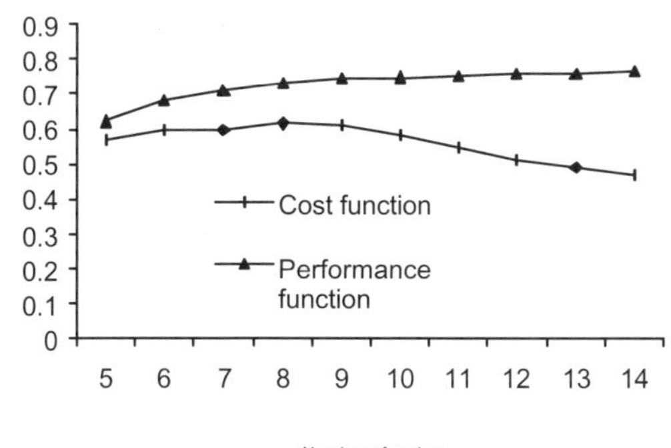

Fig. 3 Shows the cost function and performance function of subnet network with respect to conventional WDM network for different nodal configurations.

The simulation algorithm performed extensive iterations with various configurations and load patterns. We were typically interested in configurations with number of nodes from 6-12, which were most widely deployed in metro networks. The number of subnets ranged from 1-4 with the 2-subnet and 3-subnet configuration generating most interest. Fig. 2 shows the blocking probability of a 3-subnet model compared to a conventional WDM OADM configuration for the same number of nodes (12 in this case). Due to the complete wavelength reuse available at each WDM OADM node the conventional WDM model performs slightly better in terms of blocking probability. The performance for low loads (<0.5) is almost the same, while for loads from 0.5-0.7 the performance is marginally better in favor of the conventional network. The first passage time so critical in WDM networks is almost the same for both network types.

Fig. 3 shows the cost and performance functions of a subnet-based network in comparison to a conventional network. As can be seen performance in terms of throughput increases with the number of nodes (and hence subnets). The throughput of a subnet-based network is never less than 65 % of the throughput of a conventional network; however at the same time the cost is never greater than 60 % of the OADM network cost. In fact the cost decreases with number of nodes. The main factor to cost in subnet-based network is the transponder cards while the main factor responsible for cost in a conventional WDM network are the OADMs as well as the transponders. If we disassociate transponder cost from both networks we observed that the cost of subnet-based network is lesser than that of a conventional network by a factor of 3. Throughput performance of a subnet based network stabilizes between 78-88 % of conventional network for rings with 6<N<14. This result is also analytically verified in section III

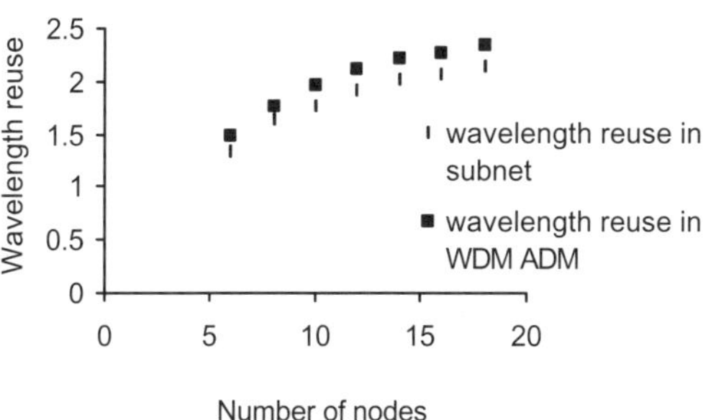

Fig.4. Wavelength reuse as a function of number of nodes

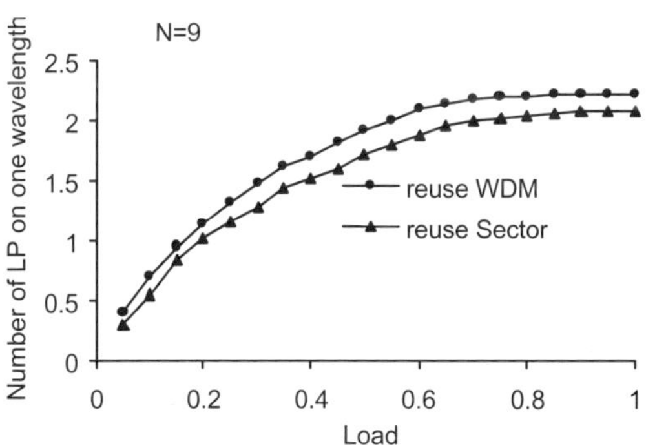

Fig. 5. Wavelength reuse as a function of load

Fig. 4 and Fig. 5 show the wavelength reuse factor as a function of the number of nodes as well as load in both network types. Wavelength reuse factor is calculated as the average number of lightpaths per wavelength in the network. The variation of wavelength reuse factor with number of nodes for both network types is also shown. The conventional network performs better by 10 %. This difference is minimal and can easily be discarded for systems having low-to-average loads. Also both networks have a similar response in terms of wavelength reuse.

IV Comparison with Conventional Network

The low cost subnet based network comprising of passive nodes divided by wavelength gateways performs almost as good as a conventional network. The main comp arison characteristic for performance is the blocking of lightpaths which states the mean difference in throughput between the two networks. The total flexibility to reuse wavelengths in a conventional network allows it to carry more traffic than the subnet based network by a certain fraction.

This fractional difference between performances is quite minimal. Let the number of nodes within a given subnet be k, and let this subnet be named S. Consider a lightpath whose destination node is one of the k nodes in the subnet (Fig. 7). If the destination node was the kth node (which is a wavelength gateway device) we could reuse the same wavelength immediately, and the subnet system would perform equally as efficient as the conventional system. To estimate the difference in efficiency between a subnet system and a conventional system, we need to find the probability that an arbitrary lightpath would be destined for a node 1…k-1, for a particular subnet S but not node 'k' (wavelength gateway). The probability that a particular node within a subnet could be a destination node is 1/k. Further the probability that the destination node would be in subnet S is the probability of the subnet itself given by $\lambda_{max}/[(\text{Ó}Tr_i)_{max}+\lambda_{max}]$. Thus the probability that the destination node would be 1…k-1 is given by

$$\Pr(d = 1,2,3,...,k-1) = \frac{\lambda_{\max}}{(\sum Tr_i)_{\max} + \lambda_{\max}}\left[1-\frac{1}{k}\right] \quad (1)$$

The probability in equation 1 compares the performances of conventional and subnet based networks in terms of blocking and throughput. Higher the probability of an intermediate node (other than a gateway) as a destination node, lower the efficiency of the subnet-based network as compared to a conventional WDM network. This result is verified through a simulation model (Fig. 6). The efficiency in terms of throughput of the system as compared to a conventional network is given by this probability expressed as a percentage.

Fig. 6 shows the comparative performance of a subnet-based network with conventional WDM OADM based network. The analytical as well as the simulated performance is shown. The simulation model assumes random lightpath requirements for various source-destination pairs. The efficiency is theoretically calculated using the analytical result in equation 1

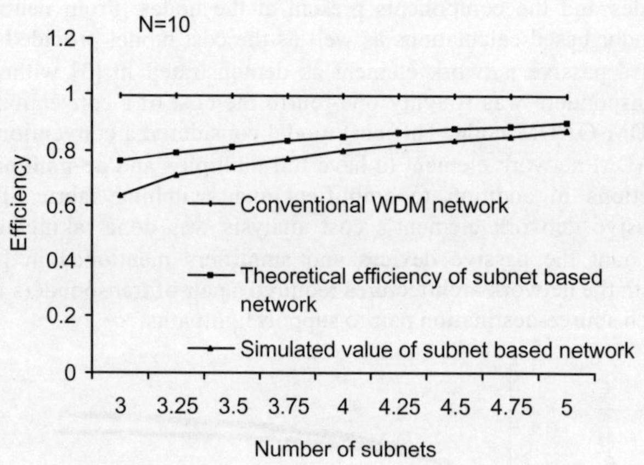

Fig. 6. Comparative analysis of WDM ADM based netw ork and subnet-based network for theoretical and simulated values.

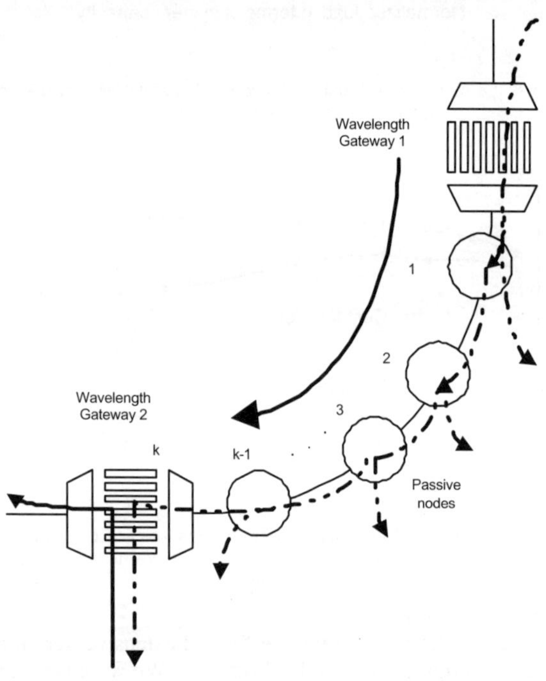

Fig. 7. The model used to perform analysis of inefficiency of subnet as compared to conventional network

V Discussion

In this paper we have introduced the concept of optical subnets. Optical subnets in metro rings can insure system performance comparable with conventional WDM networks. For low-to-moderate loads the subnet system performance is at par with the conventional system, while the cost is always lesser than the latter. The subnet system utilizes the drawback in WDM networks of multiple nodes providing pure transit (pass-through) function due to hop lengths greater than one hop. Using passive nodes divided by a few OADMs the network is a highly scalable and effective architecture in terms of performance. The analytical result for loss in efficiency between the two network architectures have been verified through simulation. For most networks the performance difference would not cross 10 % for low loads and 15 % for moderate loads. It is quite unlikely that rings would exploit the full load potential, and even if they do so, the difference in efficiency would not be greater than 25 %. At the same time there always would be a sizable cost reduction factor. Including transponder costs, the subnet configuration would conservatively save approximately 40-50 % of conventional network costs.

We have thus introduced a new class of hybrid networks for metropolitan area applications, which provides a cost effective and performance oriented solution for optical WDM networking

Acknowledgement: The authors wish to thank Seemant Choudhary and Cechan Tian for the discussions as well as Dr. Kuwahara for his support during this work.

REFERENCES

1. I.Chlamtac, A. Ganz and G. Karmi,'Lightpath Communications: An Approach to High Bandwidth Optical WAN's', *IEEE Trans on Commun* Vol. 40, No. 7, July 1992,pp 1171-1182
2. A. Narula and E. Modiano,'Efficient Routing and Wavelength Assignment for Reconfigurable WDM Networks', *IEEE Journal on Select. Areas in Commun.* Vol. 20, No. 1, Jan. 2002,pp 75-88
3. N. Slauze et al,' A novel, low cost optical packet metropolitan ring architecture' Proceedings of ECOC, Sept. 2001
4. K. Takeguchi, R. Pecquer and S. Kinoshita,' A novel dynamic optical add/drop multiplexer with cascaded coupler based distributing amplifier and combiner' submitted to OECC '02, Yokohama
5. A. Gumaste et al,'BITCA: Bifurcated Interconnection to Traffic and Channel Assignment in Metro Ring Networks', *Proceedings of OFC TuG Anaheim, CA March 2002*
6. O. Gerstel, R.Ramaswami and G.H.Sasaki,' Fault Tolerant Multiwavelength Optical Rings with Minimal Wavelength Conversion,' *IEEE J. Select. Areas in Commun.*, Special Issue on High-Capacity Optical Transport Networks, Vol. 16 Nov 1998

Bidirectional Optical Crossconnects Based on FBGs and Optical Circulators

Hai YUAN and Wen-De ZHONG
Network Technology Research Centre
School of Electrical & Electronic Engineering,
Nanyang Technological University, Singapore 639798
Fax: +65 6790 4540, Email: ewdzhong@ntu.edu.sg

Abstract: We propose a novel bidirectional optical crossconnect (BOXC) using fiber Bragg gratings (FBGs) and optical circulators. Dynamic control for wavelength routing is achieved by employing tunable FBGs. Because only a few optical circulators are used, the proposed BOXC is more cost-effective than previously reported BOXCs. Our preliminary study shows that the proposed BOXC has potentially small insertion loss and low crosstalk. It requires a small tuning range equal to wavelength channel spacing for tunable FBGs. We show that a large BOXC can be built based on the Benes network structure and present a method to reduce the complexity.

Key Words: wavelength division multiplexing, fiber Bragg grating, optical circulator, bidirectional optical crossconnect.

1. INTRODUCTION

FBG-based OXC's are considered to have the potential of being employed at nodes in WDM networks. However, most of the reported FBG-based OXCs are based on a unidirectional transmission where data is transmitted only in one direction in the fiber links.

Bidirectional OXCs and optical add–drop multiplexers (ADMs) are required, since single-fiber bidirectional WDM ring networks (SFBRNs) have emerged as cost-effective and reliable topologies for metropolitan-area networks (MANs). At a node where multiple SFBRNs are interconnected, we need a bidirectional OXC (BOXC), which could selectively interchange wavelength channels among the SFBRNs in both directions. Several types of the BOXCs were reported in [1]–[4]. In [1], a micromirror switch was reported for bidirectional connection, but such a connection is not independent in two directions, that is, bidirectional connectivity has to be offered in pairs. In [2], polarization beam splitters (PBSs) and gratings were used for the BOXC where the polarization state of the input signals needs to be controlled, and the position of gratings in each arm must be precisely controlled to avoid induced polarization mode dispersion (PMD) due to the propagation time difference between two arms. The BOXCs reported in [3] and [4] suffer a high insertion loss and crosstalk since AWGs were used.

In this paper, we propose a novel BOXC using reciprocal optical circulators (ROCs) and fiber Bragg gratings (FBGs) for interconnecting SFBRNs. Dynamic control for wavelength routing is achieved by employing tunable FBGs. Because only a few optical circulators are used, the proposed BOXC is more cost-effective than previously reported BOXCs. In addition, the proposed BOXC has small insertion loss and low crosstalk. It requires a small tuning range equal to wavelength channel spacing for tunable FBGs when it operates in the interleaving wavelength mode. As such, the port count and wavelength count of the proposed BOXC could be increased easily without imposing constraint on the tuning range of FBGs. We show that a large BOXC can be built based on the Benes network structure and present a method to reduce the complexity.

2. CONFIGURATIONS OF THE PROPOSED BOXC

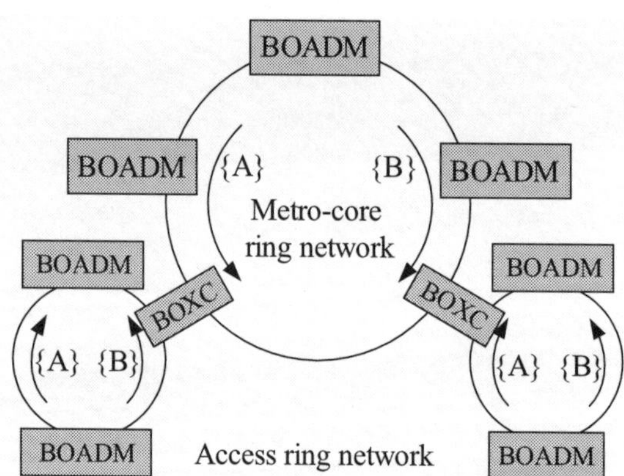

Fig. 1 Bidirectional WDM ring networks interconnected by bidirectional OXCs. In each ring network, two different sets of wavelength channels propagate in opposite directions.

Fig. 1 shows a schematic diagram of a feasible MAN composed of SFBRNs, in which several small access ring networks are connected to a big metro-core ring network. In such a scheme, each ring network contains nodes within a local area, and it is not frequent that more than two SFBRNs are required in such interconnections. Therefore, 2×2 BOXC is mainly considered here. Extension to an 2×2 BOXC to an N×N BOXC will be discussed in Section 3. As shown in Fig. 1, in each SFBRN, two different sets of wavelength channels propagate in opposite directions. We may use two separate wavelength bands for their respective directions or we may adopt the interleaved wavelength scheme in which two sets of wavelength channels are interleaved to each other. The latter is adopted here. This means that in the SFBRNs of Fig. 1, λ_1, λ_3, λ_5, ..., λ_{2n-1} (wavelength group A) and λ_2, λ_4, λ_6,...,λ_{2n} (wavelength group B) are assumed to propagate in opposite directions.

Fig. 2(a) shows the configuration of the proposed BOXC using FBGs and four 4-port optical circulators. There are four sets of fixed FBGs and two set of tunable FBGs whose function will be explained late. The optical circulators used here are reciprocal ones. A reciprocal circulator (ROC) can route the input signal from the last port back to the first port. The WDM signals at wavelengths (λ_1, λ_3, λ_5,..., λ_{2n-1}) enter at port 1 of ROC1 and be reflected by one set of fixed FBGs (the wavelength of these FBGs cannot be changed). Then the WDM signals will go through port 3 of ROC1, port 3 of ROC2, port 4 of ROC2 and encounter one set of tunable FBGs (Set A). By changing the Bragg wavelength of each tunable FBG, the incoming WDM signals will either be passed through or reflected back by the set of tunble FBGs. The reflected signals will be coupled back to port 4 of ROC2 and exit at port 1 of ROC2, which will be dubbed a bar state. On the other hand, the through signals will be coupled into port 4 of ROC4 and exit at port 1 of ROC4, which corresponds to a cross state. Similarly, for the signals (λ_2, λ_4, λ_6,..., λ_{2n}) launched into port 1 of ROC2, we can achieve the same independent switching by tuning the other set of tunable FBGs (Set B).

Fig. 2(b) illustrates an improved structure of the BOXC using 6-port ROCs. The light paths between port 3 of ROC1 and port 3 of ROC2, port 3 of ROC3 and port 3 of ROC4 could be removed if using 6-port ROCs instead of 4-port ROCs.

In a general bidirectional transmission system, Raleigh backscattering will be generated by signals in both directions due to the inherent density fluctuations in the fiber core. Optical reflections will also be produced by fiber connectors and splices in the transmission path. They will degrade the signal-to-noise ratio (SNR) and thus reduce the receiver sensitivity significantly. In addition, the maximum gain of the bidirectional optical amplifier used in a bidirectional transmission system is limited by the multiple path reflection noise caused by the Raleigh backscattering and optical reflections. Meanwhile, the bidirectional amplifiers used may cause lasing at a relatively high gain [5].

In our proposed BOXC, unwanted backscattered signals, including Raleigh scattering in fiber and reflections at connectors, could be removed by the four sets of fixed FBGs with light absorbers. For example, the wavelength channel λ_1 entering at port 1 of ROC1 is switched to port 1 of ROC4. Some backscattered signals at wavelength λ_1 will reenter port 1of ROC4, and go through port 2 of ROC4 and be absorbed

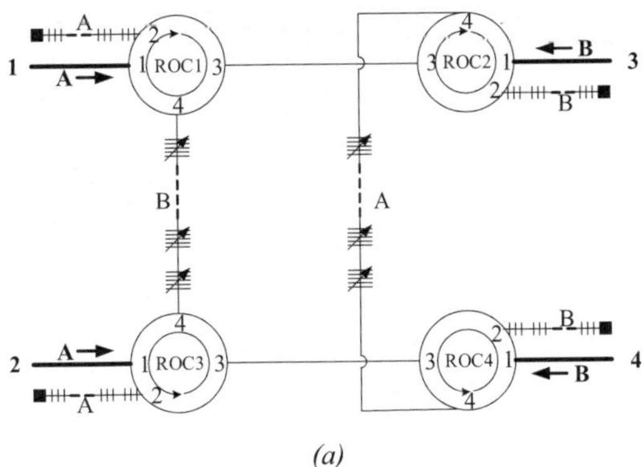

(a)

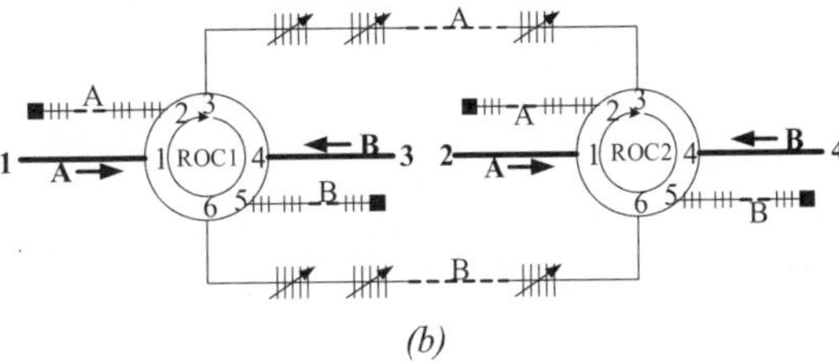

(b)

Fig. 2 BOXCs based on FBGs and OCs (a) BOXC using four 4-port reciprocal OCs and FBGs; (b) BOXC using two 6-port reciprocal OCs and FBGs.

by light absorber connected to port 2 of ROC4. But if there are no fixed FBGs with light absorber connected to port 2 of ROC4, the backscattered signals at λ_1 will be switched back to port 1 of ROC1 and interfere with signal of the same wavelength causing intensity noise.

We now examine the insertion loss for the WDM channels in both bar and cross states. Let L_{oc} be the insertion loss between two adjacent ports of an OC. Let L_{FT} and L_{FR} be the transmission and reflection losses of an FBG. With reference to Fig. 2(b), the insertion loss for the channels in the bar state, denoted as L_{bar}, is given by

$$L_{bar}= 3Loc + 2L_{FR} + (m-1)L_{FT} \quad (1)$$

The insertion loss for a channel in the cross state depends on the position of the corresponding FBG. Let $L_{cross-min}$ and $L_{cross-max}$ be the minimum and maximum insertion losses for the channels in the cross state, respectively. Then we have

$$L_{cross-min}=3L_{oc} + L_{FR} + mL_{FT} \quad (2)$$

$$L_{cross-max}=3L_{oc} + L_{FR} + (2m-1)L_{FT} \quad (3)$$

The typical values of L_{oc}, L_{FR} and L_{FT} from commercial devices are 0.6, 0.5 and 0.1 dB, respectively. Fig. 3 illustrates the insertion losses for the channels in bar state and cross state. As can be seen, the insertion loss for the channels in the bar state is very low, since they only pass though thrice of circulator ports. The power difference among WDM channels arising from the BOXC is very small. For example, if the number of WDM channels is 8, then the

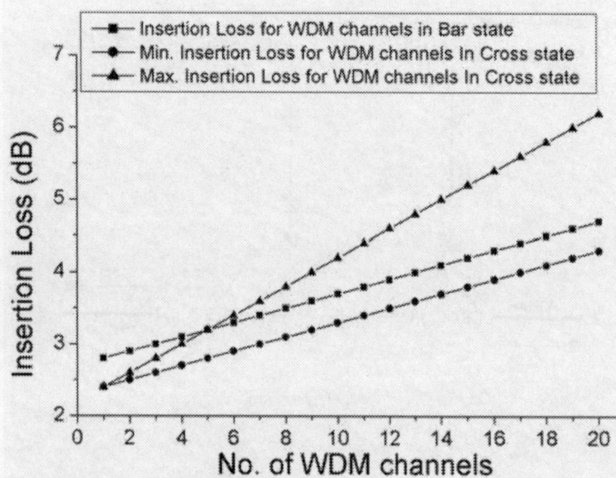

Fig. 3 Insertion Loss for through channels in bar and cross states.

insertion loss for the channels in the bar state, the min. and max. insertion losses for the channels in the cross state are 3.5, 3.1 and 3.8 dB, respectively. That is, the max. power difference among the channels arising from the BOXC is 0.7dB.

Both of the structures shown in Fig. 2 have several homodyne crosstalk sources, which include imperfect reflections of the FBGs, reflections at interfaces between connectors, and the crosstalk of the ROCs. The crosstalk due to the FBGs and ROCs is probably lower than -35 dB and -45dB respectively, and the reflection of connectors is negligible if all fiber ends are connected by fusion splicing. Thus the total homodyne crosstalk could be very low.

3. $N\times N$ BOXC BASED ON 2×2 BOXCS

A large NxN BOXC can be constructed based on the Benes network structure [6] that uses 2×2 BOXCs as the building blocks. The Fig. 4 show a 4×4 BOXC which consists of six 2×2 BOXCs. Because the main source of backscattering does not occur within the BOXC, we only need to place one set of fixed FBGs at the entry of the BOXC to remove the unwanted signals, i.e. the backscattered signals will be removed at the first stage when it enters the $N\times N$ BOXC. Therefore, only $2N$ sets of fixed FBGs are required.

We now show that the BOXC in Fig. 4 can be further simplified by compressing ports of ROCs. Let $ROC_{i,j}$ denote the jth port of the ith ROC in Fig. 4. With reference to Fig. 4, it can be seen that the light paths $ROC_{1,4}\rightarrow ROC_{3,1}$ and $ROC_{3,4}\rightarrow ROC_{5,1}$ that connect two ROCs can be compressed into one point. After removing the unnecessary sets of fixed FBGs within the BOXC and compressing some ROC ports, the three 6-port ROCs, i.e. ROC1, ROC3 and ROC5, can be replaced by a single 10-port ROC. Similarly, ROC2, ROC9 and ROC6 can also be replaced by a single 10-port ROC. The simplified 4x4 BOXC is illustrated in Fig. 5, which only requires 4 10-port ROCs instead of 12 6-port ROCs.

4. CONCLUSION

We have proposed a novel BOXC using reciprocal optical circulators and fiber Bragg gratings (FBGs) for interconnecting SFBRNs. Wavelength switching and routing is achieved by employing tunable FBGs. Because only a few optical circulators are used, the proposed BOXC is more cost-effective than previously reported BOXCs. Our preliminary investigation has shown that the proposed BOXC has potentially small insertion loss and low crosstalk. It only requires a small tuning range equal to wavelength channel spacing for tunable FBGs if it operates in the interleaving wavelength mode. We have illustrated that a large NxN BOXC can be built based on the Benes network structure and shown how to reduce the complexity by removing unnecessary sets of fixed FBGs and compressing ROC ports.

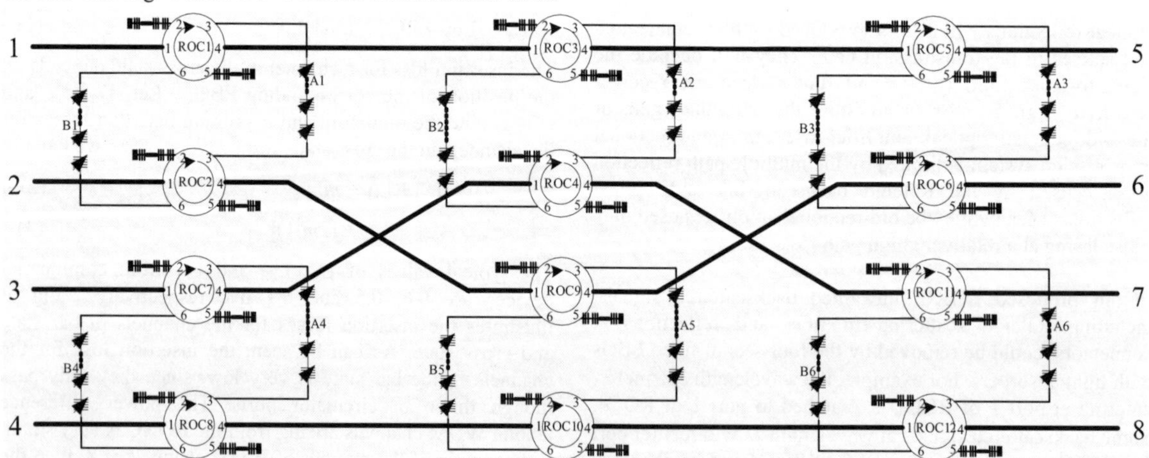

Fig. 4 4×4 rearrangable nonblocking BOXC using twelve 6-port ROCs

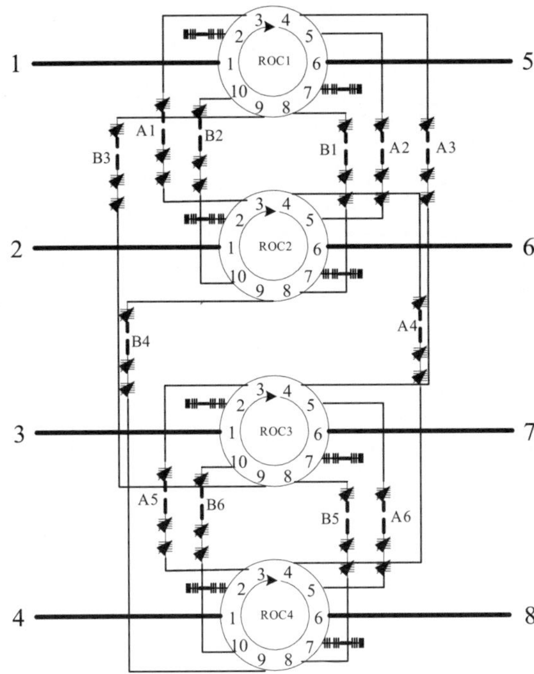

Fig. 5 4×4 rearrangable nonblocking BOXC using four 10-port ROCs

REFERENCES

[1] J. M. Simmons, A. A. M. Saleh, *et al*, "Optical crossconnects of reduced complexity for WDM networks with bidirectional symmetry," *IEEE Photon. Technol. Lett.*, vol. 10, pp. 819–821, June 1998.

[2] S. K. Park, J. W. Park, *et al,* "Multiwavelength bidirectional optical crossconnect using fiber Bragg gratings and polarization beam splitters," *IEEE Photon. Technol. Lett.*, vol. 12, pp. 888–890, July 2000.

[3] J. Kim and B. Lee, "Independently switchable bidirectional optical crossconnects," *IEEE Photon. Technol. Lett.*, vol. 12, pp. 693–695, June 2000.

[4] Sangin Kim, "Bidirectional Optical Cross Connects for Multiwavelength Ring Networks Using Single Arrayed Waveguide Grating Router," *J. Lightwave* Technol., vol. 20, pp. 188–194, 2002.

[5] Zhao, Y., *et al*, "A novel bidirectional add/drop module using waveguide grating routers and wavelength channel matched fiber gratings," IEEE Photon. Technol. Lett., vol. 11, pp. 1180–1182, Sep. 1999.

[6] Xiangnong Wu, et al, "Fiber Bragg Grating-Based Rearrangeable Nonblocking Optical Cross Connects Using Multiport Optical circulators" IEEE Photon. Technol. Lett., vol. 12, pp. 696–698, June 2000.

Session 13B

Fibers and Fiber-Based Devices

Biomimetic optimisation and co-design of microstructured optical fibres

L. Poladian, S. Manos, N. A. Issa and W. E. Padden
Optical Fibre Technology Centre,
Australian Photonics Cooperative Research Centre,
University of Sydney, Australia.
l.poladian@oftc.usyd.edu.au.

Abstract: Microstructured optical fibres have increased degrees of flexibility that allow tailoring of optical properties for various applications. Three different algorithms for calculating mode properties and structural or confinement loss in these fibres are compared. A number of biologically-inspired approaches to optimising the fibre design (hole size, shape and arrangement) within the context and constraints of a particular fabrication technology are presented.

Key Words: confinement, microstructure, optimisation, co-design, evolutionary strategy

1. INTRODUCTION

Microstructured optical fibres (MOFs) have generated significant interest since their development in 1996 [1]. Their ability to guide light through the appropriate arrangement of holes in a uniform host material has been used to create endlessly single moded fibres [2], guidance in air [3] and the potential to arbitrarily control polarisation, dispersion and nonlinear properties. The very large number of degrees of freedom associated with the size, shapes and arrangement of holes provides this flexibility but also requires new algorithms: both to calculate the optical properties of the fibres, and to carry out the design and optimisation. Biomimetic optimisation exploits the ideas behind naturally occurring optimisation phenomena in biology, such as Darwinian evolution. Algorithms based on such principles have been shown to perform efficiently in difficult multi-parameter and multi-criteria optimisation problems [4]. However, the design and optimisation of the fibres cannot proceed without incorporating constraints that arise from material properties and the fabrication process. Until recently, the dominant fabrication process for MOFs was the stacking of glass capillaries [5] which severely restricted both the variety of hole structures and the fabrication/processing techniques. Last year, two groups [6], [7] succeeded in producing MOFs in polymer. This breakthrough certainly immediately removed the constraint of the conventional hexagonal arrangement of holes and allowed new classes of designs to be investigated. Of course, the fabrication of microstructured polymer optical fibres (MPOF)introduces its own set of specific fabrication issues and constraints. This simultaneous development and evolution of fibre designs and fabrication techniques is referred to as co-design. The philosophy of co-design is consistent with re-emerging views within the wider photonics community on the necessity of design for manufacture and design for automation. However, we aim to take this philosophy one step further. The development of algorithms which can calculate the mode properties of this newer class of structures both efficiently and in a way that is compatible with the optimisation algorithms has also occurred over the last year [8], [9], [10]. In the context of optimisation, the ability of an algorithm to run autonomously without human intervention, whilst analysing vastly different designs, may become more important than speed (or even accuracy in initial design phases). Thus, even the development of different algorithms should be done simultaneously with the development of the fabrication and the design, thus adding yet another layer to the co-design philosophy. In the remainder of this paper we briefly review the current state of MPOF design, the various algorithms suited to analysing their modes and a particular class of optimisation algorithms.

2. MICROSTRUCTURED vs. CONVENTIONAL OPTICAL FIBRES

Conventional fibres have a central region (core) which has a higher refractive index than the surrounding region (cladding). Variations include multiple cores and inner claddings. Regardless, light propagating in such structures is confined to the regions of higher index and a fibre mode can in principle propagate indefinitely without attenuation. (In practice, losses arise from material properties and surface irregularities or microbending.)

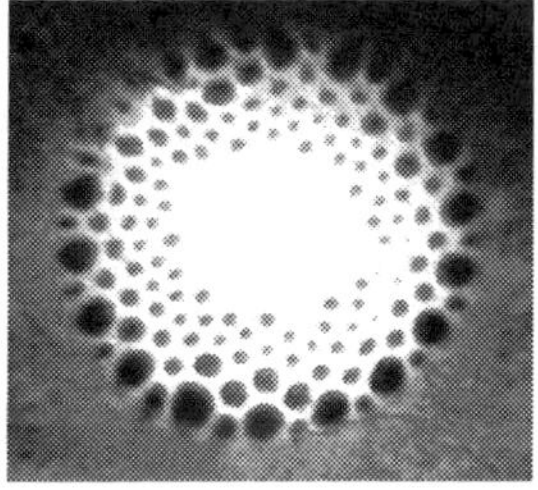
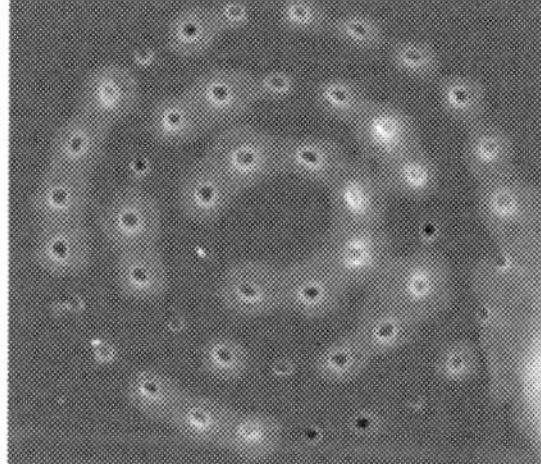

Fig. 1. Examples of microstructured polymer optical fibres with hole arrangements not constrained by hexagonal symmetry.

In almost all microstructured fibres, a single material is used and the holes provide the guidance mechanism. In effective index guidance, the holes are used to lower the effective or average index in the outer region below that in the core thus emulating conventional guidance. In bandgap guidance the periodic arrangement of holes creates resonant Bragg-like reflections which are used to confine the light. In both cases, for finite structures, light can still escape from the core region. Thus, even in principle, for perfect structures there will be a structural or confinement loss that needs to be considered. In practice, once the confinement loss is reduced to acceptable levels to all intents and purposes the leaky modes in such structures behave like bound modes in conventional structures. Some examples of recently fabricated microstructured polymer optical fibres (MPOFs) are shown in Fig. 1.

3. ALGORITHMS FOR LEAKY MODES

There are many algorithms for studying the modes of conventional fibres and in certain circumstances these can also be directly applied to the study of MOFs. However, since these algorithms were designed primarily for bound modes they are unable to calculate the structural or confinement loss that occurs for leaky modes. Direct simulation techniques, such as the beam propagation method (BPM), can model such phenomena since they follow the entire field as it propagates, but are extremely numerically intensive and do not provide direct insight. There are now three different algorithms that can be effectively used to model MOFs and leaky modes.

The earliest algorithm is the multipole method [8] which begins with an expansion of the field in the vicinity of each hole. The longitudinal field components are expanded in a series of multipole fields of the form

$$\sum_m [A_m J_m(k_\perp r) + B_m H_m^{(1)}(k_\perp r)] e^{im\phi} \tag{1}$$

where $k_\perp = (2\pi/\lambda)\sqrt{n_{\text{eff}}^2 - n^2}$. Each hole has its own expansion. One then constructs a matrix equation for the consistency of all these expansions using a technique [8] dating back to Lord Rayleigh. The method then requires searching for values where the determinant is zero (or minimised). The complexity of the calculation scales with the third power of the number of multipole moments and the number of symmetry-distinct holes. All the methods can account for symmetries of the structure and only the number of symmetry-distinct holes is relevant. The method is extremely efficient but is currently restricted to circular holes. Some example structures for which the multipole method can be used are shown in Fig. 2. In addition, the method cannot take advantage of the orthogonality of modes when calculating multiple modes of the same structure.

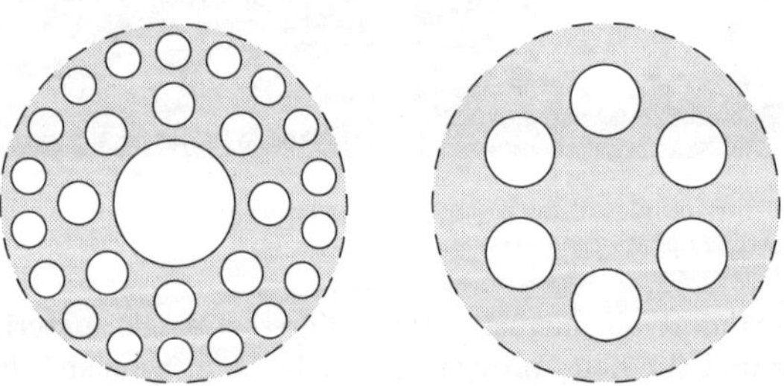

Fig. 2. MOF designs with exclusively circular holes suitable for analysis by the multipole method.

The second algorithm is the Fourier Decomposition Method with Adjustable Boundary Conditions which was initially developed within the scalar approximation [9] and subsequently generalised to the full vector analysis [10]. This is a generalisation of a well-known technique in waveguide analysis to take into account a correct matching of the fields to the outward leaking radiation. The transverse fields throughout the entire structure are expanded in a modified Fourier series:

$$\sum_{n,m} [A_{m,n} \sin(n\pi r/R) + \alpha_{m,n} + \beta_{m,n} r] e^{im\phi} \tag{2}$$

where α and β are the adjustable boundary condition parameters [9]. The coefficients are found by requiring the expansion to satisfy the wave equation. Whilst the method is computationally slow, it is applicable to arbitrary structures. Typical structures are shown in Fig. 3. The complexity of the calculation scales with the third power of the total number of Fourier components used and is independent of the number of holes. The expansions of different modes of the same structure satisfy orthogonality conditions which can be exploited when systematically analysing several modes of the structure. Some modes are shown in Fig. 4.

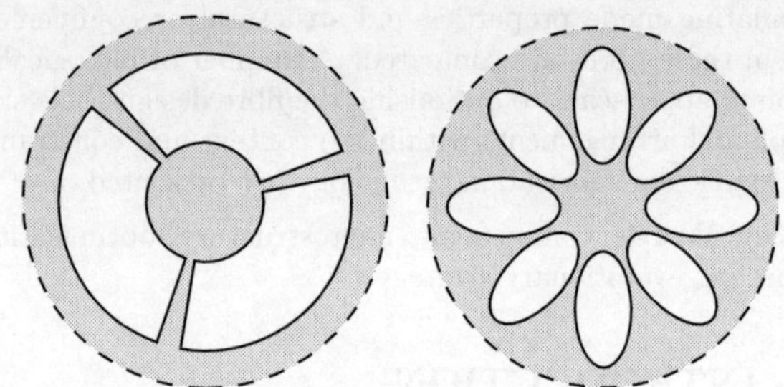

Fig. 3. MOF designs with non-circular holes that can be analysed with the Fourier decomposition method.

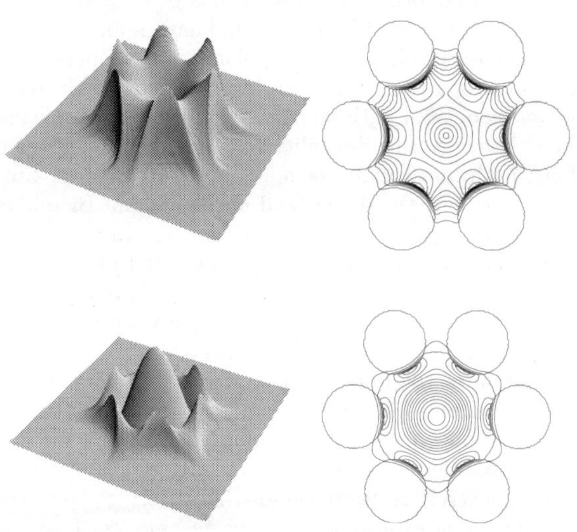

Fig. 4. Example of the longitudinal magnetic field distribution of some modes found with the vector FDM method.

The third method is intermediate between the first two and decomposes the structure into its azimuthal Fourier components:

$$n^2(r,\phi) = \sum_m n_m^2(r) e^{im\phi} \tag{3}$$

with similar azimuthal expansions performed for other quantities. Then one finds the modes by solving coupled ordinary differential equations and shooting towards the correct external boundary conditions. The method is an adaptation of an earlier scheme [11] that was only applied to bound modes. The complexity of the calculation scales linearly with the number of Fourier components used and inversely with the step-size used in the ODEs. Like the multipole method, this algorithm cannot easily exploit the orthogonality of modes. The first order approximation to this technique is essentially the Effective Azimuthal Index Method which replaces a complex structure with a circularly symmetric one. An example of a design which exploits this idea is shown in Fig. 5 and corresponds to the graded index microstructured polymer (GIMP) fibre shown in Fig. 1, which emulates a circularly symmetric parabolic index fibre

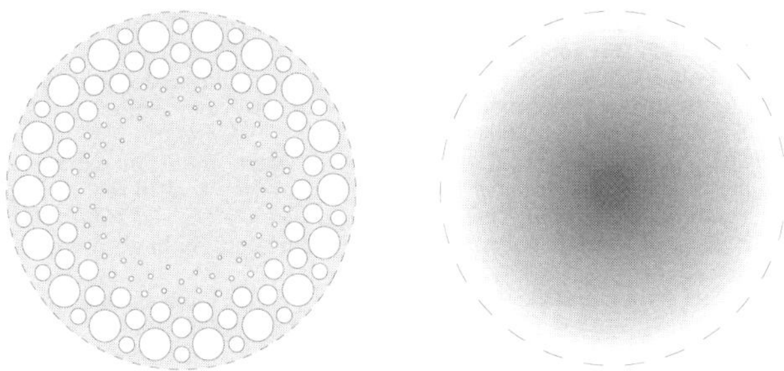

Fig. 5. MOF design (left) that emulates a circularly symmetric parabolic graded index fibre (right).

The different attributes and relative merits of the algorithms discussed above are summarized in Table I.

Criteria	Multipole Method	Fourier Method	Shooting Method
Hole Shape	Circles	Arbitrary	Arbitrary
Speed	Fast	Slow	Moderate
Orthogonal	No	Yes	No
Search	Yes	No	Yes

TABLE I
COMPARISON OF DIFFERENT ALGORITHMS FOR ANALYSING MODES OF MOFS. THE SPEED COMPARISON IS FOR TYPICAL DESIGNS AND MAY NOT BE TRUE IN ALL CASES.

4. EVOLUTIONARY STRATEGIES

Evolutionary strategies attempt to optimise designs by mimicking the three processes of mutation, recombination and selection that occur in Darwinian evolution. The analogy between biological parameters and waveguide parameters in shown in Table II. Each potential fibre design is considered as an individual within a population of diverse designs. The fitness or figure of merit for each design is the degree to which the performance (phenome or traits) of the fibre matches the desired characteristics (e.g. cut-off wavelength, loss, dispersion, birefringence). The design parameters (e.g. hole size, shape, arrangement) make up the genome of any particular design. Starting with an initial population of designs (parents), the genomes of individual designs are recombined and mutated to produce new designs (children). Those designs with the best properties (fitness) are selected and used as the parents for the next generation. After several generations, designs with high figures of merit emerge. A schematic of the evolutionary cycle is shown in Fig. 6. Early applications of evolutionary strategies to fibre designs concentrated on conventional fibres [12], [13] more recently microstructured designs have also been optimised. The improvement of the figure of merit (FOM) with successive generations for a typical fibre design problem is shown in Fig. 7. The FOM has been scaled such that a 'perfect' design that meets all objectives has a FOM equal to one.

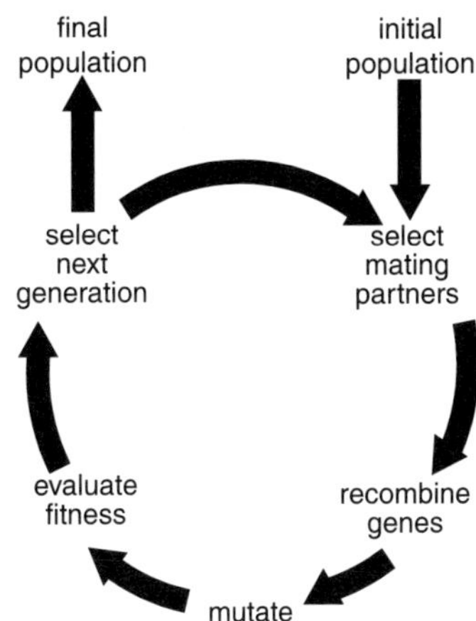

Fig. 6. The major steps in the evolutionary design cycle for optimising waveguide structures using evolutionary strategies.

	Biology	Waveguides
Genotype (genes)	DNA sequence	Waveguide structure, hole dimensions
Phenotype (traits)	Colour, Size	Loss, Dispersion, Spot size, etc.
Objective	Fitness	Figure of Merit
Genotype to Phenotype	Protein synthesis	Mode solving algorithms
Selection process	Survival of the fittest	Artificial (deterministic or stochastic)
Population progression	Sexual reproduction and mutation	Design recombination and variation

TABLE II
ANALOGY BETWEEN BIOLOGICAL EVOLUTION AND OPTIMISATION OF WAVEGUIDE DESIGNS.

5. DISCUSSION

Unconstrained optimisation will inevitably lead to designs which are difficult or, more usually, impossible to fabricate (let alone suitable for manufacture or large scale automation). External constraints can be incorporated into the evolutionary strategy in a number of ways. Firstly, the entire genome can be encoded in such a way that only fibres which meet all the relevant constraints can even exist. This is a very application specific method (sometimes impossible) and needs to be implemented anew for each problem. Alternately, designs which do not satisfy the constraints can be

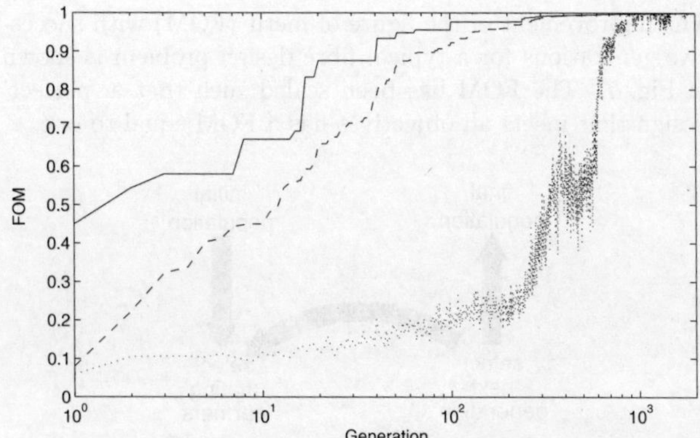

Fig. 7. Improvement of Figure of Merit (FOM) over two thousand generations. The solid curve shows the FOM for the best individual in each generation, the dashed curve shows the individual with the worst FOM that survives (selection) from one generation to the next, the dotted curve shows the average FOM for the entire population.

rejected (culled) at each generation. This is a rather drastic and wasteful process, which slows down the evolution and does not take into account that even individuals who currently fall outside the constraints may nevertheless have something worthwhile to contribute to future generations. An intermediate approach is to penalise individuals who do not satisfy the constraints by reducing their figure of merit or fitness. This requires quantifying the constraints and the desired characteristics on the same scale, which though possible is difficult.

A novel approach [14] is to copy yet another process found in nature: interaction between different species. The fibre designs and the fabrication constraints are considered different species that interact with each other but also evolve at the same time. Having a population of constraints that also evolves acknowledges that not all constraints are equally important at all stages of the design process. The encounters between specific designs and specific constraints can be stochastic and designs which satisfy constraints emerge with an enhanced fitness, while constraints that are easily satisfied fade into the background while more pressing constraints take centre stage.

ACKNOWLEDGMENTS

This work has benefited from the combined efforts of a very large number of individuals associated with the MPOF group within the Optical Fibre Technology Centre: Alex Argyros, Geoff Barton, Ian Bassett, Martijn van Eijkelenborg, Geoff Henry, Maryanne Large, Barry Reed and Joseph Zagari.

References

[1] J. C. Knight, T. A. Birks and P. St.-J. Russell, 'Holey Silica Fibers' in *Optics of Nanostructured Materials* V. A. Markel and T. F. George, (eds.) John Wiley, 2001.

[2] T. A. Birks, J. C. Knight and P. St.-J. Russell, 'Endlessly single-mode photonic crystal fiber,' *Opt. Lett.* **22** 961–963, 1997.

[3] R. F. Cregan, B. J. Mangan, J. C. Knight, T. A. Birks, P. St. J. Russell, P. J. Roberts and D. C. Allen, 'Single mode photonic band gap guidance of light in air,' *Science* **285** 1537–1539, 1999.

[4] D. Corne, M. Dorigo and F. Glover (eds.) *New Ideas in Optimisation*, McGraw Hill, London, 1999.

[5] J.C. Knight, T.A. Birks, P. St. J. Russell and D.M. Atkin, 'All-silica single mode optical fiber with photonic crystal cladding,' *Opt. Lett.* **21** 1547–1549, 1996.

[6] M. A. van Eijkelenborg, M. C. Large, A. Argyros, J. Zagari, S. Manos, N. A. Issa, I. Bassett, S. Fleming, R. C. McPhedran, C. M. de Sterke, N. A. P. Nicorovici, ' Microstructured Polymer Optical Fibre', *Opt. Expr.* **9**[7] 319–327, Sep. 2001.

[7] J. Choi, D. Y. Kim, U. C. Peak, 'Fabrication of Properties of Polymer Photonic Crystal Fibres' in *Proc. of the Plastic Optical Fibres Conference*, Vrije Universiteit, Amsterdam, 27–30 Sept 2001, pp. 355–360.

[8] T. P. White, R. C. McPhedran, C. M. de Sterke, L. C. Botten and M. J. Steel,'Confinement losses in microstructured optical fibres,' *Opt. Lett.* **26**, 1660–1662, 2001.

[9] L. Poladian, N. A. Issa, and T. M. Monro 'Fourier decomposition algorithm for leaky modes of fibres with arbitrary geometry,' *Opt. Expr.* **10**, 449–454, 2002.

[10] N. A. Issa, and L. Poladian 'Vector wave equation expansion method for leaky modes in microstructured optical fibres' in *Proceedings of the 27th Australian Conference on Optical Fibre Technology*, 8th –11th July 2002, Darling Harbour, Sydney, Australia, pp. 72–74.

[11] Y.-H. Wang and C. Vassallo, 'Circular Fourier analysis of arbitrarily shaped optical fibers,' *Opt. Lett.* **14** 1377–1379, 1989

[12] P. A. Krug, L. Poladian and M. I. Large. 'Advanced fibre design by evolutionary computation', in *Proceedings of the 25th Australian Conference on Optical Fibre Technology*, 26th June–28th June 2000, Australian National University, ACT, Australia, pp. 74–76.

[13] S. Manos and L. Poladian (2001) 'Fibre design with evolutionary strategies', Australasian Conference on Optics, Lasers and Spectroscopy, Brisbane, Australia, p.86.

[14] J. Paredis 'Coevolutionary algorithms' in *The Handbook of Evolutionary Computation* T. Bäck, D. B. Fogel and Z. Michalewicz (eds.) Oxford University Press, 1998.

Large mode area single-mode fiber: a modified segmented cladding fiber

Vipul Rastogi* and Kin Seng Chiang
Optoelectronics Research Centre and Department of Electronic Engineering
City University of Hong Kong, 83 Tat Chee Avenue, Kowloon, Hong Kong
P. R. China
*Tel: (852) 2788 9740; Fax: (852) 2784 4674; E-mail: rastogi@ee.cityu.edu.hk

Abstract: A modified design of segmented cladding fiber is presented. The fiber can provide effective single-mode operation in the wavelength range 1000 – 1700 nm with a core diameter as large as 20 μm. The mode area of the fiber is 4 times larger than that of a conventional single-mode fiber and can thus suppress the nonlinear effects effectively. The fiber should find applications in DWDM systems, fiber amplifiers, and lasers.

Key Words: Single-mode fiber; large mode area fiber; novel fibers; fiber optic communication

1. INTRODUCTION

Recently, we proposed a segmented cladding fiber (SCF) design for extended single-mode operation [1],[2]. An SCF has a uniform core of high refractive index and a radially invariant cladding with alternate regions of high and low refractive indices in the angular direction. The SCF can offer single-mode operation over an extended range of wavelengths with a large core area. In this aspect, the SCF behaves like a photonic crystal fiber or a holey fiber (see, for example, [3],[4]). Unlike a holey fiber, however, the SCF uses a small index contrast (same as that of a conventional fiber) and, hence, has potentially low polarization mode dispersion, which is essential for high-bit-rate transmission. The SCF and the holey fiber with infinitely extended cladding are essentially leaky structures. It is important to evaluate their leakage losses for long-haul communication applications. In this paper, we study the leakage loss of the SCF of Ref.[2] and present an improved design. The new design has a radially varying duty cycle of segmentation and outward tapered low-index segments (instead of inward tapered segments as in the original design [1],[2]). The modified design provides a tighter confinement of the mode in the core region and, hence, a much smaller leakage loss, compared with the previous design. The fiber is studied with the radial effective-index method (REIM) [5] in combination with the matrix method [6]. We demonstrate a design capable of effective single-mode operation from 1000 to 1700 nm using a core diameter of 20 μm. The leakage loss of the fundamental mode is as low as 0.3 dB/km. The proposed fiber is useful not only for DWDM systems but also for fiber amplifiers and lasers.

2. MODIFIED SCF

The cross section of the modified SCF (MSCF) is shown in Fig.1. The fiber comprises a uniform core region ($0 < r < a$) of refractive index n_1 and a segmented cladding region ($a < r < b$). The segmented cladding consists of 8 periodic outward

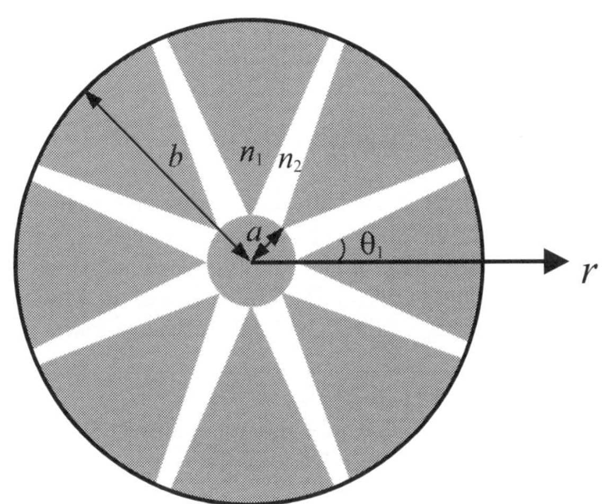

Fig.1. Cross section of the modified segmented cladding fiber, where a and b are the core radius and the cladding radius, respectively, n_1 and n_2 are the refractive indices of the segments, and θ_1 is a measure of the high-index regions in the cladding.

tapered segments of low-index (n_2) medium and inward tapered segments of high-index (n_1) medium. In other words, the fiber has a radially varying duty cycle of segmentation. The angle θ_1 is a measure of the high-index regions in the cladding and can lie between 0 and $\theta_{1max} = \tan^{-1}\{b(2^{1/2}-1)^{1/2}/[b(2^{1/2}+1)^{1/2}-2^{3/4}a]\}$. The fiber is uniform in the axial direction, which is the direction of wave propagation.

3. ANALYSIS

We analyse the fiber with the radial effective-index method (RIEM) [5]. In the RIEM, the field in the fiber is separated into the radial part, $\phi_r(r)$, and the azimuthal part, $\phi_\theta(r,\theta)$. By assuming that ϕ_θ varies slowly with r as compared with ϕ_r, the problem is reduced to solving the following two ordinary differential equations, which are satisfied by ϕ_θ and ϕ_r, respectively, using the appropriate boundary conditions [5]:

$$\frac{d^2\phi_\theta(r_i,\theta)}{d\theta^2}+k^2\left[n^2(r_i,\theta)-n_{\mathrm{effr}}^2(r_i)\right]r^2\phi_\theta(r_i,\theta)=0 \tag{1}$$

$$\frac{d^2\phi_r}{dr^2}+\frac{1}{r}\frac{d\phi_r}{dr}+k^2\left[n_{\mathrm{effr}}^2(r)-n_{\mathrm{eff}}^2\right]\phi_r=0 \tag{2}$$

where $n_{\mathrm{effr}}(r)$ is the effective refractive-index profile and n_{eff} is the effective index of the mode. Eq.(2) can be solved by the matrix method [6] to calculate the propagation constants and the losses of the modes.

4. NUMERICAL RESULTS AND DISCUSSION

As an example, we consider a silica MSCF with the following parameters:

$$\Delta = 0.007,\ a = 10\ \mu\text{m},\ b = 62.5\ \mu\text{m} \tag{3}$$

where Δ is the relative difference between n_1 and n_2. The value of θ_1 for the fiber is chosen to be 10 degrees. This fiber has a core size twice as that of a conventional single-mode fiber. According to Eq.(2), the REIM reduces the MSCF to an equivalent circular fiber with an effective-index profile

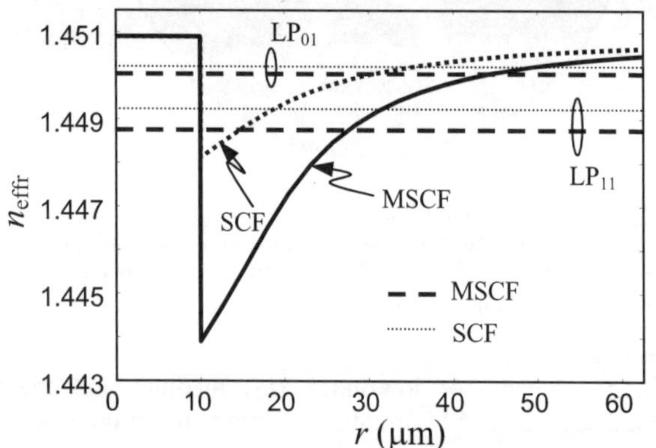

Fig. 2. Effective-index profile of the MSCF (solid curve) at the wavelength 1550 nm. The fiber parameters are given by Eq.(3). The dotted curve shows the effective-index profile of the SCF. The dashed and dotted lines show the mode indices of the LP_{01} and LP_{11} modes of the two fibers.

$n_{\mathrm{effr}}(r)$, which is shown in Fig.2. The effective-index profile is non-uniform in the cladding and varies monotonically with r. We should mention here that the effective-index profile varies with the mode type. For example, the effective-index profile for the LP_{0m} modes is different from that of the LP_{1m} modes. In our example, the difference is so small that it does not show up in Fig.2. The effective-index profile clearly shows that these kinds of fibers are essentially leaky in nature. The modes of such fibers suffer from finite leakage losses and should be considered as leaky modes. The first two leaky modes (LP_{01} and LP_{11}) along with the effective-index profile of the MSCF at the wavelength 1550 nm are shown in Fig. 2. To compare the performance of the MSCF with that of the SCF, we also plot in Fig.2 the effective-index profile and the mode indices of the SCF. The SCF used here has the same values of parameters as given by Eq.(3) and the duty cycle of segmentation is chosen to be 50%. The effective-index profiles of the fibers show clearly that both structures are leaky. However, at $r = a$, the MSCF has a much smaller value of cladding index and, therefore, is expected to have a lower leakage loss, compared with the SCF. Since the structures are leaky, the appropriate criterion to determine which modes are effectively guided is to compare their leakage losses. We should mention that photonic crystal fibers and holey fibers are also leaky structures and support only leaky modes. In all these structures, the higher modes are expected to suffer from significantly higher leakage losses and therefore be stripped off after propagating a certain distance, giving rise to effective single-mode operation. For the calculation of the leakage loss, we put an infinitely extended high-index jacket beyond the cladding of the fiber and use the transverse matrix method to estimate the losses of the LP_{01} and LP_{11} modes.

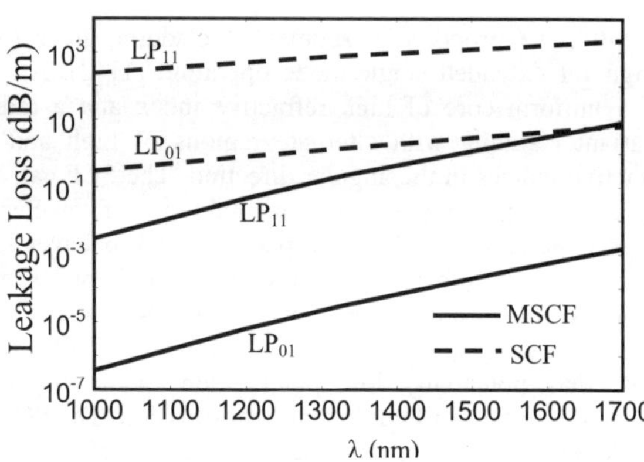

Fig. 3. The leakage losses of the fundamental and first higher-order modes of the MSCF and SCF as a function of wavelength.

Figure 3 shows the variation of the leakage loss with the wavelength for the first two modes of the two structures. Our results show that, for the MSCF, at any wavelength lying between 1000 and 1700 nm, the losses of the LP_{01} and LP_{11} modes differ by four orders of magnitude. Such a large difference in the leakage loss between the fundamental mode and the first higher-order mode makes an effective single-mode fiber. Moreover, the leakage loss of the LP_{01} mode is extremely low (~0.3 dB/km at 1550 nm), which suggests potential applications for long-distance optical communication. On the other hand, the SCF provides a loss difference of almost two orders of magnitude between the two modes. The SCF can therefore offer single-mode operation over a wide range of wavelengths including the communication window at 1550 nm. However, the relatively large leakage loss of 3.63 dB/m at 1550 nm in the SCF restricts its use only for short-length applications like fiber amplifiers and lasers. The leakage losses of the first two modes of the MSCF and the SCF at different wavelengths are listed in Table I.

Table I. Leakage losses of the first two modes of the MSCF and the SCF at different wavelengths.

λ (nm)	Leakage Loss (dB/m)			
	MSCF		SCF	
	LP_{01}	LP_{11}	LP_{01}	LP_{11}
1550	3.33×10^{-4}	1.90	3.63	1.42×10^{3}
1300	2.47×10^{-5}	0.17	1.25	7.30×10^{2}
1000	3.63×10^{-7}	2.88×10^{-3}	0.30	2.37×10^{2}

The intensity profile of the LP_{01} mode of the MSCF at 1550 nm is shown in Fig.4. The profile in the core is comparable to that of a conventional single-mode fiber, which makes it easy to excite with the use of standard Gaussian-like laser beams. Most importantly, the MSCF has the core diameter of 20 μm, which is about two times larger than that of a conventional single-mode fiber. Such a large mode area single-mode fiber can suppress the nonlinear effects effectively and should be useful in DWDM systems.

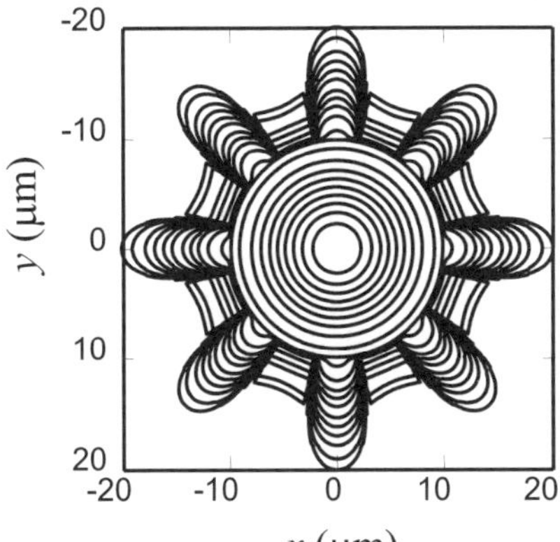

Fig. 4. The intensity profile of the LP_{01} mode of the MSCF at 1550 nm.

5. CONCLUSION

We show that all the segmented cladding structures are leaky in nature. However, by modifying the cladding structure, it is possible to adjust both the leakage loss of the fundamental mode and the loss difference between the fundamental mode and the higher-order mode. The MSCF proposed in this paper has a leakage loss that is four orders of magnitude smaller than that of the SCF. Both the MSCF and the SCF show effective single-mode operation in the wavelength range 1000 nm – 1700 nm. The MSCF should be useful for long-haul DWDM communication systems while the SCF is suitable for short-length applications like fiber amplifiers and lasers.

ACKNOWLEDGEMENT

This work was supported by a grant from the Research Grants Council of the Hong Kong Special Administrative Region, China [Project No. CityU 1034/02E].

REFERENCES

[1] V. Rastogi, and K. S. Chiang, Propagation characteristics of a segmented cladding fiber, Opt. Lett., vol. 26, page 491-493, 2001.

[2] K. S. Chiang, and V. Rastogi, "Ultra-large-core single-mode fiber for optical communications: The segmented cladding fiber," Optical Fiber Communication Conference–OFC 2002, Anaheim, CA, pp. 620-621, 2002.

[3] T. A. Birks, J. C. Knight, and P. St. J. Russell, Endlessly single-mode photonic crystal fiber, Opt. Lett., vol. 22, page 961-963, 1997.

[4] T. M. Monro, P. J. Bennett, N. G. R. Broderick, and D. J. Richardson, Holey fibers with random cladding distributions, Opt. Lett., vol. 25, page 206-208, 2000.

[5] K. S. Chiang, Radial effective-index method for the analysis of optical fibers, Appl. Opt., vol. 26, page 2969-2973, 1987.

[6] K. Thyagarajan, S. Diggavi, A. Taneja, and A. K. Ghatak, Simple numerical technique for the analysis of cylindrically symmetric refractive-index profile optical fibers, Appl. Opt., vol. 30, page 3877-3879, 1991.

A Study of Asymmetric Fiber Coupler by Fourier Series Analysis Technique

M. Liu, P. Shum and H. Ghafouri- Shiraz

Network Technology Research Center

Nanyang Technological University, Singapore

Tel: +65-67904217

Fax:+65-67926894

Email: epshum@ntu.edu.sg

Abstract: We have developed a new coupled wave equation to analyze the pulse switching dynamics of a two-unequal-core fiber coupler under different radii ratio. The analysis is based on a flexible and simple method called Fourier series analysis technique (FSAT).

Key Words: Fiber coupler, two cores, soliton switching

1. INTRODUCTION

The two-core fiber coupler has been extensively studied both experimentally and theoretically since the success of producing a two-core fiber functioning as a directional coupler in 1980. The nonlinear properties of the two-core fiber coupler were also investigated with the realization of the ultra-fast all-optical switch. However, most of the previously reported works in this area have been focused on the two-identical-core coupler where both fibers of the coupler have the same radius and refractive index.

In this paper, we have extended the study to two-unequal-core couplers in which the two cores have different radii. Therefore the commonly used symmetric nonlinear Schrödinger equations (NLSE) cannot be used to analyze the pulse switching characteristics of this unequal-core coupler. Hence, in order to analyze the characteristics of this proposed new fiber coupler we have modified the coupled wave equations so that to include different core radii of the fiber coupler. The new coupled wave equations can be solved numerically by applying the FSAT developed by the authors [1].

2. ANALYSIS

We consider weakly guiding evanescent-field coupler that consists of two parallel unequal single-mode fiber cores with Kerr nonlinearity. The radii of the two parallel fibre cores, embedded in a common cladding, are denoted by ρ_1 and ρ_2. The two cores have the same refractive index n_1 and their common cladding refractive index is n_2. The refractive index difference between cores and cladding is defined as:

$$\Delta = (1 - n_2^2 / n_1^2) / 2 \tag{1}$$

The center-to-center separation of the two cores is d.

Based on the coupled-mode equations in [2] and the analog deriving procedures presented in [3] which describes the pulse propagation in twin-core couplers, we have derived the following new set of equations which describes pulse switching dynamics in a two-unequal-core coupler.

$$i(\frac{\partial a_1}{\partial z} + \frac{1}{v_{g1}}\frac{\partial a_1}{\partial t} + C_{12}^{'}\frac{\partial a_2}{\partial t}) - \frac{k_1^{"}}{2}\frac{\partial^2 a_1}{\partial t^2} + C_{12}a_2 - \frac{C_{12}^{"}}{2}\frac{\partial^2 a_2}{\partial t^2} = 0 \tag{2}$$

$$i(\frac{\partial a_2}{\partial z} + \frac{1}{v_{g2}}\frac{\partial a_2}{\partial t} + C_{21}^{'}\frac{\partial a_1}{\partial t}) - \frac{k_2^{"}}{2}\frac{\partial^2 a_2}{\partial t^2} + C_{21}a_1 - \frac{C_{21}^{"}}{2}\frac{\partial^2 a_1}{\partial t^2} = 0 \tag{3}$$

where subscripts 1 and 2 denote the large and small core respectively ($\rho_1 > \rho_2$). $a_1(Z,T)$ and $a_2(Z,T)$ are the pulse varying envelopes carried by the fundamental modes of the two cores in isolation. $k_{1(2)}^{"}$ denotes the group velocity dispersion of both fiber cores, respectively and $v_{g1(2)}$ represents their group velocities. It should be noted that in Eqs.(2)-(3), we have also considered the coupling coefficient dispersion.

C_{12} and C_{21} are the coupling coefficients between the two cores. The parameters $C_{12}^{'}$ and $C_{21}^{'}$, which account for the dispersion properties of the coupling coefficient, represent the 1st order coupling coefficient dispersion whereas $C_{12}^{"}$ and $C_{21}^{"}$ are the 2nd order coupling coefficient dispersion.

To simplify equations (2) and (3), a normalized coordinate system based on small core is adopted, that is:

$$Z = -\frac{k_2^{"}}{t_0^2}z \tag{4}$$

$$T = \frac{1}{t_0}(t - \frac{z}{v_{g2}}) \tag{5}$$

where t_0 is the width of the input pulse. Using (4) and (5), equations (2) and (3) can be rewritten as:

$$i(\frac{\partial a_1}{\partial Z} - \frac{t_0}{k_2^{"}}(\frac{1}{v_{g1}} - \frac{1}{v_{g2}})\frac{\partial a_1}{\partial T} - t_0\frac{C_{12}^{'}}{k_2^{"}}\frac{\partial a_2}{\partial T}) + \frac{1}{2}\frac{k_1^{"}}{k_2^{"}}\frac{\partial^2 a_1}{\partial T^2} - t_0^2\frac{C_{12}}{k_2^{"}}a_2 + \frac{C_{12}^{"}}{2k_2^{"}}\frac{\partial^2 a_2}{\partial T^2} = 0 \tag{6}$$

$$i(\frac{\partial a_2}{\partial Z} - t_0\frac{C_{21}^{'}}{k_2^{"}}\frac{\partial a_1}{\partial T}) + \frac{1}{2}\frac{\partial^2 a_2}{\partial T^2} - t_0^2\frac{C_{21}}{k_2^{"}}a_1 + \frac{C_{21}^{"}\partial^2 a_1}{2k_2^{"}\partial T^2} = 0 \tag{7}$$

Adding the nonlinear terms introduced by Kerr nonlinearity, the new coupled-mode equations which describe pulse propagation in this two-unequal-core coupler, taking into account the effect of the coupling coefficient dispersion, can be expressed as:

$$i\left(\frac{\partial a_1}{\partial Z}-\frac{t_0}{k_2^{''}}\left(\frac{1}{v_{g1}}-\frac{1}{v_{g2}}\right)\frac{\partial a_1}{\partial T}-t_0\frac{C_{12}^{'}}{k_2^{''}}\frac{\partial a_2}{\partial T}\right)+\frac{1}{2}\frac{k_1^{''}}{k_2^{''}}\frac{\partial^2 a_1}{\partial T^2}-t_0^2\frac{C_{12}}{k_2^{''}}a_2+\frac{C_{12}^{''}}{2k_2^{''}}\frac{\partial^2 a_2}{\partial T^2}+|a_1|^2 a_1=0 \tag{8}$$

$$i\left(\frac{\partial a_2}{\partial Z}-t_0\frac{C_{21}^{'}}{k_2^{''}}\frac{\partial a_1}{\partial T}\right)+\frac{1}{2}\frac{\partial^2 a_2}{\partial T^2}-t_0^2\frac{C_{21}}{k_2^{''}}a_1+\frac{C_{21}^{''}\partial^2 a_1}{2k_2^{''}\partial T^2}+|a_2|^2 a_2=0 \tag{9}$$

where the coupling coefficients are given as:

$$C_{12}=\frac{1}{2}\frac{U_1U_2}{V_1}\left(\frac{V_2}{V_1}\right)^{1/2}\frac{K_0(W_1d/\rho_1)+K_0(W_1d/\rho_1)}{[K_{-1}(W_1)K_1(W_1)K_{-1}(W_2)K_1(W_2)]^{1/2}}\times$$
$$\frac{[(\rho_2/\rho_1)W_1I_1(\frac{\rho_2}{\rho_1}W_1)K_0(W_2)+W_2I_0(\frac{\rho_2}{\rho_1}W)K_1(W_2)]}{[U_2^2+(\rho_2/\rho_1)^2W_1^2]}\times\frac{(\delta_1\delta_2)^{\frac{1}{4}}}{(\rho_1\rho_2)^{\frac{1}{2}}} \tag{10}$$

The coefficient C_{21} can be obtained by interchanging the subscripts 1 and 2 in all terms. The definition of the parameters U, V, W, δ are given in [2].

Since different core diameters result in different effective refractive indices hence, the group velocity and group velocity dispersion of the two cores are different. We have [4]:

$$\frac{1}{v_{g1(2)}}=\frac{1}{c}\left(n_{eff1(2)}+\omega\frac{\partial n_{eff1(2)}}{\partial\omega}\right) \tag{11}$$

$$k_{1(2)}^{''}=\frac{1}{c}\left(2\frac{\partial n_{eff1(2)}}{\partial\omega}+\omega\frac{\partial^2 n_{eff1(2)}}{\partial\omega^2}\right)\approx\frac{\lambda^3}{2\pi c^2}\frac{\partial^2 n_{eff1(2)}}{\partial\lambda^2} \tag{12}$$

The relations between the coupling coefficient, the intermodal dispersion, and the 2nd order coupling coefficient dispersion are given as:

$$C_{12(21)}^{'}=\frac{\partial C_{12(21)}}{\partial\omega} \tag{13}$$

$$C_{12(21)}^{''}=\frac{\partial^2 C_{12(21)}}{\partial\omega^2} \tag{14}$$

To solve the coupled wave equations (8) and (9), first we obtain the parameters listed in Eqs. (10) to (14) and then we apply the Fourier series analysis technique. In doing so, a_1 and a_2 are expanded in Fourier series and substituted into equations (8) and (9). Then a set of first-order partial differential equations can be obtained, which are solved by Runge-Kutta method in the frequency domain. Employing the inverse Fourier transform, we can get the time domain solution.

We have analyzed a typical two-unequal-core coupler, which is operated at 1.55μm wavelength. Other parameters include pulse width $t_0=10\,fs$, $\rho_1=3.25\mu m$ and $d/\rho_1=2.5$. The refractive index difference between the coupler fiber cores and the common cladding is $\Delta=0.0055$, which corresponds to the weakly guiding case.

We have assumed that the pulse is initially launched into the input port of the larger core with the initial conditions of $a_1(0,T)=\mathrm{sech}(T)$ and $a_2(0,T)=0$.

We have considered the following two cases: $\rho_2/\rho_1=0.95$ and $\rho_2/\rho_1=0.9$. The switching dynamics of the two cases are shown in Figs. 1 and 2, respectively. In the figures, the parameters U and V are defined as: $U(Z,T)=|a_1(Z,T)|^2$ and $V(Z,T)=|a_2(Z,T)|^2$. Z and T are the normalized distance and time coordinates, respectively. It is well known that a nonlinear coupler can function as an all-optical switch. In the two figures shown, it can be observed that the pulse can couple back and forth between the two cores for different radii ratios. When we decrease the ratio (i.e., the difference between the two core diameters becomes greater), the power can be transferred to the larger fiber core much more easily and less power is coupled to smaller fiber core. For distinctly different cores, little power can be transferred between the two cores and the pulse mainly propagates in the input core. It can also be found that the presence of intermodal dispersion, $C_{12}^{'}$ and $C_{21}^{'}$, leads to the breakup of the input pulse and causes severe pulse distortion. It should not be surprising in that the two-non-identical-core coupler is, in fact, a two-mode device and intermodal dispersion is an important component of dispersion as explained in [5].

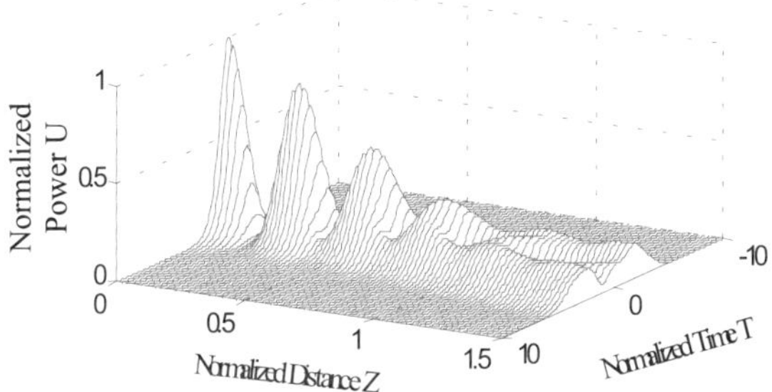

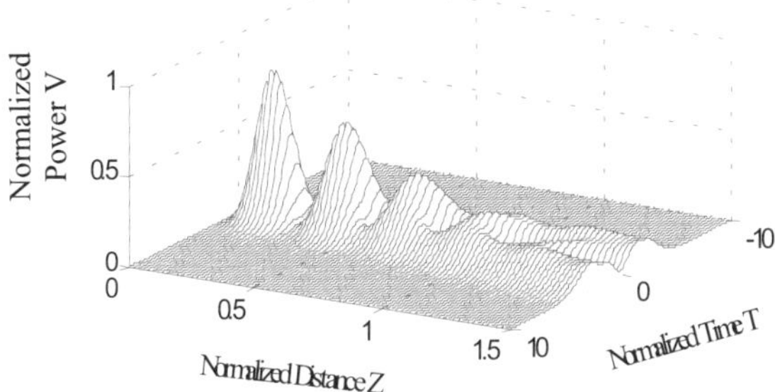

Fig.1 Propagation of optical pulse for ρ_2/ρ_1=**0.95**

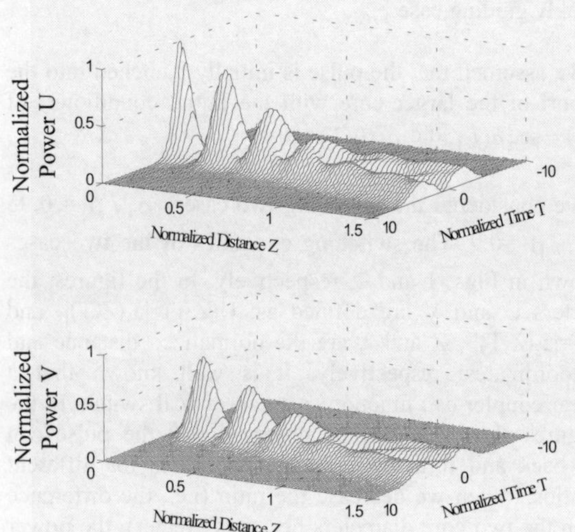

Fig.2 Propagation of optical pulse for ρ_2/ρ_1=**0.9**

In the following, the analytical and simulation results of maximum power transfer and coupling length are compared. Analytically, the maximum power transfer $F_{\_max}$ from large core to small core is described as [2]:

$$F_{\_max} = [1+(\frac{\beta_1-\beta_2}{2C_{12}})^2]^{-1} \tag{15}$$

where $\beta_{1(2)}$ is the propagation constants of the two cores respectively and C_{12} is the coupling coefficient.

For the two-unequal-core coupler, the maximum power transfer occurs at a distance [2], i.e.,

$$L = \frac{\pi}{2\beta_b} \tag{16}$$

where $\beta_b = (\frac{C_{12}C_{21}}{F_{\_max}})^{1/2}$.

Let us denote $F'_{\_max}$ and L' as the simulation maximum power transfer and coupling length. These parameters can be obtained from the numerical results as shown in Figs.3 and 4.

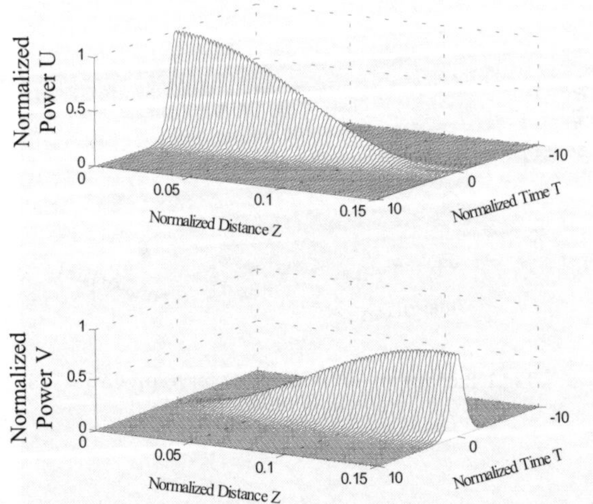

Fig.3 Propagation of optical pulse for ρ_2/ρ_1=**0.95** for one coupling length

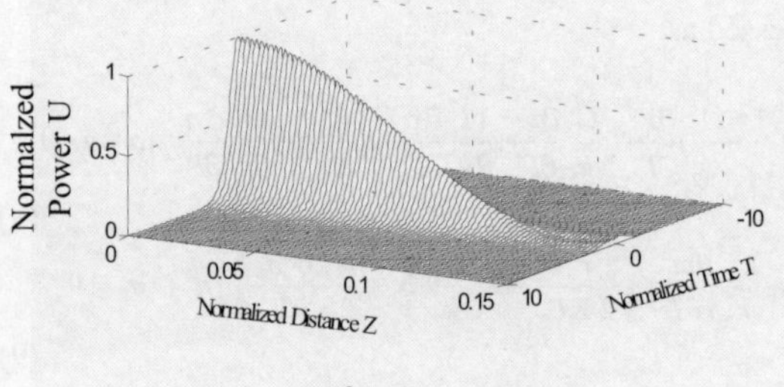

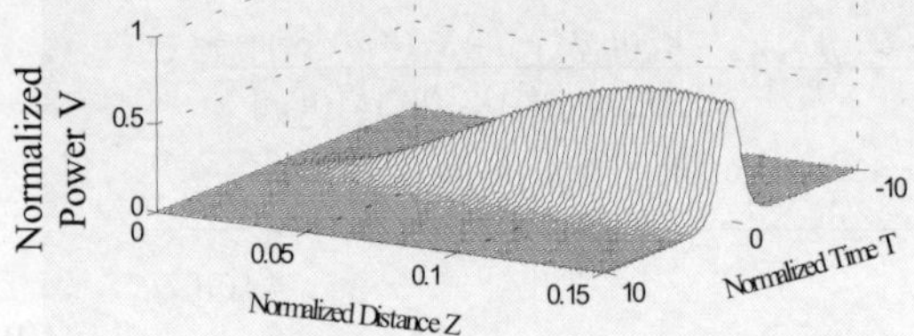

Fig.4 Propagation of optical pulse for ρ_2/ρ_1=**0.9** for one coupling length

We have listed the calculated and simulated maximum power transfer $F_{\_max}$ and $F'_{\_max}$, as well as those of coupling length L and L', in table 1 for comparison.

Table1 Comparison of power transfer and coupling length

ρ_2/ρ_1	$F_{\_max}$	$F'_{\_max}$	L	L′
0.95	0.92	0.87	0.14	0.14
0.9	0.75	0.72	0.12	0.13

Owing to the presence of intermodal dispersion, i.e., C'_{12} and C'_{21}, a complete power transfer is not attainable, i.e., $F'_{\_max} \neq F_{\_max}$. It can be observed that the numerical results are almost consistent with the analytical results.

As mentioned the total power in the coupler field is given by:

$$P = P_1(z)+P_2(z)+P_{12}(z) \tag{17}$$

where $P_i(i=1,2)$ is the power of the mode of core i, and P_{12} is the cross power to account for the modal interaction which cannot be exhibited in the figures. For identical cores, P_{12} is negligible. However, it can't be ignored for nonidentical cores. By decreasing the radii ratio of the two cores, the cross power increases [6].

3. CONCLUSIONS

The analysis of pulse switching dynamics in a two-core fiber coupler is extended to the case of unequal cores. By employing the Fourier series analysis technique, we solved the newly derived equations to describe pulse switching in a two-unequal-core coupler.

REFERENCES

[1] H.Ghafouri-Shiraz, P.Shum, M.Nagata, A Novel Method for Analysis of Soliton Propagation in Optical Fibers, IEEE J.Quantum Electron., vol 31, pp.190-200,1995

[2] P. D. McIntyre and A.W. Snyder, Power transfer between optical fibers, J. Opt.Soc. of Amer., vol.63, pp.1518-1527, 1973

[3] K. S. Chiang, Propagation of short optical pulses in directional couplers with Kerr nonlinearity, Opt. Soc. Am. B, vol. 14, pp. 1437-1443, 1997

[4] G. P. Agrawal, *Nonlinear Fiber Optics*, 1989

[5] K. S. Chiang, Coupled-Mode Equation for Pulse Switching in Parallel Waveguides, IEEE J. Quantum Electron., vol. 33, pp.950-954, 1997

[6] A. W. Snyder and A. Ankiewicz, Optical Fiber Couplers-Optimum Solution for Unequal Cores, IEEE J. Lightwave Technol., vol. 6, pp.463-473, 1988

Hydrogen loading in tin-phosphosilicate fibres: a method to achieve enhanced thermal stability in fibre-Bragg-grating based devices

Gilberto Brambilla
Optoelectronics Research Centre
University of Southampton, Southampton, SO17 1BJ, United Kingdom
Fax:+44 (0)23 8059 3149, e-mail: gb2@orc.soton.ac.uk

Abstract: Enhanced photosensitivity has been observed in hydrogen-loaded tin-phosphosilicate fibres by using a 248 nm excimer laser. Isothermal measurements up to 860 K demonstrated significant advantages over fibre gratings written in conventional H-loaded fibres. Gratings written in this fibre require a considerably shorter post-fabrication thermal annealing in order to satisfy the stability requirements of telecom components.

Key Words: Gratings, Thermal stability, Hydrogen loading, Photosensitivity.

1. INTRODUCTION

Devices based on fibre Bragg gratings require high photosensitivity and extremely small tolerance on the optical properties over a long time scale in order to be assembled in reliable WDM components. High photosensitivity in standard telecom optical fibres is achieved with the so-called hydrogen (or deuterium) loading process, which can increase the photosensitivity by more than two orders of magnitude [1]. Since its first use in germanosilicate fibres, hydrogen loading has been successfully applied to many silicate fibres [2], including phosphorus, aluminium or cerium doped or codoped fibres. No improvement has been observed when the dopant is nitrogen [3]. It has to be remarked that high photosensitivity at 248 nm has been observed in phosphorus-doped silica fibres only when gratings were written in heated fibres [4]. A weak effect has been observed in unheated H-loaded phosphosilicate fibres when exposed to the KrF laser [5]. A drawback of all the gratings written in hydrogen-loaded fibres is the extremely poor temperature stability. In fact in most fibres the photo-induced refractive index change starts to be erased at 100 ^{0}C in one hour. Tin doping has been shown to be a powerful method to achieve high photosensitivity [6]. Moreover, the enhanced stability of gratings written in tin-silicate fibres does not require post-fabrication annealing to satisfy the requirements of telecom components [7].

In this paper we study the effect of hydrogen loading on tin-doped fibres, with particular stress on the excellent thermal stability.

2. HYDROGEN LOADING

The tin-codoped phosphosilicate fibre used in the experiments was fabricated by modified chemical vapour deposition (MCVD). $SnCl_4$ was the precursor used to introduce Sn in the optical fibre preform. The fibre external diameter, numerical aperture and cut-off were d=110μm, NA~0.14 and λ_c~1.42 μm respectively. The fibre was placed in a hydrogen-loading cell at 165 bars and 25 ^{0}C for 14 days. The photosensitivity of samples taken from the vessel was tested by writing gratings at ~1.55μm with a "Lambda Physik" line-narrowed KrF-laser (mod. EMG150) and a phase-mask. The exposure time, laser repetition rate and pulse fluence were t_{exp}=30 minutes, RR=20 Hz and I_p~300 mJ/cm^2 respectively. In order to let the hydrogen outdiffuse, the fibre samples were left at room temperature for a period of 7-14 days before being spliced to a fibre-coupled diode. Spectra were collected using an optical spectrum analyser. Figure 1 compares the reflectivity spectrum of the grating written both in the hydrogen-loaded and in the pristine fibres.

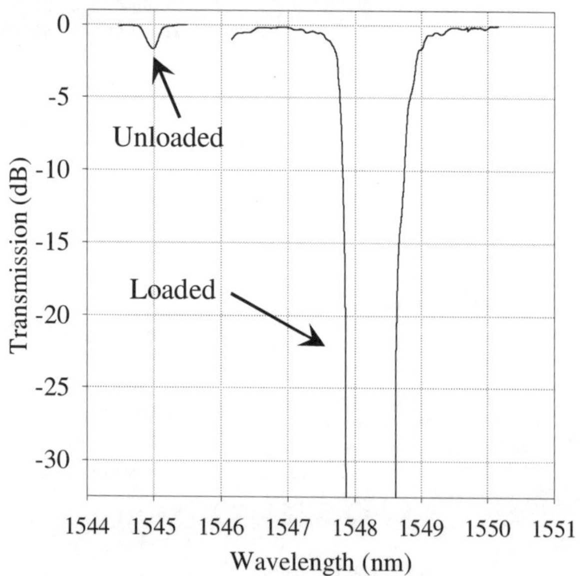

Fig. 1: Transmission spectra of gratings written in loaded and unloaded tin-phosphosilicate fibres. Grating length, exposure time, laser repetition rate and pulse fluence are L=2mm, t_{exp}=30 minutes, RR=20 Hz and I_p~0.3 J/cm^2 respectively. The loading process was carried out leaving the fibre in the hydrogen-loading cell at 165 bars and 25 ^{0}C for 14 days.

Similarly to what has previously been observed in fibres doped with Ge, Al and Ce, hydrogenation enhances photosensitivity to UV laser radiation in the tin-doped phosphosilicate fibre. While in the unloaded fibre the maximum reflectivity was <2dB, in the H-loaded fibre it was > 35 dB. The induced refractive index modulation (Δn_{mod}) was evaluated by fitting the reflectivity curve $R(\lambda)$ with [8]:

$$R(\lambda)=\frac{1}{1+\frac{1}{\kappa^2 L^2 \text{sinc}^2(\gamma L)}} \qquad (1)$$

where L is the grating length, $\kappa=\pi\cdot\Delta n_{mod}/\lambda$ the "ac" coupling constant, λ the wavelength, $\gamma=(\kappa^2-(\sigma+\delta)^2)^{1/2}$ the propagation constant inside the grating, $\sigma=2\pi/\lambda\cdot\Delta n_{ave}$ the "dc" coupling constant, $\delta=\beta-\pi/\Lambda$ the detuning, β the propagation constant in vacuum, Λ the grating pitch and Δn_{ave} the induced average refractive index change. Δn_{mod} was estimated to be ~1.4 10^{-3} and ~1.3 10^{-4} in the loaded and unloaded fibres.

Figure 2 compares the Δn_{mod} growth of gratings written in a fibre loaded for 3 days and in a pristine fibre.

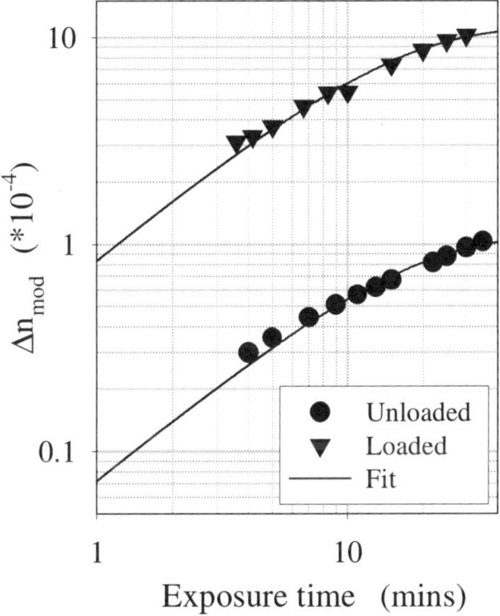

Fig. 2: Temporal evolution of Δn_{mod} in hydrogen-loaded (3 days at 165 bars and 25 ^{0}C) and pristine tin-phosphosilicate fibres.

The experimental data have been fitted with an exponential rising to the maximum:

$\Delta n_{mod}=\Delta n_{sat}\ (1-e^{-t/\tau})$ (2)

where Δn_{sat} is the asymptotic value of Δn_{mod} for long exposure times, τ is the time constant of the refractive index growth and t the exposure time. The fit parameters are reported in table 2. While Δn_{sat} differs by an order of magnitude, the τ does not show any significant modification.

	Δn_{sat}	τ (min)
Unloaded	1.10 10^{-4}	14.7
Loaded	1.11 10^{-3}	12.8

Tab. 1: Fit parameters for the data of fig. 2.

3. THERMAL STABILITY

The grating thermal stability was studied using the so-called master curve method [9]. It has been shown that this accelerating aging technique is a more general and preferable method as compared to the parallel power-law method for gratings written in hydrogen-loaded fibres [10]. In this approach the grating decay is recorded at different temperatures and all the data are combined to give a master curve from which it is possible to predict the grating reliability in different conditions. The aging parameter E_d (called demarcation energy) is defined as [9]:

$E_d=k_B T\ln(\nu t)$ (3)

where k_B is the Boltzmann's constant and T the temperature. ν represents an attempt frequency and is determined during the data fit.

In our experiment, gratings were written as previously explained in samples of fibre hydrogen loaded for 6 days at 165 bars. Real time spectra of the gratings placed in the furnace, were recorded by means of a white light source, a coupler and an optical spectrum analyser. The time evolution of the integrated coupling coefficient η (defined as $\eta=\Delta n_{mod}/\Delta n_{mod}(0)$, where $\Delta n_{mod}(0)$ is the initial value of Δn_{mod}) is shown in figure 3 for four different temperatures.

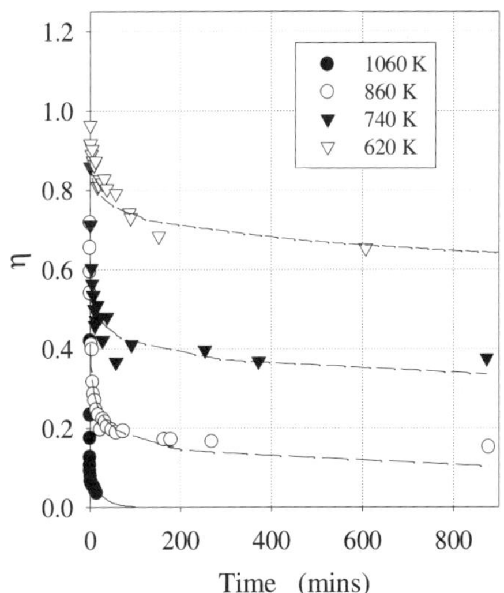

Fig. 3: Decay of the gratings written in H-loaded tin-phosphosilicate fibre for different furnace temperatures. The fibres were hydrogen-loaded at 165 bars and 25 ^{0}C for 6 days. Gratings were written by exposing the fibre for 30 mins at I_p=0.3J/cm^2.

It is worth stressing the enhanced thermal stability of these gratings. While gratings written in unloaded germanosilicate fibres had ~27% erased in 60 mins at 350 ^{0}C [9], in the same conditions the gratings written in the hydrogen-loaded tin-phosphosilicate fibre had a change η~24%. The experimental data were fitted using $\nu=2.5\cdot10^{10}$ min^{-1}, considerable smaller than the value found for hydrogen-loaded germanosilicate fibres [10]. The master curve was approximated by the equation $\eta=(1+e^{Ed-1.65)/0.1258})^{-1}$ small decays. From the curve it is possible to evaluate what is the thermal decay of a grating after 25 years at 80 ^{0}C (E_d=1.23eV): η=0.966, meaning that ~3.4% of the original grating has been wiped away. Table 2 compares the values obtained in this paper to the results published on the stability of gratings written in germanosilicate [9], in boro-germanosilicate [10] and in tin-silicate [7] fibres.

Fibre	$1-\eta$	Ref.
HL-SPS	0.034	This paper
GS	0.0782	[9]
BGS	0.138	[10]
SS	0.0014	[7]

Tab. 2: Estimation of the fraction of the initial grating erased after 25 years at 80 ^{0}C. HL-SPS, GS, BGS and SS represent hydrogen-loaded tin-phosphosilicate, germanosilicate, boro-germanosilicate and tin-silicate, respectively.

As expected, the decay is considerable higher than the one observed in unloaded tin-doped fibres (<1%); nevertheless it is better than the values of degradation reported in literature for unloaded traditional fibres.

4. CONCLUSIONS

In summary, hydrogen loading has been shown to enhance the photosensitivity of tin-doped phosphosilicate fibres. The photo-induced refractive index change of the gratings written in hydrogen-loaded fibres was one order of magnitude stronger than the gratings written in the unloaded fibres. The thermal stability of gratings written in hydrogen-loaded tin-phosphosilicate fibre has been tested with iso-thermal annealing and proved that >15 % of the initial photosensitivity survived a treatment at 860 K for 14 hours. An estimation of the grating fraction surviving 25 years at 80 ^{0}C shows that hydrogen loading of tin-codoped phosphosilicate fibres has a significant advantage over conventional fibres. Gratings written in this fibre require a less-severe post-fabrication thermal annealing in order to satisfy the stability requirements of telecom components.

The author is grateful to E. J. Tarbox and D.J. Richardson for their helpful discussions.

REFERENCES

[1] P.J. Lemaire, R.M. Atkins, V. Mizrahi and W.A. Reed, High pressure H_2 loading as a technique for achieving ultrahigh UV photosensitivity and thermal sensitivity in GeO_2 doped optical fibres, Electron. Lett., vol. 29, pp. 1191-3, 1993.

[2] M. Douay, W.X. Xie, B. Leconte, T. Taunay, P. Bernage, P. Niay, P. Cordier, J. Bayon, H. Poignant and E. Delevaque, Progress in silica optical fibre photosensitivity, Ann. Télécommun., vol. 52, pp. 543-56, 1997.

[3] E.M. Dianov, K.M. Golant, R.R. Khrapko, A.S. Kurkov, B. Leconte, M. Douay, P. Bernage and P. Niay, Grating formation in germanium free silicon oxynitride fibre, Electron. Lett. vol. 33, pp. 236-8, 1997.

[4] P.J. Lemaire, A.M. Vengsarkar, W.A. Reed and D.J. DiGiovanni, Thermally enhanced ultraviolet photosensitivity in GeO_2 and P_2O_5 doped optical fibres, Appl. Phys. Lett., vol. 66, pp. 2034 -6, 1995.

[5] Strasser, "Photosensitivity in phosphorus fibers", Proc. Conf. on Optical Fiber Communication, San Jose, CA, pp. 81-2, 1996.

[6] G. Brambilla, V. Pruneri and L. Reekie, Photorefractive index gratings in SnO_2:SiO_2 optical fibres, Appl. Phys. Lett., vol. 76, pp. 807-9, 2000.

[7] G. Brambilla and H. Rutt, "Fiber Bragg gratings with ultra-high temperature stability", Proc. Conf. on Optical Fiber Communication, Anaheim, CA, pp. 660-662, 2002.

[8] T. Erdogan, Fiber Grating Spectra, J. Lightwave Tech., vol. 15, pp. 1277-94, 1997.

[9] T. Erdogan, V. Mizrahi, P. J. Lemaire and D. Monroe, Decay of ultraviolet-induced fiber Bragg gratings, J. Appl. Phys., vol. 76, pp. 73-80, 1994.

[10] S. Kannan, J.Z.Y. Guo and P. Lemaire, Thermal stability analysis of UV-induced fiber Bragg gratings, J. Lightwave Tech., vol. 15, pp. 1478-83, 1997.

[11] S.R. Baker, H.N. Rourke, V. Baker and D. Goodchild, Thermal decay of fiber Bragg gratings written in Boron and Germanium codoped silica fiber, J. Lightwave Tech., vol. 15, pp. 1470-77, 1997.

Transfer characteristics of DI-NOLM

Y.L. Guan, Y.D. Gong, P. Shum, X.H. Xiao
Network technology research center, Nanyang technological university, Singapore
guanyali@hotmail.com

Abstract: By changing the construction of NOLM, it can possess many different functions. In this paper we analyzed the effects of input power on the transfer characteristic, and the effect of SMF or DSF length on transfer characteristic.

Key Words: NOLM, nonlinear, switching

1. INTRODUCTION

Self-switching in Sagnac interferometers is achieved by breaking the loop symmetry for counter-propagating pulses in the optical fiber. The phase changes in the loop were obtained by the interaction of SPM and group velocity dispersion (GVD). Due to its self-switching ability the nonlinear optical loop mirror has been recognized as a potentially valuable tool in optical signal processing application. The many different constructions of NOLM have been submitted to achieve self-switching function. According to the variation on construction, Nonlinear optical fiber loop mirror (NOLM) can handle the ultra-short pulses[1][3], which have been shown to be useful in a number of applications, such as soliton switching controlled by signal power, all-optical demultiplexing, pedestal suppression of short pulses, compression of pulses and noise filter in all-optical fiber communication systems.

Here we mainly discuss the DI-NOLM [2] [5], dispersion imbalanced NOLM, which can act as a nonlinear filter to remove the pedestal of narrow pulse and the pulse compressor. Since the GVD varied asymmetrically around the loop then it was possible to obtain self-switching with a symmetric coupler. This use of two fibers is a basic way of introducing an asymmetric GVD around the loop. Here discussion is mainly concentrated on the effect of the loop with different combination of fibers length and arranged order on transfer characteristics.

2. NUMERICAL ANALYSIS

The construction of DI-NOLM is showed as Figure1.

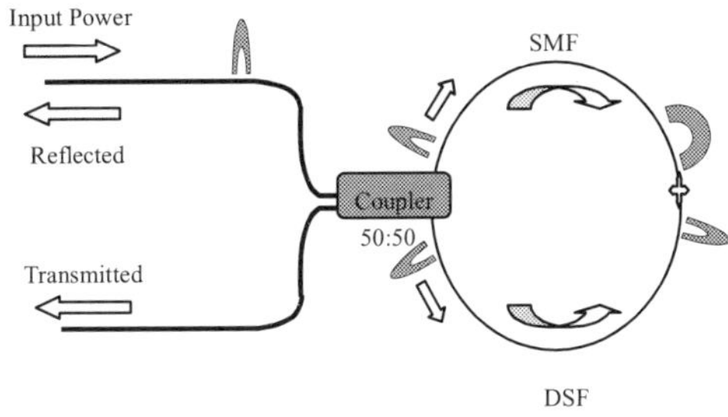

Figure 1 the schematic of the DI-NOLM

The loop consists of two kinds of fibers, SMF and DSF. The transmittivity is analyzed for different input power and the combination of fibers (SMF+DSF or DSF+SMF).

Pulse propagation in a fiber is described by the nonlinear Schrődinger equation [4], which in dimensionless form is

$$\frac{\partial u}{\partial z}+\frac{j}{2}D_i\frac{\partial^2 u}{\partial t^2}+j|u|^2 u=0 \qquad (1)$$

The normalized distance z is in units of the dispersion length $z_c=\tau^2/|\beta_2|, \tau\approx 0.567\tau_{FWHM}$, t is measured in units of τ, the complex electric field $u=u(z,t)$ is in units of for a dispersion-imbalanced loop mirror computed numerically from equation with a split-step Fourier algorithm with 1024 temporal grid points. We calculated the transfer characteristics of NOLM against the input power and the order of fiber which is decided by the clockwise pulse propagating through DSF or SMF first in loop.

the smaller input power and longer fiber, the switching characteristic of NOLM is improved.

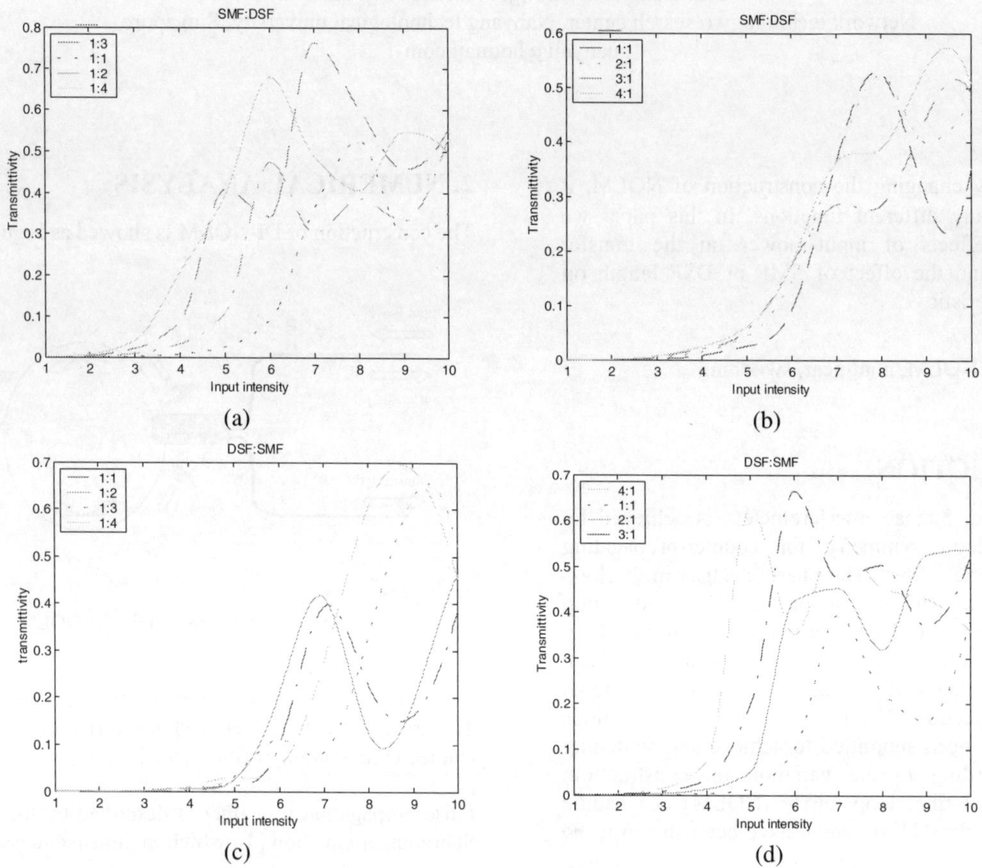

Figure2. The variations of transfer characteristics against the constructions of loop.

(The insets denote the length ratio of two fibers according to the order showed in the titles of figure)

For simplicity, we call the loop in which the clockwise pulse propagates through the SMF first as scheme 1, the transfer characteristics are showed in figure2.(a)(b), and the loop in which the counterclockwise pulse propagate through the DSF first as scheme 2, the transfer characteristics are showed in figure 2(c)(d). In figure2, the figure2.(b) and (c) give the variation of transfer characteristic against the increase of SMF length, the figure2.(a) and (d) give the variation of transfer characteristic against the increase of DSF length. Through the numerical simulation, we compared these two schemes with each other. From the figure 2, we can see that the maximum peak transmittivity shifts toward the small input power as the increasing of the ratio of length for all four changes of loop. This is because that for longer fiber length, the pulse acquires larger phase shift, therefore for

Let us observe the four figures in detail, we can find the shift speed toward small input power of the maximum peak transmittivity is different. For these two schemes, when the length of DSF increases, the shift speed is faster. In other word, the characteristics of switching are better. Under smaller input power, the transmittivity has large value. These results show that by increasing the length of DSF the pulse can get the larger phase shift than by increasing the length of SMF. This situation seems more obvious in scheme 2. , the shift of the maximum peak transmittivity has larger speed than in scheme 1. We can know that if the clockwise pulse propagates through the DSF first then into the SMF, the NOLM has better transfer characteristic. Under this situation, we can use the minimum input power to acquire the better switching characteristics among the four situations, just as showed from the Figure 2.

Through the above analysis, we can get a conclusion that for the best switching property can be acquired in the scheme 2 with longer length of DSF. That is to say, for such DI-NOLM, the pulse at the clockwise direction propagating through the fiber with larger nonlinearity first can acquire large phase shift, just so the DI-NOLM will possess better the characteristic of switching.

3. CONCLUSION

Here we gave the results of numerical simulation. We show the difference of switching characteristics of DI-NOLM in two schemes. The scheme 2 is the better choice for the acquisition of better switching characteristics of DI-NOLM. The large phase shift is important for the pulse propagating along clockwise direction. All these analysis are vital for the function of pulse compression, pedestal filter-out of pulse, and switching of NOLM.

REFERENCES

Journals:

[1] WONG, W.S., Manikin, S., Margalit, M., Haus, H.A., and Ippen, E.P., Self-switching of optical pulses in dispersion-imbalanced nonlinear loop mirrors, Optical Letter, vol.22, No.15, page 1150-1152, 1997

[2] M. Matsumoto and T. Ohishi, Dispersion-imbalanced nonlinear optical loop mirror with lumped dispersive elements, Electronics Letter, vol.34, No.11, page 1140-1142, 1998

[3] A.L. Steele, Pulse compression by an optical fiber loop mirror constructed from two different fibers, Electronics Letter, vol. 29, No.22, page 1971-1972. 1993.

[4] A. L. Steel, J. P. Hemingway, Nonlinear optical loop mirror constructed from dispersion decreasing fiber, Optical Communication, vol.123, page 487-491, 1996.

Book:

[5] Govind P. Agrawal, "Nonlinear fiber optics", 1995

Session 13C

Optical Switching

Solid Free Space Optical Fast Switches

Zeev Zalevsky, Gal Shabtay, Roy Appelman and Amir Shemer

Civcom Devices and Systems Inc., 2 Hagranit, Petah-Tikva, 49514 Israel
zeev.z@civcom.com

Abstract: This paper presents innovative concepts related to the realization of ultra-fast optical switches, which are to be used in next-generation optical networks. Experimental measurements, which characterize the performance and reliability of such switches, are presented.

1. Introduction

Optical switches are an essential part of next-generation optical networks, which carry high-bandwidth DWDM signals, where optical switches may be part of either the local access, the metro or the long-haul for realizing protection, service provisioning and routing of optical signals. All-optical switches that are aimed to take place in optical networks need to be reliable, fast (msec for protection, μsec for service provisioning and sub-μsec for burst packet switching) and should exhibit high functionality such as variable optical attenuation and multicast. The rest of this paper characterizes an all-optical switch that is based on the electro-optic effect [1,2] and exhibit these desired properties.

2. All-Optical Multi-Functional 1X2 Switch

One of the most crucial properties that are required from an all-optical switch is reliability. Being free of OEO (optical-electronic-optical) conversions, and lacking monitoring at each switch one should make sure that the switch maintain its characteristics over time and switching cycles. The 1X2 all-optical switch that is discussed in this paper, went through a series of tests for confirming its reliability. As an example, both insertion-loss and cross-talk of the proposed switch maintain their values over more than 10^{10} switching cycles.

The true relation between the rotation of the polarization angle α and the applied voltage V corresponds to the electro optic effect (Kerr effect) as expressed in Eq. 1:

$$\alpha = \frac{\pi V^2}{2V_\pi(\lambda)^2} \tag{1}$$

where λ is the illuminating wavelength, V is the applied voltage and $V_\pi(\lambda)$ is the voltage (which is wavelength dependent) required for rotating the polarization angle in angle of π. This effect itself results from the imposed electric field that causes the asymmetric molecules of the material to align with the field. This causes the material to become anisotropic and birefringent. The change in the refraction index is given by:

$$n_o - n_e = \lambda K \frac{V^2}{d^2} \tag{2}$$

where n_o and n_e are the ordinary and the extraordinary refraction index, K is the Kerr constant (which is dependent upon the wavelength), V is the applied voltage, d is the distance between the electrodes and λ is the wavelength.

The basic theory related to the physics of the Kerr effect may be obtained using Maxwell's equations. Using the following relation:

$$P = \chi^{(1)}E + \chi^{(2)}E \cdot E + \chi^{(3)}E \cdot E \cdot E + \ldots \tag{3}$$

Where, P is electric polarization. For linear case, P can be expressed as,

$$\begin{aligned} P(r,t) &= \int_{-\infty}^{\infty}\int_{-\infty}^{\infty} \chi^{(1)}(r-r',t-t')E(r',t')dr'dt' \\ P(k,\omega) &= \chi^{(1)}(k,\omega)E(k,\omega) \\ \chi^{(1)}(k,\omega) &= \int_{-\infty}^{\infty}\int_{-\infty}^{\infty} \chi^{(1)}(r,t)e^{(ik.r+i\omega t)}drdt \end{aligned} \tag{4}$$

Where the linear dielectric constant is given by,

$$\varepsilon(k,\omega) = 1 + 4\pi\chi^{(1)}(k,\omega) \tag{5}$$

For nonlinear case, P can be expressed in power series of E.

$$
\begin{aligned}
P(r,t) = & \int_{-\infty}^{\infty}\int_{-\infty}^{\infty} \chi^{(1)}(r-r_1',t-t_1')E(r_1',t_1')dr_1'dt_1' + \\
& \int_{-\infty}^{\infty}\int_{-\infty}^{\infty}\int_{-\infty}^{\infty}\int_{-\infty}^{\infty} \chi^{(2)}(r-r_1',t-t_1';r-r_2',t-t_2')E(r_1',t_1')E(r_2',t_2')dr_1'dt_1'dr_2'dt_2' + \\
& \int_{-\infty}^{\infty}\int_{-\infty}^{\infty}\int_{-\infty}^{\infty}\int_{-\infty}^{\infty}\int_{-\infty}^{\infty}\int_{-\infty}^{\infty} \chi^{(3)}(r-r_1',t-t_1';r-r_2',t-t_2';r-r_3',t-t_3')\cdot \\
& \cdot E(r_1',t_1')E(r_2',t_2')E(r_3',t_3')dr_1'dt_1'dr_2'dt_2'dr_3'dt_3' + \ldots
\end{aligned}
\tag{6}
$$

Within electric dipole approximation $\chi^{(n)}$ (**r**,t) is independent of r and hence $\chi^{(n)}$ (**k**, t) is independent of k. P can also be written as,

$$
P(\omega) = P^{(1)}(\omega) + P^{(2)}(\omega) + P^{(3)}(\omega) \tag{7}
$$

Where superscripts denotes the order of non-linearity in the polarization vector.

Fig. 1 presents the switch transition in time from one state to the other. The upper curve represents the applied voltage whereas the lower curves represent the optical response at both of the output ports. It shows that the switching time is approximately 250ns.

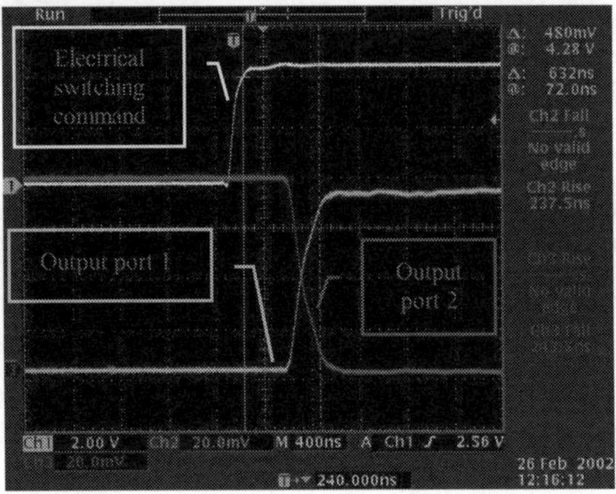

Fig. 1. The electronic and the optic signals in a switching command.

A prototype of an all-optical switch that exhibits both switching and weighted multicast functionality is displayed in Fig. 2. A first version of the prototype is displayed in Fig. 2(a,b) and a more advanced version is seen in Fig. 2(c). Fig. 2 further shows that the footprint of the switch is approximately $1cm^2$.

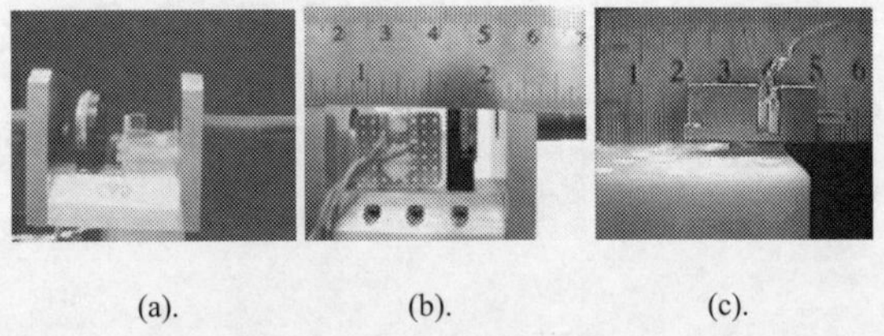

(a). (b). (c).

Fig. 2. The switching prototype.

Experimental results concerning this prototype are presented in Fig. 3. In Fig. 3(a) light is directed towards output "A" and in Fig. 3(b) the light is directed towards output "B". Fig. 3(c) shows the multicast capability case in which the light is directed towards both output ports. The measured cross-talk in these measurements was found to be better than -45dB.

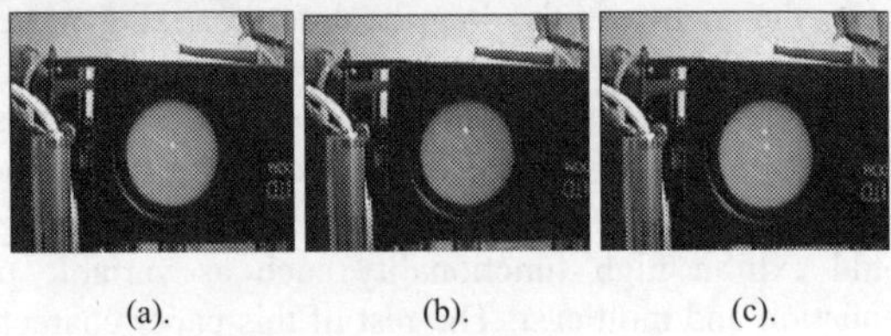

(a). (b). (c).

Fig. 3. Switching measurements.

Fig. 4 presents the improved beta prototype of the 1 by 2 all optical switches.

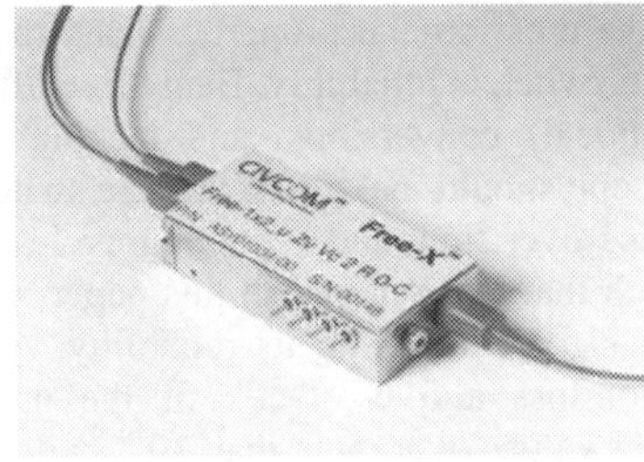

Fig. 4. The switching beta prototype.

Its optical performances such as insertion loss, cross talk, polarization dependent losses (PDL) and polarization mode dispersion (PMD) were measured and investigated as function of the wavelength (see Fig. 5).

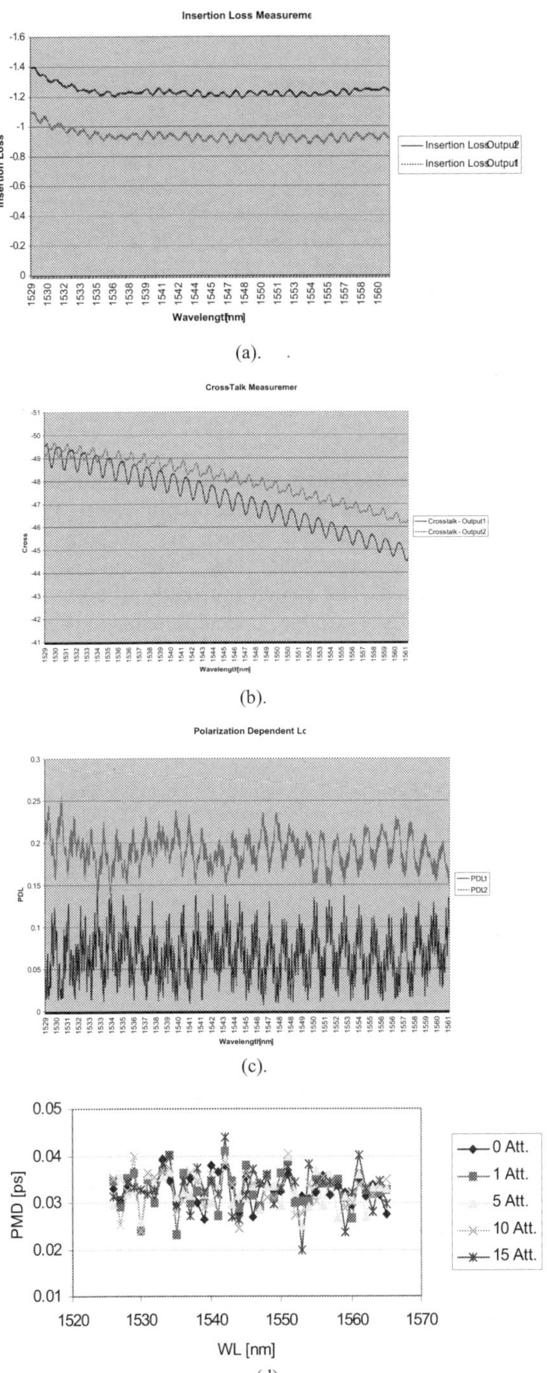

Fig. 5. Optical performance of the 1 by 2 switch.
a). The insertion loss as function of C band wavelengths.
b). The cross talk vs. wavelength . c). PDL. d). The PMD vs. wavelength for various values of attenuation.

Additional optical performance such as PDL versus attenuation, wavelength flatness versus attenuation, chromatic dispersion versus wavelength and attenuation and the multicast ability versus wavelength are respectively presented in Fig. 6.

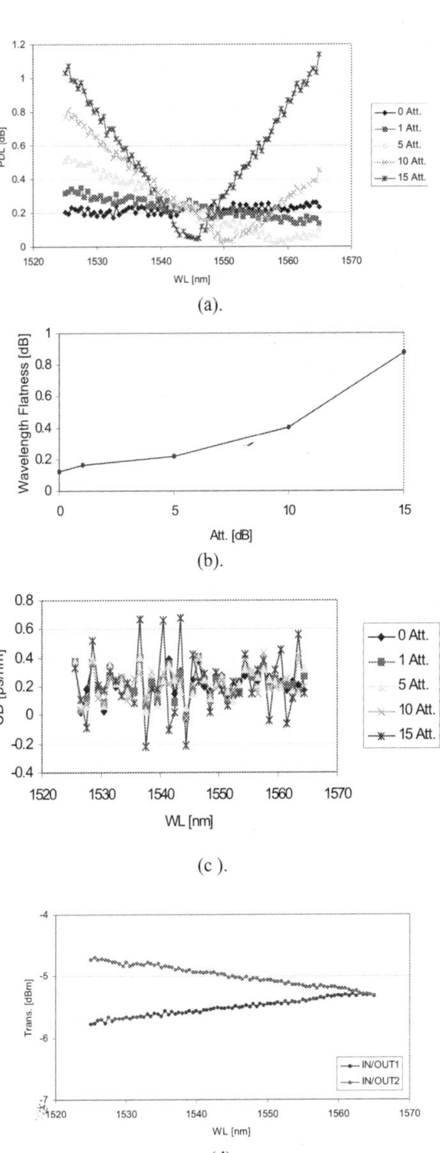

Fig. 6. Optical performance of the 1 by 2 switch.
a). PDL vs. attenuation. b). wavelength flatness vs. attenuation. (c). Chromatic dispersion vs. wavelength and attenuation. (d). multicast vs. wavelength.

The values presented in Fig. 6 are currently being significantly improved by modifying the alignment process. Anticipated improvements, for instance, for the PDL are that a PDL of 0.4dB is obtained for attenuation values of –40dB.

The switch prototype successfully passed a set of environmental tests as defined in the Telcordia specifications that include thermal cycling, shaking and shock tests in GR 1221 (see Fig. 7).

(a).

(b).

Fig. 7. (a) Thermal and (b) shaking facilities in which the switch was tested.

The performed vibration, shocks and thermal cycles are demonstrated in Fig. 8(a) (b) and (c) respectively.

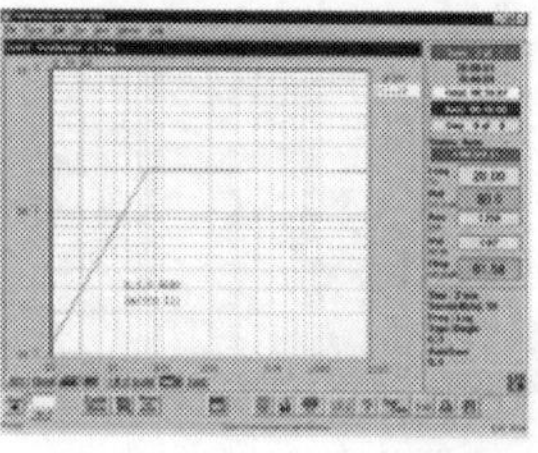

(a).

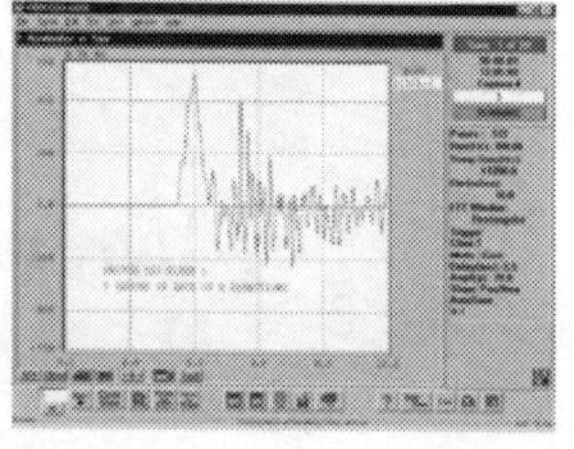

(b).

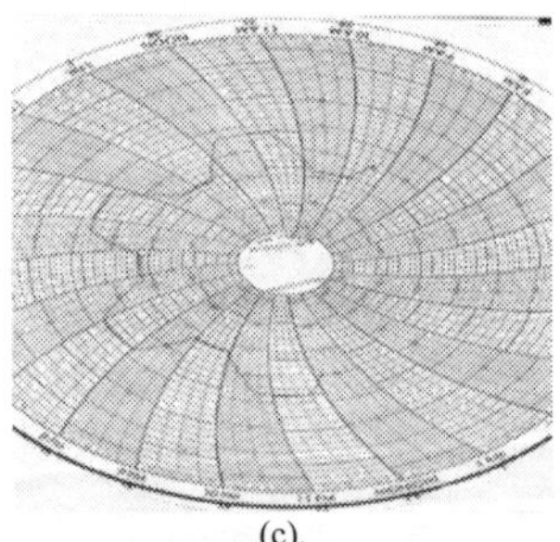

(c).

Fig. 8. (a) Vibration (b) Shocks and (c). Thermal cycles.

Finally, the entire specifications list of the above-mentioned prototype are displayed in Fig. 9.

• Insertion loss	1.2-1.6 dB
• Switching time	100 ns
• Wavelength window	1530-1625 nm
• Cross Talk	-45 dB
• Back Reflection	-50 dB
• PDL (Polarization Dependent Loss)	0.1 dB
• Max. optical power input	36 dBm (4 Watt)
• Dimensions	40x30x16 mm
• Reliability	>10*10^9 Cycles
• VOA Granularity	< 1 dB
• Input voltages	5VDC, 12VDC

Fig. 9. Spec. of the proposed switch.

3. Conclusions

This paper presented an ultra fast all-optical free space switch that is based on the electro-optic effect. The obtained results indicate that the switch has a potentially small footprint, low cross-talk, low insertion-loss and high switching speed. The switch inherently offers multicast and ultra-fast VOA (variable optical attenuation) capabilities, yielding high functionality.

4. References

1. D. Marom and D. Mendlovic, "All-optical reduced state 4 by 4 switch," Optics and Photonics News, **7**, 43 (1996).
2. D. Mendlovic, B. Leibner and N. Cohen, "Multistage optical system for broadcasting and switching information," Appl. Opt. **38**, 6103-6110 (1999).

A Novel Delay-prearranged Asymmetric Optical Burst Switching*

Huang anpeng Xie linzhen Xu anshi Wu deming Li jingcong Yin hongxi
National Laboratory on Local Optic-fiber Communication Networks & Advanced Optical Communication Systems, Department of Electronics,
Peking University, Beijing 100871, P. R. China
Tel: +6276 3334, Fax: +6275 1763
E-mail: hapku@ele.pku.edu.cn

Abstract: This paper introduces a prearranged delay asymmetric structure of optical burst switching. In this paper, system logical structure, design of concomitant spare exports and contention-resolving mechanism are investigated in detail.

Key Words: Prearranged Delay (PD), Asymmetric, Optic Burst Switch, Spare Exports.

1. INTRODUCTION

Optical burst switching is fit for developing of optical network. With data business is increasing, data traffic is dominating in future networks. On account of self-similar property of data traffic, optical burst switching is suitable for transmitting this traffic. Effective bandwidth utilization and flexible burst length can be obtained by optical burst switching [1]. Practically useable and very efficient optical burst switching is vital step to all optical networks. Delay-prearranged asymmetric optical burst switching is provided with functions of routing and switching. Goal of this project is in researching an advanced, reliable and patent switching node. In this project, burst headers don't need be rewritten and payload don't need optic-electrical conversion, thus burst could detour to get at its destination in networks instead of discarding this. Furthermore all optical transmission of payload is very advantageous. This project is of not only simply structure but also steady performance and little cost, therefore, its ratio of cost and efficiency is very preponderant.

2. THE STRUCTURE OF NOVEL PREARRANGED DELAY ASYMMETRIC OPTICAL BURST SWITCHING

In this fabrication, concomitant spare exports structure is designed as figure 1. This structure can support more wavelength links by adding switching matrix modules. All optical disposal is applied in this structure. Optical preamplifiers (OPA) and optical booster amplifiers (OBA) are used for compensating for optical energy consuming in switching, multiplexing and demultiplexing. The incoming M WDM signals on each fiber are demultiplexed into individual wavelength signals by demultiplexers. After these signals passing through a 9:1 couplers, header signals are be sent to node control center (NCC), and payload signals are delayed in anticipation according to time of header signal disposition, tunable filter response time and switching time. Payload transmission and header signal disposal should be worked in phase by prearranged delay. After prearranged delay, payload signals are amplified by OPA and are routed by tunable filters according to their header information. Here, tunable filter have router function. Switching matrix is space division optical switching. After routing, payload signals are amplified by OBAs so that they can reach destinations in adequate power. In the end these signals are multiplexed into fibers. We can upgrade this structure by adding wavelength conversions after switching. In this way, this structure can support virtual wavelength path. Wavelength conversion can be implemented with tunable lasers. Switching matrix is not symmetry. Each import is corresponding to double exports. This two type exports are named as preferential export and spare export respectively. That is to say, each export has another concomitant spare output port. We will explain concomitant spare output port in next session. The switching matrix could consist of N switches and (N+m) combiners. But this matrix can also be implemented using some 1x2 switching units, but the structure becomes complex and accumulative loss becomes very high. Designing scale is four links and four wavelengths in each link. This structure can be expanded to 32 wavelengths. Each signal is modulated at 2.5Gb/s and follows ITU-T. Thus this burst structure can attain 40Gb/s system throughput (2.5Gb/s×4×4) with full implementation. We adopt the channel space of 0.8 nm in view of tunable filter response time. Despite that big channel space can make for settling chirp, adjacent channel cross talk and the aging-induced wavelength shift and so on [2], the design of system forepart is affected by system increasing delay

*This paper is supported by the NSFC project No. 69990540-2.

due to tuning time increasing. We adopt wavelengths in 1550 wavelength band. The signal window falls in the flat-gain region of the erbium-doped fiber amplifiers. Thus optical signals can be amplified by OPAs and OBAs in this structure.

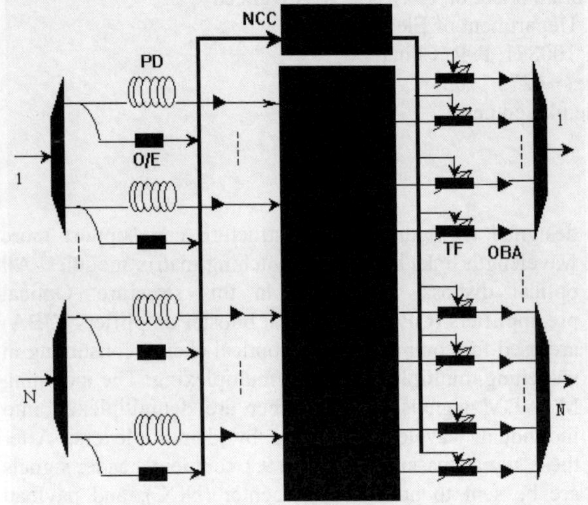

PA: prearranged delay; NCC: node control center;
O/E: optic-electrical conversion; TF: tunable filter.
OPA: optical pre-amplifier; OBA: optical booster amplifier.

Figure 1 Concomitant spare exports structure

3. CONCOMITANT SPARE EXPORT DESIGN SCHEME

This structure is not symmetry because of more exports than input ports. Purpose of this design is to decrease packet loss probability introduced contention. Reasonable design of concomitant spare export is very important, because number of spare export corresponding to each input port is tightly associated with performance and cost of this structure.

Simulation and calculation are implemented for design of concomitant spare export. We obtain figure 2. From the figure, two exports corresponding to each input port is optimal at performance and cost. This is to say, there is another one concomitant export corresponding to each output port. The two exports are preferential export and spare export respectively. Bursts will be transmitted generally through preferential export. Concomitant spare export will be used only when contentions happen.

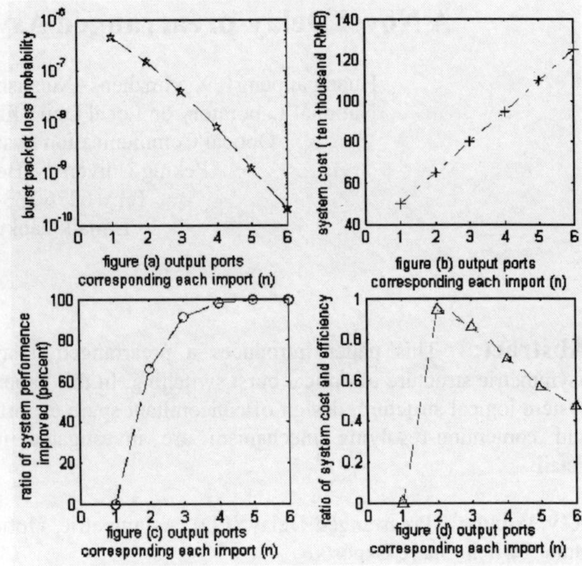

Figure 2 Relation of system performance, cost and number of export corresponding to each input port.

4. CONTENTION RESOLVING SCHEME

We utilize wavelength, space and priority dimensions to resolve contention in this scheme. Corresponding measures are wavelength band input, spare export and priority respectively. Primary principle is that we should avoid contention in most effort.

In input port, we use wavelength band so that wavelengths of each demultiplexer is different from ones of another demultiplexer. That is to say, each demultiplexer matches with one wavelength band. Thus we can completely avoid contention introduced by same wavelength in exports.

In this project, there are two exports in corresponding each import such as preferential export and spare export. Aim of concomitant spare output is to more reduce packet loss probability due to contention of two bursts in the same export. If there are not contentions, bursts will be transmitted through preferential export. If there are contentions, then contention information will be transmitted to NCC, then one of both bursts will be forwarded through spare export controlled by NCC. Precondition is that concomitant spare export is also idle status.

Because we use different wavelength band in different input fiber, same wavelength contention is avoided absolutely.

Bursts of entering switches system are divided into two priority classes. Super priority is marked in 0 and low priority is marked in 1. If same priority bursts compete, the short bursts would be forwarded through spare export basing on burst size. When two different priority bursts compete the same export, comparing their priority and size, low priority and short bursts would be transferred by spare export. If high priority bursts are corresponding to short burst, exports selecting of bursts are made according to actual value of burst size. We compare these burst size with ratio of 0 class priority and 1 class priority burst service time. According to simulation result, at high priority and short size condition, when size of the burst is less 0.4 times than another burst size, these bursts would be transmitted through spare export.

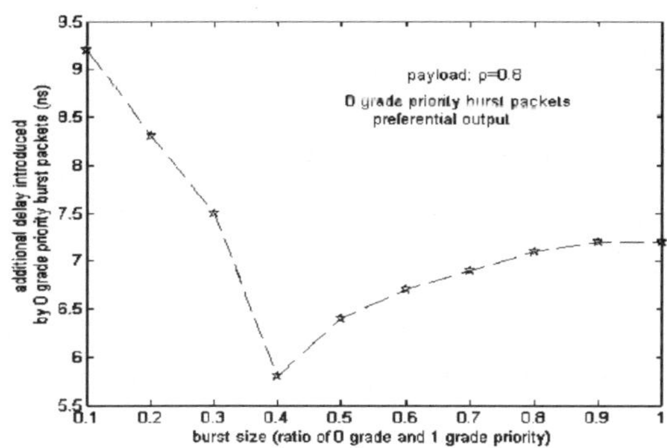

Figure 3 In different burst size condition, additional delay introduced by 0 class priority bursts preferential output

Follow as these measures, contention introduced by different priority and burst size conditions can be settled effectively 3-4]. Furthermore these strategies can more reduce networks delay and improve throughout capability of networks. If all measures don't yet prevent from contention, we should discard these conflicting bursts. According to above-mentioned measures, bursts discarded are very too little. So this structure can satisfy demand of networks high quality of service.

5. CONCLUSION

When this structure is applied to burst switching, routing function of tunable filter could be performed well. Asymmetry means that number of output ports is more than input ports. Concomitant spare exports are necessary for resolving contention. Prearranged delay synchronizes routing and switching. Another character is that headers of packets are only read instead of rewriting these. Tunable filters select wavelength according to headers information. Contention bursts keep up through detouring because of remaining header signals. Prearranged delay and asymmetry and so on are new characters in this structure. In backbone networks, this structure can adequately implement its many functions and take on advantages of new technologies. We may attain more market quotient when we make use of its good ratio of cost and performance. A Test-bed is under construction implementation.

REFERENCES

[1] Vinod M. Vokkarane, and Jason P. Jue,"Poritized Routing and Burst Segmentation for QoS in Optical Burst-Switched Networks," Proceedings,OFC 2002, WG6, MARCH 2002.

[2] S U Lee and W S Seo,"Design of An Optical Cross-connect Architecture," Telecommuncations 1998, Conference Publication No.451.

[3] D.Manjunath and Biplab Sikdar, "Variable Length Packet Switches: Delay Analysis of Crossbar Switches under Poisson and Self Similar Traffic," IEEE INFOCOM, pp.1055-1064,2000.

[4] Hai Le Vu and Moshe Zukerman, "Blocking Probability foı Priority Classes in Optical Burst Switching Networks," IEEE Communications Letters, pp.214-216, MAY, 2002.

Performance Optimisation of AWG - Based Optical Add-drop Multiplexers for Transparent Optical Networks

Mário Leiria and Adolfo V. T. Cartaxo
Optical Communications Group, Instituto de Telecomunicações,
Department of Electrical and Computers Engineering, Instituto Superior Técnico
Av. Rovisco Pais, 1, 1049 – 001 Lisboa, Portugal
Tel. +351 21 8418480; Fax: +351 21 8418172; e-mail: mario.leiria@lx.it.pt
Tel. +351 21 8418476; Fax: +351 21 8417164; e-mail: adolfo.cartaxo@lx.it.pt

Abstract: An arrayed waveguide grating (AWG) router is optimised for use in an AWG-based optical add-drop multiplexer (OADM) of a WDM transparent optical network with 10 Gbit/s per channel and channel spacing of 25 GHz. Two different techniques to flatten the AWG frequency response are investigated. It is shown, by simulation, that the optimised AWG router of an OADM with fold-back configuration should have bandwidth similar to the channel spacing, good isolation to adjacent channels and low or moderate additional losses. Numerical results reveal that AWG routers with frequency response between the flat and conventional AWG responses are the best candidates.

Key Words: arrayed waveguide grating, transparent optical network, *Q*-factor, OADM, AWG router, crosstalk.

1. Introduction

Optical add-drop multiplexers (OADM) and optical cross-connects (OXC) are key components in transparent optical networks (TON). OADMs and OXCs can be built on wavelength division demultiplexers (DEMUX) and multiplexers (MUX). The arrayed waveguide grating (AWG) MUX/DEMUX has been shown to operate as add-drop MUX and wavelength router [1]. The conventional AWG frequency response and discussion of its properties have been presented in several works [1]-[4]. Various techniques have been proposed to broaden and flatten the transfer function of an AWG [4]-[5]. Several works have studied the cascadability of optical MUX/DEMUX and OADM [6]-[8]. However, high channel spacing and an approximated AWG transfer function have been considered [7]. Also, to our knowledge, the impact of using an OADM with flat-response AWG on the TON performance has not been assessed so far.

In this work, the AWG-based OADM is optimised for a WDM TON with 10 Gbit/s per channel and channel spacing of 25 GHz. Various AWG parameters are optimised for cascades of 2 and 6 OADMs. The fold-back OADM structure is considered, because it uses only one AWG and it has low crosstalk penalties [9]. Two different principles are used to design the AWG. With the first one, various AWG filter shapes can be obtained, including the conventional and flat response, while with the second one only flat and similar to flat responses are obtained. The main purpose of this work is to realize the AWG features which optimise the performance of a cascade of AWG-based OADMs.

2. Methods of AWG design

In the first design principle, some of the arrayed waveguides are set in phase opposition to the remaining waveguides (*spatial filtering* AWG) [4]. The AWG frequency response is given by [4]

$$H_{i,k}(\upsilon)=\sum_{m=-P}^{P} C_m \cdot e^{j\cdot 2\pi\cdot(t_m - d_m(i,k)+\alpha_m)\cdot\upsilon} \tag{1}$$

where υ is the optical frequency, i is the ith input port, k is the kth output port and $2P+1$ is the number of grating waveguides. The parameter t_m is imposed by the length of the mth grating waveguide. Since the *spatial filtering* AWG is obtained from a conventional AWG, the values of t_m are as described in [4]. The parameter d_m is the length difference in the free propagation regions (FPR) when the input or output ports are not the central input or output waveguide, respectively. So, the parameter d_m is approximated by

$$d_m(i,k)=m\cdot(i-k)\cdot N_{awg}\cdot\Delta\upsilon_{ch}/(\upsilon_o\cdot(\upsilon_o-\Delta\upsilon_{ch})) \tag{2}$$

where $\Delta\upsilon_{ch}$ is the channel spacing, υ_o is the central optical frequency and N_{awg} is the order of the grating [4].

A Gaussian approximation for the waveguide modes is assumed [4]. So, the spatial light in the FPR is also Gaussian distributed [2]. Then, the spatial light distribution that illuminates the output waveguides is obtained by multiplication of the spatial light distributions of the two FPRs [3]. Therefore, for the *spatial filtering* AWG the normalized optical power of the mth grating waveguide is given by

$$C_m = e^{-(b_{awg}\cdot m)^2} \Big/ \sum_{l=-P}^{P} e^{-(b_{awg}\cdot l)^2} \tag{3}$$

where b_{awg} controls the distribution of optical power over the waveguides [4]. The parameter α_m is used to obtain phase opposition in the mth grating waveguide or not. It is given by $\alpha_m = \mathrm{sgn}(m)/(2\cdot f_o)$ for $|m|>M$ and $\alpha_m = 0$ for $|m|\le M$, where $\mathrm{sgn}(x)$ is the sign function. The sign function is required to obtain a real AWG frequency response [4]. In [4], only half of the arrayed waveguides in phase opposition to the remaining ones has been proposed to achieve a flat top response. However, it is questionable if that number of arrayed waveguides leads to the optimum performance. Changing the parameter M, shapes of *spatial filtering* AWG frequency response between the flat-top and conventional AWG response are achieved, as shown in Fig. 1 (a). For high M parameter, the *spatial filtering* AWG response is similar to a conventional AWG. Decreasing *M*, the *spatial filtering* AWG response approaches a flat response, increasing the additional loss and decreasing the sidelobes suppression. Decreasing *M* even more, the *spatial filtering* AWG response shows ripple in the passband.

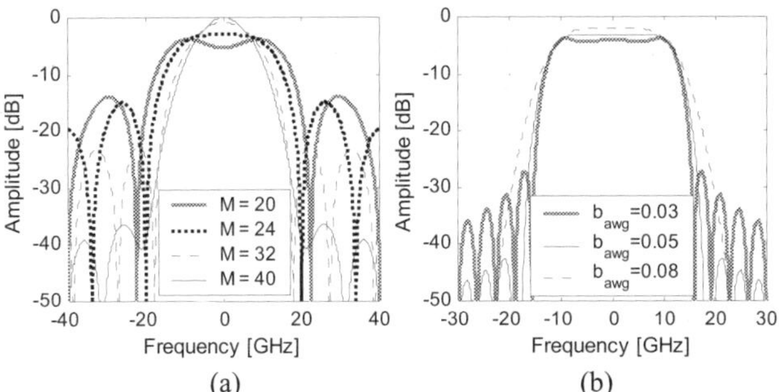

Fig. 1 – AWG amplitude response for the two designs and for P = 40. In (a), the *spatial filtering* AWG amplitude response is shown for various values of the parameter M and FSR = 800 GHz. In (b), the *sinc* AWG amplitude response is shown for various values of the parameter b_{awg}, FSR = 300 GHz and target bandwidth of 20 GHz.

As can be seen in Fig. 1 (a), the main drawback of *spatial filtering* AWG is that the flat AWG amplitude response is associated with moderate sidelobes suppression. So, other design is also considered to obtain flat AWG amplitude response based on the design principle proposed in [10] that leads to high sidelobes suppression. In this design, the spatial light distribution in the first FPR is a *sinc* distribution (*sinc* AWG). It can be physically implemented using parabolic waveguide horns in the input waveguides [5]. The key idea is to obtain the *sinc* distribution from the target bandwidth of the flat AWG (AWG desired bandwidth), the number of arrayed waveguides and free spectral range (*FSR*). The problem with this design is that the *sinc* AWG bandwidth has to be multiple of FSR/P and not all target bandwidths can be obtained. In [10], the beam propagation method has been used to obtain the AWG frequency response. Here, the AWG frequency response is assessed by taking into account the Gaussian light distribution in the second FPR through a simple expression. Numerical results show that this approximation is acceptable since the experimental results of AWG frequency response presented in [5] are approached quite well.

The *sinc* AWG frequency response is given by (1), considering the same t_m used for the *spatial filtering* AWG, with the normalized optical power of the *m*th grating waveguide given by

$$C_m = |G_m| \times e^{-1/2\times(b_{awg}\cdot m)^2} \Bigg/ \left[\sum_{l=-P}^{P} |G_l| \times e^{-1/2\times(b_{awg}\cdot l)^2}\right] \qquad (4)$$

where G_m is the target spatial distribution in the *m*th grating waveguide described in [10]. To achieve the negatives values of G_m, it is necessary to have arrayed waveguides with phase opposition, and that is obtained by properly setting α_m. There is no central waveguide and therefore, the number of arrayed waveguide is $2P$. Fig. 1 (b) shows that, for low parameter b_{awg}, the *sinc* AWG frequency response is almost flat with ripple in its passband. Increasing the parameter b_{awg}, the *sinc* AWG response first approaches the flat shape and then approaches the conventional AWG, increases the sidelobes suppression and decreases the additional loss. Also, Fig. 1 shows that, for the two designs, the AWG filters with additional loss between approximately 2.5 and 5 dB are usually flat-top filters. Notice that, in the two designs only the additional AWG loss resulting from obtaining a flat response is taken into account. Other contributions to the AWG insertion loss are not considered, because the purpose of this work is to compare different shapes of AWG frequency response, and those contributions are assumed to affect in the same manner both AWG designs.

3. AWG CHARACTERISTICS

The *spatial filtering* AWG characteristics are plotted in Fig. 2. Three different *FSR* (*FSR* = 500, 1000 and 2000 GHz) are considered, so that various AWG filter bandwidths can be obtained. In Fig. 2, the *spatial filtering* AWG characteristic are only shown for one *FSR*, because the sidelobes suppression and additional losses are independent of the *FSR*. The *spatial filtering* AWG bandwidth for other *FSR* can be obtained by multiplying the bandwidth by the *FSR*s ratio. Fig. 2 (a) shows that the sidelobes suppression increases with the parameter b_{awg}. This behaviour is due to the fact that most of the power is located in the central arrayed waveguide. So, the power in the sidelobes is reduced. The additional loss results from the arrayed waveguides with phase opposition. Using the *spatial filtering* AWG, the isolation to the adjacent channels is higher that 10 dB for *FSR* = 500 and 1000 GHz. Only for a *FSR* = 2000 GHz, the isolation is lower than 10 dB for most of the *spatial filtering* AWG parameters. This behaviour results from the AWG response bandwidth is equal or higher than the channel spacing. For most of the *sinc* AWG parameters, the *sinc* AWG frequency response has a flat response, high sidelobes suppression and high isolation. When the additional loss is lower than 2.5 dB, the amplitude response is in between a flat-top and a conventional AWG. This can be seen in Fig. 1 (b) and a similar behaviour has been reported in [11].

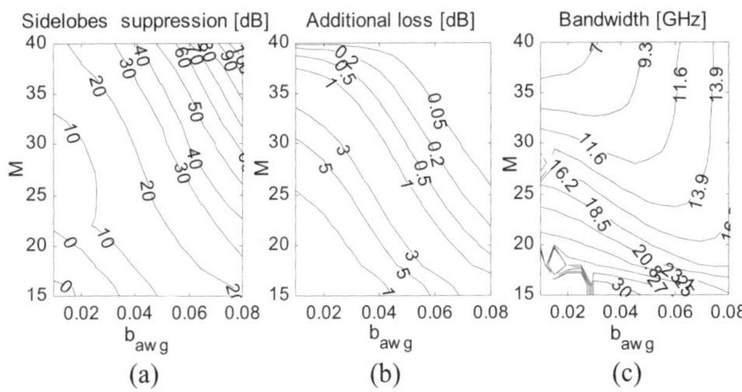

Fig. 2 – *Spatial filtering* AWG characteristics against the parameters b_{awg} and M, for P = 40 and FSR = 500 GHz.

4. SYSTEM CONSIDERATIONS

The OADM configuration considered is the fold-back configuration. But, because the *sinc* design is only for DEMUX or MUX, the fold-back configuration cannot be used. Therefore, the loop-back configuration is used in *sinc* AWG based OADMs. The structure of OADM based on AWG fold-back or loop-back configuration is reported in [9]. The TON is composed by a chain of equidistant OADM nodes connected by a span of distortionless optical fibre. A total loss of 20 dB due to fibre loss and OADM insertion loss except for the addition loss is assumed. The OADM node is composed by an optical amplifier (EDFA) followed by the AWG with fold-back or loop-back configuration. The optical amplifier gain compensates for 20 dB of loss and the additional loss of the AWG for a signal that is demultiplexed and multiplexed. The same AWG frequency response is considered for the MUX/DEMUX.

To assess the system margin the Gaussian approximation (GA) described in [12] is considered, with one signal equivalent optical filter different from the noise equivalent optical filter. To take into account the channel crosstalk, the WDM signal is composed by three 10 Gbit/s channels, with a channel spacing of 25 GHz. Three channels are enough to assess the system margin, because of the high sidelobes suppression of the AWG. The system margin is assessed for the central channel whose is at 1.55 μm. To consider the worst-case situation, the adjacent channels are dropped in all OADM and then new channels added. In each channel a NRZ signal with an extinction ratio of 10 is considered. A deBruijn sequence is transmitted in each channel and the adjacent channels have complementary sequences. Other parameters are the spontaneous emission noise factor, $n_{sp} = 2$, the p-i-n responsivity, R_s = 1 A/W, absence of a polariser, p = 2, an average emission power of 0 dBm per channel and the electrical receiver filter is a 2nd order Bessel filter with a -3 dB bandwidth of 6.5 GHz.

5. RESULTS AND DISCUSSION

The system margin, defined in [12], is assessed for a reference BER of 10^{-12} corresponding to a Q-factor of 7. In Fig. 3 the system margin is plotted for 6 OADMs based on *spatial filtering* AWG, for three values of *FSR*. For each *FSR* there is an optimum AWG. The optimum filter results from a compromise between crosstalk from adjacent channels, noise accumulation and signal distortion due to AWG cascade. The additional loss, the isolation and the bandwidth are the AWG characteristics which impair the performance. High additional loss increases the amplified spontaneous emission (ASE) noise accumulation, because the optical amplifiers compensate for the OADM loss, increasing the system margin penalty due to the ASE noise. Therefore, lower additional loss is recommended. To reduce the crosstalk, higher isolation is required and, to reduce the system margin degradation due to the cascade of filter, high bandwidth is also required.

Only for the *FSR* of 500 GHz, the AWG filter with a flat response has the best performance. This behaviour is due the tight bandwidths obtained with this *FSR*. Therefore, it is necessary a flat response to reduce the penalty due to the cascade of filters. Since flat filters have higher additional loss, the noise accumulation along the path of OADM reduces the optimum system margin in comparison to other *FSR*s. Also, the flat response increases the noise equivalent bandwidth of the equivalent TON model, described in [12], decreasing the system margin. For *FSR* of 1000 GHz, the optimum AWG filter possesses features between the ones of conventional AWG and flat AWG. These optimum filters have good isolation, moderate additional loss and bandwidth similar to the channel spacing. As can be seen in Fig. 2, there are many filters with similar bandwidths to the optimum filters but with different additional losses. Results show that, when the isolation is high, the additional loss and the shape of the filters are the important properties of the AWG to have better performance. For *FSR* of 2000 GHz, the optimum AWG filter is a conventional AWG filter. This optimum filter has also a bandwidth similar to the channel spacing, and a high isolation. This situation shows that, although the conventional AWG introduces more signal distortion, the penalty due to signal distortion is compensated for the reduction of the noise penalty due to the low additional loss and the reduction of the noise equivalent bandwidth of the filter cascade. The high differences between the obtained system margin and the experimental system margin presented in [6] are due to the high loss.

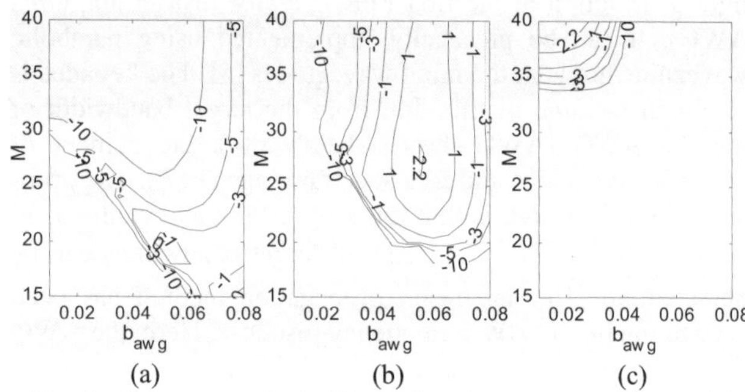

Fig. 3 – System margin, in dB, against the parameters b_{awg} and M, for 6 OADMs based on *spatial filtering* AWG and P = 40. (a) *FSR* = 500 GHz, (b) 1000 GHz and (c) 2000 GHz.

Fig. 4 (a) and (b) shows the system margin for two OADMs. In this case, a larger range of AWG parameters leads to the optimum filters than for six OADMs. For two OADMs, the optimum filters are essentially determined by the compromise between the signal distortion and the ASE noise penalty, because the adjacent channel crosstalk is only added in one OADM. Therefore, the additional loss plays an important role. The better system margin is obtained for lower additional loss, to reduce the noise power. So, the filters with a shape similar to the conventional AWG have better performance.

Fig. 4 (c) shows the system margin for six OADMs based on *sinc* AWG. As for *spatial filtering* AWG, the AWG filters with lower additional loss and a bandwidth similar to the channel spacing are the optimum filters. These optimum filters have a frequency response similar to a flat response, as can be seen in Fig. 1 (b). The system margin for the *sinc* AWG is similar to the one *spatial filtering* AWG. However, the additional loss is higher with the *sinc* AWG, but the isolation and suppression of sidelobes are also higher, compensating for the increase of the ASE noise power by the crosstalk penalty reduction. The results obtained with *sinc* AWG confirm the results obtained with the *spatial filtering* AWG, showing that the flat-top filter is not the optimum solution.

Comparing the system margin results in Fig. 3 (b) and (c) with the results in Fig. 4 (a) and (b), there are two possible choices for the optimum solution. However, because the filter obtained for a *FSR* of 2000 has a lower b_{awg} parameter and, therefore, it has low non-uniformity loss, this filter is the best choice.

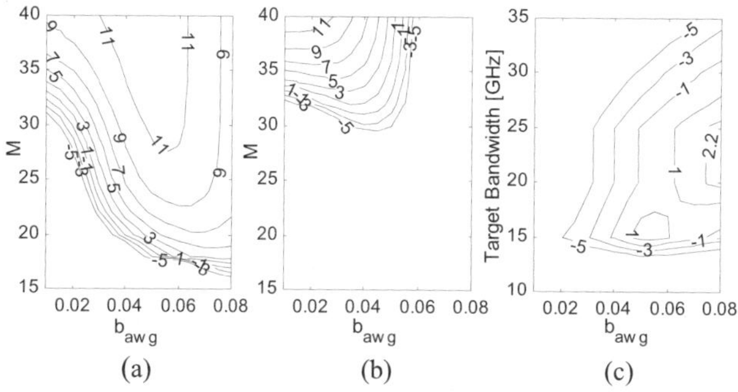

Fig. 4 — System margin, in dB, against the parameters b_{awg} and M or target bandwidth, for $P = 40$. (a), (b) 2 OADMs based on *spatial filtering* AWG and (c) 6 OADMs based on *sinc* AWG. (a) *FSR* = 1000 GHz, (b) 2000 GHz and (c) 300 GHz.

6. Conclusion

The optimisation of an OADM based on AWG optical filters has been performed for TON with a 10 Gbit/s per channel and 25 GHz of channel spacing. An optical path composed by 2 and 6 OADMs has been considered. Two designs are used for the AWG: using the *spatial filtering* AWG, different shapes of AWG filters can be obtained; using *sinc* AWG, only flat and similar to flat responses are achieved. The *sinc* design permits to obtain better sidelobes reduction and isolation to the adjacent channel. It has been observed that the best system margin is achieved for AWG filters with good isolation, bandwidth similar to the channel spacing and low additional loss. For two OADMs, a larger range of AWG optimum parameters is obtained. In this case, the importance of the crosstalk on the system margin is reduced. The optimum filters are imposed essentially by the trade-off between the signal distortion and the noise accumulation.

It has been shown that, for similar bandwidth and isolation, AWG filters with low or moderate additional loss have better performance. Therefore, the AWG with flat response is not the optimum choice for an OADM. Instead, numerical results indicate that AWG with shapes between flat and conventional responses are more suitable.

References

[1] Smit, M. K., and Dam, C., "PHASAR-based WDM-devices: principles, design and applications", *IEEE J. Quantum Electron.*, vol. 2, pp. 236-250, Jun. 1996.

[2] Muñoz, P., Pastor, D., and Capmany, J., "Modeling and design of arrayed waveguide gratings", *J. Lightwave Technol.*, vol. 20, pp. 661-674, Feb. 2002.

[3] Menezo, S., Rigny, A., Talneau, A., Delorme, F., Grosmaire, S. Nakajima, H., Vergnol, E., Alexandre, F., and Gaborit, F., "Design, realization and characterization of a ten-wavelength monolithic source for WDM applications integrating DBR lasers with a PHASAR", *IEEE J. Quantum Electron.*, vol. 6, pp. 185-190, Jan./Feb. 2000.

[4] Kamalakis, T., and Sphocopoulos, T., "An efficient technique for design of an arrayed-waveguide grating with flat spectral response", *J. Lightwave Technol*, vol. 19, pp. 1716-1725, Nov. 2001.

[5] Okamoto, K., and Sugita, A., "Flat spectral response arrayed-waveguide grating multiplexer with parabolic waveguide horns", *Electron. Lett.*, vol. 33, pp. 1661-1662, Aug. 1997.

[6] Horiuchi, Y., and Yamamoto, S., "1 Tbit/s (25×40 Gbit/s) transmission over 100km transparent network with 5 OADM nodes using AWG filters", *Europ. Conf. Optical Comm. 2000*, pp. 263-265, Munchen, 2000.

[7] Antoniades, N., Roudas, I., Wagner, R. E., and Habiby, S. F., "Simulation of ASE noise accumulation in a wavelength add-drop multiplexer cascade", *IEEE Photon. Technol. Lett.*, vol. 9, pp. 1274-1276, Sep. 1997.

[8] Roudas, I., Antoniades, N., Otani, T., Stern, T. E., Wagner, R. E., and Chowdhury, D. Q., "Error probability of transparent optical networks with optical multiplexers/demultiplexers", *IEEE Photon. Technol. Lett.*, vol. 13, pp. 1254-1256, Nov. 2001.

[9] Dods, S. D., and Tucker, R. S., "A comparison of the homodyne crosstalk characteristics of optical add-drop multiplexers", *J. Lightwave Technol*, vol. 19, pp. 1829-1838, Dec. 2001.

[10] Okamoto, K., and Yamada, H., "Arrayed-waveguide grating multiplexer with flat spectral response", *Optics Lett.*, vol. 20, pp. 43-45, Jan. 1995.

[11] Okawa, M., Maru, K. Uetsuka, H., Hakuta, T., Okano, H., and Matsumoto, K., "Low loss and wide passband arrayed waveguide grating demultiplexer", *Europ. Conf. Optical Comm. 1998*, pp. 323-324, Madrid, 1998.

[12] Leiria, M., and Cartaxo, A., "Simple expression to evaluate the margin of transparent optical networks with signal-dependent noise dominance", *Network Optical Comm. 2002*, Darmstadt, pp. 425-432, 2002.

Acknowledgments: This work was supported by Fundação para a Ciência e a Tecnologia (FCT) from Portugal and POSI within project POSI/35576/CPS/2000 - DWDM/ODC.

An Efficient Method in Optical Packet Switching for Contention Resolution

Li Yi and H. Ghafouri-Shiraz
Division of Communication Engineering
School of Electrical & Electronic Engineering
Nanyang Technological University
South Spine, Block S1
Singapore 639798

Abstract: Optical packet switching is a promising technology for future all-optical networks. The contention problem in an optical packet switch can not be easily handled because the optical random access memories are not available with present technology. Fiber delay lines and tunable wavelength converters have been investigated to emulate the optical buffering, however, the hardware volume of the optical components limits the buffer capacity. In this paper, we have proposed an efficient method for contention resolution with very limited amount of wavelength converters and fiber delay lines. The potential of this method is analyzed by simulation.

Key Words: optical packet switch, contention resolution.

1. INTRODUCTION

Present telecommunication networks are experiencing a dramatic increase in demand for bandwidth capacity. The packet switching system has an advantage of its bandwidth efficiency and hence present research is focusing on optical packet switching (OPS) [1]. In an OPS system, packets remain in the optical domain during the process of buffering and switching. All-optical header processing has been investigated in [2] and optical buffering can be realized by fiber delay lines (FDLs).

Contention problem exists in all the packet switching systems. When two or more packets are destined to the same switch output at the same time, only one packet can be successfully sent out. For an electrical packet switch, the contending packets can be easily buffered in the electrical random access memories (RAMs), however, optical RAMs are not available with present techniques. An alternative for optical buffering is the use of FDLs to delay those contending packets. Unfortunately, the large hardware volume of FDLs limits the buffer capacity. With wavelength division multiplexing (WDM) technique, FDL can contain multiple wavelength channels and the buffer capacity of FDLs is enlarged [3]. Deflection routing is another technique for contention resolution, which may become appealing in metropolitan network [4]. The contending packets in a switch node are dispersed to the available output ports except to the one which is successfully sent out to its destined output port. Therefore, optical buffers are greatly reduced or even removed. However, the network topology has great impact on the efficiency of deflection routing. Wavelength dimension which is provided by WDM technique can be viewed as a form of logic buffers for contention resolution [5]. Since each fiber contains a number of wavelengths, more than one packet can be send out simultaneously in different wavelength channels.

Based on the available techniques, we proposed a new method which can efficiently utilize the wavelength dimension and optical buffers. In the switch architecture, all tunable wavelength converters (TOWCs) and FDLs are shared by the input/output ports. The wavelength dimension is firstly exploited and then the FDLs provide optical buffering for those unsuccessfully transferred packets. The performance of the optical packet switch is analyzed and investigated.

2. CONTENTION RESOLUTION METHOD

In the contention resolution techniques, optical buffering is constrained by the hardware volume of FDLs. On the other hand, wavelength dimension can be used as logic buffers in which the TOWCs are the key components. The amount of TOWCs is proportional to both the number of input fibers and the number of wavelengths when each wavelength is associated with one TOWC. It has been verified in [5] that only a small part of TOWCs are simultaneously utilized. This is because (i) some wavelength channels do not contain incoming packets and (ii) some contending packets are already transmitted on different wavelengths, and hence they do not need wavelength conversion. We extend this principle to sharing of FDLs which can also improve the performance of OPS with limited amount.

Figure 1 shows the architecture of the OPS. It is assumed that the packet switching system is time slotted and synchronized. A time slot is equal to one packet traveling duration. The optical packet switch has N input/output fibers and each fiber contains M wavelengths ($\lambda_1 \ldots \lambda_M$).

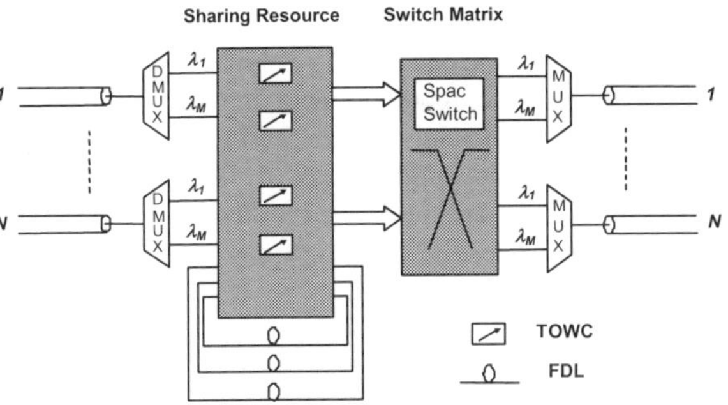

Fig. 1. Architecture of OPS.

For each input fiber, the incoming packets are demultiplexed from WDM siganals and the control unit schedules these packets to their destined output fibers according to their routing information. When contention happens, the TOWCs convert the wavelengths of the contending packets and the FDLs buffer those unsuccessfully transferred packets. TOWCs and FDLs are implemented to be shared by all input fibers. FDLs are supposed to be the same length and each FDL provides a delay of one packet traveling duration. The switch control unit drives all the optical components such as TOWCs and optical gates etc., according to the scheduling.

In the performance analysis, we assumed the packet arrival on the input fiber of the switch to follow Bernoulli process. The maximum rate of a packet arrival (ρ) is normalized. Since the input and output fibers are unbiased, the control unit randomly select the incoming packets among all input fibers. As each packet is destined to one output, we assume that the probability of the packet destination is uniformly distributed. The number of shared TOWCs is given by W and the switch has f shared FDLs. Packet scheduling of the switch is subject to the following rules.

1. Select one incoming packet and search for an available wavelength ($\lambda_1 \ldots \lambda_M$) of the destined output fiber K. If there is no contention, switch the packet directly.
2. If contention happens, select one available TOWC to convert the wavelength of the packet to a free wavelength of the output fiber K.
3. If there are no more available converters or wavelengths are all occupied, send the packet to a FDL for delaying of one time slot. When FDLs are not enough, drop the packet

The proposed contention resolution method firstly exploits the wavelength dimension for contention resolution so that the TOWCs have higher priority to be utilized. FDLs perform optical buffering for the contending packets in order to reduce the packet loss probability. However, the amount of FDLs is very small because most contentions have been resolved in wavelength dimension.

3. SIMULATION RESULT

The performance of the optical packet switch has been investigated with respect to packet loss probability. The wavelength number for an OPS is usually determined by the transmission system so that the maximum throughput capacity of the switch is subject to the fixed number of wavelengths. By using shared TOWCs, the OPS can exploit the capacity of wavelength dimension efficiently. When the requirement of the packet loss probability is determined (e.g.,10^{-6}), it is expected that the OPS can provide acceptable performance with minimum number of optical components. On the other hand, the traffic characteristic also has an impact on the packet loss. If the traffic is unbalanced, some output fibers may have much more packets than the other outputs, therefore these overloaded packets cannot be sent out successfully even with more TOWCs. In the simulations, we have assumed uniformed distributed traffic on all the input fibers.

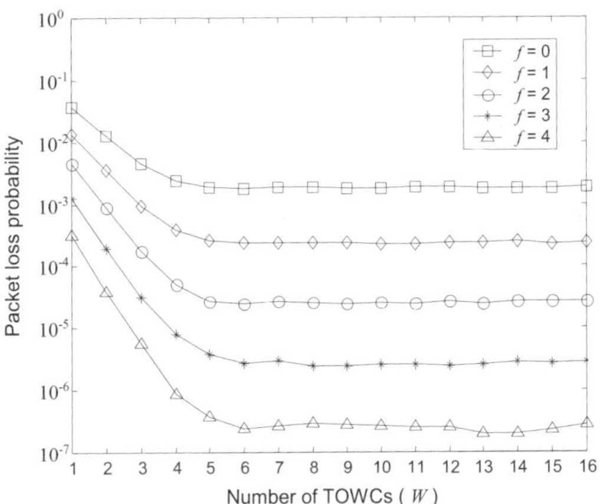

Fig. 2. Packet loss probability with parameters $N = 16$, $M = 4$, $\rho = 0.8$. The number of TOWCs (W) and FDLs (f) are used as variables for evaluation of switch performance.

Figure 2. shows the packet loss probability as a function of TOWCs. The switch size is set to $N = 16$ and the wavelength number is $M = 4$. We use the packet arrival rate $\rho = 0.8$ on each input fiber and the number of FDLs (f) increases from 0 to 4. With increasing number of converters, the packet loss probability decreases quickly and then saturates. It is noticeable that the performance of the OPS with less TOWCs is very similar to the switch with a larger number of TOWCs. For example, when the switch is equipped with 6 to 16 converters (f = 0,1,2,3,4), the packet loss probability remains at the same level. In other words, the sharing resource method can provide the same performance with much less converters (e.g., W = 6), compared with the full equipped switch where each wavelength channel is associated with one TOWC ($N \times M$ =

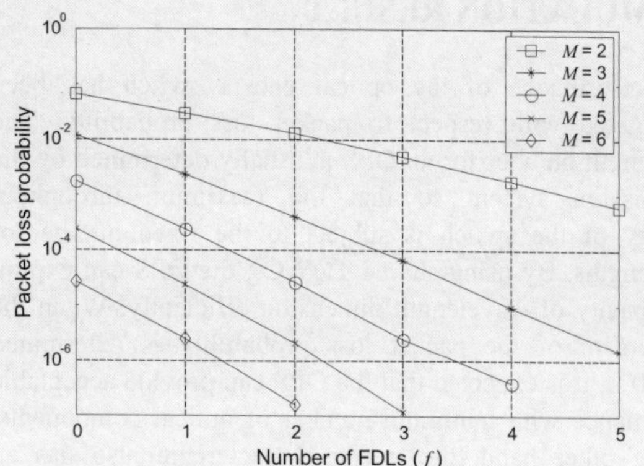

Fig. 3. Performance improvement by using shared FDLs.

64). On the other hand, the FDLs buffer the unsuccessfully switched packets for one packet traveling delay. As the number of shared FDLs' increases, the packet loss probability decreases by about one order of magnitude with each more FDL (Fig. 2.). When f = 4, the packet loss probability can reach to a level lower than 10^{-6}. More FDLs can certainly provide better performance because of larger optical buffer capacity for contending packets; but this also increases the hardware volume.

The OPS performance can be improved by shared FDLs which is shown in Fig. 3. We set the number of TOWCs W = 8 based on the result in Fig. 2. When M = 4,5,6, the packet loss probability drops below 10^{-6} with only f = 2,3,4, respectively. More wavelengths and FDLs can decrease of packet loss probability because of more buffer capacity.

4. CONCLUSION

We have proposed an efficient method for contention resolution in optical packet switching. The OPS performance has also been investigated with respect to packet loss probability. The contention problem is handled by both of the wavelength dimension and optical buffering. Since the hardware volume is an important factor in future OPS design, it is necessary to efficiently utilize the optical components. Our proposed method employs shared TOWCs and FDLs which can provide acceptable performance and may be implemented in future OPS design.

REFERENCES

[1] M. J. O'Mahony, D. Simeonidou, D. K. Hunter, and A. Tzanakaki, The application of optical packet switching in future communication networks, IEEE Communications. Magazine, pp.128-135, Mar. 2001.

[2] M. T. Hill et al., 1×2 optical packet switch using all-optical header processing, IEEE Electronics Letters, vol. 37, pp.774-775, 2001.

[3] S. L. Danielsen, P. B. Hansen, and K. E. Stubkjaer, Wavelength conversion in optical packet switching, Journal of Lightwave Technology, vol. 16, pp.2095-2108, Sept.1998.

[4] T. Chich, F. Clerot, "Wavelength converters for deflection routing networks", Proc. Fifth IEEE Symposium on Computers and Communications, pp. 140-145, 2000.

[5] V. Eramo and M. Listani, Packet loss in a bufferless optical WDM switch employing shared tunable wavelength converters, IEEE Journal of Lightwave Technology, vol.18, pp.1818-1833, Dec. 2000.

FBG-based Reconfigurable Optical Add-Drop Multiplexers

Hai YUAN and Wen-De ZHONG
Network Technology Research Centre
School of Electrical & Electronic Engineering,
Nanyang Technological University, Singapore 639798
Fax: +65 6790 4540, Email: ewdzhong@ntu.edu.sg

Abstract: We propose a new reconfigurable add-drop multiplexer using tunable fiber Bragg gratings (FBGs) and optical circulators. The new OADM allows dropping several fixed wavelength channels and at least one tunable wavelength channel. By properly assigning wavelengths channels that are add-dropped at an OADM node, the tuning range required for tunable FBGs can be made as small as the channel spacing. The proposed OADMs have potentially low insertion loss. The operation and performance of this OADM are experimentally investigated.

Key Words: wavelength division multiplexing, fiber Bragg grating, optical circulator, optical add-drop multiplexer.

1. INTRODUCTION

Reconfigurable optical add-drop multiplexers (OADMs) are key switching and routing components in wavelength-division multiplexed (WDM) optical networks. Several types of OADMs, based on different technologies, have been proposed and demonstrated. These include arrayed waveguide grating multiplexers (AWGMs) [1], Mach-Zehnder interferometers with fiber Bragg gratings (MZ-FBG) [2, 3], and optical circulators (OCs) with fiber Bragg gratings (FBGs) [4-7]. Among these, the OC-FBG-based OADM is most practical because of its low insertion loss, low crosstalk, and temperature and polarization insensitivity. In addition, OC-FBG-based OADMs do not cause bandwidth-narrowing when WDM signals pass through many OADM nodes.

A high-capacity WDM ring network may contain more than ten OADM nodes and carry tens of wavelength channels. In such a WDM network, each OADM node practically only need to drop a few channels from a large number of incoming channels. Among these dropped channels, some of them will be dropped definitely; while the others may be dropped or not, depending on the network traffic situation. A fully dynamically reconfigurable OADM can fulfil this function [7-9], but this is not a cost effective solution, because it requires many optical space switches or tunable FBGs [7-9]. In this paper, we propose a new design of reconfigurable OADMs, using tunable FBGs and optical circulators. The new OADM allows dropping several fixed wavelength channels and at least one tunable wavelength channel. By properly assigning wavelengths channels that are add-dropped at an OADM node, the tuning range required for tunable FBGs can be made as small as the channel spacing. The operation and performance of this OADM are experimentally investigated.

2. NEW OADM ARCHITECTURE

In principle, a reconfigurable multichannel OADM can be built by simply cascading multiple conventional single-channel tunable OADMs, as shown in Fig. 1.

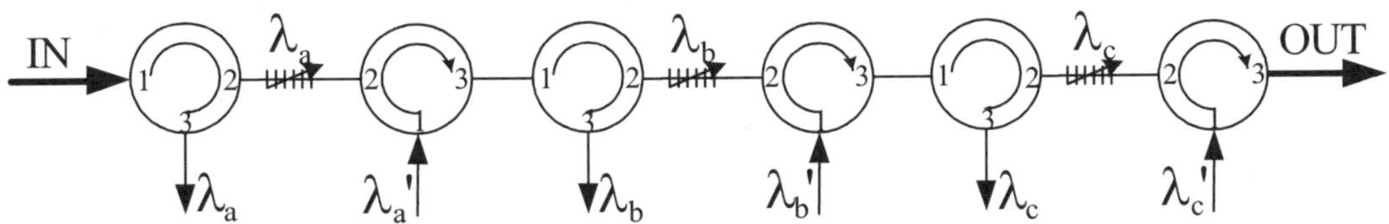

Fig. 1 Reconfigurable Multichannel OADM consisting of multiple single-channel OADMs

However, this approach is not cost-effective. Firstly it requires a large number of 3-port OCs, and wavelength channels that are not dropped pass through every circulator, leading to a high insertion loss. Secondly, it requires a large tuning range for tunable FBGs. To address the above problems, we here propose a new design using either 4-port circulators or 8-port circulators.

Fig. 2 shows a schematic diagram of the new OADM, using 4-port OCs. In this example, two channels at λ_a and λ_c are dropped definitely, and a third channel at λ_b can be either dropped or pass through the OADM. Two set of FBGs are used to perform the add-drop function. The first set of FBGs, i.e. FBG_a, FBG_b, FBG_c with center wavelengths λ_a, λ_b and λ_c are connected between port 2 of the first OC and port 3 of the last OC. This set of FBGs is used to separate the dropped channels from the through-channels, so that the through-channels will not suffer a high insertion loss [7]. The second set of FBGs (FBG'_a, FBG'_b, FBG'_c) are used to separate individual add/drop channels. Note that the FBG'_b is tunable. As shown in Fig. 2, when incoming WDM signals enter the OADM, channels at wavelengths λ_a, λ_b and λ_c will be reflected to port 3 of the first OC, and the remaining channels will pass through the first set of FBGs and emerge at port 4 of the last OC. The reflected channels are dropped at port 4 of OC1, OC2 and OC3, respectively, new channels are added at port 1 of OC2, OC3 and OC4, respectively. Since the FBG'_b is tunable, the incoming channel at λ_b can be either dropped or bypass the OADM. If the FBG'_b is detuned away from λ_b, then the incoming channel at λ_b will not be dropped. By properly assigning wavelengths to the drop channels so that at each OADM node, no two drop channels are adjacent to a tunable channel, then the tunable FBG will only need a tuning range equal to the channel spacing. In the example of Fig. 2, this means that the FBG'_b can be either detuned to λ_{b-1} or λ_{b+1} as long as λ_a and λ_c are not simultaneously adjacent to λ_b. In general, if the number of add/drop channels is m, then the number of 4-port OCs required is (m+1).

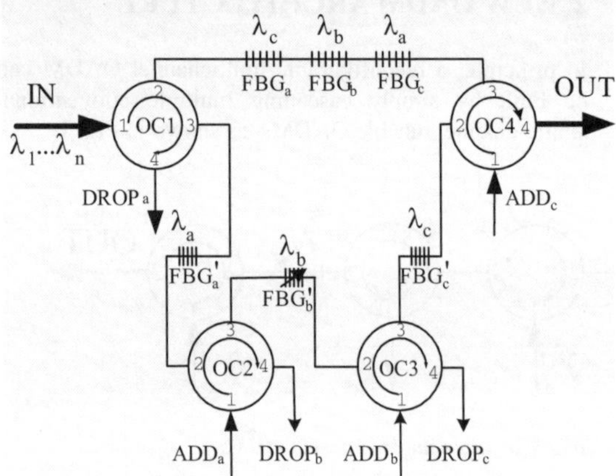

Fig. 2 New reconfigurable multichannel OADM using 4-port OCs

To reduce the number of OCs, we can use 8-port OCs to replace 4-port OCs [6-7]. Fig. 3 shows the same OADM design using 8-port OCs.

The number of OCs is reduced by half. Note that a single 8-port OC is more compact than two 4-port OCs and the cost for a commercial 8-port OC is cheaper than two 4-port OCs. Therefore, it is more economical to use 8-port OCs rather than 4-port OCs for the implementation of the above OADM.

We now examine the insertion loss for the through-channels, drop channels and tunable channels. Let L_{oc} be the insertion loss between two adjacent ports of an OC. Let L_{FT} and L_{FR} be the transmission and reflection losses of an

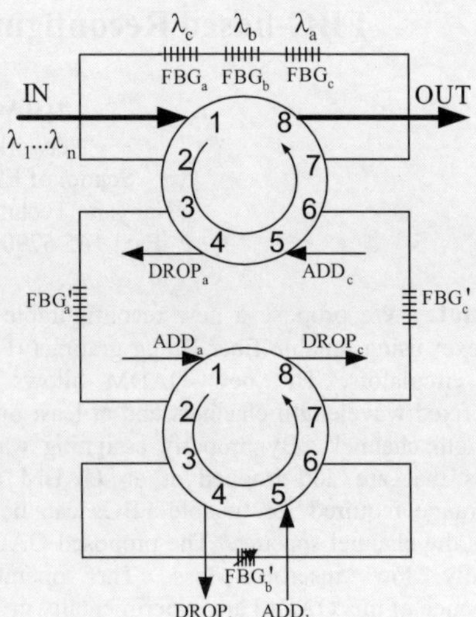

Fig. 3 *New reconfigurable multichannel OADM using 8-port OCs*

FBG. The insertion loss for the through-channels, denoted as L_{th}, is given by

$$L_{th} = 2L_{oc} + mL_{FT} \tag{1}$$

The insertion loss of a fixed drop channel depends on the position of the corresponding FBG in the OADM. It can be seen from Fig. 2 that the lastly dropped channel has the higher insertion loss, denoted as L_{drop}, which is given by

$$L_{drop} = (m+2)L_{oc} + 2L_{FR} + (m-1)L_{FT} \tag{2}$$

The insertion loss of tunable channel depends on the state of tunable FBGs. When the tunable channel bypasses the OADM, it can be seen from Fig. 2 that the tunable channel has highest insertion loss, denoted as $L_{tunable}$, which is given by

$$L_{tunable} = (m+3)L_{oc} + 2L_{FR} + (3m-2)L_{FT} \tag{3}$$

The typical values of L_{oc}, L_{FR} and L_{FT} from commercial devices are 0.6, 0.5 and 0.1 dB, respectively. Fig. 4 illustrates the insertion losses for through channels, the lastly dropped channel and the bypassing tunable channel. As can be seen, the insertion loss for through-channels is very low, since the through-channels only pass though twice of circulator ports. But the insertion losses for the lastly dropped channel and the bypassing tunable channel increase significantly in proportion to the number of add/drop channels. For example, if the number of add/drop channels is 5, then the insertion losses for the through channels, the lastly dropped channel, the bypassing tunable channel are 1.7, 5.6 and 7.1 dB, respectively.

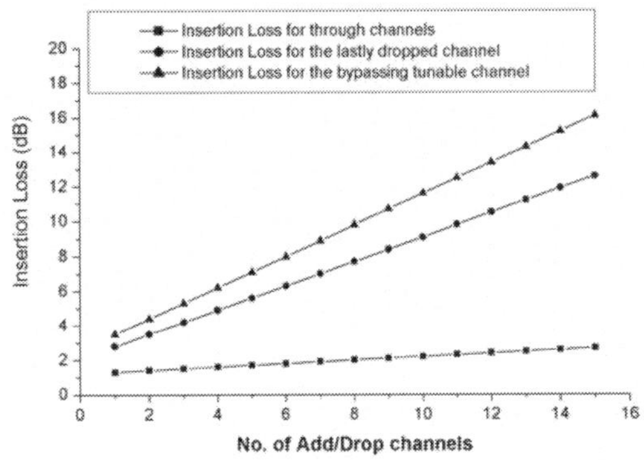

Fig. 4 *Insertion losses for the through channels, the lastly dropped channel and the bypassing tunable channel.*

3. EXPERIMENTAL SETUP AND RESULTS

The operation and performance of the new reconfigurable OADM are experimentally investigated.

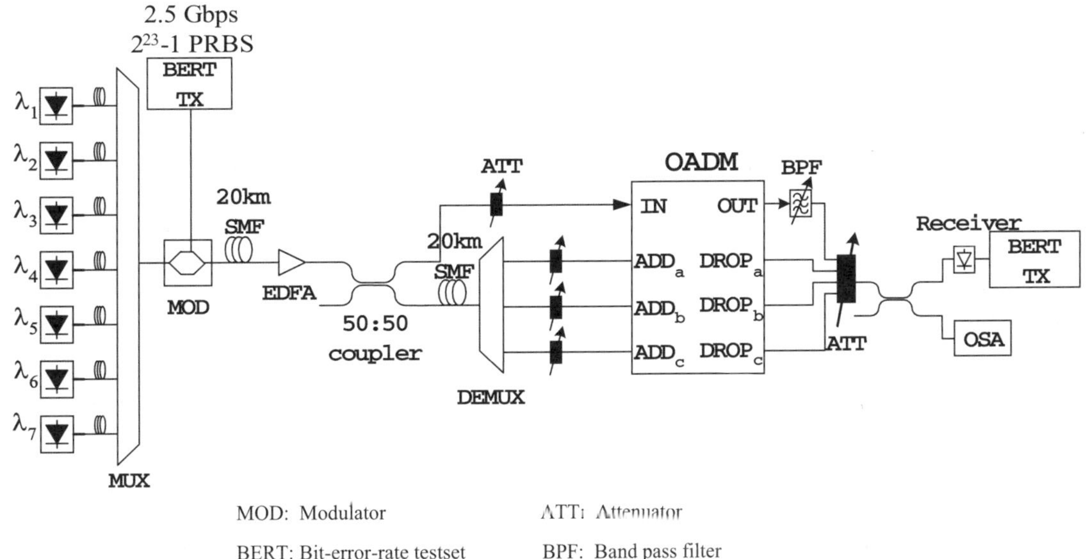

Fig. 5 Experimental setup

Fig. 5 shows the experimental setup. The OADM consists of four 4-port OCs, and 3 pairs of apodized FBGs whose center wavelengths are 1549.3, 1551.7 and 1554.1nm, respectively. One of the FBG with center wavelength 1551.7nm is tunable. 7 laser sources of ITU wavelengths from 1549.3 nm to 1554.1 nm with 0.8 nm channel spacing are externally modulated with a 2^{23}-1 nonreturn-to-zero pseudorandom bit sequence (PRBS) at 2.5 Gb/s. The modulated sources are amplified by an EDFA after being transmitted for 20 Km of standard single-mode fiber (SMF) for bit decorrelation, and then are split into two parts. One part is transmitted over another 20 Km of SMF fibre before it is demultiplexed. The demultiplexed signals are used for the add channels. The other part is connected to the input port of the OADM.

The measured optical spectra at different ports of OADM are shown in Fig. 6. The spectrum at the output port of OADM without any add channels (Fig. 6(b)) shows small in-band crosstalk of less than -30dB from the three drop channels. The spectra at the DROP$_a$, DROP$_b$ and DROP$_c$ ports of OADM with all add channels present (Fig. 6(d)(e)(f)) show very small out-of-band crosstalk of less than -30dB from the through channels and adjacent add channels.

The measured insertion loss for through channels is about 4 dB. In this experiment, fiber connectors are used for connection. If all fiber connectors are replaced by splicing, the insertion loss could be reduced by around 2dB for through channels.

CONCLUSION

We have proposed a new reconfigurable add-drop multiplexer using tunable fiber Bragg gratings (FBGs) and optical circulators. The new OADM allows dropping a few fixed wavelength channels and at least one tunable wavelength channel. The tuning range required for tunable FBGs can be made as small as the channel spacing by properly assigning wavelengths channels that are add-dropped at an OADM node. We have experimentally demonstrated the operation of the proposed OADM.

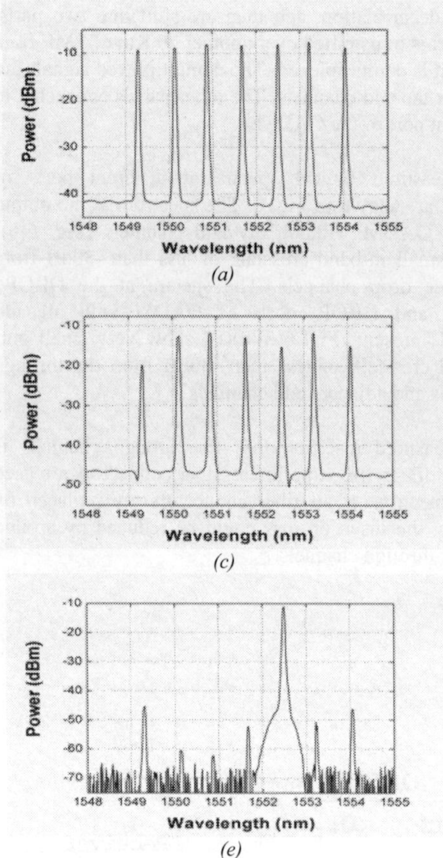

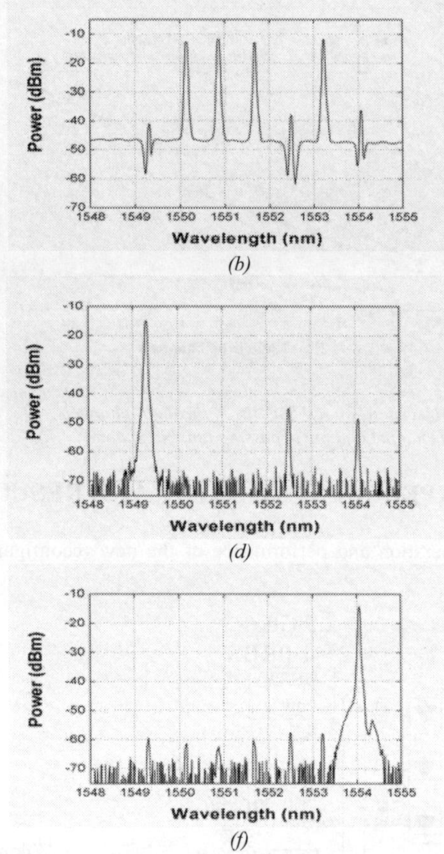

Fig. 6 Optical spectra measured at: (a) IN port, (b) OUT port without any add channels present, (c) OUT port with all add channels present, (d) $DROP_a$ port with all add channels present, (e) $DROP_b$ with all add channels present, (f) $DROP_c$ with all add channels present.

ACKNOWLEDGEMENT

The authors wish to thank Dr X. F. Yang, Ms X. Guo and Ms Y. Xu for their help in fabrication of FBGs.

REFERENCES

[1] Y. Tachikawa, *et al.*, "Arrayed-waveguide grating add-drop multiplexers with loop-back optical paths," *Electron. Lett.*, vol. 29, pp. 2133-2134, 1993.

[2] F. Bilodeau *et al.*, "An all-fiber dense-wavelength division multiplexer /demultiplexer using photoimprinted Bragg gratings," *IEEE Photon. Technol. Lett.*, vol. 7, pp. 388-390, 1995.

[3] T. Mizuochi, et al., "Interferometric crosstalk-free optical add/drop multiplexer using Mach-Zehnder-based fiber gratings," *J. Lightwave Technol.*, vol. 16, pp. 265-276, 1998.

[4] C. R. Giles and V. Mizrahi, "Low-loss add/drop multiplexers for WDM lightwave networks," *IOOC'95*, paper, ThC2-1, Hong Kong, June 1995.

[5] A. V. Tran, W. D. Zhong, R. S. Tucker, R. Lauder., "Optical add-drop multiplexers with low crosstalk," *IEEE Photon. Technol. Lett.*, vol 13, pp. 582-584, June 2001.

[6] W. D. Zhong, A. V. Tran, K. Song and R. S. Tucker, "Multichannel Optical Add-Drop Multiplexers using optical fibre Bragg gratings and circulators," Proc. ICICS, paper 1B3.5, Oct. 2001.

[7] A. V. Tran, W. D. Zhong, R. S. Tucker and Kai Song, "Reconfigurable Multichannel Optical Add-Drop Multiplexers Incorporating Eight-Port Optical Circulators and Fiber Bragg Gratings" *IEEE Photon. Technol. Lett.*, vol 13, pp. 1100-1102, October 2001.

[8] W. D. Zhong, S. Dods, J. P. R. Lacey and R. S. Tucker, "Reconfigurable multichannel add-drop multiplexer with improved performance," *Electron. Lett.*, vol. 32, pp. 1477-1478, 1996.

[9] S. Y. Kim, *et al*, "Channel-switching active add/drop multiplexer with tunable gratings", Electronics Letters, vol. 34 No.1, pp. 104-105, Jan. 1998.

Analysis of Crosstalk Generated in Synchronization Module of Photonic Packet Switches

W. Fang, W. D. Zhong, H. Ghafouri-Shiraz, X. Yi and C. Lu*
Network Technology Research Center, School of EEE, Nanyang Technological University, Singapore 639798
Tel: 65-67904540 Fax: 65-67926894 Email: pg00029463@ntu.edu.sg; ewdzhong@ntu.edu.sg
**Institute for Communications Research, Singapore*

Abstract: We investigate the inband crosstalk generated at the synchronization module of photonic packet switches and found that the inband crosstalk increases in proportion to the number of cascaded optical switches and delay lines. We show that the impact of this crosstalk on the BER is significant when the number of cascaded switches and delay lines is greater than 10.

Key words: Inband crosstalk, synchronization, photonic packet switch, power penalty

1. INTRODUCTION

Extensive work has been done in the development of photonic packet switching systems where synchronization of optical packets at the inputs of the switch fabric is necessary to simplify the operations such as packet contention resolution and buffering [1-2]. A general implementation of packet synchronization consists of (i) a packet start recognizer, which identify the packet start and (ii) a programmable delay line module (PDLM), which adjusts the arrival time of packets [1-2, 10-11]. As shown in Fig. 1, the PDLM is typically implemented by a series of n fiber delay lines (FDLs), each of which is preceded and followed by a 2x2 optical switch. The lengths of FDLs are arranged in an exponential sequence, i.e., the first FDL is equal to T/2 and the nth FDL is equal to $T/2^n$, where T is the duration of one time slot. The resolution of the PDLM is $\tau=T/2^n$. By controlling the switching states of the 2x2 switches, the arrival time of packets can be adjusted in the range of 0 to $(2^n-1)\tau$. A higher resolution (i.e. smaller τ) requires a larger number of FDLs. When a packet goes through the PDLM, inband crosstalk will arise at each 2x2 switch and will accumulate at the subsequent switches. The accumulated inband crosstalk will significantly degrade the desired signal, particularly when the value of n is large. Previously reported works have mainly focused on implementation and experimental demonstration of the synchronization module but only with a small number of FDLs [10-11]. However, the impact of inband crosstalk has not been well understood. The purpose of this paper is to give a detailed analysis of the inband crosstalk caused by the PDLM and to provide a guideline for the requirement of the optical switches. It has been found that the 2nd order inband crosstalk is dominant and its value is proportional to n^2.

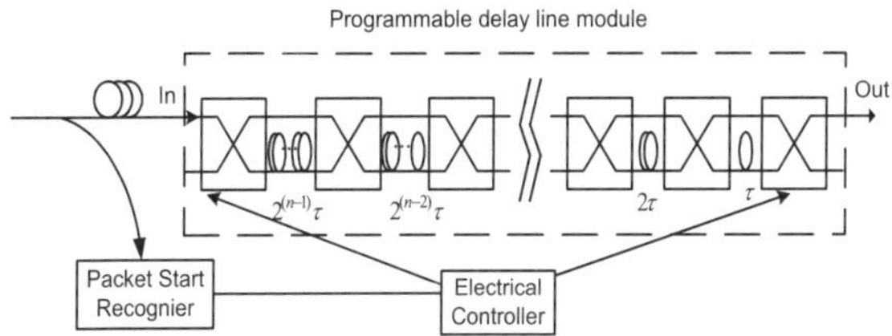

Fig. 1 The programmable delay line module (PDLM) for photonic packet switching

2. INBAND CROSSTALK IN THE PDLM

When a packet goes through a 2x2 optical switch, the main signal will appear at the designated output of the switch, but a small portion of the signal will leak to the other output of the switch due to imperfection of switching process. The ratio of the leakage to the signal at the input of the switch is defined as crosstalk of the switch and is denoted as ε. The typical value of ε for fast photonic packet switching is in the range of –20 to –25 dB [3]. When this crosstalk propagates to a down-stream switch in the PDLM, a 2nd order crosstalk whose power level is proportional to ε^2 will be generated and combined with the main signal, causing interference between the crosstalk and the main signal. There are other higher even-order crosstalks generated, but they are negligible compared with the 2nd order crosstalk. It has been noted that odd-order crosstalk will also be generated, but they do not propagate with the main signal and hence will not interfere with the main signal. It can be shown that the number of 2nd crosstalk at the output of the PDLM is $N=n(n+1)/2$, which could significantly interfere with the main signal when n is large.

3. NOISE CAUSED BY THE INBAND CROSSTALK

Neglecting the possible delay arising at each optical switch, a packet experiences a delay of $d_s\tau(0 \le d_s \le 2^n - 1)$, when it passes through the PDLM. Each of the 2nd order crosstalk follows a different path and hence has a different delay. Let $d_i\tau$ be the delay of the ith 2nd order crosstalk, where $0 \le d_i \le 2^n - 1$. It can be shown that, by ignoring the higher order terms, the intensity noise caused by the interference between the main signal and the 2nd order inband crosstalk at the output of the PDLM can be expressed as:

$$i_{noise} = 0.5E_{in}^2[\sum_{i=1}^{N} 2b_s b_i \varepsilon(1-\varepsilon)^n \cos((d_s - d_i)\omega\tau + \phi(t - d_s\tau) - \phi(t - d_i\tau))] \quad (1)$$

where E_{in} is the input electrical field, b_s and b_i are the corresponding "0" or "1" modulation states of the signal and crosstalk respectively, ω is the carrier frequency and $\phi(t)$ represents the laser phase fluctuation.

The phase fluctuation of a DFB laser is commonly modeled as Wiener-Levy random process [4-7]. According to the Wiener-Levy model, the DFB laser's phase variation is a non-stationary zero-mean Gaussian random process. However, its phase difference $\Delta\phi$ between two different times is stationary and has independent zero-mean Gaussian random distribution with a variance given by $\sigma_{\Delta\phi}^2 = 2\pi\Delta\nu\Delta t$ for a laser with power spectral density function of Lorentzian line-shape, where $\Delta\nu$ is the 3 dB line-width of the laser source and Δt is the time difference[6].

Applying the Wiener-Levy model, the variance of the phase difference of the beating items in eq. (1) varies from $\sigma_{\Delta\phi}^2 = 2\pi\Delta\nu\tau$ to $\sigma_{\Delta\phi}^2 = 2\pi\Delta\nu(2^n - 1)\tau$. As an example, for a time slot duration T=1.5μs, which was adopted in the KEOPS project [8], and a laser linewidth of 1MHz, the variance of the phase difference varies from 0.0368 to 9.351rad for n=8, covering both the coherent and incoherent regimes[4].

We now discuss each cosine term in eq. (1). The variance of the phase difference between the signal and the ith crosstalk is: $\sigma_{\Delta\varphi}^2 = 2\pi\Delta\nu\,|d_s - d_i|\,\tau$. Let $\phi_i = (d_s - d_i)\omega\tau$ be the delay difference between the signal and the ith second order crosstalk. The probability distribution function $f_i(y)$ of $y = \cos(\phi_i + \Delta\phi)$ $(-1 \le y \le 1)$ is given by eq. (2) [5]:

$$f_i(y) = \frac{1}{\sqrt{2\pi\sigma_{\Delta\phi}^2(1-y^2)}} \times \left[\sum_{n=-\infty}^{+\infty} \exp\left\{-\frac{[\cos^{-1}(y) - \phi_i + 2\pi n]^2}{2\sigma_{\Delta\phi}^2}\right\} + \sum_{n=-\infty}^{+\infty} \exp\left\{-\frac{[-\cos^{-1}(y) - \phi_i + 2\pi n]^2}{2\sigma_{\Delta\phi}^2}\right\}\right] \quad (2)$$

The mean and variance of each cosine item can be obtained from the following eq. (3).

$$\mu_i = r\int f_i(y)\,y\,dy \quad (-1 \le y \le 1), \qquad \sigma_i^2 = r^2\int y^2 f_i(y)\,dy - \mu_i^2 \quad (-1 \le y \le 1) \quad (3)$$

where $r = 0.5E_{in}^2\varepsilon(1-\varepsilon)^2$ is the coefficient, assuming that crosstalk level of the switches is ε, and the probability of the data "marks" and "spaces" is equal.

It is reasonable to assume that each cosine item to be independent because the laser phase noise difference is considered to be a stationary and independent zero mean Gaussian process. Hence, according to the central limited theory, with a reasonably large value of N (e.g. n ≥4, N≥10), the sum of the cosine items will be a Gaussian distribution with the mean and variance given as:

$$\mu_{sum} = (\sum_{i=1}^{N} \mu_i), \quad \sigma_{sum}^2 = \sum_{i=1}^{N} \sigma_i^2 \quad (4)$$

Therefore, the bit error rate (BER) $p_{ber}(d_s)$ for a packet that is delayed by a duration of $d_s\tau$ ($0<d_s<2^n-1$) in the PDLM can be obtained following the analysis of BER deteriorated by Gaussian noise [6]. Since a packet is likely to experience any of the 2^n possible different delays in the PDLM, the average BER can be expressed by eq. (5), assuming equal probability for each packet delay.

$$P_{ber} = \frac{1}{2^n}\sum_{d_s=0}^{2^n-1} p_{ber}(d_s) \quad (5)$$

4. NUMERICAL RESULTS OF BER AND POWER PENALTY

In order to get a better understanding of how the BER is affected by the 2nd order inband crosstalk, the BER and power penalty of the signal are numerically calculated, based on the analysis given above. In the following discussions we have assumed that BER=10^{-12} when there is no crosstalk (i.e. BER caused by the receiver's noise alone).

Fig. 2 shows the numerical results of BER versus different value of $\omega\tau$ at different crosstalk level and different number of serial delay lines n, with $\Delta\nu$=1MHz and T=1.5μs The value of $\omega\tau$ is normalized to the range of [0, 2π]. From Fig 2 (a) it is discovered that when the number of serial delay lines is small (e.g. n=4, 6), $\omega\tau$ has negligible effect on the BER. However when the number of serial delay line is increased to 8, $\omega\tau$ will influence the BER obviously. The figure also shows that the BER will be worst when $\omega\tau = \pi/2$ or $3\pi/2$, and achieve minimum value when $\omega\tau$ is 0, π or 2π. Also Fig 2(b) clearly shows that when the number of serial delay lines is 9, the values of BER at crosstalk levels varying from as low as -30dB to -20dB. This phenomena can be explained as since the laser phase difference obeys Gaussian distribution, the PDF of the cosine function of the total phase difference (the sum of the propagation delay difference and the laser phase noise difference) in eq. (1) follows different distributions when $\omega\tau$ is different. This discrepancy results in smaller deterioration of BER for $\omega\tau = 0, \pi$ or 2π than for $\omega\tau = \pi/2$ or $3\pi/2$. In the following analysis of BER versus other factors, $\omega\tau$ is assumed to be 0.

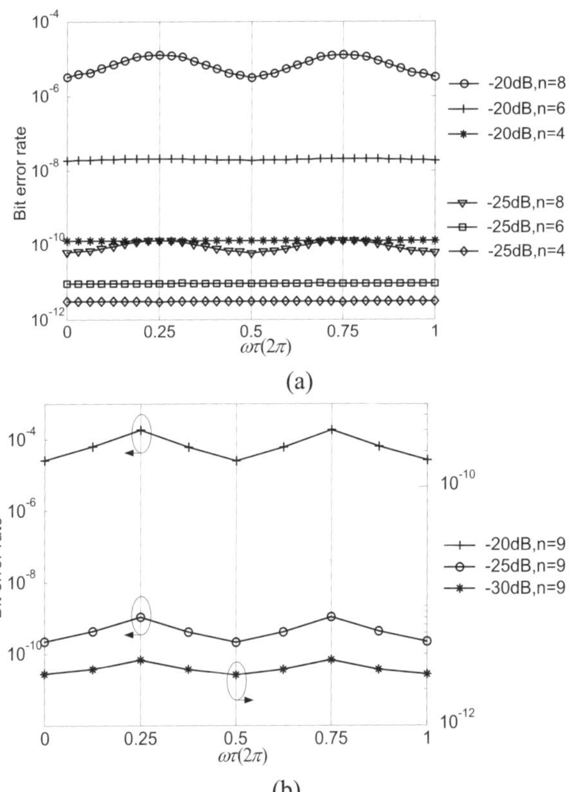

Fig 2. BER versus different value of $\omega\tau$ (a). for different crosstalk level and different n , (b). for different crosstalk level but fixed n=9.

Fig. 3 shows the numerical results of the BER versus different crosstalk levels for various values of n. Note that $\Delta\nu$=1MHz and T=1.5µs . For a given value of n, the BER is significantly increased when the crosstalk level increases from -30 dB to -20 dB, but the BER is almost not affected when the crosstalk level is less than -30 dB. When crosstalk level is greater than -30 dB, the BER rapidly deteriorates as the value of n increases. This is because the overall 2nd order inband crosstalk increases in proportion to n^2. Fig. 3 shows the relation between the BER and the laser linewidth for various values of n with crosstalk level of -25 dB. This figure clearly shows that, the BER reaches its maximum at a certain laser linewidth. Further increase or decrease in the linewidth will reduce the BER. For a large value of n (e.g. n=10), a larger laser linewidth significantly deteriorates the BER.

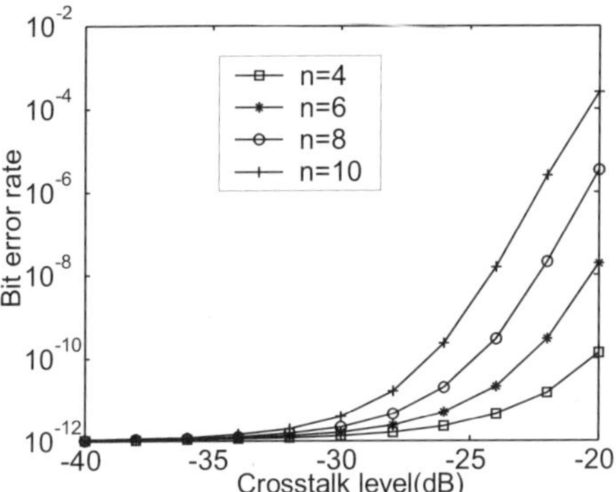

Fig 3 BER vs. different crosstalk level for various values of n

We have also investigated the power penalty caused by the 2nd order inband crosstalk. The power penalty is calculated at BER of 10^{-9}. Fig. 4 shows the power penalty versus the value of n for different crosstalk levels. The power penalty is less than 1 dB, when the crosstalk level is less than -25 dB and $n \leq 10$. The power penalty is negligible for crosstalk levels less than -30dB. However, when the crosstalk is approached to -20dB, the power penalty increases rapidly as the value of n increases. A power penalty of 1dB only allows for $n \leq 6$ with the crosstalk level of -20dB. When the crosstalk level is reduced to -25dB, n can be increased to 10 for a power penalty of 1dB.

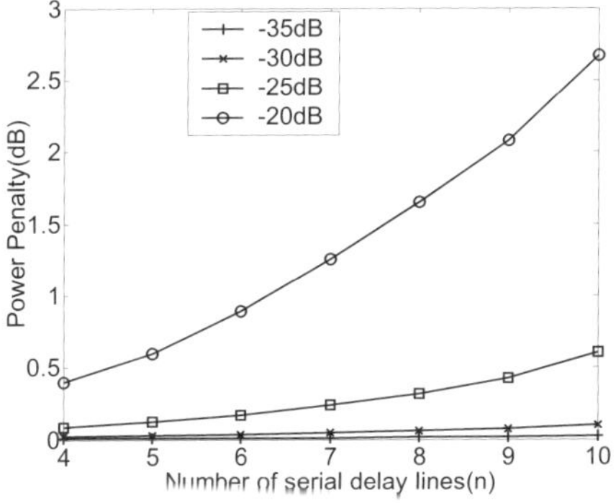

Fig.4 Power penalty versus the value of n for different crosstalk levels

5. CONCLUSION

We have analyzed the inband crosstalk caused by the synchronization module of photonic packet switches. It has been found that the 2nd order inband crosstalk is dominant but its contribution increases with the number of delay lines, which can cause significant impact on the BER performance. The numerical study has shown that the BER is considerably affected by the increase in the number of delay lines. For a reasonable time resolution (i.e. n > 10), the crosstalk level has to be less than -25 dB so as to ensure the power penalty is below 1 dB.

6. REFERENCES

[1] M. J. O'Mahony, et al, "The Application of optical packet switching in future communication networks," *IEEE Commun. Mag.* pp. 128-135, 2001

[2] R. S. Tucker and W. D. Zhong, "Photonic packet switching: an overview," *IEICE Trans. Commu.*, vol E82-B, pp. 254-264 1999

[3] M. P. Earnshaw, et al. "Semiconductor space switches based on multimode interference couplers," *IEEE J. Lightwave Technol.* vol 20 643-650 (2002)

[4] B. Moslehi, "Analysis of optical phase noise in fiber-optic systems employing a laser source with arbitrary coherence time," *IEEE J. Lightwave Technol.* vol. LT-4 pp. 1334-1351 1986

[5] Arie. A., et al., "Probability-density function of noise at the output of a two beam interferometer," *J. Opt. Soc. Am.*,A . pp. 1936-1942 1991

[6] J. L.Gimlett et al., "Effects of phase-to-intensity noise conversion by multiple reflections on Gigabit-per-second DFB laser transmission systems," *IEEE J. Lightwave Technol.* vol. 7 pp. 888-895 1989

[7] J. W. Goodman "Statistical optics," John Wiley & Sons, Inc. pp.157-170, 1985

[8] Guillemot, et al. "Transparent Optical Packet Switching: The European ACTS KEOPS Project Approach," *IEEE J. Lightwave Technol.* vol.12 pp. 2117-2133 1998

[9] H. Takahashi, et al, "Impact of crosstalk in an arrayed-waveguide multiplexer on NxN optical interconnection," *IEEE J. Lightwave Technol.* vol. 14 pp. 1097-1105, 1996

[10] L. Zucchelli, M. Burzio, and P. Gambini, "New solutions for optical packet delineation and synchronization in optical packet switched networks," Proc. ECOC'96, pp. 3.301-3.304, 1996

[11] M. Burzio, et al., "An optical cell synchronizer for packet switched nodes," Proc. Photonics in Switching, pp. 12-17, USA, Mar. 1995

Session 13D

WDM Devices II

Recent Advances in Board-Band Orthogonal Pumps Four-Wave Mixing Technique for All-Optical Wavelength Conversion

C. Shu, K. K. Chow, M. W. K. Mak and H. K. Tsang
Photonic Packaging Laboratory and Department of Electronic Engineering,
The Chinese University of Hong Kong, Shatin, N.T., Hong Kong

F. K. Tong
Applied Science and Technology Research Institute (ASTRI), Tsimshatsui, Kowloon, Hong Kong

Tel: 852-2609 8258 Fax: 852-2603 5558 E-mail: ctshu@ee.cuhk.edu.hk

***Abstract*—We report recent results on all-optical wavelength conversion using broad-band orthogonal pumps four-wave mixing. The different approaches are developed based on a semiconductor optical amplifier placed in a fiber-ring cavity. Using a single external pump laser, a conversion efficiency with a variation smaller than 3 dB has been achieved over a 40-nm operating range. We further demonstrate polarization-insensitive widely tunable wavelength conversion with a polarization diversity scheme. The power penalty is determined to be 3 dB in a 10 Gb/s bit-error-rate measurement. In a modified approach employing a double-ring SOA fiber cavity, all external pumps have been successfully eliminated.**

***Index Terms*—Four-wave mixing; Wavelength conversion; Semiconductor optical amplifier (SOA)**

I. INTRODUCTION

In a wavelength-routed optical network, all-optical wavelength conversion can be used frequently to avoid wavelength blocking. Many different methods have been developed for the conversion that are based on different mechanisms including cross-gain modulation, cross-phase modulation, wave mixing, and difference frequency generation. Four-wave mixing (FWM) in a semiconductor optical amplifier (SOA) is a promising technique for wavelength conversion in optical networks. Much research has been focused on the investigation of single-pump FWM in a SOA [1, 2]. However, the conversion efficiency of single-pump FWM drops rapidly with wavelength detuning, causing a reduction in the intensity of the converted signal and a degradation in the optical signal-to-noise ratio (OSNR) [3]. An improved scheme with a wider wavelength tuning range and a more uniform conversion efficiency has been demonstrated using broad-band orthogonal pumps (BOP) FWM [4, 5]. Different approaches are also developed to achieve polarization-insensitive wavelength conversion [6, 7]. In the setups, two wavelength-tunable external pump sources are generally needed. Recently, a technique to eliminate one of the external pumps is successfully developed by different groups [8, 9] using a fiber-ring cavity. In this paper, we investigate the performance of the BOP FWM wavelength converter using SOA fiber rings. A 40 nm tuning range is obtained with smaller than 3 dB variation in the conversion efficiency. However, the output is subjected to fluctuations in the input signal polarization. In the following section, we further described an improved approach for polarization-independent all-optical wavelength conversion. Using the scheme of polarization diversity, the converted signal power variation is found to be smaller than 1.5 dB with arbitrary input signal polarization. A 10 Gb/s bit-error-rate measurement shows a power penalty of 3 dB in the conversion process. In another modified setup, all external pump lasers have been successfully eliminated based on a double-ring configuration.

II. PRINCIPLE

A. Wavelength Conversion by Four-Wave Mixing in SOA

In the gain medium of the SOA, wave-mixing arises from a nonlinear optical response when more than one wave (signal and pump) are present. The outcome of this effect is a generation of another wave with the amplitude proportional to the product of the signal and the square of the pump amplitudes. The frequency of the generated waves is determined by the wavelength detuning between the signal and the pump [10].

The conversion efficiency of FWM in a SOA varies rapidly with wavelength detuning. The gain dynamics of the conversion depend on both inter-band and intra-band transitions. The former is dominant when the detuning is small (< 5 nm), whereas the latter becomes significant when the detuning increases (> 10 nm). Thus, both the OSNR and the conversion efficiency degrade with wavelength detuning [3].

B. Broad-Band Orthogonal Pumps Four-Wave Mixing in SOA

A relatively uniform conversion efficiency can be obtained in FWM using the scheme of BOP with two pump beams and a signal beam as shown in Fig. 1, provided that the pump beams are orthogonally polarized. In the simple lumped SOA model [9], FWM can be regarded as a lumped

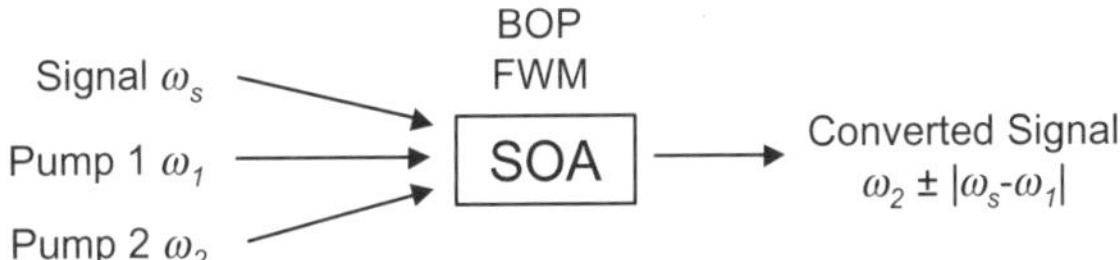

Fig. 1 Schematic illustration of broad-Band orthogonal pumps four-wave mixing.

element providing saturable gain followed by third-order nonlinearity and amplified spontaneous emission (ASE) noise. The three input beams undergo FWM in the gain medium of the SOA and produce wave-mixed signals at several wavelengths. The converted wavelength $\lambda_c = 1/\omega_c$ satisfies the condition $\omega_c = \omega_2 \pm | \omega_s - \omega_l |$ and the corresponding optical field Ec is given by [6, 10]

$$E_c = E_2 (E_1 \cdot E_s) \gamma(\omega_s - \omega_l) \exp[i(\omega_c t + \Delta\phi)] + E_s (E_1 \cdot E_2) \gamma(\omega_2 - \omega_l) \exp[i(\omega_c t + \Delta\phi)] \quad (1)$$

where ω_1, ω_2, ω_s, ω_c, E_1, E_2, E_s, and E_c are the frequencies and the field amplitudes of the first pump (P1), the second pump (P2), the input signal (S), and the converted signal, respectively. $\Delta\phi = \phi_s - \phi_1 + \phi_2$ is the phase difference between the interacting waves. The complex coupling coefficient $\gamma(\omega_s\text{-}\omega_l)$ is produced by the same carrier grating as in conventional FWM and decreases rapidly as $|\omega_s\text{-}\omega_l|$ increases [11]. In BOP FWM, the two pump beams are orthogonally polarized. Hence the second term in Eq. (1) containing the dot product of the two fields vanishes. During the tuning process, ω_2 is adjusted for a desired ω_c but $(\omega_s\text{-}\omega_l)$ remains unchanged in the first term. Therefore, a more uniform OSNR and conversion efficiency can be obtained over a large wavelength detuning range. It is worth noting that the converted signal will have the same polarization as P2.

III. EXPERIMENT

A. Single External Pump BOP FWM

Among various techniques of wavelength conversion, BOP FWM is comparatively costly owing to the use of two tunable external pump lasers. Here, we demonstrate BOP FWM wavelength conversion in a SOA fiber ring cavity where only one external pump is required. The experimental setup is shown in Fig. 2. A commercial SOA is placed in a fiber ring cavity and is biased at 198 mA. The typical fiber-to-fiber small-signal gain of the SOA is 25 dB, with 9 dBm saturation output power. To ensure unidirectional operation and to prevent back reflections, two polarization-independent optical isolators are employed. The SOA ring laser acts as one pump (P1) in the BOP FWM scheme. The operating wavelength of P1 is determined by the tunable fiber Fabry-Perot (FFP) filter. The input signal (S) is a 2^{31}-1 bits pseudorandom signal at 2.5 Gb/s. It is coupled into the ring cavity through a 3 dB coupler. The second pump (P2) for BOP FWM is derived from an external wavelength-tunable laser and is coupled into the ring cavity through a

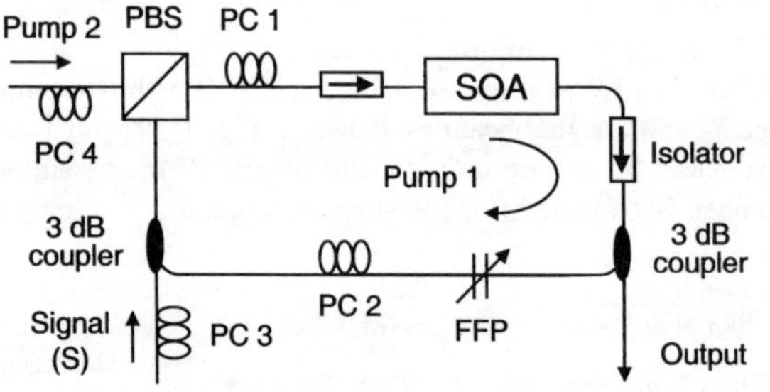

Fig. 2 Experimental setup on wavelength conversion using broad-band orthogonal pumps four-wave mixing. SOA: semiconductor optical amplifier; PC: polarization controller; FFP: tunable optical filter; PBS: polarizing beam splitter.

polarizing beam splitter (PBS).

To maximize the effect of FWM, the polarization of the input signal should match with P1 by adjusting the polarization controllers PC2 and PC3. The PC1 is used to ensure that the signal and P1 are coupled to the TE mode in the SOA. The polarization of P2 is adjusted by PC4 such that it is orthogonal with that of S and P1. The BOP four-wave mixed output is obtained from the ring cavity through another 3 dB coupler.

Fig. 3 shows the optical spectrum obtained from the output port. The input signal S (1557.5 nm), P1 (1555.9 nm), P2 (1550.1 nm) and the converted output (1548.5 nm) can be clearly identified. The channel separation between S and P1 is set to be about 200 GHz in the experiment. The input power of S and P2 are 0 dBm and 6 dBm, respectively. The output power of the converted signal is measured to be –24 dBm, corresponding to a conversion efficiency of –24 dB.

The relation between the conversion efficiency and the output wavelength is shown in Fig. 4. The overall

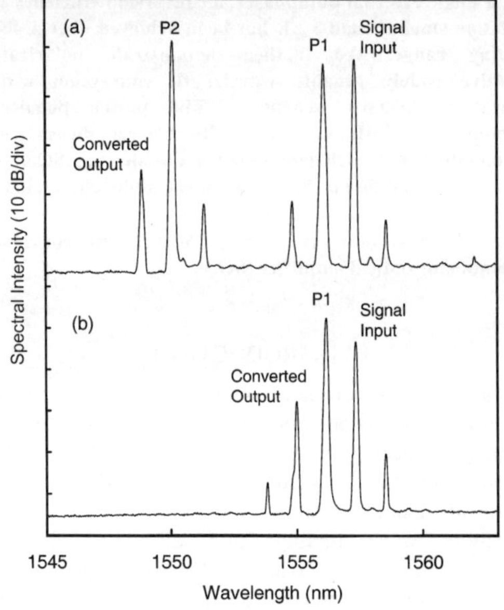

Fig. 3 Output spectra of (a) orthogonal pumps FWM and (b) conventional single-pump FWM when pump 2 is switched off.

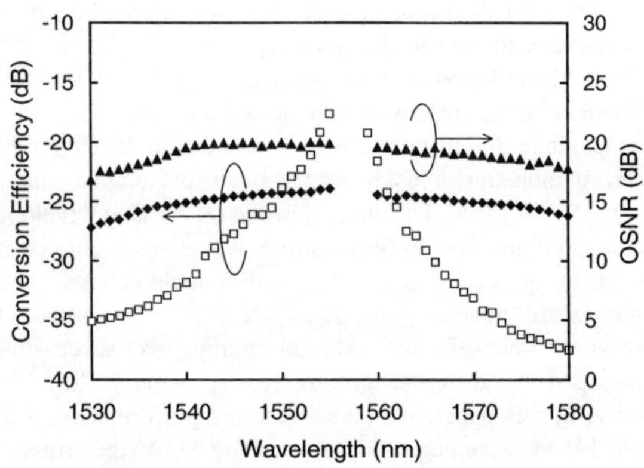

Fig. 4 Plot of the output conversion efficiency and the optical signal-to-noise-ratio against the converted wavelength. □ conventional single-pump FWM; ♦ and ▲ BOP FWM.

conversion efficiency is defined as the power ratio between the converted signal and the input signal. In the case of BOP FWM, tuning is achieved by changing the wavelength of P2. A range of 1530 to 1580 nm is covered for the converted wavelength. The 3-dB tuning range is found to be over 40 nm. The figure also shows the OSNR against the converted output wavelength. The result shows that the OSNR is over 18 dB in the 40-nm tuning range. In the case of conventional single-pump FWM, wavelength selection is performed by tuning the FFP filter in the ring cavity with P2 switched off. The conversion efficiency reaches a peak of about –17 dB near the source wavelength. However, as the converted wavelength is tuned away from the peak, the efficiency drops rapidly and results in a 3-dB tuning range smaller than 10 nm. In comparison to conventional single-pump FWM, the BOP FWM provides a more uniform conversion efficiency over the whole tuning range.

B. Polarization-Insensitive BOP FWM

In the previous section, one external pump of BOP FWM has been successfully eliminated. Another issue to be solved is that the wavelength converting process is polarization dependent. Hence, the conversion efficiency will vary with different input polarizations. In this section, we demonstrate polarization-insensitive BOP FWM using the scheme of polarization diversity. It is worth noting only one external pump is needed in the scheme. The experimental setup is shown in Fig. 5. The input signal (S) is a 2^{31}-1 bits pseudorandom signal at 10 Gb/s. It is coupled into the ring cavity through a 3 dB coupler and is combined with the fiber laser pump 1 (P1). The operating wavelength of P1 is determined by the tunable fiber Fabry-Perot (FFP) filter. An EDFA is used to increase the power of P1, which will then improve the wavelength conversion efficiency. The light beams will pass through the optical circulator (OCR1) and enter port A of the polarizing beam splitter (PBS). The pump 2 (P2) is introduced through another optical circulator (OCR2) to port D of the PBS. The SOA is placed in a loop connecting port B and port C of the PBS. The dc bias is set at 198 mA. In this scheme of polarization diversity, S is resolved into two orthogonal polarizations. Each of which undergoes separate FWM wavelength conversion before recombining to produce the converted signal. To maximize the effect of FWM, PC4 and PC5 are adjusted to ensure that the signal polarizations are TE and TM at the SOA input facets for the clockwise and the counterclockwise propagation branches, respectively. Both P1 and P2 are polarized at 45° with respect to the PBS axis so that P1 and

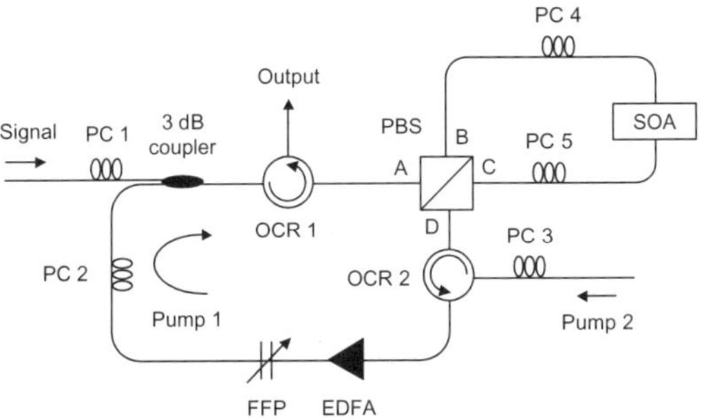

Fig. 5 Experimental setup on polarization-insensitive wavelength conversion using polarization diversity. SOA: semiconductor optical amplifier; PC: polarization controller; FFP: tunable fiber Fabry-Perot filter; PBS: polarizing beam splitter; OCR: optical circulator.

P2 will pass through the SOA loop and exit at port D and port A, respectively. Since the wavelength-converted signal generated by the BOP FWM has the same polarization as P2, the output will exit from port A of the PBS and is obtained through OCR1.

In the experiment, the channel separation between S and P1 is set to be about 200 GHz. The input power of S and P2 are –3 dBm and 5 dBm, respectively. The wavelength tuning is obtained by changing the wavelength of P2. The 3-dB tuning range is found to be over 40 nm. The OSNR is observed to be over 16 dB in the 40-nm tuning range.

To demonstrate polarization insensitivity of the converter, the wavelength-converted output is analyzed at different input signal polarizations. Fig. 6 plots the measured power penalties of the converted signal against the input signal polarization. The power variation of the converted output is found to be smaller than 1.5 dB as the polarization is rotated through 180°.

The power penalty of the wavelength converter has also been studied by performing BER measurements at 10 Gb/s. Fig. 7 plots the output BER against the received optical power. The result shows that the wavelength converter has about 3-dB power penalty at 10^{-9} BER level. The inset shows the eye-diagram of the converted signal at –16.5 dBm received power from the output port.

The above experiment demonstrate a BOP FWM wavelength converter that is polarization insensitive and eliminates an external pump laser. In our further development, the converter has been modified to a double-

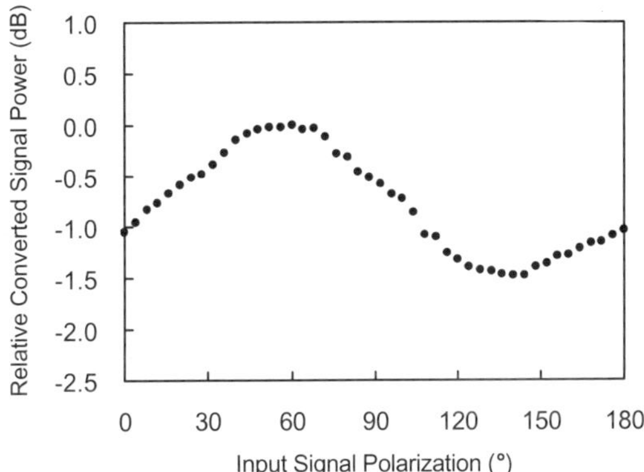

Fig. 6 The relative converted signal power against the input signal polarization.

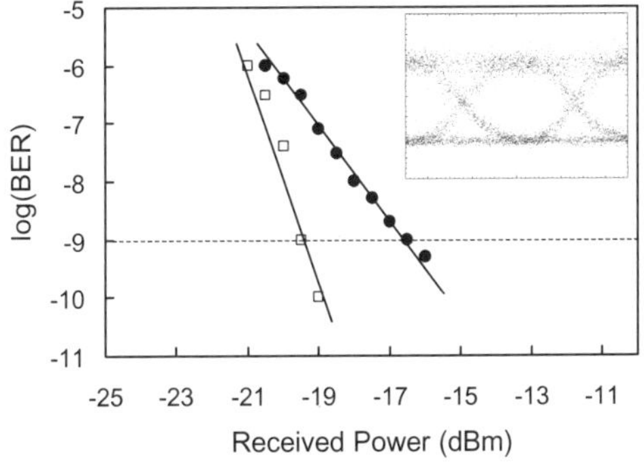

Fig. 7 Plot of the bit-error-rate against the received power in a 10 Gb/s BER measurement. • converted signal and □ back to back. The inset shows the eye-diagram of the converted signal at –16.5 dBm received power.

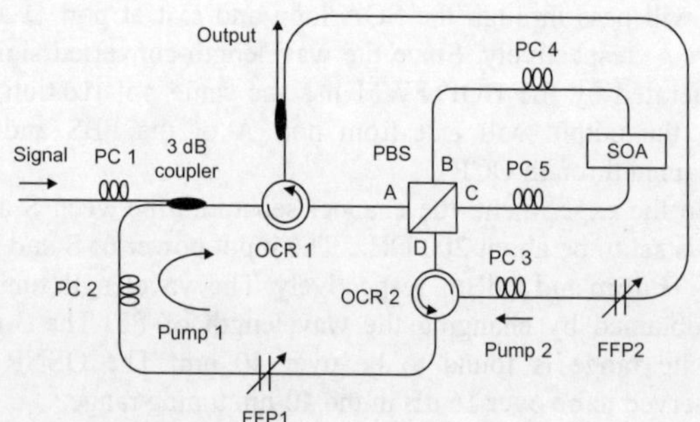

Fig. 8 Experimental setup on polarization-insensitive wavelength conversion using a double-ring configuration. SOA: semiconductor optical amplifier; PC: polarization controller; FFP: tunable fiber Fabry-Perot filter; PBS: polarizing beam splitter; OCR: optical circulator.

ring configuration such that no external pump laser is needed. The experimental setup is shown in Fig. 8. The external pump P2 is now replaced by another cw ring laser. The wavelength of P2 is determined by the tunable FFP2 in the cavity. As is similar to the configuration in Fig. 5, both P1 and P2 are polarized at 45° with respect to the PBS axis so that P1 and P2 will pass through the SOA loop and exit at port D and port A, respectively. Therefore, another ring laser for P2 can be obtained by connecting the output port of OCR1 and the input port of OCR2. Since the converted signal has the same polarization as P2, it can be coupled out from OCR1.

In this experiment, tuning of the converted wavelength is obtained by adjusting FFP2. Similar results with the previous setup are observed, where the 3-dB tuning range of the converted signal is about 40 nm. Also, the converted output power has a variation smaller than 1.5 dB with arbitrary input polarizations. The 10 Gb/s BER measurement for this setup is under investigation and will be reported. A summary on the comparison of different FWM schemes is presented in Table I.

	external pump	polarization insensitive	tuning range (3-dB)	power penalty @ 2.5Gb/s
Conventional FWM	1	NO	< 10 nm	< 0.6 dB
BOP FWM	2	NO	> 40 nm	< 1 dB
BOP FWM with PD	2	YES	> 40 nm	< 1 dB
BOP FWM with SOA ring	1	NO	> 40 nm	< 1.5 dB
BOP FWM with PD SOA ring	1	YES	> 40 nm	< 1.5 dB
BOP FWM with dual SOA ring	0	NO	> 40 nm	< 1.5 dB
BOP FWM with PD dual SOA ring	0	YES	>40 nm	< 1.5 dB

Table I Comparison of different FWM schemes. BOP FWM: broad-band orthogonal pumps four-wave mixing; PD: polarization diversity

IV. CONCLUSION

In conclusion, different schemes of widely tunable all-optical wavelength conversion have been successfully demonstrated. The converters are based on SOA fiber ring cavities. Using broadband orthogonal pumps four-wave mixing with single external pump, a nearly constant (< 3-dB variation) conversion efficiency over a 40-nm wavelength range has been obtained. In our further work, polarization sensitivity to the input signal is solved by using the polarization diversity scheme. A 10 Gb/s BER measurement is performed and a 10^{-9} BER level is obtained with 3-dB power penalty. An improved scheme of polarization insensitive BOP FWM is also demonstrated without using any external pump. The results promise a potential for wide-band applications in all-optical networks.

REFERENCES

[1] A. D'Ottavi, et al., "Wavelength conversion at 10 Gb/s by four-wave mixing over a 30-nm interval," *IEEE Photon. Technol. Lett.*, vol. 10, pp. 952-954, July 1998.

[2] Z. G. Lu, S. A. Boothroyd and J. Chrostowski, "Tunable wavelength conversion in a semiconductor-fiber ring laser," *IEEE Photon. Technol. Lett.*, vol. 11, pp. 806-808, July 1999.

[3] G. P. Agrawal, "Population pulsations and nondegenerate four-wave mixing in semicondcutor lasers and amplifiers," *J. Opt. Soc. Amer. B*, vol. 5, pp. 147-159, Jan. 1988.

[4] G. Contestabile, F. Martelli, A. Mecozzi, L. Graziani, A. D'Ottavi, P. Spano, G. Guekos, R. Dall'Ara and J. Eckner, 'Efficiency flattening and equalization of frequency up- and down-conversion using four-wave mixing in semiconductor optical amplifiers,' *IEEE Photon. Technol. Lett.*, Vol. 10, pp.1398-1400, Oct. 1998.

[5] T. J. Morgan, J. P. R. Lacey, and R. S. Tucker, "Widely tunable four-wave mixing in semiconductor optical amplifiers with constant conversion efficiency," *IEEE Photon. Technol. Lett.*, vol. 10, pp. 1401-1403, Oct. 1998.

[6] M. W. K. Mak, H. K. Tsang, and K. Chan, "Widely tunable polarization-independent all-optical wavelength converter using a semiconductor optical amplifier," *IEEE Photon. Technol. Lett.*, vol. 12, pp. 525-527, May 2000.

[7] G. Contestabile, A. D'Ottavi, F. Martelli, P. Spano, and J. Eckner, "Broad-band polarization-insensitive wavelength conversion at 10 Gb/s," *IEEE Photon. Technol. Lett.*, vol. 14, pp. 666-668, May. 2002.

[8] C. Nan, Q. Jiang, X. Lin, Z. Yuan, K. J. Guan and S. Q. Chen, "Wavelength conversion based on FWM with double orthogonal pump in a semiconductor-fiber ring laser," *IEEE Photon. Technol. Lett.*, vol. 13, pp. 499-501, May 2001.

[9] K. K. Chow, M. W. K. Mak, C. Shu and H. K. Tsang, "Widely tunable wavelength converter using a fiber ring laser with semiconductor optical amplifier," paper OR.TueA.2, *OECC/IOOC 2001*, July, 2001, Sydney, Australia.

[10] J. P. R. Lacey, M. A. Summerfield, and S. J. Madden, " Tunability of polarization-insensitive wavelength converters based on four-wave mixing in semiconductor optical amplifiers," *IEEE J. Lightwave Technol.*, vol. 16, pp. 2419-2427, Dec. 1998.

[11] A. Uskov, J. Mork, and J. Mark, "Wave mixing in semiconductor laser amplifiers due to carrier heating and spectral-hole burning," *IEEE J. Quantum Electron.*, vol. 30, pp. 1769-1781, Aug. 1994.

Design and Optimization of Photonic Devices by use of the Finite Element Method

B M A Rahman, T Wongcharoen, S Haxha, C Themistos, M Rajarajan, and K T V Grattan

School of Engineering, City University, London EC1V 0HB, UK
Tel: +44-20-7040-8123 Telefax: +44-20-7040-8568
Email: B.M.A.Rahman@city.ac.uk

Abstract: *A review of the characterisation of optoelectronic guided-wave devices using the modal solution, junction analysis and beam propagation methods based on the numerically efficient finite element method is presented. Numerically simulated results for various photonic devices such as optical modulators, spot-size converters, compact bends, power splitters, and Bragg gratings are presented.*

1. INTRODUCTION

With the advent of semiconductor lasers and low-loss optical fibres, the progress of optical technology has been significant in the field of optical communications. As optical technology has reached maturity, the associated devices have themselves become more complex. The optimization of such advanced devices requires an accurate knowledge of their lightwave propagation characteristics and their dependence on the system fabrication parameters. Unfortunately, analytical techniques are not adequate to model lightwave devices without significant approximations. Therefore, the optimization of existing realistic designs or the evaluation of new designs for optoelectronic devices and sub-systems has created significant interest in the development and use of effective numerical methods.

2. NUMERICAL METHODS

Of the different numerical approaches for modal solutions reported so far, the finite element method (FEM) has been established as one of the most powerful and versatile methods. Over the last two decades this technique has been used to model optical waveguides [1,2], devices based on directional couplers [3], or multimode interference couplers [4], spot-size converters [5], electrooptic modulators [6], polarization rotators [7] and second-order [8] or third-order [9] nonlinear optics, for example. In the finite element approach, the problem domain is suitably divided into a patchwork of a finite number of subregions called elements. Each element can have a different shape and size and using many elements, a complex problem can be accurately represented. A wide range of optical devices can be modeled as each element can be considered to have different optical parameters such as refractive index, anisotropic tensors, nonlinearity, and loss or gain factors.

Many important photonic devices, such as optical modulators, filters, polarization splitters, polarization rotators, power splitters, etc may be fabricated by combining several butt-coupled uniform waveguide sections. To design and analyse such photonic devices, it is important to use a junction analysis program in association with a modal analysis program. One of the most rigorous approaches, the least squares boundary residual (LSBR) method [10] has been developed by the authors. In this approach the continuity of the tangential electric and magnetic fields is imposed globally over the interface to obtain the scattering coefficients at the junction.

On the other hand, to simulate the propagation of optical waves through a z-dependent linear or nonlinear structure, the finite element-based beam propagation method (BPM) has been developed [11] using a fully vectorial approach with a difference scheme along the axial directions. Such an approach is particularly useful in the characterization of tapered sections, Y and X junctions and nonlinear optical devices. Being fully vectorial, this approach can be used to study polarization dependent issues in optoelectronic systems.

3. RESULTS

The optical waveguide is the basic building block of all guided wave optical systems. The finite element method (FEM) has been used to characterize accurately lightwave propagation through a wide range of waveguides, such as semiconductor ridge [2], titanium diffused $LiNbO_3$[1,12], stress induced highly-birefringent silica fibre [13] and ion exchanged glass waveguides. This method is particularly advantageous because of its applicability to waveguides with arbitrary shapes and arbitrary refractive index profiles. The finite element formulation, using accurate vector **H**-field formulation [1] which is "*exact-in-the-limit*", is also valid for anisotropic materials. Sometimes waveguides with the dimensions in the nanometer range, such as multiple quantum well structures [14] or surface plasmon structures [15] show enhanced physical activities, have also been simulated by using the FEM. Figure 1 shows the dispersion curves of a semiconductor laser [14] incorporating 70 periods of a InGaAs quantum well sandwiched between InP barrier layers with their individual layer heights of 4 nm and 10 nm respectively. The dispersion curves, given as a function of the rib width, *W*, for the H^y_{11} and H^y_{12} (quasi-TE) modes, are shown. It can be seen that for a rib width of up to 6.5 μm, only a single lateral mode is possible.

The optical directional couplers also play a major role in optical systems and they may be used as power dividers, power combiners, modulators, switches and filters. The isolated and supermodes of such coupled waveguides, their propagation constants and field profiles can be easily obtained by using the finite element method. The coupling length and the supermodes can be calculated for identical or nonidentical coupled waveguides for the design of directional coupler-based devices.

The finite element program has also been developed to model electrooptic devices such as titanium diffused $LiNbO_3$ or GaAs modulators or switches [16,6]. Here, the Laplace equation is solved by using the FEM to obtain the two transverse components of the modulating electric field (vector), which is subsequently used to calculate the electrooptic contribution to the tensor refractive indices. Then the vector **H**-field based FEM can be used to model the electrooptic modulators more accurately. For a directional coupler-based electrooptic modulator, with the increase of the modulating voltage, the refractive index of one of the guides increases and for the other it decreases. As the guides lose their phase matching with an increased modulating signal, two supermodes become highly asymmetrical. The dominant field profiles of the even- and odd-like supermodes are shown in Fig. 2.

Optical modulators are key components to transfer data to high-bandwidth optical fibres. Most of the high-speed modulators use the Mach-Zehnder-based [6] approach and recently it has been shown that etched Z-cut lithium niobate provides a lower switching voltage (V_π) and a shorter device length, L. The variations of $V_\pi L$, the key Figure-of-Merit, with the etch depth and buffer layer thickness are shown in Fig.3. The band-width of a high-speed modulator depends primarily on the phase matching between the microwave and optical wave. The variation of the optical bandwidth with the buffer later is shown in Fig.4, which demonstrates that the bandwidth reaches its maximum value when phase matching condition is achieved for a given electrode thickness, and this maximum value is limited by the conduction loss of the metal electrodes.

The design of many important photonic devices can be based on the directional coupler or multimode interference (MMI) [4] based approaches. The design of a compact power splitter can often be achieved by using the MMI-based technique. The variation of the splitter length for the General Interference (GI), Restricted Interference (RI), and Symmetric Inferference (SI) approaches and the polarization dependence is shown in Fig.5.

The functionality of monolithically integrated sub-systems critically depends on the design of optical interconnections. To reduce the chip-size and manufacturing costs and also to improve the system reliability, it is important to design low-loss compact bends [17]. The optical loss for compact bends can be reduced by using deeply-etched semiconductor waveguides, or by using low index buffer layers. The variation of the optical loss for a 90 degree bend is shown in Fig.6.

In most cases, when a photonic device is butt coupled to a standard single mode fibre (SMF) nearly 80 to 90% power is lost at the interface due to their highly unequal spot-sizes. Only recently, the approach of designing a monolithically integrated spot-size converter is gaining considerable interest as the coupling efficiency can be significantly increased without deteriorating the alignment tolerance. However, instead of traditional tapered spot-size converters (SSC), a novel approach [5] has been developed, where two uniform coupled waveguides with nonidentical spot-size but with identical propagation constants, are considered. The numerically simulated results are shown in Fig.7, which reveal more than 9 dB coupling efficiency improvement along with a relaxed vertical alignment tolerance, Δy.

Bragg gratings are important components for many optical communications and sensing applications. The coupled mode approach (CMA) is often adequate for the analysis of fibre Bragg gratings as refractive index change due to UV exposure is quite small. However, in integrated optical semiconductor structures, the discontinuity could be significantly larger and thus the CMA may not offer the accuracy required for the design of such structures. Devices with strong gratings, such as photonic crystals, could be analysed more effectively by using a combination of the FEM, the LSBR and the transfer matrix method (TMM) [18]. The numerically simulated reflection spectrum of a multi-wavelength apodised narrow-band filter is shown in Fig. 8.

4. CONCLUSIONS

The finite element method is a powerful and versatile method suitable to model accurately the optical, thermal, elasto-optic, electro-optic, and electronic effects in passive or active opto-electronic devices. The characterisation of important optical devices and the identification of important fabrication parameters by performing numerical experiments are invaluable in the optimization of such advanced photonic devices.

5. REFERENCES

[1] B M A Rahman and J B Davies, Finite element solution of integrated optical waveguides, J. Lightwave Tech., **2**, pp.682-688, Oct. 1984.

[2] B M A Rahman and J B Davies, Vector-H finite element solution of GaAs/GaAlAs rib waveguides, Proc. IEE-J Optoelectronics, **132**, pp.349-353, Dec. 1985.

[3] T Wongcharoen, B M A Rahman and K T V Grattan, Electro-optic directional coupler switch characterization, J. Lightwave Technol., pp.377-382, 1997.
[4] M Rajarajan, B M A Rahman, T Wongcharoen, and K T V Grattan, Accurate characterization of MMI devices with 2-dimensional confinement, J Lightwave Technol., pp.2078-2084, Sept. 1996.
[5] T Wongcharoen, B M A Rahman, M Rajarajan, and K T V Grattan, Spot-size conversion using uniform waveguide sections for efficient laser-fiber coupling, J. Lightwave Technol., pp.708-716, May 2001.
[6] N Anwar, S S A Obayya, S Haxsa, C Themistos, B M A Rahman, and K T V Grattan, The effect of fabrication parameters on a ridge Mach-Zehnder interferometric modulator, J Lighwave Technol., pp.826-833, May 2002.
[7] N Somasiri and B M A Rahman, Fabrication tolerance study oof a compact passive polarization rotator, J Lightwave Technol. pp.751-757, April 2002.
[8] F A Katsriku, B M A Rahman and K T V Grattan, Finite element analysis of second harmonic generation in GaAs and AlGaAs waveguides, IEEE J. Quantum Electron., pp.282-289, March 2000.
[9] S S A Obayya, B M A Rahman and H A El-Mikati, Full vectorial finite element beam propagation method for nonlinear directional coupler devices, IEEE J Quantum Electron., p.556-562, May 2000
[10] B M A Rahman and J B Davies, Analyses of optical waveguide discontinuities, J. Lightwave Tech., **6**, pp.52-57, 1988.
[11] S S A Obayya, B M A Rahman and H A El-Mikati, New full-vectorial numerically efficient propagation algorithm based on the finite element method, J. Lightwave Technol., pp.409-415, March 2000.
[12] F A Katsriku, B M A Rahman, and K T V Grattan, Finite element analysis of diffused anisotropic channel waveguides, J. Lightwave Technol., pp.780-786, May 1996.
[13] Y Liu, B M A Rahman and K T V Grattan, Thermal stress induced birefringence in bow-tie optical fibers using the finite element method, Applied Optics, pp.5611-5616, 1994.
[14] B M A Rahman, Y Liu and K T V Grattan, Finite element modelling of one- and two-dimensional MQW semiconductor optical waveguides, IEEE Photonics Technology Letters, **5**, pp.928-931, 1993
[15] C Themistos, B M A Rahman, and K T V Grattan, Finite element analysis of lossy surface plasmon modes using perturbation technique, Applied Optics, pp.7695-7701, 1995.
[16] N Anwar, C Themistos, B M A Rahman, and K T V Grattan, Design considerations for an electrooptic directional coupler modulator, J. Lightwave Technol., pp.598-605, April 1999.
[17] M Rajarajan, S S A Obayya, B M A Rahman, K T V Grattan, and H A El-Mikati, Design of compact optical bends with a trench by use of finite element and beam propagation methods, Applied Optics, pp.4946-4953, Sept. 2000.
[18] N Mahmud, B M A Rahman and K T V Grattan, Analysis of Bragg reflectors pairs using transfer matrix, overalp and LSBR methods, Meas. Sci. Tech., pp.1059-1064, Oct. 1997.

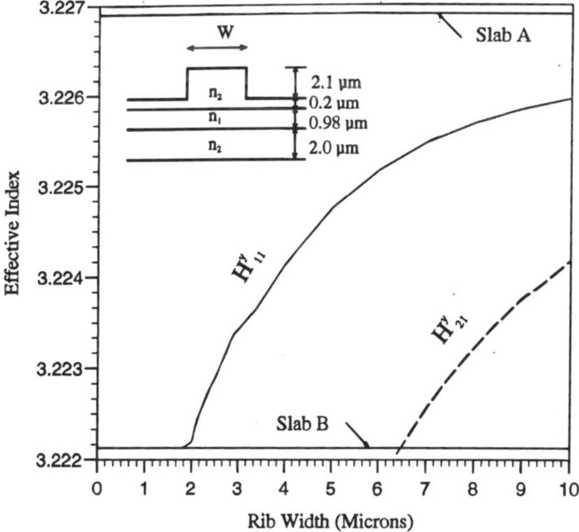

Fig.1 Dispersion curves of a MQW semiconductor laser waveguide.

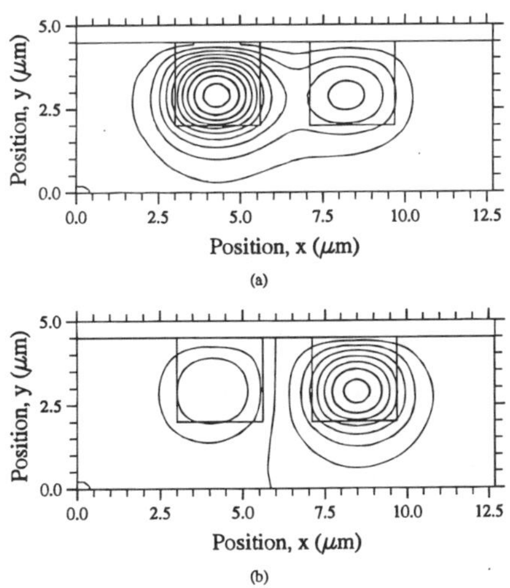

Fig.2 Even- and odd-like supermodes for nonidentical coupled waveguides

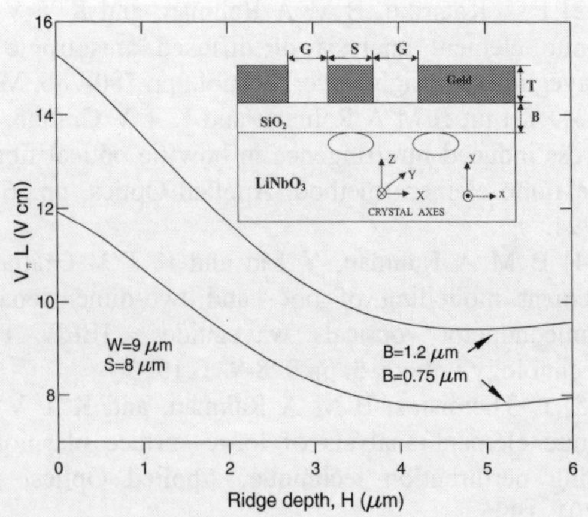

Fig.3 Variation of the $V_{\pi}L$ with the ridge depth

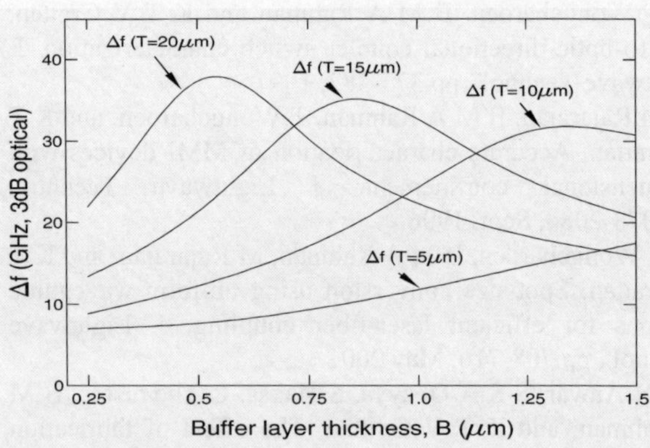

Fig.4 Variation of the optical bandwidth with buffer layer thickness.

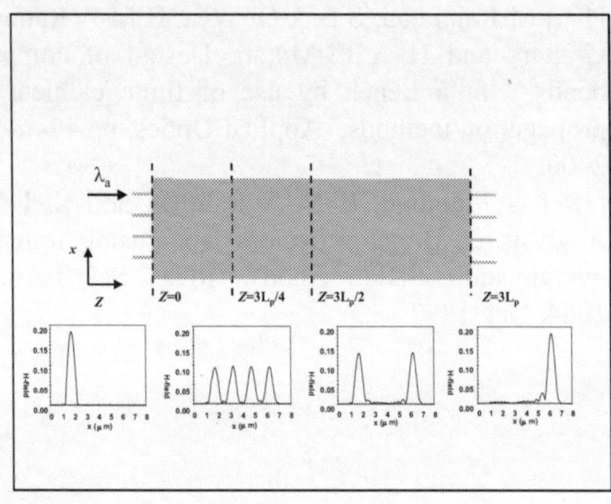

Fig.5 MMI-based optical power splitters.

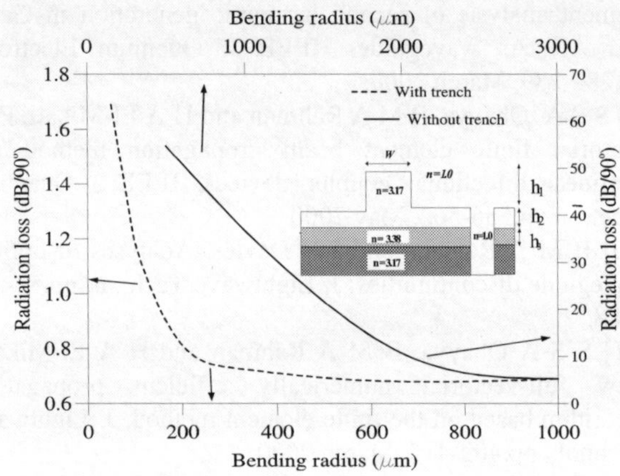

Fig.6 Radiation loss of curved rib waveguides with and without an isolation trench..

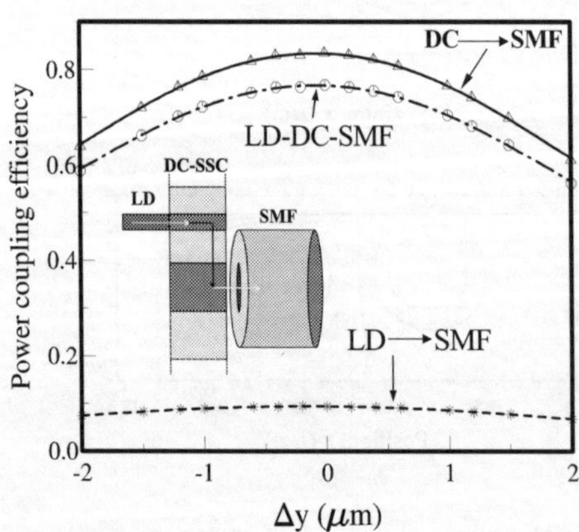

Fig.7 Variation of the coupling efficiency with the alignment tolerance, Δy.

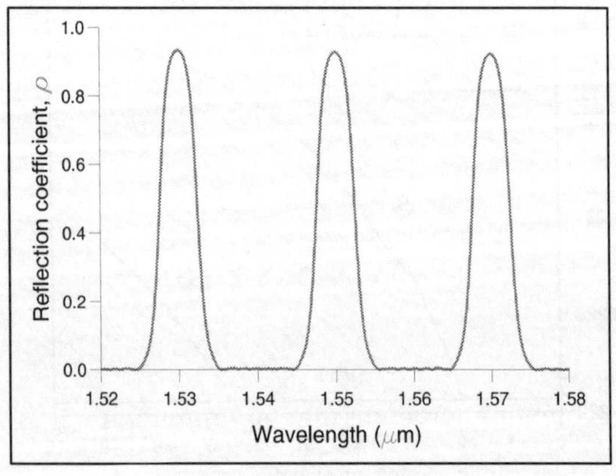

Fig.8 Gaussian apodised multiple grating with 1000 layers per section.

Optical Multichannel Flattop Filter with Complementary Outputs for DWDM Applications

Yicheng Lai, *W. Zhang, J. A. R. Williams, I. Bennion
Photonics Research Group, School of Engineering and Applied Science, Aston University, Birmingham B4 7ET, UK
Phone: (44) 0121 3593611, Fax: (44) 0121 3590156, Email: laiy@aston.ac.uk
**Indigo Photonics Ltd, Faraday Wharf, Holt Street, Birmingham, B7 4BB, UK*
Phone: (44) 0121 2606111, Fax: (44) 0121 2606064, Email: W.Zhang@Indigo-photonics.com

Abstract: A technique to implement an optical multichannel flattop filter with simultaneous complementary outputs is proposed and demonstrated. Using a bipolar delay line structure based on cascaded high birefringence (Hi-Bi) fibres, flattened optical passbands with ripples <0.1dB and sidelobe suppression >30dB is achieved. Due to the single-line cascaded Hi-Bi fibre configuration, the filter exhibits stable wavelength characteristics with low insertion loss. The configuration is not limited to optical domain applications and can be easily adapted as a microwave transversal filter that overcomes limitations imposed by the coherence of the source. Microwave flattop bandpass filtering is further demonstrated based on this technique.

Key words: Flattop filter, High Birefringence fibre, Microwave.

1. INTRODUCTION

Wavelength division multiplexing technology offers cost effective transmission network. Ever increasing demand on transmission capacity has led to Dense Wavelength Division Multiplexed (DWDM) technique that reduces channel spacing to increase the channel numbers. Optical filters such as multi/demultiplexer, optical add/drop, and interleaver are of special interest for DWDM systems [e.g. 1]. In these optical devices various parameters like flatness in passband, channel crosstalk, insertion loss and stable wavelength spectral characteristics are always of concern. Due to narrower channel spacing, fabrication of planar devices to realize such filters face greater difficulties [2]. For all-fiber based structures [e.g. 3], it is difficult to realize broad, flattened passbands under environmental perturbations and interference noise.

In this paper, we propose and demonstrate an optical multichannel flattop transversal filter suitable for DWDM applications. The use of non-interfering orthogonal polarized modes in high birefringence (Hi-Bi) fibre to realize a delay line notch filter has already been demonstrated [4]. By using further Hi-Bi fibre in sequence, cascaded Mach-Zehnder interferometers are realized in the proposed filter. Owing to the single-line cascaded Hi-Bi fibre structure, the environmental sensitivity typical of fibre-based optically coherent devices is greatly reduced. Furthermore, bipolar taps are realized in the transversal filter to yield flattened optical bandpass response with low loss, low ripples and high sidelobe suppression. The filter delivers two complementary outputs simultaneously with stable wavelength spectral characteristics. Based on the same principle, a microwave transversal filter not limited by source coherence and delivers flattop bandpass microwave responses is further realized.

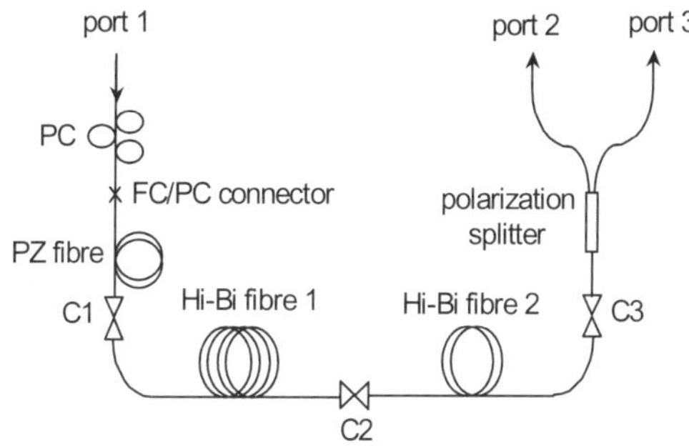

Fig.1 Experimental setup of the optical transversal filter.

2. EXPERIMENT

The configuration of proposed optical transversal filter is as shown in Fig.1. The light is optimized and aligned to Hi-Bi fibre 1 at port 1 through a combination of a fibre polarization controller (PC) and a polarizing fibre (PZ). As shown in Fig. 2, the linearly polarized light launched into Hi-Bi fibre 1 at an azimuth θ_1 with respect to the birefringence axes, degenerates into two orthogonal polarized modes propagating down the fibre, then coupled into Hi-Bi fibre 2 at an azimuth θ_2 with respect to its birefringence axes. These degenerated polarized modes (x_s, x_p, y_s and y_p) travelling down the Hi-Bi fibre constitute the 4 optical signal taps.

By making the length of Hi-Bi fibre 1 to be twice that of Hi-Bi fibre 2, the 4 taps will be spaced evenly to yield a 4-tap optical bandpass transversal filter response. The time delay between the adjacent optical signal taps is simply determined by the birefringence and the length of the shortest Hi-Bi fibre section and can be expressed as

$$\tau = \frac{\Delta n L_s}{c} \tag{1}$$

where L_s is the length of the shortest Hi-Bi fibre section, Δn is the birefringence and c is the light velocity in vacuum.

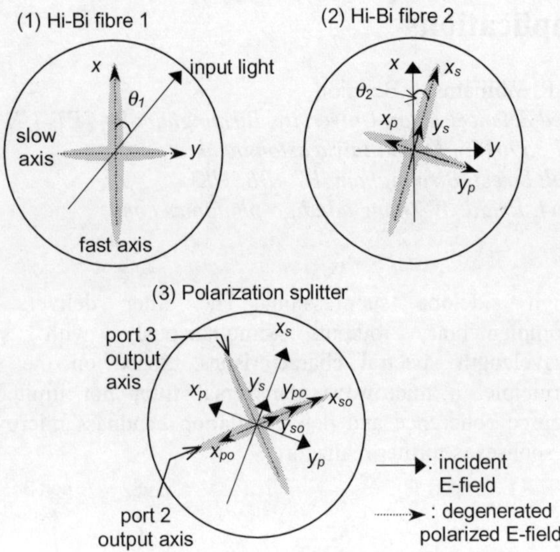

Fig.2 Diagrams showing degeneration of light in the cascaded Hi-Bi fibre transversal filter.

It is evident that the amplitudes of all polarized modes at the exit end of the second Hi-Bi fibre section can be set by varying angles θ_1 and θ_2. With the birefringence axes of the fibre polarization splitter at 45° azimuth to that of Hi-Bi fibre 2, the normalized magnitude of the electric fields of the 4 taps (x_{so}, x_{po}, y_{so} and y_{po}) at port 2 and port 3 are given by,

$$\begin{bmatrix} \cos\theta_1 \cdot \cos\theta_2 \\ -\cos\theta_1 \cdot \sin\theta_2 \\ \sin\theta_1 \cdot \sin\theta_2 \\ \sin\theta_1 \cdot \cos\theta_2 \end{bmatrix} \quad and \quad \begin{bmatrix} \cos\theta_1 \cdot \cos\theta_2 \\ \cos\theta_1 \cdot \sin\theta_2 \\ \sin\theta_1 \cdot \sin\theta_2 \\ -\sin\theta_1 \cdot \cos\theta_2 \end{bmatrix} \qquad (2)$$

respectively. Due to the degeneration of the electric fields propagating through the cascaded structure, there is, in this instance, 1 tap negated with respect to the rest in each output group. The optical response of the filter registered is the magnitude of the coherent sum of the constituents' electric fields.

Two pieces of Hi-Bi fibre with lengths ~4.53m and ~2.265m were first used in the experiment. The beat length of the Hi-Bi fibre is ~2.3mm corresponding to a birefringence of ~6.7 x 10^{-4}. The cleaved ends of the fibres are butted together (at C1, C2 and C3) such that by rotating the fibre ends with respect to each other, the orientation of the input light with respect to the birefringence axes can be varied. No special attention was required to isolate the set-up from surrounding environmental perturbations.

Based on (2), under two special cases: θ_1~45°; θ_2~0° and θ_1~45°; θ_2~90°, the relative orientation between the birefringence axes of Hi-Bi fibre 1 and 2 are aligned and orthogonal respectively. This reconfigures the structure to a 2-tap optical notch filter with maximum tap time interval for the former and minimum tap time interval for the latter. Furthermore, taking note that the output axis of port 2 is orthogonal to that of port 3, the set of output optical taps differs in the order of occurrence and the magnitude of the negated tap, based on (2). This results in complementary output response between port 2 and port 3. The measured and simulated (dotted) normalized complementary optical notch responses from the two output ports at θ_1~45°; θ_2~90° is as shown in Fig. 3. The measured free spectral range (FSR) is ~1.6nm with notch depth measured >27dB.

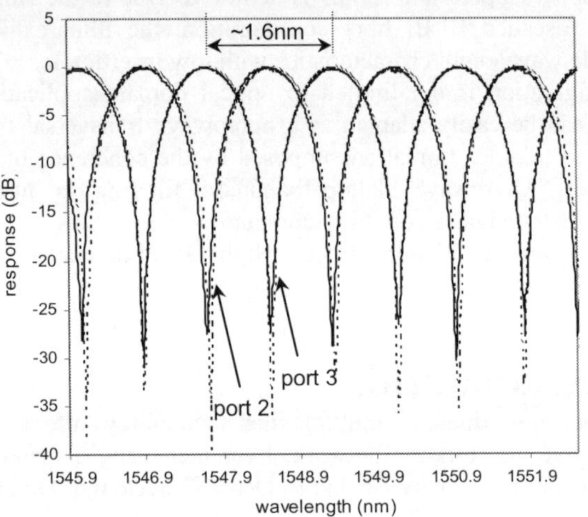

Fig.3 Measured and simulated (dotted) optical notch responses from output port 2 and port 3 at θ_1~45^0;θ_2~90^0.

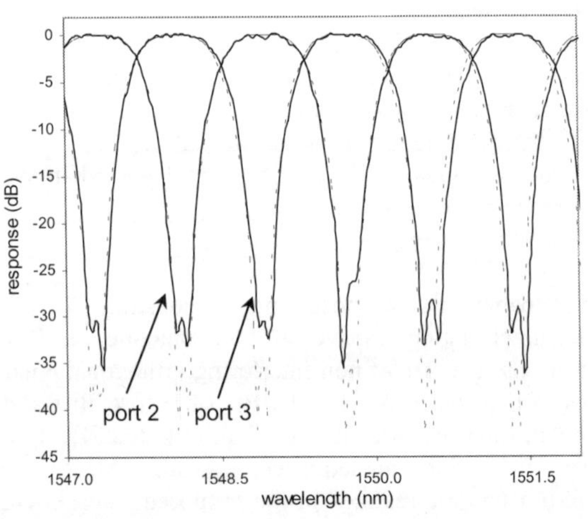

Fig.4 Measured and simulated (dotted) optical responses from output port 2 and port 3 at θ_1~14^0; θ_2~58^0. Response shows flattened passband width ~0.35nm with FSR ~1.6nm.

By setting θ_1~14^0; θ_2~58^0, a filter response with flattened passband can be realized, based on (2). The corresponding normalized tap weightings for port 2 are [0.514, -0.823, 0.205, 0.128] and the obtained spectrum agrees well with the superimposed simulation result, as shown in Fig. 4. The measured response exhibits flattened passband width ~0.35nm with FSR ~1.6nm. The measured sidelobe suppression is ~30dB with ripples in the passband <0.1dB. More importantly the output response from port 3, on the orthogonal axis of the polarization splitter, is the complement of port 2. The corresponding normalized tap weightings [0.514, 0.823, 0.205, -0.128] enabled an identical flattop bandpass response, offset by half a FSR from that of port 2, as a result of the difference in the order of occurrence and magnitude of the negated tap. This response is a characteristic feature of a coherent delay line filter consisting of bipolar taps. The normalized measured response from port 3 with the superimposed simulation result is as shown in Fig. 4.

The concept of this filter scheme applies equally for larger number of cascaded sections. The sidelobe suppression improves with increased number of cascaded sections due to the increase in number of optical taps. Correspondingly, the bandwidth of the flattened passband and FSR also increases. In the experiment, a filter consisting of 3 cascaded Hi-Bi fibre sections measuring ~4.53m, ~2.265m and ~1.133m were used. The normalized tap weightings at the output were [0.049, -0.049, -0.35, -0.35, 2.491, -2.491, 0.35, 0.35]. Measured filter response shows a FSR of ~3.2nm and flattened passband width of ~0.87nm with ripples <0.1dB. The normalized measured response superimposed with the simulation result is as shown in Fig. 5.

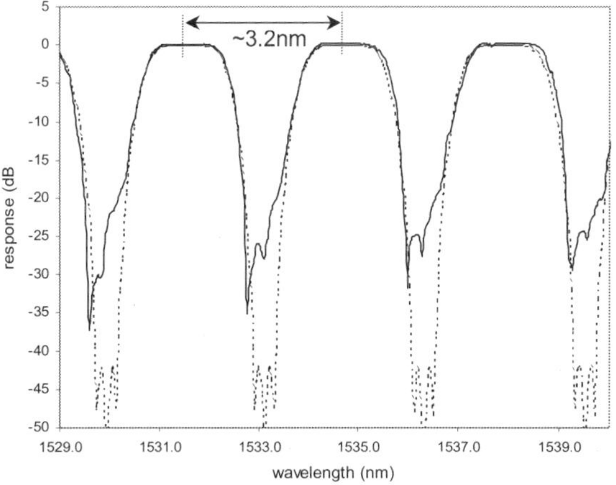

Fig.5 Measure and simulated (dotted) optical response from a 3-section cascaded Hi-Bi fibre transversal filter. Response shows a FSR of ~3.2nm and flattened passband width of ~0.87nm with ripples <0.1dB.

Due to mismatch in the length ratios of the Hi-Bi fibre sections and angular misalignment between the fibre ends, in particular for larger number of sections, the sidelobe suppression and passband ripples obtained experimentally is less ideal than the simulation results. The length ratios mismatch however can be compensated through temperature over a small, isolated section of the Hi-Bi fibre. Optionally, using general athermal packaging, the rest of the Hi-Bi fibre sections can easily be isolated from large surrounding temperature variations. Precise angular alignment of the fibre ends, on the other hand, can be achieved through a fusion splicer for polarization maintaining fibres.

The FSR of the filter depends on the time delay defined by (1). By using fibre of higher birefringence, the length of the structure can be effectively shortened. On the other hand, larger tap intervals can be obtained by increasing the lengths of Hi-Bi fibres, thus achieving FSR within the microwave regime. In fact, using Hi-Bi fibres of lengths ~64.978m and ~32.489m, a flattop bandpass microwave response as shown in Fig. 6 is realized, suitable for use in fibre-radio systems. The measured filter response through the Lightwave Component Analyzer (LCA) shows flattened passband width of ~4GHz with FSR ~18GHz.

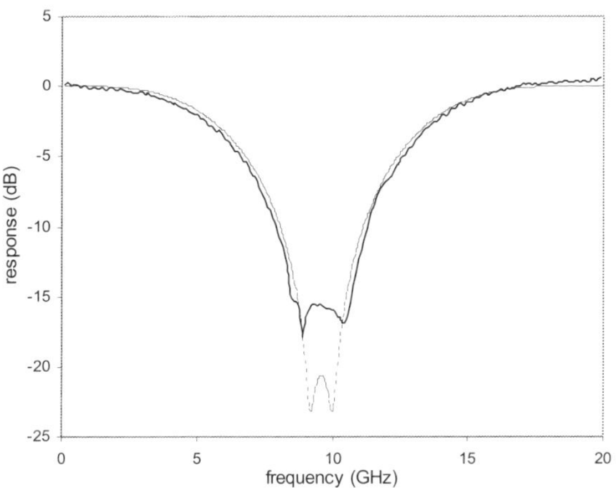

Fig.6 Measured and simulated (dotted) microwave response from the cascaded Hi-Bi fibre transversal filter scheme. The lengths of the Hi-Bi fibre sections used are ~64.978m and ~32.489m with θ_1~14^0; θ_2~58^0.

Since all the optical taps travel within a single line of Hi-Bi fibre, environmental perturbations imposed on all signal taps within the fibre are identical. The large birefringence within the Hi-Bi fibre ensures negligible coupling between the two polarization modes, thus removing any interferometric intensity noise arising from surrounding disturbances. The filter hence exhibits stable long-term operation without stringent measures required to isolate it from its surroundings.

Excess loss within the filter will be largely attributed to splice loss between the fibre section ends. Since the optical taps are the key to the operation of this transversal filter scheme, a variety of output responses and their complements can be flexibly set by changing the orientations between the fibre ends and lengths of the cascaded sections.

3. CONCLUSION

An optical multichannel flattop transversal filter using cascaded Hi-Bi fibres structure has been proposed and demonstrated. Bipolar optical signal taps are realized through the single-line cascaded structure and can be flexibly set by changing the orientation between the fibre ends and the lengths of the cascaded sections. The filter delivers two complementary outputs simultaneously and optical responses with flattened passbands, low ripples and large side mode suppressions have been achieved. Based on the same concept, a flattop bandpass microwave transversal filter not limited by the coherence of the source is further demonstrated. Due to common mode suppression to environmental perturbation, the filter exhibits stable long-term operation without stringent measures required to suppress interferometric intensity noise arising from surrounding disturbances. Results illustrate that the device offers a simple and effective means for signal processing in DWDM applications as well as fibre-radio systems.

REFERENCES

[1] Joseph Ip, "Component technologies for DWDM demultiplexers", OFC, TuH3-1, 1999.

[2] M. Kohtoku, et al., 200-GHz FSR periodic multi/demultiplexer with flattened transmission and rejection band by using a Mach-Zehnder interferometer with a ring resonator, IEEE Photonics Technology Letters, vol. 12, pp. 1174-1176, 2000.

[3] C. Huang, et al, “Ultra-low loss, temperature-insensitive 16-channel 100GHz dense wavelength division multiplexers based on cascaded all-fibre unbalanced Mach-Zehnder structure”, OFC/IOOC, TuH2-1, 1999.

[4] W. Zhang, et al, Optical Fibre Delay Line Filter Free of the Limitation Imposed by Optical Coherence, Electronics Letters, vol. 35, pp. 2133-2134, 1999.

Power Stabilization of Erbium-Doped Fibre Laser by External Injection

N. S. Mohd. Shah, T. C. Teyo, and H. Ahmad
Photonics Laboratory, Physics Department,
University of Malaya, 50603 Kuala Lumpur,
Malaysia.
Tel: +603-79676770
Fax: +603-79674146
E-mail: nshahida@hotmail.comAddress

Abstract: Injection locking technique has been commonly studied and can be used to achieve high power stable laser. This work is done to investigate such a phenomenon in an erbium-doped fiber laser (EDFL). The injection locking phenomenon occurred at high injected power. A stable output power has been achieved under locking condition.

Key Words: Erbium-doped fibre laser; Injection locking; Power stabilization

1. INTRODUCTION

Power and frequency stability is necessary in laser application, such as optical communication, medicine and remote sensing [1-2]. Stabilization of the laser power through frequency locking can be achieved by injecting an external signal into the slave laser. This technique is called injection locking. Laser frequency stabilization was first demonstrated by Stover and Steier in 1966 [3]. They directly injected the beam of one He-Ne laser into the resonator of a second He-Ne laser. Since that, injection locking has been studied in many kinds of laser systems. Injection locked EDFL was first presented by Jones and Urquhart [4], where the EDFL was in a linear configuration. In this study, power stabilization by using injection-locking technique was used. Throughout this paper, the EDFL cavity is called slave laser, while the external injected laser from a tunable laser source (TLS) is called master laser.

2. EXPERIMENT

A schematic diagram of the EDFL subject to external injection is illustrated in Fig. 1. The slave laser system consists of two 980/1550 nm wavelength division multiplexers (WDMs), three couplers: C1 and C2, with the output coupling ratio of 95%, and C3 for monitoring the spectrum in anti-clockwise direction from 1% port, and two optical isolators at the input and output ends to avoid back-reflection. Another isolator, ISO 1, is placed in the cavity to ensure a unidirectional oscillation of the laser. A 15 m long erbium-doped fibre (EDF) with a cut-off wavelength of 950 nm, a refractive index of 1.473 and an Er^{3+} concentration of +440 ppm is used as an active medium. Pump power is provided by a laser diode from the 980 nm port of WDM I. A master laser is provided from a tuneable laser source (TLS) and is coupled to the system from the 95% arm of the coupler C1. The master laser wavelength is set to be at the centre of the lasing wavelength of the slave laser, which in this case is 1557.7 nm. For all the results demonstrated here, the power of slave laser is fixed and the power of master laser is varied.

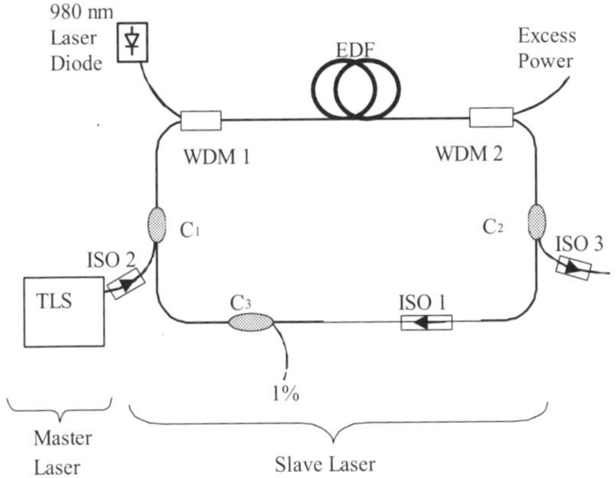

Fig.1 Schematic diagram of EDFL subject to external injection. (EDF: erbium-doped fibre; WDM: wavelength division multiplexer; C: coupler; ISO: isolator; TLS: tuneable laser source).

3. RESULTS AND DISCUSSION

Fig. 2 shows the output spectrum when the pump power of the slave laser is fixed at twice above the oscillation threshold, which is 58.7 mW. Fig. 2 (a) shows the power spectrum of a free-running oscillator. The dynamic behaviour in the laser makes the lasing output power fluctuates severely. Fig. 2 (b) shows that all the side modes have been locked to the high power master laser, i.e. –27.6 dBm. This is the results of saturation effects and the competition between the master and the slave laser. The master laser quenches the gain of the slave laser. Thus, the slave laser is entirely suppressed [5]. The slave laser system is then dominated by the injected laser, signifying the condition of injection locking.

The result of output laser power versus injection power is depicted in Fig. 3. The output laser powers are measured five times for each injected input power. Only the mean of the measurements are shown in the graph and the error bars show the deviation of the results. It can be seen clearly that the output powers fluctuate severely at low injected powers: the values of output powers diverse even at the same condition. The powers become stable at the injected power of about -25 dBm. Above this injected power, the slave laser is completely locked to the external signal, and it oscillates at the wavelength of the injected signal.

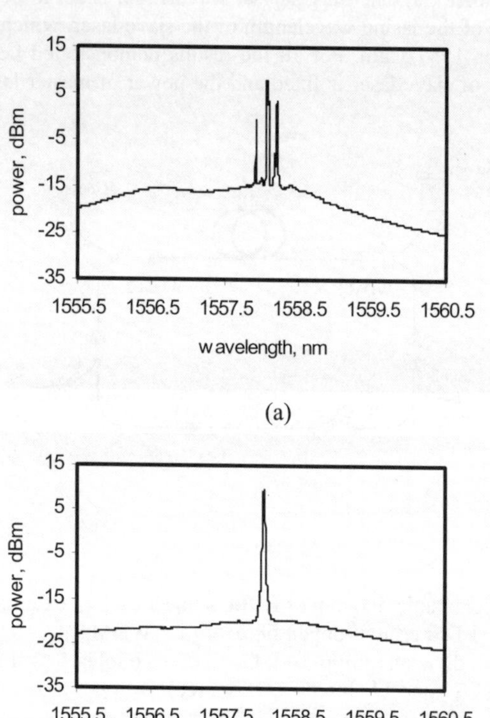

Fig. 2 Output spectrum of slave EDFL pumped at twice above oscillation threshold.
(a). Without external injection.
(b). With injection power of –27.6 dBm.

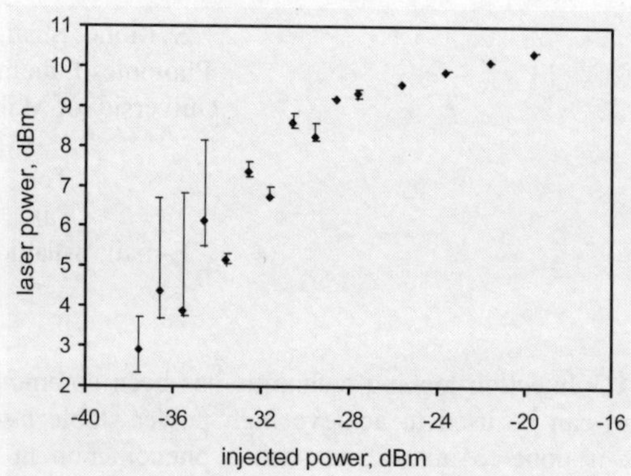

Fig. 3 Laser peak power versus injection power. The mean of five measurements is shown with their deviation. The pump power is 58.7 mW.

4. CONCLUSION

We have demonstrated the phenomenon of injection locking in an EDFL system subjected to external injection. Frequency locking was observed and stabilization of the output power was established. It was found that the stability of the EDFL power was dependent on the level of the external injection.

REFERENCES

[1] I. D. Avramov, Microwave oscillators stabilized with surface transverse wave resonant devices, Proc. of the 46th IEEE Frequency Control Symposium, pp. 391-408, 1992.

[2] A. L. Valerio, S. Alessadro, S. Marc, and D. Silvano, Dynamic behavior and Locking of a semiconductor laser subjected to external injection," Journal of Quantum Electronics, 34, pp. 2350-2357, 1998.

[3] H. L. Stover, and W. H. Steier, Locking of laser oscillator by light injection, Applied Physics Letters, vol. 8, pp. 91-93, 1996.

[4] J. D. C. Jones, and P. Urquhart, An injection-locked erbium-doped fiber laser," Optics Communications, vol .76, pp. 42-46, 1990.

[5] C. J. Buczek, and R. J. Freiberg, Hybrid Injection Locking of Higher Power CO2 Lasers, Journal of Quantum Electronics vol 8, pp. 641-650, 1972.

Multiwavelength L-Band Brillouin / Erbium Fiber Laser

Nizam Tamchek, Sulaiman Wadi Harun and Harith Ahmad
Physic Department, University of Malaya, 50603 Kuala Lumpur, Malaysia
E-mail address: *nizamtm@lycos.com*

Abstract: Twin cavity multiwavelength Brillouin/erbium fiber laser (BEFL) demonstrates multiwavelength output in long wavelength band (L-band) region. The system consists of two identical EDFLs that share a length of single-mode fiber to produce interdependently alternate bi-directional multiple Stokes-shift frequencies. The twin cavity BEFL generates a 22GHz comb of 2 x 11 output channels at 1600nm that has potential application in future wavelength division multiplexing (WDM) system.

Key Words: Optical fibre devices; optical fibre lasers; L-band; Brillouin scattering; optical fibre amplifiers.

1. INTRODUCTION

The rapid development of dense wavelength division multiplexing (DWDM) technology in current optical communication system has prompted researchers to investigate multiwavelength laser sources operating in the conventional band (C-band) and long wavelength band (L-band) communication windows. Besides the deployment of laser diodes, fiber lasers are a promising choice of light source due to their various advantages. One of which is their ability to provide tuneability features. Multiwavelength erbium-doped fiber lasers (EDFLs) have been demonstrated by cooling the gain medium to prevent mode competition between adjacent laser lines [1]. Brillouin/erbium fiber lasers (BEFLs) have recently been demonstrated as a densely (10GHz) spaced and robust multiwavelengths laser source, employing both the nonlinear Brillouin gain in single mode fiber (SMF) and linear gain in the erbium-doped fiber (EDF). Several techniques were introduced to generate multiwavelength BEFL in the C-band region, such as using an intra-cavity DFB laser in the ring cavity [2] and using injection-locked seeding in the ring resonator [3].

In this paper, we demonstrate a distinctive configuration for a multiwavelength twin cavity BEFL that could generate 2 x 11 channels with 22-GHz wavelength spacing at the 1600nm (L-band) region. Alternatively, the system could output 22-lines with 11GHz spacing. The all fiber BEFL system consists of two ring cavities that share the same SMF and both free running EDFL (with no Brillouin pump) peak mode is made to be superimposed. The SMF that served as a common component to the two resonant cavities would generate interdependently alternate bidirectional multiple Stokes-shift frequencies. The Stokes signal generated in a ring at one end of the SMF, which is initiated from an external Brillouin pump (BP), will become the Brillouin pump to the other ring at the other end of the same SMF to produce the next Stokes signal. This Stokes signal generates another and the operation iterates, consequently producing multiple Stokes to a limit within the superimposed spectral gain bandwidth of the EDFLs in the system.

2. EXPERIMENTAL PROCEDURES

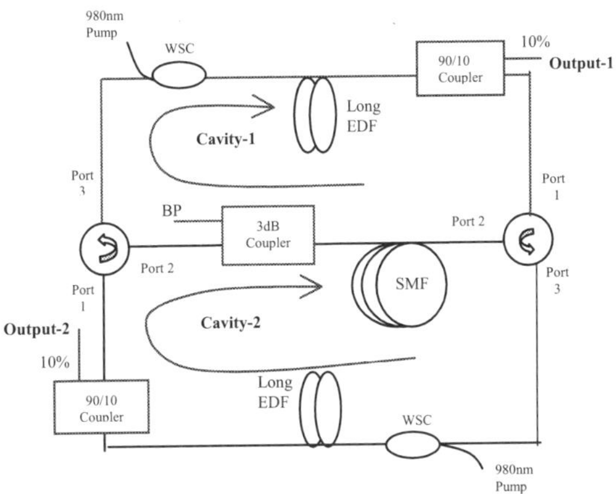

Fig. 1. Schematic of dual cavity L-band BEFL

The scheme of the twin cavity L-band BEFL is depicted in Fig. 1. Cavity-1 passes through the SMF in a clockwise direction while cavity-2 in anticlockwise. The system can be viewed as two separate resonant cavities that operate in tandem as a BEFL system. Both rings consist of similar components of a 50m long EDF with 400ppm Er^{3+} concentration, wavelength selective coupler (WSC), two circulators, a 90/10 coupler, a 3dB coupler and a 490m long communication grade single mode fiber (SMF). The two cavities share the same SMF, 3dB coupler and two circulators. The free running EDFL (with no Brillouin pump) of the two rings is shown in Fig. 2, where the peak gains are close to one another. This is due to the both rings are composed of two identical cavities and experience almost similar cavity loss.

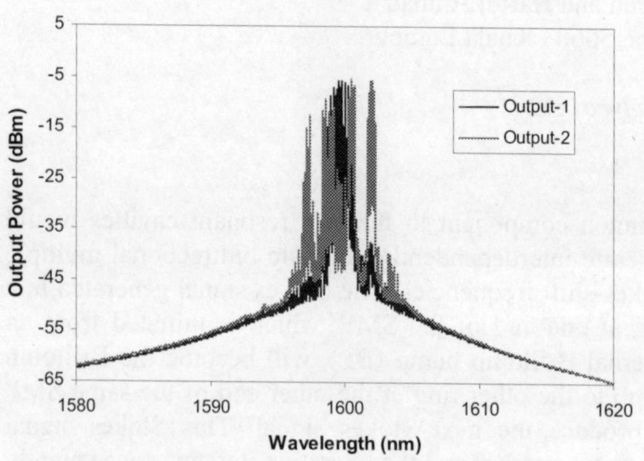

Fig. 2: Free running EDFL spectrum of the two cavities showing peak gain is closed to one another.

Brillouin pump signal from a tunable laser source (TLS) is launched into the SMF at a wavelength near the EDFL peak gain of both cavities. The Stokes scattered signal from the BP, experiences gain in the EDF which itself becomes a BEFL provided the gain equals to the cavity loss. This Stokes signal is counter direction to the BP and when strong enough induces a Stokes scattering from itself, in the same direction as the external BP. The process of gain in EDF, in cavity-2 in this case, is repeated and subsequent scatterings are output at both output ports. The two laser outputs are monitored by an optical spectrum analyzer (OSA) with 0.015nm spectral resolution.

3. RESULT and DISCUSSION

With sufficient BP power, cascaded stimulated Brillouin scattering (SBS) is initiated by increasing the 980nm pump power. Fig. 3 shows the output spectra of the odd order Stokes wave at output-1 in cavity-1, and the Brillouin pump wave and the even order Stokes wave at output-2 in cavity-2. The Brillouin pump was 1599.020 nm and the 1st Stokes and 2nd Stokes were 1599.108 nm and 1599.198 nm, respectively. The whole spectrum of output-2 is similar to output-1 but it is downshifted by 11GHz due to the Brillouin Stokes-shift. The line spacing is approximately 0.18nm in the wavelength domain and 22GHz in the frequency domain.

The combination of the output power from both couplers produces 22 lines with 11GHz line spacing at L-band region. The 3dB spectral width of each line is about 0.02nm, limited by the OSA bandwidth resolution of 0.015nm.

The proposed BEFL cavities permit the simultaneous presence of two oppositely propagating pump and Stokes components in the loops. This will initiate the four wave mixing (FWM) process, which produces both red-shifted Stokes lines and blue-shifted anti-Stokes lines as denoted by 1 ~ 15 and -1 ~ -6, respectively, in Fig. 3. These anti-Stokes signals are obvious when the power of the 980nm pumps and Brillouin pump are increased. No spatial hole burning is expected to occur in the system because the travelling wave signal propagates in a single direction in separate cavity at different wavelengths. An attractive feature of the twin cavity system is the flatness of the peak power levels covering more lines compared to a single cavity multiwavelength BEFL [4].

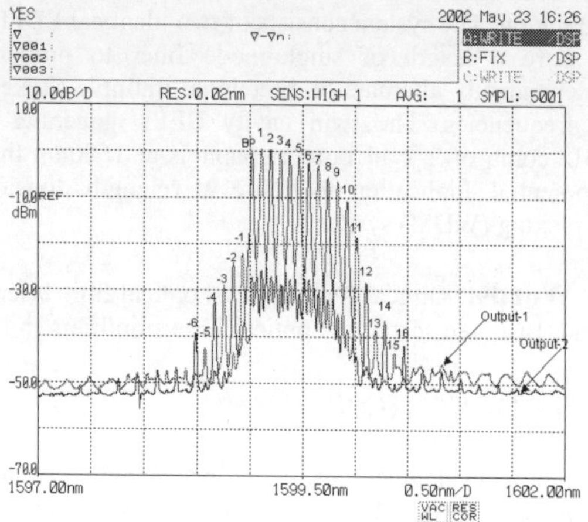

Fig. 3: The BEFL output spectra

The power characteristics of the first Stokes and second Stokes lines is shown in Fig. 4 as a function of the launched 980nm EDF pump power. The BP power is fixed at 4.8mW (before 3dB coupler) and both 980nm-pump powers were varied from 0 to 92mW. It can be seen that the BEFL power increases gradually as the pump power is increased. Also, it has been confirmed that operation of the BEFL shows a threshold behavior. The 980nm-pump power threshold is 34.7mW when a Brillouin pump power of 4.8mW is used in the BEFL cavity. The power characteristics of the lines are non linear, since the nth order Stokes is generated with depletion of (n-1)th order Stokes power. As soon as the second-order Stokes starts to appear, the first-order Stokes shows a distinct reduction in its output power.

4. CONCLUSION

Multiwavelength generation in 1600nm region was realized using the twin cavity L-band BEFL. A laser comb up to 2 x 11 lines with 22 GHz line spacing has been obtained by Brillouin pump of 4.8mW and 980nm pump of 92mW. The laser operates at multiple Stokes lines mainly caused by cascaded SBS process as well as at anti Stokes lines caused by four waves mixing. Some work has been investigated to increase the number of lasing by widening the superimposed spectrum free running EDFL from the two cavities. Further work on equalizing gain for individual channels will enhance the laser potential in future DWDM communication system.

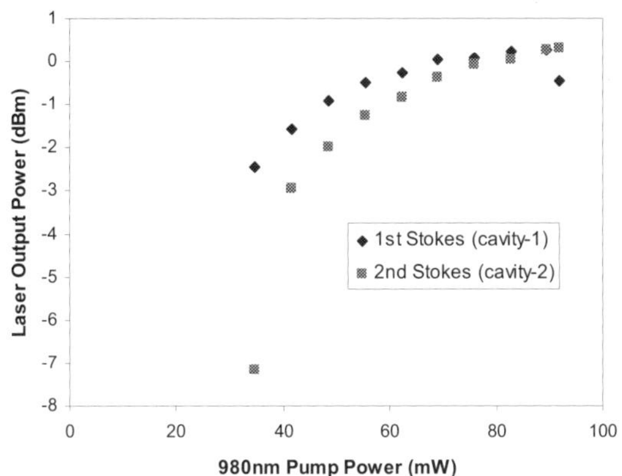

Fig. 4: Power characteristics as a function of the launched 980nm pump power.

REFERENCES

[1] N. Park and P. F. Wyscoki: IEEE Photon. Technol. Lett. **8** (1996) 1459.

[2] S. Yamashita and G. J. Cowle: In Proc. Conf. on Lasers and Electro-Optics **paper CthO46** (1998) 417.

[3] G. J. Cowle, W. H. Loh, R. I. Laming and D. Y. Stepanov: Opt. Fiber Conf. OFC'97 Technical Digest **paper TuH7** (1997) 34.

[4] S. W. Harun, N. Tamchek, M. K. Abd-Rahman, P. Poopalan and H. Ahmad: to be published in IEICE Trans. on Commun., July 2002.

Session 13E

Network Management and Optimization

MIB Modeling for the Control & Management of OADM/OXC Optical Devices

Gee-Swee Poo and Tee-Hiang Cheng
School of EEE, Nanyang Technological University, Nanyang Avenue, Singapore 639798.
{egspoo, ethcheng}@ntu.edu.sg

Abstract: The Management Information Base (MIB) provides network-level configuration information for the management and control. Numerous MIBs have been defined by IETF for the IP over WDM development. However, there is no MIB specified for the management and control of optical devices such as Optical Add-drop Multiplexer (OADM) and Optical CrossConnect (OXC). This paper intends to fill in the gap by presenting the development of a MIB for OADM/OXC optical devices. The logical basis of the MIB is provided by the development of a network-level information model using TINA-C NRIM specifications. The information model is made to portray managed object classes relevant to network resource management, leading to the construction of the MIB for OADM/OXC using the SNMP SMIv2 standard. The MIBs have been used to develop a network management system prototype for a WDM Ring network with appropriate GUIs using Java and the AdventNet SNMP APIs. The usefulness of the MIBs is demonstrated by developing the capabilities for controlling the devices and managing the faults.

Key Words: MIB, Management, Optical devices

1. INTRODUCTION

The Optical Add-Drop Multiplexer (OADM) and Optical Cross Connect (OXC) constitute the two key building blocks of a WDM network [1]. Typically, OADMs are organized into ring configurations and OXCs are structured into meshes. A combination of the two produces a mesh of rings – a reconfigurable WDM network [2]. At the heart of the WDM network is the network control and management software that dynamically reconfigures the network in response to changing traffic demands and patterns.

The management system can be structured along the multi-tiered TMN (Telecommunications Management Network) logical layered architecture [3]. In the architecture, the Element Management Layer (EML) is used to manage different subnetworks such as WDM rings, and Network Management Layer (NML) interacting with EML is used to provide end-to-end management. Internally, a management information model is structured along the TMN layers using TINA-C (Telecommunication Information Networking Architecture Consortium) [4,5]. TINA-C provides a computing and management architecture for network management development. Its modeling principles are based on the NRIM (Network Resource Information Model) [5]. The model contains specifications of the information elements, such as trail, links, and network termination points.

The information to be managed for each network element is represented in the form of a MIB (Management Information Base). The MIB provides network-level configuration information for the management and control. Numerous MIBs have been defined by IETF for the IP over WDM development [6]. However, there is no MIB specified for the management and control of optical devices. This paper intends to fill in the gap by presenting the development of a MIB for OADM/OXC optical devices. The logical basis of the MIB is provided by the development of a network-level information model using TINA-C NRIM specifications. The information model is made to portray managed object classes relevant to network resource management.

We adopt the Simple Network Management Protocol (SNMP) [7] platform for implementation. We construct the MIB for OADM/OXC using the SNMP SMIv2 standard. Then, using the MIB, we develop a network management system prototype for a WDM Ring network with appropriate GUIs using Java and the AdventNet SNMP APIs. We have also developed the capabilities for controlling and managing the OADM/OXC devices. The usefulness of the MIBs is demonstrated by these capabilities.

This paper is structured in the following manner. Section 2 outlines the WDM network management architecture. Section 3 provides the WDM network-level information model. Section 4 describes the WDM Manager MIB. Section 5 describes the OADM Agent MIB. Section 6 describes the OADM Device Manager. Section 7 describes the fault management. Section 8 outlines the implementation and section 9 gives the conclusions.

2. WDM NETWORK MANAGEMENT ARCHITECTURE

We consider a WDM Manager (Network Management System (NMS)) to manage a WDM ring network that consists of OADMs as nodes as shown in Figure 1. The number of OADMs is scalable. For illustration, we consider only four OADMs operating in a unidirectional ring network. The optical fiber carries four channels. Each OADM is connected to its subnetwork via an access node (eg. switch, router, etc). Each OADM is also connected to a PC that acts as its controller agent. All the PC agents are connected to the WDM Manager through an Ethernet network. In this case, the data plane (WDM ring network) is separated from the control plane (Ethernet network).

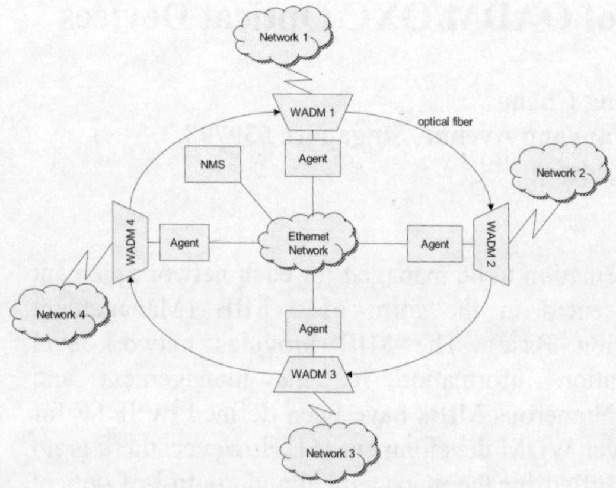

Figure 1: A 4-Node WDM Ring Network (WADM=OADM)

The WDM Manager resides at the Network Management Layer (NML) and provides two main management functions. Firstly, it performs resource management of the network by keeping track of the elements (i.e. fiber, wavelength, OADM, etc) in the network. Secondly, it performs fault and log management for the network. The WDM Manager consists of three main components, namely, Graphical User Interface (GUI), the Management Kernel and the SNMP Engine. These components provide network management functions and SNMP services for network administrators.

The OADM Agent also consists of three main components, namely, Graphical User Interface (GUI), the Management Kernel and the SNMP Engine. These components provide network management functions and SNMP services in response to the commands sent by the network administrator or any operations between itself and the Device Manager.

3. WDM NETWORK-LEVEL INFORMATION MODEL

We derive the WDM network-level information model according to the TINA NRIM specifications [5]. The model represents a global view of the transport network and considers how individual network elements are interconnected and configured to provide end-to-end connectivity. They contain information objects and their relationships.

3.1 A 4-Node WDM Ring Network

We apply the above technique to the WDM ring network with 4 OADMs. The information model is depicted in Figure 2. The WDM layer network consists of a ring subnetwork. The ring subnetwork consists of OADM 1, 2, 3 and 4. Every OADM is connected to its neighboring OADMs by a link each. Each OADM has 3 linkTPs denoting 3 connecting fibers - 4 bidirectional incoming access links from the access stations (shown as one), one incoming from and one outgoing to neighbouring OADMs. Each linkTP of a fiber link (i.e. link A, B, C, D) contains 4 nwCTPs representing 4 wavelengths supported on the fiber. A single nwCTP denotes an endpoint of a particular wavelength connection. It can either be a laser transmitter or receiver depending on its linkTP.

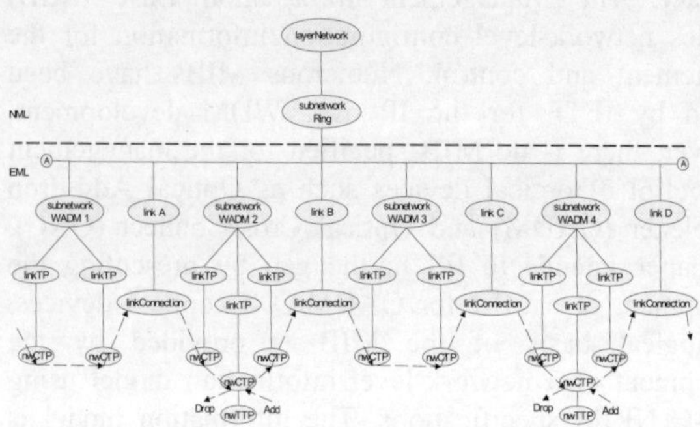

Figure 2: Network-level Information model for a 4-Node WDM Ring Network. A connection between OADM2 and OADM4 is illustrated.

Figure 2 also shows a connection between OADM 2 and 4. The direction of the links is from left to right. OADM 1 and 2 are interconnected by link A, 2 and 3 by link B, 3 and 4 by link C, and 4 and 1 by link D. The Figure only shows the nwCTPs that are involved in the connection. The nwCTPs that are adapted to be nwTTPs are the endpoints of the WDM network. Their linkTPs are connected to one access station each to another network layer. The dotted line with arrow traces the trail from the source (OADM 2) to the destination (OADM 4) and vice versa. Note that NML refers to the WDM manager, while EML refers to the agents.

4. WDM MANAGER MIB

The WDM Manager MIB is designed to represent objects in a generic and abstract manner. Part of the MIB structure reflects the OADM Agent MIB. However, other parts are for internal use. Figure 3 depicts the different classes maintained by the Manager. It shows only the main attributes.

The wdmAgents group maintains the agents that have registered with the manager. Each wdmAgent created from this class will have 5 attributes as shown. The wdmPhysicalTopology maintains the network links and their physical connection to agents. The wdmVirtualToplogy maintains the network connections. These connections are circuit-switched and use lightpaths.

Figure 3: WDM Manager MIB

5. OADM AGENT MIB

The design of OADM Agent MIB requires the detailed mapping with the OADM device architecture so as to discover the various hardware parts and parameters for the control and monitor.

5.1 OADM Device

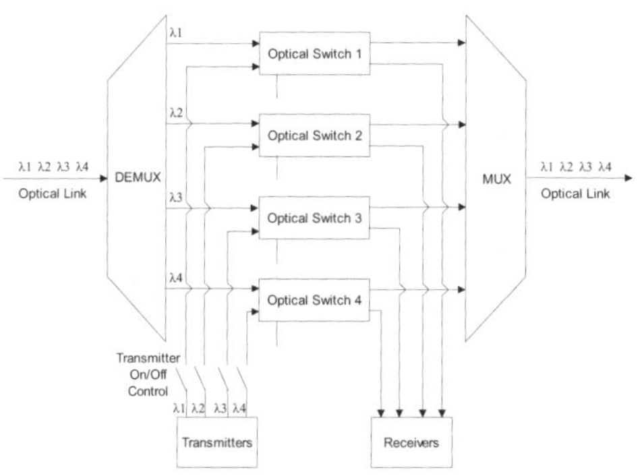

Figure 4: The Architecture of OADM device

Figure 4 shows the architecture of OADM device with the capacity of adding and dropping 4 wavelengths. An incoming fiber link is demultiplexed into 4 wavelengths. Each wavelength goes through an optical switch that can be set via a status control line to either cross or bar state. At bar state, the wavelength will pass through the switch to the multiplexer. At cross state, the wavelength will drop to the receivers while a wavelength from the transmitter will be directed to the multiplexer.

5.2 OADM Agent MIB

Figure 5 shows the OADM Agent MIB groups and tables. There are four groups: wdmObjects, wdmNode, wdmLink, wdmWavelength, and wdmTraps. The numbers indicate the Object Identifiers (OIDs) for the groups and tables. The number gap between wdmWavelength and wdmTraps is for future expansion. The wdmObjects consist of wdmProtectionType that indicates the adopted protection type. A wdmNode refers to the OADM and its property. It consists of the switch table, Laser table and receiver table for the node. The wdmLinkTable gives the property of each link. The wdmWavelengthTableGroup gives the characteristics of each wavelength.

Figure 5: OADM Agent MIB

The information objects for fault management consist of AlarmRecord, FaultManageable and FaultManagementDomain. The FaultManageable objects defined in the TINA specifications will send a communication alarm when a failure caused by loss of signals, degraded signals or communication subsystem failure occurs. The FaultManageable objects in the OADM device are objects that are prone to failure and will send an event notification when a fault occurs. With all the information objects defined for fault management, a MIB group under the name of wdmTraps is formed in the OADM Agent.

6. OADM DEVICE MANAGER

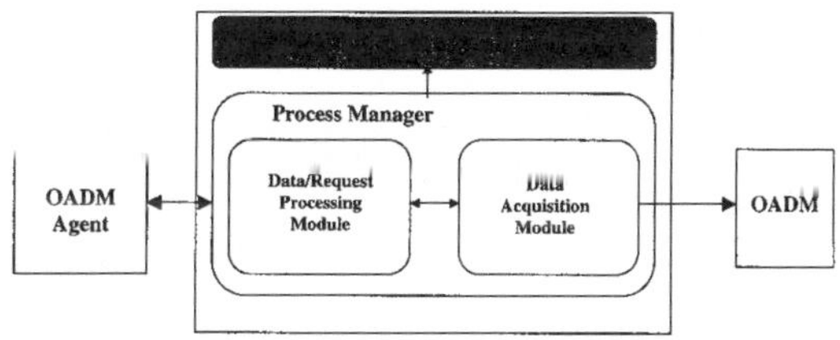

Figure 6 Overview of Device Manager

The OADM hardware device is connected to the controlling PC via a DAQ Card and is managed by the Device Manager (Figure 6). The Device Manager and Agent are software modules co-located at the PC. The Device Manager manages the OADM device by obtaining the signals read by the DAQ Card and sending signals to the DAQ Card. Thus, it forms the bridge between the low-layer physical optical equipment (OADM) and the high-layer network management routine (OADM Agent). The data acquisition module constantly polls the OADM to obtain signal values from the OADM via DAQ card. The Data/Request processing module processes data sent from the Data Acquisition module and the requests from the OADM Agent.

7. FAULT MANAGEMENT

Fault management consists of fault detection, event notification, fault localization and fault recovery. Fault management is modelled based on the TINA-C specification. FaultManageableObjects are the components that are prone to failure and will send an event notification whenever there is a change of state in the components. Hence, the FaultManageableObjects in the WDM Network are namely the optical switches, the transmitter, the receiver, the node temperature and the links. The FaultManageableDomain will be the WDM Manager since it manages all faults reported by OADM Agents and it contains a fault manager that holds all fault management functions. The log manager maintains a fault list that serves as the AlarmRecord.

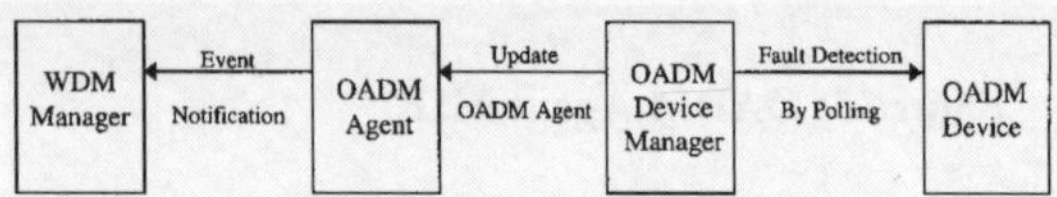

Figure 7 Process of Fault Detection and Event Notification

Figure 7 shows that the Device Manager performs fault detection by constantly polling the OADM Device to detect any change in the OADM Device. The Device Manager then updates the OADM Agent the change detected in the OADM device. Since not all changes are critical, there is a need to set criterion to define the criticality of the change so that the Agent will know when to send an event notification to the WDM Manager. When the WDM Manager receives the notification, the Fault Manager will perform the necessary actions.

8. IMPLEMENTATION

The WDM manager and OADM agent are implemented using Java and the AdventNet SNMP API [8]. The programs are coded on Windows98 platform and tested on Windows98, WindowsNT and RedHat Linux. The Device manager serves as a device driver to communicate with the OADM device via DAQ Card and RS232 port.

We have implemented appropriate Graphical User Interfaces (GUIs) for WDM Manager, OADM Agent and the Device Manager, the description of which is omitted here. We have tested the MIBs, the fault management and the log function with the OADM hardware and found that we are able to control the device, trigger the fault events and log the event records. However, not all the hardware signals (eg. temperature signal) are implemented due to the limitation of monitoring equipment.

9. CONCLUSIONS

This paper presents the development of the MIB for OADM/OXC optical devices. A network-level information model has been constructed using TINA-C NRIM specifications to provide a logical foundation of the MIB. The information model is made to portray managed object classes relevant to network resource management, leading to the construction of the MIB for OADM/OXC using the SNMP SMIv2 standard. Using the MIBs, we have developed a network management system prototype for a WDM Ring network with appropriate GUIs using Java and the AdventNet SNMP APIs. We have also developed the capabilities for controlling the configuration of the devices and managing the faults of the devices. The usefulness of the MIBs is demonstrated by these capabilities.

The WDM Manager is currently limited to one ring network. To support multiple ring networks, one needs to provide interfaces to integrate with higher-level managers to achieve scalability. Such interfaces should follow the TINA-C specifications in [9]. Manager to manager communications can be performed using the SNMPv2 Inform command.

REFERENCES

[1] C. A. Brackett, Dense wavelength division multiplexing networks: principles and applications, IEEE Journal on Selected Areas in Communication, vol. 8, no. 6, August, 1990, pp. 948-964.

[2] L. Wuttisittikulkij and M. J. O'Mahony, "Design of a WDM network using a multiple ring approach," IEEE Global Communications Conference, vol. 1, pp. 551-555, 1997

[3] ITU-T, "Principles for a Telecommunications Management Network", M3010, 1993.

[4] F. Steegmans, "Network resource architecture, Version 3.0". TINA-C, February 1997.

[5] N. Natarajan, H. Flinck, R.M. Rosli, "Network Resource Information Model Specification, Version 3.0". TINA-C, December, 1997.

[6] Various IETF drafts for example: draft-ietf-ccamp-lmp-mib-01-txt; draft-ietf-mpls-ldp-mib-08-txt; draft-ietf-mpls-te-mib-08-txt; draft-ietf-mpls-lsr-mib-08-txt; draft-ietf-mpls-ftn-mib-08-txt; draft-ietf-mpls-bundle-mib-01-txt.

[7] IETF RFC 1901-1908, Simple Network Management Protocol (SNMPv2), SNMPv2 Working Group, 1996.

[8] AdventNet, see http://www.adventnet.com.

[9] C. Abarca, P. Farley, J. C. Garcia, T. Hamada, P. F. Hansen, P. Hellemans, C. A. Licciardi, K. Nakashiro, M. Yates, "Service components specifications, computational model and dynamics", TINA-C, January 1998.

Integrated Control Plane for IP Enabled Optical Network

Li Zhengbin, Wu Deming and Xu Anshi
National Laboratory on Local Fiber-Optic Communication Networks & Advanced Optical Communication Systems, Department of Electronics, Peking University, Beijing, 100871 P. R. of China, Lizhb@ele.pku.edu.cn

Abstract: As control plane and signaling seems to be a critical part in next generation networks, this paper proposes an integrated control plane (ICP) for IP enabled optical networks. Signaling procedure for ensuring dynamic end-to-end lightpath setup has also been presented. Path provisioning comprises a string of operations like service and neighbor discovery, route computation, signaling requests, and path setup.

Key Words: Control Plane, Signaling, Route, Optical Network, IP/GMPLS.

1. INTRODUCTION

The developments in WDM technology have dramatically increased the traffic capacities to satisfy increasing demand for bandwidth. Common control plane for data and optical layers has been introduced to provide optical and IP layer more intelligence, survivable, flexible, controllable and open for traffic engineering. On the other hand, clients connect to the optical networks requires an appropriate interface at each point. As generalized MPLS (GMPLS) creates frontiers for traffic engineering and binds well to WDM by using ports in optical layer as labels. With the flexibility of IP itself, IP/GMPLS may achieve the minimum possible complexity for the networks. Existing packet oriented control planes as IP/MPLS, were designed to encompass devices recognizing packet and or cell boundaries, where the header is processed to create the label switched path (LSP) and make forwarding decision. However, in optical transport networks (OTN), devices and their constituent protocols make forwarding decisions based on GMPLS, which forwards data not directly based on the information carried in packet or cell. By providing protocol to support the OTN devices, the MPLS and GMPLS are combined to provide a signaling interface between the optical layer and upper layers such as IP, ATM, and SONET/SDH. This control plane is referred to as integrated control plane (ICP), which is an open structure for control and processing of optical transport elements and capacity for not only IP + optical but also multi-services networks over optical.

2. IP ENABLED OPTICAL NETWORK ARCHITECTURE

The interconnected OXCs make the network model in figure 1, the clients may be the devices providing services directly to users, and nodes in the network have IP addresses to be used in control plane. The control plane is realized through user-network Interface (UNI) and network-network interface (NNI). By IP-based protocols for optical networks, optical control plane can be an IP-centered based on the practical view that signaling and routing mechanisms developed for IP traffic engineering applications could be re-used in optical networks to support a seamless connection. As most nodes consist of IP functionality and OXCs, the IP module is responsible for all non-local management functions.

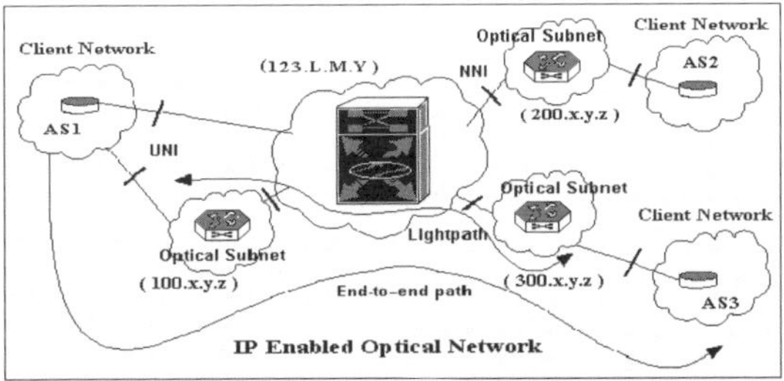

Fig.1. Generic model of IP Enabled Optical networks

By bi-directional control channels in-band or out-of-band, UNI & NNI interfaces are realized between common controllers. There are three kind of interactions between the optical network and client networks, peer model where the IP/GMPLS layers act as peers of the optical transport network, overlay model separates the IP/GMPLS routing, topology distribution, and signaling protocols from that at the optical layer, and augmented model actually separates routing instances in the clients and optical domains, while information from one routing instance is passed through the other routing instance.

2.1 LABEL HIERARCHY FOR INTEGRATED CONTROL PLANE

The MPLS architecture has been defined to support the forwarding of data based on a label in Label Switching Routers (LSRs) capable of recognizing packet or cell boundaries and processing headers. However, there are LSRs devices forwarding data based labels in GMPLS, which can be subdivided into the following classes: Packet-Switch Capable (PSC) LSR, Time-Division Multiplex Capable (TDM) LSR, Lambda Switch Capable (LSC) LSR or Waveband Switch Capable (WSC) LSR and Fiber-Switch Capable (FSC) LSR. Then the labels corresponded are combined into an integrated control plane described in Fig.2, a forwarding hierarchy is constructed by the concept of nested LSPs. LSP that starts and ends on a LSC interface, which in turn can be nested (together with other LSPs) into an LSP that starts and ends on a FSC interface.

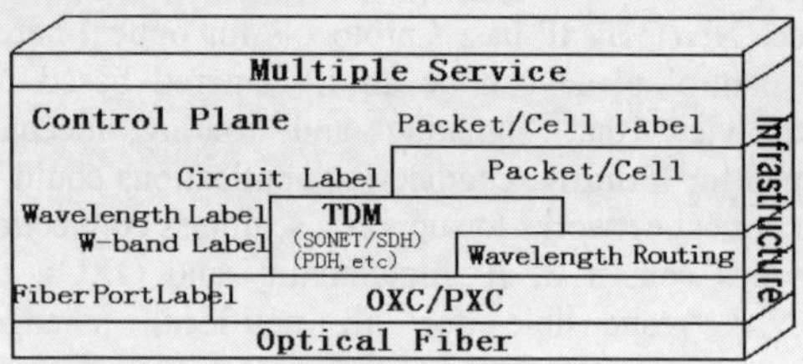

Fig.2. The hierarchy of labels in integrated control plane

2.2 INTEGRATED CONTROL PLANE REQUIREMENTS

The ICP may includes addressing scheme for devices; a signaling network for communication between entities requesting service and entities that provision those services; a signaling protocol for the set-up, maintenance and tear-down of trails such as LSP; a routing process to handle topology resource use and route calculation etc..

Addressing: The ICP can be expected to receive requests to set up traffic trails and support large amount of connection requests at very short intervals for IP packet flows and non-IP packet services by a scalable naming and addressing scheme, capable

1	2	3	4	5	6	7	8
General IP Head (V.4/V.6)							
(Node/Device Address)							
....							
Fiber/Physical Port ID/number							
Wavelength Start ID/number							
Wavelength End ID/number (option)							
TDM Time Slot Start ID/number (option)							
TDM Time Slot End ID/number (option)							

Fig.3. Message of a node resource required

of meeting expected demand for new names and addresses, such a network provides tools for capacity management, utilization, statistics on failed call attempts, plan and build facilities, providing ordering triggers, and dealing with long term localized congestion. The same type of addressing can be used in both client and optical layer control planes, which make the ICP protocol independent of hardware medium and decouple the addressing mechanism from the devices. So devices in the clouds of whole networks in Fig. 1 should be IP addressing enabled, especially the interfaces have an ID hierarchy in OTN, this kind of OTN can be referred to as IP enabled OTN. That means that the nodes addressing and interface ID hierarchy attached may be used for the establishment of all kinds of LSPs. By extension of IP protocols as OSPF and IS-IS and RSVP, node resources can be advertised to neighbors, used for LSP establishment of request and protection. As GMPLS supports the nested G.LSP, i.e. the G.LSP within G.LSP, the address of an interface determines which level of G.LSP can be established, as Fig. 3. The example is shown in Fig.1. There are three messages received by AS1 from (100.x.y.z), (123.L.M.Y) and (300.x.y.z) in Fig.1, which indicate the fiber port ID and wavelength ID are (6, 5), (9, 5) and (4, 5), then an LSP can be established through nodes with addresses (100.x.y.z), (123.L.M.Y) and (300.x.y.z) with out wavelength conversion if all the wavelengths are numbered in the same way.

Signaling: Signaling is the exchange of information involving each connection sets up, controls, and terminates. This bi-directional channel is essential for implementing the ICP. The control channel among provides connectivity between nodes. The choice of transmission media as control channel must be appropriate to optical networks and any other physical networks. However the signaling network does not have to have the same physical topology as the traffic network and indeed a separate diverse topology has both functional advantages and can be made more reliable. In a word, the ICP network must always be more reliable than the user plane traffic network. There are two ways used in ICP to carry the signaling, channel associated and common channel. Common channel signaling is usually carried out-band for consideration of transparent optical networks as well as

Fig.4 Summary of protocols functioning in a LSR

possible, while channel associated signaling is not.

The signaling protocol used for associating labels to individual paths including MPLS-based Resource reSerVation Protocol (RSVP) and Constraint Routed Label Distribution Protocol (CR-LDP), which are capable of providing quality of service (QoS), traffic engineering, and support for bi-directional paths, switching without wavelength conversion, establishing shared backup paths, and fault tolerance. The corresponding services model of ICP can bc described as figure 4.

Routing: Routing establishes an end-end connection which involves the establishment of path segments in each sub-network sequentially. Inside the optical domain, a path segment is established from the source LSR (OXC) to a border LSR (OXC) in the source sub-network. From this border LSR (OXC), signaling across the NNI is performed to establish a path segment to a border LSR (OXC) in the next sub-network. Provisioning continues this way until the destination LSR (OXC) is reached. During this, the procedure involves following steps: resource discovery, route calculation, UNI operation and NNI operation.

Resource Discovery: Routing relies on knowledge of network topology and resource availability. Each LSR (OXC) must determine the aliveness of each link, the bandwidth and other parameters of the link, and the identity of the remote end of the link. Each network node goes through neighbor discovery following which it creates an inventory of local resources and resource hierarchies, namely, channels, channel capacity, wavelengths, and links. The end result is that each LSR (OXC) creates a port state database. All of the required information is obtained via routing protocol updates, and is propagated throughout the network by extension OSPF/IS-IS.

Route Calculation: In ICP the IP and optical layer are assumed to run the same instance of an IP routing protocol, e.g., OSPF, which permits a LSR to compute an end-to-end path to another LSR across the network. In domain specific routing, the protocol allows the separation of routing within the optical and IP domains, with a standard routing protocol running between domains. The routing information exchange within the domain is performed using an interior gateway protocol with traffic engineering attribute. Then the route computation reduces to a mathematical problem as solving a problem of Routing and Wavelength Assignment (RWA) for the new connection from an IP router (over the UNI), a border node (over the NNI), or a management system. After computing the route, the actual path establishment commences.

UNI Operation: After performing service & neighbor discovery, the client network will make a handshake with backbone to register. The procedure, including Client Registration and De-Registration, aids in verifying local port connectivity between the optical and client devices, and allows each device to learn the IP address of the other to establish a UNI control channel. The UNI signaling messages are structured as requests and responses.

NNI Operation: The routing inside the optical networks is done as follows: The source LSR looks up its routing information corresponding to the specified destination IP address to determine a path; Then the computed path is signaled from the source to the destination LSR within the source sub-network.

3. CONCLUSION

An Integrated Control Plane for IP enabled optical network for IP and non-IP packet services is proposed to give the interfaces hierarchy for labels and highlights some requirements of procedures and options for ICP towards dynamic bandwidth on demand. The proposed ICP can use protocols such as service and neighbor discovery, route computation for LSP setup at UNI and NNI, Link Management Protocol (LMP), Optical Service Level Agreements, Link bundling, Signaling for Survivability All the individual protocols used can be found at the OIF, ODIS and IETF, which are developing on the way.

REFERENCES

[1]. B. Rajagopalan, "IP Over Optical Networks –A Framework," Internet Draft draft- many-ip-optical-framework-01.txt, July 2000.

[2]. Z.B. Tang et al. "Extensions to CR-LDP for Path Establishment in Optical Networks," Internet Draft draft-tang-crldp-optical-00.txt, Work in progress, March 2000.

[3]. J.P. Lang, K. Mitra, J. Drake,, "Extensions to RSVP for Optical Networking," Internet Draft draft-lang-mpls-rsvp-oxc-00.txt, Work in Progress, March 2000.

[4]. R.R.Stewart et al., "Stream Control Transmission Protocol," RFC 2960.

[5]. N.Chandhok et al., "IP over MPLS – A Summary of Issues," Internet Draft draft-osu-ipo-mpls-issues-00.txt, Work in progress, July 2000.

[6]. O. S. Aboul-Magd, "Signaling Requirements at the Optical UNI," Internet Draft draft-bala-mpls-optical-uni-signaling-00.txt, Work in Progress, July 2000.

[7]. D. Pendarakis, et al, "Routing Information Exchange in Optical Networks," Internet Draft draft-prs-optical-routing-00.txt, Work in progress, April 2000.

[8]. S. Chaudhuri et al., "Control of Lightpaths in an Optical Network," Internet Draft draft-chaudhuri-ip-olxc-control-00.txt, Work in progress, February 2000.

[9]. J.P. Lang, Link Management Protocol (LMP) Internet Draft draft-lang-mpls-lmp-01. txt, Work in progress, July 2000.

[10]. B. Rajagopalan, D. Saha., "Link Bundling Considerations in Optical Networks," draft-rs-optical-bundling-00.txt, July 2000.

[11]. P. Ashwood-Smith et al., "Generalized MPLS–Signaling Functional Description", Internet Draft draft–ash wood –generalized-mpls -signaling-00.txt, Work in progress, June 2000.

A Novel Heuristic Algorithm for Multi-layer Traffic Engineering in WDM Optical Networks

Tao Qin, Hanyi Zhang, Yili Guo and Xiaoping Zheng
Department of Electronic Engineering, Tsinghua University, Beijing
Tel: 86-10-62773377, Email: qintao00@mails.tsinghua.edu.cn
P. R. China

Abstract: A novel heuristic algorithm for multi-layer traffic engineering is proposed in this paper, with the objective of minimizing network congestion. The basic idea of this algorithm is to optimize network resource utilization by lowering the cost of setting up connections for fulfilling the traffic demands. This algorithm can be applied to arbitrary network topologies with various OXC structures. Another advantage of this algorithm is that it is designed for nonbifurcated traffic routing, making it more suitable for practical implementation. Computational results show that it can achieve congestion close to the LP lower bound.

Key Words: traffic engineering, logical topology design, nonbifurcated routing, WDM optical network

1. INTRODUCTION

WDM optical networks are the most promising candidates for backbone and metropolitan high-speed networks. In WDM optical networks, optical connection (lightpath) can be established and tear down dynamically to build up various logical topologies on top of its physical topology. Various access nodes can be connected to optical networks, grooming lower-rate (sub-wavelength) traffic demands onto wavelength channels. All these optical switch nodes and electrical access nodes can be configured by a GMPLS based control plane. In such a network environment, multi-lay traffic engineering can optimize network performance by jointly reconfiguring the optical layer and the electrical layer.

The multi-layer configuration problem is somewhat similar to the logical topology design problem studied in [1][2][3]. This problem can be decomposed into two parts: the lightpath connectivity problem and the traffic routing problem. The traffic routing problem falls into two categories: 1) traffic demand between a node pair can be split over multiple explicit routes (bifurcation), and 2) traffic can not be split (nonbifurcacation). The nonbifurcation case is preferred in practical networks because arbitrary traffic split is hard to implement. But because the nonbifurcation problem is much harder, most of the previous works only consider the bifurcation case [1][2]. Reference [3] proposes a scheme based on mixed integer linear programming (MILP), but it also assumes that bifurcated routing is allowed.

In this paper, we present a novel heuristic algorithm for multi-layer traffic engineering, with the objective of minimizing the network congestion, which is defined as the utilization rate of the most heavily loaded link in the network. This algorithm differs from previous works in two aspects: 1) it assumes that only nonbifurcated routing is permitted, 2) it can take into consideration of different OXC structures. For example, OXC may support different wavelength conversion capabilities: full wavelength conversion, limited wavelength conversion or no wavelength conversion. All these wavelength conversion capabilities can easily be modeled in our algorithm.

2. THE BASIC IDEA

Consider a network consisted of OXCs and electrical access nodes. The access node may be router, ATM switch or SDH equipment, and here we take the MPLS router, called label-switching router (LSR), as example. OXC can support high capacity, coarse granularity optical switching. The granularity of LSR is flexible, but its total capacity is limited and is much less than that of optical switching. Multi-layer traffic engineering can optimize network performance by combining advantages of both wavelength switching and packet switching [4].

Typically, this multi-layer configuration problem is formulated as a MILP problem [2][3]. In this paper, however, we formulate this problem as a minimum cost multi-commodity problem, by assigning suitable cost functions to network resources occupied by traffic demands. The objective is to minimize the total cost of setting up connections for fulfilling all the traffic demands, that is,

$$\min \Sigma C_{LSP} \quad (1)$$

where LSP (label switching path) is the explicit route established to satisfy the traffic demand. C_{LSP} is defined as follow:

$$\begin{aligned} C_{LSP}(\Delta v) &= \Sigma C_{Link}(\Delta v)+\Sigma C_{OXC\text{-}Port}(\Delta v)+\Sigma C_{LSR\text{-}Port} \\ &= \Sigma C_{Lightpath}(\Delta v)+\Sigma C_{LSR\text{-}Port}(\Delta v) \quad (2) \end{aligned}$$

$C_{LSP}(\Delta v)$ denotes the cost of setting up a LSP for traffic demand with bandwidth Δv. This cost is made up of three parts, representing three kinds of network resources occupied by the LSP. $\Sigma C_{Link}(\Delta v)$ is the cost of optical links, which should be proportional to their total length. $\Sigma C_{OXC\text{-}Port}(\Delta v)$ and $\Sigma C_{LSR\text{-}Port}(\Delta v)$ are the cost of OXC ports and LSR ports respectively. The summation of $\Sigma C_{Link}(\Delta v)$ and $\Sigma C_{OXC\text{-}Port}(\Delta v)$ can also be denoted as $\Sigma C_{Lightpath}(\Delta v)$, the cost of lightpaths

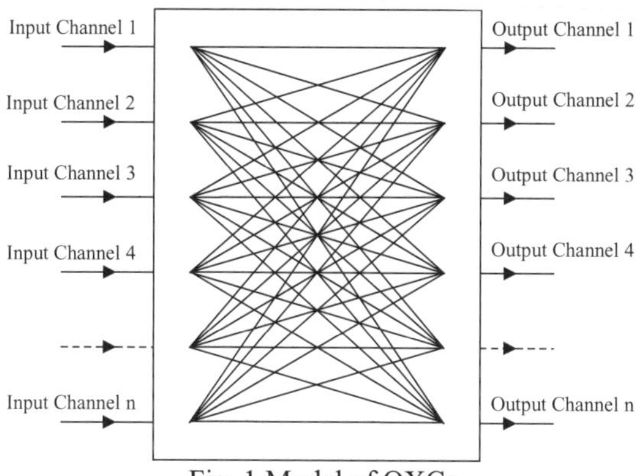

Fig. 1 Model of OXCs

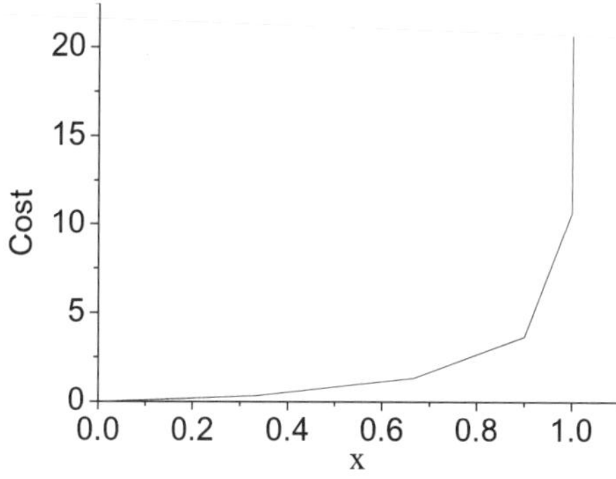

Fig. 2 LSR port cost function $F_T(x)$, assuming that the congestion threshold T=1.

that the LSP traverses.

OXC is modeled as a complete bipartite graph $K_{n,n}$ shown in fig.1, where n is the number of wavelength channels the OXC supports. Denote (i, j) as the arc connecting input channel i and output channel j. $C(i,j)$ is the weight associated with (i, j), i.e. the cost of switching optical connection from input channel i to output channel j. If signal from some input channel i can not be switched to output channel j, either because the switch fabric inside the OXC is not nonblocking or because wavelength conversion from input i to output j is not supported, then $C(i,j)=\infty$. Fig. 1 can describe different types of OXC structures with various wavelength conversion capabilities, including full wavelength conversion, no wavelength conversion or even limited wavelength conversion.

A lightpath may be occupied by a LSP exclusively, or be shared among multiple LSPs. If the lightpath is shared, its cost is also apportioned among these LSPs, according to their bandwidths. Denote the total bandwidth of all LSPs sharing a common lighpath as TB, then we define:

$$C_{Lightpath}(\Delta v)= C_{Lightpath}\times\Delta v\ /\text{TB} \qquad (3)$$

Thus unit cost of bandwidth is $C_{Lightpath}/\text{TB}$. The idea behind the cost function $\Sigma C_{Link}(\Delta v)$ is that: for traffic demand with very small bandwidth, assigning dedicated lightpath for it is not cost effective because the unit cost of bandwidth will be large. In this scenario, a lightpath should be shared by multiple traffic demands. Doing so will reduce per bandwidth cost, and increase the utilization efficiency of the lightpath.

If multiple LSPs share a common lightpath, the intermediate LSR nodes are responsible for grooming lower-rate traffic onto wavelength channels. This will aggravate electronic processing burden of these LSRs, and may lead to congestion. To avoid congestion, suitable cost function should be assigned to LSR port. Denote the cost of routing a LSP through a LSR port as $C_{LSR\text{-}Port}(v, \Delta v)$, where Δv is the bandwidth of the LSP to be routed and v is the current utilized bandwidth of that port. It is defined that:

$$C_{LSR\text{-}Port}(v, \Delta v)=F_T(v+\Delta v)-F_T(v) \qquad (4)$$

$F_T(\cdot)$ is the cost function associated with the LSR port, T is the threshold of congestion. $F_T(\cdot)$ is defined as:

$F_T(0)=0$, and

$$F_T'(x)=\begin{cases} 1 & for \quad 0\le x/T<1/3 \\ 3 & for \quad 1/3\le x/T<2/3 \\ 10 & for \quad 2/3\le x/T<9/10 \\ 70 & for \quad 9/10\le x/T<1 \\ 10^7 & for \quad 1\le x/T<\infty \end{cases} \qquad (5)$$

The cost function $F_T(\cdot)$ is shown in fig. 2. The idea behind $F_T(\cdot)$ is that: it is cheap to route a LSP over a port with small utilization, but when the utilization increases, the cost will get more expensive, so the heavily utilized port will be penalized. If the total traffic goes above the preset threshold T, the penalty will go so high that it should never happen [5]. It should be noted that exact definition of the cost function is not so important, as long as it is a piece-wise linear increasing and convex function.

3. THE AWSP-RR ALGORITHM

Given a network consisted of OXCs and LSRs, and some traffic demands between LSR s-d pairs, the problem we try to solve is to place explicit nonbifurcated routes between LSR nodes such that all traffic demands are fulfilled, and the maximum of link utilization is minimized.

We present an adaptive weighted shortest path based re-routing (AWSP-RR) algorithm to solve this problem. The starting point of AWSP-RR is a solution obtained by fixed weighted shortest path routing. Then network configuration is moved from this starting point to the optimal configuration step by step, by iterations of re-routing operations based on the cost function defined in section II. Two re-routing operations are defined:

Operation 1: *Single LSP re-routing* is made up of two steps. Firstly, the given LSP is removed from the network, and all resources it occupies are released. If the removed LSP is the only traffic passing through a lightpath, then that lightpath should also be deleted. Secondly, a new LSP is established for the traffic demand removed in the first step, through its shortest path. Denote α as the current congestion, we set the congestion threshold $T=0.99\alpha$ at the beginning of each re-routing iteration, hoping that network congestion will be lowered below T after re-routing. Another purpose of single LSP re-routing is to remove under-utilized lightpaths, freeing network resources occupied by them for later use.

If the network configuration is trapped in some local optimization points, single LSP re-routing may not be able to further reduce congestion. Under this scenario, bottleneck LSR re-routing should be performed to enable the network configuration to escape from local minimum. A LSR is called bottleneck LSR if one of its ports is the most heavily utilized port in the network.

Operation 2: *Bottleneck LSR re-routing* is also made up of two steps. In the first step, all LSPs passing through the bottleneck LSR, including those originating from or terminating at it, are removed from the network, and network resources occupied by them are all released. In the second step, new LSPs are established to restore all the traffic demands deleted in the first step.

Now the AWSP-RR algorithm is described in table 1:

TABLE 1
THE AWSP-RR ALGORITHM

Step 1.	(Initialization) Setup LSPs for all the traffic demands one by one based on fixed weighted shortest path routing, and get a feasible solution. Denote congestion of this solution as α_0. Set $\alpha_{min}=\alpha_0$, $T=0.99\alpha_0$.
Step 2.	Do the following circle *l* times:
2.1	For each LSP in the network, perform the operation of single LSP re-routing one by one.
2.2	Calculate congestion of the network α, and set $T=0.99\alpha$. If $\alpha<\alpha_{min}$, set $\alpha_{min}=\alpha$.
2.3	If $\alpha\geq\alpha_0$, which indicates that the solution may be trapped in a local minimum, then Find the bottleneck LSR, and perform bottleneck LSR re-routing operation. Calculate the network congestion α, and set $T=0.99\alpha$. If $\alpha<\alpha_{min}$, set $\alpha_{min}=\alpha$.
2.4	Set $\alpha_0=\alpha$.
Step 3.	Show the minimum congestion α_{min}, and the algorithm terminates.

4. NUMERICAL RESULTS

The performance of the AWSP-RR algorithm is evaluated by comparing it with previous works [2]. The network topology is the 14-node NSFNET backbone network shown in fig.3. It is assumed that each node in fig. 3 represents an OXC node, and an electrical access node connected to it. The number of wavelength channels that can be transmitted in a fiber is denoted as W, and the logical degree (ports number of electrical access nodes) is denoted as D. Two traffic matrices P_1 and P_2 are considered, and the computational results are shown in fig. 4.

In fig.4, HLDA, MILP and LP are results presented in [2], assuming that the number of wavelength is not a constraint ($W=\infty$), and traffic bifurcation is allowed. LP is the lower bound, obtained by LP-relaxation of the original problem. MILP indicates congestion obtained by solving the MILP problem directly using a branch and bound routine. HLDA is the heuristic algorithm proposed in [2]. AWSP-RR shows our results, assuming that W=8 and OXC is strictly nonblocking.

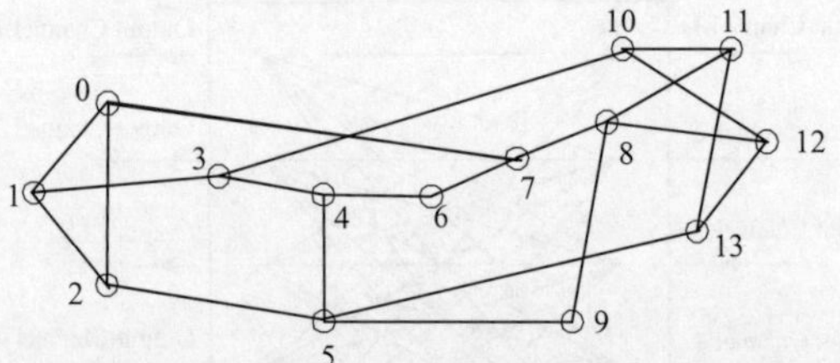

Fig. 3. The 14-node NSFNET network.

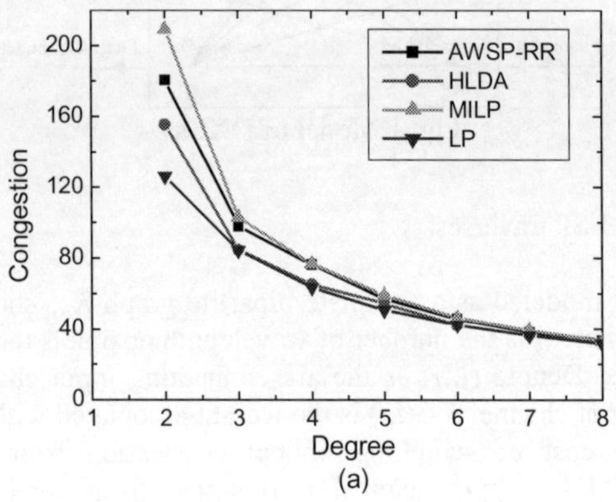

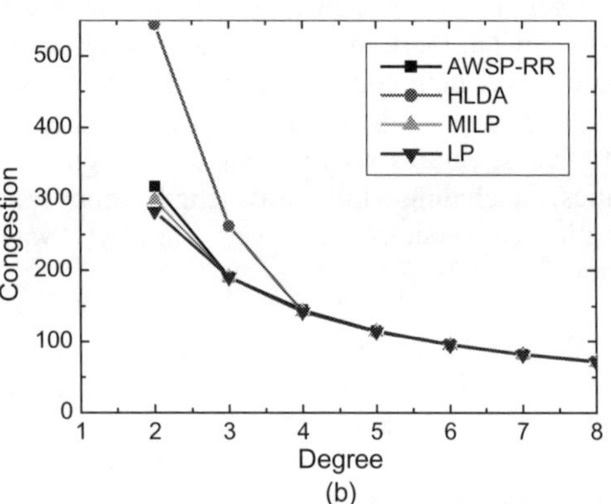

Fig.4 Congestion versus logical degree (ports number of electrical access nodes) obtained by various algorithm. (a) For traffic P_1. (b) For traffic P_2.

AWSP-RR is an algorithm designed for nonbifurcated traffic routing. When comparing it with bifurcated traffic routing algorithm, in order to be fair, the traffic demand from a s-d pair is divided into multiple sub-demands if the bandwidth is larger than MAX_BB, and traffic bifurcation of sub-demand is not permitted. MAX_BB is set to 3.0 in this paper.

As can be seen in fig.4: for traffic matrix P_1, HLDA is better than AWSP-RR, but the difference is not significant; but for traffic P_2, AWSP-RR is much better than HLDA for logical degree 2 and 3. When the logical degree is larger than 5, both AWSP-RR and HLDA can obtain congestion very close to the LP lower bound.

5 CONCLUSION

A novel heuristic algorithm AWSP-RR is proposed in this paper to minimize network congestion. This algorithm can be applied to various OXC structures with different wavelength conversion capabilities. It is designed for nonbifurcated traffic routing, and is more suitable for practical implementation. Computational results show that it can achieve congestion very close to the LP lower bound.

REFERENCES

[1] J. P. Labourdette and A. S. Acampora, logical rearrangeable multihop lightwave networks, IEEE Transactions on Communication, vol. 39, page 1223-1230, August 1991.

[2] R. Ramaswami and K. N. Sivarajan, Design of logical topologies for wavelength-routed optical networks, IEEE Journal of Selected Areas in Communications, vol. 14, page 840-850, June, 1996.

[3] Rajesh M. Krishnaswamy and Kumar N. Sivarajan, Design of logical topologies: a linear formulation for wavelength routed optical networks with no wavelength changers, IEEE/ACM Transactions on Networking, vol. 9, page 186-198, April 2001.

[4] Eiji Oki, Kohei Shiomoto, Satoru Okamoto, Wataru Imajuku, and Naoaki Yamanaka, A heuristic multi-layer optimum topology design scheme based on traffic measurement for IP + Photonic networks, *OFC* 2002, pp. 104-105, 2002.

[5] Bernard Fortz and Mikkel Thorup, Internet traffic engineering by optimizing OSPF weights, IEEE *INFOCOM* 2000, pp. 519-pp528, 2000.

MPλS vs LOBS in an Optical Burst Switching Node: Impact of Long-Range Dependent Traffic on Multiclass Services *

A. Brandani, M. Casoni, *Member IEEE*, and M. L. Merani, *Member IEEE*
Department of Information Engineering
University of Modena and Reggio Emilia
Via Vignolese, 905 - 41100 - Modena - Italy
email: casoni@ieee.org, merani.marialuisa@unimo.it; Fax:+39-059-2056129

Abstract

This work investigates the behavior of a reservation mechanism known as JET, Just Enough Time, highly promising in WDM networks adopting the optical burst switching (OBS) paradigm. In this article we assess the performance, in terms of burst blocking probability, that an optical cross-connect achieves when employing JET in conjunction with the MultiProtocol Lambda Switching in the control plane on top of the optical layer, and compare it with the Labeled Optical Burst Switching, LOBS, solution. The burst blocking probabilities of three service classes with different QoS requirements are determined, demonstrating that the LOBS alternative guarantees a remarkable improvement with respect to the MultiProtocol Lambda Switching - OBS combination.

Key Words: Optical Burst Switching, Self-Similarity, Quality of Service

1 Introduction

During the last twenty years optical communications have effectively evolved leading to very powerful transmission systems. At the same time, Internet has definitely become the solution for internetworking, connecting wireline and wireless systems and reaching a world-wide diffusion. Designing networks capable of supporting the ever increasing amount of information that Internet will deal with therefore becomes more and more compelling. In addition, as current and future applications require different network performance in terms of, e.g., loss, delay, jitter, network design becomes more complex. Optical networking has been boosted by the advances in coherent optical transmission, which led to Dense Wavelength Division Multiplexing (DWDM) systems able to accommodate up to hundreds of wavelengths per fiber, and in integrated optics, for passive and active optical components design. The ultimate goal is the development of the so-called optical transparent network, where the signals carried within the network will never leave the optical domain. A first important step in this direction is to have optical networks transparent at least for data, being the control part converted and processed in electronics. In this paper the focus is on the performance of an optical switching technology which offers a dynamic mechanism to set up high capacity end-to-end optical data connections in DWDM networks, the Optical Burst Switching (OBS) solution [1]-[2]. In OBS networks data never leave the optical domain: for each data burst assembled at the network edge, a reservation request is sent as a separate control packet, well in advance. The reservation mechanism known as Just Enough Time, JET, advantageously exploits this and through the assignment of longer offset times to higher priority bursts, JET can also enforce QoS differentiation among distinct classes of applications, directly at the optical layer [2]. [2][3]address the issue of providing basic quality of service in OBS networks and also proposes the use of LOBS, Label Optical Burst Switching, as the control solution under the IP MPLS, MultiProtocol Label Switching, framework [4]. In this paper the performance of an OBS node which implements the JET reservation approach is investigated, with three distinct traffic classes mapped in either two or three Forward Equivalence Classes (FEC). Each of them is characterized by a different statistical traffic description and features a different QoS requirement, i.e., asks for a specific value of burst blocking probability. The impact on the OBS node behavior of different traffic classes is examined, as well as the adoption of MPλS, MultiProtocol Lambda Switching [5], as the forwarding technique within the optical network, or of the alternative LOBS paradigm [3]. In detail, Section 2 explains the OBS approach and discusses the aforementioned solutions, MPλS vs. LOBS. In Section 3 the investigated switching system and the different traffic profiles examined are described. In Section 4 the OBS node performance are reported with reference to different operating conditions.

*This work has been partially supported by the University of Modena and Reggio Emilia within the framework of the reasearch project "Planning of Optical Networks with Protection Mechanisms".

2 IP Over WDM

Optical burst switching can be seen as a middle term solution towards all optical packet switching, whose goal is to improve wavelength utilization and sharing by introducing a dynamic wavelength management. The basic idea in OBS is to dynamically set up a wavelength path whenever a *large* data flow is identified and needs to traverse the network: a separate control packet therefore precedes each burst by a basic offset time, carrying relevant forwarding information. Resource allocation in OBS networks does not require a two-way end-to-end signaling like circuit switching, as data is transferred across the network without waiting the set-up of the entire lightpath: bursts therefore transparently travel across the network, but might incur in blocking if contentions arise at the OXCs which lead to destination. One relevant degree of freedom in the design of OBS networks is the dimensioning of the offset time interval. For the reservation mechanism known as Just-Enough-Time (JET), the OXC allocates a specific input-output path to a burst, only for the burst duration; furthermore, the offset might also account for an additional contribution which allows to differentiate the performance of various classes of service and grant them different priorities. The principle is that the longer the extra-offset, the higher the priority, as the corresponding flow can reserve very in advance the required resource. Several algorithms are clearly necessary to properly manage and assign the optical resources: in OBS they must be reserved along the path, and a mechanism has to be developed to complete the connection setup. One of the possible approaches is to integrate the IP control plane with the optical control plane following the MPLS paradigm. In MPLS the forwarding operation is realized by means of a label, associated with each packet and used as entry of the forwarding table. Each label identifiesthe Forward Equivalent Class (FEC) which a packet belongs to. Packets belonging to the same FEC are, from a forwarding point of view, indistinguishable and are forwarded in a connection oriented fashion from source to destination along an Label Switched Path (LSP). The most significant example of standard proposed to integrate MPLS with wavelength routed all-optical networks is Multiprotocol Lambda Switching MPλS [5]. This approach considers the wavelength as the unit of capacity within the optical network. MPλS maps LSPs into wavelengths, a possible approach for the design of a common control plane for optical cross-connects (OXCs) and Label Switching Routers (LSRs). The generality of the MPλS proposal also makes it suitable within an Optical Burst Switching architecture. Each OBS node can be equipped with an IP/MPLS controller which makes it similar to a conventional electronic label switch router. Core nodes will perform burst switching, as an incoming burst on a given wavelength is switched through an optical fabric to a wavelength (the same or different) of an outgoing fiber. Another possible approach is Labeled Optical Burst Switching, LOBS [3]. The basic idea is to exploit the control packet for labeling the corresponding data burst. Rather than associating the data flow on an input (fiber, wavelength) pair to a specific output pair, dictated by the MPλS control plane, the label is carried by the control packet itself, freeing the wavelength resource for increasing the capacity of outgoing links. As a matter of fact, now the control packet, being processed by electronics, can carry any kind of label which identifies the next hop towards the destination, which means to select one of the outgoing fibers. Any wavelength within the selected fiber can then be employed to transfer data to the next OCX/ LSP router.

3 System Model and Traffic Analysis

The OBS node investigated in this paper is equipped with $M \times M$ optical interfaces capable of supporting N wavelengths each. Over each wavelength optical transmission devices allow for transmissions at 2.5 Gbit/s. We suppose that wavelength converters are available and that the system is bufferless, i.e., no fiber delay lines are available in order to resolve contention for an output fiber, output wavelength. When the JET reservation scheme is employed in the node, one way to reduce the blocking of higher priority users is to grant them a longer extra-offset time [2]. Node behavior is examined when traffic on each input wavelength is generated by an ON/OFF source. In order to relate our results to existing literature [2] and to maintain some degree of analytical tractability, we start our analysis under the conventional assumption that the ON/OFF periods of each of the $M \times N$ traffic sources are exponentially distributed: extra-offset values which guarantee the class isolation condition are also considered. Traffic exclusively consisting of IP datagrams is examined: according to [6], medium and large-size datagrams are 576 and 1500 byte long; a third IP datagram size, equal to 200 bytes, is introduced to account for Voice over IP applications (20 bytes of IP header, 8 bytes of UDP header, 12 bytes of RTP header and 160 bytes of G.711 payload). As regards small packets, $40-44$ byte long, that represent a relevant percentage of Internet backbone traffic [6], it is assumed that they are not managed optically in the core routers, but either undergo a "grooming" process at the network edges, or cross the core network over separate channels. Burst (datagram) classes are termed 1, 2 and 3, ranging from the smallest to the largest in size. Class 1 is attributed the highest priority and the longest extra-offset time. On each input wavelength let λ represent the mean overall arrival rate of bursts. Given that $\overline{x}_1$, $\overline{x}_2$ and $\overline{x}_3$ are the mean duration times of the ON periods due to bursts of short, medium and long-sized datagrams, respectively, on each wavelength the offered load ρ is the sum of three distinct contributions,

$$\rho = \sum_{i=1}^{3} \rho_i = \sum_{i=1}^{3} p_i \lambda \overline{x}_i \,, \tag{1}$$

where p_1, p_2 and p_3 define the percentages of the three traffic classes. When MPλS is adopted, an incoming data flow has a label which identifies the wavelength over which the burst is traveling. In order to create an LSP, the control plane preliminary associates a well-defined output label, i.e., wavelength, to this input label in each optical node. In a symmetric $(M \times N) \times (M \times N)$ switching system the outputs can therefore be seen as $M \times N$ distinct $M/M/1$ queueing systems. On the other hand, when the LOBS approach is employed, the label carried by the control packet associated to the incoming data flow simply indicates the next hop that data has to be forwarded to. Equivalently, bursts are switched to the desired output interface – fiber – regardless of the output wavelength. From a modeling perspective, the system describing each output fiber can be interpreted as an $M/M/N$ queue, with M queues in total. It is evident that this approach provides better results in terms of blocking probability than the previous one. It is also worth underlining that wavelength converters are necessary in both cases, in order to allow an incoming data flow on any input wavelength to be switched to any wavelength of a generic outgoing fiber. Unfortunately, in either architecture the blocking probability can be determined by the well-known Erlang's B formula only in some very specific circumstances [2]. To this regard, observe that for the JET-MPλS approach the blocking probability P_{b1} of the highest priority bursts in class isolation is given by Erlang's B formula, so that

$$P_{b1} \simeq B(\rho_1, m), \qquad m = 1 \tag{2}$$

whereas for LOBS this blocking modifies into

$$P_{b1} \simeq B(8 \cdot \rho_1, m), \qquad m = 8\,, \tag{3}$$

where in either case m represents the number of available servers (wavelenghts). Aiming to model self-similar traffic [7], Pareto-distributed ON/OFF periods are also considered. Indicating by X_{on_i} the random variable which represents the duration on the ON period of class i bursts, its cumulative distribution function is

$$F(x) = P[X_{on_i} \leq x] = 1 - \left(\frac{k_{on_i}}{x}\right)^{\alpha_{on_i}}, 0 < k_{on_i} \leq x\,, \tag{4}$$

where the constant k_{on_i} is the smallest possible value of the random variable X_{on_i}, and $0 < \alpha_{on_i} < 2$ in order to have a heavy-tailed distribution. The corresponding mean value is

$$E[X_{on_i}] = \frac{\alpha_{on_i} k_{on_i}}{\alpha_{on_i} - 1}\,. \tag{5}$$

Note that under the hypothesis $0 < \alpha_{on_i} < 2$, X_{on_i} has infinite variance; if in addition $\alpha_{on_i} \leq 1$, $E[X_{on_i}]$ is infinite as well. Equivalently, one can affirm that the probability of very large random sampling of X_{on_i} is not negligible. A measure of the degree of self-similarity is the Hurst parameter, $H(\alpha) = \frac{3-\alpha}{2}$, a dimensionless quantity which ranges between 0.5 (no self-similarity, $\alpha = 2$) and 1 ($\alpha = 1$). An equivalent distribution is also assumed for the OFF period, X_{off_i}, with parameters k_{off_i} and α_{off_i}:

$$F(x) = P[X_{off_i} \leq x] = 1 - \left(\frac{k_{off_i}}{x}\right)^{\alpha_{off_i}}, 0 < k_{off_i} \leq x\,, \tag{6}$$

$$E[X_{off_i}] = \frac{\alpha_{off_i} k_{off_i}}{\alpha_{off_i} - 1}\,. \tag{7}$$

The load ρ_i that each class generates on the generic wavelength is therefore

$$\rho_i = \frac{E[X_{on_i}]}{E[X_{on_i}] + E[X_{off_i}]}\,. \tag{8}$$

4 Results and Conclusions

In this Section we present the results referring to an optical node with $M = 4$ incoming and $M = 4$ outgoing fibers, with $N = 8$ wavelengths per fiber. Each of the outgoing fibers is chosen with equal probability $p_F = 1/M = 1/4$. The examined service classes have mean burst lengths 200, 576 and 1500 bytes: as the channel transmission rate is $B_r = 2.5$ Gbit/s per wavelength, for Markovian sources their mean duration times are $\overline{x}_1 = .6976$ μs, $\overline{x}_2 = 1.8432$ μs and $\overline{x}_3 = 4.8$ μs, respectively. The current study refers to a single node, hence the base offset is equal to 0; in order to differentiate service among classes, class 1 bursts are assigned an extra-offset equal to $5 \times \overline{x}_3 = 24$ μs, class 2 and 3 bursts employ no extra-offset. Figs. 1 and 2 refer to the MPλS approach: for a given output fiber, an output wavelength is chosen with probability $p_\lambda = 1/N = 1/8$. The following values for p_1, p_2 and p_3 are considered, $(0.2, 0.5, 0.3)$; in both figures the case of Markovian traffic sources (exponentially distributed ON/OFF periods) is compared to the case of heavy-tailed sources (Pareto distributed ON/OFF periods). For the latter circumstance, the adopted parameters are $\alpha_{on_i} = 1.2$ and $\alpha_{off_i} = 1.5$ $\forall i$, $i = 1, 2, 3$; $k_{on\_1} = E[X_{on_1}]$, $k_{on\_2} = E[X_{on_2}]$ and $k_{on\_3} = E[X_{on_3}]$; for a given input load ρ and a specific (p_1, p_2, p_3) set, the actual values of k_{off_1}, k_{off_2} and k_{off_3} are derived from eqs.(8) and (7). Fig. 1 shows that the difference in performance that class 1 experiences in the two cases is striking, whereas class 2 and 3 do not undergo any significant variation; note however that their blocking probability is relevant, i.e., above 0.1, for both the exponential and the Pareto case. Figure 2 further elaborates on the influence that the Pareto distribution has on performance: for a given input load, $\rho = 0.6$, it shows how the blocking probability of highest priority users P_{b1} modifies as a function of the extra-offset time, and compares it with the conventional Markov-based results. The convergence towards the class isolation condition is now far slower, and would imply the adoption of an unbearable high value of the extra-offset time. Last, Fig.3 compares the blocking probability of class 1, 2 and 3 bursts (exponential vs. Pareto distribution), assuming that the LOBS paradigm is implemented and $t_{extra-offset-1} = 24$ μs: provided an output fiber is chosen, *any* output wavelength on that fiber can therefore be used. In order to avoid the $\rho > 1$ condition, unacceptable for an ON/OFF type of source, this figure refers to different p_1, p_2 and p_3 probabilities with respect to the previous case, namely, $(0.5, 0.3, 0.2)$. Nevertheless, performance noticeably improves with respect to the MPλS (single server) approach: as an example, at $\rho = 0.8$ JET in conjunction with MPλS would guarantee a blocking of high priority bursts equal to $P_{b1} = 0.1303$ for the Markovian sources, which reduces to less than 10^{-5} when the LOBS solution is adopted. In conclusion, it has been demonstrated that the LOBS alternative guarantees a remarkable improvement with respect to the MPλS - OBS combination. Different traffic source descriptions have been employed to compare the achievable node performance under Markov as well as heavy-tailed distributed traffic. The latter approach has highlighted how notably long-range dependent traffic deteriorates blocking, for any of the examined solutions and traffic classes.

References

[1] J. S. Turner, "Terabit Burst Switching," Journal of High Speed Networks, No.8, pp.3-16, 1999.

[2] M. Yoo, C. Qiao, S. Dixit, "QoS Performance of Optical Burst Switching in IP-Over-WDM Networks," IEEE Journal on Selected Areas in Communications, Vol.18, No.10, pp.2062-2071, October 2000.

[3] C. Qiao, "Labeled Optical Burst Switching for IP over WDM integration," IEEE Communication Magazine, September 2000, pp.104-114.

[4] E. Rosen, A. Viswanathan, R. Callon, "Multiprotocol Label Switching Architecture," IETF RFC 3031, January 2001.

[5] D. O. Awduche, Y. Rekhter, J. Drake, R. Coltun, "Multi-Protocol Lambda Switching: Combining MPLS Traffic Engineering Control With Optical Crossconnects," Internet Draft, Work in Progress, 2001.

[6] K. Thompson, G. J. Miller, R. Wilder, "Wide-Area Internet Traffic Patterns and Characteristics," IEEE Network, November/December 1997, pp.10-23.

[7] K. Park, W. Willinger, *Self-Similar Network Traffic and Performance Evaluation*, J.Wiley & Sons: New York, 2000.

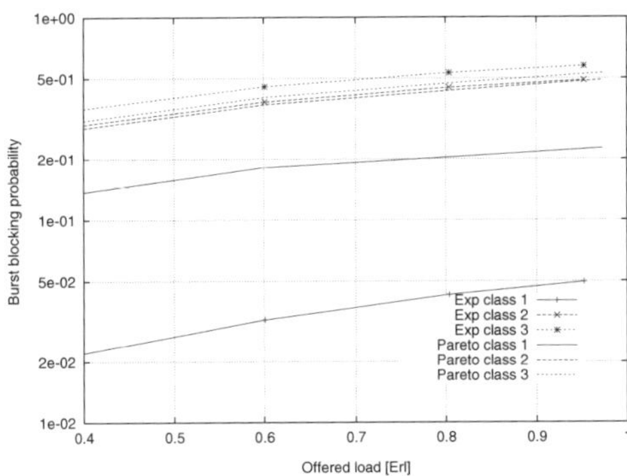

Figure 1: Blocking probability as a function of the overall offered load ρ: comparison between exponential and Pareto distributed ON-OFF periods. MPλS approach.

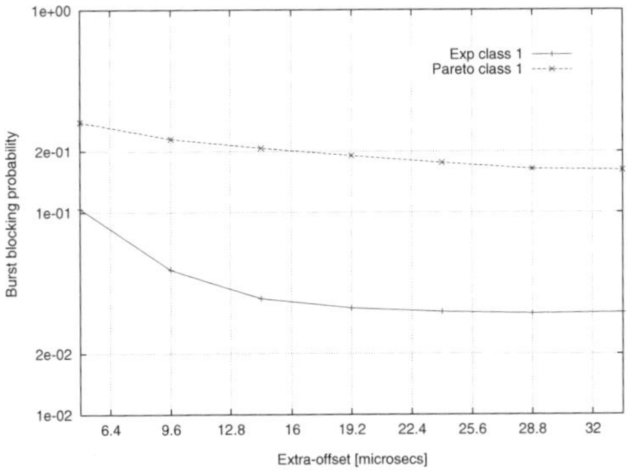

Figure 2: Blocking probability of the highest priority bursts (class 1) as a funtion of the extra-offset time, at $\rho = 0.6$. Exponential vs. Pareto distributed ON-OFF periods. MPλS approach.

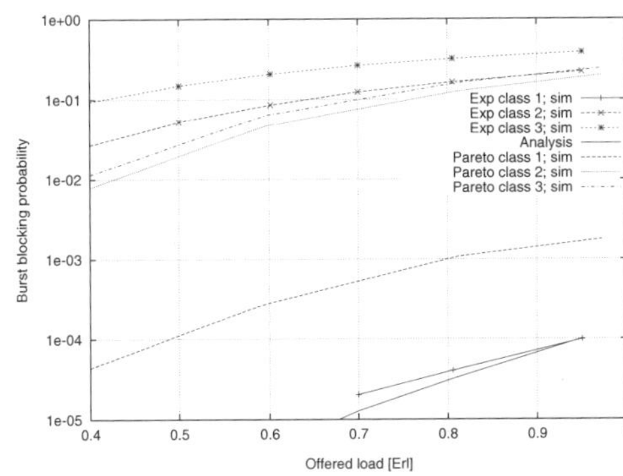

Figure 3: Comparison between blocking probabilities for the exponential and Pareto distributed ON-OFF periods: LOBS approach.

A Scheduling Procedure for Control Signaling in an Optical Burst Switched Network

Ashwin Gumaste and Jason Jue
Center for Advance Telecommunication Studies and Services
The University of Texas at Dallas, TX 75083 USA
+1(214) 476 1034 email ashwing@ieee.org

Abstract: In this paper we examine a scheduling algorithm for burst switched networks, based on a policy to aggregate bursts at the edge routers of a network. We also consider aggregation as well as scheduling of bursts from a QoS perspective. We guarantee delay based QoS to individual packets for end-to-end flows. We also describe a new aggregation process that assembles bursts from packets and determines both the appropriate size of the burst as well as the optimum time at which a control packet needs to be sent to reserve bandwidth.

I INTRODUCTION

An algorithm like JET assumes that a control packet is sent to subsequent nodes along a particular route only after a burst has been received completely at the optical network element (NE). This assumption is based upon the layered approach to networking whereby the optical layer performs only the connectivity functions. The effect of this assumption on system performance is that burst aggregation is entirely independent of transmission heuristics. A degradation of system capacity can be expected especially if there are multiple QoS requirements to be provided in the network. In this paper we introduce an approach to integrate packets into bursts in the electrical domain at the network edge, assuming multiple interfaces containing various bit-streams feeding into the edge network element. For each arriving packet into an NE buffer such that this buffer is named after a particular port, there is a QoS requirement associated with the packet. We assign a local label depicting a decrementing counter whose value is proportional to the stringency of the service requirement. This label is compared with the similar requirement of every additional packet. If the service requirement of the latest packet is more stringent than the decremented value of the label then the required service value of the new packet becomes the value of the label. When the value of the label reaches a decremented value equal to the summation of net propagation and processing delays of the burst along the entire multi-nodal path the control packet is injected into the network, while the burst is injected after the offset time. By use of this hierarchy we hope to achieve more stringent QoS requirement through optical layer performance.

II SYSTEM DESIGN

Consider a 2-fiber ring; we choose the ring topology due to vast deployment of rings in today's networks. Assume there are 'w' wavelengths in the system and further each wavelength is supported at each node by a transponder card which is fed bursts from an electronic buffer. A virtual queue (VQ) is created for each wavelength. Packets arrive from higher network interfaces at various intervals characteristic of their distributions. For simplicity we consider two main classes of packets C_1, and C_2, such that they have different QoS requirements associated in terms of delays. Packets are buffered up in the electronic buffer. Though there are w buffers corresponding to the w wavelengths there is no fixed mapping of wavelengths and buffers. Buffers are mapped to a destination node. For example at epoch we can visualize a buffer b_x containing packets each destined for a node n_y. The buffers are of fixed size usually given by the maximum size of the burst. In this paper since we are considering purely QoS based networks, buffer sizes are dictated by the lowest QoS (delay) requirement. If we assume that C_1 has the highest stringency (lowest allowable delay) requirement and C_2 to have the lowest stringency (highest allowable delay) requirement, then if the transfer rate is L Gb/s then the maximum buffer size needed is $L*C_2$ Since we are working with electronic buffers these buffers can be easily implemented.

III BURST SCHEDULING ALGORITHM

The algorithm used is a derivative of the basic JET [1] algorithm, and uses the same feature of delayed reservation as well as offset time. However, in JET there is no discrete mention of burst assembly, which we will call burst aggregation here, as well as scheduling of control information. In this paper we discuss end-to-end procedure on how to assemble a burst, send the control packet, reserve bandwidth, send the burst and in the process attempt to avoid contention. We guarantee QoS to the two classes of service. Packets destined for a particular node are made to assemble in a randomly selected VQ. VQs can be imagined to be fixed blocks of memory allocated within a main memory matrix. In our simulation we use Poisson distribution for the highest priority packets and self-similar or Pareto distribution for the lowest priority packets. The bursty nature of data traffic is characterized by the self-similar traffic flow [2]. In general we can visualize a particular burst consisting of many low priority packets and very few high priority packets. This is quite in accordance with the way Internet traffic is today [3]. The above procedure highlights burst assembly based on QoS. The objective of maximizing throughput is accomplished by the burst assembly and scheduling algorithm itself. At a given time assume that a particular node 's' has no packets in its buffer destined for a particular destination node 'd'. Now packets arrive from multiple interface lines (from the line interfaces) and those destined for 'd' are buffered up in a virtual queue. Upon arrival of the first packet in this virtual queue, the maximum permissible delay of this first packet is allotted to the VQ. This is the maximum time the packet can stay in the queue before transmission. At the arrival of every packet following the first one the QoS (delay) requirement of the newly arrived packet is compared to the delay requirement of the queue (buffer). If the delay requirement of the new packet is less than the delay requirement of the queue then the delay requirement of the queue is replaced by the delay requirement of the newly arrived packet. If on the other hand if the delay requirement of the newly arrived packet is greater than the delay requirement of the queue the queue holds on to its existing value. In the simulation program the delay requirement of the queue is decremented by one time unit at every iteration. The delay requirement for a given QoS is calculated by subtracting the mean delay requirement from the cumulative propagation delay to the destination node. The maximum allowable delay at any given time is stored in a counter and this also gives a metric of the approximate burst size (depending upon the arrival process of the packets). For example, if the first packet in the queue has the most severeQoS requirement then the maximum size of the buffer is quite small on account of the maximum delay this packet can wait. At a given time if the delay counter value equals the maximum permissible delay, a control packet is sent along the propagation path. At this time a wavelength is also selected for burst transmission. The control packet carries information such as destination node, route, mean burst size, reservation initiation time and reservation duration. The first two variables are quite easy to determine; while burst size is calculated by the normal approximation of the average rate of occupancy of packets in the VQ and the propagation delay. The burst size at the transmission of a control packet is not the final burst size, as more packets are likely to arrive in the interval between the transmission of the control packet and the actual transmission of the burst. We use a simple normal approximation to calculate the amount of packets that may appear between the intervals. The control packet also carries the information pertaining to the reservation of bandwidth. Assuming equal granularity of lightpaths, the control channel puts in a metric where by each intermediate node en-route to the destination knows the time the corresponding burst would arrive as well as the duration of the burst. If the intermediate node cannot give resources at that time (due to contention), the bust is temporarily dropped, and switched on to another wavelength by complete O-E-O feature. In this case the intermediate node fully detects the burst and sends it on a different wavelength. At the same time it stops the control packet from going any further. Contention is thus resolved by retransmission function, which is very easily implemented by the O-E-O feature of the nodes. The only issue here is that the offset time is calculated as the sum of the propagation delay plus twice the processing time at each node. The reason we assume a factor of two is to facilitate contention resolution using retransmission. When contention does occur, this contention is actually between two bursts one currently traveling and one lagging or both lagging, but never a case when both are traveling. Hence this resolution too is quite futuristic, in the sense the control packets basically help a node predict if at all there would be contention. Once it is predicted that contention would occur, it is resolved ahead in time making full use of the O-E-O as well as the multi-hopping feature of the nodes. When the initial node chooses a wavelength the assignment is done based on a lookup table. An out-of-band (OOB) optical service channel carries such a lookup table, and the source node tries to choose a wavelength on the route, which is currently not being used by any of the intermediate nodes on that route. However, being a packet like dynamic system there is still possibility of contention to occur and we have already discussed the method to deal with such contention. Another way to solve contention is by creating a smaller bust with only high QoS requirement packets and delaying the remaining low QoS packet part of the burst. This would be showcased in a future study.

Finally, despite retransmission bursts may be dropped due to wavelength unavailability. We have to note that at intermediate nodes, the bursts are stored only momentarily as we have to guarantee the delay based QoS requirement of the overall burst. Also if all wavelengths are used then the burst will in all probability be dropped without any buffering attempt, as we have assumed that the total memory available at each node is proportional to the number of wavelengths (times of course the lowest stringency buffer size). In such cases deflection [4] or loop-less [5] routing may be a good alternative but the utilization would suffer. One more impairment is the burst size approximation, whereby the source node approximates the size the burst would build, upon the propagation delay of the control packet. It is beyond the scope of this paper to work on a model for Poisson or Pareto distributions to predict the size in interval (0,delay) though such work has been done in [6]. However, by assuming that the stringency level is much greater than propagation delay this discrepancy can be nullified to a great extent.

IV SIMULATION MODEL

Based on the system description above we built a simulation model to calculate the heuristics of the system. In the simulation model we assumed a ring topology with N nodes and w wavelengths such that w>=N. We further assumed each arc of the N arcs to have the same propagation delay for simplicity. The algorithm first initializes the buffer values of the VQ to zero and each node has a compliment of two packet interfaces corresponding to the two classes of packets which are expected to be transported through the network. The first interface is modeled as a Pareto interface and this has the least priority but maximum volumes of packets come from this interface. The model assumes 1000 Pareto generators (sources) superimposing the intervals of their values on each other. By choosing 1000 generators we try to make the output stream as bursty as possible (depending on Hurst parameter too). The maximum delay of this class (C_2) is given as 250 milliseconds. The second interface is the high priority Poisson interface and its maximum permissible delay requirement is 90 msec end-to-end. For simulation purpose we assumed 10 nodes and 32 wavelengths. Bursts were statistically multiplexed by efficiently using the control channel mechanism. We observed that a large portion of the bursts consisted of the lower class packets. Due to the close resemblance of the lower class packets to data packets we measured performance for various degrees of burstiness. Shown in Fig. 1 is the blocking probability versus the load. Load was measured, as the amount of time the system was busy as compared to the maximum capacity of the system. We observed that despite statistical multiplexing approach we still were facing heavy blocking beyond the load of 0.8. This could be explained due to the retransmission strategy used in the algorithm. As a burst faced contention, it was re-transmitted on another λ, by using multi-hop like feature. However as load increased the available wavelengths decreased and hence it was not possible to find a free wavelength to send the contended burst on. Since QoS was our basic priority and we did not have buffers at intermediate nodes for regenerated traffic as this would degrade the end-to-end delay and hence this burst had to be dropped. In Fig. 2 we observe burst size and network utilization as a function of the burstiness parameter H (Hurst parameter, typically>0.5). We see that the algorithm does an additional job of traffic shaping. As burstiness increases we should intuitively observe a very sharp increase in blocking, but here we see that instead network utilization instead improves. This is due to the shaping feature of burst assembly. Note that we had assumed 10 Gb/s line rate and keeping in conformity with that line rate, traffic streams are shaped. For burstiness above H=0.95 we observe that the burst size far exceeds the normalized utilization. If we correlate this to Fig. 1 we can see that the sudden spurt in blocking probability is the reason. In Fig. 3 we plot the error in measurement of burst size as a function of Hurst parameter. In the algorithm we mentioned that a control packet does not carry the full information about the burst heuristics but only carries an approximate burst size based on the present size, and the arrival rate (average) of the packets in the buffer since inception. This is no full proof measure of the exact expected size; hence we calculate the error in the discrepancy as a function of burstiness and observe that the error is quite linear till H-0.. In Fig. 4 we compare the performance of such a scheme with JET such that for JET we assume a contention resolution based on deflection routing. We see that the number of retransmissions in our proposed scheme is far less than the number of deflected bursts in JET. This is due to the bust aggregation policy. Further JET cannot accord any kind of QoS on its own, but in our scheme, we can accord QoS to bursts. We show here that burst aggregation at the edge of a network play a crucial role in determining efficient burst switched communication.

The control packet is scheduled not after the entire burst is aggregated but just in time before the packet with the most severe delay requirement is now exhausting its allowable delay. In Appendix 1 we show the assembly feature of the bursts for the two arrival processes. A more analytical model is beyond the scope of this paper.

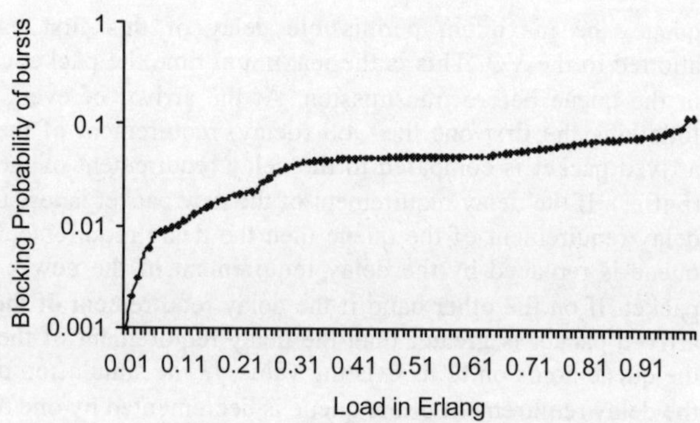

Fig. 1 Blocking Probability versus load for the proposed scheme.

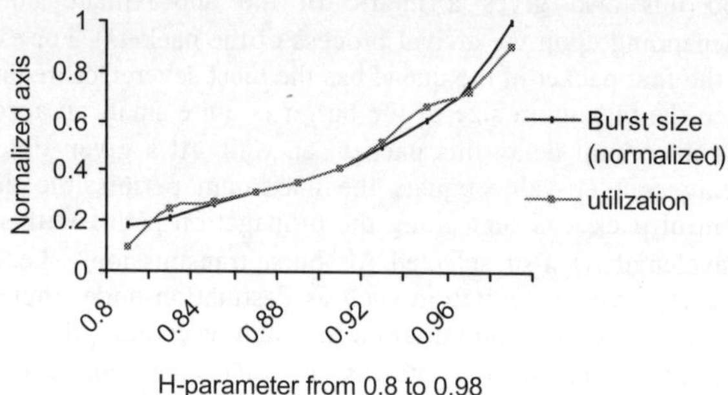

Fig. 2 Variation of Burst size and Network utilization for various degrees of burstiness.

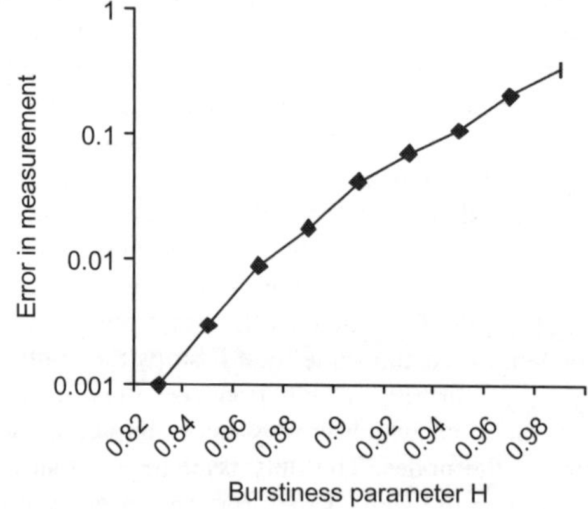

Fig. 3 Error in measuring burst size due to control packet scheduling.

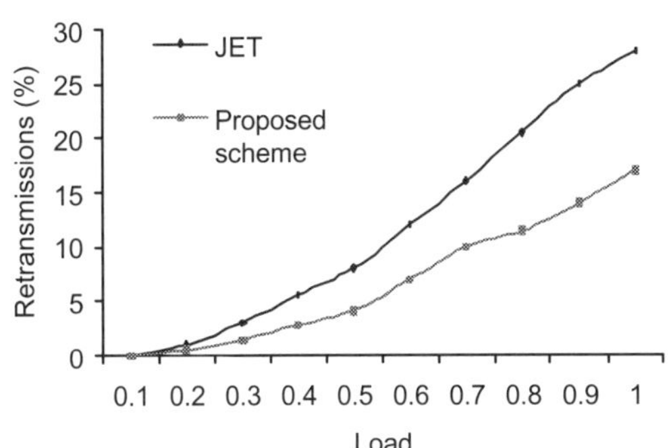

Fig. 4 Comparison with JET and proposed scheme for retransmission of bursts (statistical multiplexing).

DISCUSSION

In this paper we have proposed a new scheduling protocol, which assembles bursts as well as guarantees QoS parameters. The protocol is simple and intuitive yet at the same time it performs the complex role of traffic shaping. Traffic shaping is a very important issue in burst switched networks and this protocol to a great extent addresses the shaping issue keeping in mind end-to-end service guarantee. The protocol also determines a rule to optimally schedule control packets based on QoS delays for their bursts. Also we have coalesced the role of QoS, burst assembly, as well as scheduling all in one algorithm.

The protocol by deploying burst aggregation techniques, sends bursts of optimum sizes and from our simulation we observed that bursts had a large number of data packets and a small number of voice packets. Yet both data and voice packets were able to get delay based QoS. This can be a good motivating factor for Voice over IP kind of schemes and we are currently investigating VOIP over burst switched networks.

REFERENCES

[1] I.Chlamtac, A. Ganz, G. Karmi, 'Lightpath Communications: A Novel method to increasing bandwidth in all optical WANs' IEEE Trans on Commun. Vol.. 40 No. 7 July 1992 pp-1152

[2] M.Yoo and C. Qiao, ' Just-Enough Time (JET) : A High Speed Protocol for Bursty Traffic in Optical Networks,' Proceedings of the OFC, 2000, Baltimore TuB2

[3] W. Leland, M. Taqqu, and W. Willinger, ' On the Self-Similar Nature of Ethernet traffic,' IEEE?ACM Trans. on Networking, Vol. 2 No. 1, pp 37-48 Feb 1993

[4] E. Nahum, T. Barzilai and D. Kandlur,' WWW Servers and Internet Traffic, IEEE/ACM Trans on Networking, Feb 2002, Vol. 10, No. 1 pp1-11

[5]G. Castanon, Bononi, 'Deflection Routing in Manhatten Street Network'

[6] J.Jue,' An Efficient Algorithm for Loopless deflection Routing,' Proceedings of ICC NY 2002

APPENDIX-1 BURST AGGREGATION POLICY

For maximum burst size we consider the two distributions namely Poisson and Pareto and label the two arrival processes as $C_1(t)$ and $C_2(t)$, where $C_i(t)$ represents the number of packets that arrived in interval (0,t) due to process C_1. If d_1 and d_2 are the maximum permissible delays for each packet type such that $d_2<d_1$. Then the max burst size can be given as:

$B=C_1(d_1)$ if only Pareto packets arrive
$B=C_2(d_2)$ is only Poisson packets arrive
$B=C_1(k+d_2)+C_2(d_2)$ if first Poisson packet arrives k interval after 1st Pareto packet arrives
$B=C_1(d_2-k)+C_2(d_2)$ if 1st Pareto packet arrives after k interval and $k<d_2$
$B=C_1(d_1)+C_2(d_1-k)$ if 1st Poisson packet arrives after k interval of 1st Pareto packet and $d_1-k<d_2$

Session 13F

Fiber Gratings

Recent Advances in the Analysis of Long-Period Fibre Gratings

Kin Seng Chiang
Optoelectronics Research Centre and Department of Electronic Engineering
City University of Hong Kong, 83 Tat Chee Ave., Kowloon, Hong Kong
P. R. China
Tel: (852) 2788 9605; Fax: (852) 2788 7791; E-mail: eeksc@cityu.edu.hk

Abstract: Analytical formulas are presented to describe the transmission characteristics of a long-period fibre grating (LPFG) in response to changes in temperature, strain, pressure, external refractive index, etc. The effects that arise from the modal dispersion in the fibre are highlighted. The results are useful for the evaluation and the design of LPFG-based devices and sensors.

Key Words: Long-period fibre grating; fibre device; fibre sensor; fibre dispersion

1. INTRODUCTION

A long-period fibre grating (LPFG) is usually formed by introduction of a periodic modulation in the refractive index with a pitch of the order of 100μm along the core of a photosensitive single-mode optical fibre. The transmission spectrum of a typical LPFG consists of a number of rejection bands (resonance wavelengths), which arise from light coupling from the fundamental mode to the forward-propagating cladding modes of the fibre [1]. LPFGs have found many applications as filters [1]-[4], mode converters [5], EDFA gain equalisers [6]-[8], sensors [9]-[12], and add/drop multiplexers [13],[14]. In general, the transmission spectrum of an LPFG is sensitive to environmental changes, such as temperature, strain, pressure, and surrounding index changes, and this phenomenon has been explored for the realization of LPFG-based sensors [9]-[12]. However, a high sensitivity to environmental changes can be a serious problem in the development of stable LPFG-based devices. On the other hand, one can take advantage of the high sensitivity to a specific parameter to realize wavelength-tunable devices.

For the design and packaging of LPFG filters and sensors, it is necessary to understand how the transmission characteristics of the LPFG respond to changes in the environmental and the fibre parameters. In this paper, analytical formulas are presented to describe the sensitivities of an LPFG to a large number of parameters, including temperature, strain, hydrostatic pressure, high-frequency acoustic pressure, external refractive index, and fibre radius. The effects that arise from the modal dispersion in the fibre are highlighted and discussed.

2. SENSITIVITY FORMULAS

The resonance wavelength of an LPFG λ_o is determined by the phase-matching condition [1]:

$$\lambda_o = (N_{01} - N_{0m})\Lambda \tag{1}$$

where N_{01} and N_{0m} are the effective indices of the LP_{01} and LP_{0m} modes, respectively, and Λ is the pitch of the grating. To obtain the sensitivity of the resonance wavelength to a particular parameter, one can differentiate Eq.(1) with respect to that parameter, bearing in mind that the effective indices are functions of the wavelength [15].

2.1. Temperature

From Eq.(1), the temperature sensitivity of the resonance wavelength can be expressed as

$$\frac{d\lambda_o}{dT} = \Lambda\left(\frac{dN_{01}}{dT} - \frac{dN_{0m}}{dT}\right) + (N_{01} - N_{0m})\frac{d\Lambda}{dT} \tag{2}$$

where T is the temperature. Because the thermal expansion coefficient of silica is small (~$0.5\text{x}10^{-6}$/°C), the second term in Eq.(2) due to $d\Lambda/dT$ (which gives a shift in λ_o of less than 0.001nm/°C) can normally be neglected. Therefore, it is the temperature dependence of the effective indices that determines mainly the temperature sensitivity of λ_o. Taking into account the wavelength dependence of the effective indices, Eq.(2) with the second term ignored becomes

$$\frac{d\lambda_o}{dT} \cong \Lambda\left[\left(C_{co}\frac{\partial N_{01}}{\partial n_{co}} + C_{cl}\frac{\partial N_{01}}{\partial n_{cl}} - C_{co}\frac{\partial N_{01m}}{\partial n_{co}} - C_{cl}\frac{\partial N_{0m}}{\partial n_{cl}}\right) + \left(\frac{dN_{01}}{d\lambda} - \frac{dN_{0m}}{d\lambda}\right)\frac{d\lambda_o}{dT}\right] \tag{3}$$

where n_{co} and n_{cl} are the refractive indices of the core and the cladding, respectively, and C_{co} and C_{cl} are the corresponding thermal-optic coefficients. According to a standard perturbation theory [16], $\partial N_{01}/\partial n_{co}$ is equal to the fractional power of the LP_{01} mode in the core and $\partial N_{01}/\partial n_{cl}$ is equal to the fractional power of the LP_{01} mode in the cladding. Similarly, $\partial N_{0m}/\partial n_{co}$ is equal to the fractional power of the LP_{0m} mode in the core (which is negligible) and $\partial N_{0m}/\partial n_{cl}$ is equal to the fractional power of the LP_{0m} mode in the cladding. Using these relations and neglecting small terms, Eq.(3) can be simplified to [17]

$$\frac{d\lambda_o}{dT} \cong \gamma\Lambda\eta_{01}(C_{co} - C_{cl}) \tag{4}$$

with

$$\gamma = \frac{1}{1 - \Lambda\left(\frac{dN_{01}}{d\lambda_o} - \frac{dN_{0m}}{d\lambda_o}\right)} \tag{5}$$

where η_{01} denotes the fractional power in the core for the LP_{01} mode, and γ is the modal dispersion factor [15],[17],[18],[19]. As shown explicitly by Eq.(4), the temperature sensitivity of λ_o can be positive or negative, depending on the sign of the modal dispersion factor γ and whether the thermo-optic coefficient of the core is larger or smaller than that of the cladding.

2.2. Strain

When an axial strain ε is applied along an LPFG, both the effective indices and the pitch of the grating are changed. By following the same procedure as described in the previous section, the strain sensitivity of the resonance wavelength is obtained as

$$\frac{d\lambda_o}{d\varepsilon} \cong \gamma[\Lambda\eta_{01}(Z_{co} - Z_{cl}) + \lambda_o] \tag{6}$$

with

$$Z_{co} = \frac{dn_{co}}{d\varepsilon} = -\tfrac{1}{2} n_{co}^3 p_{12}^{co} \tag{7}$$

$$Z_{cl} = \frac{dn_{cl}}{d\varepsilon} = -\tfrac{1}{2} n_{cl}^3 p_{12}^{cl} \tag{8}$$

where p_{12}^{co} and p_{12}^{cl} are the cross-coupling strain-optic coefficients of the core and the cladding, respectively [20].

2.3. Hydrostatic Pressure

When a hydrostatic pressure p is applied to a bare LPFG, both the lateral strain and the axial strain are given by $-(1-2\nu)p/E$, where ν and E are Poisson's ratio and Young's modulus of the glass fibre. By applying the same procedure to Eq.(1), the sensitivity of the resonance wavelength is found as

$$\frac{d\lambda_o}{dp} \cong -\gamma[\Lambda\eta_{01}(P_{co} - P_{cl}) + \lambda_o](1-2\nu)/E \tag{9}$$

with

$$P_{co} = -\tfrac{1}{2} n_{co}^3 (p_{11}^{co} + 2p_{12}^{co}) \tag{10}$$

$$P_{cl} = -\tfrac{1}{2} n_{cl}^3 (p_{11}^{cl} + 2p_{12}^{cl}) \tag{11}$$

where p_{11}^{co} and p_{11}^{cl} are the self-coupling strain-optic coefficients of the core and the cladding, respectively [20].

2.4. High-Frequency Acoustic Wave

When an acoustic wave is incident upon an LPFG, a time-varying pressure is produced on the grating. At a low acoustic frequency (below 1kHz), the pressure is hydrostatic. At a sufficiently high frequency (from 10kHz to 1MHz in liquid), however, the axial strain can be assumed zero because of the fibre inertia, and the pressure can be regarded as an axially constrained radial pressure, which is denoted here as q. The lateral strain in this case is given by $-(1-\nu-2\nu^2)q/E$. The sensitivity of the resonance wavelength to q is thus obtained as

$$\frac{d\lambda_o}{dq} \cong -\gamma\Lambda\eta_{01}(Q_{co} - Q_{cl})(1-\nu-2\nu^2)/E \tag{12}$$

with

$$Q_{co} = -\tfrac{1}{2} n_{co}^3 (p_{11}^{co} + p_{12}^{co}) \tag{13}$$

$$Q_{cl} = -\tfrac{1}{2} n_{cl}^3 (p_{11}^{cl} + p_{12}^{cl}) \tag{14}$$

It can be seen that $P_{co} = Q_{co} + Z_{co}$ and $P_{cl} = Q_{cl} + Z_{cl}$.

2.5. External Refractive Index

When the refractive index that surrounds an LPFG (external refractive index) changes, the effective index of the cladding mode, and hence, the resonance wavelength of the LPFG changes accordingly. For the calculation of the effective index of the cladding mode, the fibre core can be ignored and the effective index can be obtained from a simple asymptotic expression [11]. The sensitivity of the resonance wavelength to the external refractive index is thus found as

$$\frac{d\lambda_o}{dn_{ex}} \cong -\frac{\gamma\Lambda n_{ex} u_m^2 \lambda_o^3}{8\pi^3 \rho^3 n_{cl} (n_{cl}^2 - n_{ex}^2)^{3/2}} \tag{15}$$

where ρ is the radius of the fibre cladding and u_m is the m-th root of the Bessel function J_0 for the LP_{0m} mode.

2.6. Fibre Radius

It has been demonstrated that the resonance wavelength of an LPFG can be shifted to a longer wavelength by etching the LPFG with HF acid, i.e., by reducing the fibre radius [11],[21]. As shown by Eq.(15), an etched LPFG is more sensitive to the external index [11]. The sensitivity of the resonance wavelength to the fibre radius is given by

$$\frac{d\lambda_o}{d\rho} \cong -\frac{\gamma\Lambda u_m^2 \lambda_o^2}{4\pi^2 \rho^3 n_{cl}} \tag{16}$$

2.7. Effects on the Grating Strength

In general, the strength of the LPFG, i.e., the contrast at the resonance wavelength, changes together with the shift in the resonance wavelength. According to the coupled-mode theory [1], the transmission at the resonance wavelength, denoted as P_o, is given by

$$P_o = \cos^2(\kappa L) \tag{17}$$

where L is the length of the grating and κ is the coupling coefficient, which arises from the index modulation along the core of the fibre. Assuming a sinusoidal index modulation with amplitude Δn_{co}, the coupling coefficient is given by

$$\kappa = \frac{\pi\Delta n_{co} I}{\lambda_o} \tag{18}$$

where I is the overlap integral over the core area of the fibre between the fields of the LP_{01} and LP_{0m} modes. The variation of P_o with respect to a parameter X ($X = T$, ε, p, q, n_{ex}, or ρ) can be expressed as

$$\frac{dP_o}{dX} = -\kappa L \sin(2\kappa L)\left(\frac{1}{I}\frac{dI}{dX} - \frac{1}{\lambda_o}\frac{d\lambda_o}{dX} + \frac{1}{L}\frac{dL}{dX}\right) \quad (19)$$

In general, the variation dI/dX can be expanded as

$$\frac{dI}{dX} = \frac{\partial I}{\partial n_{co}}\frac{\partial n_{co}}{\partial X} + \frac{\partial I}{\partial n_{cl}}\frac{\partial n_{cl}}{\partial X} + \frac{\partial I}{\partial n_{ex}}\frac{\partial n_{ex}}{\partial X} + \frac{\partial I}{\partial \lambda_o}\frac{\partial \lambda_o}{\partial X} \quad (20)$$

Of course, for a specific parameter X, not every term in Eqs.(19) and (20) is significant or even exists. In most cases (except for $X = n_{ex}$), only the last term in Eq.(20) is important. With all the insignificant terms neglected, the sensitivity expressions for various parameters are obtained as

$$\frac{dP_o}{dT} \cong -\kappa L \sin(2\kappa L)\left(\frac{1}{I}\frac{dI}{d\lambda_o} - \frac{1}{\lambda_o}\right)\frac{d\lambda_o}{dT} \quad (21)$$

$$\frac{dP_o}{d\varepsilon} \cong -\kappa L \sin(2\kappa L)\left[\left(\frac{1}{I}\frac{dI}{d\lambda_o} - \frac{1}{\lambda_o}\right)\frac{d\lambda_o}{d\varepsilon} + 1\right] \quad (22)$$

$$\frac{dP_o}{dp} \cong -\kappa L \sin(2\kappa L)\left[\left(\frac{1}{I}\frac{dI}{d\lambda_o} - \frac{1}{\lambda_o}\right)\frac{d\lambda_o}{dp} - \frac{1-2\nu}{E}\right] \quad (23)$$

$$\frac{dP_o}{dq} \cong -\kappa L \sin(2\kappa L)\left(\frac{1}{I}\frac{dI}{d\lambda_o} - \frac{1}{\lambda_o}\right)\frac{d\lambda_o}{dq} \quad (24)$$

$$\frac{dP_o}{dn_{ex}} \cong -\kappa L \sin(2\kappa L)\left[\left(\frac{1}{I}\frac{dI}{d\lambda_o} - \frac{1}{\lambda_o}\right)\frac{d\lambda_o}{dn_{ex}} + \frac{1}{I}\frac{dI}{dn_{ex}}\right] \quad (25)$$

$$\frac{dP_o}{d\rho} \cong -\kappa L \sin(2\kappa L)\left(\frac{1}{I}\frac{dI}{d\lambda_o} - \frac{1}{\lambda_o}\right)\frac{d\lambda_o}{d\rho} \quad (26)$$

Clearly, the sensitivity of the grating strength depends on, among other things, the sensitivity of the resonance wavelength. Its sign depends also on the range of κL because of the periodic function $\sin(2\kappa L)$. It is important to note that the magnitude of dP_o/dX is amplified by the factor κL. While the same grating strength can be achieved with many possible values of κL according to Eq.(17), a smaller value of κL gives a smaller sensitivity in the grating strength. When $\kappa L = n\pi/2$ (n = 1,2,3,...), the sensitivity of the grating strength vanishes.

3. MODAL DISPERSION FACTOR

As shown by the formulas given in the previous section, the sensitivities of the resonance wavelength are proportional to the modal dispersion factor, which is given by Eq.(5). In this section, an accurate analytical formula for γ is presented to illustrate its characteristics. To derive the formula from Eq.(5), the effective index of the LP_{01} mode is obtained from the well-known empirical dispersion formula for a single-mode step-index fibre [22], while the effective index of the LP_{0m} cladding mode is obtained from an asymptotic dispersion expression for a multimode step-index fibre [11],[23]. The result is

$$\gamma = \frac{1}{1 - \Lambda(D_M + D_W)} \quad (27)$$

with D_M, the material dispersion, given by

$$D_M \cong \eta_{01}\left(\frac{dn_{co}}{d\lambda_o} - \frac{dn_{cl}}{d\lambda_o}\right) \quad (28)$$

and D_W, the waveguide dispersion, given by

$$D_W \cong \frac{u_m^2 \lambda_o}{4\pi^2\rho^2 n_{cl}} - \frac{0.996}{2\pi a n_{cl}}\left[1.1428(n_{co}^2 - n_{cl}^2)^{1/2} - \frac{0.996\lambda_o}{2\pi a}\right] \quad (29)$$

where a is the radius of the fibre core. The factional power in the core η_{01} in Eq.(28) can be approximated by

$$\eta_{01} \cong 1.1428\left[1.1428 - \frac{0.996\lambda_o}{2\pi a(n_{co}^2 - n_{cl}^2)^{1/2}}\right] \quad (30)$$

When the materials of the core and the cladding are known, the material dispersion D_M can be computed from the Sellmeier equations of the materials. With the knowledge of the fibre parameters, the waveguide dispersion D_W can be calculated directly from Eq.(29).

As an example, consider a B-Ge co-doped fibre, which has a silica core of 3.8 μm in radius, doped with 9.7Mol% B_2O_3 and 4.03Mol% GeO_2, and a pure silica cladding of 62.5 μm in radius [19]. The inverse of the modal dispersion factor $1/\gamma$ is plotted in Fig.1 as a function of the mode order m at the resonance wavelength 1550 nm, where two curves are shown, one for the waveguide dispersion only and the other including both waveguide dispersion and material dispersion. It is important to note that the factor $1/\gamma$ changes sign (from positive to negative) as the mode order increases, which suggests that large values of γ are expected for the modes near the zero-crossing point. The sensitivity of the resonance wavelength to any of the parameters considered in this paper should be particularly large for these modes. This phenomenon has in fact been characterized in great detail for the temperature, strain, and external index sensitivities [19]. In principle, it is possible to achieve an infinite sensitivity by locating a mode order right at the zero-crossing point of the $1/\gamma$ curve. This could be done by solving the equation $\Lambda(D_M + D_W) = 1$ for a specific cladding mode (i.e., by fixing a proper value of u_m). In practice, however, this zero-crossing point is difficult to determine accurately, because of its high sensitivity to the fibre parameters. As shown in Fig.1, neglecting the apparently unimportant material dispersion can shift the zero-crossing point quite significantly. In Fig.2, the factor γ is shown for three different resonance wavelengths. For this particular fibre, the peak value of γ moves towards a lower mode order as the resonance wavelength increases, but this should not be considered as a general trend for all fibres. It is clear from these results that an exceptionally high sensitivity in the resonance wavelength can be achieved with only a few high-order modes. On the other hand, to achieve a low sensitivity, one should use low-order modes. For this fibre, at the resonance wavelength 1550 nm, the highest sensitivity of λ_o is obtained with the LP_{011} mode. The sensitivities measured for this mode are –1.854 nm/°C for the temperature, 30.31 nm/1000$\mu\varepsilon$ for the strain, and –1.481 nm/10^{-3} for the external index [19]. On the other hand, the corresponding sensitivities for the LP_{02} mode are only –0.327 nm/°C, 0.204 nm/1000$\mu\varepsilon$, and –0.003 nm/10^{-3} [19]. The consideration of the sensitivity in the strength of the grating is more complicated because of the influence by the value of κL. In general, however, a highly sensitive resonance wavelength goes along with a highly sensitive grating contrast, as shown by the recent experimental results on the temperature effects [17].

4. CONCLUSION

Analytical formulas have been presented to describe the sensitivities of the resonance wavelength and the strength of an LPFG to a wide range of parameters. The role of the modal dispersion factor of the fibre in affecting these sensitivities has been highlighted. It has been shown that an exceptionally high sensitivity is possible with a proper choice of a high-order cladding mode. On the other hand, to achieve a low sensitivity, one should use low-order modes. These formulas are useful for the evaluation and the design of LPFG-based devices and sensors.

This work was supported by a grant from the Research Grants Council of the Hong Kong Special Administrative Region, China [Project No. CityU 1041/99E]. Miss M. N. Ng's assistance in computer simulations is gratefully acknowledged.

REFERENCES

[1] A. M. Vengsarkar, J. P. Lemaire, J. B. Judkins, V. Bhatia, T. Erdogan, and J. E. Sipe, *J. Lightwave Technol.*, LT-14 (1996) 58.

[2] L. Tallone, L. Boschis, L. Cognolato, E. Emelli, E. Riccardi, O. Rossotto, *OFC'97* (1997) 175.

[3] H. Ke, K. S. Chiang, and J. H. Peng, *IEEE Photonics Technol. Lett.*, 10 (1998) 1596.

[4] A. A. Abramov, A. Hale, R. S. Windeler, and T. A. Strasser, *Electron. Lett.*, 35 (1999) 81.

[5] D. Ostling and H. E. Engan, *Opt. Lett.*, 21 (1996) 192.

[6] A. M. Vengsarkar, J. R. Pedrazzani, J. B. Judkins, P. J. Lemaire, N. S. Bergano, and C. R. Davidson, *Opt. Lett.*, 21 (1996) 336.

[7] M. Akiyama, K. Nishide, K. Shima, T. Sakai, A. Wada, and R. Yamauchi, *OECC'98* (1998) 150.

[8] M. K. Pandit, K. S. Chiang, Z. Chen, and S. Li, *Microwave Opt. Technol. Lett.*, 25 (2000) 181.

[9] V. Bhatia and A. M. Vengsarkar, *Opt. Lett.*, 21 (1996) 692.

[10] H. J. Patrick, A. D. Kersey, and F. Bucholtz, *J. Lightwave Technol.*, 16 (1998) 1606.

[11] K. S. Chiang, Y. Liu, M. N. Ng, and X. Dong, *Electron. Lett.*, 36 (2000) 966.

[12] M. N. Ng, Z. Chen, and K. S. Chiang, *IEEE Photon. Technol. Lett.*, 14 (2002) 361.

[13] K. S. Chiang, Y. Liu, M. N. Ng, and S. Li, *Electron. Lett.*, 36 (2000) 1408.

[14] K. S. Chiang, M. N. Ng, Y. Liu, and S. Li, *LEOS 2000*, (2000) 836.

[15] T. W. MacDougall, S. Pilevar, C. W. Haggans, and M. A. Jackson, *IEEE Photon. Technol. Lett.*, 10 (1998) 1449.

[16] K. S. Chiang, *Electron. Lett.*, 23 (1987) 717.

[17] M. N. Ng and K. S. Chiang, *Optics Commun.*, (2002) (to appear).

[18] X. Shu, T. Allsop, B. Gwandu, L. Zhang, and I. Bennion, *IEEE Photon. Technol. Lett.*, 13 (2001) 818.

[19] X. Shu, L. Zhang, and I. Bennion, *J. Lightwave Technol.*, 20 (2002) 255.

[20] G. B. Hocker, *Appl. Opt.*, 18 (1979) 1445.

[21] S. A. Vasiliev, E. M. Dianov, D. Varelas, H. G. Limberger, and R. P. Salathe, *Opt. Lett.*, 21 (1996) 1830.

[22] H. D. Rudolph and E. G. Neumann, *Nachrichtentech. Z.*, 4 (1976) 328.

[23] M. Miyagi and S. Nishida, *J. Opt. Soc. Am.*, 69 (1979) 291.

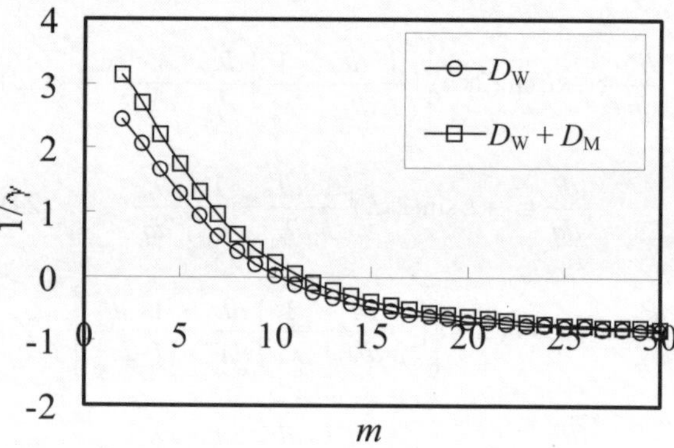

Fig. 1. The inverse of the modal dispersion factor, $1/\gamma$, as a function of the mode order m at λ_o = 1550 nm.

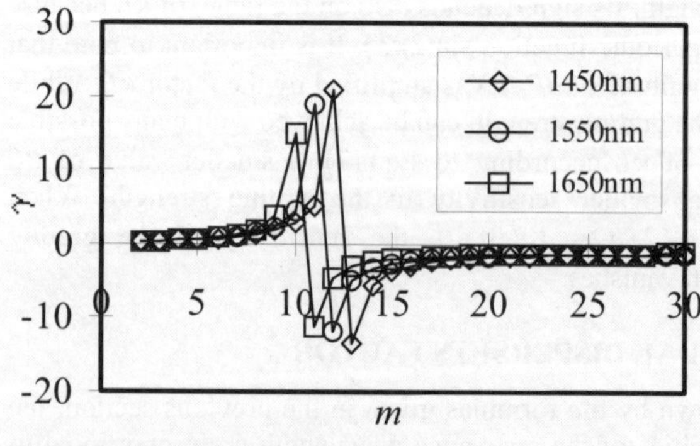

Fig. 2. The modal dispersion factor γ as a function of the mode order m at three different resonance wavelengths.

Theoretical investigation of wavelength response characteristics of Fiber Bragg grating under radial compression

Jinu Paul *, B.K.A.Ngoi
School of Mechanical and Production Engineering, 50 Nanyang Avenue, Nanyang Technological University, Singapore 639798, E-mail*: pa8669753@ntu.edu.sg Tel: (++65) 6793 2072
Zhao Liping, Fang Zhong Ping
Singapore Institute of Manufacturing Technology, 71 Nanyang Drive, Singapore 638075

Abstract

Theoretical investigation of fiber Bragg grating (FBG) under two cases of lateral compression has been carried out. Lateral pressure is applied over two different angles of 20^0 and 100^0 along the circumference of the fiber. The shift in the center wavelength of FBG with respect to the radial pressure is analysed. Pressure sensitivity is more than 4 times bigger if the pressure is distributed over a large angle. The strain-induced birefringence created due to the refractive index difference in the radial directions has also been presented. The simulation results show that distributing the pressure can reduce the birefringence effects.

Key Words: Fiber Bragg gratings, Radial Compression, Birefringence.

1. INTRODUCTION

Fiber Bragg gratings (FBGs) are extensively used in optical communication and sensing systems. Many of these applications make use of the advantage of Bragg center wavelength shift due to an induced gradient in effective refractive index or the period of the grating. Various techniques have been developed in order to accomplish this shift in the center wavelength. Some methods for tuning the gratings include mechanical stretching [1], compression tuning [2] and thermal tuning [3]. Bragg gratings can also be tuned by applying a lateral stress. T. Komukai et al. experimentally evaluated the case of FBG compressed between two metal rods enclosed in a plastic sleeve [4]. Lateral compression of FBG considering the load as line force has also been reported [5].

Here we theoretically investigate the spectral response of FBG under two cases of lateral compression. In case 1, the lateral force is applied over an angle of 20^0 along the circumference of the fiber, which is analogous to a FBG compressed between two plates. In case 2, the pressure is distributed over an angle of 100^0 producing a nearly hydrostatic effect. The wavelength response and birefringence effects in both cases have been analysed and compared.

2. THEORY

2.1 Strain dependence of Bragg center wavelength

The center wavelength λ_b of a Bragg grating is given by [7] the condition

$$\lambda_b = 2n_{eff}.\Lambda \qquad (1)$$

where n_{eff} is the effective refractive index of the fiber core at free space center wavelength and Λ is the spacing of the grating. The above equation can be differentiated to find the relative shift in Bragg center wavelength with respect to effective refractive index and grating period.

$$d\lambda_b = 2n_{eff}.d\Lambda + 2\Lambda.dn_{eff} \qquad (2)$$

Dividing equation (2) by equation (1) gives the relative wavelength gradient.

$$\frac{d\lambda_b}{\lambda_b} = \frac{dn_{eff}}{n_{eff}} + \frac{d\Lambda}{\Lambda} \qquad (3)$$

The strain dependence of the Bragg center wavelength can be formulated as [6]

$$\frac{\Delta\lambda_b}{\lambda_b} = \varepsilon_z - \frac{n_{co}^2}{2}\left[\varepsilon_r\left(P_{11} + P_{12}\right) + P_{12}\varepsilon_z\right] \qquad (4)$$

where $\frac{\Delta\lambda_b}{\lambda}$ is the relative shift in the center wavelength, ε_z is the axial strain, ε_r is the radial strain, n_{co} is the refractive index of the fiber core P_{11} and P_{12} are the photoelastic coefficients of silica in the axial and radial directions respectively. For an isotropic fiber, the radial strain can be written in terms of axial strain using Poisson's law as

$$\varepsilon_r = -\nu\varepsilon_z \qquad (5)$$

where ν is the Poisson's ratio of the fiber material.

Taking the values of photoelastic coefficients as P_{11}=0.113 and P_{12}=0.252, refractive index n_{co} = 1.51 and ν=0.16, the relative wavelength shift

$$\frac{\Delta\lambda_b}{\lambda} = 0.78\varepsilon_z \qquad (6)$$

2.2 Birefringence

This strain induced birefringence B can be calculated as [5]

$$B = \Delta n_x - \Delta n_y \cong \frac{n^3}{2}(P_{12} - P_{11})(\varepsilon_x - \varepsilon_y) \quad (7)$$

where Δn_x-Δn_y is the difference in refractive index in the radial X and Y directions and ε_x, ε_y are the radial strains in the X and Y directions respectively.

3. MODELS FOR FINITE ELEMENT SIMULATION

A finite element model of the fiber has been developed in order to analyse various cases of lateral compression. The fiber is considered as a homogeneous, isotropic, axis symmetric material. The properties of the fiber used for simulation are, Young's modulus E = 72.5 Gpa, Poisson's ratio ν = 0.16. Since the fiber core and cladding is made of same material and their elastic properties are almost same, the entire fiber is modelled as an isotropic material. Plane stress conditions are assumed where the longitudinal stress along the axis of the fiber is zero.

Fig.1b. and Fig.1c. shows the two cases of loading geometry of the cylindrical fiber.

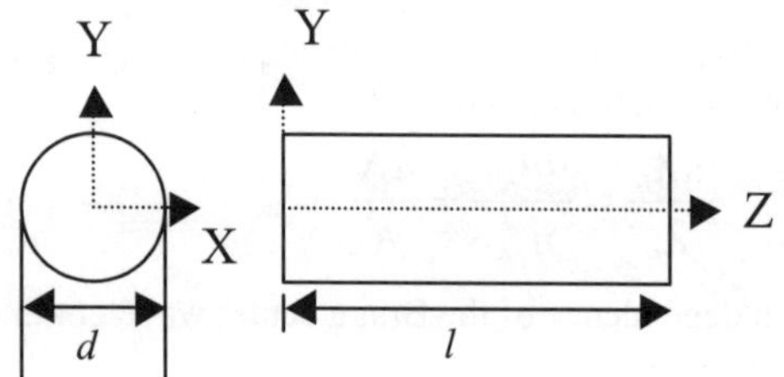

Fig. 1a. Co-ordinate system

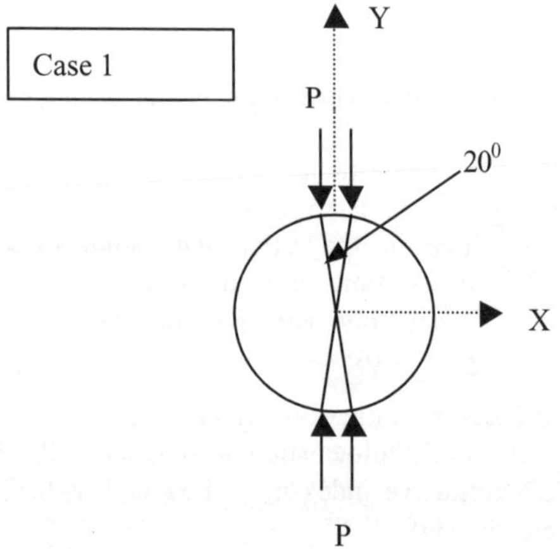

Fig.1b. Loading geometry in which the pressure acts over an angle of 20^0

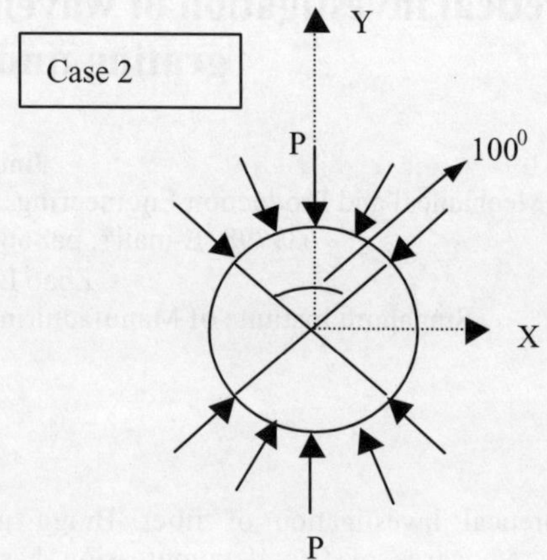

Fig.1c. Loading geometry in which the pressure acts over an angle of 100^0

The simplest case of lateral compression is the case of a FBG compressed between two plates. In this case, the contact angle is very small and the lateral load can be considered as a line force. But our interest is to analyse the strain characteristics when the applied load is distributed along the circumference of the fiber. In case 1, we apply the lateral load over a very small angle of 20^0. The applied pressure varies from 0.1 to 0.9 kN/mm^2. The pressure is applied in the Y-direction, perpendicular to the axis of the fiber.

In case 2, the applied pressure is distributed over an angle of 100^0. Thus an almost hydrostatic loading condition is maintained. Same pressure range as in case 1 is maintained for comparison purposes.

When a cylindrical elemental volume is laterally loaded [8], the stress strain relations can be written in the matrix form as,

$$\begin{bmatrix} \sigma_r \\ \sigma_z \\ \tau_{rz} \\ \sigma_\theta \end{bmatrix} = \frac{E(1-\nu)}{(1+\nu)(1-2\nu)} \begin{bmatrix} 1 & \frac{\nu}{1-\nu} & 0 & \frac{\nu}{1-\nu} \\ \frac{\nu}{1-\nu} & 1 & 0 & \frac{\nu}{1-\nu} \\ 0 & 0 & \frac{1-2\nu}{2(1-\nu)} & 0 \\ \frac{\nu}{1-\nu} & \frac{\nu}{1-\nu} & 0 & 1 \end{bmatrix} \begin{bmatrix} \epsilon_r \\ \epsilon_z \\ \gamma_{rz} \\ \epsilon_\theta \end{bmatrix} \quad (8)$$

where σ_r, σ_z, σ_θ, and τ_{rz} are the stress components in the radial, axial, circumferential and shear directions and ε_r, ε_z, ε_θ, γ_{rz} are the respective strains. E is the modulus of elasticity and ν is the Poisson's ratio.

4. RESULTS AND DISCUSSION

The finite element model of the original shape and the deformed shape of the fiber after loading are as shown in Fig. 2a. and Fig. 2b. The dotted line represents the pre-loading condition and the firm lines represent the deformed shape.

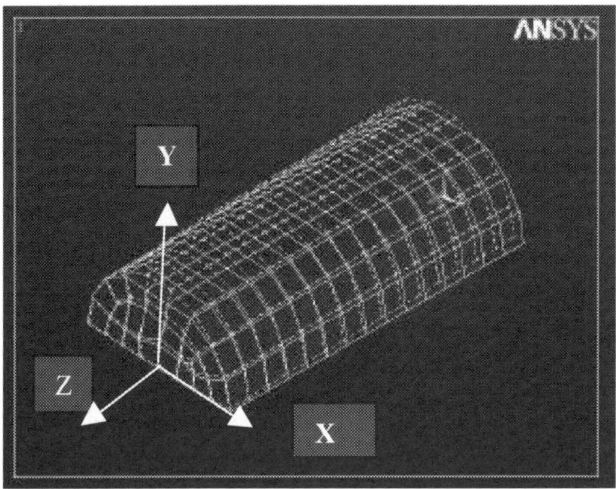

Fig.2a. Isometric view of finite element model

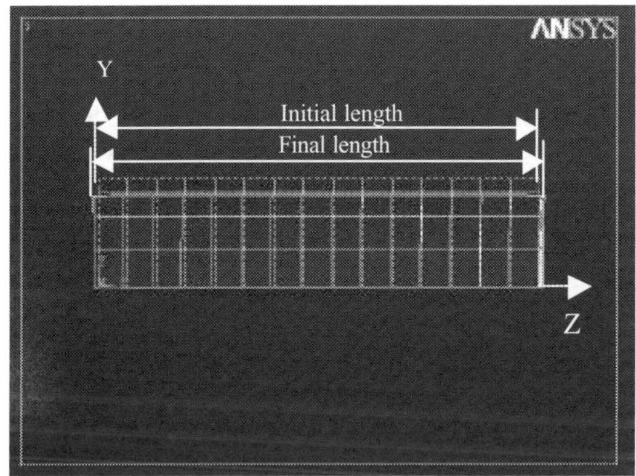

Fig.2b. Side view of finite element model

4.1 Wavelength shift

The wavelength shift with respect to the radial pressure is calculated as per equation (4) for both the cases and shown in Fig.3. The absolute value of wavelength shift is calculated from the relative shift at a center wavelength of 1550 nm. It can be seen that the shift in the Bragg center wavelength is directly proportional to the axial strain.

The sensitivity of wavelength shift to applied pressure in case 2 is much more (more than four times) than that in case 1. A radial pressure of 0.5 kN/mm^2 produces only a 0.3 nm shift in the wavelength in case 1 where as in case 2, the same pressure gives a shift of about 1.48 nm. So the pressure sensitivity is 0.6 nm per kN/mm^2 in case 1 and 2.9 nm per kN/mm^2 in case 2. In case 1, since the pressure is acting on a small contact angle, the ability to produce an axial strain is low.

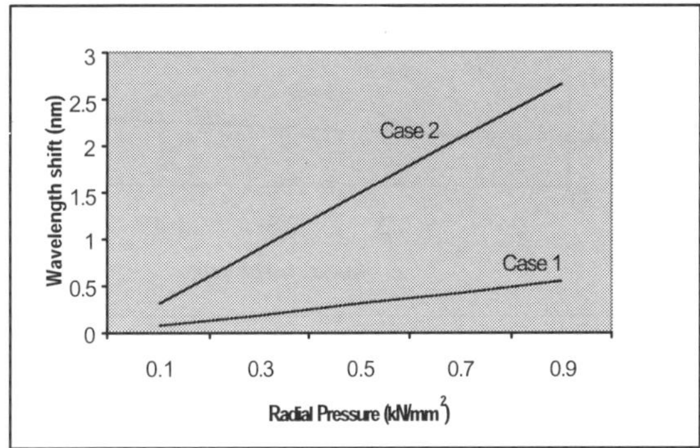

Fig.3. Radial Pressure Vs Wavelength shift

Pressure applied over a small angle induces a high radial strain in the Y-direction and a smaller axial displacement compared to the radial displacement. Whereas in case 2, the axial strain produced is much bigger because the pressure is acting on more contact area and hence the total load acting per unit length of the fiber is high compared to the first case.

4.2 Birefringence

When a fiber is laterally compressed in the y direction (Figure1), the radial strain in the vertical direction (Y-direction) and horizontal direction (X-direction) is not same. The radial strain in Y-direction ε_y is compressive in nature whereas the radial strain in the X-direction ε_x is tensile (Fig.4.).

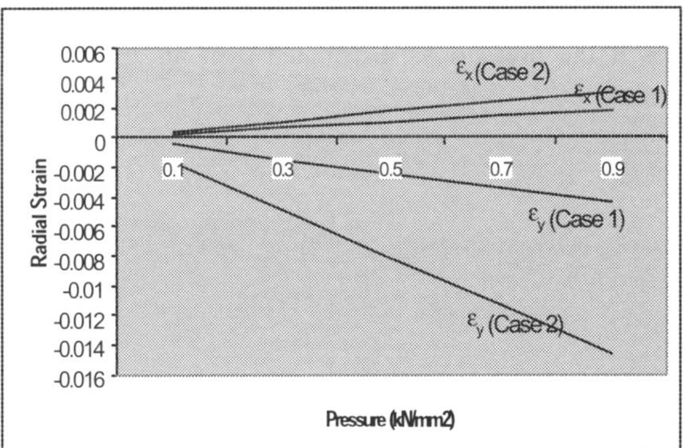

Fig.4. Radial Strain

This differential strain causes a gradient in the refractive indices in the two perpendicular radial directions (X and Y directions). This strain-induced birefringence is calculated according to equation (7) and is shown in Fig.5.

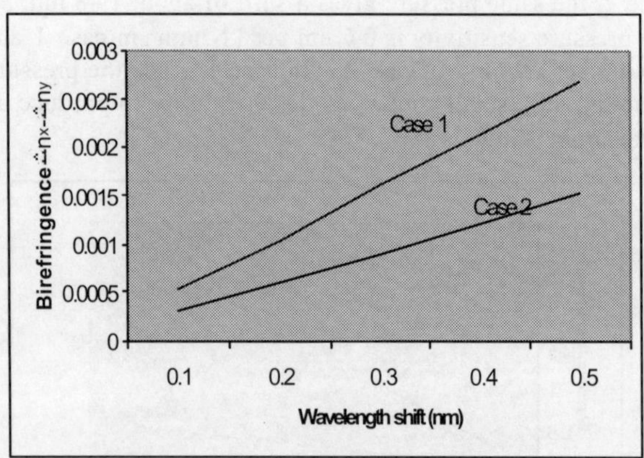

Fig.5. Wavelength shift Vs Birefringence

It can be seen from the graph that the birefringence produced in the case of concentrated pressure is much more than that in the case of distributed pressure. A wavelength shift of 0.5 nm is associated with a birefringence of 0.0015 in case 2. Whereas the same wavelength shift in case 1 induces a bifringence, which is almost double as that in case 2. In case 1, almost all the components of pressure are acting in Y-direction only. This produces a high compressive strain in the Y direction and tensile strain in the X-direction. In case 2, the applied pressure gets resolved into horizontal X and vertical Y components. So the difference in the strain in X and Y-directions is less compared to the first case.

4. CONCLUSION

Finite element simulation of fiber Bragg grating under two cases of lateral loading is done. Wavelength response characteristics of these cases are studied. It is found that the sensitivity of wavelength shift to lateral pressure is more significant in the case of distributed pressure. It can be concluded that the pressure sensitivity is more than four times if the pressure is distributed over an angle of 100^0. Also the birefringence effects are analysed. The birefringence produced per unit shift in the Bragg center wavelength is almost double in the case of concentrated pressure when compared to distributed pressure.

REFERENCES

1. Alessandro Iocco, Hans Georg Limberger, Rene Paul Salathe, et al., Bragg grating fast tunable filter for wave length division multiplexing, Journal of light wave technology, Vol.17, No.7, page 1217-1221, 1999.
2. G.A.Ball and W.W.Morey, Compression tuned single frequency Bragg grating fiber laser, Optics letters, Vol.19, No.23, page 1979-1981, 1994
3. H.Mavoori, S.Jin, R.P.Espindola, and T.A.Strasser, Enhanced thermal and magnetic actuations for broad range tuning of fiber Bragg grating-based reconfigurable add-drop devices, Optics letters, Vol.24, No.11, page 714-716, 1999.
4. T.Komukai, Y.Miyajima, M.Nakazawa, In-line fiber grating type optical band pass filter tuned by applying lateral stress, Jpn. J. Appl.Physics, Vol.34, Part 2, page 306-308, 1995.
5. A-Ping Zhang, Bai-Ou Guan, Xiao-Mng Tao, Hwa-Yaw Tam, Experimental and theoretical analysis of fiber bragg gratings under lateral compression, Optics Communications, 206, page 81-87, 2002.
6. Sylvain Magne, Stephane Rougeault, Manuel Vilela and Pierre Ferdinand, State-of-strain evaluation with fiber Bragg grating rosettes: application to discrimination between strain and temperature effects in fiber sensors, Applied optics, Vol.36, No.36, pp 9437-9447.
7. Andreas Othonos, Kyriacos Kalli, *Fiber Bragg gratings, fundamentals and applications in telecommunication and sensing*, Artech House, Boston, 1999.
8. S.P.Timoshenko, J.N.Goodier, *Theory of Elasticity*, McGraw Hill, New York, 1970.

Optical fiber Bragg grating accelerometer with temperature insensitivity

Yinian Zhu, Ping Shum, and Chao Lu
Network Technology Research Centre, Nanyang Technological University, Singapore 639798
Phone: +65-6790 5363 Fax: +65-6792 6894 E-mail: eynzhu@ntu.edu.sg

B.M. Lacquet, P.L. Swart, A.A. Chtcherbakov
Center for Optical Communications, Rand Afrikaans University, Auckland Park 2006, South Africa

S.J. Spammer
Science and Technology Division, Corning Inc., New York 14831, USA

Abstract: A cantilever beam and fiber Bragg grating is used to measure acceleration. The cantilever induces strain on the grating resulting in a Bragg wavelength modification that is subsequently detected. The output signal is insensitive to temperature variations and for a temperature change from $-20\,^0C$ to $40\,^0C$, the output signal fluctuated less than 5% without any temperature compensation schemes. Because the sensor does not utilize expensive and complex demodulation techniques it is potentially inexpensive. For the experimental system a linear output range of 8 g could be detected.

Key Words: Fiber Bragg grating, accelerometer, insensitivity, sensor, temperature.

1. INTRODUCTION

The first fiber Bragg gratings were reported approximately two decades ago and since then applications of fiber Bragg gratings increased [1]. Because the optical line-width of the reflected Bragg wavelength can be in the order of 0.1 nm, the main area of interest has been in optical fiber communication systems such as amplifiers and dispersion compensation elements [2,3]. However, many optical fiber sensor applications are also based on this technology. Typically these sensors are embedded or epoxy onto a transducer to provide measurement of a variable. Any change in the gratings such as strain, pressure or temperature, will change the reflected Bragg wavelength [4-8].

Because acceleration will result in a wavelength shift of the grating, previously reported sensors make use of wavelength demodulation schemes such as scanning Fabry-Perot filters, spectrum analyzers or interferometric techniques [9,10]. In additional to the acceleration induced wavelength shift, the Bragg shift is also temperature dependent. Additional temperature compensation techniques are therefore necessary adding to system cost and complexity. This paper reports on the measurement of acceleration utilizing fiber Bragg grating that is not based on a wavelength detection scheme. This novel sensor configuration ensures automatic temperature insensitivity without any additional components and circuitry.

2. THEORY

Fiber Bragg gratings are formed by generating a permanent periodic modulation of the fiber's core refractive index using an ultraviolet laser with a wavelength around 248 nm. This periodic index modulation with period Λ can be written into Ge-doped single mode fiber core using various methods such as interferometry [11] or phase masks [12]. According to Bragg's law, the grating will only reflect a specific wavelength namely the Bragg wavelength that is given by

$$\lambda_B = 2n\Lambda \qquad (1)$$

where n is the core index of refraction. Any change in the refraction or the index modulation pitch will result in a shift of the Bragg wavelength. The strain and temperature dependent variations on the Bragg wavelength as a result of the photo-elastic and thermo-optic effect respectively, is given by [4]

$$\lambda_B = 2n\Lambda\left(\left\{1-\left(\frac{n^2}{2}\right)\left[P_{12}-\nu\left(P_{11}+P_{12}\right)\right]\right\}\Delta\varepsilon + \left[a+\frac{\left(\frac{dn}{dT}\right)}{n}\right]\Delta T\right) \qquad (2)$$

with P_{ij} the Pockel's coefficients of the stress-optic tensor, ν Poisson's ratio, a the coefficient of thermal expansion of the material, ΔT the temperature change, and $\Delta\varepsilon$ the strain variation. The second term in equation (2) incorporating the refractive index, Pockel's coefficients and Poisson's ratio, is known as the photo-elastic coefficient. Assuming bulk values for the strain-optic coefficients and low levels of germanium doping in the core, the photo-elastic coefficient is approximately equal to 0.22 [3]. This results in a theoretical strain dependent wavelength shift of approximately 1.2 pm / $\mu\varepsilon$ at a wavelength of 1550 nm.

The sensor is based on a cantilever structure consisting of a mass (m) at the end of a rectangular shaped beam with length L, width b, thickness d and Young's modulus E. To measure acceleration the grating is attached to the surface of the beam. If the structure is subjected to an acceleration a, the beam will bend resulting in a strain gradient along the length of the beam. The resulting strain at a given position x on the beam is given by $(0 < x < L)$ [13]

$$\varepsilon(x) = \frac{6(L-x)}{mbd^2E}a \tag{3}$$

From equation (3) it is clear that the strain will vary linearly along the length of the cantilever beam for an applied acceleration. If this linear strain gradient is applied to a uniform period grating it will result in a linear chirped grating with an FWHM line-width proportional to the strain gradient. The line-width difference between a chirped grating and a unchirped grating shown in Fig. 1 [14].

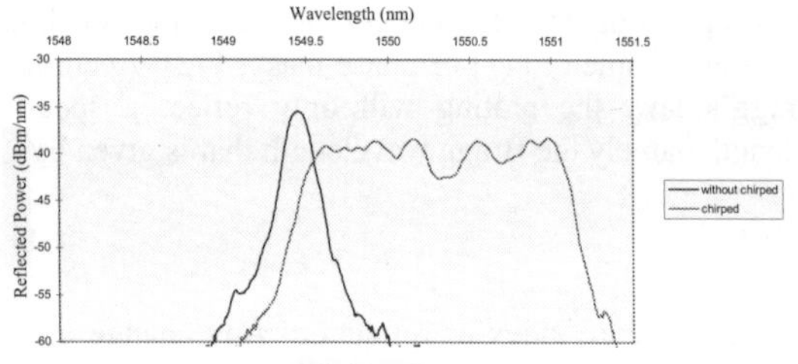

Fig.1 . Reflection spectra of both the uniform and the chirped fiber Bragg gratings.

Considering Fig.1 it can be seen that the total amount of reflected power for the chirped grating is more compare to the uniform period grating. By simply detecting this amount of reflected power the acceleration could be measured.

3. EXPERIMENTAL

The fiber Brag grating used in our experimental setup is 5 cm long with a center wavelength of 1549.7 nm, and an FWHM line-width of 0.13 nm. It was epoxy onto the surface of a stainless steel beam with a length of 5 cm, width of 5 mm and thickness 1 mm. A super luminescent diode (SLD) was used to illuminate the fiber via an optical circulator while the output was measured with an optical spectrum analyser as well as a pin-photodiode (PD). The experimental arrangement is shown in Fig. 2.

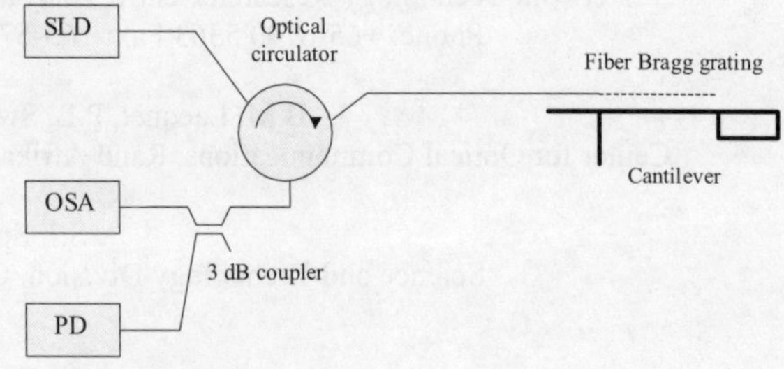

Fig. 2. Experimental setup.

Firstly the grating spectra and output optical power was measured for different static displacements of the cantilever end. For every increase in displacement of the beam tip, the line-width of the grating spectrum widens as a result of an increasing strain gradient on the grating. The resulting detector output signal for displacements varying from + 0.9 mm to – 0.6 mm is depicted in Fig. 3. It can be seen that the output is linear except in the case of large displacements. This is attributed to the non-linearity of the stress-strain relationship of the beam material above approximately 1200 $\mu\varepsilon$. The fact that zero deflection does not correspond to the minimum output voltage is as a result of non-uniform curing of the epoxy grating. However, a pre-strain on the grating in the zero deflection position is necessary in order to ensure that directional information of the acceleration could be obtain.

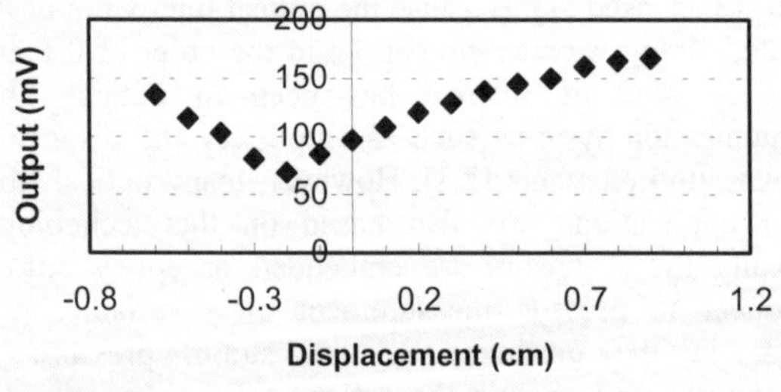

Fig. 3. Fiber Bragg grating output for beam tip displacement.

Because the sensor output is not dependent on an absolute wavelength detection technique, it should be insensitive to temperature fluctuations. From equation (2) it is clearly shown that the wavelength shift is temperature and strain dependent. If the sensor is subjected to temperature variations the Bragg wavelength will therefore shift as a result of the thermo-optic effect and the strain induced thermal expansion of the beam. However, this wavelength shift will not affect the output signal because the photodiode detects the total reflected power. Measurements were taken for zero beam deflection while the temperature was varied between $-20^{0}C$ to 40 ^{0}C. The corresponding wavelength shift as function of temperature is shown in Fig. 4.

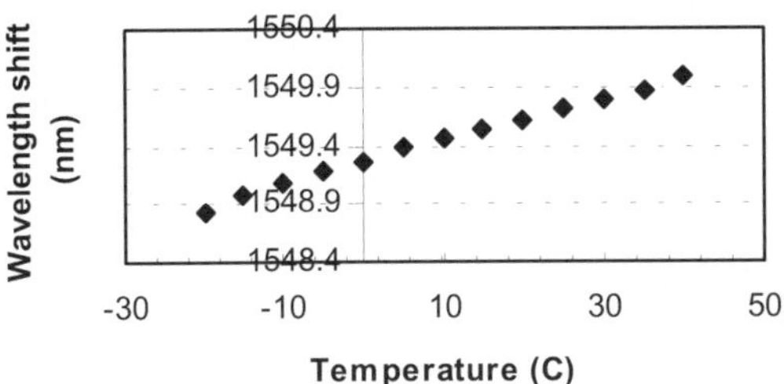

Fig. 4. Bragg wavelength shift vs temperature.

The sensor output for a constant beam deflection as function of temperature is depicted in Fig. 5. The output variation over the whole temperature range was less than 5 mV. Finally the output of the Bragg accelerometer was compared to the output of a piezo-resistive accelerometer (ICSensors 3021, sensitivity: 8.5 mV / g, maximum acceleration: 2 g). The amplified output signals from both sensors were obtained in our experiments, in which the maximum output on the piezo-resistive accelerometer is – 6.8 V. this corresponds to an acceleration of 1.48 g for a given an amplifier gain of 540.

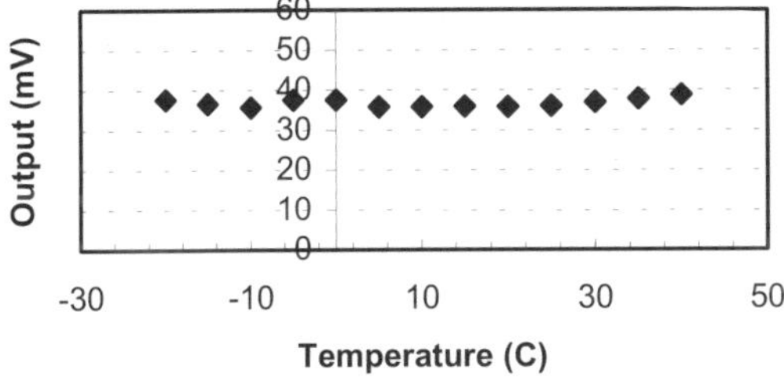

Fig. 5. Sensor output signal vs temperature.

4. CONCLUSION

The feasibility of a fiber Bragg grating accelerometer has been investigated by measuring the reflected optical power of a strain induced chirped fiber Bragg grating. The grating which was attached on the surface of a cantilever beam have a varying linear strain distribution along the beam length. Because the detection of the bending induced strain on the grating is not measured with any wavelength monitoring techniques, the sensor is insensitive to wavelength shift dependent temperature variations. For a temperature range from $-20^{0}C$ to $40^{0}C$, the output signal variation did not exceed 5 mV.

To obtain a linear output signal for a positive and negative acceleration, it is necessary to use a chirped fiber Bragg grating or a pre-strained uniform fiber Bragg grating with a linear strain distribution. Although the linear elastic range of the beam material used is approximately 1200 με, other material could be used resulting in a higher linear acceleration range of the sensor. Considering Fig. 4, the linear output range is approximately 100 mV for this specific experimental arrangement. This would imply a maximum linear acceleration range of approximately 8 g with a dynamic range of 26 dB. This by no means the optimum system design and these figures could improve significantly if a more thorough accelerometer design is conducted.

REFERENCES

[1] K. O. Hill, F. Fujii, D. C. Johnson, B. S. Kawasaki, Photosensitivity on optical fiber waveguides: application to reflection filter fabrication, Appl. Physics Lett., vol. 32, pp. 647-649, 1978.

[2] V. Mizrahi, Components and devices for optical communications based on UV-written-fiber phase gratings, Digest of OFC'93, San Jose, CA, pp. 243-244, 1993.

[3] J. A. R. Williams, I. Bennion, K. Sugden, N. J. Doran, Fiber dispersion compensation using a chirped in-fiber Bragg grating, Electron, Lett., vol. 30, pp. 985-987, 1994.

[4] A. D. Kersey, M. A. Davis, H. J. Patrick, M. LeBlanc, K. P. Koo, C. G. Askins, M. A. Putnam, E. J. Friebele, Fiber grating sensors, J. Lightwave Technol., vol. 15, pp. 1442-1462, 1997.

[5] S. M. Melle, T. Alavic, S. Karr, T. coroy, K. Lui, R. M. Measures, A Bragg grating-tuned fiber laser strain sensor system, IEEE Photon. Technol. Lett., vol. 5, pp. 516-518, 1992.

[6] Y. J. Rao, In-fiber Bragg grating sensors, Meas. Sci. Technol., vol. 8, pp. 366-374, 1997.

[7] W. W. Morey, G. Meltz, W. H. Glenn, Fiber optic Bragg grating sensors, SPIE-2718, pp. 98-107, 1989.

[8] R. T. Jones, T. A. Berkoff, D. G. Bellemore, D. A. Early, J. S. Sirkis, M. A. Putnam, E. J. Friebele, A. D. Kersey, Cantilever plate deformation monitoring using

wavelength division multiplexed fiber Bragg grating sensors, SPIE-2718, pp. 258-268, 1996.

[9] T. A. Beroff, A. D. Kersey, Experimental demonstration of a fiber Bragg grating accelerometer, IEEE Photon. Technol. Lett., vol. 8, pp. 1677-1679, 1996.

[10] M. D. Todd, G. A. Johnson, B. A. Althouse, S. T. Vohra, Flexural beam-based fiber Bragg grating accelerometers, IEEE Photon. Technol. Lett., vol. 10 pp. 1605-1607, 1998.

[11] G. Meltz, W. W. Morey, W. H. Glenn, Formation of Bragg gratings in optical fiber by a transverse holographic method, Opt. Lett., vol. 14, pp. 823-825, 1989.

[12] K. O. Hill, B. Malo, F. Bilodeau, D. C. Johnson, J. Albert, Bragg gratings fabricated in mono-mode photosensitive optical fiber by UV exposure through a phase mask, Appl. Physics Lett., vol. 62 pp. 1035-1037, 1993.

[13] P. L. Fuhr, S. J. Spammer, Y. Zhu, A novel signal demoduation technique for chirped Bragg grating strain sensors, Smart Mater. Struct., vol. 1 pp. 85-94, 2000.

[14] Y. Zhu, B. M. Lacquet, P. L. Swart, S. J, Spammer, Realization of chirped fiber Bragg gratings by using differently tapered transducers and loading procedures, Meas. Sci. Technol., vol. 12 pp. 922-926, 2001.

Strain and Temperature Response of Polymer Fiber Bragg Gratings

H. B. Liu, H. Y. Liu, R. Wang, and G. D. Peng
School of Electrical Engineering & Telecommunications
University of New South Wales, Sydney 2052, Australia
Phone: +61-2-9385 4014, Fax: +61-9385 5993
Email: g.peng@unsw.edu.au

Abstract-Strain and temperature response of polymer fiber gratings have been measured and analyzed. A rather large tuning range with excellent reversibility is obtained by simple tensile strain on polymer fiber Bragg gratings. Furthermore, polymer fiber Bragg gratings show much higher temperature sensitivity compared with silica fiber Bragg gratings. Very high linear response of Bragg wavelength to temperature with no hysteresis effect is also revealed. Because of these two linear responses of polymer fiber gratings, its future application in strain and temperature sensing has also been proposed.

I. Introduction

Silica fiber Bragg grating (FBG) has been widely investigated since its photosensitivity was revealed by Hill et al. in 1978 [1]. The applications such as sensing, dispersion compensation, wavelength demul/multiplexing etc. by using silica FBG in optical communication systems have been paid lots of attention in past decades. However, the tunability of silica FBG is always a big concern. Typically, the Bragg wavelength shift for silica FBG is a few nanometers [2] due to strain and temperature variation; thus this dramatically disfavors its applications. Because of its large Young's modulus and small thermal-optic effect, silica FBG doesn't show high strain and temperature sensitivities.

Lots of interests have been put on polymer fiber because of its ease of handling. Due to the vastly different material system from silica, polymer fiber has very large breakdown strain and high stress sensitivity owing to its small Youngs' modulus. Therefore, we have reasons to expect that polymer FBG would have high strain and thermal tunabilities and be promising in fiber sensor applications. We have found quite useful photosensitivities of PMMA based POF and gratings can be written in the PMMA based POF. In order to take use of this advantage of polymer FBG, it makes understanding the strain and temperature response of polymer FBG important and worthwhile.

II. Polymer FBG Fabrication

A polymer FBG is fabricated and a series of experiments are carried on it in order that we are able to comprehensively understand the thermal and strain responses of polymer FBG. Figure 1 is the setting of fabrication. The polymer fiber we use for grating fabrication is PMMA based. The core diameter is 6 μm and the outer diameter is 125 μm. The difference in the refractive index between the core and the cladding is 0.0086. The fiber is single-moded in the 1550 nm window. The Polymer FBG was prepared using the setup with the introduction of a static ring interferometer based on the patent invented by Ouellettee [3]. The UV writing beam was obtained from a frequency-doubled MOPO pumped by a frequency-tripled Nd:YAG laser. The output UV light is in short pulses of 5 ns in width and the repetition rate of 10 Hz. The wavelength of UV light for polymer fiber gratings fabrication is 325 nm.

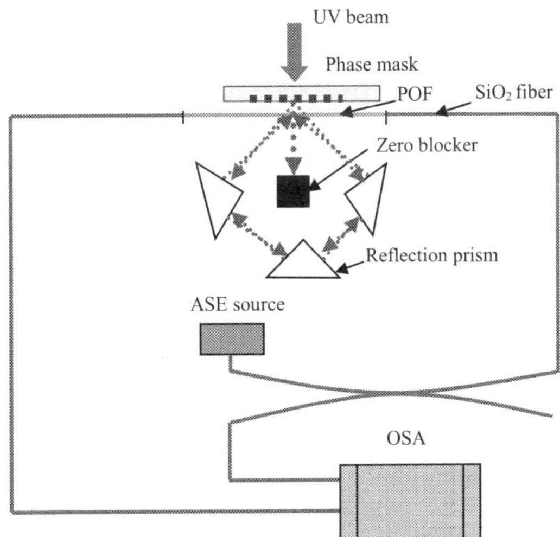

Fig. 1. Setup for polymer FBG fabrication.

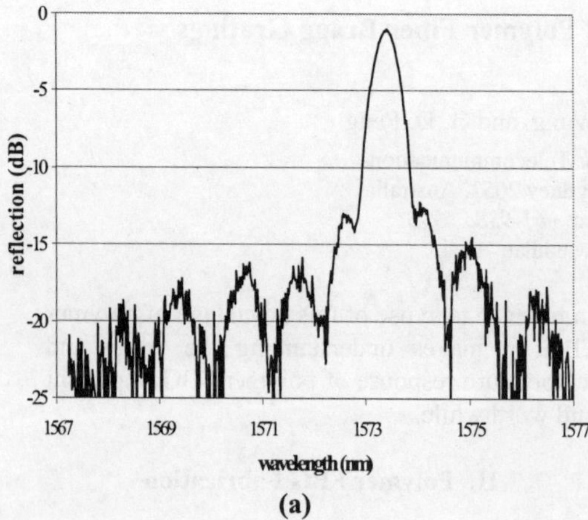

(a)

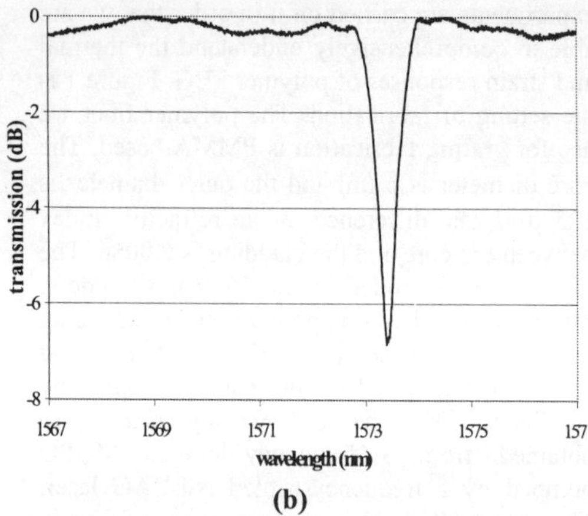

(b)

Fig. 2. Reflection spectrum (a) and transmission spectrum (b) of polymer FBG.

The reflection and transmission spectra of the polymer fiber grating at ambient temperature are shown in Figures 2(a) and 2(b) respectively. The grating has reflection about 80% with the linewidth less than 0.5 nm.

III. Strain and thermal response of polymer FBG

The strain response of this polymer FBG has been shown in Figure 3. This polymer FBG is stretched up first, and then is released back. Polymer FBG shows clearly linear relation of Bragg wavelength with the applied strain. Furthermore, the loaded curve matches perfectly with the unloaded one, which shows the reversibility for polymer FBG strain tuning is excellent. In our test, the maximum strain is only 3.75 milli-strain and the Bragg wavelength is tuned within around 10 nm. Generally, polymer optical fiber can stand up to 3% without decay. Therefore, polymer FBG could be tuned within a rather large tuning range with perfect reversibility, while the maximum tuning range for silica FBG by tensile strain is only 2 nm [4]. By using the least-square fitting method, the strain response for polymer FBG shown in Figure 4 can be expressed as,

$$\lambda_B = 1574.5(1+1.71\varepsilon) \qquad (1)$$

The strain sensitivity of polymer FBG is found to be 2.7 pm/με. This value clearly shows more than two times higher strain sensitivity in comparison with silica FBG whose strain sensitivity is approximately 1.2 pm/με at wavelength 1570 nm.

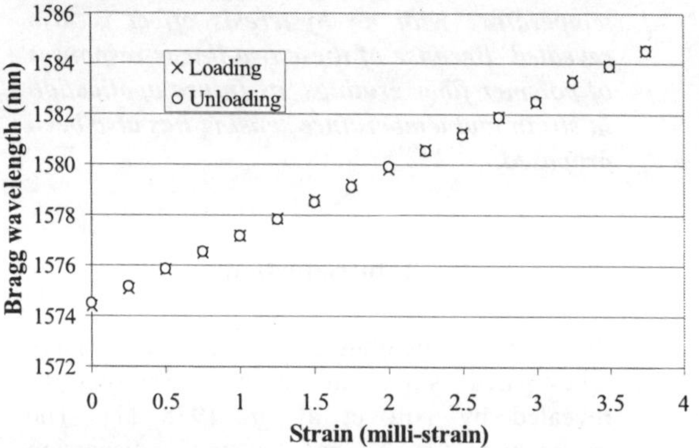

Fig. 3. Strain response of polymer FBG.

Figure 4 shows the shift of Bragg wavelength as a function of temperature of polymer FBG. There are two curves in this figure. One is obtained when the grating was heated up from room temperature to 65 °C while the other is obtained while the grating was cooled down. Both curves are quite linear and coincide with each other. Here, we noticed that polymer FBG had total different temperature response curve from silica FBG. The Bragg wavelength of polymer FGB is moving down with the increase of temperature. However, it is shifted up for silica FBG. At the same time, the coincidence of heating up and cooling down curves indicates the absence of hysteresis and a sign of usage by taking advantage of polymer FBG's thermal response. The expression for the temperature response of polymer FBG shown in Figure 4 is found to be,

$$\lambda_B = 1577(1-9.64\times10^{-5}T) \qquad (2)$$

Thus, the temperature sensitivity of polymer FBG at 1570 nm is about 152 pm/°C. This means polymer FBG is about 10 times more sensitive than silica FBG, whose temperature sensitivity is only 13.7 pm/°C [5].

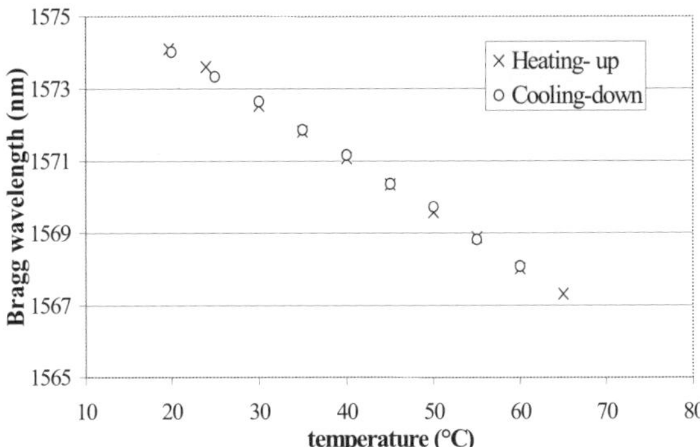

Fig. 4. Temperature response of polymer FBG.

IV. Polymer FBG strain and temperature sensing application

Since the strain and thermal sensitivity for polymer FBGs is very high and the tuning reversibility is rather excellent, polymer FBGs are quite promising for fiber sensor systems.

One of the very popular topics in fiber sensing field is the discrimination of stain and temperature effects. The widely accepted proposal is to use two silica FBGs with different Bragg wavelength λ_1 ανδ λ_2 το φορμ τηε σενσινγ ηεαδ. Ονε σιμπλε σχηεμε ισ σηοων ασ φολλοωσ:

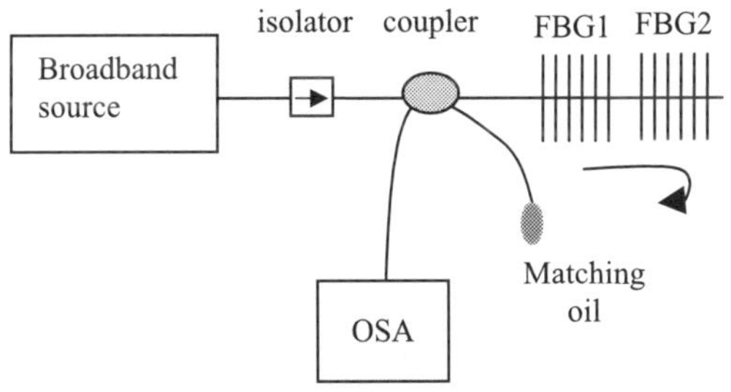

Fig. 5. Discrimination sensing system

The shift of Bragg wavelength as the reason of combined strain and temperature factors is given by

$$\begin{bmatrix} \Delta\lambda_1 \\ \Delta\lambda_2 \end{bmatrix} = \begin{bmatrix} K_{1T} & K_{1\varepsilon} \\ K_{2T} & K_{2\varepsilon} \end{bmatrix} \begin{bmatrix} \Delta T \\ \Delta\varepsilon \end{bmatrix} \quad (3)$$

where K_{1T} and K_{2T} are the temperature sensitivities while $K_{1\varepsilon}$ and $K_{2\varepsilon}$ are the strain sensitivities. By calculating the inverse transfer matrix in Eq. (3), the information of T and ε can be recovered,

$$\begin{bmatrix} \Delta T \\ \Delta\varepsilon \end{bmatrix} = \begin{bmatrix} K_{1T}' & K_{1\varepsilon}' \\ K_{2T}' & K_{2\varepsilon}' \end{bmatrix} \begin{bmatrix} \Delta\lambda_1 \\ \Delta\lambda_2 \end{bmatrix} \quad (4)$$

But because of the similar strain and temperature response of the two silica FBGs, the inversion operation of the transfer matrix is always troublesome.

A number of other techniques have been proposed for the discrimination of temperature and strain sensing. These include hybrid fiber Bragg grating/long period gratings [6], dual-diameter FBGs [7], FBGs superimposed with polarisation-rocking filter [8], FBGs combined with EDFAs [9], long period grating method [10], and FBGs written at splice joint [11]. The long period grating method utilises the multiple resonance bands in a single long period grating to measure strain and temperature simultaneously. However, it is difficult to accurately measure small shifts in the resonant wavelength in long period grating by using broad bandwidth spectrum. The splicing joint method complicates the sensor system and might limit the maximum strain, which can be applied to sensor head. All of these methods involve complex gratings or other optical elements. It adds up the system cost.

As we found out previously, strain and temperature sensitivity of Polymer FBGs is quite different from silica FBGs. If we combine one polymer FBG (FBG1) and one silica FBG (FBG2) together as a sensor head to improve the proposal in Figure 5, the trouble for the transfer matrix operation won't exist any more. From our previous results, the value for K_{1T} and $K_{1\varepsilon}$ are –152 pm/°C and 2.68 pm/με respectively for polymer FBG. K_{2T} and $K_{2\varepsilon}$ are for silica FBG, which are 10.47 pm/°C and 1.2 pm/με [12]. Thus, the inversed transfer matrix is given as

$$\begin{bmatrix} 0.0495 & 0.7229 \\ -0.0057 & 0.0127 \end{bmatrix}$$

Therefore, the temperature and strain can be easily figured out if the wavelength shift is known.

$$\begin{bmatrix} T \\ \varepsilon \end{bmatrix} = \begin{bmatrix} 0.0495 & 0.7229 \\ -0.0057 & 0.0127 \end{bmatrix} \begin{bmatrix} \Delta\lambda_1 \\ \Delta\lambda_2 \end{bmatrix}$$

Furthermore, because of the higher sensitivity of polymer FBG, it improves the sensitivity of the whole sensing system at the same time.

V. Conclusion

Polymer FBG's strain and thermal responses have been investigated. A large strain tuning range with excellent reversibility and temperature tuning without hysteresis have been discovered. The strain and temperature sensitivities for polymer FBGs is also found to be much higher than that of silica FBGs. This has made a combined sensing system, which includes one polymer FBG and silica FBG possible with low cost and large dynamic range.

References

[1] K. O. Hill, Y. Fuji, D. C. Johnson, and B. S. Kawasaki, "Photosensitivity in optical fiber waveguides: Application to reflection filter fabricaton," *Appl. Phy. Lett.*, 1978, **32**, (10), pp. 647-649.

[2] K. O. Hill and G. Meltz, "Fiber Bragg grating technology: Fundamentals and overview," *J. Lightwave Technol.*, 1997, **15**, (8), pp. 1263-1276.

[3] F. Ouellette, Univ. Sydney Patent Cooperation Treaty AL/96/00 782.

[4] P. L. Chu, G. D. Peng, and Z. Xiong, "Highly-tunable polymer optical fiber gratings," in Proc. 8th Intl. POF Conf.'99, July 1999, pp. 242-245.

[5] O. Andreas and K. Kyriacos, "Fiber Bragg Gratings Fundamentals and Applications in Telecommunications and Sensing," Chapter 3, pp. 98-99.

[6] H. J. Patrick, G. M. Williams, A. D. Kersey, J. R. Pedrazzani, and A. M. Vengsarkar, "Hybrid fiber Bragg grating/long period fiber grating sensor for strain/temperature discrimination," *IEEE Photonics Technol. Lett.,*1996, **8**, (9), pp. 1223-1225.

[7] S. W. James, M. L. Docknhy, and R. P. Tatam, "Simultaneous independent temperature and strain measurement using in-fiber Bragg grating sensors," *Electron. Lett.,* 1996, **32**, (12), pp. 1133-1134.

[8] S. E. Kanellopoulos, V. A. Handerek, and A. J. Rogers, "Simultaneous strain and temperature sensing with photogenerated in-*fiber gratings," Opt. Lett*., 1995, **20**, (3), pp. 333-335.

[9] J. Jung, H. Nam, J. H. Lee, N. Park, and B. Lee, "Simultaneous measurement of strain and temperature by using of a single-fiber Bragg grating and an erbium-doped fiber amplifier," *Appl. Opt*., 1999, **38**, (13), pp. 2749-2751.

[10] V. Bhatia, D. Campbell, R. O. Claus, and A. M. Vengsarkar, "Simultaneous strain and temperature measurement with long period gratings," *Opt. Lett.,* 1997, **22**, (9), pp. 648-650

[11] B. O. Guan, H. Y. Tam, S. L. Ho, W. H. Chung, and X. Y. Dong, "Simultaneous strain and temperature measurement using a single fiber Bragg grating," *Electron. Lett*., 2000, **36**, (12), pp. 1018-1019.

[12] A. D. Kersey, M. A. Davis, H. J. Patrick, M. Leblanc, K. P. Koo, C. G. Askins, M. A. Putnam, and E. J. Friebele "Fiber Grating Sensors" *IEEE J. Lightwave Technol. Lett*., 1997, **15**, (8), pp. 1442-1463.

Trapping of Solitons in Fibre Bragg Gratings

William C.K. Mak[1], Boris A. Malomed[2], and Pak L Chu[1]
[1]Optoelectronics Research Centre
City University of Hong Kong
[2]Department of Interdisciplinary Studies
Tel-Aviv University, Israel

Abstract--We show that solitons can be trapped in a fibre Bragg grating by a local defect provided that the soliton energy and defect strength are within a certain range of values.

Index Terms--Soliton, fibre Bragg gratings, Optical memory

A. INTRODUCTION

Optical fibre Bragg gratings have attracted a great deal of attention for various reasons. They form the basis for the design of many devices used in optical telecommunications such as filters [1], add-drop WDM multiplexers [2], dispersion compensators [3], etc. They also find important application as sensors, since the optical properties of the fibre-based gratings are highly sensitive to local stresses [4,5]. Furthermore, fibre Bragg grating is interesting in its own right because it is a unique physical medium supporting various optical excitations. In particular, strong dispersion induced by the grating opens a way to observe slow light in this medium, which is a topic of great current interest [6].

Gap solitons in fibre Bragg gratings have been studied theoretically [7,8] and observed experimentally [9]. They attracted a lot of attention because of their vastly different properties from those of conventional solitons propagating in transmission optical fibres. Attempts to slow the solitons down in the Bragg grating are being researched intensively [8,10].

Here we report results on trapping of the soliton by a local defect in a fibre Bragg grating. We obtain an analytical solution for two coupled-mode equations describing the propagation of resonant electromagnetic waves in a grating in the presence of a local defect, which changes the local Bragg reflection coefficient and local refractive index. We then examine the stability of the trapped soliton as a function of its energy and the defect strength, solving the coupled-mode equations numerically.

Earlier, it was reported that light can be trapped by a local defect in a fibre Bragg grating by means of four-wave mixing without forming a soliton [11]. In fact, the soliton trapping is a different kind of a problem. Very recently, it was considered in [12], where attention was focused on effects produced by a finite size of the defect. Here, we consider the case of a very narrow defect, obtaining results which are essentially new in comparison with those reported in [12].

B. THE MODEL

We consider a configuration shown in Figure 1. The trapping defect, generally, combines a local change of the refractive index, λ, and a change of the grating reflectivity, κ. The defect is assumed to be localized as the delta-function around x=0.

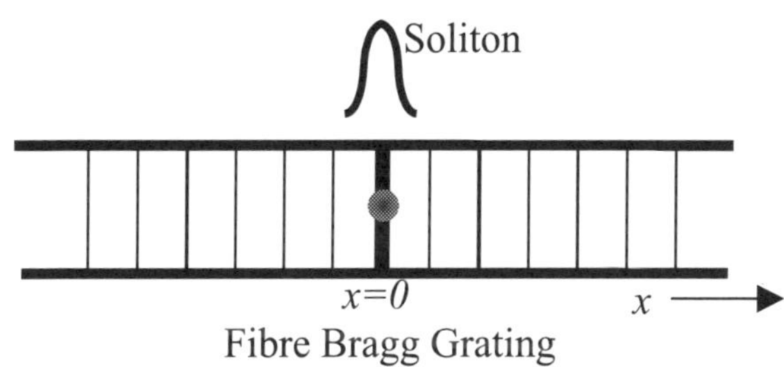

Figure 1. The soliton trapped by the defect in fibre Bragg grating

The light propagation in the grating is represented by the coupled-mode equations for the right- and left-traveling waves $u(x,t)$ and $v(x,t)$:

$$i\frac{\partial u}{\partial t}+i\frac{\partial u}{\partial x}+v+\left(\frac{1}{2}|u|^2+|v|^2\right)u=\kappa\delta(x)v+\lambda\delta(x)u \quad (1)$$

$$i\frac{\partial v}{\partial t}-i\frac{\partial v}{\partial x}+u+\left(\frac{1}{2}|v|^2+|u|^2\right)v=\kappa\delta(x)u+\lambda\delta(x)v, \quad (2)$$

where the usual ratio between the self-phase modulation (SPM) and cross-phase modulation (XPM) coefficient, 1:2, is adopted.

To find *exact* solutions to Eqs.(1) and (2), we substitute into Eqs.(1) and (2)

$$u(x,t)=U(x)\exp(i\omega t), \quad and \quad v(x,t)=V(x)\exp(i\omega t), \tag{3}$$

which leads to

$$\omega U + iU' + V + \left(\sigma |U|^2 + |V|^2\right)U = \kappa\delta(x)V + \lambda\delta(x)U \tag{4}$$

$$\omega V - iV' + U + \left(\sigma |V|^2 + |U|^2\right)V = \kappa\delta(x)U + \lambda\delta(x)V \tag{5}$$

where the prime stands for d/dx. In the case $\kappa=\lambda=0$, a solution to Eqs. (4) and (5) takes the form:

$$U(x)=\sqrt{\frac{2}{3}}\sin\theta \,\mathrm{sech}\left((x-\xi)\sin\theta-\frac{i}{2}\theta\right), \quad V=-U^* \tag{6}$$

where ξ is the coordinate of the soliton's center, and θ is a real parameter taking values

$$0<\theta<\pi \tag{7}$$

which also determines the soliton's frequency:

$$\omega=\cos\theta. \tag{8}$$

This solution is called a gap soliton.

If $\kappa,\lambda\neq 0$, Eqs.(4) and (5) admit the following exact solution:

$$U=\sqrt{\frac{2}{3}}\sin\theta \,\mathrm{sech}\left[(x+a\,\mathrm{sgn}\,x)\sin\theta-\frac{i}{2}\theta\right], \quad V=-U^* \tag{9}$$

where the parameter a can be determined by substituting Eq.(9) into Eqs.(4) and (5) and integrating in an infinitesimal vicinity of the point $x=0$:

$$\tanh(a\sin\theta)=\sqrt{\frac{\kappa-\lambda}{\kappa+\lambda}}\,\frac{\tanh\left(\sqrt{\kappa^2-\lambda^2}/2\right)}{\tan(\theta/2)} \tag{10}$$

where we assume $\kappa^2>\lambda^2$. For the special case when there is no defects in the refractive index, i.e. $\lambda=0$, Eq.(10) takes the form

$$\tanh(a\sin\theta)=\frac{\tanh(\kappa/2)}{\tan(\theta/2)}. \tag{11}$$

If $\kappa<0$, i.e., the local defect represents an enhancement of Bragg reflection, Eqs.(10) and (11) yield $a<0$. This means that $|U(x)|$ has, according to Eq.(9), two symmetric local maxima at $x=\pm a$, and a local minimum at $x=0$.

If $\kappa>0$, i.e., the local defect represents a suppression of the Bragg reflection, we have $a>0$, i.e., the solution has a single local maximum at $x=0$.

In the case $\kappa^2<\lambda^2$, a can be obtained from a straightforward analytical continuation of (10):

$$\tanh(a\sin\theta)=-\sqrt{\frac{\lambda-\kappa}{\lambda+\kappa}}\,\frac{\tan\left(\sqrt{\lambda^2-\kappa^2}/2\right)}{\tan(\theta/2)} \tag{12}$$

In the special case $\kappa=-\lambda$, Eqs.(10)&(12) give

$$\tanh(a\sin\theta)=\frac{\kappa}{\tan(\theta/2)}. \tag{13}$$

Obviously, the expressions (10) through (13) make sense if they yield $|\tanh(a\sin\theta)|<1$, otherwise the solution for the trapped soliton does not exist. This translates into the following trapping condition:

$$\theta_{min}<\theta<\pi, \tag{14}$$

$$\theta_{min}\equiv 2\tan^{-1}\left(\sqrt{\frac{\kappa-\lambda}{\kappa+\lambda}}\tanh\left(\sqrt{\kappa^2-\lambda^2}/2\right)\right) \tag{15}$$

In the further consideration, we will make use of the field energy, which is a dynamical invariant of Eqs. (1) and (2), and is given by

$$E=\int_{-\infty}^{\infty}\left(|u(x)|^2+|v(x)|^2\right)dx. \tag{16}$$

An analytical expression of E for the general case of the trapped soliton (9) is cumbersome. However, in the case $\lambda=0$, the expression is simple:

$$E=\frac{8}{3}\left\{\theta-(\mathrm{sgn}\,\kappa)\left[\frac{\pi}{2}-\sin^{-1}(\mathrm{sech}\,\kappa)\right]\right\}. \tag{17}$$

C. STABILITY OF THE TRAPPED SOLITON

The stability of the free gap soliton has been studied in detail on the basis of variational approximation, and then by accurate numerical method. It was found that the soliton is stable only within the range

$$0<\theta<\theta_{cr},\ \theta_{cr}\approx 1.011(\pi/2). \tag{18}$$

Note that θ_{cr} is just slightly larger than $\pi/2$, the value at which the gap soliton has the largest amplitude and smallest width.

We carried out a numerical study of the stability of the trapped soliton, directly solving Eqs. (4) and (5) with $\kappa>0$, $\lambda=0$. Basic results depicting the stability of the trapping are displayed in Figure 2 showing the evolution of the soliton for $\kappa=0.08$, and $\theta=0.4\pi$, 0.5π, 0.7π, and 0.9π. The left part of each panel in Fig. 2 shows a "side view" of the evolution of the wave field in time. The right parts of the panels show a "top view" of the evolution.

As is seen in Fig. 2, only a single value $\theta=\theta_{stab}$ directly gives rise to a stable soliton. θ_{stab} is equal or very close to $\pi/2$. Figure 2c shows that if the initial value $\theta=\theta_{in}$ is somewhat larger than θ_{stab}, the soliton evolves until θ becomes equal to θ_{stab}, shedding off excess energy in the form of radiation. It is noted that, as per Eq.(17), θ is directly related to the energy of the soliton. If $\theta_{in}\geq 0.8\pi$, the soliton does not relax into a stable one, but instead it decays into radiation, sometimes also generating a small-amplitude residual soliton not trapped by the defect, see Figure (2d). On the other hand, if $\theta_{in}<\theta_{stab}$, the soliton cannot be stable, directly decaying into radiation (Fig. 2a).

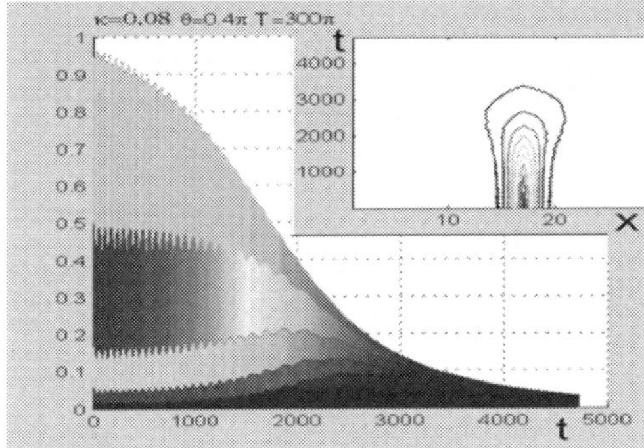

(a) θ=0.4π

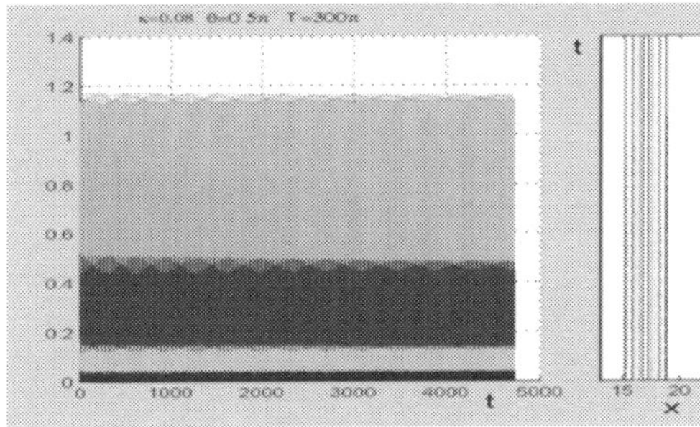

(b) θ=0.5π

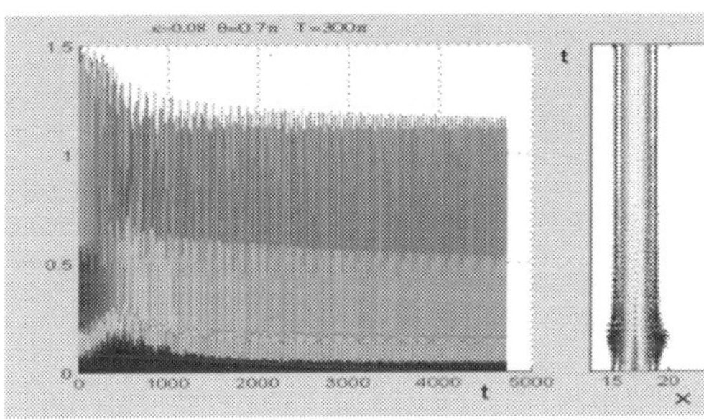

(c) θ=0.7π

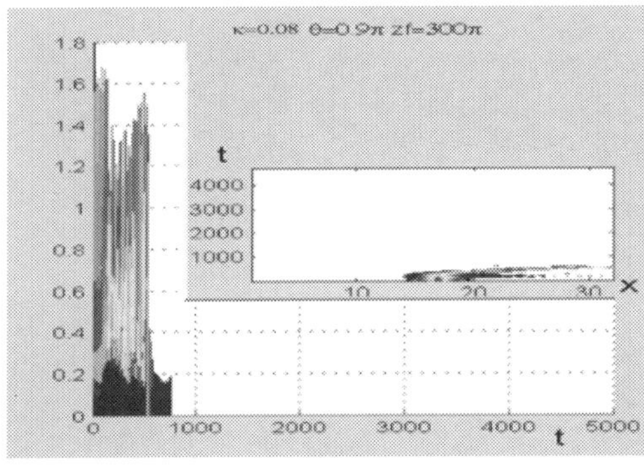

(c) θ=0.9π

Figure 2 Stability of trapped solitons λ=0,κ=0.6

Figure 3 summarizes the stability results. In this figure, two borders define a region in the plane (κ, θ_{in}), marked as "relaxation to stable solitons", in which the soliton is stably trapped. It is noticeable that the trapped soliton is stable for a whole range of κ as long as κ>0.

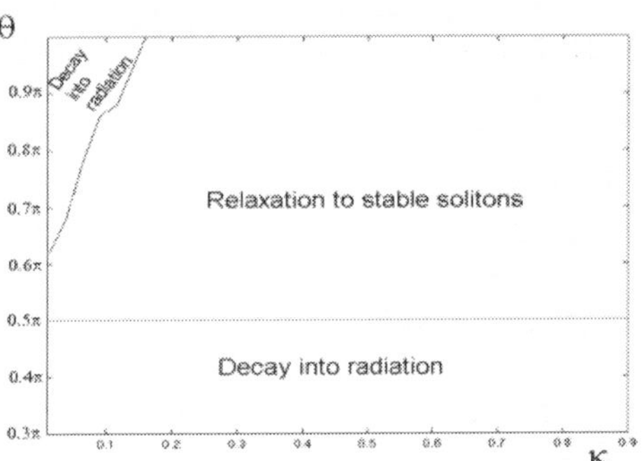

Figure 3 Region of Stability of trapped soliton

D. CONCLUSIONS

We have shown that a soliton can be stably trapped by a defect in a fibre Bragg grating, provided that the soliton's intrinsic parameter θ has a value close to π/2. It noted that the defect strength κ does not impose a severe condition for the trapping, if it is positive. The trapping of the soliton in the grating can be used to design optical memory devices, logic circuits, and optical delay lines.

E. REFERENCES

[1] T. Inui, T. Komukai, and M. Nakazawa: "Highly efficient tunable fibre Bragg grating filters using multiplayer piezoelectric transducers", Opt. Comm. Vol.190, pp.1-4, 2001

[2] A.S. Kewitsch, G.A. Rakuljic, P.A. Willems, and A. Yariv: "All-fibre zero-insertion-loss add-drop filter for wavelength-division multiplexing", Opt. Lett., Vol.23, pp.106-108, 1998

[3] B.J. Eggleton, T. Stephens, P.A. Krug, G. Dhosi, Z. Brodzeli, and F. Ouellete: "Dispersion compensation using a fibre grating in transmission", Electron. Lett., Vol.32, pp.1610-1611, 1996

[4] K. Pran, G.B. Havsgard, G. Sagvolden, Ø. Farsund, and G. Wang: "Wavelength multiplexed fibre Bragg grating system for high-strain helth monitoring applications", Meas. Sci. Technol., Vol.13, pp.471-476, 2002

[5] J.M. Senior, S.E. Moss, and S.D. Cusworth: "Multiplexing techniques for noninterferometric optical point-sensor networks: A review", Fibre Int. Opt., Vol.17, pp.3-20, 1998

[6] J. Marangos: "Slow light in cool atoms", Nature, Vol.397, pp.559-560, 1999

[7] A.B. Aceves and S. Wabnitz: "Self-induced transparency solitons in nonlinear refractive periodic media", Phys. Lett. A, Vol.141, pp.37-42, 1989

[8] D.N. Christodoulides and R.I. Joseph: "Slow Bragg solitons in nonlinear periodic structures", Phys. Rev. Letter, A, pp.141, pp.37-42, 1989

[9] B.J. Eggleton, R.E. Slusher, C.M. de Sterke, P.A. Krug, and J.E. Sipe, Phy. Rev. Lett., Vol.76, pp.1627, 1996

[10] J.E. Heebner, R.W. Boyd, and Q.H. Park:"Slow light, induced dispersion, enhanced nonlinearity, and optical solitons in a resonator-array waveguide", Phys. Rev. E, Vol.65, 6 March 2002

[11]C.M. de Sterke, E.N. Tsoy, and J.E. Sipe: "Light trapping in a fibre grating defect by four-wave mixing", Opt. Lett., Vol.27, pp.485-487, 2002

[12] R.H. Goodman, R.E.Slusher,and M.I.Weinstein, "Stopping light on a defect", J. Opt. Soc. Am. B. Vol. 19, pp. 1635-1652, 2002.

[13] B.A. Malomed and R.S. Tasgal, "Vibration modes of a gap soliton in a nonlinear optical medium", Phys. Rev. E , Vol. 49, pp. 5787-5796, 1994.

[14] I.V. Barashenkov, D.E. Pelinovsky, and E.V. Zemlyanaya, "Vibrations and oscillatory iInstabilities of gap solitons", Phys. Rev. Lett., Vol. 80, pp. 5117-5120, 1998.

Session 13G

Optical Transmission I

Optical Fibre-Fed Radio System for Broadband Services

Invited Paper

H. Ghafouri-Shiraz

School of Electrical and Electronic Engineering
South Spine, Block S1
Nanyang Technological University
Nanyang Avenue
Singapore 939798

The use of fibre-fed millimetre-wave radio systems seems to attract the attention for future broadband services, such as Intelligent Vehicle Highway Systems, high speed Wireless Local Area Networks, Interactive Multimedia Services, and Mobile Broadband Services. Optical fibre and millimetre-wave radio are capable of supporting the large bandwidth requirements of these systems, and on the other hand the low attenuation of the optical fibre provides the best method for the distribution of millimetre-waves at present.

In this paper, new configurations for the fibre-fed radio system are proposed. The advantage of these configurations is that they offer minimum complexity of the system central site where a dual-mode laser source generates the optical millimetre-wave signals, and also they offer the feasibility of a compact and low cost implementation of the radio node. The planar circuit technology with the minimum number of active devices, the remote site can be simple with low power requirements.

A diplexer-antenna integration is designed and implemented with promising results and a first attempt of a radio node prototype is constructed operating at 6GHz. Furthermore, the concept of optically injection-locked patch oscillators is investigated and the results showed that the mutually coupled patch oscillators can satisfy the power requirements of the fibre-fed radio system, where power combining can take place in free-space.

Subcarrier-multiplexed fiber-optic mm-wave radio using sideband injection locking of semiconductor lasers

Madhumita Bhattacharya, Bipul Sarkar and Taraprasad Chattopadhyay, MIEEE
Department of Physics , Visva-Bharati University, P.O. Santiniketan(731235),W.B., India
Fax : 91-3463-52672 ; E-mail : tara_vb@hotmail.com

Abstract

A scheme for subcarrier-multiplexed fiber- optic mm-wave radio is proposed in this paper. Using an overdriven phase modulator, higher order optical sidebands can be generated. Three coherent lightwaves are produced from three laser diodes injection-locked to the optical carrier and the two optical modulation sidebands. The two laser diodes locked to the upper optical sidebands are bias current modulated by a group of subcarriers. When these waves are mixed in a PD, two mm-waves are obtained which contain the same amplitude modulations that are present in two of the lightwaves.

Key Words: fiber-optic radio, mm-wave, injection-locking, subcarrier multiplexting

1. INTRODUCTION

The mm-wave frequency band covering the range 30- 70 GHz is useful for the fiber-optic mm-wave radio [1-6]. High atmospheric attenuation around 60 GHz makes the mm-wave frequencies reusable. Fiber-optic mm-wave radio is a modern scheme of communication where low loss of optical fiber makes the mm-wave modulated lightwaves propagate over long distances from the central station to several base stations.

In this paper, we propose a fiber-optic mm-wave radio where the optical carrier and optical sideband locking generate three coherent lightwaves. Two of these contain subcarrier - modulation in amplitude. Mixing these three lightwaves we obtain subcarrier-modulated mm-waves which are radiated over a cell. The proposed scheme finds application in wireless CATV.

2. SYSTEM DESCRIPTION

The schematic circuit diagram of the subcarrier-multiplexed fiber-optic mm-wave radio is shown in Fig.1. The lightwave from a laser diode (LD) having power P_{in} and frequency f_c is phase modulated in an overdriven phase modulator by a microwave signal having frequency f_m . The voltage amplitude of the microwave signal, V_{mo}, is greater than V_π , the half-wave voltage of the phase modulator. In the overdriven case , $V_{mo} = kV_\pi$ where $k > 1$.The output PM lightwave consists of frequency components f_c , and the modulation sideband frequencies , $f_c \pm nf_m$ where $n = 1, 2, 3, ...etc.$ The amplitude of the optical carrier is proportional to $J_0(\beta)$, while that of the sidebands are proportional to $J_n(\beta)$, where $\beta = \pi V_{mo}/V\pi = k\pi$. The PM lightwave is split into three lightwaves by a 1X3 optical coupler. Each PM lightwave is injected to a laser diode. The laser diode (LD1) in port 1 is tuned so that its free-running oscillation frequency, f_{01} , is almost equal to f_c , the frequency of the optical carrier.The LDs in port 2 and 3 are so chosen that their free-running oscillation frequencies are close to upper side modulation frequencies $f_c + nf_m$ and $f_c + (n-1)f_m$ respectively. LD2 and LD3 are directly intensity modulated by a group of baseband-modulated subcarriers, SC1 and SC2 respectively. The laser diodes LD1, LD2 and LD3 are injection-locked to the optical carrier and the optical sidebands having frequencies $f_c + nf_m$ and $f_c + (n-1)f_m$ respectively. Here, we consider the Q- factor of the LDs high enough so that the lockbands of the LDs are much smaller [7] than f_m. The intensity modulated lightwaves from LD2 and LD3 are combined with the CW lightwave from LD1 in a 3X1 optical coupler and the combined lightwave propagates down the optical fiber to a base station , BS. In the BS, at the output of the PD one gets the mm-waves having frequencies nf_m and $(n-1)f_m$ along with the frequency component f_m

due to the mixing of the three coherent lightwaves. These mm-wave signals are amplitude modulated by the group of subcarriers SC1 and SC2 respectively. Since the mm-waves are generated from coherent lightwaves, the phase noise of the generated mm-waves are greatly reduced. The other rf signal generated due to the mixing of the lightwaves in the PD having frequency f_m is filtered out by the mm-wave bandpass filter, BPF. The mm-wave signals are amplified and fed to an antenna and are radiated over a cell.

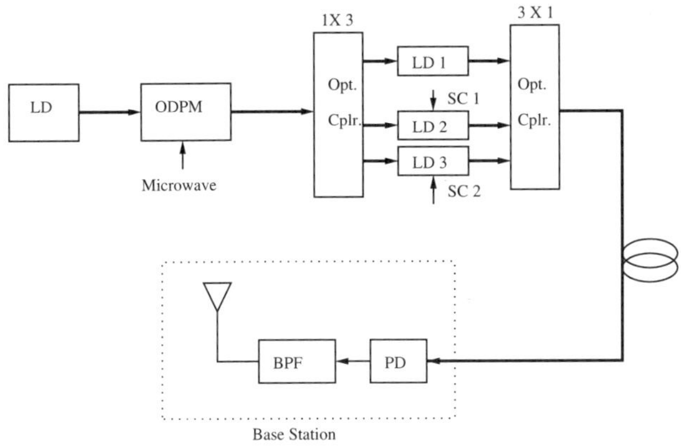

Fig. 1. Schematic diagram of the proposed fiber-optic mm-wave radio.

3. ANALYSIS

The PM lightwave at the output of the ODPM biased at $V_{dc} = V_\pi/2$ can be written as

$$\begin{aligned} E_{pm}(t) = \sqrt{P_{in}}\, [J_0(\beta) sin\omega_c t + \textstyle\sum_{p=1}^{\infty} J_{2p}(\beta) \\ [sin(\omega_c + 2p\omega_m)t + sin(\omega_c - 2p\omega_m)t] \\ - \textstyle\sum_{p=1}^{\infty} J_{2p-1}(\beta)\, [cos(\omega_c + (2p-1)\omega_m)t \\ + cos(\omega_c - (2p-1)\omega_m)t]] \end{aligned} \quad (1)$$

The free-running oscillation frequencies of the laser diodes LD1, LD2 and LD3 are chosen as f_c , $f_c + nf_m$ and $f_c + (n-1)f_m$ respectively. 'n' is an integer corresponding to near maximum value of $J_n(\beta)$ and $J_{n-1}(\beta)$. The subcarriers SC1 and SC2 carrying the baseband signal/information may lie in the CATV band (50-750 MHz).

The output lightwaves from the LD1, LD2 and LD3 can be written as

$$a_1 = a_{10} exp\{j(\alpha_1 + \omega_c t)\} \quad (2)$$

$$a_2 = a_{20} m_1(t) exp\{j(\alpha_2 + \omega_c t + n\omega_m t)\} \quad (3)$$

and

$$a_3 = a_{30} m_2(t) exp\{j(\alpha_3 + \omega_c t + (n-1)\omega_m t)\} \quad (4)$$

respectively. a_{10}^2 , a_{20}^2 and a_{30}^2 are the free-running output powers of the laser diodes LD1, LD2 and LD3 respectively. $m_1(t)$ and $m_2(t)$ are the subcarrier modulation functions. Neglecting the coupler loss and the optical fiber loss, the output photocurrent can be expressed as

$$\begin{aligned} I_{pd} = \eta\, [a_{10}^2 + a_{20}^2 m_1^2(t) + a_{30}^2 m_2^2(t) \\ +2a_{10}a_{20}m_1(t) cos(n\omega_m t + \alpha_2 - \alpha_1) \\ +2a_{10}a_{30}m_2(t) cos\,((n-1)\omega_m t + \alpha_3 - \alpha_1) \\ + 2a_{20}a_{30}m_1(t)m_2(t) cos(\omega_m t + \alpha_2 - \alpha_3)] \end{aligned} \quad (5)$$

Where η is the responsivity of the PD. The mm-wave signals at the output of the PD having frequencies nf_m and $(n-1)f_m$ and being amplitude-modulated by the corresponding subcarrier signals are passed by the BPF. While the other signals in (5) fall outside the passband of the BPF. The mm-wave signals are fed to the base station antenna. Each mm-wave signal carries a group of subcarrier channels. The system can find application in CATV.

Taking $V_\pi = 2$ volt and $k = 2.54$, $\beta = k\pi \cong 8$. If the microwave frequency, $f_m = 8\,GHz$, the mm-wave signals at the output of the PD have frequencies $64\,GHz$ and $56\,GHz$ assuming that LD2 and LD3 are injection-locked to the 8th and 7th modulation sideband frequencies which are $f_c + 8f_m$ and $f_c + 7f_m$ respectively. Taking free-running output power of each LD equal to 10 mW, $R_L = 50\,\Omega$, and $\eta = 0.8\,A/W$, the mm-wave power at the output of the BPF is calculated as 12.8 mW.

4. CONCLUSION

A new scheme for fiber-optic mm-wave radio, where each mm-wave can carry a group of SCM signals, is proposed. The mm-waves are optically generated by mixing three coherent lightwaves in a photodiode. If two of the lightwaves are bias current modulated by group of subcarriers, the amplitude modulations are transferred to the mm-waves. Typical value of mm-wave power available in this scheme without post-detection amplification is 12 mW.

Acknowledgement

T. Chattopadhyay acknowledges CSIR, New Delhi, India for financial assistance under project No. 03(0950)/02/EMR-II.

REFERENCES

[1] D. Wake, C. R. Lima, and P. A. Davies, "Optical generation of millimeter-wave signals for fiber radio systems using a dual mode DFB semiconductor laser, " IEEE Trans. Microwave theory and Techniques, vol.43, page 2270-2276, no.9, Sept. 1995.

[2] J. O'Reilly and P. Lane, "Remote delivery of video service using mm-waves and optics," IEEE Journal of Lightwave Technology, vol.12, no. 2, page 369-375, Feb. 1994.

[3] T. Kuri, K, Kitayama, A. Stohr and Y. Ogawa, "Fiber-optic millimeter-wave downlink system using 60 GHz band external modulation," IEEE Journal of Lightwave Technology, vol. 17, no. 5, page 799-806, May 1999.

[4] T. Chattopadhayay and M. Bhattacharya, " A scheme for fiber-optic mm-wave radio," Procedings of the International Conference in Fiber-optic and Photonics, PHOTONICS-2000, Vol.2, pp. 578-580, 2000, Calcutta, India.

[5] T. Chattopadhayay and M. Bhattacharya, "Subcarrier multiple access fiber-optic mm-wave radio using Cascaded Mach-Zehnder interferometric square law modulators," Journal of Optical Communications vol. 23, page 48-51,April 2002.

[6] C. Lim, A. Nirmalathas, D. Novak, R. Waterhouse and G. Yoffe, "Millimeter-wave broad-band fiber-wireless system incorporating baseband data transmission over fiber and remote LO delivery," IEEE Journal of Lightwave Technology, vol. 18, no. 10, page 1355-1362, October 2000.

[7] M. Bhattacharya, B. Sarkar and T. Chattopadhyay, " Detection of optical FM-SCM signal through coherent carrier reconstruction," Journal of Optical Communications, vol. 23, 2002 (in press).

Optical Generation of Millimetre Wave Signal Using a Co-located Dual Distributed Feedback Fibre Laser

Yicheng Lai, *W. Zhang, and J. A. R. Williams
Photonics Research Group, School of Engineering and Applied Science, Aston University, Birmingham B4 7ET, UK
Phone: (44) 0121 3593611, Fax: (44) 0121 3590156, Email: laiy@aston.ac.uk
**Indigo Photonics Ltd, Faraday Wharf, Holt Street, Birmingham, B7 4BB, UK*
Phone: (44) 0121 2606111, Fax: (44) 0121 2606064, Email: W.Zhang@Indigo-photonics.com

Abstract: A co-located dual DFB fibre laser configuration is realized for optical generation of a millimetre wave signal. Dual polarization mode operation is obtained and the heterodyne signal measured at ~32.5GHz has a linewidth of < 1kHz. The two operating laser modes, originate from co-located cavities, exhibit high common-mode suppression of phase noise to produce a beat signal of high spectral purity. No complex feedback scheme is required and the beat frequency achievable is determined by grating parameters only. Dual mode operation with wavelength spacing measured ~0.75nm is also demonstrated.

Key words: Distributed Feedback fibre laser, dual mode, polarization.

1. INTRODUCTION

Optical generation and delivery of microwave/millimetre signals over fibre has potential applications in various fields like future broadband wireless access networks [1], spectroscopy [2] and phase array antennas [3]. The generation of high frequency signals ranging from gigahertz (GHz) to terahertz (THz) directly in optical domain has been focus of much research. Among various techniques currently employed, an effective and favoured technique is to use the heterodyning between two lasers whose frequencies differ by the required millimetre-wave frequency, and which are mixed at a photodetector to produce the desired electrical beat signal. Optical mixing of two laser sources has the advantage over other generation techniques of giving higher RF power [4]. It is particularly advantageous in fibre microwave network applications since it reduces base station hardware complexity and overcomes limitations on transmission imposed by fibre chromatic dispersion. Usually an extremely narrow linewidth microwave/millimetre wave signal is required for those applications. Implementations of this technique include frequency locking [4, 5], injection locking [6] as well as the optical phase locked loop (OPLL) [7]. Generally, these schemes rely on unique solid-state laser configurations, with external high-frequency oscillator injection and/or feedback control to produce phase locked, narrow linewidth millimetre wave signals. Furthermore, feedback loops like OPLL, though effective for correlating low frequency phase noise between two laser cavities, may face limitations in overcoming high frequency phase noise in the beat signal unless very narrow linewidth (kHz) optical sources [8] are used.

Apart from their obvious fibre-compatibility and fabrication simplicity, Distributed Feedback (DFB) fibre lasers typically exhibit extremely narrow optical linewidth of ~10kHz [9] with good frequency stability against temperature fluctuation and without mode-hopping. We have previously reported optical generation of very narrow linewidth microwave signal from a dual polarization mode fibre DFB laser [10]. The common mode suppression of phase noise of the lasing modes from the same cavity allowed very narrow linewidth (<1kHz) beat signal without feedback control. In this work, we describe an alternative fibre laser configuration for generating signals at millimetre wave regime and beyond based on a dual co-located phase shifted gratings structure. We emphasise the simplicity of the fabrication process and the flexibility. Compared to other methods [e.g. 10,11] the range of frequency generated is not determined by fibre parameters. The laser generates stable dual polarization output. A narrow linewidth (<1kHz, limited by measurement setup) millimetre wave signal is achieved without complex feedback control.

2. BACKGROUND

The total optical field of two laser modes can be expressed by,

$$E = E_1 e^{i(\omega_1 t+\phi_1)} + E_2 e^{i(\omega_2 t+\phi_2)} \quad (1)$$

where $\omega_{1,2}$ represent the two optical frequencies, $\phi_{1,2}$ the phase noises for the corresponding modes. Photocurrent generated by direct detection of the optical field defined by (1) is proportional to $E \cdot E^*$, given by,

$$I \propto E_1^2 + E_2^2 + 2E_1E_2\cos[(\omega_1 - \omega_2)t + (\phi_1 - \phi_2)] \quad (2)$$

The difference of the two optical frequencies produced a beat signal locates in microwave/millimetre wave domain. Generally $\phi_{1,2}$ are high frequency noise and uncorrelated to each other if the two modes originate from different lasing cavity. The difference between them is a high frequency noise source, which consequently leads to large linewidth of the generated beat signal. This phase noise, however, is very small provided the two modes are from the same lasing cavity.

By taking advantage of the linear birefringence in a practical DFB fibre laser cavity, a dual polarization modes fibre laser source can be realized to generate a beat note of very low phase noise in the microwave regime [10]. This leads to an effective means of generating very narrow linewidth microwave signal without feedback control. Based on a similar context, by co-locating two DFB fibre laser cavities within the same fibre, a dual mode optical source can be achieved to generate a beat note at any frequency depending on the wavelength spacing between the two laser cavities during fabrication.

3. EXPERIMENT

It has been illustrated [12] that for a dual overwritten fibre Bragg grating (FBG), the spectrum of individual grating will be independent of each other under condition,

$$\Delta\Lambda L \gg \Lambda^2 / \pi \qquad (3)$$

where Λ is the grating period, $\Delta\Lambda$ is the period difference and L is the grating length.

A single phase-shifted Bragg grating was first inscribed into a 40mm long Er/Yb doped fibre by a focused UV beam through a phase mask with Λ ~1069nm. The Er/Yb doped fibre has a small signal absorption ~230dB/m at 980nm and ~25dB/m at 1535nm. The fibre was adhered firmly on its ends to a translation stage to allow a strain to be applied to the fibre. By real time monitoring the spectrum of the first grating, the fibre was strained ~0.06% so that a wavelength shift of ~0.75nm, corresponding to $\Delta\Lambda$ ~0.52nm, was observed in the Optical Spectrum Analyzer (OSA). A second phase-shifted grating was then produced under the exact same conditions. The total fabrication time of the device was less than 5min and the resultant profile of the laser cavity is as shown in Fig. 1.

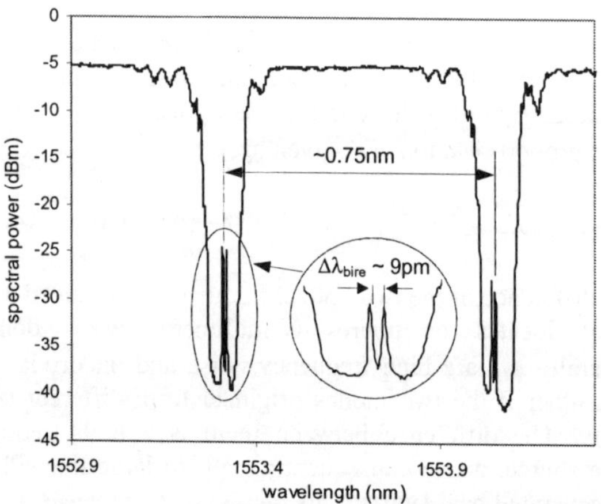

Fig. 1 Spectral profile of the co-located dual DFB fibre laser. Inset highlights the birefringence-induced polarization mode splitting within the transmission peak of the phase-shifted grating.

It is evident that the spectra profile of each phase-shifted grating in the dual overwritten structure remains independent. The wavelength spacing between the transmission peaks from each grating was measured ~0.75nm and birefringence induced polarization mode splitting ($\Delta\lambda_{bire}$ ~ 9pm i.e. birefringence ~ $1\text{x}10^{-5}$) can be observed for both gratings.

The experimental setup for the operation of the co-located dual DFB fibre laser is as shown in Fig. 2. A 980nm laser module was used as the pump source and isolators were used for both the 980nm pump laser and the output of the fibre laser structure. A high birefringence (Hi-Bi) fibre grating was used for polarization discriminating self-injection feedback. To reduce the effects of environmental temperature fluctuations, the fibre laser was placed in a temperature controller consisting of a Peltier device, with temperature control better than 0.1°C. The output of the laser was observed on a 0.1nm resolution OSA and the obtained spectrum is as shown in Fig. 3.

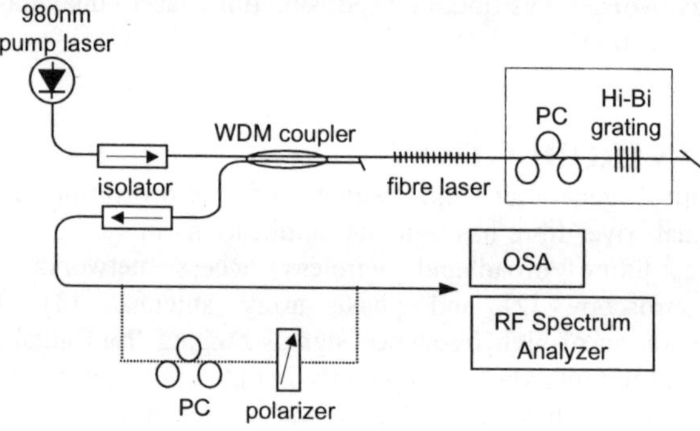

Fig. 2 Experimental setup arrangement. Hi-Bi fibre grating spliced to the fibre laser provides polarization discriminating self-injection feedback during operation.

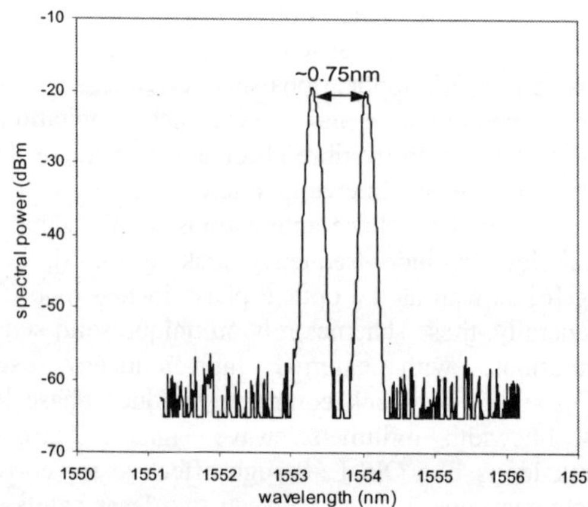

Fig. 3 Optical output spectrum from the co-located dual DFB fibre laser. The wavelength spacing between the two optical modes measured ~0.75nm. Linewidths are broadened by the limited instrumental resolution.

The optical output spectrum reveals two optical frequencies with a wavelength spacing measured ~0.75nm, corresponding closely to the passive measurement. Feeding the output through a polarization controller (PC) and a polarizer enabled one or other of the two optical frequencies to be eliminated, indicating that they correspond to orthogonal polarisation states. Furthermore, the absence of any polarization mode-beating (PMB) signal within the measurement range of the RF spectrum analyser (40GHz) indicates that one polarization mode from each DFB laser constitutes the dual mode output. The dual mode operation was stable over a measurement period of 2 hours with maximum differential power of < 0.25dBm due to polarization dependent gain.

To evaluate the operation and to allow direct measurement of the PMB signal generated through such a fibre laser configuration, a separate co-located dual DFB fibre laser with mode spacing of ~0.26nm was fabricated. The desired Hi-Bi grating can be effectively realized by a dual overwritten Hi-Bi uniform FBG. Each DFB fibre laser cavity will match to one reflection wavelength, along orthogonal axes, of the Hi-Bi grating. But by using Hi-Bi fibre of appropriate birefringence, a single Hi-Bi grating with peak reflection wavelengths, along the fast and slow axis, that matches each of the DFB fibre laser cavity for polarization discriminating self-injection feedback can be realized in a one-step process. The 15mm long Hi-Bi grating was spliced to the end of the fibre laser and a polarization controller (PC) was placed in the feedback path to help the polarization alignment between the fibre laser and the grating. The Hi-Bi grating ensures the two oscillating wavelengths, one from each cavity, are of orthogonal polarization states. Since the feedback paths for both laser cavities are within the same length of fibre, the configuration does not compromise common mode suppression of phase noise in the beat signal generated. The superimposed spectral profile of the co-located dual DFB fibre laser with the Hi-Bi grating is as shown in Fig. 4.

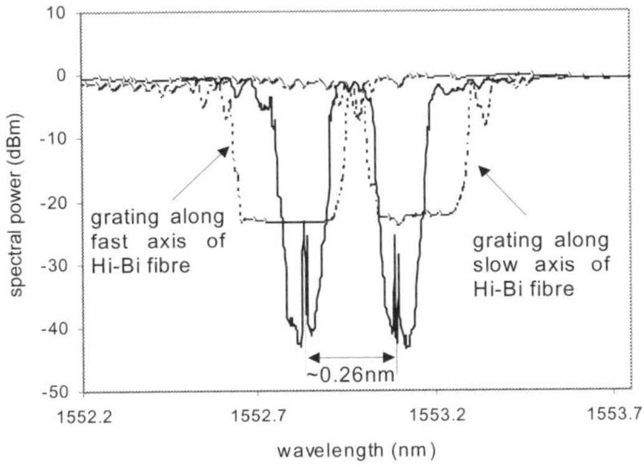

Fig. 4 Superimposed spectral profile of co-located dual DFB fibre laser with Hi-Bi grating. Note that the profile along the fast and slow axis of the Hi-Bi grating was measured separately.

The obtained optical output spectrum of the fibre laser reveals stable dual mode operation with mode spacing measured ~0.26nm as shown in Fig. 5. The output signal after the polarizer was then fed directly to a high-speed photodetector (New Focus 1014) and the resulting electrical spectrum was monitored on the RF spectrum analyser. The recorded electrical spectrum of the beat signal is given in Fig. 6. The millimetre wave signal is measured at ~32.5GHz, and an estimated linewidth of less than 1kHz.

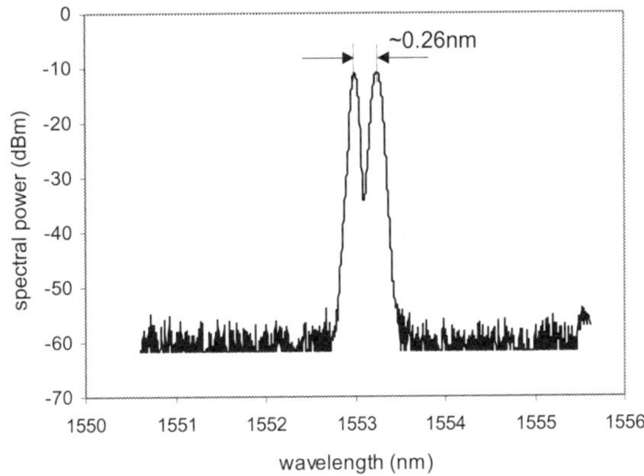

Fig. 5 Optical output spectrum from the co-located dual DFB fibre laser with mode spacing measured ~ 0.26nm.

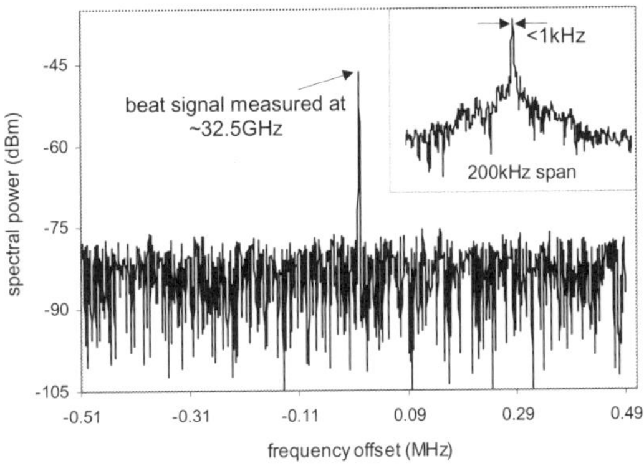

Fig. 6 Millimetre wave beat signal generated from the co-located dual DFB fibre laser. Measured beat signal linewidth is <1kHz.

The narrow linewidth millimetre wave beat signal obtained highlight the low phase noise incurred during operation. The two laser cavities, inclusive of the self-injection polarization discriminating feedback path, are co-located. Hence there is a high common mode suppression of phase noise between the dual optical frequencies generated. No complex circuitry or high-frequency synthesizer is required for the generation of the beat note with a linewidth of <1kHz. With better stabilization of the laser, possibly with active feedback control, 3dB beat linewidth in the order of a few Hertz should be possible.

As compared to [11], the frequency of the millimetre wave signal is not determined or limited by fibre parameters but essentially controlled by the wavelength spacing between the two phase-shifted gratings. This allows commercially available conventional rare-earth doped fibres to be adopted. The fabrication involves straightforward inscription of phase-shifted gratings and does not require slow, complex interferometric control. It is however important to note that the wavelength spacing between the two phase-shift gratings in the experiment is achieved through straining the fibre. Hence the frequency range attainable will be limited by the mechanical strength of glass fibre (typically ~3%). This nevertheless implies beat signal up to 3THz could be generated. To achieve THz-waves based solely on fibre birefringence instead would require birefringence in order of 10^{-2}. As a means to fine-tune the desired frequency of the millimetre wave signal generated, the low temperature sensitivity of the fibre birefringence measured ~ -1.3MHz/^{0}C can be utilized.

Due to its dual polarization output, the co-located dual DFB fibre laser has a potential advantage in the relative ease by which one polarization state can be selectively phase modulated before heterodyning. Phase shifts impose on the millimetre wave signal through such means can be attractive for phased array antenna applications. Furthermore, as opposed to dual longitudinal mode fibre grating laser [13], the power equalization issue can be readily solved through a single polarizer at the output.

4. CONCLUSION

A co-located dual DFB fibre laser configuration is proposed and demonstrated. Stable dual polarization mode operation is obtained from which a 32.5GHz millimetre wave beat signal is generated. Due to strong common mode rejection of phase noise from the dual optical frequencies, the linewidth of the beat signal measured <1kHz without extensive feedback control or external frequency synthesiser. The simplicity of the fabrication and the ease of operation of this fibre laser configuration offer a potential alternative to optical generation of very narrow linewidth millimetre wave signal.

REFERENCES

[1] A. J. Seeds, “Broadband wireless access using millimetre-wave over fibre systems,” IEEE MTT-S Int. Microwave Symp. Dig., Ch. 430, 1997.

[2] S. Matsura, M. Tani, H. Abe, K. Sakai, H. Ozeki and S. Saito, High resolution terahertz spectroscopy by a compact radiation source based on photomixing with diode lasers in a photoconductive antenna, Journal of Molecular Spectroscopy, vol. 187, 1998.

[3] W. H. Loh and C. L. Tang, Polarisation modulation for optical microwave applications in phased array antennas, IEEE Photonics Technology Letters, vol. 3, 1991.

[4] C. R. Lima, D. Wake, and P. A. Davies, Compact optical millimetre-wave source using a dual-mode semiconductor laser, Electronics Letters, vol. 31, no.5, 1995.

[5] A. C. Bordonalli, B. Cai, A. J. Seeds, and P. J. Williams, Generation of microwave signals by active mode locking in a gain bandwidth restricted laser structure, IEEE Photonics Technology Letters, vol. 8, no. 1, 1996.

[6] L. Goldberg, H. F. Taylor, J. F. Weller and D. M. Bloom, Microwave signal generation with injection-locked laser diodes, Electronics Letters, vol. 19, no. 13, 1983.

[7] R. T. Ramos and A. J. Seeds, Fast heterodyne optical phase lock loop using double quantum well laser diodes, Electronics Letters, vol. 28, no. 3, 1992.

[8] F. N. Timofeev et al, High spectral purity millimetre-wave modulated optical signal generation using fibre grating lasers, Electronics Letters, vol. 34, no. 7, 1998.

[9] M. Sejka, P. Varming, J. Hubner and M. Kristensen, Distributed feedback Er^{3+}-doped fibre laser, Electronics Letters, vol. 31, no. 17, 1995.

[10] Y. C. Lai, W. Zhang, J. A. R. Williams, “Optical generation of tunable microwave signal from a dual mode fiber grating DFB laser”, IEEE Int. Top. Meet. Microwave Photonics, 2001.

[11] W. H. Loh, J. P. de Sandro, G. J. Cowle, B. N. Samson and A. D. Ellis, 40GHz optical-millimetre wave generation with a dual polarisation distributed feedback fibre laser, Electronics Letters, vol. 33, no. 7,1992.

[12] J. Bao, X. Zhang, K. Chen and W. Zhou, Spectra of dual overwritten fiber Bragg grating, Optics Communication, vol. 188, (1-4), 2001.

[13] S. Taccheo et al, Generation of micro-and THz-waves at 1.5μm by dual frequency Er:Yb laser, Electronics Letters, vol. 37, no. 24, 2001.

USE OF UNEQUAL CHANNEL SPACING AND NONUNIFORM CHROMATIC DISPERSION IN REDUCING THE FWM LIMITATIONS ON OPTICAL WDM SYSTEM

Muhammad Zulfiker Alam[*], *Mohammed Nazrul Islam, Md. Forkan Uddin, A. B. M. Nasirud Doulah, Abul Bashar Mohammad Ishteak Hossain and Md. Ataur Rahman Sarkar*

Department of Electrical and Electronic Engineering, Bangladesh University of Engineering and Technology, Dhaka-1000, Bangladesh. [*]E-mail: zalamker@eee.buet.ec.bd, zulfiker93@yahoo.com

Abstract: *The performance of an optical wavelength division multiplexed system is analyzed including the effect of four wave mixing (FWM) in presence of chromatic dispersion. FWM process in the single mode fiber as well as in the fiber amplifier is taken into account. It is observed that the efficiency of FWM generation can be greatly reduced by simultaneously making the spacing between channels unequal and by employing nonuniform chromatic dispersion along the fibers in the communication link. The power limitations are evaluated under various conditions of system parameters. The power depletion is compared under different channel spacing schemes.*

Keywords: *Bit error rate, four wave mixing, erbium doped fiber amplifier, nonuniform chromatic dispersion, power depletion, power penalty, unequal channel spacing, wavelength division multiplexing.*

1. INTRODUCTION

With the increase in demand for high data rates, optical transmission system, especially, wavelength division multiplexed (WDM) system is becoming more and more popular day by day. However, a number of nonlinear effects can cause serious performance degradation of such a system [1]. Of these effects, the four wave mixing (FWM) effect is very severe. Unequal channel spacing may be employed to reduce the effect of FWM, however, at the expense of increased bandwidth [2]. Even with unequal channel spacing, the system may suffer from power depletion of the transmitted channels because of the generation of FWM signals [3]. Another solution to the FWM problem is to utilize a number of fiber sections with nonuniform chromatic dispersion so that the phase matching conditions for FWM generated waves are not satisfied [4].

The objective of this paper is to compare the performance of a long distance optical WDM communication system in presence of FWM with various dispersion management and channel spacing schemes. A system with multiple-fiber sections having different chromatic dispersion, maintaining a minimum overall dispersion, is studied. Now-a-days WDM systems frequently make use of erbium doped fiber amplifiers. However, in the literature, the FWM process in EDFAs has not yet been taken into account [5]. In this paper, the FWM effect in the fiber amplifier is also included in the performance analysis. System limitations are depicted under various conditions, to find the best system configuration.

2. THEORETICAL ANALYSIS

In an optical transmission link, each repeater span is composed of a number of short-length fibers as shown in figure 1. There are also a number of amplifiers with equal repeater span throughout the system to compensate for the power loss of the section just before the amplifier. Thus the power input at the beginning of each section is essentially equal. It is assumed that the polarization states of all optical sources are matched throughout the transmission system. For simplicity polarization mode dispersion is ignored.

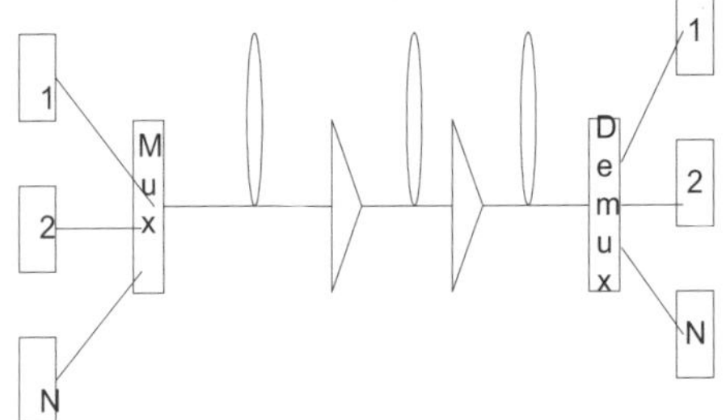

Figure 1: Optical WDM system

Now if there are three channels of frequencies f_a, f_b and f_c, then the frequency components generated by FWM process are

$$f_F = f_i + f_j - f_k, \quad i, j, k = a, b, c \quad \text{and} \quad i, j \neq k \tag{1}$$

The power of FWM signal for general case of chromatic dispersion can be found as [6]

$$P_{FWM} = \left(\frac{D_{abc}}{3}\right) P_i P_j P_k \left[\frac{\sin(N_{amp}(\Delta\beta l + \Delta BL)/2)}{\sin((\Delta\beta l + \Delta BL)/2)}\right]^2$$

$$\times \Big\{ \left|\gamma_{EDF}\sqrt{G(l)} \int_0^l G(z) e^{i\Delta\beta z} dz\right|^2 + \gamma_{TF}^2 G(l)^2 \frac{(1-e^{-\alpha L})^2 + 4e^{-\alpha L}\sin^2(\Delta BL/2)}{\alpha^2 + \Delta B^2} + 2\gamma_{EDF}\gamma_{TF} G(l)^{3/2}$$

$$\times \Big[\int_0^l G(z)\cos(\Delta\beta z)\Big] dz . \int_0^L e^{-\alpha z} \times \cos(\Delta\beta l$$

$$+ \Delta Bz) dz + \int_0^l G(z)\sin(\Delta\beta z) dz \times \int_0^L e^{-\alpha z} \sin(\Delta\beta l$$

$$+ \Delta Bz) dz \Big] \Big\} \qquad (2)$$

where, D_{abc} is the degeneracy factor, P_a, P_b and P_c are the input powers of channels a, b, c, respectively, N_{amp} is the number of fiber amplifiers, γ is the nonlinearity coefficient, $\Delta\beta$ and ΔB are phase mismatching parameters, L and l are length of fiber section and EDFA section, respectively, and $G(z)$ is the gain evolution function of EDFA. The expression for phase mismatching is given by [7]

$$\Delta\beta^{mn} = \beta_a^{mn} + \beta_b^{mn} - \beta_c^{mn} - \beta_F^{mn} = \frac{\pi\lambda^4}{c^2}\frac{dD_c}{d\lambda}\left\{\left(f_a - f_0^{mn}\right) + \left(f_b - f_0^{mn}\right)\right\} \times \left(f_a - f_c\right)\left(f_b - f_c\right) \qquad (3)$$

where, β_a^{mn}, β_b^{mn}, β_c^{mn} and β_F^{mn} are the propagation constants for channel a, b, c and FWM, respectively, of the nth fiber in the mth section, f_0^{mn} is zero-dispersion frequency of nth fiber in mth section, D_c is the chromatic dispersion coefficient.

The above analysis assumes that power depletion due to FWM is negligible. However, the effect of power depletion can also be estimated [8]. Suppose three channels are launched with powers $P_1(0)$, $P_2(0)$ and $P_3(0)$. For finding power after transmission of length z the following relations can be used

$$Q_1 = P_1(0)y$$
$$Q_2 = 0.5P_1(0)(1-y) + P_2(0) \qquad (4)$$
$$Q_3 = 0.5P_1(0)(1-y) + P_3(0)$$

Here $P_j = Q_j \exp^2(-\alpha z/2)$. For unequal channel spacing

$$y = \frac{r+1}{r\cosh^2(\sqrt{r+1}f) + 1} \qquad (5)$$

$$f = \gamma P_1(0)[1 - \exp(-\alpha z)]/\alpha \qquad (6)$$

$$r = 2P_2(0)/P_1(0) \qquad (7)$$

For equal channel spacing

$$y = \frac{4\bar{r}^{-2}\exp(2\bar{r}f)}{r_0^2\exp(4\bar{r}f) + 2(2 + r_2 + r_3)\exp(2\bar{r}f) + r_0^{-2}(r_3 - r_2)^2} \qquad (8)$$

where, $r_2 = 2P_2(0)/P_1(0)$, $r_3 = 2P_3(0)/P_1(0)$, $\bar{r} = \sqrt{(1+r_2)(1+r_3)}$, $r_0 = \sqrt{r_3(r_2+1)} + \sqrt{r_2(r_3+1)}$.

3. RESULTS AND DISCUSSION

The performance of an optical WDM system is evaluated considering the effect of FWM using the theoretical formulation presented in earlier section. A multi-amplifier optical system with ten sections each consisting twenty different fibers is studied. The typical system parameters assumed in calculation are n_o = 1.45, operating guide wavelength λ = 1.55 μm, A_{eff} = 50 μm^2, $\chi = 4\text{x}10^{-15}$ esu, α = 0.21 dB/km, length of each fiber L_o = 2.5 km, $dD_c/d\lambda$ = 0.07 ps/(km-nm^2), $\gamma_{tf} = 1.2\text{x}10^{-3}$ and $\gamma_{edf} = 3.692\text{x}10^{-3}$ 1/mw-km.

The bit error rate (BER) characteristics of the system are estimated for various cases of channel spacing and chromatic dispersion. In the equal channel spacing case the spacing is assumed to be 100 GHz between channels. For unequal channel spacing the minimum separation between channels is taken as 100 GHz so that no FWM frequency will interfere with any of the transmitting channels.

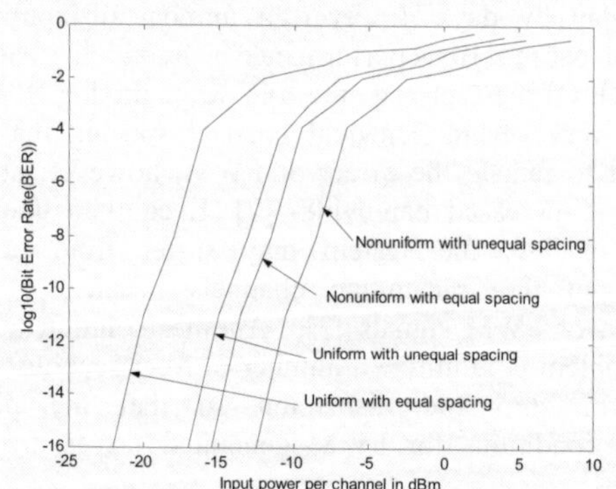

Figure 2: Bit error rate with input power

Figure 2 shows the BER characteristics of the optical system for various cases of channel spacing and chromatic dispersion for a power penalty of 0.7 dB. The BER is found to increase with the input power because of increasing

FWM effect. The system with equal channel spacing and uniform dispersion is found to suffer the most. As seen in the figure, the BER can be reduced by employing unequal channel spacing. The performance can also be improved by utilizing nonuniform chromatic dispersion where each fiber in a section is assumed to have a different dispersion. The zero-dispersion wavelengths of the fibers are assumed to be distributed over an 8 nm width around 1550 nm such that the average zero-dispersion wavelength is maintained at 1.55 μm. The system with unequal channel spacing and nonuniform chromatic dispersion is observed to be the least sufferer.

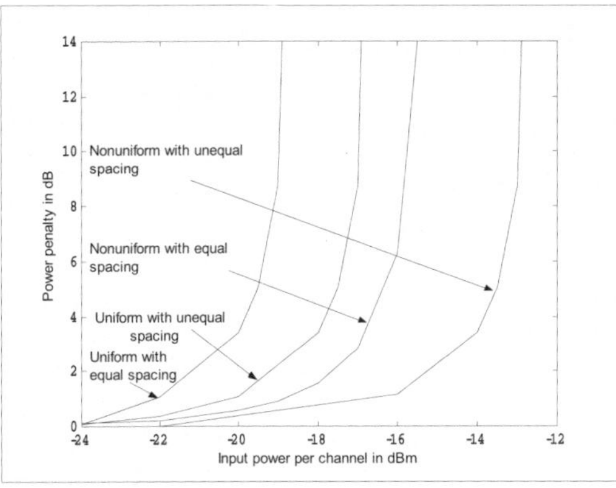

Figure 3: Power penalty with input power

The penalty suffered by the system to achieve a BER of 10^{-9} is estimated and presented in figure 3. It can be observed that the penalty increases with the input power. The worst penalty is suffered by the system with uniform dispersion and equal channel spacing.

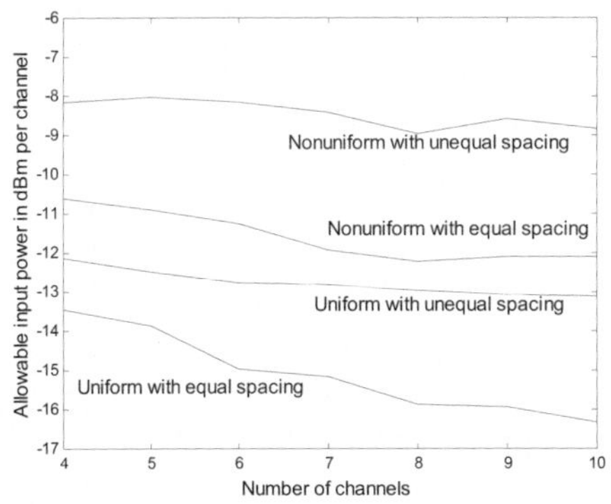

Figure 4: Allowable power with number of channels

The allowable input power for a given power penalty of 0.7 dB at BER $=10^{-9}$ is calculated and shown in figure 4 for various number of channels. As seen in the figure, the allowable input power for nonuniform dispersion is always higher than that for uniform dispersion. Similarly, unequal channel spacing permits much higher allowable input power as compared to equal channel spacing. Thus the maximum possible input power can be obtained by utilizing unequal channel spacing and nonuniform dispersion.

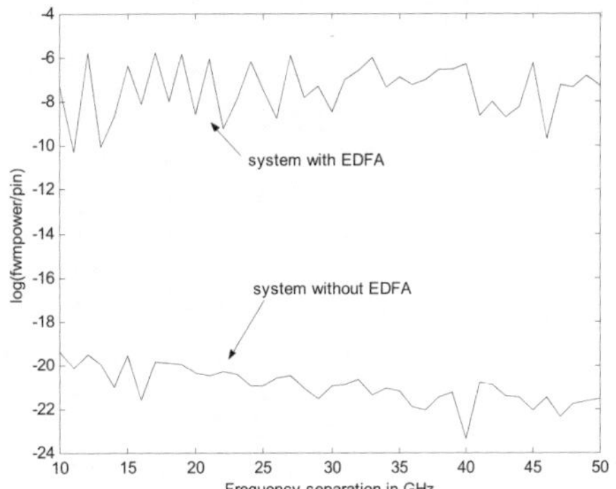

Figure 5: FWM power with and without EDFA

The analysis presented here has taken the FWM process in EDFAs into account. To emphasize the importance of this consideration, the relative FWM power has been plotted for two systems as a function of channel spacing. The first system uses only single mode fiber (SMF) and the second one uses both SMF and EDFA. It is observed that FWM power is much higher for the second system. This clearly demonstrates the importance of FWM process in EDFA while evaluating system performance.

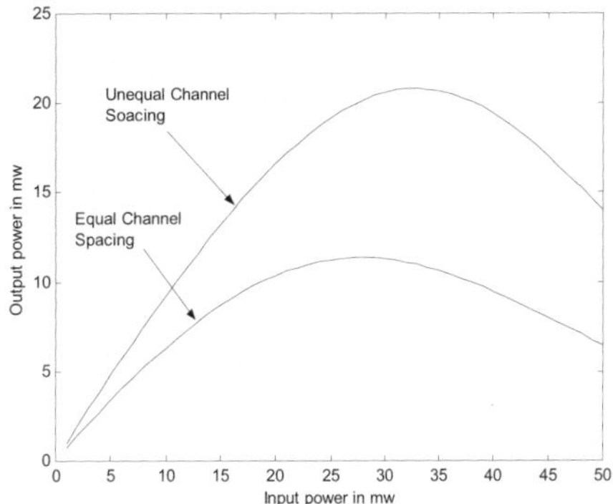

Figure 6: Output power with input power

The effect of power depletion was also studied. We have considered a three-channel system for simplicity and calculate the depletion of one channel. We have used $\gamma = 1.853\times10^{-5} cm^{-1}W^{-1}$ and A_{eff}=50 μm^{-2} and a transmission length of 30 km. With the increase in input power, the output power at first increases. But with high level of input power output power start to decrease because depletion due to FWM becomes severe. But the power depletion is much prominent in case of equal channel spacing compared to unequal channel spacing. This also supports the view that unequal channel spacing scheme should be the choice for WDM systems [9].

4. CONCLUSION

An analysis is presented to evaluate the limitations imposed by FWM on an optical WDM system. The system performance can be improved by employing varying dispersion along the fibers while maintaining a low average chromatic dispersion. Also unequal channel spacing is found to reduce the FWM efficiency appreciably. Therefore, a high bit rate long haul optical WDM system can be designed by controlling the channel spacing and chromatic dispersion to overcome the effect of FWM. The result also suggest that FWM effect in EDFA s should be taken into account while evaluating system performance.

ACKNOWLEDGEMENT

The authors would like to thank the Department of Electrical Engineering, Bangladesh University of Engineering and Technology for providing necessary financial and technical support in the research work.

REFERENCES

[1] G. P. Agarwal, *Nonlinear Fiber Optics*, Academic Press, 1989.

[2] F. Forghieri, R. W. Tkach, A. R. Chraplyvy, "WDM systems with unequally spaced channels," *J. Lightwave Technol.,* vol 13, no. 5, pp. 889-897, 1995.

[3] H. Zheng, S. Xie, Z. Zhou and B. Zhou, "A simple formula on allowable fiber input power for unequal channel spaced WDM transmission system," *J. Optical Commun.*, vol 20, no. 4, pp. 153-155, 1999.

[4] K. Inoue and H. Toba, "Fiber four-wave mixing in multi-amplifier system with Non-uniform Chromatic Dispersion," *J. Lightwave Technol.*, vol. 13, no.1, 1995.

[5] M. N. Islam and M. Z. Alam, "Performance of a long distance optical WDM communication network in presence of four wave mixing," *IEB J. Elect. Engg.*, vol. EE29, no. 1, 2001.

[6] Kai Song, Malin Paramratne, " Effect of SPM,XPM and Four-Wave-Mixing in L-Band EDFAs on Fiber-Optic Signal Transmission", IEEE Photonics Technology Letters, vol.12,no.12,December 2000.

[7] K. Inoue, "Fiber four-wave mixing in the zero dispersion wavelength region," *J. Lightwave Technol.*, vol. 10, pp 1553-1561, 1992.

[8] Weishu Wu and Pochi Yeh, " Energy Coupling by Partially Degenearte Four Wave Mixing in Multichannel Lightwave Systems", IEEE Photonics Technology Letters, Vol 7, No.5, May 1995

[9] M.Z.Alam,"Performance of an optical WDM Network in presence of four wave mixing", M.Sc.Engineering Thesis to be disserted, Department of Electrical and Electronic Engineering, Bangladesh University of Engineering and Technology.

Experimental investigation of Raman Amplifier with dynamic gain control

Y. X. Wang, Y. D. Gong, P. Shum, Y. L. Guan, N. Q. Ngo, X. C. Yuan
Network Technology Research Center, Nanyang Technological University,
Singapore 639798

Abstract: A gain-clamped fiber Raman amplifier is proposed and builds up. The result shows that gain fluctuation will be champed to a lower level with the dynamic gain control. And the dynamic gain control method is also fit for the Raman amplifiers with multi-wavelength pump.

Keywords: Raman Amplifier, Dynamic, Gain control, Feedback loop, Pump, Clamp, RFA, ASE

1. INTRODUCTION

The last research on WDM transmission systems, both distributed and lumped Raman amplifiers have been used in order to improve the noise performance and to expand the signal wavelength bandwidth, respectively. [1] Most of them optimize the gain profile of Raman Amplifiers through adjust the parameters of the Pumps. But, as we know such a gain level of Raman amplifiers is produced by the interaction of pumps and signals. Further more, in the design of optical amplified link for DWDM application in a network scenario, the number and the power level of the input channels may vary randomly in time, there for it is also essential to stabilize the amplifier gain profile

.

In this paper, we discuss a fiber Raman amplifier with dynamic gain control base on the experiments. The basic idea is similar to a gain-clamped EDFA. [2] By analyze the result of the experiments, we can found the dynamic gain control method is also fit for the Raman amplifiers with multi-wavelength pumps.

2. PRINCIPLE

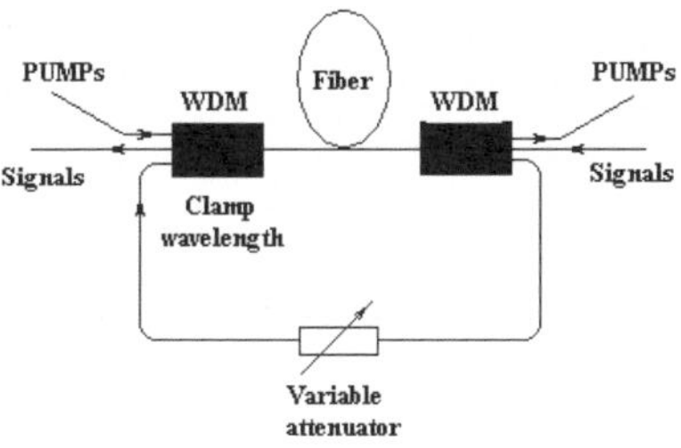

Fig. 1 Diagram of fiber Raman Amplifier

The proposed gain-clamp Raman amplifier is shown in Fig. 1. The pumps and the lasing light are all counter propagating against the signals in order to decrease the pump-induced crosstalk. The gain of the lasing light will be at a stable state because of the loss of the variable attenuator. In order to generate a wide-band and gain-flatten spectrum profile, Raman fiber amplifiers usually have more than 2 pumps. [3] At a stable state, neglecting additional various effects such as Rayleigh back scattering, temperature dependent spontaneous Raman emission, and the interaction between signals, the propagation equations governing signal lights and lasing lights power evolution can be simplified in the form as the follows.

$$\frac{dS(z,v_i)}{dz}=\frac{1}{K_{eff}A_{eff}}\left(\sum_{j=1}^{P\_num} g_r(v_i-\varsigma_j)P(z,\varsigma_j)+g_r(v_i-\xi)F(z,\xi)\right)S(z,v_i) \tag{1}$$

$$\frac{dF(z,\xi)}{dz}=\frac{1}{K_{eff}A_{eff}}\left(\sum_{j=1}^{P\_num} g_r(\xi-\varsigma_j)P(z,\varsigma_j)\right)F(z,\xi) \tag{2}$$

Here $S(z,v_i)$, $P(z,\varsigma_j)$ and $F(z,\xi)$ denote the signal power, pump power and lasing light power, respectively. $g_r(v_i-\varsigma_j)$, A_{eff} and K_{eff} represent Raman gain coefficient, effective area of the fiber and polarization-related factor. P_num means the number of the used pumps. Integrating (1) and (2) along the fiber length, we get signal and lasing light gain as

$$G_s(v_i)=\exp\left(\sum_{j=1}^{P\_num} g_r(v_i-\varsigma_j)H_p(\varsigma_j)+g_r(\xi-\varsigma_j)H_F(\xi)\right) \tag{3}$$

and

$$G_F(\xi)=\exp\left(\sum_{j=1}^{P\_num} g_r(\xi-\varsigma_i)H_p(\varsigma_j)\right) \tag{4}$$

respectively. Here $H_p(\varsigma_j)=\int_0^L P(z,\varsigma_j)dz$ (L means the fiber length) and $H_F(\xi)=\int_0^L F(z,\xi)dz$. Note that the signal gain is only dependent on the integration of pump power (including the lasing light) along the fiber length. Unlike a gain-clamped EDFA, even if $G_F(\xi)$ remains a constant, the value of $G_s(v_i)$ is still uncertain. However, observing (3) and (4) carefully, we can found that, the introduction of lasing light can still reduce the gain the lasing light. It can also be supposed that, the "gain-locking" bandwidth will be dependent on the wavelength of the lasing light. To verify these suggestions, we performed computer simulations for both the stable state performance and transient dynamics.

3. EXPERIMENT

We have done the experiments to measure the performance of the Raman amplifier with such a dynamic gain control and compared with the simulation result. The setup of our experiment is shown in Fig.2.

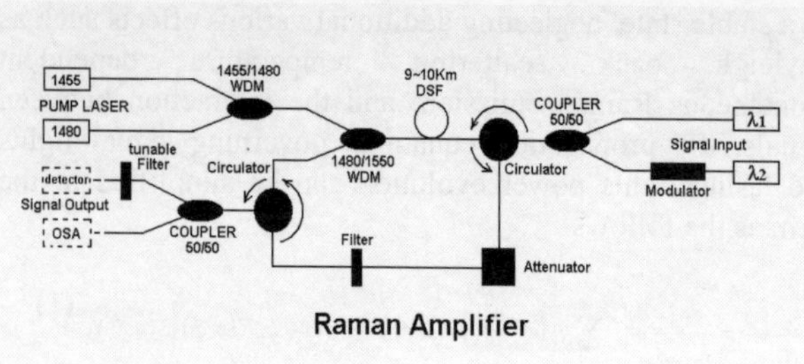

Fig.2 Experiment setup of Raman amplifier with the dynamic gain control

In this experiment setup, we use two high power pump light (one is 1455nm and another is 1480nm). Then we use two circulators and an attenuator to build a feed back loop with a filter additional. So there is a variable lasing light propagating in this loop. Our theoretical analysis have shown that, for a RFA with N pump lights, the dynamic gain spectrum profile cannot be completely stabilized even with N lasing lights (at different wavelength). This is due to the fact that the introduced lasing lights themselves can be very strong and also acts as pump lights. But we still can see the ASE spectra of RFA with the dynamic gain control or with out it. And using the all-optical feedback loop in this structure, the gain fluctuation of the RFA will be clamped to a lower level. The introduction of lasing light can reduce the gain fluctuation caused by power variation of the incoming signal light, especially for signals around the wavelength of the lasing light. The "gain-locking" bandwidth will be dependent on the wavelength of the lasing light.

As the detail of our experiment, one of the signal channels will be modulated act as a signal variation. And as we know the other signal channel will be affected by the signal variation. In our simulation, the gain fluctuation of the RFA can be clamped to a lower level. We use a detector with a filter additional to analyze the signal channel affected. The pump power will be very high in this experiment because of the two high power pumps used. But the power should not much higher than the loss of the feedback loop. It will be harm for the experiment equipments and the operators. In this experiment, when we find the lasing light propagating in the loop, we should stop increasing the pump power. After the introduction of our experiment, we will compare the result of the simulation and the experiment, and analyze the result in chapter 4.

4. RESULT

Our experimental results have revealed that, even for a 10 km length of RFA, operation is still very stable feedback systems. The Fig. 3 shows the clamping signal effect on ASE. It is clear that the ASE spectra are in different level when the feedback loop works or not. The upper line indicates the cases with no feedback loop, and the lower line indicates the cases with the lasing light propagating in the feedback loop. Then the gain spectra of the RFA with the clamping signal will be shown in Fig. 4. The constant pump powers launched into the DSF are: 0.5W for 1480nm pump and 1.2W for 1455 nm pump. The gain/ASE spectra are not optimized (for flatness) because what we concern here is the stability of the dynamic gain spectrum profiles. From these ASE spectra it is clear that the gain variations due to signal cross saturation can be suppressed in a great degree by introducing all-optical feedbacks. The lasing peaks is near the 1560.0 nm are due to Rayleigh scattering in the amplifying paths. Such reflection peaks are quite detrimental when they fall in the useful amplifier gain band. Figure. 5(a), (b) and (c) shows the measured output signals respectively. Here we always let the signal channel very close (1557.8 and 1558.6) in order to let the cross talk between the signals channels to be in the biggest level. From the result of our experiment we found that the optical feed back loop clamped the gain fluctuation directly and efficiently. The interaction between the two signal channels was decrease to a lower level. By the way, because the lasing light propagated as a pump light, the level of gain flattening will be better with the dynamic gain control. The gain will be more than 16dB in our experiment. And the efficiency of the gain clamping will not change with the change of the pump power when the power is high enough and the lasing light can propagate in the feed back loop.

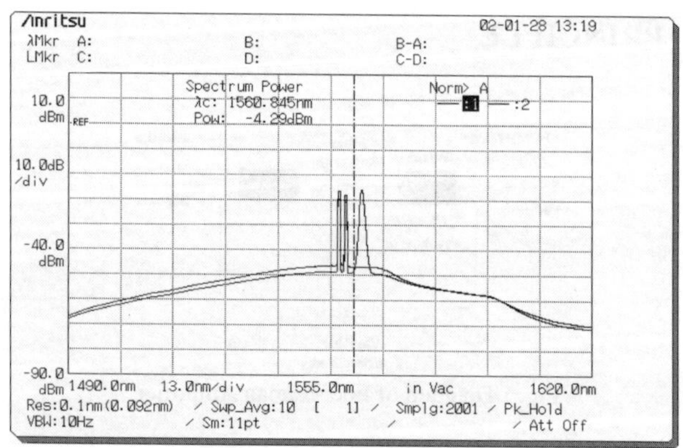

Fig. 3 clamping signal effect on ASE

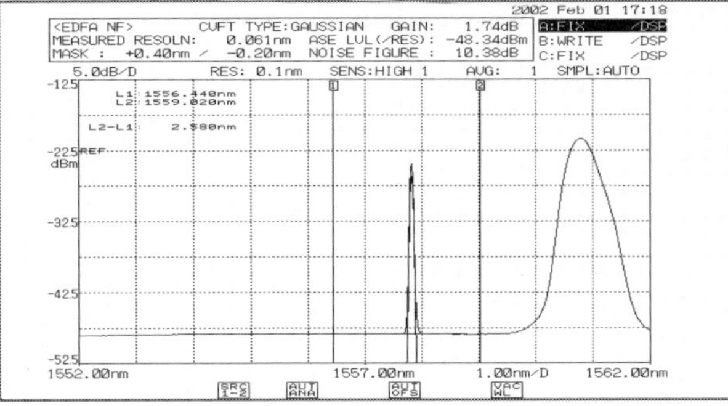

Fig. 4 gain spectra of the RFA with the clamping signal

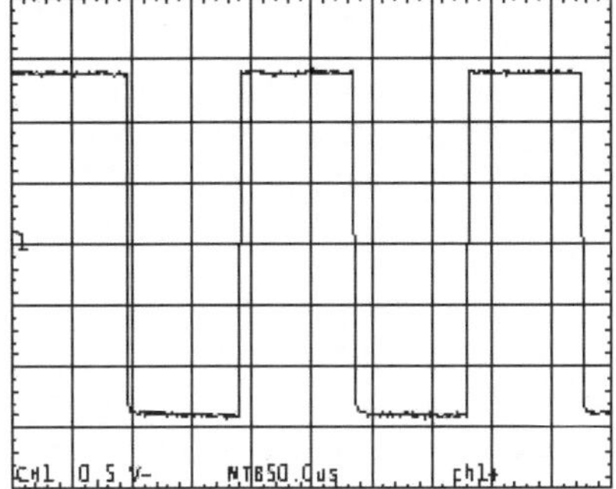

Fig. 5 (a) modulated signal #2 to emulate the channel add/drop

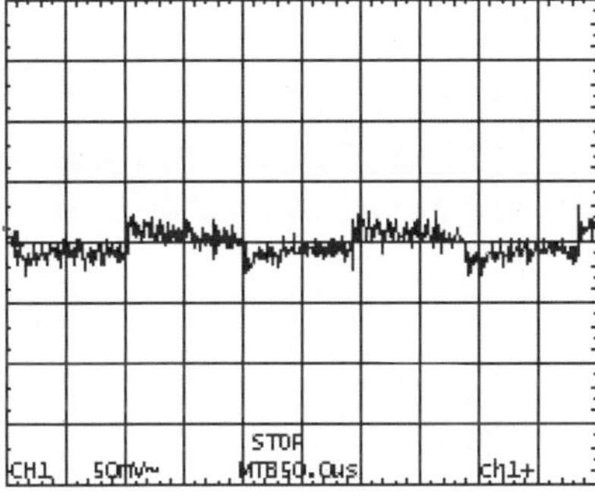

Fig. 5 (b) channel #1 result without feed- back loop

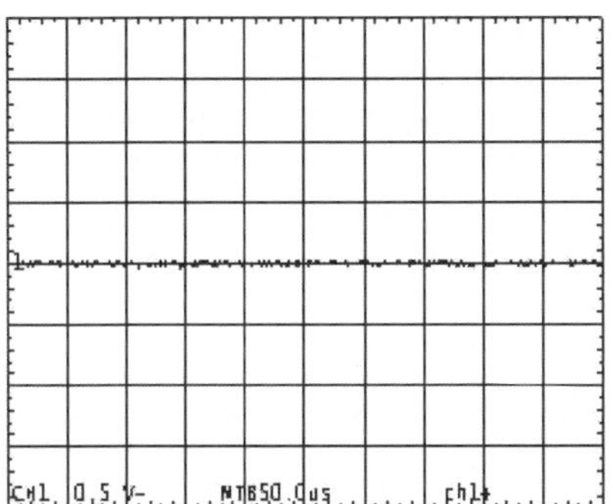

Fig. 5 (c) channel #1 result with all optical feedback loop

5. CONCLUSION

Base our setup of Raman amplifier with the dynamic gain control; we compare our result with the result of simulation. We found that the gain fluctuation can be clamped to a lower level with the optical feed back loop efficiently. With the propagation of the lasing light, the gain profile will be better on gain flattening. In addition, we found that the optimal lasing wavelength is dependent on the Raman gain profile, fiber length and pump powers in our experiment.

REFERENCE

[1] Simple gain control method for broadband Raman amplifiers gain-flattened by multi-wavelength pumping, Yoshihiro Emori, Soko Kado, Shu Namiki, Optical communication, 2001. ECOC '01. 27th European Conference on, Volume: 2 , 2001 Page(s): 158 –159.

[2] Theoretical investigation of fiber Raman amplifier with dynamic gain control, Xiang Zhou; Shalaby, H.H.M.; Chao Lu; Ping Shum; Teehiang Cheng, Optical Fiber Communication Conference and Exhibit, 2001. OFC 2001, Volume: 3, 2001 Page(s): WDD17 -WD1-3 vol.3.

[3] Howard Kidorf, Karsten Rottwitt, Morten Nissov, Matthew Ma and Eric Rabarijaona, "Pump interaction in a 100-nm bandwidth Raman amplifier", IEEE Photonics Technology Letters, Vol 11, no.5, pp. 530-532 (1999).

Contact:
mr_wangyuxiang@hotmail.com,G01534263@ntu.edu.sg; phone. 65 67905363, fax. 65 67926894, Network Technology Research Center, School of Electrical and Electronic Engineering, Nanyang Technological University, Block S2, B3c-23 50 Nanyang Avenue Singapore 639798

Launch Power Management for DWDM Transmission Systems Using Two Stage Lumped Amplification

Yang Jing Wen and Sarah Dods
Australian Photonics Cooperative Research Centre, Photonics Research Laboratory
Department of Electrical and Electronic Engineering
The University of Melbourne, Victoria 3010, Australia.
Tel: +61 3 8344 4936, Fax: +61 3 8344 6678, Email: y.wen@ee.mu.oz.au

Abstract: This paper investigates launch power management for DWDM transmission systems that use two stage lumped amplification surrounding dispersion compensating fiber (DCF). The impact of modulation format and fiber type on optical launch power is investigated. The results show that for all modulation formats and transmission fiber types, launch power into DCF should be much lower than into transmission fiber, to maintain high optical signal to noise ratio (OSNR). However, the values of the optimum launch powers vary with the modulation format and transmission fiber type.

Key Words: WDM transmission system, nonlnearity management.

I. INTRODUCTION

Continuing growth in demand for bandwidth is pushing carriers and service providers to increase optical backbone transmission capacity. Dense wavelength division multiplexed (DWDM) transmission systems are used to boost the capacity of long distance networks, by packing a single fiber with multiple wavelengths, each carrying a different data stream. In these networks, signal distortion is caused by dispersion, amplified spontaneous emission (ASE) noise, and fiber nonlinearities. To manage these effects, new concepts including dispersion management, distributed Raman amplification, and nonlinearity control have been proposed [1]. However, many WDM systems are based on already deployed standard single mode fibers (SMF), and use lumped dispersion compensation at the end of each amplifier span [2, 3].

If distributed Raman amplification is not used, two stage erbium doped fiber amplifiers (EDFA) are usually required to compensate for the total fiber loss in each span, and are placed before and after the DCF. The allocation of launch power into the DCF and into SMF is critical to transmission performance, however the optimization of these powers has generally been overlooked in work published to date. In this paper, we study the management of launch power into DCF and transmission fiber for these two stage EDFA amplified WDM systems, using computer simulations. The impacts of modulation format and different types of transmission fiber are also investigated.

II. LAUNCH POWER OPTIMIZATION FOR TWO STAGE EDFA AMPLIFIED WDM SYSTEMS

To investigate launch power management for two stage EDFA amplified WDM systems, we simulate a 40 channel 10 Gb/s based WDM transmission link over 25 fiber spans of 100 km, as shown in Fig. 1, using *VPItransmissionMaker*. The central wavelength of the 40 channels is 1550nm, and the channel spacing is 100 GHz. The simulation parameters used for the transmission fibers and DCF considered in the study are listed in Table I. Currently many new types of DCF are being announced, which adds to the complexity of the study. In order to simplify matters, we have chosen here to assume a single type of DCF to compensate each transmission fiber type. The DCF slope is customized for each type of transmission fiber, in order to give full dispersion compensation across all the total transmission bandwidth. The length of DCF used for each type of transmission fiber is chosen to fully compensate the total dispersion of one fiber span. The lumped amplification is provided by gain flattened EDFAs with a noise figure of 5.0 dB. The WDM multiplexer at the start of the link is modeled

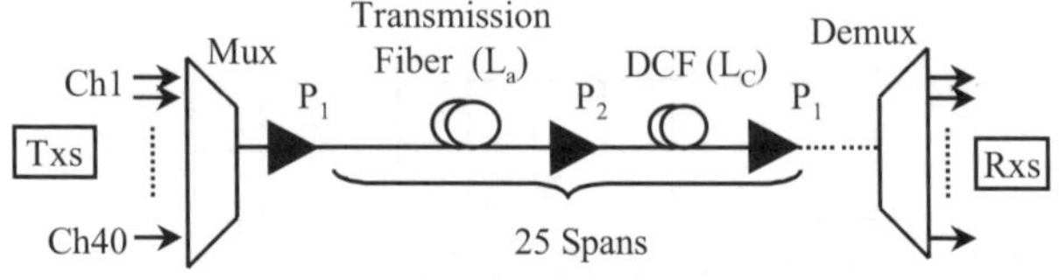

Fig. 1 Schematic used for investigation

Table I – Fiber parameters used in the simulation. TWRS – Truewave Reduced Slope, TL – Teralight

	Dispersion (ps/nm/km)	Slope (ps/nm^2/km)	α (dB/km)	A_{eff} (μm^2)	n_2 (10^{-20} m^2/W)
SMF	17	0.058	0.25	80	2.35
TWRS	4.5	0.05	0.22	50	2.35
TL	8.0	0.058	0.22	65	2.35
DCF	–90	Matching transmission fiber	0.5	20	2.7

as a combiner, and the demultiplexer as 3rd order Bessel filters with a bandwidth of 0.3 nm.

Initially, we consider SMF transmission fiber transmitting return-to-zero (RZ) data, which is generated by two cascaded modulators. Figure 2 shows the calculated Q value at the end of the link as a function of launch power into the SMF (P_1) and DCF (P_2) for different channels. Although the Q values shown are low, the BER is still acceptable if forward error correction is used. The results show that the optimal launch power into SMF for each channel is around 2 dBm, while the optimum launch power into DCF is around –7 dBm/channel. These powers give Q values of 5.20, 5.24 and 5.28 for channels 4 (191.9 THz), 20 (193.5 THz), and 37 (195.2 THz) respectively. The slightly lower Q value for channel 4 is probably due to higher local dispersion at the longer wavelength. All three channels show that the optimal launch power into SMF is much higher than into DCF, and that Q decreases rapidly away from this point. The cause of this optimum is the interaction between fiber nonlinear effects and ASE noise. We define the nonlinear phase shift due to self-phase modulation accumulated along each span as [4]:

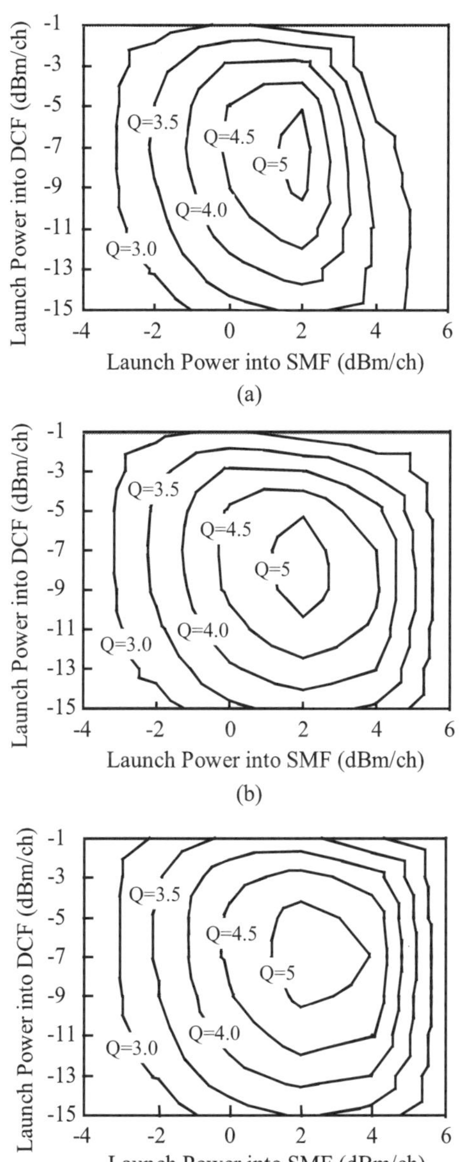

Fig.2 Q values as a function of launch power into SMF and into DCF using RZ modulation format for (a) Channel 4, (b) Channel 20 and (c) Channel 37.

$$\phi_{NL} = \frac{2\pi}{\lambda}\int_0^{L_a} \frac{P(z,\alpha^{(+)})}{A_{eff}^{(+)}} n_2^{(+)} dz + \frac{2\pi}{\lambda}\int_0^{L_a} \frac{P(z,\alpha^{(-)})}{A_{eff}^{(-)}} n_2^{(-)} dz \quad (1)$$

where λ is transmission wavelength, L_a is the span length, L_C is the length of DCF, P is the optical power at a given point along the fiber, and $\alpha^{(\pm)}$, $n_2^{(\pm)}$ and $A_{eff}^{(\pm)}$ are the attenuation coefficient, the nonlinear refractive index, and the effective area for the transmission fiber (+) and DCF (–). From (1) we can obtain

$$\phi_{NL} = \psi_1 P_1 + \psi_2 P_2 \quad (2)$$

where

$$\psi_1 = \frac{2\pi n_2^{(+)}}{\lambda A_{eff}^{(+)} \alpha^{(+)}} \left(1 - \exp\left(-\alpha^{(+)} L_a\right)\right) \quad (3)$$

$$\psi_2 = \frac{2\pi n_2^{(-)}}{\lambda A_{eff}^{(-)} \alpha^{(-)}} \left(1 - \exp\left(-\alpha^{(-)} L_C\right)\right) \quad (4)$$

are measures of the nonlinear phase shift in the transmission fiber and DCF respectively. We can also derive a simplified optical signal to noise ratio expression for a single span as:

$$OSNR^{-1} = Bh\nu n_{sp}\left(\frac{\exp\left(\alpha^{(+)} L_a\right)}{P_1} + \frac{\exp\left(\alpha^{(-)} L_C\right)}{P_2}\right) \quad (5)$$

where B is the bandwidth of optical filter and n_{sp} is the spontaneous emission factor of EDFA.

From Table I, we can see that $\exp\left(\alpha^{(+)} L_a\right) \gg \exp\left(\alpha^{(-)} L_C\right)$, which combined with Equation (5) shows that increasing P_1 increases the OSNR much more quickly than increasing P_2. Therefore the optimum launch power into the transmission fiber $P_{1,opt}$ will be much higher than $P_{2,opt}$. Applying the parameters from Table I, into Eq. (3) and (4) gives $\psi_1/\psi_2 \sim$ –3 dB, while from Fig.1, we get the optimum ratio $P_{1,opt}$ / $P_{2,opt} \sim$ 9 dB. Combining these gives $(\psi_1 P_{1,opt})/(\psi_2 P_{2,opt}) \sim$ 6 dB, which means that the nonlinear phase shift introduced by SMF is larger than that of DCF. This also explains why the Q value is more sensitive to changes in P_1 than P_2.

III. IMPACT OF MODULATION FORMATS

In this section, we present the impact of modulation format on the launch power management for two stage lumped amplified WDM systems. A variety of modulation formats have been proposed to achieve higher capacity and longer distance WDM transmission. Available formats include non-return-to-zero (NRZ), RZ, duo-binary etc. RZ formats include simple RZ, carrier suppressed RZ (CS-RZ), chirped RZ (CRZ), vestigial side band RZ (VSB-RZ), and dispersion managed soliton based RZ (DMS-RZ). The performance of VSB-RZ depends on the profile, bandwidth and center wavelength detuning of the VSB filter used for the generation of VSB-RZ data. For WDM systems using a DMS-RZ format, the pulse returns to the same shape after every period in the dispersion map, requiring additional pre-emphasis in the launch power, and much higher launch power than that of conventional RZ systems. The increased launch power leads to stronger nonlinear effects from self phase modulation (SPM), cross phase modulation (XPM) and four wave mixing (FWM). SPM can be balanced by the average dispersion over the link, however XPM and FWM will significantly increase the transmission impairment in WDM systems. Due to these limitations, DMS-RZ is better suited to very high speed single channel or sparse WDM transmission, and is not considered here. Duo-binary formats, which include conventional duo-binary and carrier-suppressed (CS) duo-binary formats, have a very narrow optical spectrum width, and so have a high dispersion tolerance allowing very high spectral efficiency. However they also have very low tolerance to nonlinearity, and so are not advantageous for ultra long haul WDM transmission, and are also not considered.

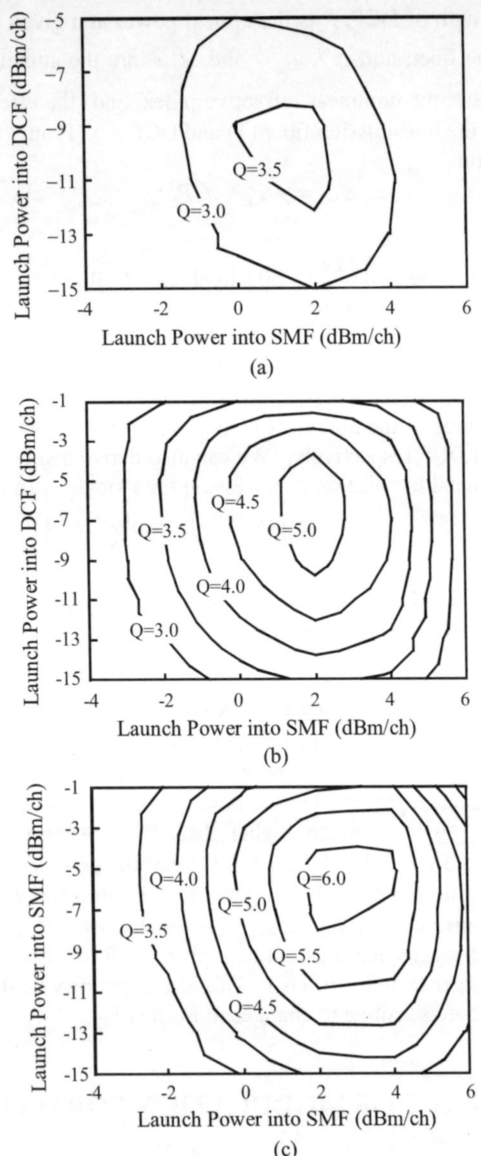

Fig.3 Q values of channel 20 as a function of launch power into SMF and DCF for (a) NRZ, (b) CS-RZ and (c) CRZ modulation formats.

In this work, we compare the optimum launch powers and performance of NRZ, CS-RZ and CRZ formats with the RZ results presented in the previous section. The CS-RZ data was generated via two cascaded Mach-Zehnder modulators, with the pulse generating modulator biased at the null point. The CRZ data was generated by passing RZ data through an additional phase modulator. The relative phase of phase modulation and data bits was optimized, with a modulation index of 0.4π.

Fig.3 shows Q as a function of the launch power into the SMF and DCF for channel 20, using the different modulation formats. The maximum Q for the NRZ format is much lower than for RZ, due to its narrower optical spectrum, which causes larger dispersion tolerance but increased nonlinear effects. The optimum value of P_1 for NRZ data decreases to 1.5 dBm/ch, while the optimum P_2 value is –10dBm/ch, which are both lower than for the RZ format. The optimal values of P_1 and P_2 for the CS-RZ format are similar to RZ format, but give a slightly higher maximum Q value. The improvement in Q is attributed to reduced intersymbol interferece for CS-RZ data due to the alternate optical phase of successive bits.

Compared with RZ, the CRZ format shows a significant increase in the maximum Q. The optimum launch power into the DCF increases to –6dBm/ch, while optimal launch power into the SMF is also increased, to 3 dBm/ch. The CRZ format shows better transmission performance and higher tolerance against nonlinearity because its initial chirp reduces the dispersive wave caused by the interaction of dispersion and nonlinearity [5], and the broader optical spectrum is less phase matched since the spectral components are lower in power, leading to reduced nonlinear effects.

IV. IMPACT OF TRANSMISSION FIBER TYPE

This section presents the impact of transmission fiber type on the launch power management for two stage lumped amplified WDM systems. Different transmission fibers have different attenuation, dispersion, effective area and nonlinear refractive index parameters, leading to different transmission performance. Some of the many commercially available transmission fibers are TruewaveRS, Teralight, LEAF, SMF, Ultrawave, and pure silica core fiber. Of these fibers, SMF, Ultrawave and pure silica core fiber have high dispersion and large effective area, TruewaveRS has low dispersion and a small effective area, while Teralight and LEAF have a trade off in dispersion and effective area. We compare the performance of a low dispersion fiber: TruewaveRS (TWRS) and a moderate dispersion fiber: Teralight, with the SMF results from Section II for RZ data modulation. DCF is used to compensate the dispersion of all fiber types, with the dispersion slope set to match the slope of each fiber type, and the length set to give full dispersion compensation.

Fig.4 shows Q value as a function of the launch powers into TWRS and DCF for channel 20. The optimal launch power into the DCF is reduced to around –8 dBm/ch, while the launch power into TWRS drops to –1 dBm/ch. The maximum Q value for the TWRS system is only slightly higher than that from SMF system, even though the fiber loss of TWRS is lower. However, the optimum operating range for P_1 and P_2 for a Q value greater than 5 is much larger, and the Q value is much less sensitive to the DCF launch power. The reduction in $P_{1,opt}$ can be attributed to the reduced effective area of TWRS. Since $(\psi_1 P_{1,opt})/(\psi_2 P_{2,opt}) \sim 11$ dB,

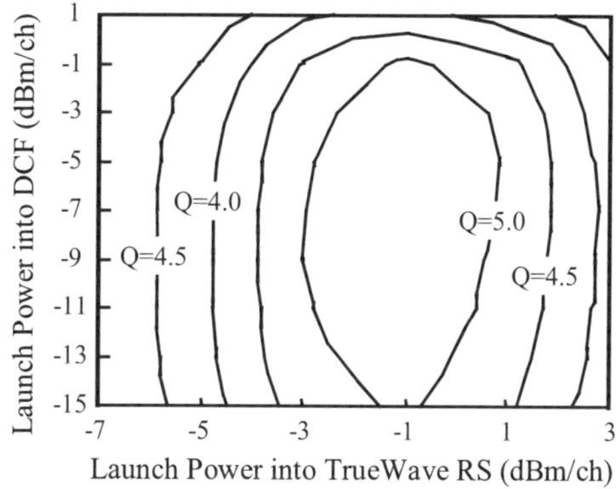

Fig.4 Q value as a function of launch power into TruewaveRS and DCF for channel 20 using RZ modulated data.

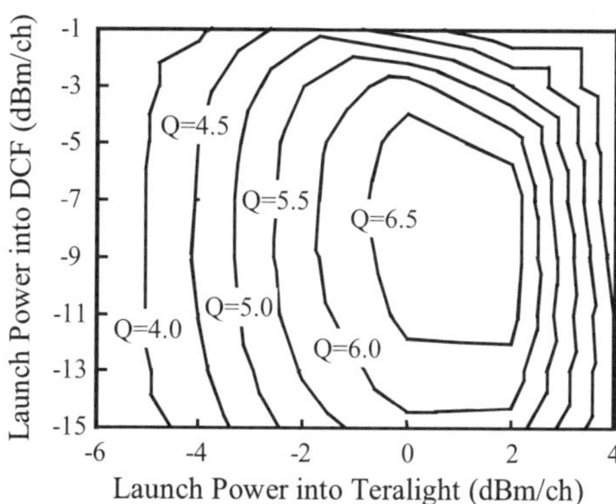

Fig.5 Q value as a function of launch power into Teralight and DCF for channel 20 using RZ modulated data.

the nonlinear phase shift introduced by TWRS is dominant over the DCF effects, leading to the insensitivity of Q value to P_2.

Fig.5 shows Q as a function of the launch powers into Teralight and DCF for channel 20. The optimum power levels are between those of the SMF and TruewaveRS systems, with $P_{1,opt}$ of 1 dBm/ch, and $P_{2,opt}$ of –8 dBm/ch. The maximum Q is much higher than either the SMF or TruewaveRS results. The improved performance is attributed to the fiber trade-off between local dispersion and effective area, and the reduced fiber loss compared with SMF.

V. CONCLUSION

We have demonstrated that that there is an optimum combination of launch powers into transmission fiber and DCF for DWDM transmission systems using two stage lumped amplification. The results here show that launch power into DCF should be much lower than into transmission fiber to maintain high OSNR, for all the modulation formats and transmission fibers considered. However, the optimum launch power levels, and the maximum Q value achievable depend on both the modulation format and transmission fiber type used, due to the interplay of ASE noise, nonlinear effects and dispersion.

ACKNOWLEDGEMENT: The authors would like to thank Telstra for their support in this investigation.

REFERENCES

[1] R. Hainberger, and J. Kumasako, K. Nakamura, T. Terahara, H. Onaka, and T. Hoshida, "Optimum span configuration of Raman-amplifier dispersion-managed fibers", Optical Fiber Communication Conference and Exhibit, 2001, OFC 2001, vol. 1, page MI5/1- MI5/4, Anaheim, USA.

[2] R. I. Killey, H. J. Thiele, V. Mikhailov, and P. Bayvel, "Reduction of intrachannel nonlinear distortion in 40-Gb/s-based WDM transmission over standard fiber", IEEE Photonics Technology Letters, vol.12, page 1624-1626, 2000.

[3] S. Bigo, and A. Bertaina, "WDM transmission experiments at 32/spl times/10 Gb/s over nonzero dispersion-shifted fiber and standard single-mode fiber IEEE Photonics Technology Letters vol.11, page 1316-1318, 1999.

[4] J-C. Antona, S. Bigo, and J.-P. Faure, "Nonlinear cumulated phase as a criterion to assess performance of terrestrial WDM systems", Optical Fiber Communication Conference and Exhibit, 2002, OFC 2002, Technical Digest, page 365-367, Anaheim, USA.

[5] T. Georges, and B. Charbonnier, "Reduction of the dispersive wave in periodically amplified links with initially chirped solitons", IEEE Photonics Technology Letters, vol.9, page 127-129, 1997.

Session 13H

Lasers and Amplifiers I

SOME RECENT ISSUES IN FIBER OPTICS

Ajoy Ghatak
Physics Department
Indian Institute of Technology
New Delhi - 110016

ABSTRACT

The dramatic reduction of transmission loss in optical fibers coupled with equally important developments in the area of light sources and detectors has brought about a phenomenal growth of the fiber optic industry during the past twenty five years. Recent advances in the area of zero dispersion operation, dispersion compensation, rare earth doped fiber amplifiers and soliton communication systems have provided communication engineers with almost limitless bandwidth operating over millions of kilometers. Some of the recent developments in this area, including devices like Fiber Gratings will be briefly described in this lecture.

A novel structure of external cavity semiconductor lasers with suppressed thermally induced wavelength shift and mode hopping

H.G.Yu , C.Q.Xu,
Department of Engineering Physics, McMaster University, Canada
honggay@mcmaster.ca
S.F.Yu
Photonics Research Group, School of EEE, Nanyang Technological University, Singapore
L.P.Zhao, Z.P.Fang
Photonics Systems Group, Singapore Institute of Manufacturing Technology, Singapore

Abstract

A novel structure of external cavity semiconductor lasers with suppressed thermally induced wavelength shift and mode hopping is proposed in this paper. The calculation results have shown that the wavelength stability with temperature is dramatically improved and the side mode suppression ratio (SMSR) remains a higher value (>35dB) when temperature change from 20°C to 50°C.

Key Words: waveguide grating external cavity semiconductor lasers (WG-ECL), temperature insensitive fiber Bragg grating (FBG).

1. Introduction

Integrated waveguide grating-external cavity semiconductor lasers (WG-ECL) is a promising light source for the future DWDM systems and access area network. WG-ECL is composed of a UV written Bragg grating on silica waveguide and a laser diode (LD) with HR coating on one facet and AR coating on the other facet. The oscillation wavelength of WG-ECL is less sensitive to temperature than that of the conventional distributed feedback lasers due to the relatively low temperature coefficient of the silica waveguide. However, thermally induced mode hopping in WG-ECL has to be suppressed because it will affect the stability of output power and side mode suppression ratio (SMSR). The channel spacing of future DWDM system will be less than 0.8nm and the tolerance of emitted wavelength should be less than 10% of the channel spacing on the whole operating range. Consequently, thermally induced wavelength shift in WG-ECL (normally 1nm/100°C) has to be further reduced for the future DWDM applications.

Experimental study has shown that the thermally induced wavelength shift can be reduced in ECL by using temperature insensitive Bragg grating [1], but there is no study on the thermally induced mode hopping in ECL when using temperature insensitive Bragg grating. On the other hand, although it has been reported that thermally induced mode hopping in ECL can be suppressed by using a silicone waveguide with negative thermo-optic coefficient as a part of the external cavity [2], there is no improvement on the thermally induced wavelength shift using this structure.

In this paper, it is found for the first time that the thermally induced mode hopping will be increased by using temperature insensitive Bragg grating in WG-ECL. A novel structure of WG-ECL using temperature insensitive grating integrated with a silicone waveguide with negative thermo-optic coefficient is proposed to suppress both the thermally induced wavelength shift and the thermally induced mode hopping.

2. Theoretical study on thermally induced mode hopping and wavelength shift in WG-ECL

Thermally induced mode hopping in WG-ECL is mainly due to the fact that the external cavity modes shift more rapidly than the Bragg wavelength of grating with temperature change. Hence, the temperature interval of mode hopping ΔT is given as:

$$\Delta T = \frac{\Delta\lambda}{d\lambda_c / dT - d\lambda_B / dT} \quad (1)$$

where ΔT is related to the cavity mode spacing, $\Delta\lambda$, the thermal coefficient of cavity mode, $d\lambda_c/dT$, and the thermal coefficient of Bragg grating, $d\lambda_B/dT$.

On the other hand, thermally induced wavelength shift in WG-ECL is determined by the thermally induced Bragg wavelength shift of Bragg grating, which is given by:

$$\frac{d\lambda_B}{dT} = \lambda_B \left(\frac{1}{n_{eff}} \frac{dn_{eff}}{dT} + \frac{1}{\Lambda} \frac{d\Lambda}{dT} \right) \quad (2)$$

where λ_B is the Bragg wavelength, Λ and n_{eff} are the period and effective refractive index of the Bragg grating, respectively.

From (1) and (2), the condition of no thermally induced mode hopping and no thermally induced oscillation wavelength shift is given by:

$$d\lambda_c / dT = d\lambda_B / dT = 0 \quad (3)$$

In the following calculations, the 1.55μm InP QWs LD with HR coating ($R=95\%$) and AR coating ($R=10^{-4}$) is modeled using multimode rate equations [3]. The variation of peak optical gain with temperature is also taken into account using an empirical formula and the corresponding optical gain spectrum is approximated by a parabolic curve. The output power is maintained around 3 mW.

3. A novel structure of WG-ECL with suppressed thermally induced mode hopping and wavelength shift

We have proposed the simulation results of wavelength stability of the ECL with and without temperature insensitive fiber Bragg grating in [4]. It is found that the thermally induced wavelength shift is reduced due to the use of temperature insensitive Bragg grating. However, the temperature interval of thermally induced mode hopping is also reduced. This is because the difference between $d\lambda_B/dT$ and $d\lambda_c/dT$ is increased by the use of temperature insensitive grating.

In order to suppress both the thermally induced mode hopping and the thermally induced wavelength shift, a novel structure of integrated WG-ECL is proposed. Figure 1 is the proposed structure of integrated WG-ECL. Core and cladding regions of the waveguide are constructed using SiO_2 and SiO_2+GeO_2. A 5mm waveguide gap including a $300\mu m$ silicone ($dn_g/dT=-3.7\times10^{-4}/^oC$) and $4.7\ mm$ silica is used to suppress thermally induced mode hopping. A polymer [5] with negative thermo-optic coefficient ($dn_{eff}/dT = -1.7\times10^{-6}/^oC$) is used as waveguide cladding to decrease the temperature sensitivity of Bragg grating ($\lambda_B/dT\approx0.001\ nm/_oC$).

Figure 2 is the calculated wavelength stability with respect to temperature in (a) an ECL with a conventional uniform fiber Bragg grating and (b) the proposed WG-ECL, respectively. Compared with Figure 2 (a), it is shown in Figure 2 (b) that there is no thermally induced mode hopping in a temperature range of $30\,^oC$ ($20^oC\sim50^oC$) in our proposed structure. The thermally induced wavelength shift is also as small as $d\lambda/dT=-0.0017nm/^oC$ due to the use of temperature insensitive Bragg grating. Consequently, the thermally induced mode hopping and the thermally induced wavelength shift are both suppressed in our proposed WG-ECL.

Figure 3 are the calculated Side Mode Suppression Ratio (SMSR) (a) in the ECL with common uniform fiber Bragg grating and (b) in our proposed WG-ECL, respectively. As shown in Figure 3 (b), compared with figure 3 (a), the SMSR remains a higher value (35dB) in our proposed structure.

4. Conclusion

It is found that the thermally induced mode hopping occurs more frequently although the thermally induced wavelength shift can be reduced by using temperature insensitive Bragg grating in WG-ECL. A novel structure of WG-ECL is proposed to suppress both thermally induced wavelength shift and thermally induced mode hopping by using temperature insensitive Bragg grating and a suitable length of gap with negative thermo-optic coefficient. It is shown theoretically that there is no thermally induced mode hopping over a temperature change of 30^oC and thermally induced wavelength shift is also as small as $0.0017nm/^oC$ by using the proposed structure. This structure can be easily realized using the existing waveguide fabrication technologies.

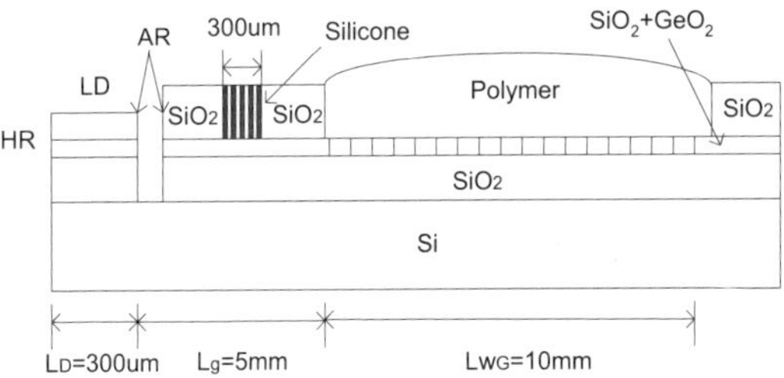

Fig.1 Proposed structure of WG-ECL to suppress both thermally induced mode hopping and wavelength shift. It is composed of a *5mm* waveguide gap including a $300\mu m$ silicone with negative thermo-optic coefficient ($dn_g/dT=-3.7\times10^{-4}/^oC$) and *4.7 mm* silica and a polymer [4] with negative thermo-optic coefficient ($dn_{eff}/dT=-1.7\times10^{-6}/^oC$).

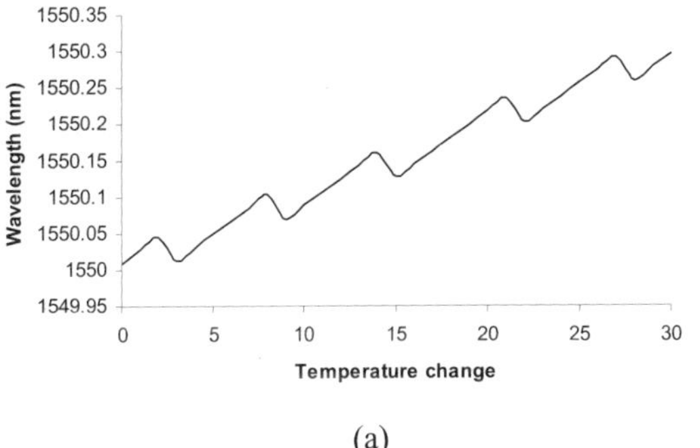

(a)

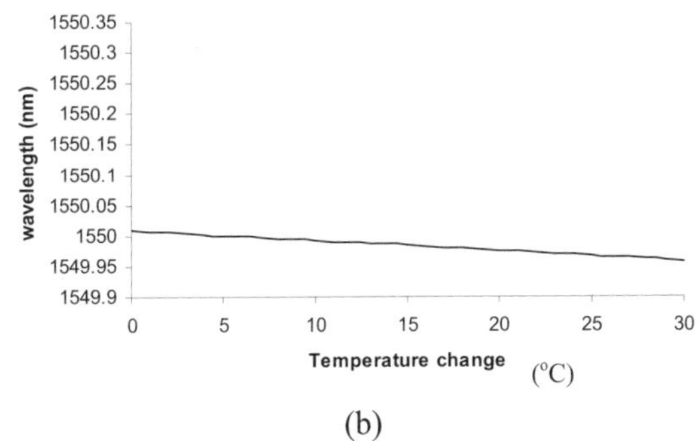

(b)

Fig.2 calculated wavelength stability with respect to temperature in (a) ECL with common uniform fiber Bragg grating; (b) proposed WG-ECL.

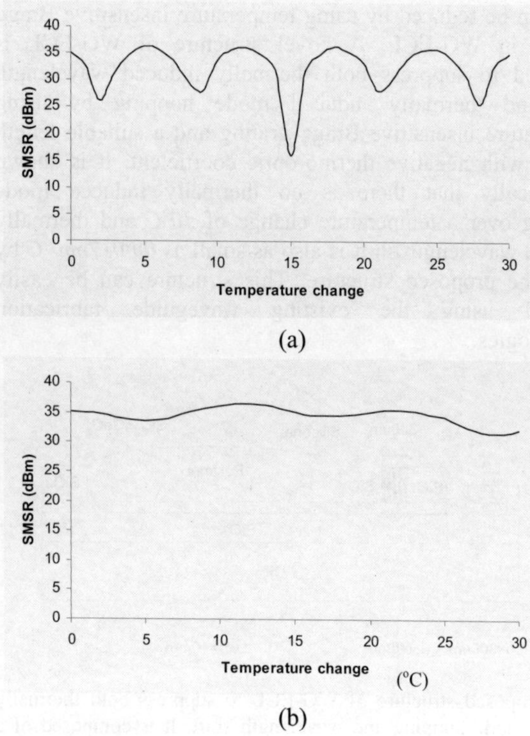

Fig.3 calculated SMSR with respect to temperature in (a) ECL with common uniform fiber Bragg grating; (b) proposed WG-ECL.

References

[1] A.Leroy, et al., "Low-cost wavelength stabilized plug and play lasers for WDM systems in future local networks", Electronics Letters, Vol.**37**, No.16, 1012-1014 (2001)

[2] T.Tanaka, et.al., "Hybrid integrated external cavity laser without temperature dependent mode hopping", Electronics Letters, Vol.**35**, No.2, 149-150 (1999)

[3] Tien-Pei Lee, et al., "Short-Cavity InGaAsP injection lasers: dependence of mode spectra and single-longitudinal-mode power on cavity length", IEEE Journal of Quantum electronics, Vol.QE-18, No.7, 1101-1112 (1982)

[4] H.G.Yu and C.Q.Xu, "Suppression of thermally induced wavelength shift and mode hopping in fiber grating external cavity semiconductor lasers", ISPA 2001, Singapore

[5] D.Bosc, et.al., "Temperature and polarisation insensitive Bragg gratings realised on silica waveguide on silicon", Electronics Letters, Vol.**33**, No.2, 134-136 (1997)

Analysis of Multiple-Phase-Shift DFB Laser Diodes Based on Transmission Line Laser Model

Liu Hairong, H. Ghafouri-Shiraz
Network Technology Research Centre
School of Electrical & Electronic Engineering,
Nanyang Technological University, Singapore 639798
Fax: +65 6790 4510, Email: ehgshiraz@ntu.edu.sg

Abstract: In this paper, we propose a novel dynamic transmission-line-laser-based model of the multiple-phase-shift distributed feedback (DFB) lasers. The dynamic and static characteristics of three phase shift (3PS) DFB structures are analyzed based on this model. The longitudinal carrier density distribution of quarterly wavelength shifted (QWS) DFB LDs and 3PS DFB LDs are compared. Our results show that 3PS DFB structure has a more uniform longitudinal carrier density distribution. Therefore, such a structure reduces the spatial hole burning effect and hence strengthens the single-longitudinal-mode stability.

Key Words: Transmission-line laser model, multiple-phase-shift DFB laser

1.INTRODUCTION

Advanced semiconductor laser diodes, such as the InGaAsP distributed feedback (DFB) lasers, are important devices in communication systems [1]. Recently, there has been a considerable effort spent on attempting to find a DFB structure for producing a laser which has a stable single longitudinal mode operation with narrow spectral linewidth and efficient power utilization. To overcome the two-mode degeneracy in uniform DFB lasers, introducing a $\lambda/4$ phase shift in the grating structure is effective for achieving single-mode operation. However, when the coupling coefficient is large, a hole-burning effect will enhance the side mode. One design is to insert more phase shifts along the laser cavity [2]. The multiple shift DFB lasers are proposed and expected to combat the hole burning and suppress the side mode and achieve single-mode operation [3].

In Transmission-Line Laser Modeling (TLLM) theory, a time-domain numerical algorithm is used for laser simulation, which was developed by Lowery [4]. TLLM models the laser cavity as a set of series-connected transmission lines. Voltages pulses traveling along the lines represent the index modulation of laser waveguide [5]. The TLLM method gives a clear physical interpretation of the laser's internal processes such as reflections, gain/loss, filtering, noise, and coupling, in terms of circuit theory.

A novel dynamic transmission-line-laser-modeling-based model of the multiple phase shift distributed feedback (DFB) lasers has been proposed. We have analyzed the dynamic and static characteristics of three phase shift (3PS) DFB lasers based on this model. Moreover, the longitudinal carrier density distribution of quarter-wavelength shifted (QWS) DFB LDs and 3PS DFB LDs are compared. The results show that 3PS DFB structure has a more uniform longitudinal carrier density distribution. Therefore, such a structure reduces the spatial hole burning effect and strengthens the stability single-longitudinal-mode operation.

2. MODELING PROCEDURE

Transmission-line laser models split the laser cavity longitudinally into a number of much smaller sections. For each of these sections, TLLM use a scattering matrix to represent the internal optical processes, such as stimulated emission, spontaneous emission and attenuation. The matrices of these sections are then connected together by transmission matrices, which represent the propagation delays of the waves. From the iteration of scattering and connecting processes we can obtain the output wave in the time domain as well as its spectra by Fourier transforms.

In the conventional DFB TLLM model [6], quarter-wave-shifted grating devices are modeled by inserting a zero-reflection interface (identity matrix) half way along the cavity. Consequently, the connecting matrices are changed. However, this model cannot work when a phase shift of other value or multiple phase shifts are applied. In our proposed model, the phase shifter stubs are inserted into the scattering matrices to model the effects of multiple phase shifts with any phase shift values without changing the connection matrices. In this way, we can model arbitrary phase shifts easily.

$$\begin{bmatrix} A(n) \\ A_C(n) \\ A_L(n) \\ A_P(n) \end{bmatrix}_k^r = \frac{1}{y(1+Z_s)} \begin{bmatrix} t(g+y)(Z_S-1) & 2tY_C(Z_S-1) & 2tY_L(Z_S-1) & 2y \\ g(Z_S+1) & (2Y_C-y)(Z_S+1) & 2Y_L(Z_S+1) & 0 \\ g(Z_S+1) & 2Y_C(Z_S+1) & (2Y_L-y)(Z_S+1) & 0 \\ t(g+y)2Z_S & 4tY_C Z_S & 4tY_L Z_S & y(1-Z_S) \end{bmatrix} * \begin{bmatrix} A(n) \\ A_C(n) \\ A_L(n) \\ A_P(n) \end{bmatrix}_k^i + \begin{bmatrix} (Z_s-1)\cdot I_S Z_P t/2 \\ 0 \\ 0 \\ 0 \end{bmatrix}_k^S \quad (1)$$

Scattering matrices, *S* in the new model

Different from the conventional scattering matrix, a phase-shifter stub is included in a section to represent an inserted phase shift. Therefore, the scattering process (represented by the scattering matrix) of this section is given by (1) at the top of this page, where $y = 1 + Y_L + Y_c$ is the sum of the gain-curve filter's stub admittances, Y_c (capacitive) and Y_L (inductive) the stubs' admittances as given in [6], g the gain and t the attenuation.

The impedance of the stub is given by $Z_s = Z_c \tan(\pi f \Delta T)/\tan(\beta l)$, where f is the laser resonant frequency, β is the propagation constant inside the cavity and l is the length of the stub which will be added to the section to produce a phase shift. For λ/4 phase shift, l=ΔL/2 will provide π/2-radian phase shift. For λ/8 phase shift, l=ΔL/4.

For the other sections without phase shift, the scattering matrix for each section remains unchanged, which is written as

$$\begin{bmatrix} A(n) \\ A_c(n) \\ A_l(n) \end{bmatrix}_k^r = \frac{1}{y} \begin{bmatrix} t(g+y) & 2tY_C & 2tY_L \\ g & 2Y_C-y & 2Y_L \\ g & 2Y_C & 2Y_L-y \end{bmatrix} \begin{bmatrix} A^i(n) \\ A_C(n) \\ A_L(n) \end{bmatrix}_k^i + \begin{bmatrix} I_s Z_P t/2 \\ 0 \\ 0 \end{bmatrix}^s \quad (2)$$

The connecting matrix will remain the same as in reference [7].

3. RESULTS AND DISCUSSION Conclusion

A 1550nm device is modeled using TLLM to prove the validity of our model. Its parameters are obtained from reference [7]. The schematic of structure of 3PS are shown in Fig.1. In the following simulation, we set L_1=L_2. θ_2=θ_3= θ_4=π/3.

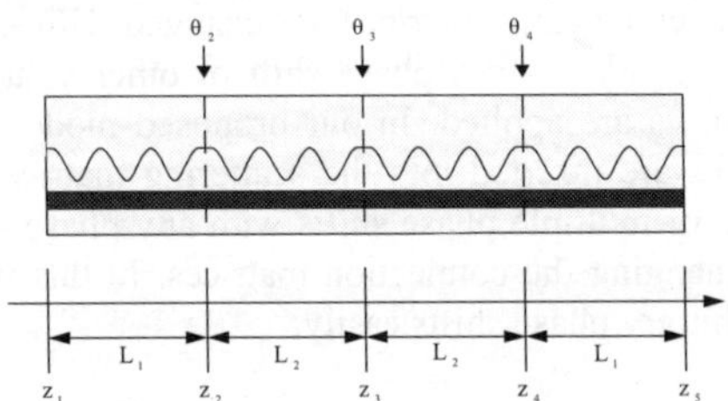

Fig.1 Scematic of 3PS DFB In the simulation

In Figure 2, the longitudinal carrier density profiles of QWS DFB structure calculated using the new model, the

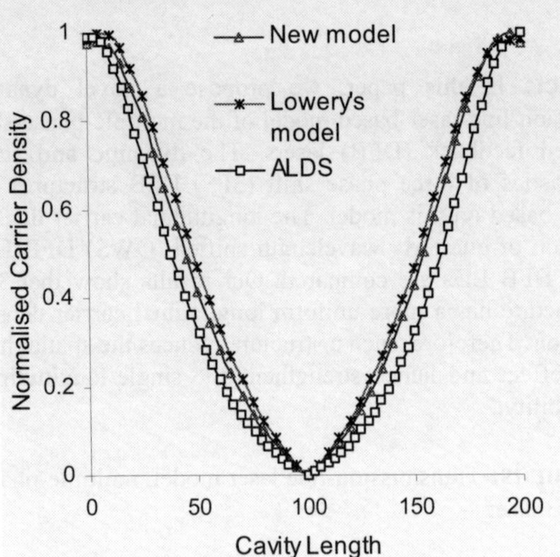

Fig.2 Longitudinal carrier density distributions of QWS DFB by different models

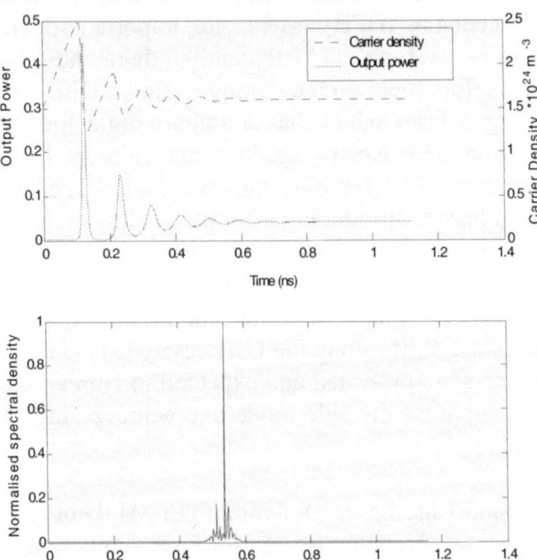

Fig.3 Dynamics response of 3PS DFB LDs

Lowery's model and the commercial software ALDS (ALDS is static model and provided by Apollo Photonics) are plotted. The results show that the new model has a good agreement with the Lowery's model and the ALDS model. The discrepancy between the new model and ALDS may be caused by the insufficient number of sections used in ALDS.

Both the transient response and the spectrum simulation results of a 3PS DFB LDS under 100 mA injection current are showed in Fig. 3. It shows that the 3PS operated at single mode.

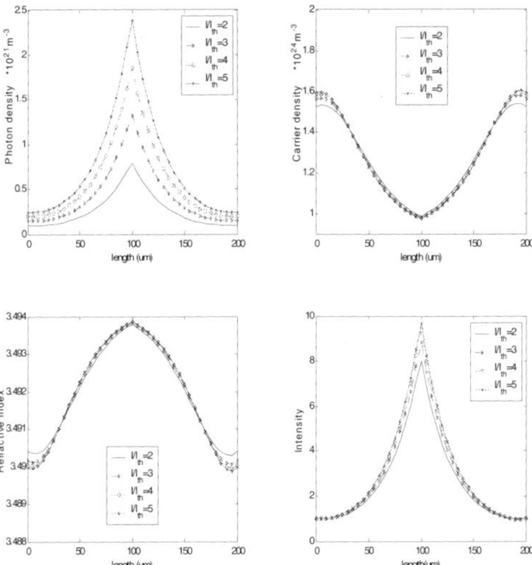

Fig. 4 Longitudinal distribution of the carrier density, photon density, refractive index and intensity in a QWS DFB LD for different current injection currents

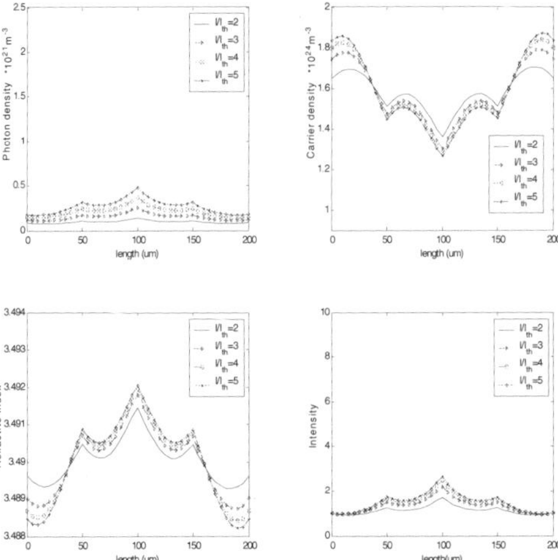

Fig. 5 Longitudinal distribution of the carrier density, photon density, refractive index and intensity in a 3PS DFB LD for different injection currents

The longitudinal variations in the carrier density, thephoton density, the refractive index and the internal field of the QWS and 3PS DFB LDs are shown in Fig 4 and Fig.5 respectively. In the analysis all the electric field distributions have been normalized so that the intensity at the laser facets is unity. The injection current threshold of these two devices is 20mA.

From the photon density profiles of both structures, it could be observed that the average photon density of these two structures increased with the bias current. However, rather than a single peak found at the center of the cavity (which is observed in QWS), local maxima can be seen corresponding to the phase-shift positions in the 3PS structures. Moreover, it is found that the difference between the central photon density and the escaping photon density at the facet reduces more in the 3PS structures.

Compared with the QWS structure, a more uniform distribution of carrier density can be seen in the case of the 3PS DFB structure. Local minima can be seen at the location of the 3PS DFB structures. Compared with the QWS DFB structure, we can observe that an increase in the bias current causes little change in the refractive index in the 3PS structures.

The variations of the internal intensity distribution along the laser cavity of these two structures are compared. It can be observed that the intensity of the QWS DFB LDs show highly non-uniformity. Compared with the QWS DFB, the internal intensity profile of the 3PS DFB LDs appears to be more uniform. The narrower span of the longitudinal intensity distribution strengthens the single-mode stability of the 3PS structure and hence reduces the threat from the SHB effects.

4. CONCLUSION

A novel phase-shift model has been proposed and incorporated into the TLLM for modeling of the multiple-phase DFB LDs. The dynamic and static characteristics of the 3PS phase shift DFB lasers are studied based on the new model. From the results, compared with the QWS structure, uniform distributions are observed in the carrier density, photon density and the refractive index and internal intensity profiles in the 3PS DFB structures. The results show that the use of an optimized 3PS laser structure could strengthen the single-mode operation stability. Furthermore, we can optimize the multiple-phase-shift laser structures by this new proposed model.

REFERENCES

[1] Xun Li, A.D.Sadovnikov, W.-P. Huang, and Toshihiko Makino, A Physics-Based Three-Dimensional model for Distributed Feedback Laser Diodes, IEEE J. Quan. Electr. Vol.34, No.9 . Sept., page1545-1553,1998

[2] Jing-Yi Wang, Michael Cada, Jin Sun, Theory for optimum Design and Analysis of Distributed-Feedback Lasers, IEEE Photonics. Tech. Lett., Vol . 11, No.1 Jan.,page24-26,1999

[3] L.M.Zhang, S.F.Yu, M.C.Nowell, D.D.Marcenac, J.E.Carroll, and R.G.S. Plumb, Dynamic Analysis of Radiation and Side-mode Suppression in a Second-order DFB Laser Using Time-domain Large-signal Traveling Wave Model, IEEE J. Quan. Electr.Vol.30 No.6,June, page 1389-1395 1994

[4] A.J.Lowery, A new dynamic semiconductor laser model based on the transmission line modeling method, IEE Proc. J. Optoelectron., vol.134, page 281-289,1987

[5] H. Ghafouri-Shiraz, B.S.K.Lo, *Distributed Feedback Laser Diodes Principles and Physical Modeling*, John wiely & Sons Ltd, 1996

[6]A.J.Lowery, dynamic modeling of distributed-feedback lasers using scattering matrices, Electronics letters, Vol. 25, No. 19,page1037-1038,1989

[7] A.J.Lowery, New dynamic model for multimode chirp in DFB semiconductor lasers, *IEE Proceedings*, Vol.137, Pt.J, No.5, October, page293-300, 1990

Optimization Design By Genetic Algorithm With A New Crossover Method For Broadband Fiber Raman Amplifiers

Fei Lin*, Yandong Gong, Ping Shum
Network Technology Research Center, Nanyang Technological University
50 Nanyang Ave., S2-B3-C23 Sch. of EEE, Singapore 639798
Email: afeilin@msn.com, G02048743@ntu.edu.sg

Abstract: An optimization design for broadband gain-flattened multi-wavelength backward-pumped fiber Raman amplifiers (MW-FRAs) was presented. We proposed a new crossover method (Bounded Gaussian Distributed Crossover) in real-coded Genetic Algorithm for the optimization design of broadband gain-flattened MW-FRAs. By applied it to the simplified inverse design problem, we achieved optimal designs with gain ripple ranged from 0.72% to 12.55% (which corresponds to 15 pumps~3 pumps). Furthermore, we also gave the partial optimization design when the wavelengths of pumps are not ideal or fixed.

Keywords: Optimization, Raman, optical amplifier, genetic algorithm, crossover, gain-flattening.

1. INTRODUCTION

Recently, broadband fiber Raman amplifiers (FRAs) arose many researchers' interests in DWDM system due to their advantages over erbium-doped fiber amplifiers (EDFAs). Raman amplifiers offer the extra-broad bandwidth (>80nm) and distributed amplification, and can operate through the low loss window (from 1.1μm to 1.7μm) under appropriate pumps compared with EDFAs which are limited in 1.55μm region with relative narrow bandwidth. More important, FRAs provide distinctive flexibility in bandwidth designs by multiple pumps arrangement [1], which are more efficient than gain-equalizing by filters for EDFAs. Although FRAs have these advantages, there are still some difficulties in design of broadband amplifiers.

Highly nonlinear interactions (which experienced by signals include stimulated Raman scattering, amplified spontaneous emission noise, Rayleigh back scattering) increased the difficulties in optimal design of gain-flattening. Given the target amplifier gain, the goal of inverse design of amplifier for MW-FRAs is to find out appropriate input pump wavelengths and powers. But the nonlinear interactions make the inverse-problem solving complicated, extremely time-exhaustive. Although the powerful genetic algorithm (GA) has been applied to gain-flattening optimization [2], only simplified analytic model for MW-FRAs instead of numerical model can be solved efficiently. So new methods in GA for optimal design are necessary.

In this work, we simulated and optimized the performance of MW-FRAs. We proposed a new crossover method BGD (Bounded Gaussian Distributed Crossover) in real-coded GA for optimization design.

2. SIMPLIFIED PROBLEM

The highly nonlinear interactions complicated inverse design, which drove us to simplify the problem. Since ASE and Rayleigh back scattering are negligible to the gain of amplifier, we simplified the model (reported in Ref. [1]), and get the system gain as

$$G_s(\nu,\varsigma_p,\mathbf{I}_p)=G_{ss}(\nu)\cdot\exp\left(\sum_{\varsigma_p=\{pump\}}g(\varsigma_p,\nu)I_p(\varsigma_p)\right) \tag{1}$$

where power integrals $I_p(\varsigma_p)=\int_0^L P(z,\varsigma_p)dz$ (L is the length of fiber). And the signal-to-signal gain is

$$G_{ss}(v)=\exp(-\alpha(v)+\sum_{\varsigma=\{signal\}} g(\varsigma,v)\int_0^L P(z,\varsigma)dz), \quad (2)$$

where $\varsigma_p=\varsigma_p(\varsigma_{p1},\varsigma_{p2,...,}\varsigma_{pn_p})$ and $\mathbf{I}_p=\mathbf{I}_p(I_{p1},I_{p2,...,}I_{pn_p})$ are vectors represented frequencies and power integrals of pumps.

Now the gain was given in two parts: gain caused by signal-to-signal (G_{ss}) and gain caused by pump-to-signal. For a given target gain G, the inverse-problem is to minimize the gain ripple R:

$$R(\varsigma_p,\mathbf{I}_p)=\frac{1}{G}\sqrt{\frac{1}{n_p}\sum_{v=\{signal\}}(G-G(v,\varsigma_p,\mathbf{I}_p))}\cdot 100\% \quad (3)$$

where n_p is the number of pumps.

In other words, the problem is to search the optimal pump power integrals $\mathbf{I}_p$ and optimal pump frequencies ς_p.

3. REAL-CODED GENETIC ALGORITHM WITH NEW CROSSOVER METHOD

For Raman amplifier with n_p pumps, the search space is very large ($2n_p$-dimension). So we applied the genetic algorithm (GA) to our simplified problem.

The GA is a strong stochastic solution-searching techniqure[3] based on the mechanics of natural genetics and survival of the fittest. The basic idea is derived from Darwin's theory of Natural Selection. For its strong search ability in complex space, the GA has been applied to optimization problem, machine learning, automatic programming, social system and so on.

Similar to natural biological system, many individuals (or candidate solutions) for the special problem construct a population. The individuals in this population will experience three kinds of genetic operations: selection, crossover and mutation. Thus the population will evolve generation by generation. The good individuals will survive and produce the children. So the best one will be selected as the final solution for our optimization problem after evolution.

Crossover operation is the most important characteristic for GA, which determines the way to combines substructures of parents to produce new individuals and performance of GA. For our optimization, we tested many crossovers such as Bend-a, 2-point, uniform, and others with crowding GA[4]. But all of these showed poor performance in searching optimal pumps. Here we developed a new crossover name Bounded Gaussian Distributed (BGD) crossover, which showed a good search ability and stable performance in optimum searching.

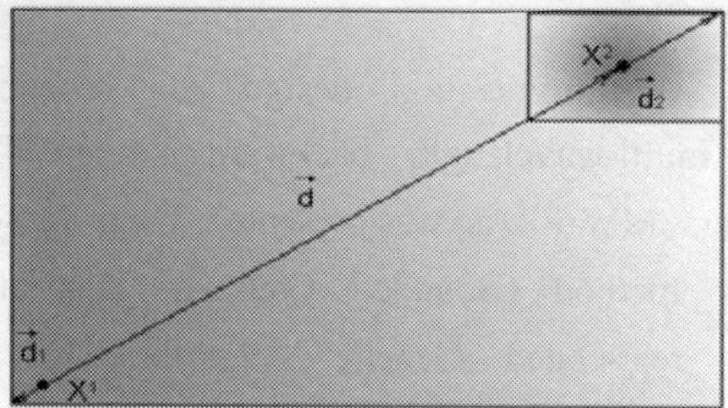

Fig. 1. Illustration of Bounded Gaussian Distributed Crossover.

The idea of BGD crossover is that new solutions will be generated in distance and fitness dependent areas with Gaussian distribution. Firstly, we select two parental solutions represented as vectors $\mathbf{X}^1$, $\mathbf{X}^2$ randomly from the population, which are the two points X^1 and X^2 in Fig. 1. Two offspring vectors $\mathbf{X}^{c1}$, $\mathbf{X}^{c2}$ (supposed the fitness of latter is higher) are chosen from bounded areas following the Gaussian distribution as

$$X_i^{c1}=X_i^{c1}+d_i^1+|G(0,1)|(-d_i^1+d_i^2+d_i) \quad (4)$$

$$X_i^{c2}=X_i^{c2}+G(0,1)d_i^2 \quad (5)$$

where the subscription i denote the i-th element of vector and

$$\mathbf{d}=\mathbf{X}^2-\mathbf{X}^1, \quad (6)$$

$$\mathbf{d}^1=-\alpha\cdot(1+F)\mathbf{d}, \quad \mathbf{d}^2=\beta\cdot(1+F)\mathbf{d}, \quad (7)$$

$$G(\mu,\sigma^2)=g(\mu,\sigma^2)(F\cdot\gamma+\eta). \quad (8)$$

$$(if \ |G|>1, \ G=rand(1)\cdot G/|G|)$$

In above equations, $g(\mu,\sigma^2)$, $rand(1)$denote Gaussian distribution function with mean value of μ and mean-square deviation σ^2, random function within range[0,1], respectively. The value of constant parameters α, β, γ, η, μ and σ for optimization are 0.00015, 0.15, 0.1, 0.41, 0, 1, correspondingly. These parameters can be changed for various optimization problems. For example, BGD crossover has good performance for optimizing sphere function when $\alpha=1$.

$F = F(\mathbf{X}^1, \mathbf{X}^2) = 1 - \exp(-\kappa \overline{f})$ ($k = 0.01$) is the function of parents' average fitness $\overline{f}$, which makes the GA to explore bigger bounded areas when fitness is low and converge to parents when fitness is high.

As we can see from Fig. 1, $\mathbf{X}^{c1}$, $\mathbf{X}^{c2}$ are bounded in a smaller and a bigger rectangle respectively, which undertake the local search (exploitation) and global search (exploration).

4. OPTIMIZATION DESIGN

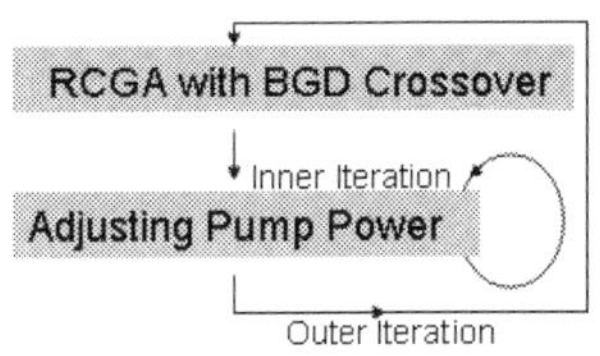

Fig. 2. Procedure of Optimization Design.

4.1 Principle

Considered a Raman amplifier with n_s Signals and n_p backward pumps, we can get the optimal frequencies ς_p and power integrals $\mathbf{I}_p$ of pumps by GA with BGD crossover in the first step (see Fig. 2). In the next step, we need to search the pump powers $\mathbf{P}_p$ according to the power integrals $\mathbf{I}_p$.

The powers-searching procedure is iterative. Before starting search, we need to assume the reasonable values for pump powers. Then in each iterative procedure, we get the new power integrals of pumps $\mathbf{I'}_p$ by simulating the Raman amplification with average power method. Thus the powers of pumps can be adjusted by ΔP:

$$\Delta P(\Delta I, \varsigma_p) = \alpha(\varsigma_p)/(1 - \exp(-\alpha(\varsigma_p) \cdot L)) \cdot \Delta I, \tag{9}$$

or

$$\Delta P(\Delta I, \varsigma_p) = P(\varsigma_p)/I'(\varsigma_p) \cdot \Delta I \tag{10}$$

(where $\Delta I(v) = I(\varsigma_p) - I'(\varsigma_p)$) until $\mathbf{I'}_p$ and $\mathbf{P}_p$ are self-consistent. This is the inner iteration.

The power integrals of signals $\mathbf{I}_s(I_{s1}, I_{s2}, \ldots, I_{sn_s})$ got from the inner iteration will be used to search optimal frequencies ς_p and power integrals $\mathbf{I}_p$ of pumps by GA in the first step again, then the whole procedure is repeated as illustrated in Fig. 2 until outer iteration gives self-consistent results of $\mathbf{I}_s$.

4.2 Results of Optimization

Using the above methods, Raman amplifiers with 80 nm bandwidth (from 1520nm to 1600nm) and 50km length were designed. The amplifiers to be optimized have 100 channels with 0.5 mW/channel. The pumps were limited in the wavelength range [1400nm, 1519nm] with maximum power of 1000mW. The other used parameters are given as follows. Effect area of fiber A_{eff} and polarization factor K_{eff} are $50\mu m^2$, 2, respectively. Raman gain coefficient and loss profile were taken from reference [5]. The amplifiers were designed with output gain of 5 dB.

We optimized the Raman amplifiers with 3, 6, 9, 12 and 15 pumps respectively. Fig. 3(a)~(e) give the output gain profiles. It can be seen that higher flatness of gain profile can be achieved for amplifiers with more pumps. The gain ripple varied from 0.72% to 12.55% which corresponds to 15 pumps~3 pumps. It should be noticed that more pumps would cost more time for optimization design because the search space has $2n_p$ dimensions.

4.3 Partial Optimization

In reality, the wavelengths of pumps are always not ideal (or pump wavelengths are fixed) as that required by previous optimal designs. So here the partial optimization means only searching optimum powers of pumps for amplifiers with fixed pump wavelengths. The size of search space decreased to n_p dimensions.

Fig. 4 gives the optimization results for 6 pumps amplifier and shows the gain ripple increased to 10.37% due to less space-freedom for searching procedure compared with Fig. 3(b). The flat output gain was got in a narrower wavelength range about 60nm (1540nm~1600nm).

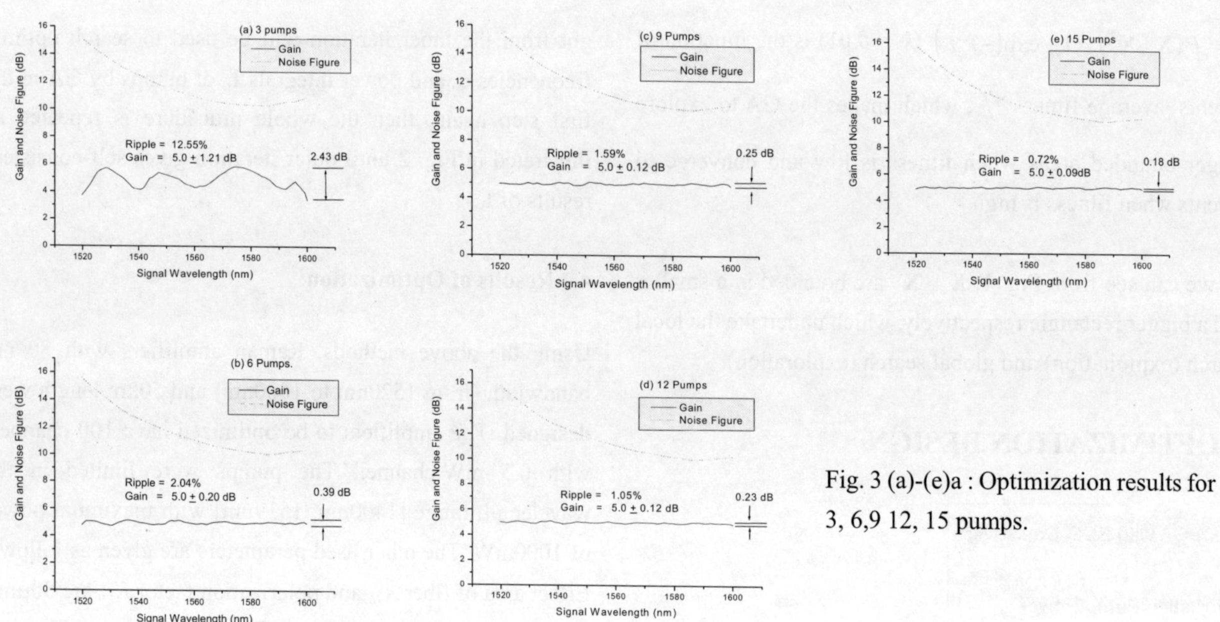

Fig. 3 (a)-(e)a : Optimization results for 3, 6,9 12, 15 pumps.

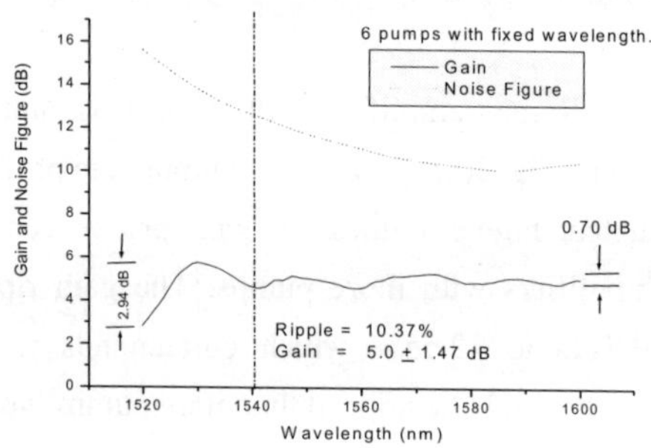

Fig. 4 Partial Optimization for 6 pumps with fixed wavelength.

5. CONCLUSION

We presented an optimization design for broadband gain-flattened multi-wavelength backward-pumped fiber Raman amplifiers. A new crossover method in GA for the optimization design was presented. By applied it to the simplified inverse design problem, we achieved optimal designs with gain ripple ranged from 0.72% to 12.55% (which corresponds to 15 pumps~3 pumps). And we also discussed the partial optimization design for fixed pumps wavelengths in this paper.

REFERENCES

Journal Papers:

[1] H. Kidorf, K. Rottwitt, M. Nissov, M. Ma, and E. Rabarijaona, Pump interactions in a 100-nm bandwidth Raman amplifiers, IEEE Photon. Technol. Lett., vol. 11, page 530-532, 1999.

[2] X. Zhou, C. Lu, P. Shum, and T. H. Cheng, A simplified model and optimal design of a multiwavelength backward-pumped fiber Raman amplifier, IEEE Photon. Technol. Lett., vol. 13, page 945-947, 2001.

Books:

[3] D. E. Goldberg, *Genetic Algorithms in Search, Optimization, and Machine Learning,* Addison-Wesley, 1989.

[4] M. Mitchell, *An introduction to genetic algorithms,* MIT press, Cambridge, MA, 1996.

[5] G. P. Agrawal, *Nonlinear Fiber Optics,* Academic Press, San Diego, 2001.

A DUAL-PUMPED TUNABLE WAVELENGTH CW RAMAN FIBER RING LASER

K.Y. Kong, Y.D. Gong and P. Shum
Nanyang Technological University, Network Technology Research Centre,
Block S2, S2-B2b-71, Nanyang Avenue, Singapore 639798.
Fax: (65) 6792 6894, Email: D-NTRC@ntu.edu.sg

Abstract: The authors have demonstrated a dual-pumped tunable Raman fiber laser. Tuning of the laser is done by adjusting a tunable bandpass filter that yields a range of about 45 nm from 1520 to 1565 nm and with peak power output of over 11 mW. Two 1.2 W fiber lasers at 1455 and 1480 nm pump the laser.

Keywords: fiber laser, Raman amplification, tunable

1. INTRODUCTION

The use of erbium-doped fiber amplifiers (EDFAs) has been dominating optical amplifier technology at the wavelength region of the C band (1530-1565 nm). However, the insatiable consumer demand for data bandwidth has resulted in saturation of the C band in dense wavelength division multiplexed (DWDM) systems [1]. Thus, in recent years, there has been a lot of attention on the nonlinear effect of stimulated Raman scattering (SRS) because of their distinctive flexibility in bandwidth designs and growing maturity of high power pump module technologies [2,3,5]. Through Raman amplification, expanding span lengths and capacity of fiber optic transmission systems will not be a problem.

Pumps used for Raman amplification may be constructed from semiconductor laser diodes [4] or all fiber lasers. In all fiber lasers, Bragg gratings [6] or fused fiber couplers [7] are generally used to cascade, via the Raman effect, through several Stokes orders from a short pump wavelength to a longer output wavelength. If we can have a fiber based, tunable Raman pump source covering the 1520 -1565 nm region, it could possibly provide tunable gain over the L and U band (1565-1625 and 1625-1675 nm respectively).

An erbium fiber ring laser with a wavelength tuning element would offer a high power, economic pump source but the short wavelength limit of ~1530 nm would restrict the tunable gain over to the U band only. Instead we propose the use of a Raman fiber laser pump with a wavelength tunable final cascade. In this letter we detail the optimization of a tunable fiber Raman ring laser with continuous wave (CW) output between 1520 and 1565 nm.

2. THEORY

Spontaneous Raman scattering can transfer a small fraction (generally ~ 10^{-6}) of power from one optical field to another field, whose frequency is downshifted by an amount determined by the vibration modes of the molecular medium. This process is called the Raman effect. It is described quantum-mechanically as scattering of a photon by one of the molecules to a lower-frequency photon, while the molecule makes transition to a higher-energy vibration state. Incident light acts as a pump for generating the frequency-shifted radiation called Stokes wave. It was observed that in 1962 that, for intense pump fields, the nonlinear phenomenon of SRS can occur in which the Stokes wave grows rapidly inside the medium such that most of the pump energy is transferred to it. Since then, SRS has been studied extensively in a variety of molecular media.

The development of fiber-based Raman lasers is an important application of the SRS phenomenon in optical fibers. Such lasers can be tuned over a wide frequency range (~10Thz). A simple example of this is when a piece of single mode fiber is placed inside a Fabry-Perot cavity formed by two partially reflecting mirrors [7]. The cavity provides wavelength-selective feedback for the Stokes light generated inside the fiber through SRS. An intracavity prism allows tuning of the laser wavelength by dispersing various Stokes wavelengths spatially, each can be selected by turning the mirror after the prism. If we assume a typical value of 10 dB for the round-trip loss, the threshold condition is

$$G = \exp(2g_R P_0 L_{eff}/A_{eff}) = 10$$

There has been tremendous progress over the years in the development of Raman lasers especially during the 1990s. In an early method, a ring cavity configuration was used to make a low-threshold, all fiber Raman laser using a fiber loop and a fiber coupler [8]. With the advent of fiber-Bragg gratings, it has become possible to replace cavity mirrors with such gratings [9].

3. EXPERIMENT

The experimental set-up of the tunable wavelength CW Raman fiber ring laser is shown in Fig. 1. Two Raman fiber lasers with a maximum output power of 1.2W (post-coupler) at 1455 and 1480 nm were used as the pump sources. The Raman active medium was the Dispersion shifted fiber (DSF) and the optimum length for highest power output was about 9 km. Counter-propagating pumped Raman amplifier configuration using optical circulators was used to combine and extract signal and pump wavelengths. This gave the advantage of removing pump radiation from the tunable laser output. The extraction of the tunable laser power and the reseeding of the ring laser were done using a broadband optical coupler. A 1.2 nm tunable band pass filter (TBPF) was chosen for low loss and extended wavelength tunability of 1520 to 1565 nm. Disrupting effects of Stimulated Brilliouin Scattering in the gain fiber can thus be reduced with the employment of this broad band pass filter.

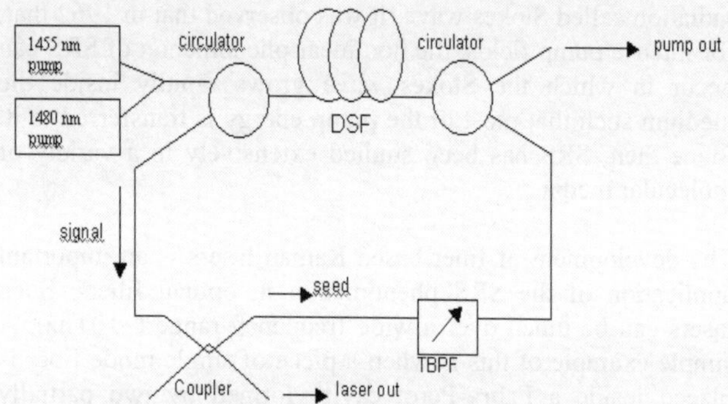

Figure 1: Experimental configuration of wavelength tunable CW Raman fiber ring laser with dual pumps.

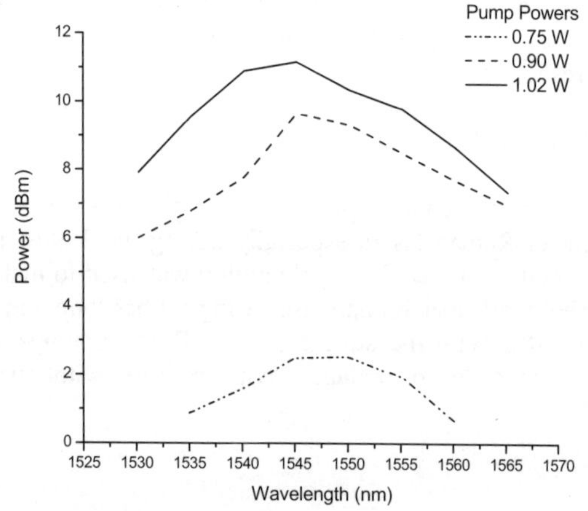

Figure 2: Laser output power against wavelength (single pump)

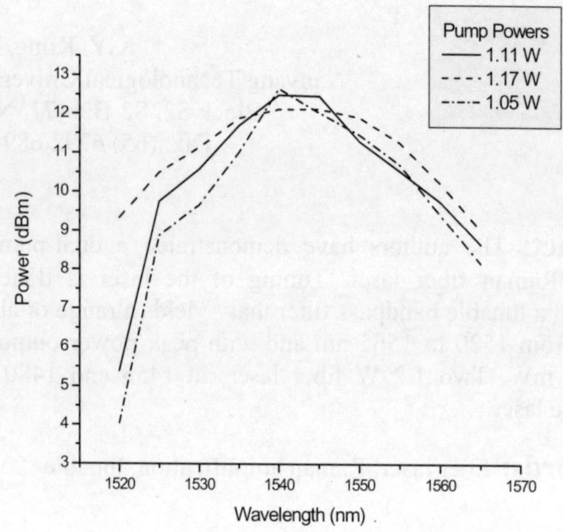

Figure 3: Laser output power against wavelength (double pump)

4. RESULTS

Initially we had used a single pump Raman fiber laser of wavelength 1455 nm and observed that when the pump power of the laser is increased, the tuning range of the ring laser increases from 25 nm (0.75 W) to about 35 nm (1.02 W) (Fig. 2). At the maximum pump power (1.02 W), the tuning range is from 1530 to 1565 nm. The CW power output is at least 5 mW and within it, it has a tuning range of 15 nm with power output of over 10 mW. It has a peak output power of 11.18 dBm at about 1545 nm. The 3 dB line width of the ring laser output was about 1.0 nm for all tunable wavelengths.

We added another pump laser of wavelength and observed the following results (Fig. 3). We find that with increasing pump powers, the output laser power is more consistent over a given range as compared to the one with single pump only. With this dual pump configuration at pump powers 1.17 W, we have a tunable range of 1520 nm to 1565 nm with maximum power of 12.13 dBm at 1545 nm. This range has at least 9 mW of power and from 1530 to 1550 nm, the laser is at least 11 mW. The 3 dB line width of the ring laser output for this configuration was about 0.5 nm at the lower wavelengths to about 1.5 nm for the higher wavelengths.

5. CONCLUSION

The wavelength tunable Raman ring laser demonstrated in this Letter has a tuning rage of 1520 to 1565 nm with a CW power output of at least 9 mW and a tuning range of 20 nm with power output of over 11 mW. Further amendments can be made to the experiment to increase the power of the ring laser. For example the optical coupler with a higher laser output : seed ratio could be used to optimize output. More powerful pump lasers could be used to give a higher power output. With these considerations, perhaps the extension of the transmission bandwidth of the WDM telecommunications systems could be explored in future.

REFERENCES

Proceedings Papers:

[1] Davidson C.R., Chen C.J., Nissov M., Pilipetskii A., Ramanujam J., Kidorf H.D., Pedersen B., Mills M.A., Lin C., Hayee M.I., Cai J.X., Puc A., Corbett P.C., MengesR., Li H., Elyamani A., Rivers C. and BerganoN.S., 1800 Gb/s Transmission of one hundred and eighty 10 Gb/s WDM channels over 7000 km using the full EDFA C-band, Optical Fiber Communication Conf., Tech. Dig. Postconference Edn., Trends in Optics and Photonics, 2000, Vol. 37, pp. 242-244.

[2] OSA IEEE: Koch F., Lewis S.A.E., Chernikov S.V., Taylor J.R.. "Broadband gain flattened Raman amplifier to extend operation in the third telecommunication window", OFC (Baltimore, USA, 2000), Paper FF3-1.

[3] OSA IEEE: Hiroji Masuda, "Review of wideband hybrid amplifier", OFC (Baltimore, USA, 2000), Paper TuA 1-1.

[4] Y. Emori and S. Mamiki, 100 nm flat Raman amplifiers pumped and gain equalized by 12-wavelength-channel WDM high power laser diodes, Optical Fiber Conf., 1999, Postdeadline Paper PD 19-1.

Journal Papers:

[5] Hansen P.B., Eskildsen L., Grubb S.G., Stentz A.J., Strasse T.A., Judkins J., DeMarco J.J., Pedrazzani R. and DiGiovanni D.J., Capacity upgrades of transmission system by Raman amplifier, IEEE Photonics Technology Lett., 1997, Vol. 9, No. 2, pp. 262-264.

[6] Grubb S.G., Rogan T., Mizrachi V., Strasser T., Cheung W.Y., Reed W.A., Lemaire P.J., Kosinkski A.E., Nykolak G., and Becker P.C., 1.3 μm cascaded Raman amplifier in germanosilicate fibres, Conf. Optical Amplifiers and their Applications, 1993, (Optical Society of America, 1993), Paper PD3.

[7] Chernikov S.V., Platonov N.S., Gapontsev D.V., Do Ill Chang, Guy M.J., and Taylor J.R., Raman fiber laser operating at 1.24 μm, Electron. Lett., 1998, 34, pp. 680-681.

[8] E. Desurvire, A. Imamoglu and H.J. Shaw, J. Lightwave Technol. 1987, LT-5, 89.

[9] P.N. Kean, K. Smith, B.D. Sinclair and W. Sibbett. J. Mod. Opt. 1988, 35, 397.

Book

[10] G.P. Agrawal, *Nonlinear Fiber Optics*, third edition, Academic Press, 2001

Session 14A

Optical Networking III

A Scheme for managing the presence of Ambient Lights in Diffuse Infrared Wireless Home/Office Network

Sanjay Jamindar, Sr. Software Engineer
Satish Jamadagni, Research Scientist
{sanjam, satishj }@sasken.com
Sasken Communication Technologies Ltd
#139/25, Amar Jyoti Layout, Ring road,
Domlur P.O, Bangalore – 560071, India

Abstract: In recent years, optical wireless communications has emerged as a suitable alternative to radio frequency transmission for short-range indoor data communications since infrared offers several advantages over radio. To utilize the advantages of diffuse IR wireless system such as very high data rate and security over the RF system in an indoor environment we need to exploit the ubiquitous computing and smart sensing mechanism which could be used to mitigate some of the drawbacks of diffuse IR such as different background light strengths at different places, in which case at some places BER gets deteriorated or receiver fails to detect the signal due to its limited sensitivity. In this paper a scheme for tackling the ambient or background lights of different strengths in different locations in a typical indoor space using smart and aware environment has been discussed.

Key Words: Diffuse Infrared Wireless, smart Environment, and ambient Light

1. INTRODUCTION

Indoor wireless access networks are one important part of today's research in communication technology. For seamless wireless extension of the fixed network a further requirement is bit rates far above 10Mbps. Possible solutions could be based either on radio frequency or optical infrared signals. Optical system has the potential to support very high data rate wireless systems. The Infrared (IR) spectrum is license-free worldwide and offers a large amount of bandwidth. In modern communication technology there is on the one hand a strong growth of bandwidth request and on the other hand an increasing demand for wireless applications. At the same time typical wireless environments are highly unreliable and the channel condition changes quite rapidly. Wireless communication system of the future will experience more dynamic channel conditions and a wider range of applications requirements than systems today.

In recent years, optical wireless communications has emerged as a suitable alternative to radio frequency transmission for short-range indoor data communications [1][2]. As a transmission medium, infrared offers several advantages over radio, including a virtually unlimited spectral region that is not regulated worldwide, relative security against eavesdropping, reuse of the same spectrum in every room of a building without interference, and immunity to multipath fading.

Wireless indoor infrared (IR) transmission systems have been used in many applications in the past few years, ranging from simple remote controllers for home appliances to more complex wireless diffuse IR LANs. The performance of IR transmission systems is impaired by several aspects such as the speed limitations of the optoelectronic devices, the high optical path loss, the multipath dispersion, the receiver noise, the shot noise induced by the background ambient light. Multipath dispersion and the background ambient light are the most important sources of degradation. The effect of the background ambient light manifests into shot noise produced in the receiver photodiode, with power proportional to the background optical power impinging the photodiode at the relevant wavelength band.

It is estimated that the dominant mode of data transfer in future enterprise wireless networks would be through optical systems. Typical application that would be using such high bandwidth optical access network would have very high quality of service (QoS) requirements. It will be a real challenge designing networks that can maintain consistent QoS in dynamically changing situation as faced by a typical optical system due to background light effects. The problem is more intense in some parts of the indoor locations, where there is temporary/permanent excessive background light created by different light sources. In these locations optical carrier fails to maintain QoS data transfer due to highly negative channel conditions and proper actions should be taken to maintain QoS data transfer. At extremely negative channel conditions when an optical system might fail, an RF system would be a choice for continued link availability at the cost of sacrificing data rate/QoS upto a certain extent, as typical RF systems are order of magnitude behind optical systems in providing bandwidth.

To design an efficient optical wireless indoor system background light effects will have to taken into consideration as it can affect the bit rates supported in an optical system drastically. Thus there is a need in the art to provide for such systems.

There are a small number of wireless network systems based on diffuse optical (infrared) currently available but they provide low data rates (up to 4 Mbps). There are also research projects, which under experimental conditions reach data rates in the order of 50 Mbps using diffuse infrared links. Despite the dominance of solutions based on radio frequency (RF) a system based on infrared (IR) communication links would offer certain advantages, which cannot be realized otherwise. IR transmissions are confined to a room, because the walls block them. Compared to RF this means an exactly defined shape and size of one cell and results in smaller cells. So the same spectrum can be re-used in the room next door for another communication link without interfering each other. This increases the overall system throughput significantly. The IR spectrum is license-free worldwide and offers a large amount of bandwidth.

IR communication is invulnerable to matters of electromagnetic interferences (EMI). IR transmission does not disturb any other electronic device and the reception of IR signals is not interfered by electromagnetic fields. Since modern offices are crowded by several kinds of RF electronic devices this is certainly advantage. Assuming the power used by the communication devices does not exceed the limits given for eye safety there is no biomedical impact to the human organism contrary to RF, which is discussed to cause health-hazards. A further non technical but also important advantages of using IR is the relatively low costs of transmitters and receivers. So the conclusion has to be that a high-performance diffuse infrared network system would be the ideal solution for indoor wireless Internet access.

Although IR system has many advantages stated above it is not without drawbacks. It has few drawbacks, which are, shadowing, limited range and vulnerable to high intensity background light, which affect it's performance in a typical indoor wireless environment, which happens to be dynamic in nature. One major problem in diffuse infrared link is the shadowing caused by obstacles like pedestrians, furniture etc which directly affects it's mobility and range. In wireless infrared communications, the effects of shadowing are in general much severe than those in radio wireless communications, because infrared signal never passes through human bodies or other obstacles because of the fundamental reason of having very small wavelength. Shadowed channels exhibit characteristics that are less predictable and highly dynamic than the unshadowed wireless optical channels. The shadowing significantly degrades the channel frequency and impulse responses. The diffuse optical indoor system is quite vulnerable to shadowing caused by humans and other obstacles, which hamper reliable wireless connectivity and sometimes even cause severe link failure.

Because infrared cannot penetrate walls, communication from one room to another requires the installation of infrared access points that are interconnected via a wired backbone. In many indoor environments there exists intense ambient infrared noise, arising from sunlight, incandescent lighting and fluorescent lighting, which induces noise in an infrared receiver. In virtually all, short range, indoor applications, IM/DD is the only practical transmission technique. The signal-to-noise ratio (SNR) of a direct-detection receiver is proportional to the square of the received optical power, implying that IM/DD links can tolerate only a comparatively limited path loss. Often, infrared links must employ relatively high transmit power levels and operate over a relatively limited range. While the transmitted power level can usually be increased without fear of interfering with other users, transmitter power may be limited by concerns of power consumption and eye safety, particularly in portable transmitters.

Radio and infrared are complementary transmission media, and different applications favor the use of one medium or the other. Radio is favored in applications where user mobility must be maximized (although diffuse infrared can provide same mobility as RF) or transmission through walls or over long ranges is required in a typical indoor shadowing condition, and may be favored when transmission power consumption must be minimized. Infrared is favored for short range applications in which per-link bit rate and aggregate system capacity must be maximized, cost must be minimized, international compatibility is required, or receiver signal processing complexity must be minimized.

2. TYPICAL BACKGROUND LIGHT SCENARIO

In a typical indoor office environment a mobile Terminal (MT) could face different background light (infrared noise at the transmission wavelength band) strengths at different locations in a room. This is due to different intensity distributions of background or ambient light in a room. The distribution pattern depends on positions of fluorescent/incandescent lamps, positions of tables, chairs, partitions etc and also the proximity of windows where sunlight can enters the room. So near windows direct sunlight would cause excessive infrared noise to the MT. This infrared background light can be captured and intensity can be measured by suitable "infrared camera" and the room can be divided into a number of "noise zones" (N) each having different levels of background light strengths. In a particular zone it can be assumed that the intensity of ambient light is uniform (I_N). The zones could have of any shape depending on the positions of lamps and shadowing caused by furniture (fig 1). The "infrared camera" will identify these noise zones, which can degrade the BER performance of the MT (IR receiver) at different levels as it crosses from one zone to another. It is assumed here that these zones are fixed as far as the background noise intensity and zone area is concerned. The fixed smart sensor network is deployed according to the positions of these already identified "noise zones" for having the full coverage.

3. SMART AND AWARE ENVIRONMENT

To utilize the advantages of diffuse IR wireless system such as very high data rate and security over the RF system in an indoor environment we need to exploit the ubiquitous computing and sensing mechanism which could be used to mitigate some of the drawbacks of diffuse IR such as different background light strengths at different places, in which case at some places BER gets deteriorated or receiver fails to detect the signal due to its limited sensitivity. So there is a need to either modulations of transmitted power level (up to eye safety limit) or sacrificing data rate or handoff from IR to RF mode of transmission. For this to achieve we need an indoor smart and aware environment which is having embedded computing infrastructure everywhere [3]. Such environment will be aware of the activities performed within it and the configurable embedded sensor network will provide high-end sensory data to the Access Points or Base Stations (BT) about the status in the environment without increasing the cognitive load on the users in the space. We expect such a ubiquitous sensing infrastructure to support different environments, with varying needs and complexities. These sensors can be mobile or static, configuring their sensing to suit the task at hand while sharing relevant information with the other available sensors. A network of sensors that is configured with a network of processing devices can provide a rich multi modal stream of sensory data. The Base Stations (BT) to determine the specifics of the environment in turn, analyze the sensory data and it provides context for interaction with mobile terminal (MT). Base Stations (IR and RF) will process the sensory data and both the BTs and the MT will perform the relevant operations, which suits with the current environmental situation and also the applications that are running on it with keeping in mind the required QoS and also how far the QoS could be sacrificed. Computing and sensing in such environments must be reliable, persistent (always remains on), easy to interact with, and transparent.

Using the ubiquitous sensors, the space senses the activities of its occupants. Using computer vision and audio processing techniques, the environment locates and identifies its occupants and recognizes their activities [4]. The sensors should be able to sense the following behaviors of the MT for the present purpose of mitigating background light variations.

Location: using visual and tactile tracking, users can be located. Where the users are will serve as a strong cue about what they are doing. Phased-array microphones are used to locate a speaking person [5].
Activity and behavior: By processing a video stream, we can track the users as they move. These movements could be over short terms or over long terms. Each type of movement provides important cues about what is happening in a scene.
Pose tracking: Recognizing where a user is looking is a significant cue about what a user is doing. This is achieved in an aware space by tracking the body and face as it moves.
Expression/gesture recognition: Visual tracking and recognition of facial expression and hand gesticulation is essential in determining the state of the users.

The main function of these ubiquitous sensors should be to find out the present MT location within a noise zone and to predict well its location in near future by tracking the user activity, behavior and pose.

4. SCHEME

The MT can be found well inside at any zone at an instant or it can be found near the boundary of a zone. Each zone will have an area inside, surrounding the boundary called "Look Ahead zone (LAZ)". In fig. 1, "A" is the LAZ and "B" is well inside the noise zone. The "location/pose tracking sensors" of the smart environment are placed in the room in such a manner that each sensor will cover a particular noise zone to sense the MT location and their behavior/activity at frequent intervals (fig 4). It is assumed that ambient light condition in each zone is not changing with time and the intensities I_N's of different zones are stored in Ambient Light Intensity (ALI) table (fig 3) inside the smart sensors management module. The sensor network in the room is linked with Smart Space Management Module (SSMM) where the actual processing of the sensed data takes place (fig 2). After processing the sensed data (location, pose, behavior etc) and using the ALI table, the SSMM notifies the IR and RF base stations to take appropriate action before the MT actually crosses it's present zone to enter into the next zone which is having different background light level. If the MT is well inside a particular zone, no action will be taken by the BTs. But if the MT is in the LAZ and approaches towards the next zone, it will be notified as a "Look Ahead Technique". The LAZ for each noise zone will be known to the SSMM and every time the sensor network sends the sensed information (location), SSMM will compare it with the location information corresponding to the LAZ. If the next zone where the MT is approaching has extremely high background light level, there will be a complete handoff between IR and RF mode of data transfer and thus link failure can be avoided in advance. Otherwise the IR mode of data transfer will continue and variation of any background intensity levels from one zone to next will be reported to the

IR Access Point, which then reports this to the mobile terminal as a link layer parameter. Both BT and MT in that case will adjust its power level/data rate or there will be switch over to more efficient modulation scheme to cope with the excess background noise.

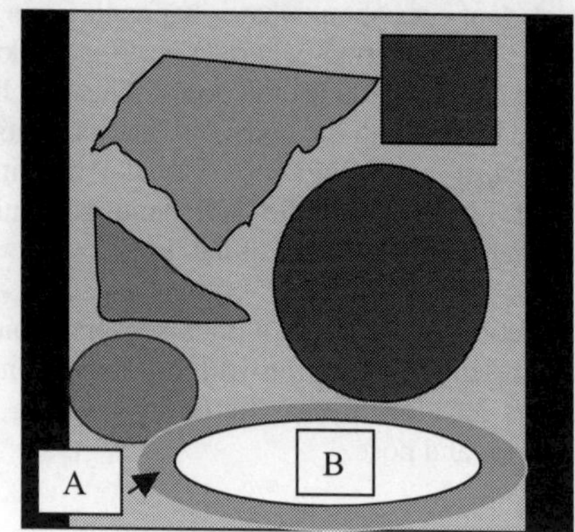

Fig. 1

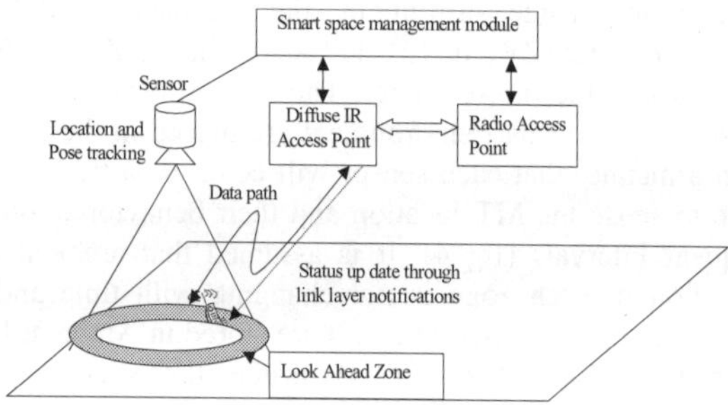

Fig. 2

Table: Ambient light intensity table

I_{11}	I_{12}	..	..
I_{21}			
			I_{kk}

Fig. 3

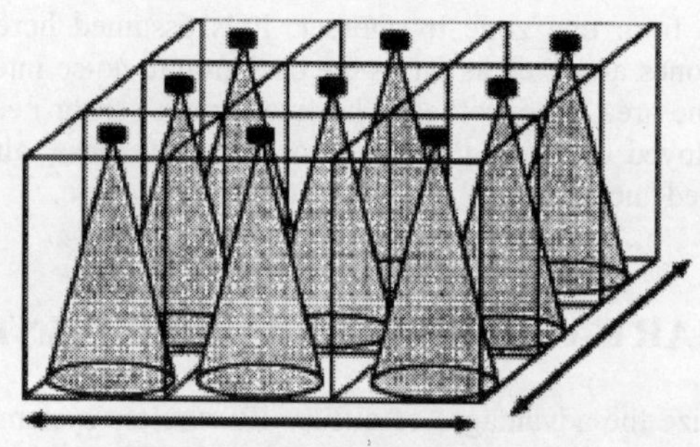

Fig. 4

5. CONCLUSION

The diffuse infrared wireless system has many advantages over RF wireless like high and unregulated bandwidth, inherent security but the performance of IR transmission systems is impaired by one aspect like shot noise induced by the background ambient light. So in an indoor space, to mitigate the degradation effects of background light adaptively, the smart and aware sensor network could be utilized to have a robust diffuse IR system without having the problem of link failure.

REFERENCES

[1] J. M. Kahn and J. R. Barry, " Wireless Infrared Communications", *Proceeding of the IEEE,* vol. 85, no. 2, 265-98, Feb. 1997.

[2] J. R. Barry, " Wireless Infrared Communication", *Kluwer Academic Publishers*, 1994.

[3] Irfan A. Essa, " Ubiquitous Sensing for Smart and Aware Environments", *IEEE Personal Communications*, October 2000.

[4] I. Essa, " Computers Seeing People*", AI Magazine*, vol. 20, no. 1, summer 1999, pp. 69-82.

[5] S. Stillman, R. Tanawongsuwan, and I. Essa, " A System for Tracking and Recognizing Multiple people with Multiple Cameras*", Proc. 2nd Int'l. Conf. Audio-Vision-based Person Authentication*, Washington, DC, Apr. 1999.

Routing with Resource Reservation

Dan Mancas
University of Craiova
dmancas@ucv.ro
13, A.I. Cuza Street, 1100, Craiova
Romania

Abstract: Multimedia applications present new challenges to the current networking technology. One of them is Quality of Service (QoS) requirement. Due to the increasing dynamics of traffic introduced by multimedia applications, dynamic and flexible QoS (Quality of Service) control is needed to ensure both QoS satisfaction and resource efficiency. Because integrated service networks are designed to support a wide range of traffic classes, which have different service requirements, developing metrics to evaluate them is complex. In this paper there are presented the results of some tests performed on Cisco routers in a quality of service environment.

Key Words: Quality of Service, routing, queues.

1. INTRODUCTION

In a QoS environment, routing algorithms must provide routes, which conform to users end-to-end requirements while minimizing the cost of a connection. A connection cost is defined as the sum of the costs of individual links participating in the connection. Link costs can be assigned in various ways. Generally, they should reflect the desirability of using a given link as part of a connection path. As such, they are often referred to as administrative weights in the ATM world. QoS requirements translate into path constraints for a routing algorithm, and impact its computation complexity. Complexity is a major concern in network scalability since the routing algorithm must sometimes be run in real time, especially in case of multicast connections. A given application may not afford to wait too long for a routing algorithm to compute a QoS suitable path, before a connection setup signaling can be initiated. Moreover, if route computation takes too long, topology information in which this computation is performed is more likely to be outdated, leading to a higher probability of having a connection setup failure.

2. TESTBED ENVIRONMENT

The QoS tools were tested in a network environment (figure 1.) set up within the Network Management Laboratory, Fachhochschule Regensburg, Germany.

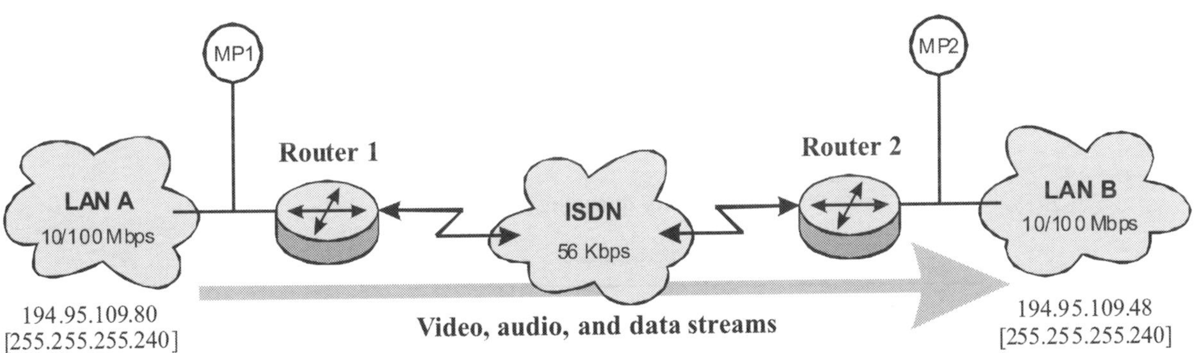

Figure 1. Cisco testbed

LAN A and ***LAN B*** are Ethernet local area networks having the specified IP addresses and sub-network masks.
Router1 and ***Router 2*** are Cisco 1600 Series Routers using CISCO IOS 11.2. They are interconnected through an ISDN line, which acts as a "bottle-neck" in the evaluation process.
MP1 is the measurement point for the transmitted packets. It is a program installed on a Windows NT workstation (the same program that it was used as traffic generator).
MP2 is the measurement point for the received packets. Here it was used a Sniffer machine - a dedicated computer with an Ethernet network card configured on promiscuous mode. On

this machine runs a program that listen the wire and displays some statistical information upon the "sniffed" packets.

3. CISCO TEST RESULTS

3.1. Without Using Qos Features (FIFO)

In its simplest form, FIFO queuing involves storing packets when the network is congested and forwarding them in order of arrival when the network is no longer congested.

FIFO is the default queuing algorithm in some devices, thus requiring no configuration, but it has several shortcomings. Most importantly, FIFO queuing makes no decision about packet priority; the order of arrival determines bandwidth, promptness, and buffer allocation. Nor does it provide protection against ill-behaved applications (sources).

Bursty sources can cause long delays in delivering time-sensitive application traffic, and potentially to network control and signaling messages.

The FIFO algorithm's only benefit is that it is very simple, so it is very fast and it is very easy to be implemented on low resources.

3.2. Using Qos Features

3.2.1. Port Address Bandwidth Distribution

3.2.1.1 Priority Queuing - PQ

PQ ensures that important traffic gets the fastest handling at each point where it is used. It was designed to give strict priority to important traffic. Priority queuing can flexibly prioritize according to network protocol (for example IP, IPX, or AppleTalk), incoming interface, packet size, source/destination address, and so on.

PQ is useful for making sure that mission-critical traffic traversing various WAN links gets priority treatment. For example, PQ might be used to ensure that important Oracle-based sales reporting data gets to its destination ahead of other, less-critical traffic. PQ currently uses static configuration and thus does not automatically adapt to changing network requirements. Although PQ can be enabled for any interface, the intended application was for low-bandwidth, congested serial interfaces. PQ introduces extra overhead that is acceptable for slow interfaces such as ISDN, but may not be acceptable for higher-speed interfaces such as Ethernet.

Port Address	traffic rate A (Kbps)		traffic rate B (Kbps)		traffic rate C (Kbps)	
	transmitted	received	transmitted	received	transmitted	received
20 (high)	30	30	35	35	60	54
80 (medium)	10	10	20	20	50	0
25 (normal)	5	5	15	0	45	0

Table 1. Priority Queuing

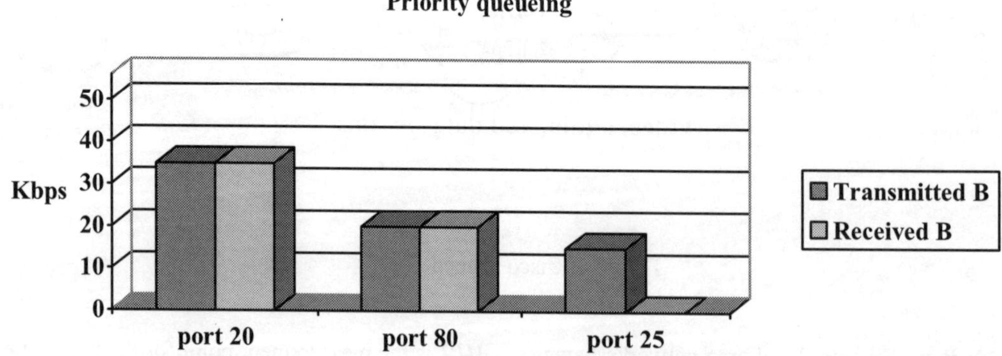

Figure 2. PQ - Protocol Distribution B.

The ports 20 (ftp) and 80 (http) have higher priority than the port 25 (e-mail). As the first two flows use the entire disponible bandwidth, any e-mail will never reach the destination. The 3rd flow starved for bandwith.

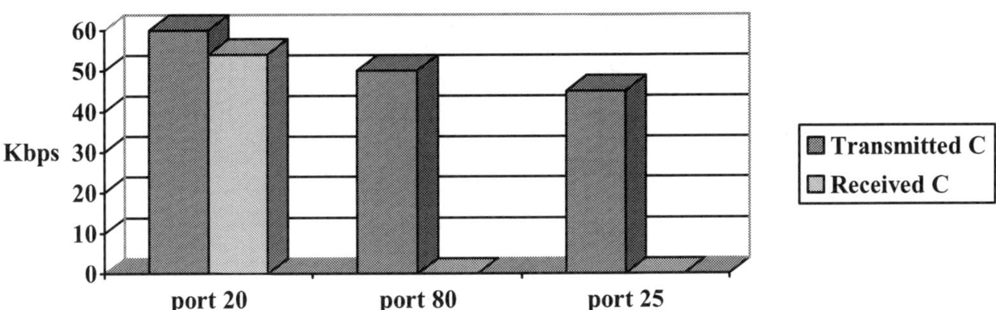

Figure 3. PQ - Protocol Distribution C.

The ftp flow is very intense, so that in this network http and e-mail will never work at all. The ftp bandwidth hits the maximum value.

3.2.1.2. Custom Queuing - CQ

CQ was designed to allow various applications or organizations to share the network among applications with specific minimum bandwidth or latency requirements. In these environments, bandwidth must be shared proportionally between applications and users.

CQ feature can be used to provide guaranteed bandwidth at a potential congestion point, ensuring the specified traffic a fixed portion of available bandwidth and leaving the remaining bandwidth to other traffic. Custom queuing handles traffic by assigning a specified amount of queue space to each class of packets and then servicing the queues in a round-robin fashion.

The queuing algorithm places the messages in one of 17 queues (queue 0 holds system messages such as keepalives, signaling, and so on), and is emptied with weighted priority. The router services queues 1 through 16 in round-robin order, dequeuing a configured byte count from each queue in each cycle. This feature ensures that no application (or specified group of applications) achieves more than a predetermined proportion of overall capacity when the line is under stress.

Like PQ, CQ is statically configured and does not automatically adapt to changing network conditions.

Although, with CQ a "weighted fair" queuing strategy is implemented for the processing of interface output queues. The percentage of an interface's available bandwidth that is used by a particular kind of traffic can be controlled.

Associated with each output queue is a configurable byte count, which specifies how many bytes of data the system should deliver from the current queue before it moves on to the next queue. When a particular queue is being processed, packets are sent until the number of bytes sent exceeded the queue byte count or the queue is empty. Bandwidth used by a particular queue can only be indirectly specified in terms of byte count and queue length.

Note: With CQ or PQ enabled, the system takes longer to switch packets because they are classified by the system's processor card.

The following values were assigned to the queues: port 20 – 70 %, port 80 – 20 %, and port 25 – 10 %.

In case B, the queuing is „weighted fair". The ftp flow passes with the highest priority. Although, e-mail will not longer starve for bandwidth, the e-mail will work (slower than http because has lower queuing priority). For each flow, a maximum bandwidth is defined, but no bandwidth remains unused.

The ftp port has the highest priority, although the bandwidth it uses cannot be greater than the maximum imposed by the CQ parameters (byte count) configuration. The rest of the bandwidth is distributed "fair" to the other flows in such a way, that no one will starv for bandwidth.

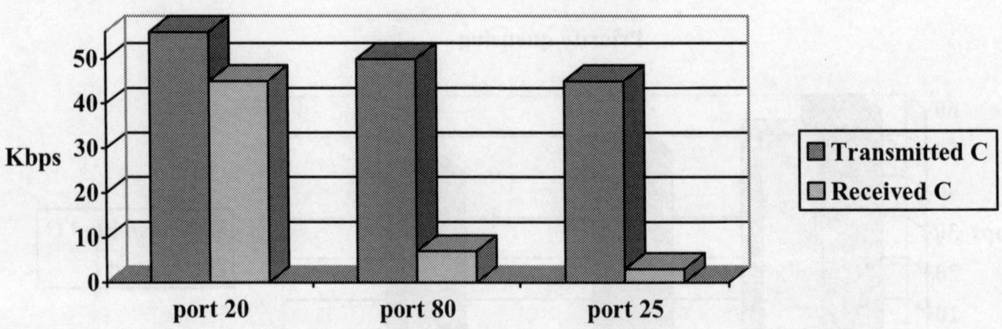

Figure 4. CQ - Protocol Distribution C.

3.2.2. IP Address Bandwidth Distribution

The priority lists can be made based on the source/destination address. The algorithms are the same, so, the behavior of the policy queuing based on addresses is the same as the policy queuing based on ports.

IP Precedence Bandwidth Distribution
Weighted Fair Queuing

WFQ is a flow-based queuing algorithm that does two things simultaneously: it schedules interactive traffic to the front of the queue to reduce response time, and it fairly shares the remaining bandwidth among high-bandwidth flows.

In case of my evaluation, the ISDN bandwidth is ~56Kbps. So the reserved bandwidths are: 56*8/13 = ~34; 56*4/13 = ~17; 56*1/13 = ~4. The conclusions of this test consists of: no flow can use more than the bandwidth that was reserved for it at the interface configuration time; no flow will starv, if the device is properly configured; the second flow (IP priority = 3) does not use the entire bandwidth that was reserved for him, so this remaining bandwidth is used by the first flow (IP priority = 7) - this was an automated WFQ decision.

4. CONCLUSION

WFQ can manage duplex data streams, such as those between pairs of applications, and simplex data streams such as voice and video. From the perspective of WFQ, there are two categories of data streams: high – volume (bandwidth) sessions and low – volume sessions. WFQ ensures that queues do not starve for bandwidth, and that traffic gets predictable service. Low-volume traffic streams - which comprise the majority of traffic - receive preferential service, transmitting their entire offered loads in a timely fashion. High-volume traffic streams share the remaining capacity proportionally between them. WFQ is efficient in that it uses whatever bandwidth is available to forward traffic from lower-priority flows if no traffic from higher-priority flows is present. This is different from time-division multiplexing (TDM), which simply carves up the bandwidth and lets it go unused if no traffic is present for a particular traffic type.

WFQ works with primary QoS signaling techniques - IP precedence and Resource Reservation Protocol (RSVP), - to help provide differentiated QoS as well as guaranteed services. When WFQ is enabled for an interface, new messages for high-bandwidth conversations are discarded after the congestive – messages threshold you set for the default one has been met. However, low-bandwidth conversations, which include control-message conversations, continue to enqueue data. As a result, the fair queue may occasionally contain more messages than are specified by the threshold number.

REFERENCES

Journal Papers:
[1] Pasquale J. C., Polyzos G. C., Xylomenos G., The Multimedia Multicasting Problem, ACM Multimedia Systems Journal, vol. 6, pages 43-59, 1998.
[2] Vogel A., Kerherve B., Von Bochmann G., Distributed Multimedia and QoS: A Survey, IEEE Multimedia, vol. 2, no. 2, pages 87-96, 1995

Books:
[3] Doyle J., DeHaven C., *Routing TCP/IP, Volume II (CCIE Professional Development)*, MacMillan Publishing Company, 2001.
[4] Huitema C., *Routing in the Internet* - second edition, Prentice Hall, 2000
[5] Minoli D., Minoli E., *Delivering Voice over IP Networks*, 2nd Edition, John Wiley & Sons, Inc, 2001
[6] Siegel E., *Designing Quality of Service Solutions for the Enterprise*, John Wiley & Sons, Inc, 1999.

Comparison of the Self-Phase Modulation Effects Tolerance Performance for NRZ, RZ and CS-RZ Formats

Yi Dong, Qi Zhang, Wendi Li, Nan Ma, Minghua Chen and Shizhong Xie
Department of Electronic Engineering, Tsinghua University,
Beijing 100084, China
Tel. 86-10-62773197, Fax. 86-10-62788161, E-mail: yidong@mail.tsinghua.edu.cn

Abstract: The self-phase modulation (SPM) effects of nonreturn-to-zero (NRZ), return-to-zero (RZ) and carrier-suppressed return-to-zero (CS-RZ) signal formats are experimentally compared for 10Gbit/s single-channel transmission in a SMF dispersion managed system at nearly zero residual dispersion. The results show that NRZ format is seriously affected by SPM, while RZ and CS-RZ formats have excellent tolerance. With the input power increasing further and due to the SPM effects, RZ and CS-RZ experience a recovery after the deterioration.

Index Terms: Self-phase modulation, stimulated Brillouin scattering, NRZ modulation, RZ modulation, CS-RZ modulation.

1. INTRODUCTION

The chromatic dispersion and the nonlinear optical phenomena are widely known as two critical limitations to 10Gbit/s transmission systems operating at 1.55 μm over standard single-mode fiber (SMF). Various dispersion compensation schemes, such as dispersion compensating fiber (DCF) and fiber Bragg gratings (FBG), have been utilized to overcome the chromatic dispersion limitation. Thus, the nonlinear optical phenomena, especially the self-phase modulation (SPM) owing to the nonlinear Kerr effect are crucial to the performance of long-haul optical fiber communication systems.

The choice of modulation format is of great interest for the management of fiber nonlinearities. Various formats including nonreturn-to-zero (NRZ), return-to-zero (RZ) and carrier suppressed return-to-zero (CS-RZ) have been studied [1]-[6]. However, to our acknowledgement, their performances of tolerance to SPM have not been clearly compared in a same experimental setup.

In this paper we experimentally compare the performance of NRZ, RZ and CS-RZ modulation formats for single-channel SMF transmissions in terms of the tolerance to SPM. The results show that RZ and CS-RZ have excellent tolerance to SPM.

2. EXPERIMANTAL AND DISCUSSION

The schematic of the experimental setup is shown in Fig.1. The transmitter consists of a continue-wave (CW) laser diode and a $LiNbO_3$ (LN) modulator. The 10Gbit/s pseudorandom bit sequence (PRBS) data of length 2^7-1 from a bit-error-rate test set (BERTS) is externally modulated onto the CW carrier in either NRZ or RZ format by the modulator. When desired, the 10Gbit/s CS-RZ format can also be obtained by adjusting the bias voltage of the modulator at the modulation null point with 5GHz clocks. The CW output is amplified to ~18dB by an Erbium-doped fiber amplifier (EDFA) to make SPM effects obviously.

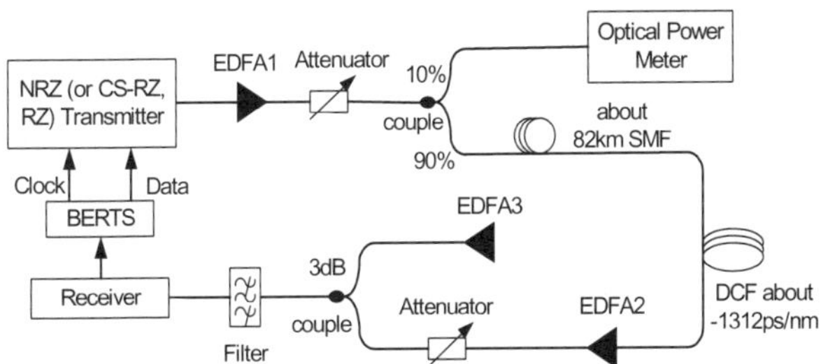

Fig.1 Experimental setup

For a single optical channel transmission system, the main sources of power penalty are the chromatic dispersion, SPM, SBS and the accumulated ASE noise. The comparison of the tolerance to SPM is reasonable only if the impact of the chromatic dispersion, SBS and the accumulated ASE noise have been eliminated.

In the transmitter, a 20KHz rectangular signal is modulated directly on the CW to suppress the effect of stimulated Brillouin scattering (SBS). The transmission line consists about 82-Km SMF with 17ps/km·nm fiber dispersion and a DCF compensation module for compensating 80-Km SMF, whose total dispersion is about –1312ps/km. The overall dispersion is compensated to nearly zero at the 1551.72nm wavelength. Thus the SPM effects will dominate the system performance. The EDFA3 operating without input signal injects its amplified spontaneous emission (ASE) noise into the system. Together with the use of the attenuator, the optical signal-to-noise ratio (OSNR) at the receiver was kept a constant to eliminate the effects of SNR to the system power penalty.

At the receiver, the bit-error-rate (BER) performance is measured by changing input average powers while keeping a constant OSNR. The system performance is evaluated through the power penalty at 1×10^{-10} BER.

Fig. 2 shows the corresponding transmission optical spectra and the eye diagrams for NRZ, RZ and CS-RZ signal.

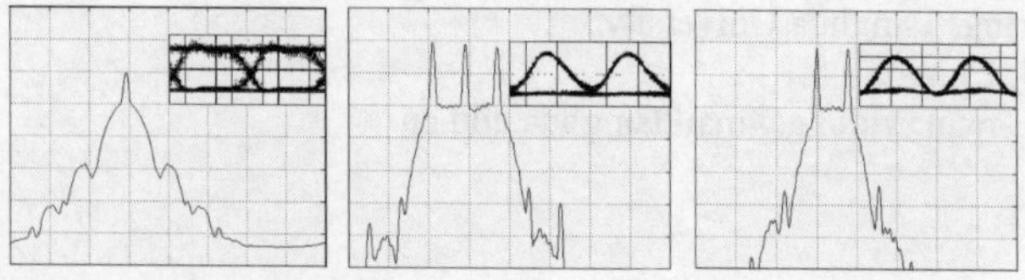

Fig. 2 Optical signal spectra and eye diagrams of NRZ (left), RZ (center) and CS-RZ (right)

Fig. 3(a, b, c) shows the measurements of SBS effects for NRZ, RZ and CS-RZ formats respectively. It is clear that the NRZ data has worse tolerance to SBS than RZ and CS-RZ. The 20KHz rectangular dither signal effectively increases the SBS threshold power. Within our measurement range, the impact of SBS is negligible.

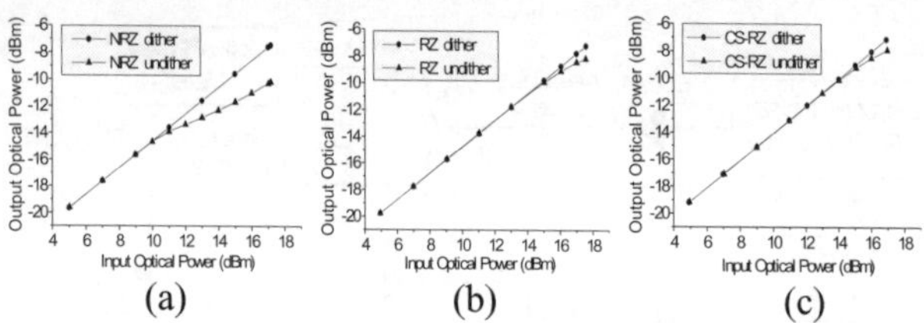

Fig. 3 SBS performance of NRZ (a), RZ (b) and CS-RZ (c)

In Fig. 4 the power penalty results at 1×10^{-10} BER is plotted versus the average input optical power for the three modulation formats.

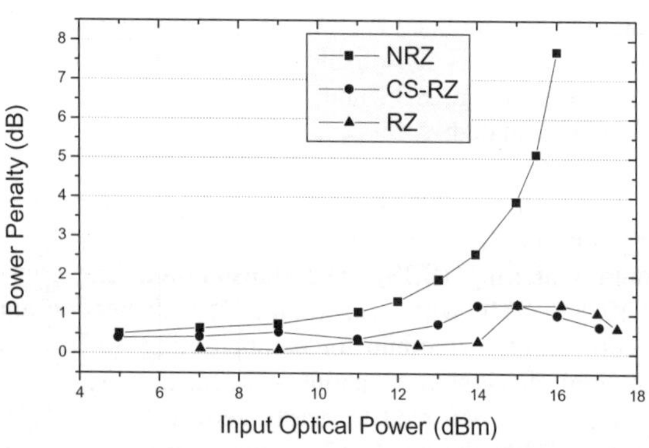

Fig. 4 Power penalty at 1×10^{-10} BER versus average input power for NRZ, RZ and CS-RZ

Fig. 5 illustrates corresponding signal waveforms of the three modulation formats at different input power. All results are obtained at nearly zero residual dispersion system and the same system OSNR of 22 dB at receiver to reduce the effects of other factors and make the comparison more reasonable.

The SPM-induced spectral broadening is a consequence of the time dependence of the nonlinear phase shift ϕ_{NL}. The induced frequency chirp $\delta\omega$ is given by [4]

$$\delta\omega = \frac{1}{2\pi}\cdot\frac{d(\phi_{NL})}{dt} = \frac{n_2 L_{eff}}{\lambda A_{eff}}\cdot\frac{dP}{dt}$$

Where n_2 is the nonlinear-index coefficient, λ is the signal wavelength, A_{eff} is the effective fiber area, P is the signal power and $L_{eff}=(1-exp(-\alpha L))/\alpha$, the effective length. The equation indicates that the SPM-induced frequency chirp $\delta\omega$ is directly proportional to dp/dt. Thus the SPM effects are considerably larger at the steeper leading and trailing edges and nearly zero at the center part of pulses.

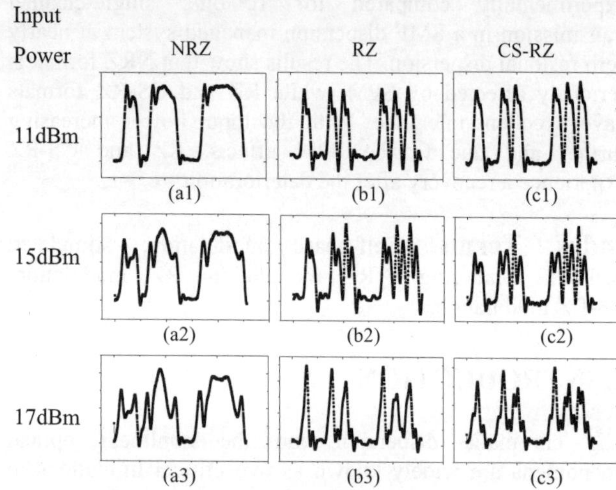

Fig. 5 Output signal waveforms of NRZ (a), RZ (b) and CS-RZ (c)

For NRZ format, the consecutive "1" codes are split at leading and trailing edges owing to the SPM. At the same time, the relative power of the consecutive "1" codes' central part increases (as shown in Fig.5 (a2) and (a3)). This waveform distortion results in great power penalty because the decision voltage is in accordance with the average optical power. Thus, in Fig. 4, the NRZ power penalty increases rapidly with the input power, which indicates NRZ codes are of the most susceptible to SPM.

Similarly, for RZ and CS-RZ formats, the "1" codes are also split at the edges for SPM effects (as shown in Fig. 5 (b2) and (c2)). However, the split will not induce significant power penalty by use of electric equalized filter in the receiver, as shown in Fig. 4. When the input power increases further more, with the continuous influence of SPM to the split codes, the codes will be further split. This makes the power of center sunken part of "1" code increasing and makes the code narrowed as shown in Fig.5 (b3) and (c3). Thus, the power penalty curve has a local extremum at 14dBm and 15dBm for RZ and CS-RZ respectively. When the input power increases, the power penalty turns out to be decreasing.

3. CONCLUSION

We have experimentally performed a comparison of NRZ, RZ and CS-RZ modulation formats in a single-channel SMF dispersion managed system at nearly zero residual dispersion. The results imply the robustness of RZ and CS-RZ format against SPM. We also find that when the input power increases, the SPM effects will give RZ and CS-RZ experience a recovery after the deterioration.

REFERENCE

Proceedings Papers:

[1] Vassilieva et al, "Numerical comparison of NRZ, CS-RZ and IM-DPSK formats in 43Gbit/s WDM transmission", *LEOS 2001*, the 14th Annual Meeting of the IEEE, ThC2, vol. 2, pp. 673-674 , 2001.

[2] Arne Rroyset, "Self phase modulation limitations in long nonrepeatered standard fibre transmission: influence from dispersion compensation scheme and modulation format", OFC '98, TuB4, 1998.

[3] B.Bakhshi et al., "comparison of CRZ, RZ and NRZ modulation formats in a 64x12.3Gb/s WDM transmission experiment over 9000 km", OFC 2001, WF4-1, 2001.

[4] Yann Frignac et al., "Numerical optimization of residual dispersion in dispersion-managed systems at 40 Gbit/s", OFC 2000, TuD3-1, 2000.

Journal Papers:

[5] Toshiya Matsuda et al., "Comparison Between NRZ and RZ signal formats for in-line amplifier transmission in the zero-dispersion regime", J. Lightwave Technol., vol. 16, pp. 340-348, Mar. 1999.

[6] C. Caspar et al., "RZ versus NRZ modulation format for dispersion compensated SMF-based 10-Gb/s transmission with more than 100-km amplifier spacing ", IEEE Photonics Technol. Lett., vol. 11, pp. 481-483, April. 1999.

ROUTING AND WAVELENGTH ASSIGNMENT OF SCHEDULED LIGHTPATH DEMANDS IN A WDM OPTICAL TRANSPORT NETWORK*

Josue KURI Nicolas PUECH Maurice GAGNAIRE Emmanuel DOTARO†
ENST Paris, Dept. of Computer Science and Networks. 46, rue Barrault. F-75634 Paris, France.
Tel: (+33) 1.45.81.75.70 Fax: (+33) 1.45.81.31.19 E-mail: {kuri|npuech|gagnaire}@enst.fr
†Alcatel R&I, Optical Systems and Networks. E-mail: Emmanuel.Dotaro@alcatel.fr

Abstract: We present RWA algorithms for a new class of traffic model where, besides the source, destination and size (number of lightpaths) of the traffic demands, their set-up and tear-down dates are known. They are called Scheduled Lightpath Demands (SLDs). We model the RWA problem as a spatio-temporal combinatorial optimization problem and provide two solution algorithms. The time disjointness that could exist among SLDs is taken into account in order to maximize the *utilization* of resources and hence, minimize the amount of globally required resources. We compare our algorithms to an on-line RWA algorithm and show that taking into account the time disjointness of demands can lead to a gain of resources of 20 % in average.

Key Words: WDM; RWA; scheduled demands; Tabu Search.

1 INTRODUCTION

The optical transport networks based on architectures like OTN [1] and control protocols like GMPLS [2] are expected to provide a set of functions that will include the set-up, maintenance and tear-down of permanent, soft-permanent and switched optical connections[1]. These functions will be used to implement services like Bandwidth on Demand (BoD) and Optical Virtual Private Networks (OVPNs). For example, a network operator offering a OVPN service can use permanent optical connections to provision the *base* capacity of the OVPN. Soft-permanent connections can be used to provision the additional capacity required at *expected* times (e.g., scheduled backups, etc.). Finally, switched connections can be used to satisfy *unexpected* bandwidth requirements. Thus, the offered services will lead to three types of connection demands: static, scheduled and unexpected.

In this paper we deal with the scheduled connection demands, that we call Scheduled Lightpath Demands (SLDs). An SLD is represented by a tuple $(s, d, n, \alpha, \omega)$ where s and d are the source and destination nodes of the demand, n is the number of requested connections (hereafter called lightpaths) and α, ω are the set-up and tear-down dates of the demand. The traffic model based on SLDs is different from the static and dynamic lightpath traffic models previously considered in the literature [3]. Static traffic means that all the demands are known in advance and do not change over time, whereas dynamic traffic means that the arrival and holding time of demands are random. In the SLD traffic model, the demands and their distribution over time are known in advance.

We address the Routing and Wavelength Assignment (RWA) problem for SLDs. The parameters of the problem are a set of SLDs and a physical network. The objective is to define for each lightpath a route[2] in the network and a wavelength to be used along this route by the lightpath. We model this RWA problem as an spatio-temporal combinatorial optimization problem and propose two solution algorithms that we call SLD-RWA algorithms.

Since all SLDs are not simultaneous, some *wavelength channels*[3] can be used by several lightpaths provided that these lightpaths do not overlap in time. The SLD-RWA algorithms take into account this property in order to maximize the utilization of wavelength channels when routing the demands and, as such, minimize the global number of required channels.

The next section provides a detailed description of the considered RWA problem. Section 3 presents the problem's spatio-temporal combinatorial optimization model and Section 4 the algorithms used to solve the model. In Section 5 we describe an online RWA algorithm described in [3] that we use as reference to evaluate the SLD-RWA algorithms. In Section 6 we experimentally compare our algorithm with the online RWA algorithm. Finally, in Section 7 we give some conclusions and describe ongoing work.

2 PROBLEM DESCRIPTION

There are two basic versions of the RWA problem investigated in the literature. If the traffic demands are static, the problem typically consists in determining a routing and a wavelength assignment that minimize the cost of network resources required to satisfy all the demands. Alternatively, if the resources are fixed, the goal is to maximize the number of satisfied demands or, when a revenue model exist, maximize the revenue generated by the satisfied demands. If the traffic demands are random, given a fixed amount of resources, the problem typically consists

*Work funded by Alcatel R&I, Marcoussis, France under Grant No. CI 578. This paper has neither been published before nor currently being submitted elsewhere.

[1] Connections described in Recommendation ITU-T G.8080.

[2] A concatenation of optical fibers (connected by OXCs) from the source node to the destination node of the lightpath.

[3] A *wavelength channel* corresponds to the use of a wavelength on an optical fiber. A fiber has as many wavelength channels as wavelengths can be multiplexed in the fiber.

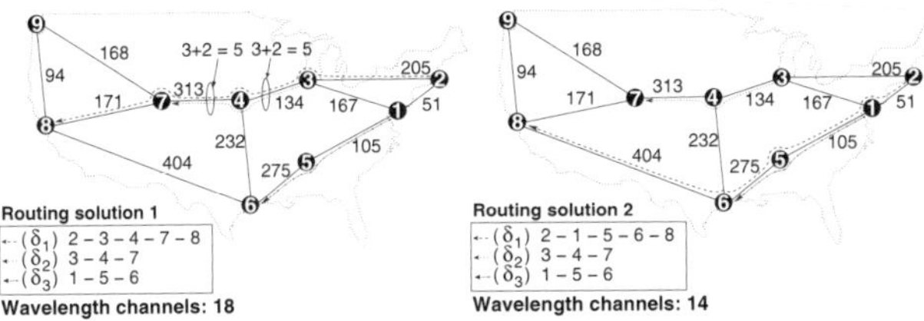

Figure 1: Two alternative routing solutions for the SLDs of Table 1.

SLD	Src	Dest	# of LPs	set-up	tear-down
δ_1	2	8	2	08:00	14:40
δ_2	3	7	3	11:00	13:00
δ_3	1	6	2	17:00	19:30

Table 1: A set of 3 SLDs.

in defining a routing and a wavelength assignment that reduce the blocking probability of demands. The RWA problem instances are usually defined in terms of either a traffic matrix (static traffic) or a statistical model (random traffic); the network characteristics (e.g., the topology); and, depending on the problem type, the capacity of links and nodes. Additionally, assumptions are defined about the directionality of links or fibers, the number of fibers per link, the functionality in nodes (e.g., wavelength conversion), etc.

Given a set of SLDs and a physical network, we want to determine a routing and a wavelength assignment that minimize the number of required wavelength channels to satisfy all the SLDs. A wavelength channel is required on each fiber traversed by a lightpath. We assume that there is one fiber on each direction of every physical link and that there is no wavelength conversion in nodes. Figure 1 illustrates how two routing solutions lead to a different number of wavelength channels. Table 1 shows the set of SLDs considered. Note that SLD δ_3 is time-disjoint with respect to SLDs δ_1 and δ_2.

3 MATHEMATICAL MODEL

We solve the routing and the wavelength assignment (WA) problems separately. The solution of the routing problem is the input for the WA problem. The spatio-temporal combinatorial optimization models for these problems are respectively presented in subsections 3.1 and 3.2

3.1 ROUTING

We use the following notations:

$G = (V, E, w)$
is an edge-weighted undirected graph with vertex set $V = \{v_1, v_2, \dots, v_N\}$, edge set $E = \{e_1, e_2, \dots, e_L\}$ and weight function $w : E \to \mathbb{R}^+$. The graph represents a physical network. The set V corresponds to the nodes, the set E to the bidirectional links and the function w to the links' cost (e.g., their physical length).

$\Delta = \{\delta_1, \delta_2, \dots, \delta_M\}$
is the set of M SLDs, where $\delta_i = (s_i,\ d_i,\ n_i,\ \alpha_i,\ \omega_i)$ is a tuple representing the SLD number i; $s_i, d_i \in V$ are the source and destination nodes, n_i is the number of requested lightpaths, and α_i and ω_i are the set-up and tear-down dates.

(G, Δ)
is a pair representing an instance of the RWA problem.

$N = |V|$, $L = |E|$, $K_{\max}$
are, respectively, the number of vertices and edges in G and the maximum number of possible alternate paths for each demand (e.g., the $K_{\max}$ shortest paths computed with the Eppstein algorithm [4]).

$P_{k,i}$, $1 \leq k \leq K_{\max}$, $1 \leq i \leq M$
represents the kth alternate path in G from s_i to d_i.

$\pi_{\rho,\Delta} = (P_{\rho_1,1}\ P_{\rho_2,2}\ \dots\ P_{\rho_M,M})$, $\rho \in \{1, \dots, K_{\max}\}^M$
is called an *admissible routing solution* for Δ. ρ is an M-dimensional vector whose elements can take a value between 1 and $K_{\max}$. An admissible routing solution is fully characterized by ρ.

$\Pi_\Delta = \{\pi_{\rho,\Delta} \,|\, \rho \in \{1, \dots, K_{\max}\}^M\}$
is the set of all admissible routing solutions for Δ. The cardinality of the set is $|\Pi_\Delta| = (K_{\max})^M$.

$C : \Pi_\Delta \to \mathbb{N}$
is the cost function that computes the number of wavelength channels required by an admissible routing solution $\pi_{\rho,\Delta}$. In such a solution, a path is defined for each SLD in Δ. Each lightpath of an SLD requires a wavelength channel on every fiber spanned by the path. The same channel can be used by several lightpaths provided that they use the channel at different times. In Figure 1, the values of C are 18 (left side) and 14 (right side).

The combinatorial optimization problem to solve is:

$$\textbf{Minimize } C(\pi_{\rho,\Delta}),\quad \textbf{subject to: } \pi_{\rho,\Delta} \in \Pi_\Delta,$$

that is, we search an admissible routing solution $\pi_{\rho,\Delta}$ that minimizes the number of required wavelength channels for the set of demands Δ. To formalize the cost function, $C : \Pi_\Delta \to \mathbb{N}$, we define the following additional notations:

$\theta = (\theta_{ij})$
is a $\{0,1\}^{M\times M}$ upper triangular matrix; θ_{ij}, $i \leq j$, indicates whether SLDs i and j overlap in time ($\theta_{ij} = 1$) or not ($\theta_{ij} = 0$). By definition $\theta_{1 \leq ii \leq M} = 1$, and $\theta_{ij} = 0$, $i > j$. The matrix expresses time dependence among SLDs.

$\beta = (\beta_{ij}) = diag(n_i)$
is the diagonal matrix; $\beta_{ii} = n_i,\ 1 \leq i \leq M$.

$\gamma^{\pi_{\rho,\Delta}} = (\gamma_{ij}^{\pi_{\rho,\Delta}})$
is a $\{0,1\}^{L\times M}$ edge-path incidence matrix; $\gamma_{ij}^{\pi_{\rho,\Delta}}$ indicates whether edge $i \in E$ is part of path $P_{\rho_j,j}$ in routing solution $\pi_{\rho,\Delta}$ ($\gamma_{ij}^{\pi_{\rho,\Delta}} = 1$) or not ($\gamma_{ij}^{\pi_{\rho,\Delta}} = 0$). For the sake of simplicity, we note γ instead of $\gamma^{\pi_{\rho,\Delta}}$. This matrix describes the physical routing of SLDs for a given ρ.

$\eta = \theta \cdot \beta \cdot \gamma^T = (\eta_{ij})$
is a $\mathbb{N}^{M\times L}$ matrix; η_{ij} indicates the number of time-overlapping lightpaths on link e_j between SLD i and SLDs k, $k > i$.

Thus, the cost function is defined by:

$$C(\pi_{\rho,\Delta}) = \sum_{j=1}^{L} \max_{1\leq i\leq M} \eta_{ij}. \qquad (1)$$

3.2 WAVELENGTH ASSIGNMENT

The solution to the routing problem is an admissible routing solution $\pi_{\rho,\Delta}$ that defines a route in the physical network for each SLD in Δ; it represents an instance of the WA problem (i.e., the input for the WA algorithm).

The solution to the WA problem defines the wavelength channels that must be assigned to each lightpath of the SLDs in Δ when the $\pi_{\rho,\Delta}$ routing is used, such that *i)* each lightpath uses the same wavelength channel on all the fibers spanned by its route (wavelength continuity constraint), *ii)* any two time-simultaneous lightpaths whose routes have at least one fiber in common use different wavelengths channels, and *iii)* the global number of *different* wavelengths channels used is minimal[4].

The WA problem in networks with one fiber per link direction assuming the wavelength continuity constraint and the static traffic model reduces to a well known graph vertex coloring problem [3]. We generalize this model for the WA problem using the SLD traffic model. The following notations are used:

$\psi = \gamma^T \cdot \gamma = (\psi_{ij})$
is an $\mathbb{N}^{M\times M}$ matrix; ψ_{ij} is the number of edges on routing solution $\pi_{\rho,\Delta}$ where SLD i overlaps with SLD j (remember that γ is a simplified notation for $\gamma^{\pi_{\rho,\Delta}}$).

μ_i
is the set of lightpaths belonging to the SLD i; μ_i^j, $1 \leq j \leq n_i$, denotes the jth lightpath of SLD i.

$G_c = (V_c, E_c)$
is a conflict graph where $V_c = \cup_{i=1}^{M} \mu_i$ and $E_c = \{(\mu_p^i, \mu_q^j) \mid \mu_p^i \neq \mu_q^j, \psi_{pq} > 0, \theta_{pq} = 1\}$.

[4]Note that the "number of *required* wavelength channels" computed by the routing algorithm and the "number of *different* wavelength channels" computed by the WA algorithm are different things.

Once the graph vertex coloring problem is solved, the color of each vertex in G_c represents the wavelength channel to be assigned to the lightpath represented by this vertex.

4 SOLUTION ALGORITHMS

We developed an exact resolution algorithm for the routing problem based on branch-and-bound that we called B&B SLD-RWA [5]. The major drawback of the algorithm is its time complexity $O(|\Pi_\Delta|)$, i.e., in the worst case, it has to explore $(K_{\max})^M$ admissible routing solutions. This means that only problem instances of small size can be solved in practice. In [6] we propose an approximate resolution algorithm based on the Tabu Search (TS) meta-heuristic that we called TS SLD-RWA. Meta-heuristics explore only a fraction of the solution space associated to a problem instance. Consequently, large size problems may be solved, though, the obtained solutions are not guaranteed to be optimal. We found in [6] that, for the considered problem instances, the average difference in cost between B&B SLD-RWA and TS SLD-RWA computed solutions was not greater than 1.18 %. The closeness to optimality of the computed solutions and the ability to solve large instances make TS SLD-RWA a viable alternative for realistic routing problem instances. We used this algorithm in the experiments of Section 6. For the WA problem we used the greedy algorithm described in [7], which computes approximate solutions.

5 ONLINE RWA ALGORITHM

We use the online RWA algorithm described in [3] as a reference to evaluate the SLD-RWA algorithms.

The algorithm solves the RWA problem for dynamic lightpath demands. This means that the route and the wavelength to be assigned to the lightpaths of a demand are computed as these demands arrive over time. In order to use the algorithm with the SLD traffic model, we use the set-up and tear-down dates of SLDs as arrival and departure times.

The algorithm uses fixed-alternate routing and a First Fit (FF) wavelength assignment strategy. In fixed-alternate routing, k alternate routes are computed in advance for each source/destination pair in the network. When a demand with destination d arrives at node s, one route among the k precomputed is selected according to the following criterion. In the FF wavelength selection strategy, the wavelengths channels in an optical fiber are numbered in ascending order. When searching for an available wavelength channel in the fibers of a route, a lower-numbered channel is considered before a higher-numbered one. A wavelength channel is available in a route if it is not currently used by any lightpath demand on the fibers spanned by the route. When a traffic demand with destination d arrives at source node s, the route (among the k routes) with the available wavelength channel whose number is the smallest one is selected. If more than one route has the same smallest number of available wave-

	Online RWA		TS SLD-RWA		
Δ set class	AWC (routing)	ADW (WA)	AWC (routing)	ADW (WA)	ACCG
1	562.90	15.12	442.82	5.32	21.33 %
2	453.12	10.64	371.64	4.76	17.98 %
3	769.00	11.36	600.20	4.73	21.95 %
4	909.70	16.26	720.45	5.37	20.80 %

Table 2: Average values obtained with the online and TS SLD-RWA algorithms ($M = 500$).

length channel, the physically shortest route is selected.

6 EXPERIMENTAL EVALUATION

In this section we evaluate the gain in terms of required wavelengths channels provided by the TS SLD-RWA algorithms when compared to the online RWA algorithm. Recall that an instance of the RWA problem is defined by a pair (G, Δ). We used as the graph G the same network considered in [6] ($N = 29$, $L = 44$)[5].

We classify the sets Δ according to the time and the space correlation of the constituent SLDs. The mathematical definitions of these correlation criteria are given in [5] and [6][5]. In the experiments, we consider sets Δ of the following classes: "strong time - strong space", "weak time - strong space", "weak time - weak space" and "strong time - weak space" correlation. We denote these classes 1, 2, 3 and 4, respectively. We generated 50 sets Δ of each class with $M = 500$ SLDs each. The SLDs were generated randomly being constrained only by the time and space correlation defined for the set they belong to. In the TS routing algorithm we used a value $K_{\max} = 4$.

Table 2 shows the average number of required wavelength channels (AWC) and the average number of *different* wavelengths channels (ADW) computed by the online and the TS SLD-RWA algorithms for the four classes of Δ sets. The column 'Avg channel count gain' (ACCG) shows the average gain, in terms of required wavelength channels, of the solutions computed with TS SLD-RWA with respect to the online algorithm solutions.

The AWC is usually smaller for the classes with strong space correlation (classes 1 and 2) than for the classes with weak space correlation because in the former the endpoints of demands are topologically closer than in the latter. The ACCG is, however, approximately the same for all the classes (around 20 %). By comparing the two classes with the same space correlation (classes 1 and 2, on one hand, and classes 3 and 4 on the other hand), we note that classes with weak time correlation lead to a smaller AWC than classes with strong time correlation for both the online and the TS SLD-RWA algorithms. This is because the time-disjointness of demands in weak time correlation classes allows for a better utilization of resources. Regarding the ADW, the classes with weak time correlation (classes 2 and 3) have a smaller average value than the classes with strong time correlation.

The combined execution time of TS SLD-RWA and the greedy graph vertex coloring algorithm on the problem instances considered in the experiments was of 6480 seconds in average (200 instances). The execution time of the online RWA algorithm was of 20 seconds in average. It is interesting to note that the higher computing capacity (illustrated by the execution time) required by the proposed RWA algorithm is worthwhile, considering the gain in channels of about 20 %. Furthermore, the algorithm is not expected to be executed in a real-time environment (as would be the case of the online RWA algorithm) but as a batch process at time intervals of several hours (remember that the SLDs correspond to planned demands).

7 CONCLUSION

In this paper we presented RWA algorithms for scheduled lightpath demands. We have shown that, if the set-up and tear-down dates of traffic demands are known in advance, these pieces of information can be used by the algorithm to achieve a gain of about 20 % in terms of required channels when compared to an RWA algorithm that does not take this information into account. The gain is achieved by exploiting the time disjointness among demands so that a same channel can be used by several lightpaths at different times.

REFERENCES

[1] ITU-T Recommendation G.872, Architecture of Optical Transport Network, revision November 2001.

[2] E. Mannie (Editor), Generalized Multi-Protocol Label Switching (GMPLS) Architecture, IETF ccamp group, Work in progress, March 2002.

[3] H. Zang, J. Jue and B. Mukherjee, "A Review of Routing and Wavelength Assignment Approaches for Wavelength-Routed Optical WDM Networks," Optical Networks Magazine, January 2000.

[4] D. Eppstein, Finding the k shortest paths, SIAM Journal of Computing, 28(2):652-673, 1998.

[5] J. Kuri, N. Puech, M. Gagnaire and E. Dotaro, Optimal Routing of Scheduled Lightpath Demands in a WDM Network Using a Branch & Bound Algorithm, Internal Report ENST Paris, May 2002.

[6] J. Kuri, N. Puech, M. Gagnaire and E. Dotaro, "Routing Foreseeable Lightpath Demands using a Tabu Search Meta-heuristic," Procs. of IEEE Globecom 2002. OPNT-04-7. Taipei, R.O.C. November 2002.

[7] D. Kirovski and M. Potkonjak, "Efficient Coloring of a Large Spectrum of Graphs," Procs. of the 35th Conference on Design Automation. San Francisco, CA, June 1998.

[5] Not reproduced here because lack of space.

Multi-layer Model for Optical Multicast over WDM Networks

Aijun Ding †, Gee-Swee Poo ‡ and Sun-Teck Tan †
† Department of Computer Science, National University of Singapore; {dingaiju, tanst}@comp.nus.edu.sg
‡ School of EEE, Nanyang Technological University; egspoo@ntu.edu.sg

Abstract: Optical multicast is usually modeled as a two-step process: multicast routing and wavelength assignment. This paper proposes a new multi-layer model, in which the WDM network is represented as a multi-layer wavelength structure, each of which has finite capabilities, to integrate routing and wavelength assignment. The algorithm makes use of the light splitting and wavelength conversion capabilities simultaneously in a deterministic manner, avoiding the re-routing procedure. It handles the finite light splitting capability, thus providing a general solution to the problem. Some simulations results are presented in the paper.

Key Words: Multicast, WDM Network, Steiner Problem

1. INTRODUCTION

Advances in optical network technologies such as wavelength division multiplex (WDM) provide enormous bandwidth to meet the future demand of next generation Internet. In a single fiber, WDM is capable of providing multiple communication channels at various wavelengths, thus, increasing the cable capacity to many folds. Future high-speed network can carry voice, data, and other multimedia information, and multimedia traffic will overwhelm pure data traffic. Efficient multicast mechanism is a fundamental issue for the success of group applications. Multicast communication involves transmitting a message from a single source node to multiple destination nodes. This naturally fits for the group applications. Besides, multicast also provides an economical way for utilizing bandwidth.

Recently, studies on multicast over WDM have been active [1]. The studies can be divided into two groups. The first group approaches the problem from high to low by defining protocols at the higher layer and integrate them with the lower WDM layer. These protocols are based either on the modification of existing multicast tree or makes use of powerful nodes as cores. The second group tackles the problem directly at the WDM layer. This group makes use of powerful modeling tools such as Graph theory and mathematical programming to handle lightpaths or light-trees provisioning. In this approach, the Steiner Minimal Tree (SMT) heuristics are used extensively to construct the multicast tree. The wavelength assignment is done either by mathematical programming techniques such as Integer Linear Programming (ILP) or by graph coloring heuristics.

However, both groups suffer from a number of drawbacks. Firstly, the multicast routing and wavelength assignment are handled separately. Secondly, although the algorithms do make use of the light-splitting (LS) capability of nodes, they consider only two extreme situations, either none or full LS capability. This may not be valid in general. Finally, the algorithms seldom consider wavelength conversion (WC) capability. Even if they do, such as the VS-routed approach, the WC is not used in a systematic and efficient way.

This paper proposes a new model to integrate the routing and wavelength assignment, and to overcome all the above limitations. In the model, the WDM network is structured as a collection of wavelength layers, each of which consists of a collection of nodes with finite light splitting and wavelength conversion capabilities. This is illustrated in Figure 1. The centralized SMT heuristics plus appropriate extensions to deal with wavelength assignment should be the best choice. We construct a framework for developing SMT heuristics in the model. By doing so, our heuristics can generate light-trees for all-optical routing in the wavelength-routed optical network.

The rest of the paper is organized as follows. Section II presents the multi-layer WDM model, and gives the genetic algorithm flow. Section III describes the simulation methodology, and gives the numerical results. A conclusion is given in Section IV.

2. MULTICAST OVER WDM ISSUES

2.1. The multi-layer WDM model

Multicast routing and wavelength assignment (MCRWA) problem can be modeled as a tri-criteria problem. The three objectives are: (1) to minimize the use of light splitters, (2) to minimize the use of wavelength converters, and (3) to minimize the cost of the multicast tree. The overall objective is to find a network of minimum cost, subject to the LS and WC constraints, and then maximize the blocking performance of the optical network. However, this kind of greedy approach may not fulfill the goal.

To address the problem, we adopt a layered approach, in accordance to the multi-layer model of WDM network (Figure 1a). It does routing one layer at one time, and moves to a new layer only when necessary. The purpose is to minimize the number of wavelengths used. It finishes routing and wavelength assignment in one step, making efficient use of the LS and WC capabilities in a more systematic manner, and avoiding the re-routing procedure all together. In addition, it is able to model the finite capacity of light-splitting as well as wavelength conversion in a node.

By modeling the WDM network as a collection of wavelength layers, one adds a distinctive dimension, i.e. wavelength to the problem. The nodes with WC capability serve as ladders connecting different layers (Figure 1c).

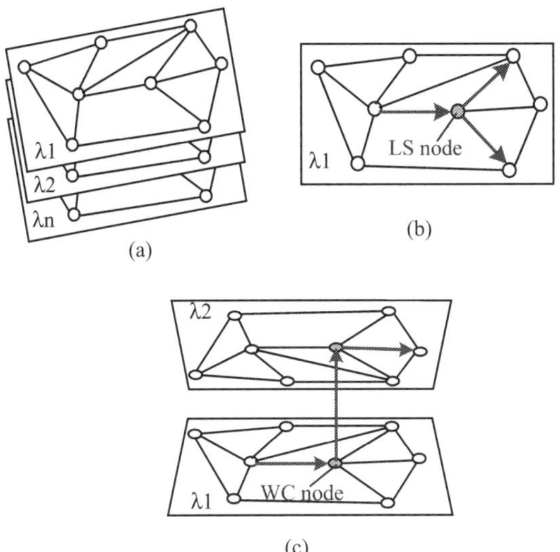

Figure 1 WDM network, light splitting and wavelength conversion

In the WDM environment, the finite LS capability can be transformed to a degree constraint. For instance, the LS node in Figure 1b has a 1:2 splitter, so its degree constraint is 3. Therefore, with appropriate extension, DCSP can be applied to solve the multicast over WDM problem.

DCSP was once used in ATM network, where some ATM switches can only make limited number, compared with its fan-out, of copies of multicast packets [3][4]. By modeling the limitation of ATM copying capability into the degree constraint, DCSP was applied with some heuristics to solve the problem. Similarly, it can also be applied to address the MCRWA problem.

2.2. Generic Algorithm Flow

Our algorithm adopts a layered approach, and considers LS and WC constraints of WDM network as well. We have developed a few heuristics under the multi-layer model, which can be divided into two groups: quasi-Prim and quasi-Kruskal heuristics. For the description of original Prim's or Kruskal's algorithm, please refer to [5]. The two groups are similar to the insertion and component connecting methods with 2Basic building block, respectively [2]. Figure 2 shows the outline of the quasi-Prim algorithm, which is an open architecture.

We have also developed other type of heuristics, such as Distance Network Heuristics (DNH), which fall into the category of tree heuristics [6]. However, due to the degree constraint, mapping the minimum spanning tree (MST) back into the original graph cannot be done in an efficient manner. We reserve this kind of heuristics for further study.

1. Start the tree with a randomly chosen node (terminal or non-terminal)
2. While there is still terminal not on the tree
   - 2.1 If there is no free layer available, the heuristic fails
   - 2.2 Add a new layer
   - 2.3 Initialize the network topology
   - 2.4 While there are shortest paths that connect a terminal and an on-tree node
     - 2.4.1 Choose one shortest path by some rule and add it to the tree
     - 2.4.2 Adjust the network topology, concerning LS and WC of on-path nodes
   - 2.5 If the set of off-tree terminals is identical to that when iteration begins, the heuristic fails
3. Output the multicast tree

Figure 2 Quasi-Prim Algorithm Flow

In Step 2.3, the network topology of current layer can be initialized, according to the usage of nodes and edges in this layer. In Step 2.4.1, different *rule* makes up different heuristic. For instance, to naïve heuristic, the rule is to randomly pick up a path, while SPH heuristic will pick the shortest path. In Step 2.4.2, the adjustment is carried out in two parts. First, the degrees of all on-path nodes are checked. For any node that reaches the upper bound, all remaining (not on-tree yet) edges incident to the node will be removed from the network. Next, one terminal (already on-tree before being connected) of the path is checked to see whether the use of wavelength converter is needed. If the terminal is connected to the tree, but in the previous layer, then the wavelength converter is necessary. In Step 2.5, if no new terminal can be added into the tree during the iteration, it means that there are no nodes available for tree expansion. Therefore, the heuristic fails. Wavelength assignment is straightforward by directly mapping layer to wavelength. Of course, a separate wavelength assignment algorithm is optional, but not recommended.

The quasi-Kruskal algorithm is similar to the quasi-Prim algorithm. However, it starts from a multicast forest, which contains all the isolated terminals. At first, every terminal forms a component. Every time, a path (not an edge as in the Kruskal's algorithm), which connects two components and meats certain criteria (e.g. not bring in a loop in the forest, pass through designated node, etc.) is picked and added to the forest. The algorithm ends when the forest becomes a tree, or all the components are connected as one.

Several algorithms have been developed in either group. In quasi-Prim group, there are naïve, Shortest Path Heuristic (SPH) and its variants, which are SPH-dV, SPH-dNN, and SPH-dZ, in particular. In the quasi-Kruskal group, there are

K-SPH, Average Distance Heuristic (ADH), and P-ADH. Their original forms (except P-ADH) can be found in [6] and [7]. The worst-case error ratio of original SPH and its variants, ADH, and K-SPH is tightly bounded to 2 [6]. In other words, the cost of tree is less than twice of the optimized tree.

P-ADH is our own algorithm. It develops from ADH, but uses a different function as below:

$$f(v) = \rho L(v) + \sigma W(v)$$

L(v) equals d(v) – deg(v), and stands for the number of new edges that v can connect. W(v) equals the number of usable converters of node v. ρ and σ reflect the importance of LS and WC. Both parameters have to be finely tuned to get best performance.

3. SIMULATION RESULTS

3.1. Simulation Methodology

Currently, the network topology is randomly created following the model described in [8], which is widely adopted for its simplicity and efficiency. The model is as this: given an area of $L \times L$; randomly pick N points inside this area; and the probability of having a link between any two nodes u & v is $\beta \exp\frac{-d(u,v)}{L\alpha}$, where $d(u,v)$ is the distance.

Table 1 Simulation parameter values

N	100
L	160
α	{0.2, 0.3, 0.4, 0.5}
β	{0.1, 0.2, 0.3, 0.4}
numWave	16
tDense	{10%, 20%, 30%}
lDense	{10%, 20%, 30%}
wDense	{10%, 20%, 30%}
numLS	16
LScap	{4, 8, 16}
numWC	16

This method may generate disconnected graphs, especially when α and β are both small. Hence, in our implementation, we added one part to check the connectivity of graph and add least number of edges to make it connected.

Besides network topology, to simulate multicast over WDM network, additional parameters are needed. At the network level, those parameters are the number of wavelengths (numWave), and the density of terminals (tDense), LS-capable nodes (lDense) and WC-capable nodes (wDense).

At the node level, if a node is LS-capable, then the number of splitters (numLS) and the capability of each splitter (LScap, i.e. 1:LScap) are necessary. All over the network, only one kind of splitters is deployed. On the contrary, we only consider 1:1 converter. So for a WC-capable node, only the number of converters (numWC) is necessary. By the way, it can be learnt from the description of heuristics that splitters and converters are not dedicated to one wavelength each, instead they are put in a pool and can be deployed in whichever wavelength. Of course, there will be no confliction of allocation of splitter or converter, because each splitter or converter is bound to different ports.

In simulation, the values of parameters are combinations of those listed in Table 1. In [3] and [4], the authors use α and β as two parameters to describe the network. Instead, we will use the *order* (|V|) and *size* (|E|) to describe the network [5]. This is more explicit and meaningful. There is a fixed relationship between the size of graph and sum of degrees of all nodes, i.e. $\sum_{v \in V} \deg(v) = 2\,|\,E\,|$. Therefore, we use the average degree (avgDeg) of all nodes to reflect the network size implicitly.

3.2. Numerical Results

The metric we measured is the Quality of Solution (QoS). We cannot get the optimum results, because the network order is prohibitive to exact algorithms. For appropriate comparison, we run an unconstrained naïve heuristic in every network instance, and compare all the other heuristics with it. The ratio generally reflects the QoS of each heuristic. If T_U and T_H denote the trees generated by the unconstrained naïve and one heuristic, respectively, then the QoS is defined as:

$$\sum_{e \in T_U} c(e) \Big/ \sum_{e \in T_H} c(e)$$

Note that unconstrained naïve heuristic usually generates the best result, so the ratio is usually less than one.

In the system, there are totally five factors: AvgDeg, tDense, lDense, wDense, and LScap. An interesting finding is that if we vary one factor and relatively fix other factors, then the result will show the exactly the same trend, no matter what the values of fixed factors may be. Therefore, when considering the impact of one factor on one system parameter (e.g. QoS), we will only show the factor and the system parameter. Also, wDense is trivial in single-session case, because only a few instances need wavelength converter.

As a whole, SPH and its variants, K-SPH, and ADH produce the most optimized multicast tree. This is expected because the error ratios of the original forms are all bound to 2. The increase of tDense slightly increases QoS (Figure 3). Also, the increase of LScap almost has no effect on QoS (Figure 4). This finding is surprising, which means that it is vain to design more powerful splitter to improve QoS. However, increase lDense will significantly improve QoS (Figure 5). This is expected, because bigger lDense results in more elements in set B, which in turn increases the chance to find a shorter path to connect terminals. Moreover, increase the network connectivity also contributes to better QoS (Figure 6), mostly because dense network will generate shorter paths.

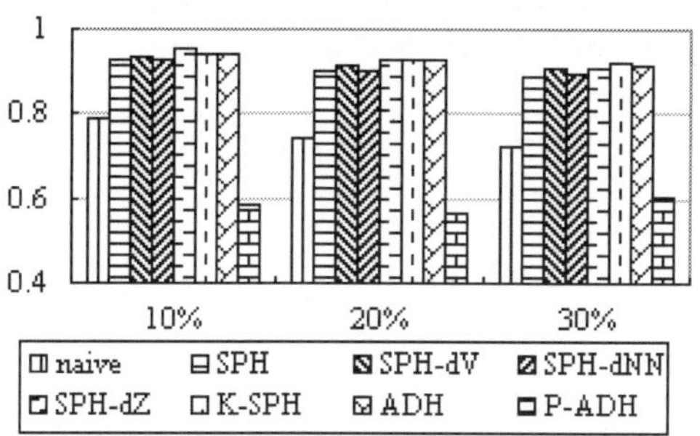

Figure 3 QoS vs. terminal density

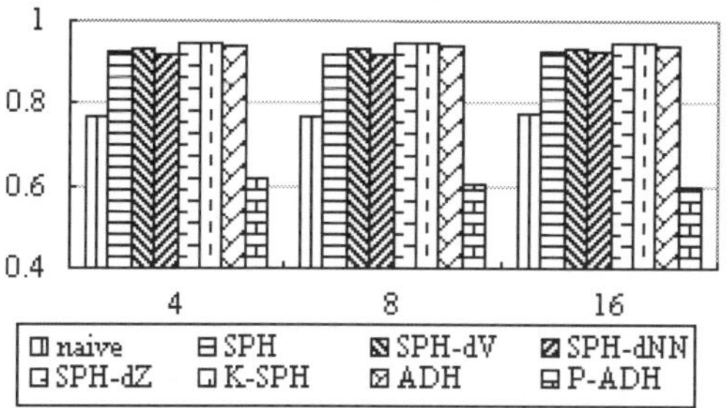

Figure 4 QoS vs. light splitter capability

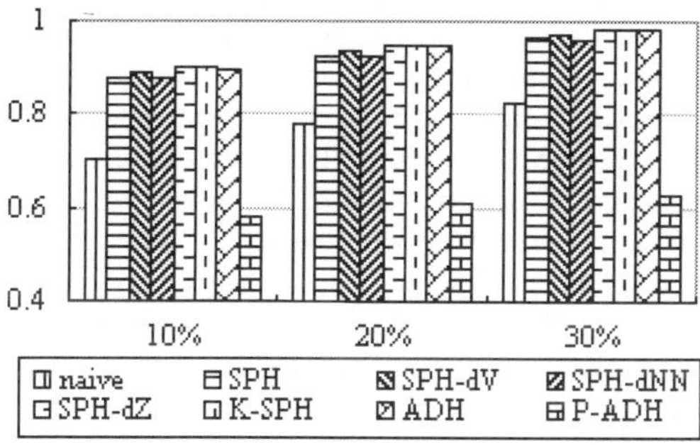

Figure 5 QoS vs. LS density

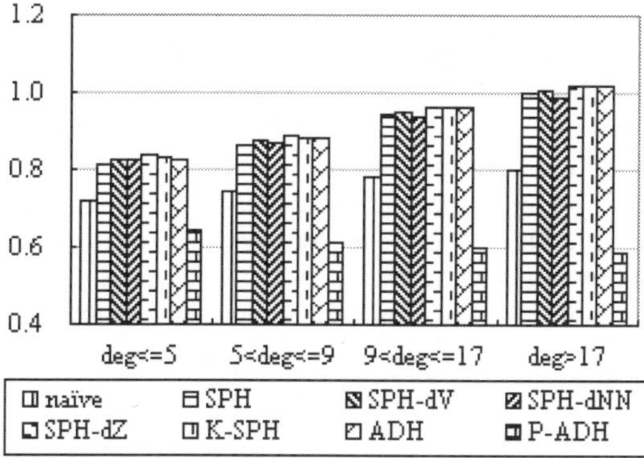

Figure 6 QoS vs. average degree

Naïve and P-ADH, especially the latter, do not perfectly match the conclusion. In conclusion, tDense is not a controllable parameter. However it has negligible impact on network cost. Splitter capability cannot save network cost, so that powerful splitters are not necessary, which helps save network construction cost, and also avoid power attenuation and single point of failure. But splitter density and network connectivity strongly affect the cost. To save network cost, it is advisable to deploy more normal splitters and increase network connectivity.

4. CONCLUSION

In this paper, we have developed a new multi-layer model to represent the WDM network, and applied the DCSP technique to the model to solve the problem. Any 2Basic heuristic can be modified to fit in the multi-layer model, and then used to solve the MCRWA problem.

We have tested all the heuristics in single-session environment. It appears that minimizing the numbers of splitters and converters used will lead to accommodate more multicast sessions. It also provides guidance on the use of LS and WC to support optical multicast. The case of multiple sessions is under study.

In the simulation, we do not consider the cost of light splitting and wavelength conversion. These two costs can be included in the node weight. This is reserved for further study.

REFERENCES

[1] A. Ding, and G.-S. Poo, *A Survey of Optical Multicast over WDM Networks*, to appear in Computer Communications
[2] S. Voss, *Steiner's problem in graphs: heuristic methods*, Discrete Applied Mathematics, vol. 40, pp. 45-72, 1992
[3] F. Bauer and A. Varma, *Degree-Constrained Multicasting in Point-to-Point Networks*, in proc. INFOCOM'95, vol. 1, pp. 369-376, 1995
[4] S.-J. Chung, S.-P. Hong, et al., *Algorithms for the Degree-Constrained Multicast Trees in Packet-Switched Networks*, in Proc. GLOBECOM'98, vol. 2, pp. 1054-1059, 1998
[5] J. A. McHugh, *Algorithmic Graph Theory*, Prentice-Hall, 1990
[6] F. K. Hwang, D. S. Richards, and P. Winter, *The Steiner Tree Problem*, Elsevier Science, 1992
[7] S. Voss, *Problems with Generalized Steiner Problems*, Algorithmica, vol. 7, pp. 333-335, 1992
[8] B. M. Waxman, *Routing of Multipoint Connections*, IEEE JSAC, vol. 6, no. 9, Dec. 1988

Incremental restoration scheme in Optical Networks

Hyeon Park *, Sun-Hee Yang*, Sang-Ha Kim**
*Next Generation Protocol Team, ETRI, Yusong P.O.Box 106, Taejon, 305-600, KOREA
**Chungnam National University, Yusong-Gu, Taejon, 305-600, KOREA
*email:{hpark, shyang}@ etri.re.kr, fax:+82-42-860-5440

Abstract

To provide the high capacity transport network facilities, with the development of the WDM technology the optical networks containing multiple OXCs are deployed in carrier networks. Furthermore, a consensus is emerging in the industry on utilizing an IP-centric control plane within optical networks to support dynamic provisioning and restoration of optical layer connection [1]. So, in this paper, we propose the fast incremental line restoration method: how to discover fast the secondary path through the incremental examination of the available resource. The proposed scheme is able to use the resource efficiently. And it is more appropriate for the strict route specified all the exact nodes explicitly rather than the loose or abstract route since the traffic information of the previous path can be mostly kept.

1. Introduction

The rapid growth of the data traffic and its diverse service demands in the Internet are induced the huge bandwidth and the need for transportation. Recently, to provide the high capacity transport network facilities, with the development of the WDM (Wavelength Division Multiplexing) technology the optical networks containing multiple OXCs (Optical Cross Connector) are deployed in carrier networks. WDM technology harnesses the enormous bandwidth of the fiber by multiplexing hundreds of optical wavelengths, each of which operates at speeds of several gigabits per second [1].

The OXC containing the WDM technology switches a wavelength to an output port with the same wavelength (for the wavelength continuity via the lightpath) or with another wavelength (just for the wavelength conversion), or to an electronic controller port for the O-E-O (Optical-Electronic-Optical) conversion.

Thus, IP layer controller (or router) requiring the O-E-O conversion is increasingly connected directly to the optical layer (i.e. WDM layer) controller within an OXC. Furthermore, a consensus is emerging in the industry on utilizing an IP-centric control plane within optical networks to support dynamic provisioning and restoration of optical layer connection [1]. However, although the two-level architecture (IP over WDM) is simple, its critical factor is the survivability. For example, in optical networks a single cable cut is surprisingly frequent and serious. In addition, a single fiber failure can lead to the failure of all lightpaths traversing the failed fiber, and loss of each lightpath, operating at up to 10Gb/s, results in significant loss of traffic [2]. Meanwhile, the multilayer architecture including the SONET/SDH layer has the capability of the fast recovery time, known as 50 ms.

In general, the survivability can be accomplished by the two types of schemes : protection and restoration from the failure. In the protection scheme, the secondary path to replace the failed primary path is predetermined and put onto service or not. On the other hand, the secondary path in the restoration scheme is dynamically recomputed using the available resources of the networks when the failure is detected.

The recent advances in GMPLS (Generalized Multiprotocol Label Switching) have provided enhanced survivability capabilities (e.g., performance monitoring, protection and restoration), supported traffic engineering functions at both the IP and WDM layers, and made it possible to achieve end-to-end IP over WDM protection [2]. Especially, using the GMPLS capabilities, extended MPLS (Multiprotocol Label Switching) routing and signaling TLV (Type Length Value) for the protection/restoration, the protection scheme is easier than that in the existing MPLS.

In general, the recovery time of the protection scheme is more rapid than the restoration scheme. However, restoration scheme has better resource utilization than the protection scheme. Recently, as the restoration scheme, the fast recovery methods including fast failure detection and re-computation of the secondary path have been proposed and researched [3][4].

So, in this paper, we propose the fast line restoration method: how to discover fast the secondary path through the incremental examination of the available resource. That is, when the failure is occurred, first the restoration is accomplished by use of the local resources at its immediate nodes. If it is not succeeded the secondary path is then computed at the immediate node. Most restoration methods include fiber-level, link-level, or channel-level restoration. This proposed scheme considers not only these levels but also node-level restoration and is thoroughly based on the GMPLS.

This paper is organized as follows. The optical network and the system architecture to process the proposed scheme is provided in Section II. And the proposed incremental restoration scheme is also provided in Section III. The conclusion and further study are described in Section IV.

2. Optical network and its System architecture

This paper considers a mesh-based optical network, which includes LERs (Label Edge Routers) as the clients attached to its own optical core nodes. Each optical node consists of optical cross connectors OXCs and IP-based controller to process the GMPLS-based control plane. The control plane provides IP-centric routing, signaling, and survivability functionality. GMPLS has been proposed to extend MPLS to support multiple switching capabilities such as packet switching capable (PSC), time-division multiplex (TDM) capable, lambda switch capable (LSC), fiber switch capable (FSC), and so on.

The proposed restoration due to its GMPLS based control scheme is suited to an opaque network with the O-E-O conversion as well as a transparency network where the wavelength assignment continuity constraint is required.

Fig. 1 shows the GMPLS based optical network consisted of LER, which classifies the IP traffic into the related LSP (Label Switched Path) (i.e. lightpath : a sequence of wavelength channels), and LSR (Label Switched Router).

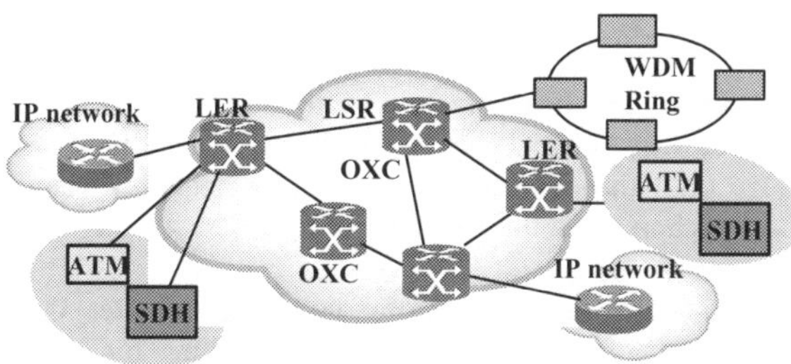

<Figure 1> The Optical Network

In the LSR as a core node, the data on a wavelength channel of an input port is switched across an optical switch fabric to another wavelength channel of an output port; the switched wavelength can be the same or different one. In other words, this is surely not only by electronic conversion but also by passing via different wavelengths.

When the failure (node, link, or wavelength channel failure) is occurred, the restoration scheme to recover its failure is needed. There are two types of restoration schemes according to the failure range to restore : link-based and path-based. In the path-based restoration case, the entire LSP is rerouted at the source LSR of the primary path. In the link-based case, however, only individual LSP hops are rerouted, and it can be considered a degenerate (single-hop) case of path-based recovery [7].

The proposed overall OXC system architecture and its control plane architecture is shown in Fig. 2.

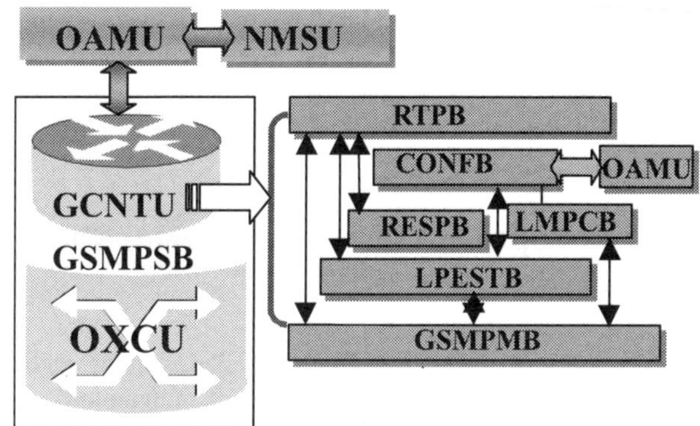

<Figure 2> The OXC System architecture

The OXC system consists of GMPLS control unit GCNTU, operation and maintenance unit OAMU, network management system unit NMSU, GSMP slave block GSMPSB, and OXC unit OXCU.

GCNTU provides the network topology information via the routing protocol block RTPB, the lightpath setup/release by lightpath establishment block LPESTB, the system configuration data via the configuration block CONFB, the link management protocol LMP (Link Management Protocol) control by the LMP control block LMPCB, and the restoration processing by the proposed restoration scheme via the restoration processing block RESPB. OAMU is used for the system operation and maintenance and NMSB is used for the network management system to control the OXC system in the distance.

In GMPLS, similar to MPLS, the input-output label associations for a path can be represented using label information base LIB in RTPB.

There are also two types of information: its own system information by LMP of GMPLS and the topology information resulted by the routing protocol processing. That is, the interior gateway protocol IGP routing information including the whole network topology information is stored in the link state information table and the physical link/fiber information between two neighbor nodes by LMP is stored in the system configuration information table. As an example, the data items by extended IGP information [5] for the optical network are summarized as follows in Table. 1.

Capacity unit			Identification
Fiber	link	Wavelength	Connection channel Id
Fiber	Data-bearing link		Interface Id
Fiber	Data-bearing link	TE link	TE link Id, Port Id
Fiber	Data-bearing link	Time Slot	TE link Id, Compon. Id

<Table 1 > The data item of extended IGP information

This information is used to get the idle channel/link to replace the failed channel/link. To select the required path for the restoration two types of information must be referred. The one is the topology information table stored by IGP (e.g., OSPF or IS-IS) and the other is the system configuration information between two neighbor nodes used by LMP in RTPB and CONFB, respectively, as described above.

3. The proposed incremental restoration scheme

There are four stages to restore the failed working path: fault detection, localization, notification, and mitigation.

The fault detection and the fault localization are processed by LMP in GMPLS. In either case, control channels (wavelengths) are pre-established. In other words, these are electronically terminated at each hop, while the data channel is transmitted along the controlled path.

In general, most proposed schemes provided fault detection and fault localization of the control channels by signaling messages (for example, Notify Message) of signaling protocol, CR_LDP (constraint-based routed Label Distribution Protocol), RSVP-TE (Resource Reservation Protocol- Traffic Engineering). In fact, it is important to note that the failure of the data channel is more essential than that of the control channel. The proposed restoration scheme is concerned to the mitigation stage for the failed data link/channel.

For the proposed restoration scheme, Fig.3 shows an example of how to restore the failed data link/channel. It illustrates the logical topology NSFNET[6]. Each optical node (shaped by oval) represents OXC system with the GMPLS-based controller of the IP layer

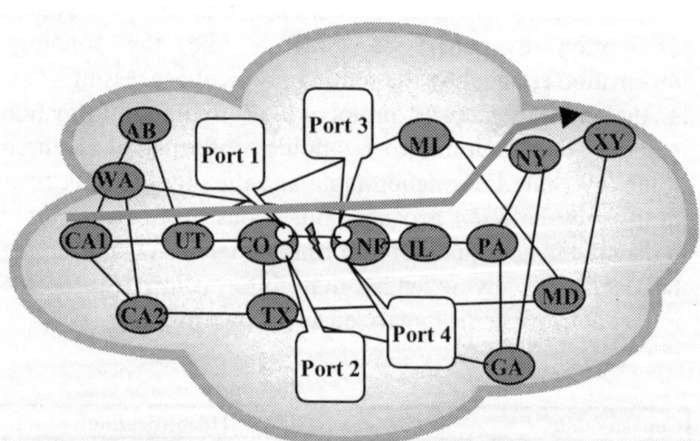

< Figure 3 > An example optical network (NSFNET)

In optical networks, most of the connection may be bi-directed due to the fast connection set-up and, moreover, GMPLS is easier to support this procedure while MPLS networks support just uni-direction. Both end nodes of the failed link are supposed to have the capable of setting up the rerouting path. As described before, the proposed restoration scheme is regard to the link restoration.

And as shown Fig. 3 we suppose that the link from port 1 on node CO to port 3 on node NE has failed. Either CO or NE is able to detect the link failure and initiates the restoration processing.

When a failure is detected, the (upstream/downstream) OXC (in this case, CO) closest to the failure refers first the configuration table (i.e., in CONFB) to get the idle channel/link on the same fiber to switch the traffic. If there is no the available channel/link, the OXC, in turn, checks if the appropriate link/channel exists in the routing table in RTPB. When the appropriate link/channel is founded, the port 1 is replaced to the new port 2 on node CO, and the port 3 is similarly replaced to port 4 on node NE. If the idle channel/link in both the tables not exists, the path to reroute the disrupted traffic between the nodes (around a failure) is computed as follows (i.e., secondary path computation algorithm).

As the explicit path (in this case, CA1->UT->CO->NE->IL->PA->NY->XY) of the primary path is known to every node over its primary path, the secondary path can be computed at the immediate node (CO).

- Shortest Path First SPF for the destination node (i.e. the downstream node (node NE) of the failed link/channel) is computed: the route into the upstream node (UT) must be excepted.
- As shown in Fig.3, if the computed path includes any node of the primary path, the next downstream node (IL) is replaced into the destination for the secondary path.
- The downstream node is replaced into the destination node until the downstream node computed previously has less hop count than the later one.

When the failure is repaired, the node CO initiates the switchback process and triggers a valid signalling for LSP. In this paper, we allow multihop paths for the local restoration of the failed span.

As described above, in case of a failure event on the working path, the basic idea of the proposed incremental restoration scheme is summarized as follows:

- Checks idle resources in CONFB to switchover the connection between end nodes of the failed link/channel
- If not exist, checks again in RTPB
- If not exist, applies the secondary path computing algorithm as described above

The proposed scheme is able to use the resource efficiently. And it is more appropriate for the strict route specified all the exact nodes explicitly rather than the loose or abstract route since the traffic information of the previous path can be mostly kept. Moreover, the switching paradigms, the optical circuit switching and the burst switching establish the wavelength-routed paths before the data can be transmitted even though there is the difference in the point of data transmission starting time. Accordingly, the explicit wavelength route is computed in advance.

In this paper, those switching paradigm can be applied to

the proposed scheme, since when a failure occurs at the intermediate node the rerouting path is computed at the span nodes of the failed link/channel.

Meanwhile, the packet switching processes establishing of the wavelength-routed path and switching of the packets synchronously; hence, the explicit route is not required. In addition, due to the technical hurdles such as the optical random access memory RAM, which is not become available in the near future, we not considered the packet switching. As a result, the proposed scheme is appropriate to the IP over WDM switching and the optical burst switching.

4. Conclusions

Although the two-level architecture (IP over WDM) is simple, its critical factor is the survivability. For example, in optical networks a single cable cut is surprisingly frequent and serious. For that purpose, we proposed the fast line restoration method: how to discover fast the secondary path through the incremental examination of the available resource.

The recovery time of the proposed scheme depends on very much on the failure detection time, the configuration/routing table searching time, rerouting computation time, connection set-up time, and the time of switching over the traffic. For the further study, we will simulate the performance of our scheme with the factors described above.

Reference

[1] Sopher De Maesschalck *et al.*, " From Network Design to Dynamic Provisioning and Restoration in Optical Cross-Connect Mesh Networks: An Architectural and Algorithm Overview," IEEE Commun. Mag., vol.39, no. 1, Jan. 2001, pp.42-49.

[2] Y. Ye, *el al.*, "A Simple Dynamic Integrated Provisiong/Protection Scheme in IP over WDM Networks," *IEEE Commun. Mag.*, Nov. 2001, pp. 174-182.

[3] Guangzhi Li, *et al.*, "Experiment in Fast Restoration using GMPLS in Optical/Electronic Mesh Networks," OFC 2001, paper PD34-1, March 2001.

[4] D. Awduche *et al.*, "Multiprotocol Lambda Switching: Combining MPLS traffic engineering control with optical crossconnects," *IEEE Commun. Mag.*, March 2001.

[5] K. Kompella *et al.*, "OSPF extensions in Support of Generalized MPLS," Internet draft, work in progress, January 2002.

[6] B. Mukherjee, Optical Communication Networks, McGraw-Hill, pp.260-265, 1997.

[7] Chunming Qiao, "Labeled Optical Burst switching for IP-over-WDM Integration," *IEEE Commun. Mag.*, Sept. 2001, pp. 104-114.

[8] C. Xin, *el al.*, "On an IP-Centric Optical Control Plane," *IEEE Commun. Mag.*, Sep. 2001, pp. 88-93.

Session 14B

Optical Signal Processing

All-optical signal processing based on nonlinear polarization rotation in a semiconductor optical amplifier

H.J.S. Dorren, Y. Liu, M.T. Hill, E. Tangdiongga, N. Calabretta, H. de Waardt,
X. Yang, R. Geldenhuys, D. Lenstra and G.D. Khoe

(Invited Paper)

Abstract—**We discuss an all-optical flip-flop memory and a wavelength converter based on nonlinear polarization rotation in a semiconductor optical amplifier. The flip-flop is realized from two coupled polarization switches. The concept is explained and experimental results are presented that demonstrate that a contrast ratio of over 20 dB and a switching power of less than –3 dBm can be obtained. The all-optical wavelength converter employs a single polarization switch. We show that error-free inverted and non-inverted wavelength conversion can be realized at a data-rate of 10 Gbit/s. Both optical signal-processing functionalities can be utilized in all-optical packet switches.**

I. INTRODUCTION

The need to produce a transparent all-optical packet switch has been well documented [1]. Key components for all-optical packet switches are all-optical flip-flop memories and wavelength converters [2,3]. In this paper, we describe an all-optical flip-flop memory that is realized out of two coupled nonlinear optical polarization switches. This all-optical flip-flop implementation has a simple structure, separate all-optical set and reset inputs and large input wavelength range. Moreover, it turns out that this all-optical flip-flop has similar properties as the one that is based on two coupled Mach-Zehnder interferometers [4]. This implementation however is easy to realize by using commercially available pigtailed components and shows stable operation without photonic integration. We demonstrate the feasibility of the concept and we show that the contrast ratio between output states of the flip-flop is over 20 dB while the switching power is less than –3 dBm.

Another essential building block for WDM networks and optical packet switches is a wavelength converter [5-9]. We discuss wavelength conversion based on nonlinear polarization rotation in a single SOA [6,8]. We demonstrate that error-free inverted and non-inverted wavelength conversion can be realized at 10 Gbit/s, using a single SOA. No error floor is observed. Furthermore, for non-inverted wavelength conversion an extinction ratio of 12.3 dB is obtained at a bit-rate of 10 Gbit/s.

This work was supported by the Netherlands Organization for Scientific Research (NWO) through the "NRC photonics" grant.

The authors are with the COBRA Research Institute, Eindhoven University of Technology, P. O. Box 513, 5600 MB Eindhoven, The Netherlands, (Telephone: +31 40 247 4362, E-mail: H.J.S.Dorren@tue.nl).

D. Lenstra is also with Vrije Universiteit, FEW, Division of Physics and Astronomy, De Boelelaan 1081, 1081 HV, Amsterdam, The Netherlands.

H.J.S. Dorren is currently on leave at the National Institute of Advanced Industrial Science and Technology (AIST), AIST Tsukuba Central 2-10, Tsukuba 305-8568 Japan

II. ALL-OPTICAL FLIP-FLOP MEMORY

The all-optical flip-flop concept is depicted in Figure 1. It consists of two coupled polarization switches (PSWs). The PSW, that acts as a logic AND gate [10], is made from a laser source, a SOA, two polarization controllers and a polarization beam splitter (PBS). A laser emits a continuous wave (CW) probe beam at wavelength λ_1 that is fed into a SOA. The SOA output is sent into a PBS. The system contains two polarization controllers. The first polarization controller is used to adjust polarization of the input signal with the orientation of the SOA layers, while the second polarization controller is used to adjust the polarization of the amplified SOA output with the orientation of the PBS. The SOA can be saturated by injection of a high intensity pump (control) signal. The solid curve in Figure 2 shows the typical (experimental) PBS output as a function of the intensity of the saturating control light. It follows from this figure that a control beam of sufficient intensity, can suppress the PSW output. This effect is caused by the additional birefringence that is introduced in the SOA by the control light [11,12], which causes the TE and the TM modes of the probe beam to experience a different refractive index. At the PBS, the two modes combine coherently. If the phase difference between the two modes is an odd multiple of π, the PSW output is suppressed. Note that the curves of Figure 2 are similar to the curves presented in [4] in which the suppressed output of an active Mach-Zehnder interferometer is discussed.

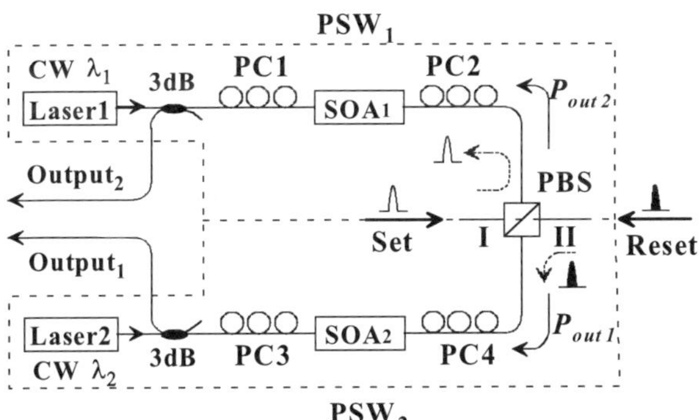

Figure 1: The configuration of the all-optical flip-flop based on two polarization switches. CW: continuous wave, PSW: polarization switch, PC: polarization controller, SOA: semiconductor optical amplifier, PBS: polarization beam splitter, $P_{out\,1}$ is the output of PSW_1 and $P_{out\,2}$ is the output of PSW_2.

Similarly as in [4], an optical flip-flop can be realized by coupling two identical PSWs as shown in Figure 1. The first PSW, hereafter to be called PSW_1, outputs light that is injected into the second PSW (to be called PSW_2). Hence, the light that outputs PSW_1 acts as a saturating control signal that can suppress PSW_2 and the light that outputs PSW_2 can act as a saturating control signal to suppress PSW_1. The solid curve in Figure 2 represents the intensity $P_{out\ 1}$ of the light that outputs PSW_1 as a function of the intensity $P_{out\ 2}$ of the light that outputs PSW_2. The system is set in such a way that the maximum output intensity $P_{out\ 1}$ equals the intensity of the control light $P_{out\ 2}$ that is required to suppress PSW_1. Since the PSWs are identical, the solid curve is complementary to the dashed curve that represents the intensity of the light that outputs PSW_2. At point A, PSW_1 is dominant and PSW_2 is suppressed while at Point B, PSW_1 is suppressed and PSW_2 is dominant. Both A and B can be shown to be stable states of the system, Point S can be shown to be an unstable point [13].

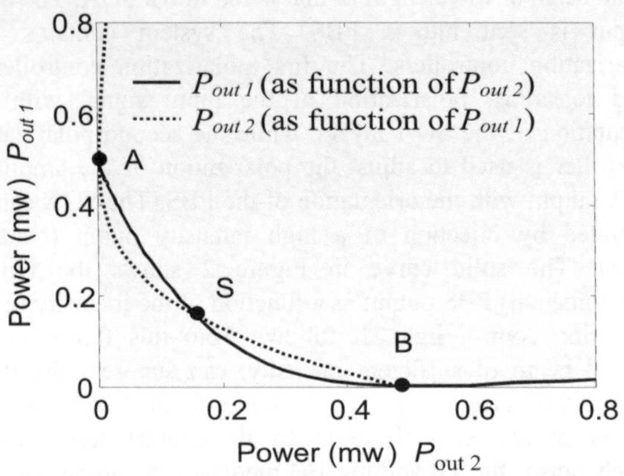

Figure 2: Intensity $P_{out\ 1}$ of the light that outputs PSW_1 as a function of the intensity $P_{out\ 2}$ of the light that outputs PSW_2 (solid curve). Intensity $P_{out\ 2}$ of the light that outputs PSW_2 as a function of the intensity $P_{out\ 1}$ of the light that outputs PSW_1 (dashed curve). It is clear that the two curves are complementary.

The system of two coupled PSWs can function as an optical flip-flop as follows. The state of the flip-flop can be determined by observing the amount of light at the PSW outputs. In State 1, PSW_1 dominates and suppresses PSW_2, while in State 2 PSW_2 dominates and suppresses PSW_1. To switch the flip-flop between the states, light can be injected in the PSW that dominates (that is the one injecting the most light into the other PSW) via the set state or the reset ports. The injected light reduces the light exiting the dominant PSW, which allows the suppressed PSW to increase its light output and become the dominant PSW.

The all-optical flip-flop is implemented as in Figure 1. Laser 1 and Laser 2 emit CW light at wavelength λ_1=1549.32 nm and λ_2=1550.92 nm respectively, however, it is not essential to bias the PSWs with light at different wavelengths. The output power is –3.34 dBm for Laser 1 and –3.05 dBm for Laser 2. The SOAs were manufactured by JDS Uniphase and employ a strained bulk active region. SOA_1 is biased with 163.97 mA of current and SOA_2 is biased with 161.86 mA of current. The PBS has four ports and an extinction ratio of 30 dB. PSW_1 and PSW_2 are coupled to each other via the PBS.

The steady state PSW output intensity versus the intensity of the input light is presented in Figure 2 as discussed in the previous section. The two states of the flip-flop are shown as point A and point B. The dynamic operation of the flip-flop is demonstrated by toggling the state of the flip-flop by injecting a regular sequence of optical pulses into the PSW that is currently the master. The pulses had a wavelength of 1552.52 nm and a duration of 150 ns. The pulses were injected in the master once every 1.85 μs through the set and reset port (see Figure 1). Figure 3 shows the oscilloscope traces of the optical output power of the flip-flop for each state. The optical peak power for the set and reset pulses were –3.91 dBm and –4.35 dBm respectively. In Figure 3, regular toggling between the flip-flop output states every 1.85 μs is visible. Furthermore, it can be observed that the flip-flop state is stable in the time between changing states. Also the contrast between State 1 and State 2 was investigated by using an optical spectrum analyzer. It turned out that the contrast ratio between the two states of the flip-flop is over 20 dB.

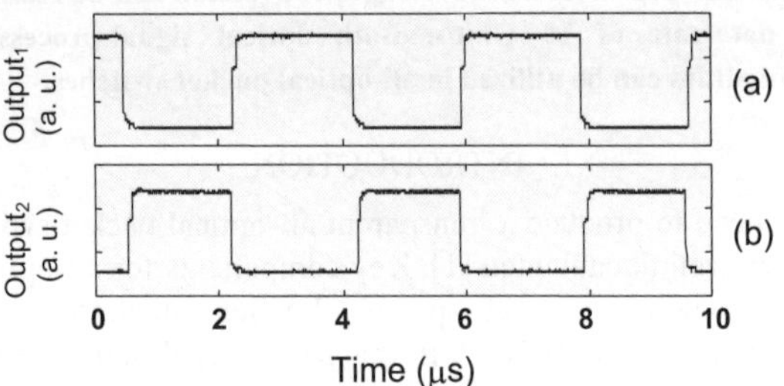

Figure 3: Regular toggling between the two flip-flop states by injecting set or reset pulses every 1.85 μs.

III. WAVELENTH CONVERSION

The all-optical wavelength converter is depicted in the dashed box of Figure 4. A similar polarization switch as used in the flip-flop memory, is employed for wavelength conversion. The laser emits a CW probe beam at λ_{probe}. The first polarization controller (PC1) is used to adjust the initial polarization of the probe beam with the SOA layers so that an optimum birefringence in the SOA is obtained [11,12]. The SOA output is then fed into a polarization beam splitter (PBS) that acts as a polarization discriminator. The second polarization controller (PC2) is used to adjust the polarization of the amplified SOA output with the orientation of the PBS so that initially the probe beam cannot pass through the PBS. If a saturating pump beam at λ_{pump} is coupled into the SOA via an optical circulator, additional birefringence in the SOA will be introduced [11,12]. This leads to a phase difference between the TE and TM modes of the probe signal, causing the polarization angle of the probe light to be rotated. As a consequence of this, some probe light can pass through the PBS. Hence, a variation of the intensity of the pump light results in a proportional variation in the intensity of the probe light that outputs the PBS, which indicates that non-inverted wavelength conversion is obtained. Conversely, PC2 can be adjusted so that initially a maximum amount of probe light can pass through the PBS. Thus, if a saturating pump beam is applied, the phase change between the TE and TM modes leads to a lowered intensity at the PBS output. Variations in the pump intensity lead to complementary variations in the probe intensity at the PBS output, which indicates that the inverted wavelength conversion is realized.

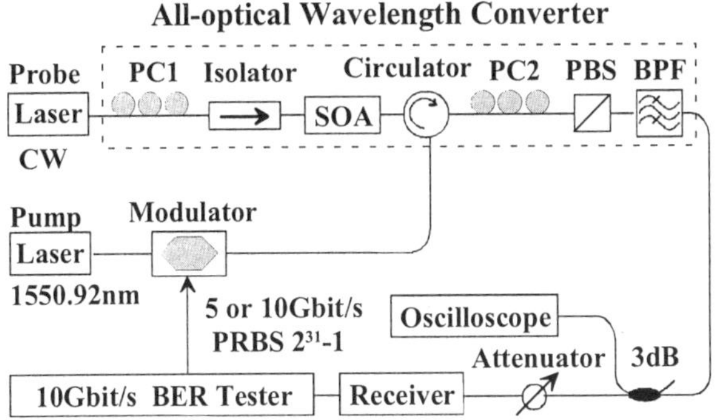

Figure 4: The configuration of the all-optical wavelength converter based on nonlinear polarization rotation. PC: polarization controller, SOA: semiconductor optical amplifier, PBS: polarization beam splitter, BPF: optical band pass filter.

All-optical wavelength conversion based on the concept described above is demonstrated using the setup that is shown in Figure 4. The CW pump and probe beams are at 1550.92 nm and 1552.52 nm respectively. The probe light is sent into the SOA via PC1 and an optical isolator. The SOA (JDS Uniphase) has an active length of 800 μm and employs a strained bulk active region. The pump light is firstly modulated by a 10-Gbit/s external modulator, and then coupled into the SOA via an optical circulator. The bit-rate was 9.95328 Gbit/s (2^{31}–1 NRZ PRBS). The PBS has an extinction ratio of 30 dB. A band pass filter (1 nm) is used to sufficiently suppress the spontaneous noise of the SOA.

In the first experiment we demonstrate the static operation of the proposed wavelength converter. In Figure 5, the (measured) power of the probe light that outputs the PBS is plotted versus the power of the pump light in the case of inverted as well as non- inverted wavelength conversion. The intensity of the probe light that enters the SOA is 2 dBm. The SOA is pumped with 237 mA of current. It follows from Figure 2 that the extinction ratio of the converted signal is more than 20 dB.

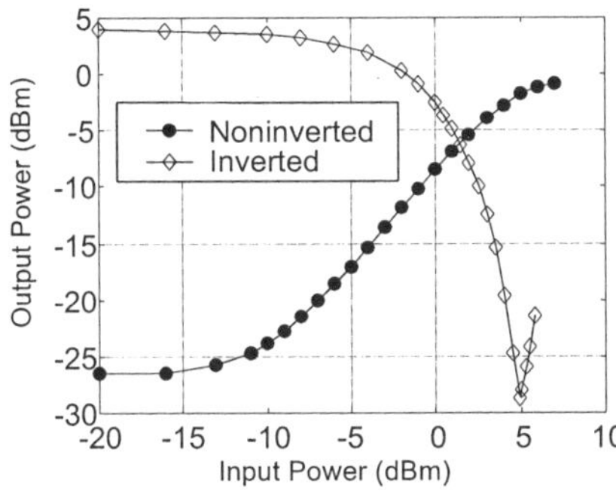

Figure 5: Measured static output power of probe light (1552.52 nm) versus input power of the pump light (1550.92 nm)

In the second experiment we demonstrate wavelength conversion at 9.95328 Gbit/s. Figure 6a shows the bit-error-rate (BER) curves for the converted signal (1552.52nm) for both inverted and non-inverted operation, together with a reference back-to-back measurement (of the modulated pump signal). It is observed that non-inverted conversion results in a 3-dB penalty at a BER of 10^{-9}. Inverted conversion leads to an additional 3-dB penalty compared to the non-inverted case. We note that no BER floor is observed up to a BER of 10^{-12}, indicating an excellent performance of the proposed scheme. The eye diagram is presented in Figure 6b. The non-inverted wavelength conversion displays an extinction ratio of 12.3 dB and a Q-factor of 11.1. The eye pattern of inverted wavelength conversion is presented in Figure 6c, and leads to an extinction ratio of 4.7 dB and a Q-factor of 7.3. Figure 6d shows the oscilloscope traces of the converted signals (non-inverted and inverted). We have observed a 0.3 dB penalty for non-inverted wavelength conversion at a bit-rate of 5 Git/s. The induced penalties at 10 Gbit/s are currently under investigation. The extinction ratios, Q-factors and eye-diagrams were measured on a HP 83480A Digital Communication Analyzer with a 30GHz O/E converter (83482A).

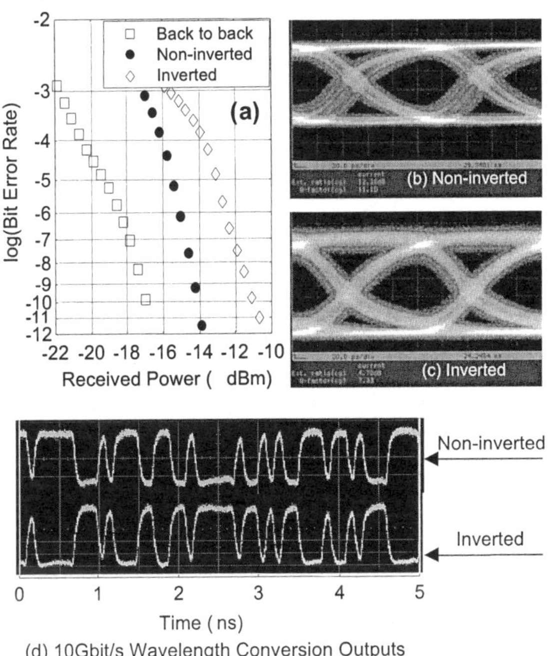

Figure 6: Performance of wavelength conversion at 10 Gbit/s

IV. CONCLUSIONS

An all-optical flip-flop memory based on two coupled polarization switches has been demonstrated. The contrast ratio between output states of the flip-flop is over 20 dB and the optical switching power is less than –3 dBm.

The speed of this flip-flop is determined by the speed of the PSW and the propagation distance between two SOAs. In the experimental setup, approximately 12 meters of fiber are used between the two SOAs, which implies that about at least 100 ns are required for the states of the flip-flop to change. However, integrated versions of the flip-flop could reduce the distance between the two SOAs to several millimeters. In this case, the speed of the flip-flop is dominated by the speed of the PSW. It has been demonstrated that the PSW can operate at 10 GHz [14], thus we expect that the flip-flop can reach similar speeds. Finally, we remark that the curves presented in Figure 2 are similar to the ones presented in [4] which reveals that a PSW acts as a Mach-Zehnder interferometer where the role of the different light paths is now realized by independently operating TE and TM modes of the optical field.

We have also shown that inverted and non-inverted wavelength conversion can be obtained by using a single SOA. The selection for inverted or non-inverted operation is achieved by using polarization controllers. The static extinction ratio of the converted signal is more than 20 dB. Error-free wavelength conversion at 10 Gbit/s is obtained. No error floors are observed. Moreover, an extinction ratio of 12.3 dB is achieved for the non-inverted case.

REFERENCES

[1] Boyraz, O., Lou, J.W., Ahn, K.H., Liang, Y., Xia, T.J., Kao, Y.H. Islam, N.: 'Demonstration and performance analysis for the off-ramp portion of an all-optical access node', *Journal of Lightwave Technology, 1999,* 17, pp. 998-101

[2] Hill, M.T., Srivatsa, A., Calabretta N., Liu, Y., Waardt, H. de, Khoe, G.D., and Dorren, H.J.S.: '1x2 optical packet switch using all-optical header processing', *Electronics Letters*, 2001, 37, pp. 774-775

[3] Liu Y., Hill M.T., Waardt, H., de, Khoe G.D., Lenstra, D. and Dorren, H.J.S., All-optical flip-flop memory based on two coupled polarization switches, Electronics Letters, Accepted for publication (2002)

[4] Hill, M.T., Waardt, H. de, Khoe G.D., and Dorren, H.J.S.: 'Fast optical flip-flop by use of Mach-Zehnder interferometers', *Microwave Optical Technology Letters*, 2001, 31, pp. 411-415

[5] T. Durhuus, B. Mikkelsen, C. Joergensen, S. L. Danielsen and K. E. Stubkjaer, "All-Optical Wavelength Conversion by Semiconductor Optical Amplifiers," *J. of Lightwave Technol.*, Vol. 14, pp. 942-954, June 1996.

[6] M. F. C. Stephens, M. Asghari, R.V. Penty, and I. H. White, "Demonstration of ultrafast all-optical wavelength conversion utilizing birefringence in semiconductor optical amplifiers", *IEEE Photonics Technology Letters*, Vol. 9, pp. 449-451, April 1997.

[7] D. Nesset, T. Kelly and D. Marcenac, "All-optical wavelength conversion using SOA nonlinearities," *IEEE Communications Magazines*, Vol. 36, pp. 56-61, December 1998.

[8] Liu Y., Hill M.T., Tangdiongga, E., Waardt, H., de, Calabretta, N., Khoe G.D., Lenstra, D. and Dorren, H.J.S., All-optical wavelength converter based on cross-polarization modulation in a single semiconductor optical amplifier, ECOC 2002 (accepted)

[9] H. Soto, D. Erasme, and G. Guekos, "Cross-polarization modulation in semiconductor optical amplifiers", *IEEE Photonics Technology Letters*, Vol. 11, pp. 970-972, August 1999..

[10] Soto, H., Diaz, C.A. Topomondzo, Erasme, D, Schares, L and Guekos, G: 'All-optical AND gate implementation using cross-polarization modulation in a semiconductor optical amplifier', *Photonics Technology Letters, 2002,* 14, pp. 498-500

[11] Manning, R.J., Antonopoulos, A., Roux R. Le, and Kelly, A.E.: 'Experimental measurement of nonlinear polarization rotation in semiconductor optical amplifiers', *Electronics Letters*, 2001, 37, pp. 229-231

[12] Dorren, H.J.S., Lenstra, D., Liu, Y., Hill, M.T., Khoe, G.D. Nonlinear polarisation rotation in semiconductor optical amplifiers: theory and application to all-optical flip-flop memories, submitted to IEEE Journal of Quantum Electronics

[13] Hill, M.T., Waardt, H. de, Khoe G.D., and Dorren, H.J.S.: 'All-optical flip-flop based on coupled laser diodes', *Journal of Quantum Electronics*, 2001, 37, pp. 405-413

[14] Stephens, M. F. C., Kelly, A. E., Nesset D., Wonfor, A., and Penty, R.V.: 'Demonstration of time demultiplexing from 40Gb/s to 10Gb/s via polarization switching in a semiconductor optical amplifier', *in Proceedings CLEO/Europe '98*, 1998, pp. 383-383

An Overhead Processing Model for OCh-level OAM

Yunhee Cho, Seungil Myong, Jyung Chan Lee and Yool Kwon
Optical Communications Department, Network Research Laboratory
Electronics and Telecommunications Research Institute (ETRI)
161 Gajeong-dong Yuseong-gu, Daejon, 305-350, KOREA
☎ +82-42-860-5196, (Fax) 042-860-6104, e-mail: yhcho@etri.re.kr

Abstract:

The widespread use of DWDM technology has presented service providers with a new challenge: how to manage increasing optical channels (OChs) to provide fast, reliable services to their end customers cost effectively. OCh-level operation, administration and maintenance (OAM) function in optical are important to effective management, and digital wrapper can provide these functions.

In this paper, we propose an ITU-T Rec. Draft G.709 and G.798 based overhead processing model for OCh-level OAM.

Keywords: Digital wrapper, Overhead processing, OAM

1. INTRODUCTION

Optical Transport Networks (OTNs) are divided into three layers as Optical Channel (OCh), Optical Multiplex Section (OMS) and Optical Transport Section (OTS) as Figure 1. OTS layer provides optical signal transmission function. OMS layer provides networking function of multi-wavelength optical signals. OCh layer provides an end-to-end networking of OCh for transparent conveying of a client signal format such as SONET/SDH, ATM, GbE/10GbE, etc. OTNs should support OCh-level operation, administration and maintenance (OAM) function to manage OChs effectively. Digital wrapper technology provides these network management functions based on ITU-T Recommendation G.709 and G.798 [1][2][3]. These include an optical layer performance monitoring and a forward error correction (FEC) on a per-wavelength (or OCh) basis, independent of the client signal format.

Adding an overhead to the OCh client signal digitally, it makes sense to use this to support all of the OCh-level OAM requirements. In particular, digitally added overhead makes it almost trivial to solve the major performance monitoring problem of the OTN, namely providing access to a bit error rate (BER) in the client-independence. The digital wrapper can significantly enhance the BER performance of the client signal and can be minimized the opto-electronic conversion requirement by using FEC [4].

In this paper, we propose the overhead processing model for OCh-level OAM that is based on ITU-T G.709 (OCh-level frame structure) and G.798 (functionality of OTN equipment).

2. DIGITAL WRAPPER OVERHEAD

Digital wrapper is composed of overhead part and payload part. The overhead assignment and payload in the OCh transport/data/payload unit signal on each part are defined in Figure2 (a). In this overhead processing model, we use the alignment overhead, the section monitoring (SM) overhead, the path monitoring (PM) overhead, the payload structure identifier overhead, and the FEC overhead as defined in Figure2 (b).

OTU*k* SM overhead is used to support section monitoring between digital wrapper processing points and contains the sub-fields as Table1. ODU*k* PM overhead is used to support end-to-end path monitoring as shown in Table1. OPU*k* overhead is associated with the client signal mapping to the OPU*k* payload.

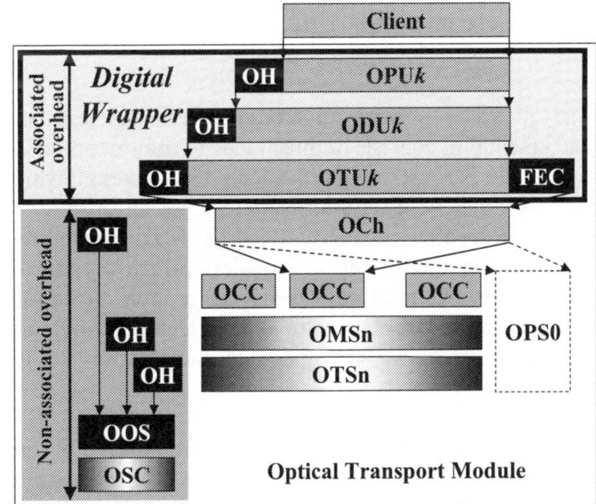

OPU*k* (OCh Payload Unit)
ODU*k* (OCh Data Unit)
OTU*k* (OCh Transport Unit)
OCh (Optical Channel)
OCC (Optical Channel Carrier)
OMS (Optical Multiplex Section)
OTS (Optical Transmission Section)
OOS (Optical Overhead Signal)
OSC (Optical Supervisory Channel)

Figure 1. Structure of OTM(optical transport module).

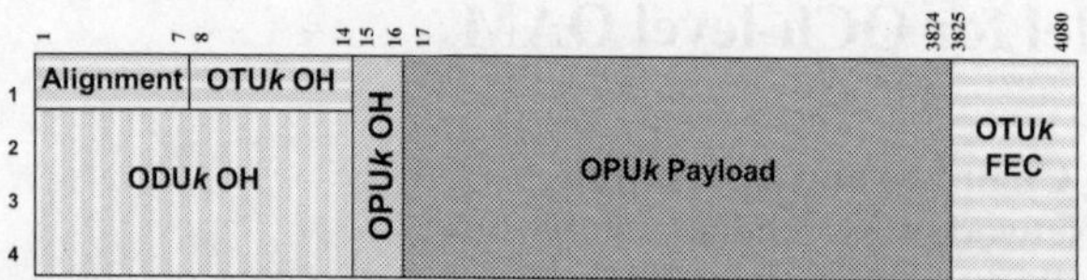

(a) OTU*k* and ODU*k* frame structure.

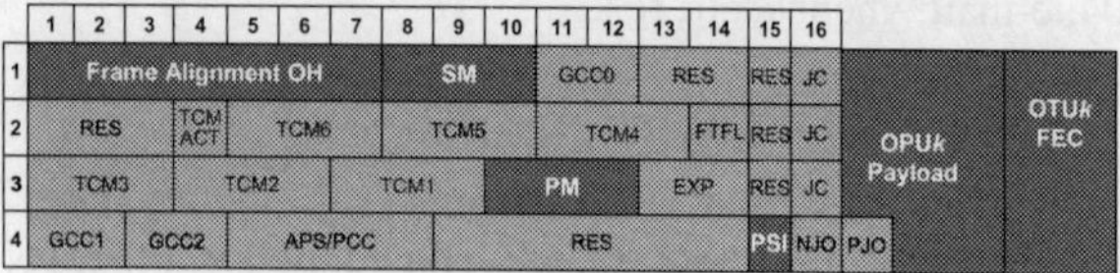

(b) OTU*k*, ODU*k* and OPU*k* overhead used in the overhead processing model.

Figure 2.

Table 1. OTU, ODU and OPU overhead used in the overhead processing model.

OTU OH (SM)	ODU OH (PM)	OPU OH (PSI)
TTI (1 byte) BIP-8 (1 byte) BEI (4 bits) BDI (1 bit)	TTI (1 byte) BIP-8 (1 byte) BEI (4 bits) BDI (1 bit) STAT (3 bits)	PSI (1 byte)

TTI (trial trace identifier), BIP (bit interleaved parity), BEI (backward error indication), BDI (backward defect indication), STAT (status field), PSI (payload structure identifier)

3. OVERHEAD PROCESSING UNIT

Figure 3 shows structure of overhead processing model. This overhead processing unit is composed of source function and sink function on each direction. OAM information in this overhead processing unit is shown in Figure 4.

The interface between overhead processing unit and management unit consists of three kinds of information flow. The first one is source/sink provisioning information, which brings periodically. This information allows an operator to control overhead processing unit and OCh. The second one is performance monitoring information, which reports the results of defect and error monitoring for section and end-to-end path and so on. The third one is Defect Correlation information, which reports probable fault causes to the Management Unit.

The interface between overhead processing unit and digital wrapper chip consists of two kinds of information flow. The first one is register control command from overhead processing unit to digital wrapper chip. This information changes the register value of digital wrapper chip for control and management. The second one is interrupt from digital wrapper chip. Digital wrapper chip receives OTU*k* frame and separates overhead field from payload field. It computes BIP-8 and extracts monitoring overhead as shown Table 1. And it sends notification information, which notifies the overhead extraction, digital wrapper state, update of performance monitoring information and so on. This information activates defect detection functions and performance monitoring functions.

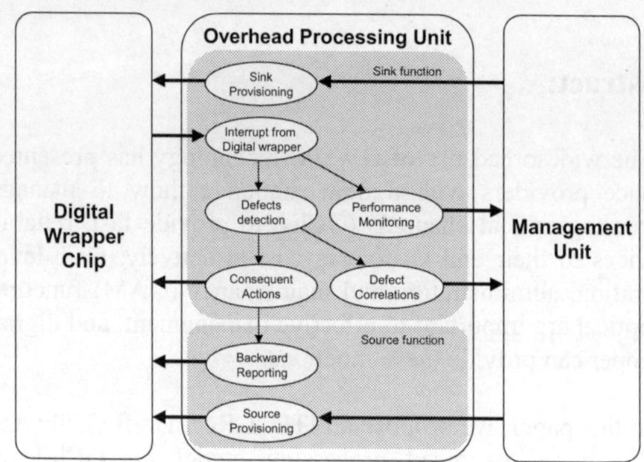

Figure 3. Structure of overhead processing model.

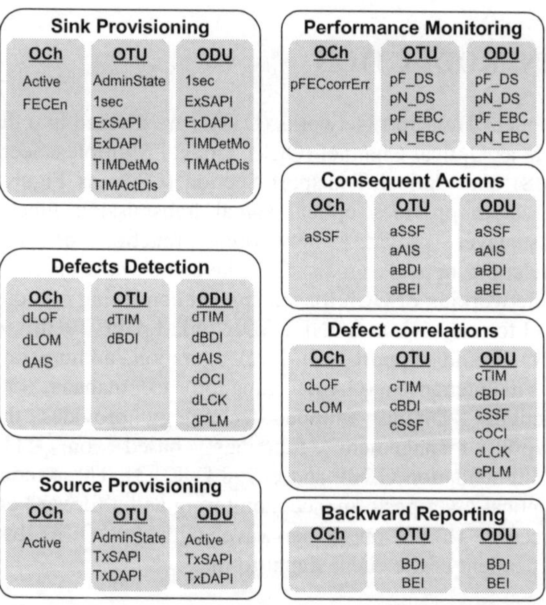

Figure 4. OAM information in overhead processing unit. (See Reference [3])

(1) Source Provisioning

Management unit sends OAM information to this function by polling. The layer function in overhead processing unit is activated with the 'Active' command in OAM information. The LCK maintenance signal is generated by operator request (AdminState =LCK) in order to lock the signal. In this case, ODU*k* signal is replaced by 01010101 patterns. And TxSAPI and TxDAPI are inserted into the TTI byte position.

(2) Sink Provisioning

This function is almost same as source provisioning. In addition, TIMDetMo (TIM detection mode) and TIMActDis (TIM consequent actions disabled) control defects detection and consequent actions in connection with TIM (trail identifier mismatch).

(3) Interrupt from Digital Wrapper

When this function receives an interrupt, it must search the corresponding action and layer as soon as possible. If the interrupt is related with defects, it calls defect detection function. If the interrupt is related with performance monitoring as accumulated error blocks, it calls performance monitoring function.

(4) Defects Detection

This function extracts SM, PM, PSI overhead, and detects for dLOF, dLOM, dAIS, dTIM, dBDI, dOCI, dLCK and dPLM defects in compliance with ITU-T Rec. G.798. And then, it calls consequent actions function.

(5) Consequent Actions

When each layer receives a server signal fail (SSF) indication from its server layer, it performs of parameters supervision pertaining to the layer, and generates a SSF indication to its client layer. The SSF state of the layer is indicated via an alarm indication signal (AIS) in the OTU and ODU layer. On declaration of aAIS, this function sends a control commendation to digital wrapper chip; in this case the current signal is replaced by AIS signal pattern. And then it calls backward reporting function.

If the ODU or OTU layer enters SSF state, all upper layer defects must be cleared and interrupts are deactivated. If the layer comes out SSF state, all upper layer interrupts must be activated.

(6) Backward Reporting

A single bit BDI signal is inserted in proper overhead position of digital wrapper chip to convey the signal fail status detected in consequent actions in sink direction. And also 4 bits BEI signal is inserted to convey the count of interleaved-bit blocks that has been detected in error by digital wrapper chip using the BIP-8 code.

(7) Defect Correlations

This function correlates the defects and incoming SSF in order to determine the probable fault causes and reports to the Management Unit.

(8) Performance Monitoring

This function fetches the values of digital wrapper chip registers containing the number of accumulated error blocks and processes the defect second (DS) information by using defects for each layer. And it reports performance monitoring information to the management unit.

4. CONCLUSIONS

In this paper, we propose the overhead processing model for OCh-level OAM is based on ITU-T G.709 and G.798. This overhead processing unit can manage effectively OChs by digital wrapper overhead processing functions. The Overhead processing unit is composed of source function and sink function. The source function manages the processing in source direction, which accomplishes two sub-functions: source provisioning and backward reporting. The sink function manages the processing in sink direction, which accomplishes five sub-functions: sink provisioning, interrupt from digital wrapper, defects detection, consequent actions and performance monitoring. Recently, a lot of digital wrapper chips comply with ITU-T G.709 but not G.798. Some functions based on ITU-T G.798 are not implemented with digital wrapper. So we expect the digital wrapper chip that supports all G.798 functions to manage OChs effectively.

REFERENCES

[1] ITU-T, http://www.itu.org

[2] ITU-T Draft Rec. G.709, Network Node Interface for Optical Transport Network

[3] ITU-T Draft Rec. G.798, Characteristics of Optical Transport Network (OTN) Equipment Functional Blocks

[4] Paul Bonenfant et al., Optical data networking, IEEE Communications Magazine, vol. 38, page 63-70, March 2000

50 GHz millimeter-wave generation by using waveguide based optical technology for Fiber-Radio Systems

Ki-Hwan Park and Wataru Chujo
Basic and Advanced Research Division, CRL,
4-2-1, Nukui-Kitamachi, Koganei-shi, Tokyo 184- 8795,
JAPAN
e-mail: phwan@crl.go.jp Phone: +81-42-327-7551, Fax: +81-42-327-7035

Abstract: We generated millimeter-wave (MMW) signals for fiber-radio systems. The frequency separation between two optical signals was approximately 30-50 GHz. The generated millimeter-wave signals were transmitted through a millimeter waveguide and detected. The spectral linewidth of the millimeter- wave was 2 kHz, and power fluctuation of the millimeter-wave was less than 2 dB over a 30-50 GHz range. This method can reduce the loss of millimeter- wave power and obtain a wider bandwidth compared with a detection method using a photodiode.

Key Words: Fiber-Radio, optical waveguide, millimeter wave generation

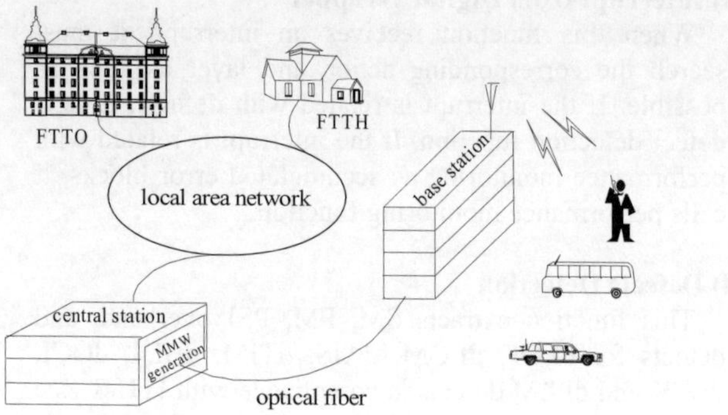

Fig.1 Photonic and fiber-radio distribution network

1. INTRODUCTION

Millimeter-waves have been studied as a favorable means of data-transmission for wireless or mobile communications. However, the available frequency range for millimeter-waves has been gradually limited because of the popularization of communication equipment. Consequently, researchers have proposed a millimeter-wave region between 30-70 GHz for new communication frequency bands[1]. These millimeter-wave frequencies easily provide high antenna gain. Recently, fiber-radio systems that generate millimeter-wave signals by optical means and transmit them through optical fiber have attracted interest as to their applicability to mobile communications[2-3]. Millimeter-wave radio on fiber should be a reliable transmission method for delivering a broadband service in mobile communications. We have concentrated on combining photonic and mobile networks(Fig.1), which convey through fiber the millimeter-waves generated by optical technology and radiates them from an antenna. This technology has low signal loss, and is immune to signal distribution interference or exposure to unfavorable environmental conditions, including sun, snow, wind, and rain.

Millimeter-waves with very narrow linewidths are required for broadband communications. They can be obtained by using an optical technology for fiber-radio systems. Generation of coherent sub-millimeter and millimeter-waves by using difference-frequency generation(DFG) has also been investigated with optically nonlinear materials[4-5]. The DFG method easily obtains the desired millimeter-wave frequency by controlling the frequency spacing of the input optical signals. However, pico- or femto-second optical pulses have been used in most studies to ensure high temporal coherence of the generated millimeter-wave. Therefore, a very high optical peak power was required for the experiments. Dual-mode distributed feedback lasers (DFB) have been developed for the optical generation of millimeter-waves for broad bandwidth mobile communications and fiber-radio systems. However, so far, their spectral linewidths have been too large for practical use. Furthermore, using a photodiode to detect millimeter-waves leads to a rapid decrease in signal power at high frequencies. That is, this method can provide only a narrow bandwidth (several GHz), as indicated by the gain characteristics at these frequencies.

In this study, we used optical technology to generate CW millimeter-waves. We detected the millimeter-wave generated in the optical waveguide by using a millimeter waveguide. This method can reduce the millimeter-wave power loss and can obtain a wider bandwidth compared with the other method, which uses a photodiode.

2. EXPERIMENT

The experimental setup is shown in Fig. 2. It

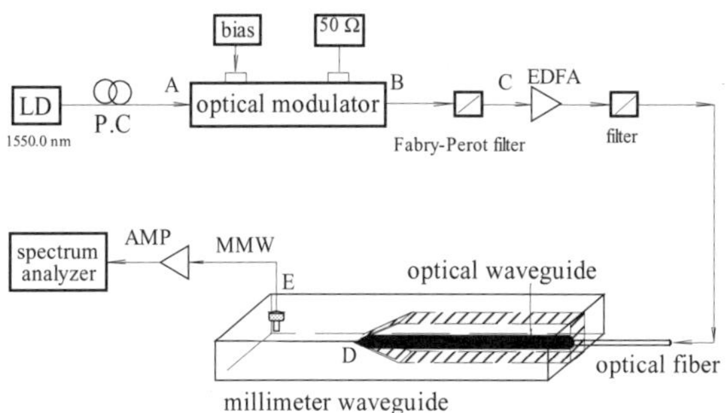

Fig. 2 Experimental setup

consists of a tunable laser diode, polarization controller, optical modulator, Fabry-Perot filter, erbium-doped fiber amplifier (EDFA), 3-nm bandpass filter, millimeter-wave amplifier, an optical waveguide inserted into a millimeter waveguide, and a spectrum analyzer. A 1550.03 nm optical signal was injected into the optical modulator through the polarization controller. The controller adjusts the polarization of the optical signal going into the optical modulator. A 15-25 GHz modulation signal was applied to the optical modulator via a bias tee from a synthesizer. The B output included two large optical signals positioned at the same frequency interval from the central optical carrier. The central optical carrier component was sufficiently suppressed by adjusting the bias of the optical modulator and by tuning the Fabry-Perot filter. Optical signal C, which included the two large signals, was amplified by the EDFA. The optical signals were input into the optical waveguide and output from port D.

The millimeter-wave generated in the optical waveguide propagates in the direction of the optical signals passing through the LN substrate. The millimeter-wave is radiated in the space inside the millimeter waveguide from the tapered cross-section of the substrate. The detector combined with the millimeter waveguide detects the millimeter-wave propagating through within the waveguide. The generated millimeter-wave was then detected at port E, boosted by the millimeter-wave amplifier, and measured by a high-resolution spectrum analyzer.

A Z-cut plate of $LiNbO_3$ was used as the substrate, and the optical waveguide was made by Ti diffusion. The length of the optical waveguide inserted inside the millimeter waveguide was 5 cm, and the cross-section was 4.2×10^{-11} (m^2). The height and width of the LN substrate were 1.8 and 1.0 millimeter, respectively. The nonlinear optical coefficient of the LN was $5.29\times10^{-23} cm^2/W/nm^2$.

Figures 3 (a) and (b) show a typical spectrum of the output signals at points B and C (Fig. 2), respectively. (a) shows the signal modulated by the optical modulator. The central signal is an optical carrier component. (b) shows the central optical carrier and other modes suppressed more than 30 dB by the Fabry-Perot filter. The large optical signals(two modes) had different wavelengths, 1549.83 and 1550.23 nm, and their frequency separation was approximately 50 GHz. The resolution of the optical spectrum analyzer used for the experiment was 0.007 nm.

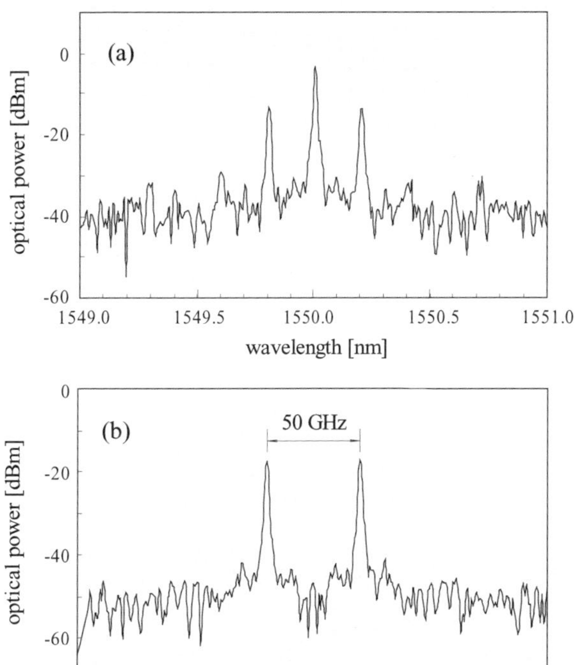

Fig.3 Typical spectrum of the measured optical siganls.
(a): the optical signals modulated by an optical modulator
(b): two modes passed the Fabry-Parot filter

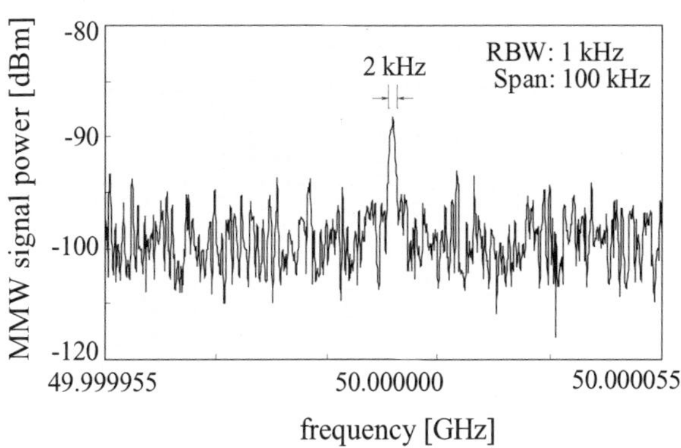

Fig. 4 Typical spectrum of millimeter-wave obtained at port E. The spectrum linewidth was 2 kHz, and the resolution of the spectrum analyzer was 1 kHz.

Figure 4 shows a typical CW millimeter-wave spectrum obtained from port E. The spectral linewidth was 2 kHz, and the resolution of the spectrum analyzer was 1 kHz. CW millimeter-waves with very narrow linewidths were obtained in this experiment.

Figure 5 shows the power fluctuation of the generated millimeter-wave as a function of the frequency difference between the two optical signals measured at port C(in Fig. 3). The power of millimeter-wave signals was about –13 dB at 30 GHz frequency difference and -12 dB at 50 GHz. The power fluctuation was less than 2 dB over a 30-50 GHz range. The proposed method has broad bandwidth gain-characteristics, thus making it superior to the photodiode detection method.

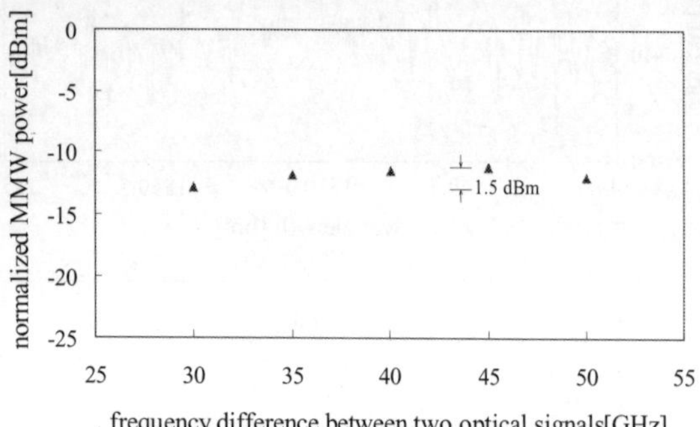

Fig. 5 Power fluctuation of the generated millimeter-wave shown as a function of frequency difference between the two optical signals measured at port C.

3. CONCLUSION

We demonstrated the generation of CW millimeter-wave by using an optical technology and successfully obtained signals with very narrow linewidths. We found that the power of the measured CW millimeter-wave was almost proportional to the power multiplication of two input optical signals, and that the power increased as the frequency difference of the two optical signals increased. The spectral linewidth of the generated millimeter-wave was 2 kHz. We expect that this method will lead to new technology for future mobile and wireless communication systems.

REFERENCES

[1] D. Wake, C. R. Lima, and P. A. Davies, "Optical generation of millimeter waves signals for fiber-radio systems using a dual-mode DFB semiconductor laser," IEEE Trans. on MTT, **43**(9), 2270-2276, (1995).

[2] R. –P. Braun, G. Grosskkopf, H. Heidrich, C. Helmolt, R. Kaiser, K. Kruger, U. Kruder, D. Rohde, F. Schmidt, R. Stenzel, and D. Trommer, "Optical microwave generation and transmission experiments in the 12- and 60-GHz region for wireless communications," IEEE Trans. on MTT, **46**(4), 320-329, (1998).

[3] D. Wake, C. R. Lima, and P. A. Davies, "Optical generation of millimeter waves signals for fiber-radio systems using a dual-mode DFB semiconductor laser," IEEE Trans. on MTT, vol 43, no. 9, pp. 2270-2276, 1995.

[4] A. S. Pine, "Doppler-limited molecular spectroscopy by difference-frequency mixing," J. Opt. Soc. Am., **64(**12), 1683-1690, (1974).

[5] K. P. Petrov, A. T. Ryan, T. L. Patterson, H. Lee, S. J. Field, and D. J. Bamford, "Spectroscopic detection of methane by use of guide-wave diode pumped difference-frequency generation," Optics Letters, **23**, 1052-1054, (1998).

MONITORING METHODS IN WDM RING NETWORKS

Wang Jianquan, Gu Wanyi

Beijing University of Posts and Telecommunications Beijing, China 100876

Wang Jianquan,
BOX: 128,
Number 10, Xitucheng road, Beijing, China.
Telephone: +86-10-62283383
Fax: +86-10-62283728
Email: jackie3003@sohu.com

Gu Wanyi,
BOX: 128,
Number 10, Xitucheng road, Beijing, China.
Telephone: +86-10-62283383
Fax: +86-10-62283728
Email: wyg@bupt.edu.cn

Abstract: Two failure monitoring methods, leaking lightwave monitoring and extra traffic monitoring, are proposed for monitoring failure of protection line in WDM ring networks in this paper, and which can be used both in two fiber unidirectional ring networks and four fiber bidirectional ring networks. The performance of the two monitoring methods are analyzed and compared in detail. Moreover application condition of the two monitoring methods is analyzed and advised.

Keywords: ring networks, protection and switching, failure monitoring, extra traffic (ET), automatic protection switching(APS), wavelength division multiplexing(WDM)

1. INTRODUCTION

The APS (automatic protection switching) protocol is a standard for implementing automatic protection switching of ring networks. The survivability of ring networks depends on the quality of APS.

Prescribing APS protocol is the first step to implement the protocol. Implementing the protocol, many new problems will be introduced in WDM ring networks. Failure monitoring which is the foundation of implementing APS protocol in optical fields is a quite new subject, and it is also very difficult to realize for two-fiber unidirectional protection ring and four-fiber bi-directional protection ring.

The subject mentioned above will be discussed and researched in this paper, and some failure monitoring methods are proposed and analyzed.

2. Failure monitoring methods in the protection line

It's clear that monitoring of fiber failures, node failures and other hardware failures in working line can be well solved because signal exist in working line. Alarm occurs when failure takes place in working line. But there is no alarm even if failure takes place in protection line. The reason is there is no signal in protection line when the network is in going order. Therefore, failure can't be found in protection line if not adopting especial method. Then if failure takes place in working line in another section, protection will be triggered because the failure can be detected. As a result, the ring network can't be protected normally on account of the failure which existed in protection line cannot be detected in time.

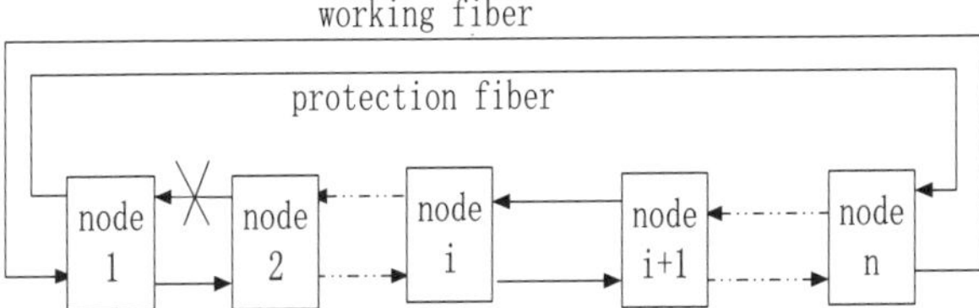

Fig. 1(a) Failure takes place in protection line in section between node 1 and node 2 at t1.

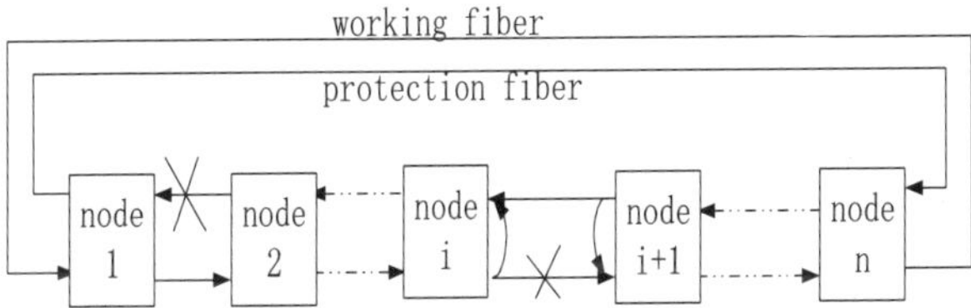

Fig. 1(b) Failure takes place in working line in section between node i and node i+1 at t2.

Take Fig.1 for example, at t1, failure takes place in protection line in section between node 1 and node 2, there is no alarm because there is no signal in the line, so failure can not be found. At t2, failure takes place in working line in section between node i and node i+1. Protection and switching will be triggered because network can detected the failure, which result in network protection out of work because the failure which has taken place at t1 in protection line. So, failure monitoring in protection line in ring networks is very important and significant.

We can conclude that failure in protection line can not be detected when network is in going order. If we can ensure that optical signal exists in protection line when the initial state of ring is in normal state, there will be no difference between failure monitoring in working line and that in protection line. Two monitoring methods are proposed to monitor failure in protection line in ring networks, which make signal exist in protection line even if network is in order.

Leaking lightwave monitoring (shown as Fig.2)

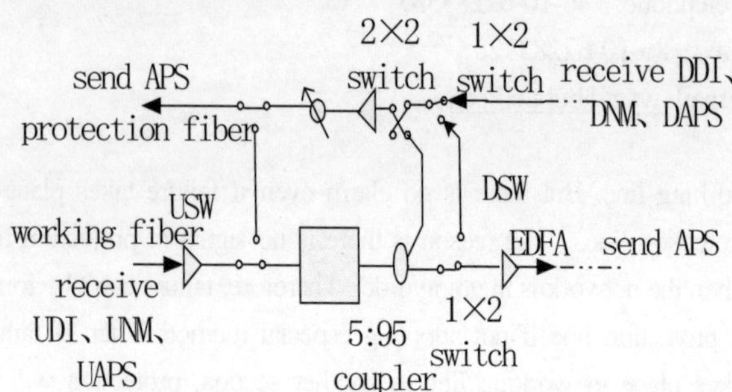

Fig.2 Node scheme of leaking lightwave monitoring

Where:

UDI : upsream defect indication

DDI : downsream defect indication

UNM : upsream network commands

DNM : downstream network commands

UAPS: upstream APS signals

DAPS: downstream APS signals

USW : upstream switch

DDI : downstream switch

As Fig.2 shown, 5:95 optical coupler introduce 5 percent optical of working line into protection line, which make signal exist in protection line. Because optical leak from working line to protection line, we call the method "leaking lightwave monitoring method". In the method, a 2×2 optical switch must be equipped at the node upstream in protection line to avoid leaking lightwave influence on working signal when the system is out of order; and a 1×2 optical coupler must be equipped at the node upstream in working line to minimize optical power loss due to leaking lightwave. Coupling coefficient of optical coupler in working line must be small. But for the purpose of keeping failure monitoring running in order, that is to say, for keeping EDFA in protection line working in gear, coupling coefficient of the optical coupler should not be too small. The coupling coefficient is commonly chosen as 5: 95.

2). Extra Traffic monitoring (shown as Fig.3)

As Fig.3 shown, adding extra traffic make signal exist in protection line. Therefore, the method can be titled "Extra Traffic monitoring method". Priority of extra traffic will be set the lowest initially in order to protect the traffic in working line in time when failures happened, i.e. remove extra traffic immediately whether it has reached its aim or not when failures occur in the system.

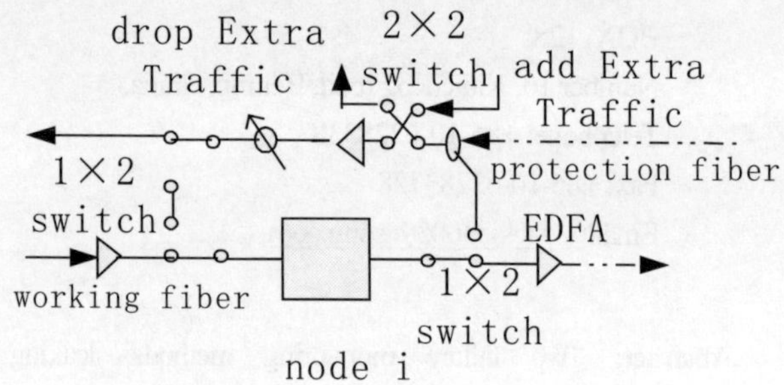

Fig.3 Node scheme of extra traffic monitoring

As Fig.3 shown, extra traffic monitoring method requires that every node must be configured as follows: a 2×2 optical switch plus a 1×2 optical coupler or a 3×3 optical switch . As far as the cost is concerned, a 2×2 optical switch plus a 1×2 optical coupler is adopted.

3. Performance analysis of failure monitoring methods

In this section, we discuss the difference of performance between the 2 failure monitoring methods, and summarize their advantages, so as to find their each application condition. Then take Fig.4 for example. We can analyze the difference of performance between the 2 monitoring methods in detail.

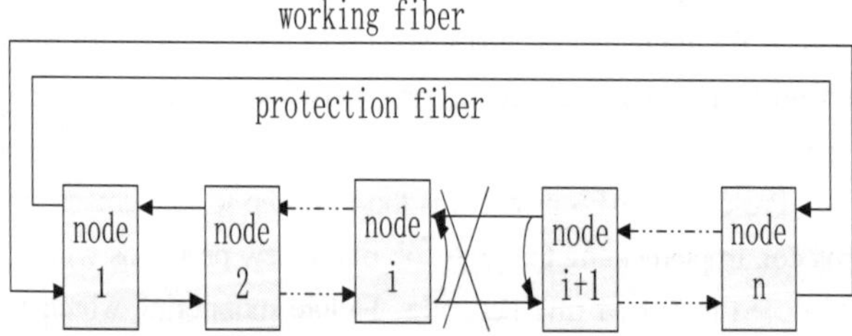

Fig.4 The sketch of protection and switching in WDM all optical networks

Only optical power, the center wavelength and optical Signal-to-Noise (OSNR) can be monitored in optical field up to now. OSNR is more precise in evaluating the quality of optical signal than other two parameters of optical signal. Because the 2 failure monitoring methods are operated in WDM ring networks, OSNR can be used to analyze the performance of these monitoring methods. For the sake of simplicity of analyzing, assuming:

1. The optical power equilibrium is perfect, i.e. when the initial state of ring networks is in the idle state, the optical power discrepancy among different channels is limited within 1dB.

2. There are n nodes in this system. A certain wavelength signal that located failures span upstream from protection line direction, whose Signal Noise Ratio (SNR) is A_0 before the failure takes place are taken into account.

3. Neglecting the fixed insert loss in power of 2×2 optical switch, but its response expression are taken into account: $\begin{pmatrix} a & 1-a \\ 1-a & a \end{pmatrix}$, where a is pass-through coefficient, and $(1-a)$ is the crosstalk coefficient of 2×2 optical switch. In general, a is no less than 0.9 because channel isolation is often more than 20dB.

4. The SNR of output signals from I-th node is SNR_I . I is a positive integer from 0 to n.

1). Leaking lightwave monitoring

In this failure monitoring method, all nodes must be configured as Fig.2, so the change in SNR of a certain wavelength signal can be expressed as equation (1).

$$SNR_i = \frac{aP\left(\frac{S}{S+N}\right)}{aP\left(\frac{N}{S+N}\right) + 0.05P(1-a)} = \frac{aSNR_{i-1}}{a + 0.05(1-a)(1+SNR_{i-1})} \quad (1)$$

$$i = 1,2,3,\cdots,n, \qquad SNR_0 = A_0$$

Where a stands for pass-through coefficient, SNR is the initial Signal Noise Ratio, P is the optical power and N is the noise power.

It is easy to find the relation between SNR and the number of nodes, and pass-through coefficient a , as shown in Fig.5.

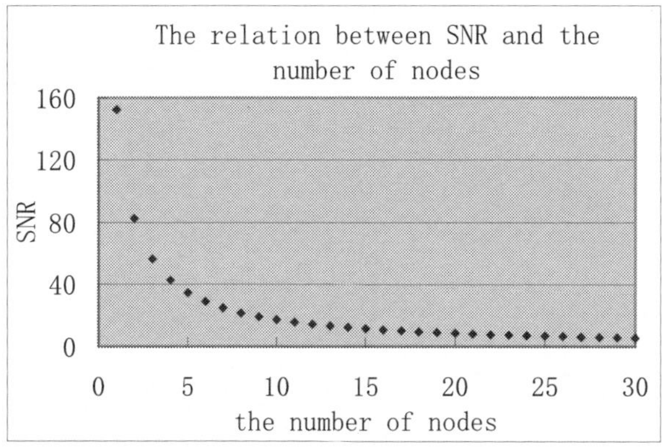

Fig5 (a) The relation between SNR and the number of nodes.

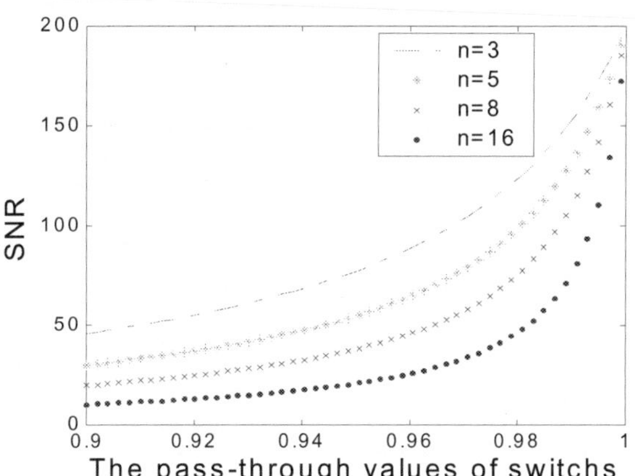

Fig5(b) The relation between SNR and pass-through coefficient of optical switch.

2). Extra Traffic monitoring

Suppose there are m wavelengths in working line and k wavelengths to carry Extra Traffic, and the mechanism of optical equilibrium in each wavelength runs very well. We may think that there is no optical power difference between different wavelengths approximately. So we can derive equation (2) as follows.

$$SNR_i = \frac{aSNR_{i-1}}{a + (1-a)SNR_{i-1}} \quad (2)$$

Where i=1, 2, …, n and $SNR_0 = A_0$.

It is easy to find the relation between SNR and the number of nodes, and pass-through coefficient a , as shown in Fig.6

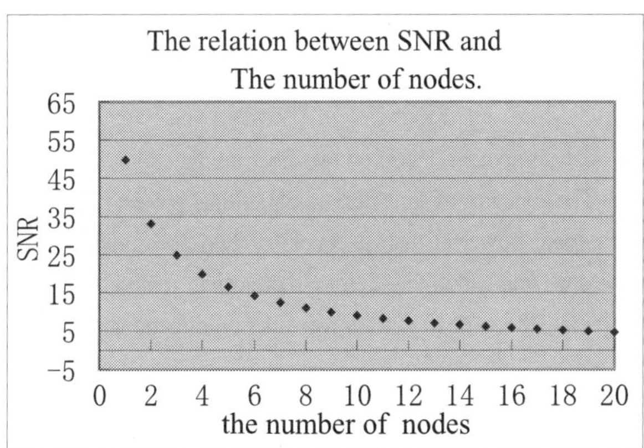

Fig.6 (a) The relation between SNR and the number of nodes.

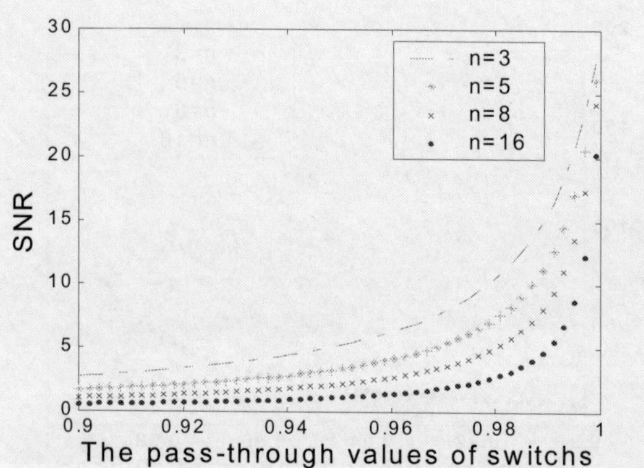

Fig.6(b) The relation between *SNR* and pass-through coefficient of optical switch.

4. CONCLUSION

Some conclusion can be drawn easily from Fig.5 and Fig.6. First, with the increase of the number of nodes and the decrease of pass-through coefficient of optical switch, SNR of signal which has been protected will decrease in spite of adopting the first monitoring method or the second method. In addition, the second method is better than the first method in network scale. However, the first method is better than the second method if the pass-through coefficient of optical switch is small. In addition, the implementation of leaking lightwave failure monitoring uses only signals in working line. It is very simple and very easy, but it leads to optical power loss and brings in crosstalk to working signal. Although Extra Traffic failure monitoring doesn't waste optical power, its crosstalk is more than that of leaking lightwave failure monitoring. It makes full use of the resource, i.e. it can transmit some traffic on protection line when the system has no failures. Therefore, each failure monitoring method has its merits and defects. One can be chosen from the two methods according to network scale and the performance of optical switch in order to monitor failure in protection line optimally.

The analysis above is based on two-fiber unidirectional protection rings The function of four-fiber bi-directional protection rings is much more powerful than a couple of two-fiber unidirectional protection rings, but the failure monitoring method for it is completely the same as that for two-fiber unidirectional ring.

5. References

Telecommunication standardization sector of ITU:

[1]. ITU-T G.841

Journal Papers:

[2]. Takatori, M.; Nakano, Y.; Ashi, Y.; Fujita, H.; Mizukami, M.; Itoh, K. " A high performance switch for OC-12 SONET self-healing ring networks ". Selected Areas in Communications, IEEE Journal on Published: Feb. 1996 Volume: 14 2 , Page(s): 353-361

Proceedings Papers:

[3] Tillerot, F.; Didelet, E.; Daviaud, A.; Claveau, G. "Efficient networks upgrade based on a WDM optical layer with automatic protection switching ". Optical Fiber Commuunication Conference and Exhibit, 1998. OFC '98., Technical Digest , 1998 , Page(s): 296-297

[4] Wen De Zhong "New bi-directional WDM ring networks with dual hub nodes ". Global Telecommunications Conference, 1997. GLOBECOM '97., IEEE 1997 Volume: 1 , Page(s): 556-560 vol.1

Optical Signal Quality Monitoring Method Based on Sub-harmonically Synchronous Sampling

Lei Ding, Wen-De Zhong, Yixin Wang, and Yaolong Sam*
Network Technology Research Center, Nanyang Technological University, Singapore 639798
Telephone: 65-67904681, fax: 65-67926894, e-mail: elding@ntu.edu.sg
** Institute for Communications Research, Singapore*

Abstract: We propose a new method for monitoring optical signal quality. The new method is based on sub-harmonic sampling technique and is experimentally verified. With the new method, the amplitude histogram of the optical signal is acquired by sub-harmonically synchronous sampling. We demonstrate experimentally that the Q-factor obtained from the amplitude histogram coincides very well with the Q-factor deduced from the bit-error-rate that is directly measured by an bit-error-rate tester. The new method does not require high-speed sampling circuits, and hence it is very cost effective for monitoring high-speed optical signals that have different bit rates and coding formats.

Key Words: amplitude histogram, Q-factor, sub-harmonically synchronous sampling, optical signal monitoring

1. INTRODUCTION

Optical networks have the ability to transport hundreds of wavelength-division multiplexed (WDM) signals that have different bit-rates and coding formats [1]. Monitoring signal quality of individual WDM signals is important for proper network operation and management, and to accomplish this the bit-error-rate (BER) has to be measured. BER is a fundamental performance parameter of any digital transmission system. However, the BER can only be correctly measured with a special BER tester, by using a pre-established test bit pattern, not the real signal. Further more, it takes a rather long time to get a precise BER.

Because the measurement of BER is very slow and very costly, Q-factor measurement based on amplitude histograms has emerged as an alternative method. Many Q-factor measurement techniques have been demonstrated recently [2]-[6]. In these methods, synchronous or asynchronous sampling is generally used to get essential information about the real digital signal. Hanik *et al* proposed a method to evaluate the optical signal by asynchronous sampling [2]. But this method can only give a qualitative evaluation, not a quantitative Q-factor. Takara *et al* reported a synchronous optical sampling method to measure the eye diagram of a 100Gbit/s optical signal using an organic crystal [3]. Shake *et al* used asynchronous optical sampling to evaluate the performance of 10 to 160Gbit/s RZ optical signals [4][5]. Bach *et al* reported a synchronous sampling method to measure the optical signal's Q-factor [6]. However these methods require either special organic crystals or complicated circuits and hence are difficult to implement.

In this paper, we propose a novel Q-factor evaluating scheme based on sub-harmonically synchronous sampling technique. With this method, optical signal is sampled at a sub-harmonic rate of the measured signal. The Q-factor is then calculated based on the sampled data. Since the signal is sampled at a rate much lower than the signal data rate, very high-speed sampling circuits are not required. The new monitoring method is experimentally demonstrated and verified.

2. PRINCIPLE

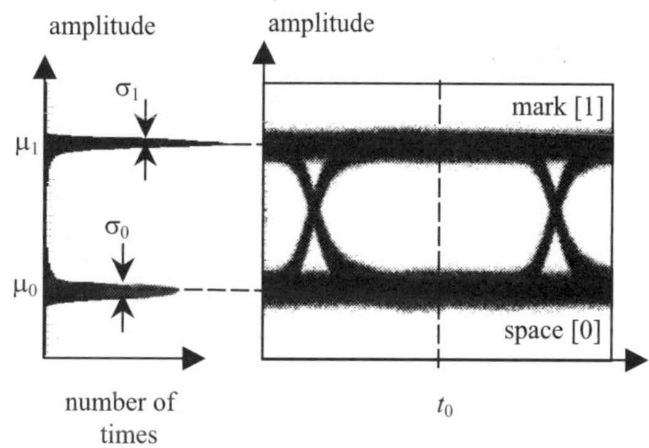

Fig. 1 Eye pattern and amplitude histogram

Fig. 1 shows a typical eye pattern of a digital signal and its amplitude histogram. The amplitude histogram shows amplitude distributions in both mark and space levels. Assuming that the signal's waveform distortion is sufficiently small and the ASE and other noise that are added to the signal can be approximated as Gaussian noise, then the Q-factor ($Q(t_0)$) estimated at a certain timing phase (t_0) from the amplitude histogram is expressed by equation (1) [7].

$$Q(t_0) = \frac{|\mu_1(t_0) - \mu_0(t_0)|}{(\sigma_1(t_0) + \sigma_0(t_0))} \quad (1)$$

where $\mu(t_0)$ and $\sigma(t_0)$ are the mean and standard deviation of the mark(1) and space(0) levels at time t_0, respectively. For the consideration of accuracy, t_0 is usually set at the center of the eye pattern. Since the eye pattern is periodic and composed of large numbers of dots, we strictly sample the dots appearing at the center of the eye to get the real Q-factor.

In modern SDH or SONET systems, transmission rates of STM-16 (2488.32Mbit/s) and STM-64 (9953.28Mbit/s) are widely used, and much higher transmission rates have emerged. However, data acquisition boards over 1GHz are either very expensive or unavailable. Our new method does not need to sample the signal for every bit, but sample at a sub-harmonic rate which is *1/N* (*N* is an integer) of the signal data rate. Though the sampling rate is low, the means and standard deviations of the amplitude histogram can be determined after sampling an enough number of bits. Based on the sampled amplitude histogram, the Q-factor is estimated, which is independent of signal data rate and coding format.

3. EXPERIMENT

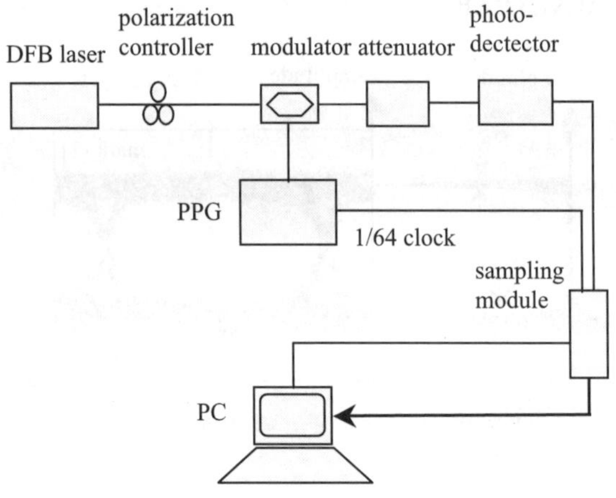

Fig. 2 Experimental setup for Q-factor monitoring

The experimental set-up for the proposed Q-factor monitoring system is shown in Fig. 2. The optical signal is a 2^{31}-1 pseudorandom NRZ signal generated by a DFB laser, a $LiNbO_3$ intensity modulator and a BERT's pulse pattern generator (PPG). The optical signal passes through an attenuator and then it is converted into an electrical signal by a photodetector. The signal's clock is divided by 64, which is then used to trigger the electrical sampling module. Once the sampling module is triggered, it samples the signal at one 64th of the signal rate. The amplitudes of all sampled dots are then transferred to a PC. The PC divides the dots into mark and space and estimates their means and standard deviations. The Q-factor is then calculated by using equation (1).

In the experiment, bit rates of 2488.32Mbit/s and 9953.28Mbit/s were used. The Q-factor for various values of signal-to-noise ratio (SNR) was obtained by changing the attenuation value of the optical attenuator. In order to verify the new method, we also measured the signal's BER using a bit-error-rate tester (BERT), from which $Q(t_0)$ can be deduced by equation (2) [7].

$$BER = \mathrm{erfc}(Q(t_0)) \quad (2)$$

where erfc(x) is a complementary error function.

Fig. 3 shows the optical signal's eye diagrams and amplitude histograms measured at the middle of the eye patterns for different SNRs. When the SNR is high (Fig. 3a), the mark and the space are sharp. Whereas when the SNR is reduced to a low level, the eye pattern's eye becomes closing, and the mark and space become much lower and wider.

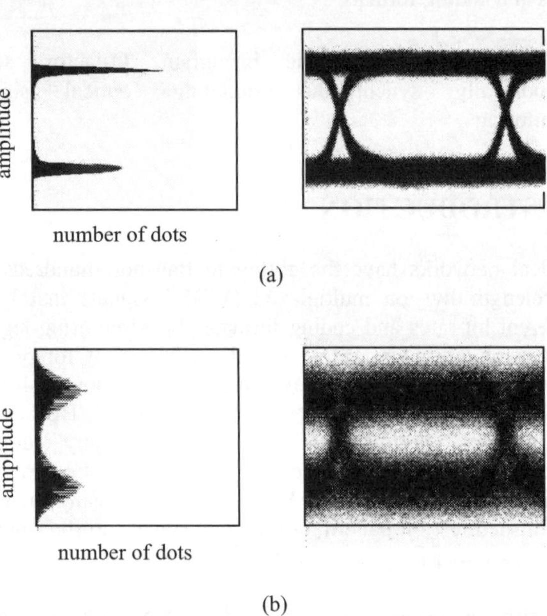

Fig. 3 Amplitude histograms of NRZ optical signal's eye pattern: (a) high SNR (BER<10^{-10}), (b) low SNR (BER=10^{-5})

Fig. 4 shows Q-factor (Q_{AH}) calculated based on the amplitude histogram obtained from the sub-harmonic sampling and Q-factor (Q_{BER}) calculated from the measured BER for the NRZ optical signal at data rates of 2488.32Mbit/s and 9953.28Mbit/s, respectively. The sampling rate is 38.88 MSample/s for the data rate of 2488.32Mbit/s and 155.52 MSample/s for the data rate of 9953.28Mbit/s.

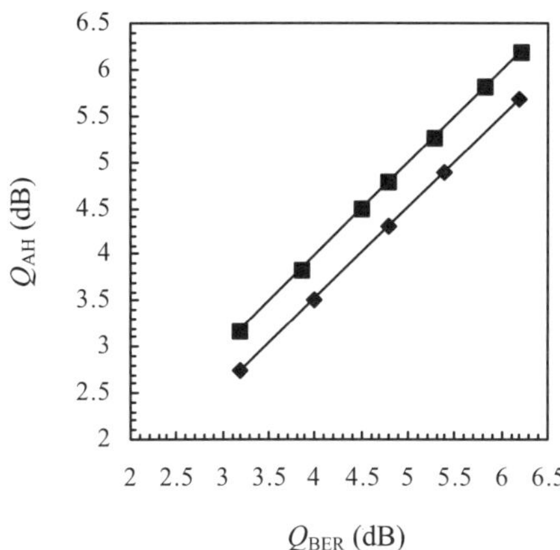

Fig. 4 Relationship between Q_{AH} and Q_{BER}
■ 2488.32Mbit/s
♦ 9953.28Mbit/s

As can be seen in Fig. 4, the slope of the linear fitting is at 45 degrees for both data rates for Q_{BER} ranging from 3.2 to 6.2. ($10^{-10}<BER<10^{-5}$). At the data rate of 9953.28Mbit/s, Q_{AH} is slightly lower than Q_{BER}. This is because the sampling module induced higher noise at the higher data rate. Fig 4 shows that Q_{AH} agrees very well with Q_{BER}. The results confirm that we can evaluate the optical signal quality by calculating the Q-factor with the new method. Since signal is sampled at a low sub-harmonic rate of the actual signal, data acquisition boards with very high speed sampling circuits are not required. Therefore, cost-effective optical signal monitoring systems can be made based on the new method.

4. CONCLUSION

We have proposed a new optical signal monitoring method based on sub-harmonic sampling technique. The new method has been experimentally demonstrated. The experimental results show that the Q-factor obtained from amplitude histogram coincides very well with the Q-factor deduced from the BER that is directly measured by a bit-error-rate tester. Since the new method does not require high-speed sampling circuits, it is very cost effective for monitoring high-speed optical signals that have different bit rates and coding formats.

REFERENCES

[1] P. E. Green Jr., Optical Networking Update, *IEEE J. Select. Areas Commun.*, vol. 5, page 764-779, 1996

[2] N. Hanik, A. Gladish, C. Caspar and B. Strebel, Application of amplitude histograms to monitor performance of optical channels, *Electron. Lett.*, vol. 35, No. 5, page 403-404, 1999

[3] H. Takara, S. Kawanishi, A. Yokoo, S. Tomaru, T. Kitoh and M. Saruwatari, “100Gbit/s optical signal eye-diagram measurement with optical sampling using organic nonlinear optical crystal”, *Electron. Lett.*, vol. 32, No. 24, page 2256-2258, 1996

[4] I. Shake, H. Takara, S. Kawanishi and Y. Yamabayashi, “Optical signal quality monitoring method based on optical sampling”, *Electron. Lett.*, vol. 34, No. 22, page 2152-2154, 1998

[5] I. Shake, E. Otani, H. Takara, K. Uchiyama, Y. Yamabayashi and T. Morioka, “Bit rate flexible quality monitoring of 10 to 160Gbit/s optical signals based on optical sampling technique”, *Electron. Lett.*, vol. 36, No. 25, page 2087-2088, 2000

[6] Roland Bach, Wolfgang Moench, Gunter Strohmaier, “In-service optical performance monitoring of high-speed transparent networks based on Q measurements”, *Proceedings of 2001 3rd International Conference on Transparent Optical Networks*, page 210-217, 2001

[7] Neal s. Bergano, F. W. Kerfoot and C. R. Davidson, “margin measurements in optical amplifier systems”, *IEEE Photonics Technology Letters*, vol. 5, No. 3, page 304-306, 1993

Session 14C

Network Protection and Restoration

Understanding *p*-Cycles, Enhanced Rings, and Oriented Cycle Covers

(invited paper)

Wayne. D. Grover, *Fellow IEEE*

TR*Labs* and Department of Electrical and Computer Engineering, University of Alberta,
c/o TR*Labs* #800, 10611-98th Avenue, Edmonton, Alberta, Canada T5K 2P7 Contact: grover@trlabs.ca

***Abstract* - This paper explains the important conceptual and technical differences between the method of *p*-cycles and two other recent advances involving a cyclic orientation to protection. These are enhanced rings and cycle double covers. The most fundamental difference that is unique to *p*-cycles is the aspect of straddling span failure protection. This enables mesh-like efficiency levels at well under 100% redundancy. In contrast enhanced rings and advanced cycle cover methods are both seeking to reduce span overlaps in what is otherwise a purely ring-like logical paradigm in which 100% redundancy remains the best that can possibly be achieved.**

***Keywords-p*-cycles, protection cycles, rings, cycle covers, network protection, restoration.**

I. Introduction and Objective

In recent years independent research efforts have been advancing a number of new approaches to survivable network architecture. One of these, the technique of *p*-cycles, has been of primary interest to the author. *p*-Cycles are remarkable and interesting because they combine the simplicity and speed of ring-like protection switching and yet, under fairly simple design optimization, they result in networks that as a whole exhibit mesh-like capacity efficiency. The key concepts and properties of *p*-cycles will shortly be reviewed. In parallel with the work on *p*-cycles, however, other advances have been made in aspects of ring-based protection itself. Notable amongst these is a recent patent by Fee on what we call "enhanced rings" and the theoretical research by Ellinas et al. on unidirectional or "oriented" cycle covers of a graph. All of these advances involve some form of cyclic layout of protection capacity on a network graph and may consequently be confused at a superficial level. Indeed, coincidentally similar names used by Grover and Ellinas has inadvertently even contributed to the confusion. So it is fair to ask: are these three advances really the same? If not, how do they relate? This paper hopes to provide a timely explanation of the important ways in which these schemes differ and, in particular, why neither of the other schemes are equivalent conceptually, functionally, or in terms of efficiency, to *p*-cycles. Such clarification is needed to ensure the industry and research community interested in *p*-cycles avoids confusion and false assumptions of equivalence arising from the superficial similarities or naming between these schemes.

II. *P*-Cycles

In protecting a network with *p*-cycles, one forms cyclic pre-connected closed paths of spare capacity while allowing working paths to take the shortest direct route over the facilities graph. *p*-Cycles are formed in advance of any failure and the switching actions required in real time are completely pre-planned and no more complex than that of a line-switched ring. Despite the similarity to rings in terms of switching function and the fact that both use a cycle on the network graph for their topology, *p*-cycles are unlike any survivable ring-based system heretofore developed (including 1+1, UPSR, BLSR and FDDI) in that they protect both *on-cycle* and *straddling span* failures. This initially seems to be a rather minor technical difference but when its implications are fully worked through it turns out that the aspect of straddling spans, and the direct routing of working paths, are the keys to obtaining mesh-like efficiency while employing ring-like protection structures.

To illustrate the *p*-cycle concept, and convey an appreciation of why such high capacity efficiency is reached, Fig.1(a) shows a typical transport network example of 11 nodes and 24 spans. In Fig. 1(b) a cycle of ten hops is configured as an example. If this cycle is used as the layout for a BLSR ring then each unit of protection capacity on the ring protects the same amount of working capacity against failure of the ten spans underlying the ring itself, shown in (c). These are called the "on-cycle" failure spans. But if we use the same unit-capacity cycle in (b) as a *p*-cycle it means that we will also access it for protection against the failure of a span such as (6-10) which is said to "straddle" the *p*-cycle. A straddling span has both its end nodes on the *p*-cycle, but is not itself part of the cycle. In the example, the *p*-cycle has eight such straddling span relationships, shown in (d).

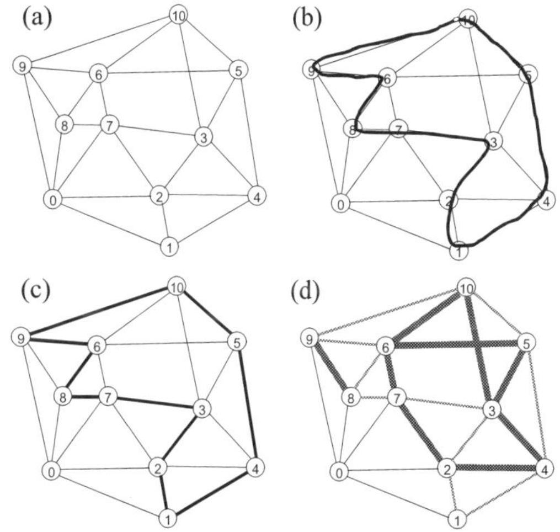

Fig. 1. (a) a network, *G*, (b) a 10-hop cycle of unit-spare capacity, *x*, on *G*, (c) the ten spans obtaining ring-like protection from *x*, (d) the eight *additional* spans each receiving protection for *two* units of working capacity if is *x* is used as a *p*-cycle.

Straddling span relationships have twice the leverage of an on-cycle span in terms of capacity efficiency because when they fail, the cycle itself remains intact and can thereby offer *two* protection paths for each of unit of protection capacity. Note, in the case of a span such as (9-8), that the straddling spans are not limited to being only inside the *p*-cycle perimeter on the graph.

Fig. 2(a) shows the resultant impact on efficiency (or conversely the redundancy). The (w, s) indications note the working and spare composition of the spans, respectively. The example *p*-cycle can achieve 38% redundancy if all its straddling spans have two units of working capacity equipped. The corresponding ring is 100% redundant.[1] In Fig. 2(b) and (c) the real time operation is illustrated. In (b) a span on the cycle breaks and the surviving arc of the cycle is used for protection. This is functionally like a unit-capacity BLSR.

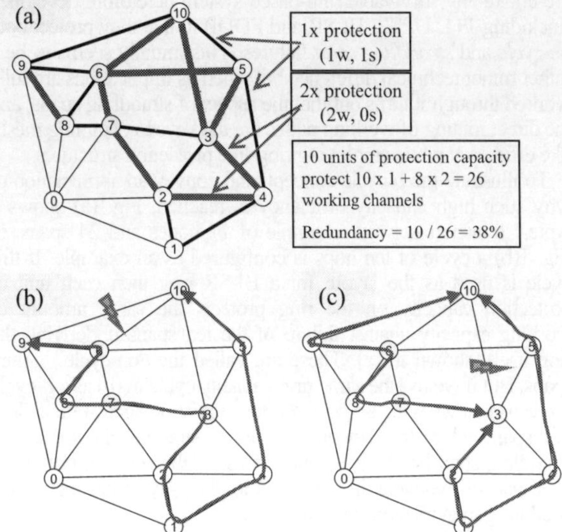

Fig. 2. (a) the compete set of protection relationships offered by *p*-cycle, *x*, and corresponding protection redundancy, (b) example of an on-cycle failure, (c) example of a straddling-span failure.

In (c), however, the same *p*-cycle is accessed to protect a *straddling* span failure. Note the significant increase in protection coverage provided, using the same investment in spare capacity as a ring, but configuring it as a *p*-cycle. In Fig. 2 the single *p*-cycle of ten hops provides protection to 18 spans and since it protects two times its own capacity on each straddling span it actually covers 26 units of working capacity, compared to the ring which protects only 10. Note also, a "signature" feature of *p*-cycles: this is the existence of spans that have working capacity but *no* spare capacity. Spans (3-4), (2-4), and all other straddling spans, may bear either one or two units of working capacity, but in either case require *zero* units of protection capacity. In a complete design of many *p*-cycles such spans may in practice bear protection capacity associated with other *p*-cycles overlying them, but the very possibility of spans with no protection capacity is a unique property of *p*-cycle based networks. In fact this gives one of the simplest "acid tests" to determine, regardless of other complexities, whether the other schemes or equivalent to *p*-cycles. Most notably, no form of enhanced ring, cycle cover, (or generalized loop-back network - to follow) will exhibit "working only" spans which bear *zero* associated protection capacity.

[1] Even this interpretation (100% redundancy) is generous to the ring if we recognize that with ring-routing we have already usually had to deviate the working paths from their shortest routes over the network graph.

There are a number of other important properties of *p*-cycles. *p*-Cycles can either be cross-connect based or based on an ADM-like "capacity slice" nodal device structure [1]. In the OXC-based approach *p*-cycles are formed from individual spare wavelength channels, offering the greatest flexibility to adapt and evolve the *p*-cycle configuration to protect evolving and patterns of dynamic demand. On the other hand, the ADM-like *p*-cycle nodal element offers the "pay as you grow" advantage of conventional rings. In either case, *p*-cycles avoid the structural association between the routing of working demands and the configuration of protection capacity. Unlike rings *p*-cycles are formed only within the spare capacity layer of the network, leaving the working paths to be routed freely on shortest paths, or any other route desired. The configuration of the *p*-cycles is adapted to the working flow, not the other way around. It is important to note that when *p*-cycle network redundancy values are stated in various studies, this is already stated with respect to the most efficient shortest-path routing of working demands. In other words, relative to rings there remains a further capacity saving due simply to replacing ring-routing with mesh-like shortest path routing. Some in the industry have attributed another 30% working capacity savings overall due to this effect alone. Note also that the average length of protection paths in a *p*-cycle is half that of the corresponding ring for straddling spans, and the same as a BLSR ring for on-cycle spans.

The literature on *p*-cycles is growing. Since its introduction in 1998 [2] the basic theory has been developed [3,4], and there have been studies on self-organization of the *p*-cycle sets [2, 5], application of *p*-cycles to the MPLS/IP layer [6], application to DWDM networking [7,13] and studies on joint optimization of working paths and spare capacity [8]. Notably in [7] it was found that full survivability against any span cut could be achieved with as little as 39% total network redundancy in the COST 239 European network model. And [13] reports *p*-cycle designs using fewer OEO conversions and fewer wavelength km's than other alternatives but without significant availability penalty. Recent work is also extending the basic concept of *p*-cycles for span protection to flow-segment protection, which is in effect a path-protection version of *p*-cycles [9].

III. Enhanced Rings

Having stressed how *p*-cycles differ from conventional rings, we now need to make the distinction between *p*-cycles and what are referred to as "enhanced rings." The basic idea of enhanced rings, in the US Patent by Fee [10], is for two otherwise conventional BLSR-type rings to be able to share a common protection channel on spans where the rings run in parallel; that is to say, if the rings were tiling the plane, where two tiles are edge to edge.

To appreciate the motivation for this enhancement to normal ring-based networking, consider first the view of a capacity cross-section along the right-of-way indicated in the conceptual example of Fig. 3, which imagines a North American IXC with two rings that have facing edges on the Salt Lake to L.A. facility route. Each ring individually comprises a 100% matching of working and protection fiber capacity. If, as is not unusual, the working fiber "fill" is itself say only 40% (it may tend to be even lower than this on a North-South mid-west US or Canadian route due to demography), then the overall redundancy of capacity on this route may be 400%.[2] The enhanced ring idea is to recognize that in a sense the situation is doubly redundant because there are two completely separate side-by-side ring systems, each with their own self-contained protection. More generally this situation is called a *span overlap* in ring network design. It arises commonly in even the most efficient ring-based network designs (especially with odd-degree nodes) and is widely known as one of the difficult aspects of ring-based networking. The example conveys its most "painful" real-world form: overlapped spans, thousands of km long, each with low working load.

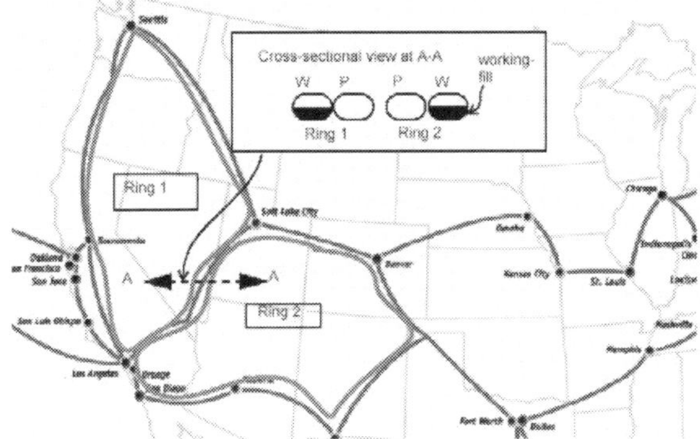

Fig. 3. - Example of two BLSR rings with a span overlap in a context aggravated by light loads and long distances.

Enhanced rings improve the situation by allowing the two facing rings to share a single allocation of protection capacity. This is done through provision of a 2:1 selector switch at each of the sites bordering the common span, and arranging an exchange of signaling between the two rings so access to the common protection span can be coordinated. See [10] for more details. In the example the net effect is that the overall redundancy is improved to 275% by elimination of one complete protection channel, and the revised cross-section view is that in Fig.4

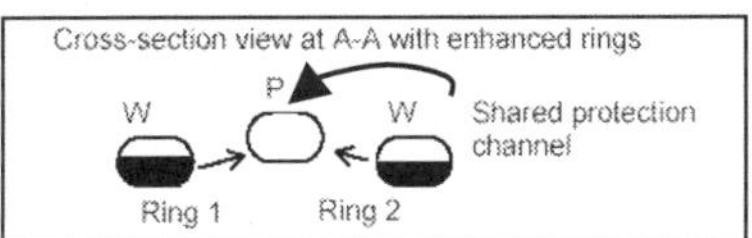

Fig. 4. Cross-section view at A-A with enhanced rings.

Some (notably patent examiners) have wondered if this amounts to being a *p*-cycle. The answer is no because these remain coupled BLSR rings with an allocation of working and protection capacity on every span and still having the ability only to protect against on-cycle failures with an on-cycle loop-back. The key difference is again the aspect of straddling spans. If a true *p*-cycle was formed it would be on the outer perimeter of the two rings in Fig. 3, and the corresponding cross-sectional view in the same example would become that in Fig. 5: two working fibers, with no protection fibers on the same span at all.

In fact, the *p*-cycle approach allows an even further efficiency increase in the example, which is categorically not available to the coupled ring scheme because each ring in the latter case must retain its complete cyclic integrity. This is the option to equip only one working fiber. Even if all the Salt Lake - L.A. traffic was groomed over to one or the other ring only, the overall redundancy would not change because the other working fiber would still have to be there, completely intact (even if empty) to complete its ring.

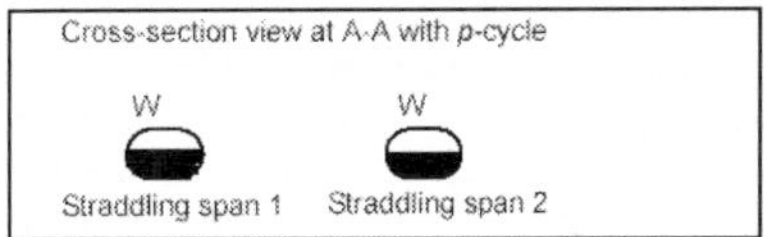

Fig. 5. Cross-section view at A-A with *p*-cycle

In contrast, while the *p*-cycle can protect *up to* two pure-working straddling spans, there is no actual need to do so if demand does not require it. Thus, in the example, the further option available on the Salt Lake - L.A. route when protected by a *p*-cycle is to have only one working line system, straddling the *p*-cycle. Fig. 6 shows the resultant cross-sectional capacity view:

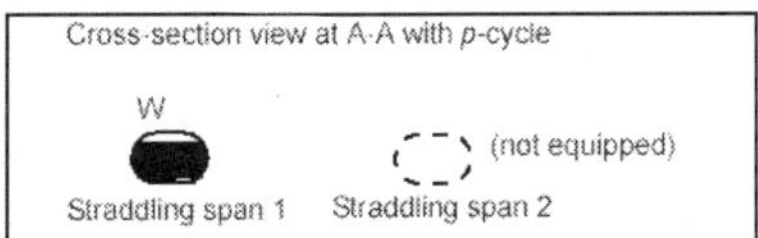

Fig. 6. Potential cross-section view at A-A with *p*-cycle

The corresponding redundancy of this span, which was initially 400% with the overlapped rings, and reduced to 275% by enhanced rings, now reaches (1.0-0.8) / 0.8 = 25% under the *p*-cycle approach.[3]

A separate argument against *p*-cycles may be that the coupled rings should exhibit better availability because each individually has a smaller perimeter distance than the single *p*-cycle which is implied in the example. However, with the sharing of access to a single protection span of the enhanced rings, the coupled rings in essence have the same total mileage exposure to a dual failure situation on the outer perimeter of themselves as a pair, as does the *p*-cycle. In general, however, there may be circumstances where a single *p*-cycle could be made so efficient, through having a large number of straddlers, that we put separate limits in design on maximum *p*-cycle circumference or number of straddling spans, to limit the dual failure availability implications.

IV. Oriented Cycle Double Covers

Let us now explain the work of Ellinas [11] on oriented cycle double covers. Given a graph G , a cycle cover is any decomposition of *G* into elementary cycles such that each edge is present in at least one cycle. To a large degree conventional ring network planning is an exercise in finding a suitably low cost cycle cover. In practice it is more than this, however, because the cycles have to take on a modular capacitated nature and some demands may be re-routed to offload an edge from the cycle-coverage requirement, a practice known as span elimination. But fundamentally every ring-based transport network comprises a specialized form of cycle-cover.

In practice where the ring systems operate at discrete OC-n or DWDM λ capacity levels the problem is not just one of finding a set of graph cycles that cover all spans of the facilities graph. One needs also to provide for different capacity requirements on each

[2] This is the ratio of total unused working plus protection capacity to used working capacity, i.e., here: [2(0.60) + 2] / 2(0.40) = 4.

[3] Note that these redundancy measures take unused working capacity into account. In other contexts one would say that the pure protection-to-working fiber-count redundancies were 100%, 50%, and zero in Fig.s 4, 5, 6, respectively

edge, and usually wants to do so with a minimum number of ring systems arranged in a near minimum-cost configuration.

But there is a technical context where the survivability design problem can be reduced to that of finding a pure logical cycle cover without further considerations of different specific capacity levels on each cycle. This is the paradigm of "four-fiber" networks (found in both [11,12]) where every span of the network is assumed to be comprised of exactly one bidirectional pair of working fibers and a completely matched pair of protection fibers. DWDM allows the consideration of such "flat" fixed-fiber-count networks, at least theoretically, by assuming that any needed number of lightwave channels can be supported on each fiber pair. When one is working in this framework, it can be appreciated that the only thing that makes a cycle cover *more* than 100 % redundant (in the sense of footnote 2) is the unavoidability of span overlaps; the same problem illustrated in Fig. 3. Somehow, if these span overlaps could be avoided, a "4-fiber" cycle-cover network design could become more efficient. In the best case it would get to where exactly 1-for-1 matching of protection and working fiber (or channel counts) are needed on every edge.

The problem which ring network planners know well, is that in general one cannot find a set of (bidirectional) covering cycles that cover every span only once. Some spans unavoidably wind up having two or more rings overlapping on them in order to support the set of cycles that cover other spans. In fact it can be fairly easily appreciated that if a graph is non-Eulerian, then more than one protection fiber will be needed on at least one span because the covering cycles are degree-2 structures and yet at least one node in a non-Eulerian network is of odd degree. Fig. 7 illustrates intuitively why odd degree nodes lead unavoidably to span overlaps in ordinary ring planning.

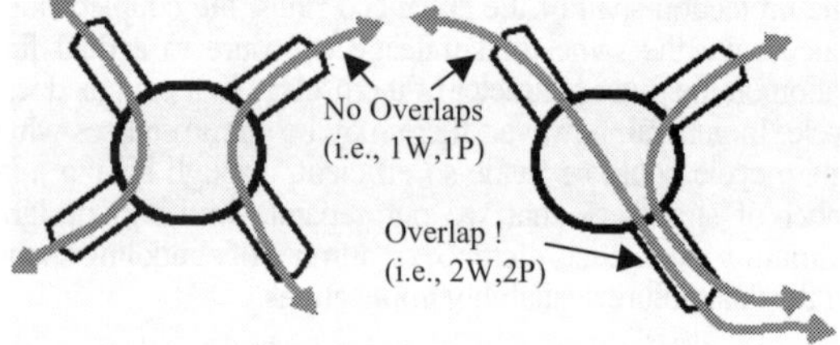

Fig. 7. With bidirectional cycles a cover without overlaps is possible at even degree nodes but not odd degree nodes.

In Fig.7 that each line represents two anti-directional fibers. Ellinas et al. [11] observe that this bi-directional nature of the actual transmission systems is almost always separable so that one could take a unidirectional view to the formation of single-fiber cycles. What [11] shows is that if one works with individual *unidirectional* cycles you can always derive a layout of such "oriented" cycles on a planar graph so that all span overlaps are avoided! Such a cycle set is called an *oriented cycle double cover* (O-CDC). Others have also exploited the same basic principle under the heading of "generalized loopback networks" [12]. Both approaches enable design solutions having *exactly* one working and one protection fibers on every span, for 100% fiber-count redundancy by avoiding span overlaps through an oriented planning paradigm. This quite interesting general mathematical result can be appreciated first-hand from the simple demonstration in Fig. 8. that with oriented cycles, it *is* possible to find a set of cycles to cover each span at the odd degree node with exactly one working and one protection fiber per span, i.e., with no span overlap, whereas this was not possible with bidirectional cycles.

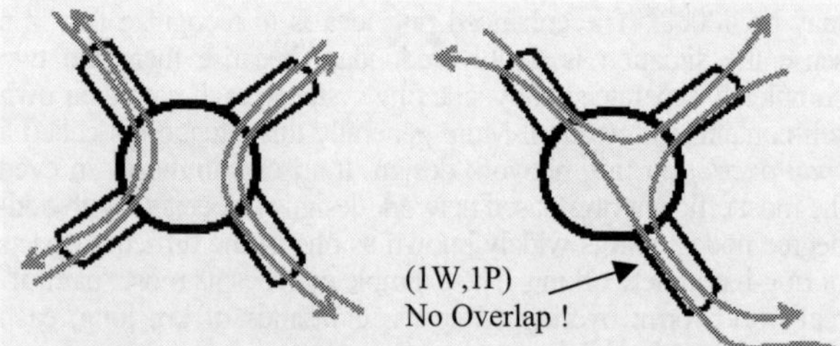

Fig. 8 With oriented cycles the even-degree node cover is equivalent but at an odd degree node the overlap can be avoided! (An identical anti-directional set of fibers is implied but not shown)

More generally [11] proves that as long as the graph is planar it is always possible to find a set of unidirectional cycles that covers every edge of the graph *exactly twice* (i.e., equivalent to only once with a bidirectional cycle). Rather coincidentally this principle for achieving 100% fiber-level protection redundancy appeared in the same issue of JSAC [11] which described *p*-cycle applications to the IP networking layer [6]. Unintentionally, both papers used the term "protection cycles"[4] in their titles, sowing the seeds for confusion. But O-CDCs are clearly not *p*-cycles. As elegant as the basic CDC cover result is, it still only achieves the goal of achieving *no worse than* 100% redundancy and protection remains logically ring-like in all cases. There are no straddling spans (which would bear zero protection) anywhere in an O-CDC design. In contrast *p*-cycles are far more efficient, involve the use of straddling spans, and can be efficiently planned on any form of arbitrarily capacitated facilities graph and are not limited to the "4-fiber network" paradigm.

REFERENCES

[1] W. D. Grover, D Stamatelakis, "Bridging the ring-mesh dichotomy with *p*-cycles", *Proc. DRCN 2000,* Munich, April 2000, pp. 92-104.

[2] W.D. Grover, D. Stamatelakis, "Cycle-oriented distributed preconfiguration: Ring-like speed with mesh-like capacity…," *Proc. ICC'98*, 1998, pp. 537-543.

[3] D. Stamatelakis, W.D. Grover, "Theoretical Underpinnings …," *IEEE Trans. Comm*, v.48, no.8, Aug. 2000, pp. 1262-65.

[4] D. Stamatelakis, …Preconfiguration of Spare Capacity in Mesh Restorable Networks, M.Sc. Thesis, Univ. of Alberta, 1997.

[5] D. Stamatelakis, W.D. Grover, "OPNET Simulation of Self-organizing Restorable SONET Mesh Transport Networks", *Proc. OPNETWORKS '98* (CD-ROM), Wash. D.C., April 1998, paper 04.

[6] D. Stamatelakis, W. D. Grover, "IP layer restoration … based on virtual protection cycles," *IEEE JSAC*, Oct. 2000, pp. 1938 - 1949.

[7] D. A. Schupke, C. G. Gruber, A. Autenrieth, "Optimal configuration of *p*-cycles in WDM networks," *Proc. ICC'02*, NYC, 2002.

[8] W. D. Grover, J. E. Doucette, "Advances in optical network design with *p*-cycles: Joint optimization and pre-selection …," in *Proc. IEEE-LEOS Topical Meetings*, Quebec, July 15-17, 2002.

[9] W.D. Grover, G. Shen, "Extending the *p*-Cycle Concept to Path-Protection," submitted to ICC 2003, Aug. 2002 (patent pending).

[10] J. A. Fee, *Method and System for Optical Restoration Tributary Switching in a Fiber Network*, U.S. Patent #5,884,017, Mar. 1999.

[11] G. Ellinas, A. G. Hailemariam, T.E. Stern, "Protection Cycles in Mesh WDM Networks," *IEEE JSAC,* Oct.2000, pp. 1924-37.

[12] M. Medard et.al., "Generalized loop-back recovery in optical mesh networks," *IEEE Trans. Networking,* Feb.2002, pp. 153-164.

[13] L.Lipes, "Understanding the trade-offs associated with sharing protection," OFC 2002, Paper ThGG121, pp.786-787.

[4] The "*p*" in *p*-cycles was chosen to suggest aspects of both protection and pre-connection concepts in general, but in early literature *p*-cycles they were variously also called "protection cycles."

Fast Restoration Mechanism through the GMPLS in Optical Networks

*Hyeon Park, *Sun-Hee Yang, **Wan-Seop Lee, **Sang-Ha Kim
*Electronics and Telecommunications Research Institute, Yusong P.O.Box 106, Taejon, 305-600, KOREA
**Chungnam National University, Yusong-Gu, Taejon, 305-600,KOREA,
*email:{hpark, shyang}@etri.re.kr, fax:+82-42-860-5440

Abstract

A single fiber cut can potentially influence a total of 1.6 Tb/s of traffic [1] and is surprisingly frequent and results in significant loss of traffic [2]. The restoration time in the optical networks consisted of optical cross connector OXCs is slower than that of the SONET/SDH networks.

For the fast link-based protection several mechanisms for the existing MPLS network are currently under discussion in IETF. However, there are some problems to apply the mechanisms to the optical network.

So, we describe the deficiencies of the existing fast re-route mechanism and then propose the fast link restoration mechanism using the extensions of LMP for the optical network and simulate the restoration time performance. The key characteristics of the proposed mechanism are the transmission of the main LSP label for the back up path by the LMP ChannelStatus message.

1. Introduction

In recent, using the wavelength division mutiplexer WDM technique to provide the high capacity transport network facilities, a single fiber can transport 100 or more different wavelengths (each 10Gb/s). Therefore a single fiber cut can potentially influence a total of 1.6 Tb/s of traffic [1] and is surprisingly frequent and serious. In other words, a single fiber failure can lead to the failure of all lightpaths traversing the failed fiber, and loss of each lightpath, operating at up to 10Gb/s, results in significant loss of traffic [2].

The restoration time in the optical networks consisted of optical cross connector OXCs is slower than that of the SONET/SDH networks. Meanwhile, because the multilayer architecture including the SONET/SDH layer has the capability of the fast recovery time, known as 50 ms, the loss of traffic in the SONET/SDH is trivial.

In general, the survivability can be accomplished by two types of schemes: protection and restoration from the failure. In the protection scheme, the backup path to replace the failed working path is predetermined and put onto service or not. On the other hand, the backup path in the restoration scheme is dynamically recomputed using the available resources of the networks when the failure is detected. There are also two types of restoration (also protection) schemes according to the failure range to restore: link-based and path-based restoration. In the path-based restoration case, the entire label switched path LSP is rerouted at the source label switched router LSR of the working path. In the link-based case, however, only individual LSP hops are rerouted, and it can be considered a degenerate (single-hop) case of path-based recovery [3].

For the fast link-based protection several mechanisms [4][5] in the existing MPLS network are currently under discussion in IETF. However, there are some problems to apply the mechanisms to the optical network. As described above, the link management in optical networks is important and the link management protocol LMP to manage the links is only one new specialized protocol required to support the operations of generalized multi-protocol label switching GMPLS.

So, we describe the deficiencies of the existing fast re-route mechanism and then propose the fast link restoration mechanism using the extensions of LMP for the optical network. The key characteristics of the proposed mechanism are the transmission of the main LSP label for the back up path by the LMP ChannelStatus message when the failure is occurred on the main LSP and the fast fault detection by the LMP fault localization.

This paper is organized as follows. The fast re-route method for the MPLS networks is described in Section 2. And the problems of the existed mechanism are described and the fast restoration mechanism for the optical network based on GMPLS control plane is also proposed in Section 3. In Section 4, the performance comparison of two mechanisms is provided. The conclusion and further study are described in Section 5.

2. Fast Re-Route in MPLS networks

2.1 Definition of the fast re-route

Fast re-route is a process where MPLS data can be directed around a link failure without the need to perform any signaling at the time that failure is detected [7]. Unlike protection switching, the repair point is the point of failure detection. Consequently there is no requirement to propagate the error to the repair point using the signaling protocol [7]. The fast re-route protection can be achieved through a parallel virtual link or the detour paths set up automatically.

2.2 Fast re-route protection with detours

In this section, we briefly describe the mechanism of the MPLS LSP fast re-route using RSVP detours. Fig. 1 shows the main LSP (mLSP) with the detour paths. If the failure between Node A and Node C is occurred, the detour path Detour 1 to replace the failure for the entire mLSP is offered.

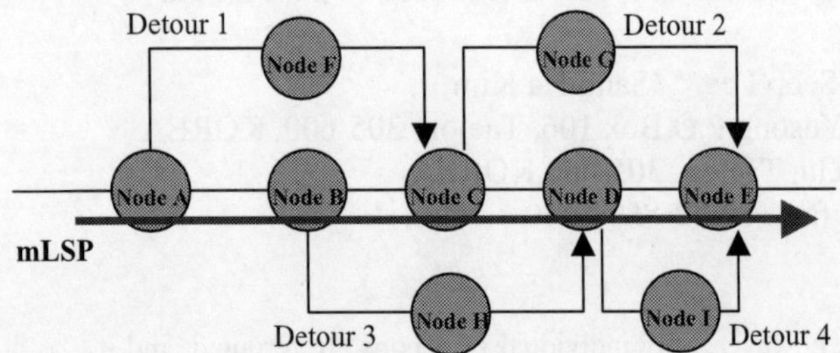

<Figure 1> The detour paths of the main LSP

To support the LSP fast re-route two new objects of RSVP are defined [5]. The one is the FAST-REROUTE object carried in the RSVP Path message. It consists of the setup priority for the session preemption, bandwidth, exclude/include colors representing a set of attribute filters associated with a detour and so on. The other is DETOUR object, which is consisted of the source address identifying the beginning point of detour and the downstream node address identifying the downstream node that source is trying to avoid. Following is the brief fast re-route procedure using the information.

- In the ingress node, to request a fast-reroute the Path message carries a FAST_REROUTE object and the object information is stored in the node for later Path refreshes.
- When a node receives FAST_REROUTE objects in the Path messages, it can trigger CSPF computation periodically to find out where to setup the detour paths.
- The detour path originates from current node and should satisfy all the requirements as specified in the FAST_REROUTE object.
- To establish a detour path the signaling using RSVP is started immediately and the branching node (for example, node A for Detour 1) must mix the messages for the main and the detour LSP's.
- The detour path must be merged back into the main LSP automatically at its destination.
- When the branching node receives ResvTear messages from the downstream for the main LSP due to error conditions the ResvTear messages must be not sent to the upstream if the detour is up.

Thus, the detour paths are set up dynamically without any operator intervention. However, there are several deficiencies of the fast re-route mechanism as described below.

2.3 The deficiencies of fast re-route mechanism

The deficiencies of the fast re-route mechanism in the existing mechanism are as follows.

- At the detour merging node, a detour LSP merges back onto a main LSP and both traffic must be going out through the same label. However, in this approach, there is no the description about them. In optical networks, the label (channel, i.e. wavelength) allocation is important because of the main part of the system cost.
- Backup path such as that of 1:1 or 1:n protection mechanisms can be used for the other working path. However, in this approach, all the detour paths are reserved for the main LSP, therefore, the resource utilization is very low. In optical networks, because the bandwidth of one wavelength is about 10G almost of the backup paths are required to reuse.
- This approach did not describe the restoration speed, which includes the failure detection time, the failure localization time, the failure mitigation time, and so on.

As a result, the main issues are the increased complexity of configuration and the amount of resources that must be reserved in the network. Furthermore, essentially, the signaling protocols in MPLS networks impose some restrictions to the extent of the local repair. For example, CR-LDP does not include facilities for routing at the node that detects the fault. In the next section, we propose the fast local repair method using the GMPLS signaling functionality, which takes quicker restoration time than the existing signaling functionality.

3. Fast restoration in GMPLS networks

3.1 The fast re-route problems in the existing networks

There are some problems to adapt the signaling functionality of the existing networks for that of the optical networks.

MPLS is designed to logically separate the control plane from the data plane [8]. However, the fault detection should be handled at the layer (i.e., the physical layer (or data plane)) closet to the failure. Thus, the fault detection time in the existing networks is not so fast. Furthermore, the fault detection by the soft state mechanism at the control plane (for example, it takes at least 15sec to detect the LSP failure) as well as fault notification by unnecessary message processing at the intermediate nodes may delay the fault restoration.

GMPLS to address some limitations of the MPLS as described above extends the signaling and the routing functionality. For example, the Notify message has been added to RSVP-TE for GMPLS to provide a mechanism informing nonadjacent nodes of LSP-related failures [8]. And LMP includes the procedure to localize failures in optical networks. In the next section, we propose the fast restoration method using the GMPLS signaling protocols in optical networks.

3.2 Failure detection and localization

In an optical network, most of the switches are photonic. How to localize the location of the failure is important and LMP provides the rapid detection of the data link failure, notification, and the mitigation. As shown in Fig. 2, if there is a link failure (i.e., LOL: Loss Of Light) on a link span between node B and E in the lightpath LP, node J will detect the link failure and report the failure to node H through the control channel. And then the node H correlates the failure locally and also reports the failure to node E.

Eventually, node B notifies the link failure to the signalling protocol. Then the signalling protocol can be used to initiate span or path protection/restoration procedures. Thus, since the failure detection is directly delegate to the physical layer instead of layer 2 or layer 3 the failure processing is achieved faster than by the existing signalling protocol.

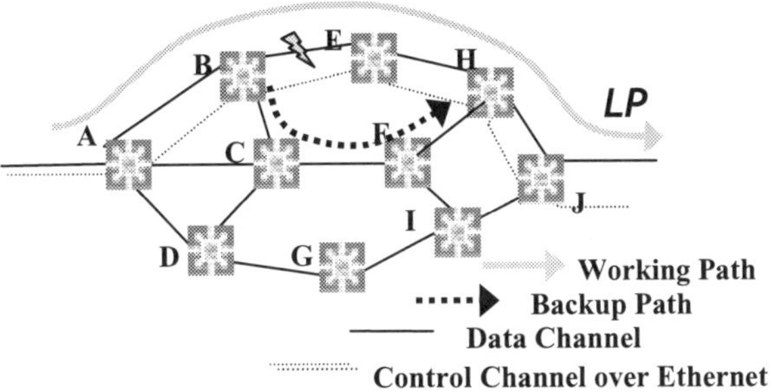

< Figure 2 > Link Restoration

LMP also provides the configuration and management of ease for the links of the optical network. The control channel can be established by discovering the IP address of the adjacent node automatically. Furthermore, thousands of data links between the adjacent nodes are aggregated into a smaller number of Traffic Engineering (TE) links using the link property correlation and verification functionality. Thus, control channels and data links are automatically established and maintained, respectively.

3.3 Fast restoration mechanism

Working LSP establishment

To establish the working path LSP ***LP*** the ingress node A generates Path message for the label request and sends it to the downstream node. Path message includes the explicit routes of the entire LSP ***LP*** and the following processing, whenever it goes through intermediate nodes, is triggered.

- Each node pre-calculates the backup path against to the failure event of the neighbor node. It means that the backup path is calculated before failure and the channels assignments of failed connections along the restoration path is also calculated, however, for resource reuse, not cross-connected the channels along the restoration path. For example, in Fig. 2 the shortest backup path B-C-F-H for the failure between B and F (link or node) is selected.
- The setup information (LSP traffic metric, LSP id, Label and so on) for the backup path establishment is stored at each node.

Backup LSP activation

- Each node H, J detects the LOL of main LSP ***LP*** when the failure between node B and node H is occurred. According to the procedure of the fault localization as described above, node J notifies the failure to the node H, node H also notifies the failure to node E.
- And then, at last, node E notifies the failure to node B, via the control channel over Ethernet separated with the data channel over the lambda. The ChannelStatus message used to notify includes the LSP id of the LSP ***LP*** and its outgoing label. And its message format and the added TLV is as shown in Fig. 3. And the RSVP-TE Path message containing the Label request TLV is also shown in Fig. 3.

<ChannelStatus Message>

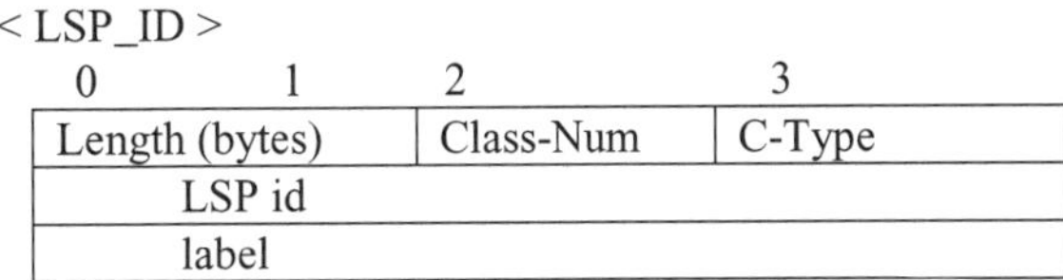

Common Header	Local Link ID	Message ID	Channel Status	LSP ID	LSP ID

< LSP_ID >

0	1	2	3
Length (bytes)		Class-Num	C-Type
LSP id			
label			

<Path message>

<Path Message> ::= <Common Header>
[<INTEGRITY>]
<SESSION> <RSVP_HOP>
<TIME_VALUES>
[<EXPLICIT_ROUTE>]
<LABEL_REQUEST>
[<SESSION_ATTRIBUTE>]
[<POLICY_DATA> ...]
<sender descriptor>

< Figure 3 > The Message Format

- Node B receives the LSP id and the label information from the E via the control channel. And node B searches for the backup path of the failed LSP using the LSP id.
- Using the information node B activates (i.e., cross-connects the lambda) the backup path B-C-F-H.
- To set up the backup path the inbound label for the node C is suggested as the outbound label 6 of node B. And the outbound label 7 for node F is suggested as the inbound label node H merging the main LSP ***LP*** and the labels of the intermediate nodes are not cared. The RSVP-TE message contains the Label Request TLV with the suggested label.

4. Performance results

In general, protection time is depends on a series of sequential events that need to take place, depending on the type of protocol used [9]. The total restoration time includes the fault detection time, the fault localization time, the fault notification time, and the fault mitigation time.

As described above, the fast re-route mechanism in the existing MPLS networks sets up the back up path before failure, in other words, the back up path is pre-established. Therefore, its mitigation time may be faster than the proposed mechanism. However, because of the fast-alive monitoring and the fault localization mechanism of LMP our mechanism will offer better fault detection time and fault localization time than the existing mechanism described in section 2.

Two mechanisms described in the previous section have been simulated through the test network model. Restoration time performance is evaluated with the number of nodes in the network. The test network is depicted in Fig.4. And the total restoration time is evaluated with the fault localization time and the fault mitigation time. In case of the proposed mechanism, the fault detection time is faster than the existing mechanism due to the fast keep-alive hello and the fault management mechanism.

The MNS 2.0(MPLS Network Simulator) aware NS 2.1b6a(Network Simulator) [10] is used to assess the performance the proposed mechanism. Traffic model is exponetial traffic and link rate is 1 Mbit/s. We assumed that the fault detection time of our mechanism is 0.15sec while the existing mechanism is about 2.1. Because our mechanism uses the LMP fault management mechanism. Fig. 5 shows the data reception rate over the data traffic flow time. It indicates that the restoration of our mechanism after the failure is faster than the existing mechanism.

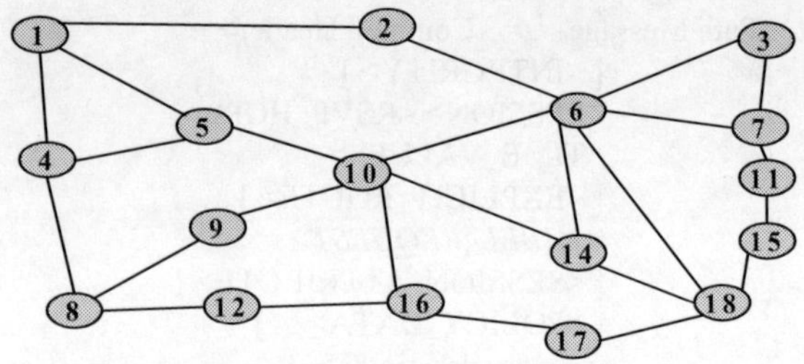

<Figure 4> The test network

Based on the result in Fig. 5 the number of nodes in a network according to the deley time of the restoration. The delay time is summarized as the fault detection time, the messaging processing time, the switch over time, and so on.

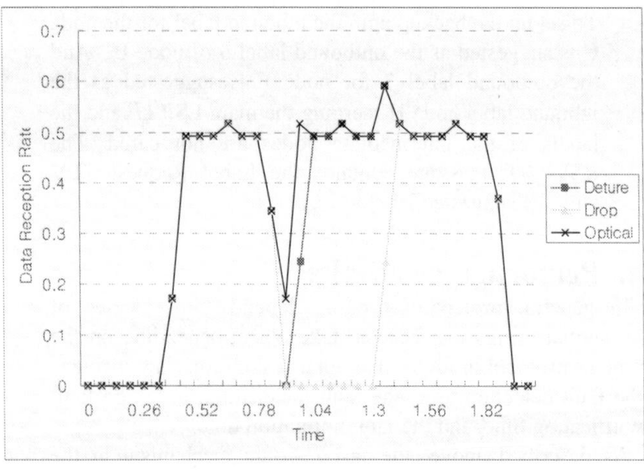

<Figure 5> The data reception rate

5. Conclusions

We described the deficiencies of the existing fast re-route and then proposed the fast link restoration mechanism using the extensions of LMP for the optical network. And we provided the performance comparison of two mechanisms.

In optical networks, most of the connection may be bi-directed due to the fast connection set-up and, moreover, GMPLS is easier to support this procedure while MPLS networks support just uni-direction. Both end nodes of the failed link are supposed to have the capable of setting up the rerouting path. The proposed approach did not consider the bi-directional LSP. For further study, we will extend our mecanism to process bi-directional signaling to reduce restoration time, for example setup latency.

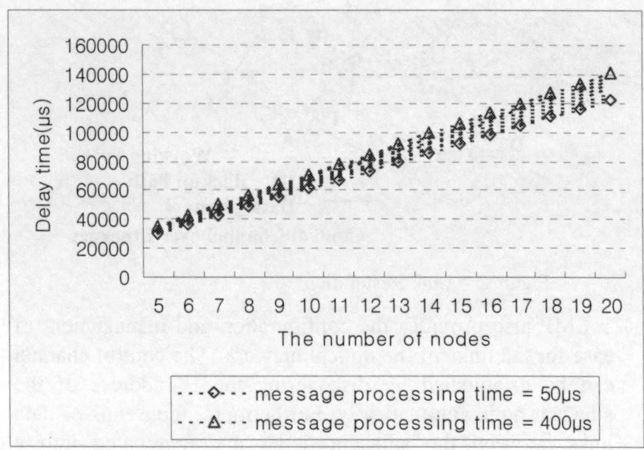

<Figure 6> The number of nodes according to the the delay time

Reference

[1] Sopher De Maesschalck *et al.*, " From Network Design to Dynamic Provisioning and Restoration in Optical Cross-Connect Mesh Networks: An Architectural and Algorithm Overview," IEEE Commun. Mag., vol.39, no. 1, Jan. 2001, pp.42-49.

[2] Y. Ye, *el al.*, "A Simple Dynamic Integrated Provisiong/Protection Scheme in IP over WDM Networks," *IEEE Commun. Mag.*, Nov. 2001, pp. 174-182.

[3] Chunming Qiao, "Labeled Optical Burst switching for IP-over-WDM Integration," *IEEE Commun. Mag.*, Sept. 2001, pp. 104-114.

[4] Data Connection White paper, Protection And Restoration in MPLS Networks, version 2, October 20001,

[5] Der-Hwa Gan *et al.*, "A Method for MPLS LSP Fast-Reroute Using RSVP Detours," Internet draft, draft-gan-fast-reroute-00.txt, work in progress, April 10, 2001

[6] Jonathan P. Lang *et al.*, "Link Management Protocol," Internet draft, draft-ietf-ccamp-lmp-03.txt, work in progress, March 2002.

[7] http://www.dataconnection.com

[8] Ayan Banerjee, *el al.*, "Generalized Multiprotocol Label Switching: An Overview of Signalling Enhancements and Recovery Techniques," *IEEE Commun. Mag.*, July. 2001, pp. 144-151.

[9] Rao, Lungampalli, *et al.*, "Performance of path protection in IP/optical core mesh networks with source-routed GMPLS protocols," OFC 2002, paper ThG4, March 2002.

[10] http://www.isi.edu/nsnam/ns/

A Design of Network Restoration and Connection Management for Optical Cross-Connect System Using Distributed Objects

Hyun-Soon Nam, Chang-Gyu Lim, Seong-Soon Joo, and Byung-Ho Ye

Electronics & Telecommunications Research Institute

namhs@etri.re.kr, ssjoo@etri.re.kr

Tel 82-042-860-5606, Fax 82-042-860-6104
KOREA

Abstraction

OXC (Optical Cross-Connect) system is on indispensable element to realize the all-optical mesh networks by supporting the increase of link channels and expansion of an OTN (optical transport network) capacity at low expense after investing initial equipment. This paper proposes a design of network restoration and connection management for OXC system based on distributed object-oriented technology. We describe the functions of NPRS, which handles UNI/NNI signaling of high layer client and manages network connection. This design of network connection and restoration can be applied in not only the layered-centralized network management but also the distributed network management.

Key Words: OTN restoration model, distributed network management

I. Introduction

Today Internet traffic is increasing exponentially and variety of users and QoS (Quality of Service) requirement demand a high speed and a large capacity of transfer network. These requirements bring the development of intelligence OTN (optical transport network) and signaling protocol for supporting high-level network layer.

OXC (Optical Cross-Connect) system becomes the most important element in OTN since it is necessary technologies that can realize all-optical mesh networks, and since IP based client network layer is required to integrate in OTN. OXC system has to provide network connection database management and network restoration management for managing the flexible and high reliable OTN.

Many official standardization bodies have made an effort to write recommendations on management of OTN and network survivability [2][3].

This paper proposes a model of network restoration and connection configuration management for OXC system in the distributed environment. This paper is organized as follows: in Section Ⅱ, trends on standardization CORBA (Common Object Request Broker Architecture) based service for OTN are described. An architecture of OXC system for a model of network restoration management and network connection configuration management is explained in Section Ⅲ, and a detail function design of proposed network restoration management and network connection configuration management is proposed in Section Ⅳ. The Section Ⅴ concludes the benefit of a proposed model.

II. Standardization efforts on CORBA based service for OTN

The distributed object-oriented design technique is used in large communication fields and several attemps to apply this technique to an optical communication network was done in. ITU-T (International Telecommunication Union - Telecommunication

Standardization Section), the representative official standards groups, defines standard CORBA based TMN (Telecom Management Network). In M.3210, CORBA Generic Network and NE Level Information Model is recommended. CORBA based TMN services are defined in Q.816 and guidelines for defining CORBS managed objects are described in X.780.

TMF (Tele-Management Forum) proposes NML-EML (Network Management Layer – Element Management Layer) interface for CORBA based network management [3]. This interface supports the Façade methodology and is composed with six mandatory managers and with three optional managers. TMF513 defines Multi-Technology Network Management Business Agreement to describe requirements and Use Cases of nine managers, TMF 608 shows classes and a class relationship (inheritance, dependency, extended etc.) of Use Cases, and TMF 814 proposes CORBA IDL (Interface Definition Language). Recently, ITU-T accepts TMF specification.

III. Architecture of OXC System

In this paper, we design a subsystem, NPRS (Network Provisioning and Restoration Subsystem), for network restoration management and network connection configuration. NPRS performs on various layers of OTN management shown as Figure 1; it works through NMS, EMS and NE level.

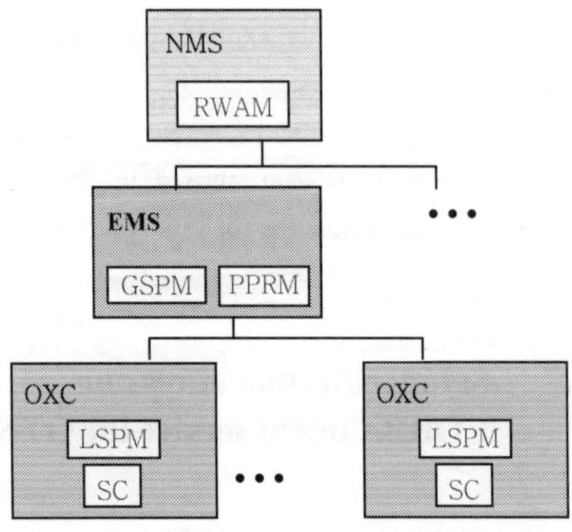

Figure 1. NPRS Architecture for layered Manager

On NMS (Network management System), RWAM (Route and Wavelength Assign Manager) selects route and assigns a wavelength for a connection. RWAM manages network resources and configures a network connection according to the request from PPRM.

On EMS (Element Network management System), GSPM (Global Switching Path Manager) and PPRM (Provisioning, Protection & Restoration Manager) are performed. GSPM manages all switching paths and connection information gathered from LSPM in OXCs, which construct physical connection path information and provide logical network connection information to RWAM. PPRM plays a major role to control the NPRS, and analyzes fault messages and performance information for protecting or restoring a connection. PPRM also handles connection requests originated by signaling protocol from high layer client router.

In OXC, LSPM (Local Switching Path Manager) and SC (Switching Controller) are located in. LSPM manages switching paths such as 2D MEMS switch configuration. LSPM controls various OXC units, switch and wavelength converter, and sends OXC switching resource information.

These managers are designed with distributed objects in CORBA architecture.

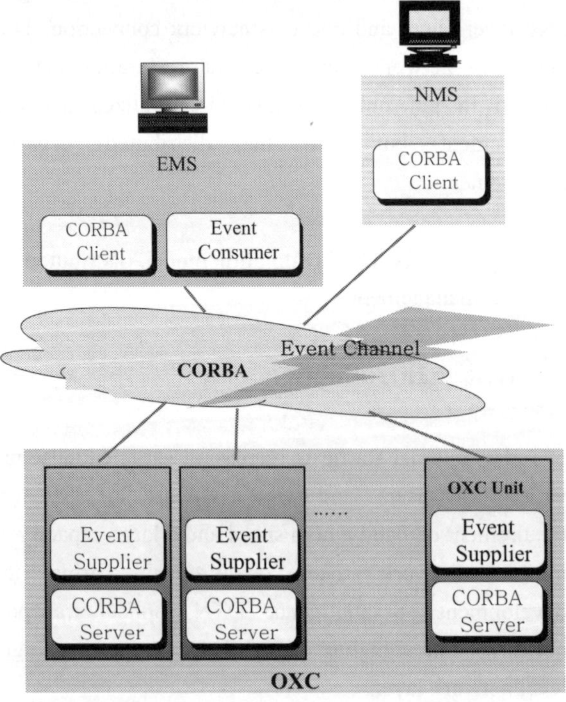

Figure 2. CORBA Architecture of OXC System

To manage fault information and performance information, network restoration management function is implemented with COS (Common Object Services) Event service. In Fiure2, The Event Consumer of EMS registers fault and performance channel types from all OXCs. When a fault manager in OXC perceives a fault or when performance manager in OXC updates a performance-information, Event Supplier of each manager sends information to Event Consumer registered its own Event Channel. Then, PPRM analyzes events and extracts network connection affected by fault or update, and requests RWAM to reroute the network connection.

To manage network connection, switch path management function is implemented using CORBA server/client model. GSPM, CORBA client, requests switching in-out port to LSPM, CORBA server.

All distributed objects in NPRS use COS Naming service that makes objects to invoke each other independent upon location and CORBA vendors.

IV. Distributed Objects Design of NPRS

NPRS supports the network restoration path with four-service level. Network connection restoration function classifies network resources according to the service level as shown table 1, and applies network resources following priority.

Table 1. Type of Network connection

Type	Description
Working	Network resource occupied for optical route
Protection	Network resource to protect a Working connection
Reserved	Network resource for pending restoration
Free	Network resource not involved in above

NPRS are composed of seven Use Cases expressed on Figure 3.

Actors of NPRS are network operator and UNI/NNI. Actor UNI/NNI client is signaling protocol handler of a high layer client router and it can communicate to all Use Case having relationship with Actor Operator by inheriting.

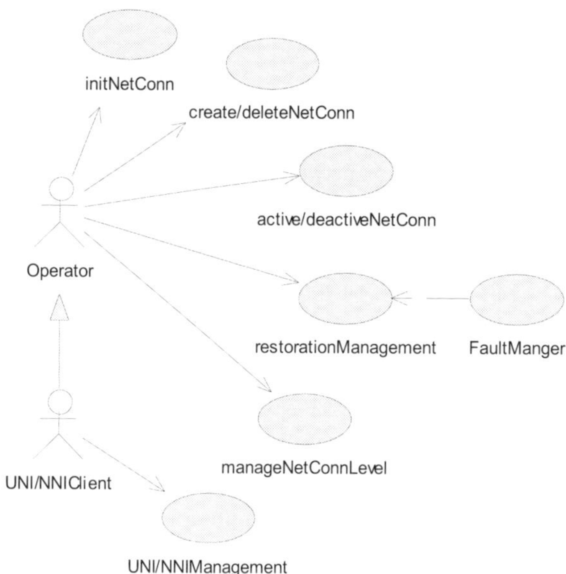

Figure 3. Use Cases Diagram of NPRS

1. Initiate Network Connection

When Actor Operator requests to initiate network connection or to starts/restarts NPRS, GSPM gathers physical path information from all of LSPMs, which are a hardware connection configuration and a resource information of OXC capacity such as number of wavelength converters, maximum number of wavelengths and maximum number of links, and so on.

2. Create and Delete Network Connection

Actor Operator or Actor UNI/NNIClient sends command to NPRS with source address of destination address of high layer client connected with ingress and egress OXC. PPRM searches network connection information. If a requested connection is already created, PPRM returns a status of network connection and reports to Actor Operator. If a requested connection is not created, PPRM sends create command to RWAM with source address, destination address and additional service level parameters. RWAM validates this command, and then selects a route for the network connection agreed with service level. PPRM requests to GSPM for creating a switch path. GSPM sends set switch command to LSPMs of OXCs within the route.

Delete network connection is performed as like create

command and Deactivated network connections set switching resource to be free.

3. Activate and Deactivate Network Connection

The created network connection is activated according to the type of network connection. When Working and Protecting network connection are created, they are activated at once, reserved type of network connection are activated when restoration is performed. If requested connection doesn't exist, PPRM returns a status of network connection and reports to Actor Operator.

Deactivate network connection command deactivates the switch operation on the network connections.

4. Analyze fault information

When a fault is happened or when performance manager updates data, Use Case Analyze fault information maps physical location information to logical switch paths and sends restoration command to Use Case restoration management. Because Use Case analyze fault information manages OXC connection configuration information tables, it can map fault information to path information.

5. Restoration management

When a fault is occurred or performance manager updates data, PPRM searches switch paths corresponding to change and gets sub-connections of network connection. PPRM supports for RWAM to reroute network connection with finding information.

6. Network Connection Level management

Service level of network connection is maintained by PPRM. Actor Operator provisions it in four levels, and RWAM maintains restoration routes corresponding to it.

7. UNI/NNI management

When PPRM receives requests from high layer client router, PPRM creates/deletes or activates/deactivates network connection corresponding to network connection level. And PPRM handles UNI/NNI signaling and maintains a status of requested network connection.

V. Conclusion

In this paper, a model of network connection and restoration management is designed based on CORBA distributed objects. We apply the CORBA naming service standards to support location independency of distributed objects and the CORBA event service standards to support asynchronous multicasting communication mechanism in distributed environment.

We propose that NPRS three managers, PPRM, RWAM and GSPM, and NPRS accomplishes seven sub-functions: initiate network connection, create and delete network connection, activate and deactivate network connection, restoration management, network connection level management, UNI/NNI management and analyze fault information.

The most popular model of OTN management has been layered-centralized architecture recommended in ITU-T. But IETF GMPLS (Generalized Multi Protocol Label Switching) requires distributed architecture for OTN management. Proposed design in this paper can support location independency using CORBA, and can be applied to both standardization bodies for OTN management.

REFERENCES

[1] S. Belotti, D. Papadimitriou, N. Larkin, Z. Lin and D. Pendarakis Generalized MPLS (GMPLS) RSVP-TE Usage and Extensions For Automatically Switched Optical Network (ASON), IETF draft, 2002

[2] ITU-T, http://www.itu.org

[3] TMF, http://www.tmforum.org

[4] ITU-T Draft Rec. G.875 "Optical transport network (OTN) management information model for the network element view"

[5] ITU-T Draft Rec. G.872 "Architecture of Optical Transport Network"

[6] TMF 513 "Multi-Technology Network Management Business Agreement"

[7] TMF 608 "Multi-Technology Network management Information Agreement"

[8] D. Benjamin, R. Trudel, and E. Kus, "Optical Services over the Intelligent Optical Network" IEEE Communication magazine, sep. 2001, pp 73-78

[9] Gardy Booch et al, *The Unified Modeling Language User Guide*, Addison-Wesley, 1999.

A Distributed Dynamic Restoration Protocol using Disjoint Path-Wavelength Grouping in WDM Networks

C. Sivakumar G. Mohan[1]
Electrical and Computer Engineering Department
National University of Singapore, Singapore 117576
Email: elegm@nus.edu.sg

Abstract— *Wavelength division multiplexing* **(WDM) networks have become a viable solution to meet the increasing bandwidth demand. These networks carry messages in the optical domain at data rates of several Gb/s. Such high data rates make it necessary for using efficient restoration protocols in an event of a failure. This paper addresses the problem of dynamic path-based restoration in WDM optical networks. Our objective is to develop a protocol that ensures rapid recovery while keeping the restoration ratio high. We propose two distributed dynamic restoration protocols called disjoint** *path-wavelength grouping protocol* **and disjoint** *weighted path-wavelength grouping protocol* **. Both these protocols group all the candidate backup paths of the failed paths into path-disjoint groups before assigning wavelengths to them. This allows the failed paths to search different sets of wavelengths on various candidate backup paths in parallel without any reservation conflicts. Also, it does not require any reservation** *retries*, **thus resulting in rapid restoration. The grouping method attempts to restore increased number of failed paths by allowing increased number of wavelengths to be searched on a path. The performance of the proposed protocol is verified through extensive simulation experiments.**

Key Words: **Wavelength division multiplexing, network survivability, protection and restoration, distributed protocol.**

I. Introduction

Wavelength division multiplexing (WDM) technology divides the huge bandwidth available on a fiber into non-overlapping wavelength channels. A communication channel, or *lightpath*, between source and destination nodes is uniquely defined by a physical path and a wavelength. Due to large data rates, network survivability is an important issue in WDM networks [1], [2], [3].

In protection-based restoration, backup lightpaths are reserved when establishing the primary lightpath. This gives good *restoration ratio* (number of lightpaths restored to the total number of lightpaths failed) and low *restoration time* (time taken to restore communication after the failure has occurred) but the resource utilization is inefficient. In a dynamic restoration method, backup paths are searched when the existing lightpath fails. There are no resource reservation, but restoration ratio is poor. Both protection and dynamic restoration allow path-based (end-to-end detouring) or link-based (local detouring) restoration. In the literature a number of work have been carried out on the WDM survivability problem [1], [2], [4], [5]. Survivability techniques for IP-MPLS/WDM networks have been studied in [6], [7]

A path-based dynamic restoration algorithm is proposed where the candidate backup paths of primary paths in a given link are grouped into link-disjoint sets. Wavelengths are distributed to each of the sets. The grouping method attempts to improve restoration ratio. The performance of the proposed protocol is verified through extensive simulation experiments. The related work is described in section II. The proposed protocol is discussed in section III. Performance study is carried out in section IV before concluding in section V.

II. Related Work

A distributed dynamic restoration methods has been proposed in [5]. The source node of a failed path searches for a backup path on the same wavelength. Any unused wavelengths are uniformly distributed among the failed paths, creating groups of wavelengths. The number of wavelength groups is equal to the number of failed paths. The source node floods the network with control message to probe for an available backup path.

Consider a network with 8 wavelengths per link. Assume that a failed link carried 5 primary paths, $p_1, \ldots, p_5$ using wavelengths $w_1, \ldots, w_5$, respectively. The source nodes probes the network on the wavelength used by the primary path. The free wavelengths, w_6, w_7, w_8 are distributed evenly, i.e, p_1, p_2, p_3 uses w_6, w_7, w_8, respectively. We call this method as *wavelength grouping* (WG) restoration.

Our work uses path-wavelength grouping rather than wavelength grouping to improve restoration ratio. Furthermore, we send control messages along pre-defined alternate paths rather than flooding to reduce signalling overhead.

III. Proposed Dynamic Restoration Protocol

We develop a disjoint *path-wavelength group* (PWG) restoration protocol. The motivation is to reduce the number of groups by tightly packing candidate backup paths into sets. This would lead to a better distribution of wavelengths for each group, hence increasing restoration ratio. Our algorithm ensures that the number of wavelength groups is at most equal to the number of failed paths.

Let L be a link in a network. Each link is associated with groups, G_1, $G_2, \ldots, G_k$, where k is the number of wavelength groups. The candidate backup paths of the paths that traverse along L are grouped such that a particular set G_i contains candidate backup paths, possibly belonging to different primary lightpaths, that are link-disjoint. Initially, all the sets G are null sets.

For example, let two paths p_1 and p_2 traverse link L. Let the candidate backup paths b_1^1 and b_1^2 of p_1 traverse links $l_1 \rightarrow l_2$ and $l_3 \rightarrow l_4$ respectively. Let the candidate backup paths b_2^1 and b_2^2 of p_2 traverse links $l_5 \rightarrow l_6$ and $l_7 \rightarrow l_8$, respectively. Since all the candidate paths are link-disjoint,

[1]Author for correspondence. This work was supported in part by the National University of Singapore Academic Research Fund under Grant R-263-000-173-112.

our algorithm keeps them in one group. If there are 8 wavelengths per link, then upon failure all 8 wavelengths are searched on the candidate backup paths by the source nodes of the failed paths. But wavelength-grouping method will form two groups. The source nodes will search only 4 wavelengths. This illustrates why path-wavelength grouping is expected to restore more paths than wavelength grouping.

We use alternate routing method to establish a path for an incoming request. When a request arrives at a source node, the possible paths are all probed and the shortest available path is chosen among them. The information about the candidate backup paths are then sent to facilitate grouping at every link that the primary path traverses. Grouping is done at the time of setting up of primary lightpaths, reducing the computational burden at the time of failures.

A. Restoration Protocol Description

Assume that link L has failed. The end nodes of L would then send a $Notify$ message to the end nodes of the affected primary paths. At the same time, the wavelengths are evenly distributed among the groups G_1 to G_k. A *wavelength assignment* indicating the wavelengths to be searched by each of the candidate backup paths for a failed path is carried by the $Notify$ message.

After receiving the $Notify$ message, the source nodes would send out $Collect$ message on the candidate backup paths. This signal contains the set of wavelengths, S_w, that it has to search for. At every link, it would find the set of free wavelengths F on the link and finds the common wavelengths between the sets. This message then traverses to the next node and the process is repeated until it reaches the destination node.

When the $Collect$ message reaches the end node, it would check whether set S_w contains some elements. If set S_w has some elements, the wavelength with the lowest index is chosen. The $Success$ control message is transmitted to the source node. At every intermediate node, this signal would configure the cross-connects for the chosen wavelength. The destination node ignores any subsequent $Collect$ messages.

If the set S_w, upon reaching a node, is a null set, then a $Fail$ message is transmitted to the source node. If the source node receives $Fail$ messages from all the candidate paths, then restoration fails.

B. Weighted Path-Wavelength Grouping

Previously, wavelengths are distributed evenly among the groups formed. This type of distribution might not be optimal. Suppose a link L has formed two groups, $G1$ and $G2$. consisting of 6 and 2 paths respectively. Assuming 8 wavelengths per link, each group would receive 4 wavelengths each. It can be seen from this example that lightly loaded groups ($G2$) command equal share of wavelength as tightly packed groups ($G1$). We propose to use a weighted distribution to rectify this problem. each group is assigned $x = (P_g/P) * MAX_n$ wavelengths where P_g is the number of paths in group G, $P = \Sigma P_g$ for a particular link L and MAX_n is the number of wavelengths per link. In this example, $G1$ would receive 6 wavelengths and $G2$ would receive 2 wavelengths. We call this protocol *weighted path-wavelength restoration* (wPWG).

C. Restoration Time

One of the objectives of this protocol is to make the restoration time close to that of a protection-based restoration scheme. Restoration time has many components. Let A be the fault detection time. Let P be the propagation delay of each link. Let B be the message processing and transmission time at each node. Let C be the time taken to process a control message and configure the cross-connect at a node. Let x be the number of hops from the end node of the failed link to the source node. Let y be the number of hops on a candidate backup path.

In a backup multiplexing based protection scheme, when source node is notified of a failed link, a message is sent from source to destination to configure the switches. Then the source node waits to receive an acknowledgement from the destination node. In this case, the restoration time would be $A + xP + (x+1)B + (y+1)B + (y+1)C + 2yP$.

In our grouping algorithm, upon failure, the source node would search all the backup paths simultaneously. It would then wait for a $Success$ message. Upon receiving a $Success$ message, it would select the backup path. The restoration time in this case would be $A + xP + (x+1)B + (y+1)B + (y+1)C + 2yP$. The above expression same as the protection case. However, the restoration time is slightly longer for this grouping algorithm due to the larger control messages and increased the transmission and message processing time. But this increase is not significant.

IV. Performance Study

A. Simulation Network

We conduct simulation experiments to verify the effectiveness of our PWG and wPWG restoration protocols. In this simulation, we assume that the link failure is detected first by the end nodes of the failed link, the control network is reliable, uses the same physical layer as the data network. Two 32-node networks are randomly generated with different degree of connectivity. Network $G1$ has 102 bi-directional links and the other network $G2$ has 81 bi-directional links. $G2$ represents a sparsely connected network, while $G1$ represents a densely connected network.

B. Simulation Parameters

We use the values for various parameters as follows. The message processing time, B, at a node is 10 microseconds. The propagation delay P on each link is randomly generated, between 100 microseconds to 1 millisecond that corresponds to approximately 20 to 200 kilometers. Time to configure the cross connects of each node C is 500 microseconds [1]. Every link is assumed to carry 16 wavelengths. We assume that the destination node is uniformly selected among all nodes except the source node. The calls arrive into the network as a Poisson process. The holding time is exponentially distributed with unit mean. We have neglected in our calculation to add the fault detection time as it is a constant value throughout and do not affect our findings when a comparison study is done. Also, the realistic values for failure detection are not known except that it is on the order of milliseconds.

C. Performance Metrics

The following metrics are used to study the performance; *blocking probability ,restoration ratio* and *restoration time*. In order to determine the restoration ratio and restoration time, we artificially failed links one at a time at a specific interval and the average of the restoration ratio and restoration time are calculated.

We have taken four different restoration scenarios.

1. Weighted Path-Wavelength grouping (wPWG)
2. Path-Wavelength grouping (PWG)
3. Wavelength grouping (WG)
4. Restoration with unlimited retries (URT)- backup paths are allowed to search on any wavelength with unlimited number of retries. This method gives the best performance in terms of restoration ratio, but it fosters conflicts for resources by the simultaneous search attempts leading to increased number of retries and increased restoration time. This method is not desirable in the perspective of restoration time. However, we use its performance in terms of restoration ratio as an upper bound for the purpose of comparison.

D. Results and Analysis

Figure 1 plots the blocking probability against various traffic load (arrival rate per node in Erlangs) for the dense network $G1$ for the proposed PWG restoration and backup multiplexing (BM)-based protection scheme. As arrival rate increases, the blocking probability increases as network resources are scarce at higher arrival rates. This graph shows the resource utilization gain is more when one adopts a restoration protocol as opposed to a protection-based method. The blocking probability of wPWG protocol is the same as PWG protocol.

In Fig. 2,the restoration ratio decreases as the arrival rate increases since network resources are used up. At lower arrival rates, both PWG protocol and WG protocol restores a significant percentage of the failed paths as free resources are available. As arrival rate increases, the restoration ratio of PWG protocol gives an improved performance over WG protocol. The wPWG distribution does better than PWG restoration. per link is high. The URT plot shows the number of paths restored if no restrictions on wavelengths searched was imposed. But this would result in undesirable reservation conflicts. The performance of the PWG protocol is closer to the URT protocol when the traffic load is light. A large deviation is noted when the traffic load is high because at high arrival rates the effect of restricted wavelength usage due to grouping becomes worse.

Figure 3 plots the restoration time in milliseconds for different arrival rates for $G1$. The restoration time increases as the arrival rate increases up to a certain point. At low arrival rates, the primary path chosen is usually the shortest path. But as the arrival rate increases, the longer paths will be chosen as primary paths. This would increase the notification time to the end nodes. After a certain arrival rate, the restoration time starts decreasing, since the network resources are scarce and only short candidate backup paths can qualify for restoration. On the whole, the weighted distribution does better for all arrival rates considered.

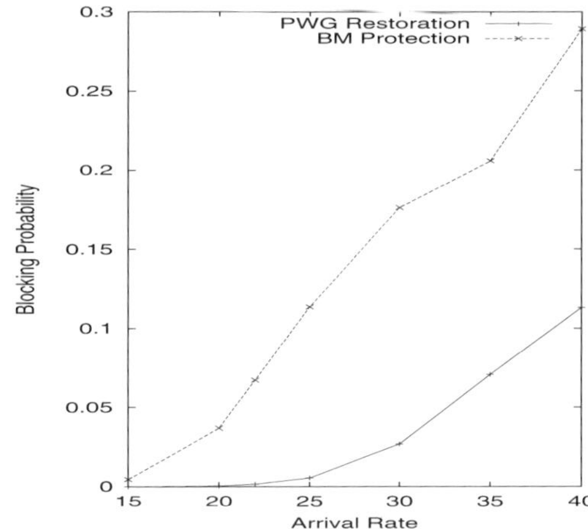

Fig. 1. Graph of Blocking Probability against Arrival Rate for network G1.

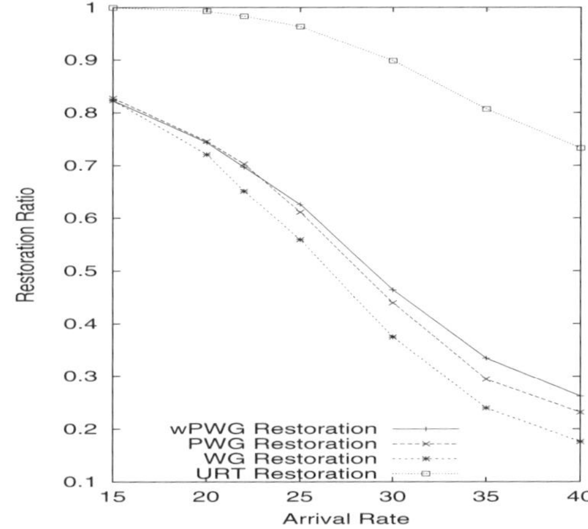

Fig. 2. Graph of Restoration Ratio against Arrival Rate for network G1.

Figures 4 through 6 plot the performance for $G2$. The trends are similar to that of $G1$. Since the degree of connectivity of $G2$ is lower than that of $G1$, for a given blocking probability, $G2$ carries lesser load than $G1$. The restoration ratio is similar at low blocking probabilities as wavelength resources are readily available. For a moderate or high blocking probability (say 0.1) the restoration ratio for $G2$ is lower than $G1$. This is because the average hop length in $G2$ is higher than $G1$. For the same reason the restoration time for $G2$ is higher than $G1$.

V. Conclusions

This paper addressed the issue of rapid dynamic failure restoration in WDM optical networks. A new distributed dynamic restoration protocol based on path-wavelength-disjoint grouping was developed. The objective was to

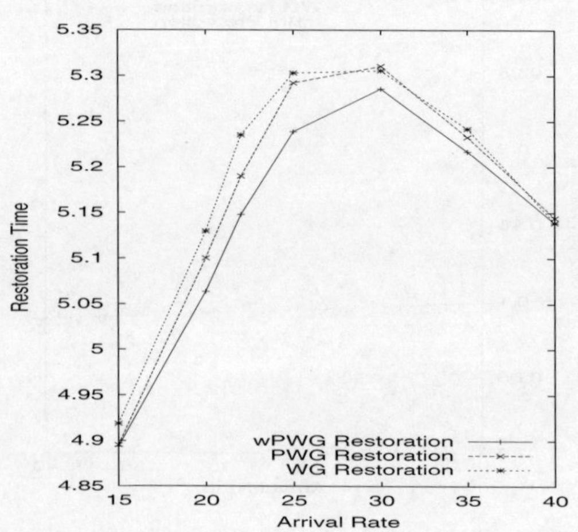

Fig. 3. Graph of Restoration Time against Arrival Rate for network G1.

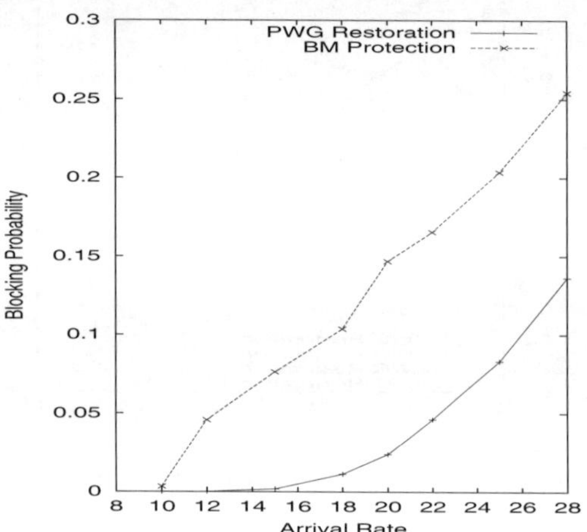

Fig. 4. Graph of Blocking Probability against Arrival Rate for network G2.

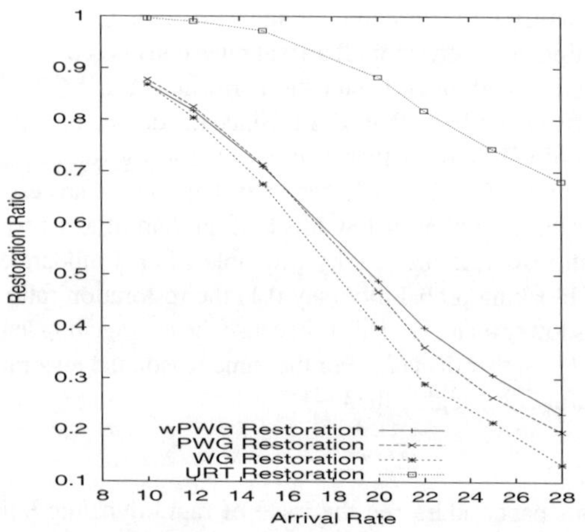

Fig. 5. Graph of Restoration Ratio against Arrival Rate for network G2.

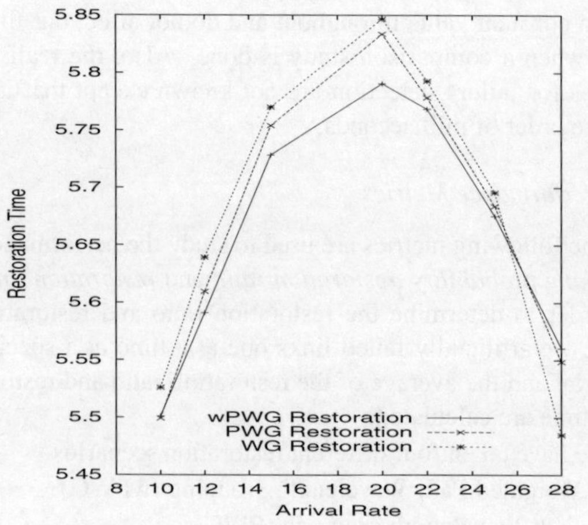

Fig. 6. Graph of Restoration Time against Arrival Rate for network G2.

restore paths rapidly while ensuring a increased number of restored paths. The proposed grouping method enables reservation-conflict-free wavelength search on different paths simultaneously. The idea was to pack the backup paths into smaller number of groups, thus having a better spread of wavelengths among groups without any conflicts for wavelength resources. This would increase the chance of a backup path finding a free wavelength, hence increasing the restoration ratio. To further improve the wavelength distribution among groups, a weighted path-wavelength distribution algorithm was developed which assigns wavelengths proportional to the group size. The effectiveness of these algorithms was evaluated by extensive simulation experiments. The results show that the restoration ratio is significantly high at light loading and it decreases with the increasing load. The results also show that the restoration time is on the order of milliseconds which is small.

REFERENCES

[1] H. Zang and B. Mukherjee, "Connection Management for Survivable Wavelength-routed WDM Mesh Networks," *Optical Networks Magazine*, pp. 17-28, July/August 2000.

[2] G. Mohan and C. Siva Ram Murthy,"Lightpath Restoration in WDM Optical Networks," *IEEE Network*, vol. 14, no. 6, pp. 24-32, November/December 2000.

[3] B. Mukherjee, "WDM Optical Communication Networks: Progress and Challenges,", *IEEE Journal of Selected Areas in Communications*, vol. 18, no. 10, pp. 1810 -1824, October 2000.

[4] G. Mohan, C. Siva Ram Murthy and Arun K. Somani, "Efficient Algorithms for Routing Dependable Connections in WDM Optical Networks," *IEEE/ACM Transactions on Networking*, vol. 9, no. 5, pp. 553-566, October 2001.

[5] S. Ramamurthy and B. Mukherjee "Survivable WDM mesh networks. II. Restoration", In Proceedings of*IEEE International Conference on Communications, ICC '99*, pp. 2023 -2030, 1999.

[6] S. D. Maesschalck *et. al.*, "Intelligent Optical Networking for Multilayer Survivability," *IEEE Communications Magazine.*, pp. 42-49, January 2002.

[7] Y. Ye, C. Assi, S. Dixit and M. A. Ali, "A Simple Dynamic Integrated Provisioning/Protection Scheme in IP over WDM Networks", *IEEE Communications Magazine*, pp. 174-182, November 2001.

Span-Restorable Mesh Network Design to Support Multiple Quality of Protection (QoP) Service-Classes

Wayne D. Grover, *IEEE Fellow*, Matthieu Clouqueur, *IEEE Student Member*

TR*Labs*, #800 10611-98th Avenue, Edmonton, Alberta, Canada T5K 2P7
and Department of Electrical and Computer Engineering, University of Alberta, Edmonton, Alberta
Contact: {grover, clouqueur}@trlabs.ca

Abstract – In the modern business environment for optical mesh networking, there is considerable interest in being able to support a range of service classes, and to charge accordingly. In this work we consider the capacity design problem for different combinations of four protection priority classes in a span-restorable mesh network. Most significantly we have test-case findings that 15% to 30% of all demand can be in the gold class and enjoy 100% restorability, *solely* through preemption of economy class service capacity. This indicates the potential to design and operate mesh-based restorable networks which have *no spare capacity at all in the conventional sense: all capacity is bearing service of some paying type*. The methods provided could be the basis for future mesh-based optical transport networks that simultaneously support multiple QoP services with optimal capacity.

Keywords – Optical networks, mesh, restoration, QoP, capacity design.

A. Introduction and Objective

In capacity-design models for almost any variety of transport architecture to date, the most commonly studied type of problem assumes one service class that must be fully restorable (or protected) against any single failure scenario. Conceptually, however, there could be at least four different policies regarding the treatment of different individual demands in a mesh-restorable network. Table 1 lists three multi-QoP service schemes defined on the four classes. In Scheme 1, traffic is either protected or un-protected. In Scheme 2, we add a best-efforts class of traffic, and in Scheme 3, we add preemptible traffic.

Table 1. Three schemes of multiple QoP service classes

Scheme 1 (2 levels)	Scheme 2 (3 levels)	Scheme 3 (4 levels)	Service-class designation
protected	protected	protected	gold (g)
n/a	best-efforts	best-effort	silver (s)
non-protected	non-protected	non-protected	bronze (b)
n/a	n/a	preemptible	economy (e)

While the proposal of a set of QoP classes is not new [1], what this paper contributes is a comprehensive model of the optimum capacity allocation problem for a generalized mix of QoP classes in a span-restorable mesh network.

Let us first touch on the operational principles that would go hand-in-hand with the multi-QoP capacity designs that we envisage. Each service path signal would be labelled as to its service class. This is easily accommodated in OC-n overheads, digital wrapper, or GFP overheads or conveyed by an optical service channel (OSC). The Optical Cross-Connects (OXC) adjacent to each span would use these designations to classify working channels on the span. Being present in the overheads or OSC for the span, this information is self-updating at every OXC. In the context of a span-restorable network (or the corresponding link-protected network [2]) this means that what has previously been a single "w_i" quantity for the working capacity to be protected on span i will now be decomposed in up to four constituent types of working capacity, w_i^g, w_i^s, w_i^b, w_i^e, which the custodial cross-connects will treat functionally as follows:

- **gold** (assured restoration): these are working channels that must be restored
- **silver** (best efforts restoration): these are working channels that should be restored if possible following full restoration of any gold-class service capacity.
- **bronze** (non-protected service): these are working channels that do not receive any restoration efforts but are also not subject to pre-emption.
- **economy** (preemptible services): these are working channels that are not protected and may also be seized and logically converted to spare capacity if needed to satisfy gold requirements for restoration.

Several variations are conceivable in the definitions. For example, economy capacity might be available to cover *either* gold or silver class restoration requirements. The only difference between gold and silver would then be that gold gets first access before silver to spare and preemptible capacity. In another alternative, economy capacity could itself also receive secondary best-efforts restoration upon its own failure, while still being preemptible for other failures. These and other variations can be represented with minor changes to the general mathematical model, which follows.

B. Capacity Design for Multi-QoP Schemes

To develop a multi-QoP capacity design model, let us start by reviewing the single-priority capacity design model for span restoration. The most general is Modular Joint Capacity Assignment (MJCA) from [3], which is an extension of the basic arc-path Integer Programming model for spare capacity allocation (SCA) [4]. The MJCA model is summarized below as a starting point for making changes that will lead to a highly general Multi-QoP design model. MJCA is:

$$\textbf{MJCA: } \text{minimize} \sum_{m\in M}\sum_{i\in S} C_j^m \cdot n_j^m \tag{1}$$

Subject to:

$$\sum_{q\in Q^r} g^{r,q} = d^r \text{ , } \forall r \in \boldsymbol{D} \tag{2}$$

$$w_j = \sum_{r\in D}\sum_{q\in Q^r} \zeta_j^{r,q} \cdot g^{r,q} \text{ , } \forall j \in \boldsymbol{S} \tag{3}$$

$$\sum_{p\in P_i} f_i^p = w_i \text{ , } \forall i \in \boldsymbol{S} \tag{4}$$

$$s_j \geq \sum_{p\in P_i} \delta_{i,j}^p \cdot f_i^p \text{ , } \forall (i,j) \in \boldsymbol{S}^2, i \neq j \tag{5}$$

$$w_j + s_j \leq \sum_{m\in M} n_j^m \cdot Z^m \text{ , } \forall j \in \boldsymbol{S} \tag{6}$$

- S *is the set of network spans,*
- M *is the set of modular capacities available,*
- D *is the set of demand pairs,*
- Q^r *is the set of eligible working routes for each demand pair* r *in* D,
- P_i *is the set of eligible restoration routes for each single-span failure scenario* i *in the set* S,
- C_j^m *is the cost of a module of type* m *on span* j,
- Z^m *is the capacity of modules of type* m,
- d^r *is the number of equivalent unit capacity demand exchanged by demand pair* r,
- n_j^m *is the number of modules of type* m *placed on span* j,
- $g^{r,q}$ *is the working flow on route* q *serving demand pair* r,
- $\zeta_j^{r,q}$ *is 1 if eligible route* q *for demand pair* r *crosses span* j *and 0 otherwise,*
- w_j *is the amount of working capacity allocated on span* j,
- f_i^p *is the flow on route* p *for restoration of span* i,
- $\delta_{i,j}^p$ *is 1 if eligible restoration route* p *for span* i *crosses span* j *and 0 otherwise, and finally*
- s_j *is the amount of spare capacity allocated on span* j.

The objective (1) minimizes the total cost of capacity required to support all working and restoration flows. Eq. (2) ensures that for each traffic demand, there is flow over all eligible working routes to fully route all traffic, and Eq. (3) places enough associated working capacity. Eq. (4) guarantees that there is enough restoration flow over all eligible restoration routes to fully restore any single physical span cut (or some other failure of a span), while Eq. (5) provisions enough associated spare capacity on each span to accommodate all restoration flows simultaneously routed over it. Finally, Eq. (6) places enough total capacity on each span to carry all working and spare capacities on it. This design model implicitly assumes only for a single class of working service priority, equivalent to what we have now re-defined as the fully restorable (i.e., gold) service class.

C. Changes to MJCA for Multi-QoP

In this section we touch on each multi-QoP scheme in Table 1 and explain how MJCP can be adapted to provide an optimal capacity design for that particular case. The changes to support Scheme 1 are the easiest: One simply replaces the ordinary span working capacity quantities (w_i) by w_i^g in Eq. 4. In effect this makes "gold" capacity the only type that is visible to the restoration process and spare capacity design process. The non-protected (bronze) capacity remains in the working capacity (Eq. (2),(3)) and modularity generating (Eq. (6)) constraints, however.

In approaching the other mixtures of QoP classes, we can first make some helpful observations about how the basic model would be altered to reflect the different treatments of working capacity. In particular, in any design problem with an economy class the corresponding capacity is:

(i) omitted from the restorability constraint,
(ii) treated as a *credit* against needed spare capacity in the spare capacity constraint, and
(ii) included in the modular capacity constraint.

Another consideration is that in any of the multi-QoP models, restorability will only ever be explicitly asserted for w_i^g quantities. In this regard the difference between best-efforts and strictly non-protected services is really only an operational distinction; there is no distinction between them from a capacity-design standpoint. In other words, neither silver or bronze classes require any assured restoration considerations in the basic design, but silver will receive operational best-efforts to exploit any available and otherwise unneeded spare capacity under the given failure scenarios, whereas non-protected bronze does not receive this effort at all.

In considering changes to the spare capacity generating constraint, Eq. (5), we note that restoration flows corresponding to gold-class working capacity need to be fully supported by the spare plus preemptible capacity dimensioning but no explicit considerations of spare capacity have to be made for silver, bronze or economy-class demands crossing the span. On the other hand preemptible working capacity is completely inter-changeable for spare capacity on the corresponding spans. Hence any w_i^e can be *subtracted* from the corresponding s_i.

Finally, all forms of working capacity, plus spare capacity required for gold-class assured restoration, must be supported by the final modular capacity placement decisions.

As a result of these considerations we can state a generalized model for joint modular capacity design with an *arbitrary mixture of demands in the four service classes.* All variables and parameters are the same as in MJCA above, with the exception that the set of demand requirements is now represented by constituent demand sets, one for each priority class. In practice, (for the capacity design problem only), this can be reduced to three by merging the best-efforts and non-protected requirements, based on the considerations above. Thus we have the new demand sub-sets:

- D^g, $D^{s,b}$, D^e *are the demand requirements for gold, silver&bronze (merged), and economy services, respectively.*

Individual values of these sets are still referred to as d^r values (integer), and index r continues for use in enumerating members of any of these sets. The sets $\mathbf{P}_i$ and $\mathbf{Q}^r$ remain unchanged because the eligible restoration routes and eligible choices for working routes remain properties of the topology alone, unaffected by the structuring of demands into different service classes. The multi-QoP capacity-design model thus becomes:

Multi QoP-MJCA: minimize $\sum_{m\in M}\sum_{i\in S} c_j^m \cdot n_j^m$ (7)

Subject to:

$$\sum_{q\in Q^r} g^{r,q} = d^r, \ \forall r \in \boldsymbol{D}^g \cup \boldsymbol{D}^{s,b} \cup \boldsymbol{D}^e \quad (8)$$

$$w_j^g = \sum_{r\in D^g}\sum_{q\in Q^r} \zeta_j^{r,q}\cdot g^{r,q}, \ \forall j \in \boldsymbol{S} \quad (9)$$

$$w_j^{s,b} = \sum_{r\in D^{s,b}}\sum_{q\in Q^r} \zeta_j^{r,q}\cdot g^{r,q}, \ \forall j \in \boldsymbol{S} \quad (10)$$

$$w_j^e = \sum_{r\in D^e}\sum_{q\in Q^r} \zeta_j^{r,q}\cdot g^{r,q}, \ \forall j \in \boldsymbol{S} \quad (11)$$

$$\sum_{p\in P_i} f_i^p = w_i^g, \ \forall i \in \boldsymbol{S} \quad (12)$$

$$s_j \ge \sum_{p\in P_i} \delta_{i,j}^p \cdot f_i^p - w_j^e, \ \forall (i,j)\in \boldsymbol{S}^2, i\ne j \quad (13)$$

$$w_j^g + w_j^{s,b} + w_j^e + s_j \le \sum_{m\in M} n_j^m \cdot Z^m, \ \forall j \in \boldsymbol{S} \quad (14)$$

$$f_i^p \ge 0 \ \forall i \in \boldsymbol{S} \ \forall p \in \boldsymbol{P}_i ; \ g^{r,q} \ge 0$$

$$\forall r \in \boldsymbol{D}^g \cup \boldsymbol{D}^{s,b} \cup \boldsymbol{D}^e ; \ \forall q \in \boldsymbol{Q}^r \quad (15)$$

A non-modular (but still jointly optimized multi-QoP model) is obtained from by dropping Eq. (14) and using the cost-weighted sum of all working and the spare capacity as the objective function. A non-joint (spare capacity only) multi-QoP model is further obtained by dropping Eq. (8) to (11) and generating working capacities through shortest-path routing or other procedural routing method, prior to optimizing only the spare capacity.

D. Experimental Multi-QoP Design Studies

We ran a set of test cases to look at some of the interesting effects and trade-offs that exist in a multi-QoP capacity-design environment. For simplicity we considered integer unit-capacity designs rather than assume a specific set of capacity modules. We compared a conventional single-priority fully-restorable joint design against the corresponding multi-QoP designs with four different profiles of multi-priority demand mix. The mixed demand scenarios are represented in the star-plots of Figure 1. The test case demand patterns are constructed by assuming a total of 20 units of demand of all types between each node-pair. For the benchmark capacity design all 20 are gold class on each node pair.

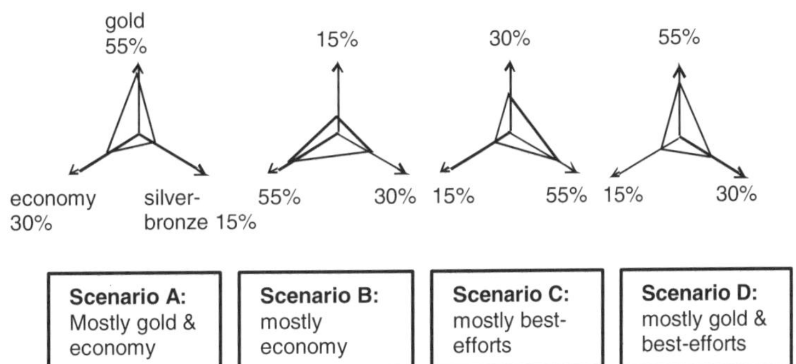

Figure 1. Four multi-QoP mixed demand scenarios for testing

The other (mixed-class) demand scenarios are generated by allocating different numbers of the 20 demand units on each pair to each class: 55% corresponds to 11 units, etc. Several random test networks were used. In each test network, edge distances times capacities are used as the measure of cost.

E. Discussion of the Results

Table 2 is a sample result illustrating the general findings. First, not surprisingly, there is a significant reduction of the spare capacity requirements because not all services need restoration. Less obvious, however, is the general finding that preemtptible economy capacity can be so efficiently used for gold restoration that in scenarios A, B and C there is literally none or almost no conventional spare capacity required at all.

Table 2. Sample results for 30-node, 60 span test network.

Scenarios	Total working	Total spare	Economy capacity
Conventional	26954	14485	n/a
A – mostly gold and economy	27705	251	8787
B – mostly economy	27385	0	14982
C – mostly best efforts	27834	52	4618
D – mostly gold & best efforts	28642	2131	5622

For the working capacity the situation is somewhat reversed, however: the multi-QoP designs use more working capacity in all four scenarios. The reason is that in a mixed scenario a gold-class demand might now deviate further from its shortest path if by doing so it can effectively pick up restorability coverage "for free" through economy class capacity. Conversely, economy class paths may also deviate from shortest paths to enhance synergies with gold. In this way it is quite possible for *working* capacity to rise noticeably so that *total* capacity is lowered even further.

Another aspect that was studied is the level of restorability of silver-class services. It was found that the restorability of these services is highly dependent on the choice whether or not to allow preemption of economy-class to support their restoration. In the case this was allowed the restorability of best-effort services was found to be very high for the first two scenarios (A and B). Generally, in inspection of failures for these demand scenarios, the average number of restoration paths that were feasible was much higher than what is needed to restore all the gold and silver service paths. This is understandable given the fact that both these scenarios have more economy class demands: the higher the proportion of economy services the higher the best effort restorability. In scenario A, this resulted in an average best-effort restorability of only about 0.55 whereas for scenario B the average best effort restorability is almost 1. The very high best-efforts restorability may not actually be that desirable from a commercial point of view if the interpretation is that gold service levels are not sufficiently superior to what customers could expect in the less costly silver class. In this regard, Scenario C, with an average best-effort restorability potential between 0.34 and 0.61, seemed to be a more preferable service-mix scenario. It is also one of the scenarios for which the total capacity requirements are the lowest.

F. Conclusion

This work has developed an optimal capacity design model for span-restorable mesh networks which may have mixtures of up to four different protection service classes. We found that the most substantive new interactions in the capacity design problem occur between the gold and economy class services. Two of the most interesting findings are that in many cases there is almost no requirement for conventional "spare" capacity: the routing of gold and economy services can be worked out so that gold restorability is almost or completely provided through pre-emption of economy-class services. Additionally relatively high levels of best-efforts restorability are observed for silver-class services if they are allowed to preempt economy class capacity after gold has been fully restored. We think this work contributes to providing a greater range of tools, strategies and knowledge that a transport network operator can use to enhance their business and provide more customer options.

G. References

[1] O. Gerstel, G. Sasaki, "Quality of Protection (QoP): A Quantitative Unifying Paradigm to Protection Service Grades," *OptiComm2001*, Aug. 2001.

[2] W. D. Grover, "Distributed restoration of the transport network," in Telecommunications Network Management into the 21st Century, S. Aidarous and T. Plevyak, Eds. New York: IEEE Press, 1994, ch. 11, pp. 337–419

[3] J. Doucette and W. D. Grover "Influence of modularity and economy-of-scale effects on design of mesh-restorable DWDM networks," *JSAC* vol. 18, no. 10, pp. 1912-1923, Oct. 2000

[4] M. Herzberg, S. Bye, "An optimal spare-capacity assignment model for survivable networks with hop limits," *GLOBECOM '94*, pp. 1601-1607, 1994.

Session 14D

WDM Devices III

High density optical data storage and fabrication of photonic crystals in photorefractive polymers for optical communications and networks

Min Gu
Centre for Micro-Photonics
School of Biophysical Sciences and Electric Engineering
Swinburne University of Technology
PO Box 218, Hawthorn, Vic 3122, Australia

Abstract-- In this presentation, we report on our recent development of erasable three-dimensional bit optical data storage, the formation of micro-cavities and micro-fabrication of waveguide structures in photorefractive polymers. These devices are important for optical communications and networks.

Index -- Two-photon excitation, 3-D optical data storage, microfabrication

A. INTRODUCTION

Research on polymer-based photonic devices such as optical data storage devices, photonic crystals and waveguides is a new field for optical communications and networks. Three-dimensional (3-D) optical data storage based on two-photon excitation provides a possibility of achieving data densities of the order of 100 to 1000 times more than conventional two-dimensional (2-D) data storage systems such as compact discs (CD) and digital versatile discs (DVD). The materials used for current research on 3-D optical data storage systems can be divided into three broad categories; photorefractive crystals (e.g. $LiNbO_3$), various types of glasses, and polymer based materials. Currently, polymer based materials can be separated into photobleaching, photochromic, photorefractive and polymer dispersed liquid crystals types. Due to the localized excitation nature, two-photon excitation is also a useful tool for fabricating 3-D structures including photonic crystals in polymers.

In this presentation, we first give a review on our work of 3-D bit optical data storage in photobleaching polymer [1] and polymer dispersed liquid crystal materials [2, 3]. Then, our recent development of 3-D optical storage and microfabrication of photonic crystals in photorefractive polymer [4-6] will be described.

B. 3-D OPTICAL DATA STORAGE

In the case of two-photon photobleaching polymer, the fluorophore is doped into a polymer block. Illuminating the sample with an appropriate laser wavelength and average power (typically < 1 mW) from an ultra-short pulsed laser will produce two-photon fluorescence. Increasing the power above the bleaching threshold will cause the fluorophore to breakdown (bleach) and stop fluorescing. By the use of this method a series of bleached patterns can be recorded in the material. As a recording medium we used a photobleaching polymer block consisting of PMMA doped with fluorescent dye AF-50 (0.3 wt %). Such a fluorescent polymer has an absorption peak at a wavelength of 400 nm. Thus an infrared laser of wavelength 800 nm can produce two-photon excitation. Fig. 1 is the multi-layered storage in the photobleaching polymer.

Fig. 1 Recorded 24x24 bit patterns at different depths in the photobleaching polymer using two-photon excitation. The spacing between adjacent layers is 5 μm, and the bit separation is 2 μm, corresponding to a density of ~50 Gbit/cm^3.

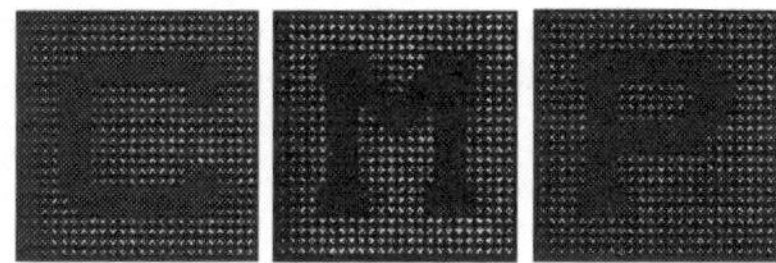

Fig. 2 Recorded 24x24 bit patterns at different depths in the PDLC material using two-photon excitation. The spacing between adjacent layers is 1.5 μm, and the bit separation is 2 μm, corresponding to a density of ~200 Gbit/cm^3.

Polymer-dispersed liquid crystals (PDLCs) are a new type of polymer based material. They consist of small micro-droplets of liquid crystals dispersed in a polymer matrix. Upon illumination of an electric field, the re-orientation of the directors of the nematic liquid crystals occurs. In the unexposed zone, the liquid crystal directors have random alignment but in the exposed zone the directors align. Such an effect leads to a field induced polarisation effect. The PDLC material was manufactured using various concentrations of 4-pentyl-4′-cyano biphenyl (E49) liquid crystal, 2,4,7-trinitro-9-flourenone (TNF) as the photosensitive material and *N*-ethylcarbazole (ECZ) as the plasticiser; all doped into the polymer poly methyl methacrylate (PMMA). The absorption of this new material is negligable at a wavelength of 900 nm. Therefore, a laser with an infrared wavelength at 900 nm can be used in the writing process to produce two-photon (2-p) excitation at wavelength 450 nm Fig. 2 shows three layers of data recorded in the PDLC.

C. ERASABLE AND NON-ERASABLE 3-D OPTICAL DATA STORAGE

Photorefractive polymer is a promising material for optical data storage and another photonic applications. Recently, we have reported on the photorefractive effect under two-photon excitation in a photorefractive polymer, which allows one to develop a low-cost, erasable, and rewritable three-dimensional (3-D) optical storage method. The photorefractive polymer we have used consists of 2,5-dimethyl-4-(*p*-nirtophenylazo)anisole (DMNPAA), 2,4,7-trinitro-9-fluorenone (TNF), 9-ethylcarbazole and poly(*N*-vinylcarbazole) (PVK). The concentration of the materials DMNPAA:PVK:ECZ:TNF was 50:33:16:1 wt% and its maximum absorption band of the material is within 380 - 600 nm. Therefore a laser beam of an infrared wavelength at 800 nm can be used in the recording process to produce two-photon excitation at 400 nm. Fig. 3 shows an example of such a storage method.

Fig. 3 Recorded 24x24 bit patterns at different depths in the photorefractive polymer using two-photon excitation. The spacing between adjacent layers is 20 μm, and the bit separation is 3.2 μm, corresponding to a density of ~4 Gbit/cm^3.

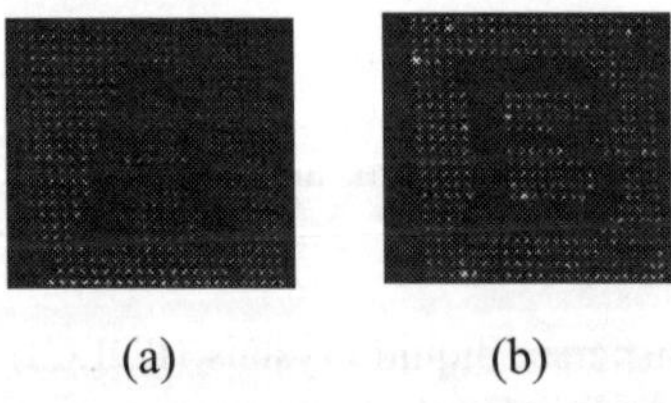

(a) (b)

Fig. 4 Multi-layered arrays of voids in the PMMA-based photorefractive polymer. (a) The first layer including the letter A is recorded near the surface. (b) The second layer including the letter B is recorded with a separation of 15 μm in the depth direction. The images are read out by confocal reflection microscopy.

The advantage of the PMMA-based photorefractive polymer is that the transition region from the erasable nature to the optical damage point is short so that it becomes a good candidate for fabricating micro-structures under multi-photon excitation. We have observed the formation of voids in the PMMA-based photorefractive polymer with a concentration of the dopants (DMNPAA:PMMA:ECZ:TNF) of 30:53:16:1, according to the micro-explosion mechanism. One of the applications of the voids is in read-only 3-D bit optical data storage, as shown in Fig. 4.

D. MICROFABRICATION

If the laser beam used for forming a void is scanning at an appropriate speed, a 3-D micro-structure can be created. Fig. 5 shows a few examples of the micro-structures formed by a scanning void. Fig. 5(a) is a waveguide structure. As shown in Fig. 5(b), this structure can be joined together, which could be used as a beamsplitter or a coupler. Fig. 5(c) is the image of a 3-layer stack formed by the waveguide structure. The line structures in the mid layer are arranged at 90 degrees with respect to those in the top and bottom layers. Thus, such a 3-D period structure may show a photonic band gap.

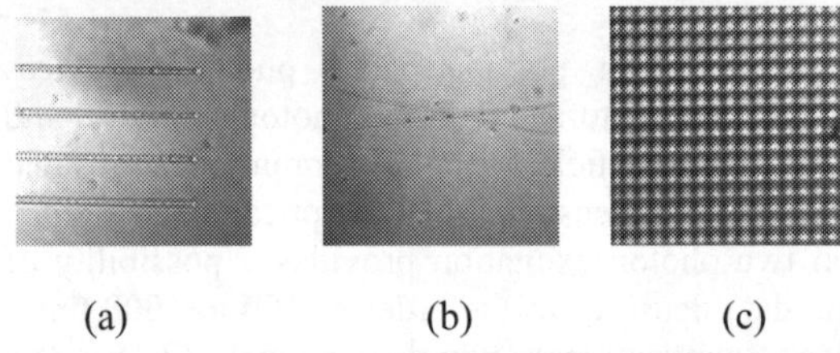

(a) (b) (c)

Fig. 5 A few examples of the micro-structures fabricated in the PMMA-based photorefractive polymer. (a) a waveguide structure; (b) a Y-junction; (c) Three layers of the waveguide structures.

ACKNOWLEDGMENTS

The author acknowledges the support from the Australian Research Council and thanks D. Day, D. McPhail and Tom Rodgers for their contributions to the topic.

References

[1] Multi-level optical data storage in a photobleaching polymer using two-photon excitation under continuous wave illumination, Djenan Ganic, Daniel Day, Min Gu, *Opt. and Laser Engineering*, (2002), in press.

[2] Multi-layered optical data storage in a polymer-dispersed liquid crystal material under two-photon excitation, Min Gu, D. K. McPhail, A. Smallridge, Technical Digest of the 2001 Pacific-Rim Conference on Lasers and Electro-Optics, p. II-756 – II-757.

[3] Use of polarisation sensitivity for three-dimensional optical data storage in polymer dispersed liquid crystals under two-photon illumination, D. McPhail, Min Gu, *Appl. Phys. Lett.* (2002), in press.

[4] Use of two-photon excitation for erasable/rewritable three-dimensional bit optical data storage in a

photorefractive polymer, D. Day, Min Gu and A. Smallridge, *Opt. Lett.* 24 (1999), 948-950.

[5] Rewritable three-dimensional bit optical data storage in a PMMA-based photorefractive polymers with continuous wave illumination, D. Day, Min Gu, A. Smallridge, *Advanced Materials,* 13 (2001), 1005-1007.

[6] Formation of voids in a doped polymethylmethacrylate polymer, Daniel Day, Min Gu, *Appl. Phys. Lett.,* 80 (2002), 2404-2406.

Scalable Liquid Crystal Optical Switches: Fiction or Reality?

Daryoush Habibi
School of Engineering and Mathematics
Edith Cowan University, Australia

Guoqiang Mao
School of Electrical, Computer and Telecommunications Engineering
University of Wollongong, Australia

Abstract

One of the most critical elements within the framework of wavelength division multiplexing optical network is the optical cross-connect. The core of an optical cross-connect is the optical switch capable of switching a large number of input and output fibers. In this paper, we report the design of a polarization independent liquid crystal optical switch. It is pointed out that liquid crystal spatial light modulator technology is a very competitive technology to realize a large-scale reconfigurable optical switch.

Keywords: optical switch, optical cross connect.

1. INTRODUCTION

The rapid rise in network applications that require high bandwidth and the ever growing internet traffic have resulted in the explosion of network traffic beyond what could possibly be predicted a decade ago. Today, many carriers are reaching close to 100% utilization across some portions of their networks. To address the rapid growth of traffic there are a number of approaches that can be taken, eg. installing more fiber, increasing the bit rate of the data link on the fiber, or increasing the number of wavelengths that are carried over the same fiber. The latter is the basis for the introduction of Dense Wavelength Division Multiplexing (DWDM). Today, we have networks that carry tens of wavelengths on a single fiber, each wavelength operating at 10 Gbps or 40 Gbps.

The introduction of DWDM has made it possible to use the existing fibers more effectively, but it has also created the need for many new devices and network components. One of these components which has a key role in the deployment of DWDM in optical networks is a cross connect, and although it has been the subject of significant research and development in recent years, still efficient and reliable cross connection of the many DWDM wavelengths at carrier central offices remains a challenge. Whilst optical-electrical-optical conversion, including cross connection in the electrical domain was a viable option for lower link rates, it is no longer satisfactory or economical when dealing with bit rates in the order of 40 Gbps. Switching these wavelengths requires all-optical switches with many tens of ports. While electrical cross connection (EXC) will still be needed to switch lower-granularity circuits, there are significant cost, power and space savings associated with the use of all-optical or photonics cross connects (OXCs). At the core of the OXC there is an optical switch, in charge of steering the light in a controllable fashion. Ideally, an optical switch should be able to switch the entire wavelength content of the fiber, selected bands of it, or single wavelengths.

Some of the factors that determine the quality of an optical switch are

- insertion loss
- insertion loss variation
- polarization-dependent loss (PDL)
- crosstalk
- wavelength range
- switching speed
- footprint
- cost per port

There are a number of technologies available to realize optical switches, including Micro-Electro-Mechanical System (MEMS) switches, Bubble-Based Waveguide Switches, Liquid Crystal on Silicon (LCOS) Switches, Electro-Optic Switches, and Thermo-Optic Switches.

With the recent announcement of the availability of 3D MEMS 64X64 optical switch from Agere and 2D MEMS 32X32 optical switch from OMM, MEMS technology is attracting the most attentions in the optical switch market. However MEMS switch is a mechanical switch, it has some inherent disadvantages, such as high power consumption, high cost, subject to mechanical wear and tear, etc. In this paper, we shall introduce a liquid crystal optical switch, which is able to overcome these disadvantages.

2. LIQUID CRYSTAL OPTICAL SWITCH

An optical cross-connect consists mainly of optical switches (OSWs), wavelength multiplexers (WMUXs) and demultiplexers (WDMXs), optical pre-amplifiers (Pre-OAs) and post-amplifiers (Post-OAs) [1]. Depending on the specific system in study, one or more components may be omitted from the general configuration.

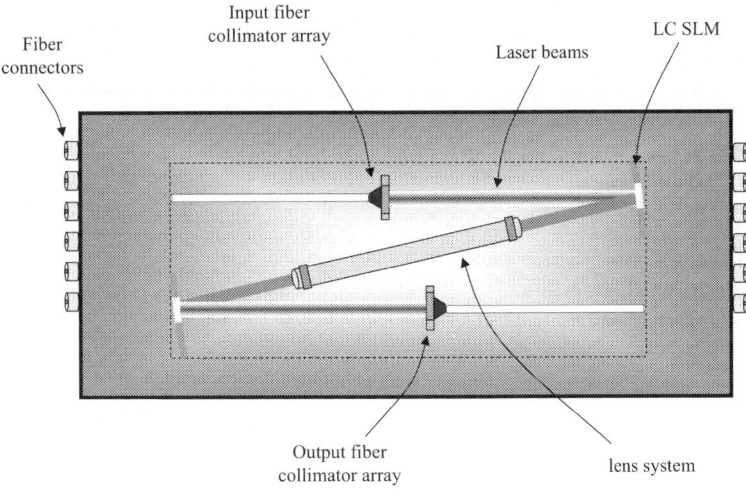

Fig. 1 Architecture of the LC optical switch in the "Z" configuration

The core of the optical cross-connect is an optical switch, which is built using polarization-independent (PI) liquid crystal spatial light modulator (SLM). The architecture of a liquid crystal optical switch can be either a "Z" configuration or a "W" configuration. Fig. 1 shows the architecture of the switch in the "Z" configuration. The basic operation of the switch in the "Z" configuration is as follows: the input fibers are arranged into an array from which the light beams are launched and subsequently collimated, either through a micro-lens array, or a Fourier lens, onto a first reflective PI LC SLM. The SLM acts as a phase-only diffraction grating with a reconfigurable pattern, which determines the deflection angle of the reflected beam.

Controlling the voltage applied on the LC pixel elements of the SLM controls the deflection angle. The deflected beam is then collimated by a lens system onto a second SLM, where it is diffracted again onto the selected output port and coupled into the output fibre. One of the most critical components of the LC optical switch is the PI LC SLM. Fig. 2 depicts the SLM.

Fig. 2 Depict of the LC SLM

The SLM contains millions of LC pixel elements. Those pixels are divided into multiple blocks, depending on the input and output fiber array. Each pixel block on the first SLM is responsible for receiving the optical beam from a specific input fiber. Each pixel block on the second SLM is responsible for switching the beam to a specific output fiber. This two-SLM architecture in either "W" or "Z" configuration ensures that optical beam from the input fiber is coupled parallel into the output fiber, therefore maximizing the coupling efficiency. Pixels in a block act as a phase-only diffraction grating with a reconfigurable hologram pattern. By changing the hologram pattern of a pixel block, one optical beam can be diffracted into a specific angle or several specific angles.

Fig. 3 is a schematic illustration of the LC optical switch in the "Z" configuration.

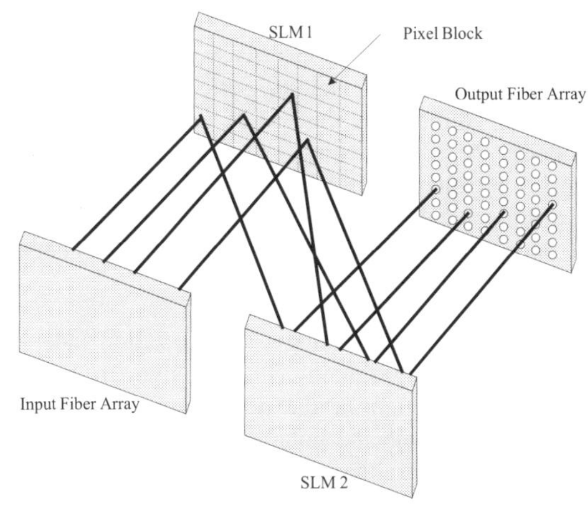

Fig. 3 A Schematic illustration of the LC optical switch

The optical beam from the input fiber can be diffracted into several diffraction angles. This implies that the optical signal from an input fiber can be switched into multiple output fibers. The intensities of the output beams into the diffraction angles can be controlled by changing the hologram pattern. Therefore the LC optical switch has the unparallel capability of realizing multicasting and channel equalization. These are very useful features in optical networking.

Furthermore the LC SLM is very flexible, which allows fine-tuning of diffraction angles as well as positions of pixel blocks to account for some minor misalignments of optical fibers.

The key to the success of the LC optical switch is making the LC SLM polarization independent. Several approaches have been proposed to design a polarization-independent liquid crystal spatial light modulator [2-5]. The proposed LC optical switch uses a proprietary technology to realize PI LC SLM.

3. PERFORMANCE OF THE LC OPTICAL SWITCH

Using the proposed design, a large-scale (64X64 or larger) LC optical switch can be realized. In this section, we shall analyse the performance of the proposed LC optical switch.

Polarization Sensitivity and Scalability

Contrary to some common belief, MEMS switch is not free from polarization dependence problem. The polarization dependent loss (PDL) of Agere's 64X64 switch is 0.2dB [6] and the PDL of OMM's 32X32 switch is as large as 1dB [7]. PDL of MEMS switch possibly arises from non-perfect micro-mirrors.

Polarization dependence is also a problem once puzzling the development of large-scale LC optical switch. LC SLM used to be polarization dependent. In order to solve the polarization dependence problem, traditionally one optical beam is divided into two orthogonal polarization components and each polarization component is switched separately by a SLM. Since each SLM handles optical beam at a specific polarization direction, the SLM can be polarization dependent. After switching, the two components are combined together. This technology potentially limits the scale of LC optical switch because it requires a perfect match of the optical paths of the two polarization components of the optical beam.

Using a proprietary technology, a polarization independent LC SLM has been designed. This revolutionary technology brings a breakthrough in the development of LC optical switch. It solves both the polarization dependence problem and the scalability problem of LC optical switch. Preliminary results indicate that a low PDL of about 0.5dB can be achieved by the LC optical switch. On-going efforts are being carried out to improve this parameter.

With the breakthrough of the polarization independent LC SLM technology, the proposed LC switch is able to achieve the same scalability as 3D MEMS technology. The proposed LC optical switch has a similar architecture as that of 3D MEMS as shown in Fig. 4. Therefore principally LC optical switch has the same scalability as 3D MEMS switch. However rather than using mechanical switching to steer the optical beam, the proposed LC optical switch uses the electro-optical properties of LC to realize beam steering, which results in much less power consumption, simpler structure and higher reliability.

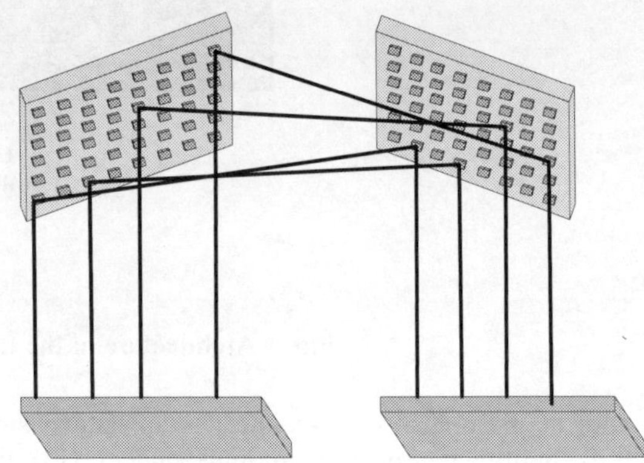

Fig. 4 A Schematic illustration of Agere's 3D MEMS switching technology

Optical Power Loss and Crosstalk

In addition to coupling loss, which accounts for the majority of the power loss of the proposed LC optical switch, another major loss of the LC optical switch comes from the LC SLM. Power loss of a LC SLM is mainly determined by the number of phase levels of the LC pixel elements, which is in turn determined by capability of the circuit underneath the LC and the LC material. Preliminary results indicate that the proposed LC optical switch, which has four phase levels, is able to achieve a power loss of less than 10dB. With the investigation and deployment of eight-phase LC SLM, as well as the investigation on LC material and VLSI circuit, this parameter expects to be improved.

The proposed LC optical switch is able to achieve a static crosstalk performance of -50dB. The switching of the LC switch involves changing the voltage of the LC pixel elements in a pixel block and resulting in the formation of another hologram pattern. During the transitional period, the hologram pattern of the pixel block is random. If a beam is present during switching, the beam will be scattered all over. Since the switching in the LC optical switch needs cooperative interaction of two SLMs, the fraction of the beam coupled into other fibers during switching, referred to as dynamic crosstalk, is negligible. Therefore the LC optical switch has a very good dynamic crosstalk performance. In comparison, during the switching of a MEMS switch, the optical beam transverses other ports, which may potentially result in large dynamic crosstalk.

Wavelength Range

Another important aspect of an optical switch is its wavelength range. The two-SLM architecture of the LC optical switch is able to offset the dispersion effect of a hologram and achieves much wider wavelength range than a single SLM architecture. Careful considerations are also given in the lens design to offset the dispersion effect. As a result, the LC optical switch can be designed to work at either 1310nm wavelength range, or, L and C band.

Stability, Reliability and Manufacturability

Stability and reliability are possibly the biggest advantages of the LC optical switch over MEMS switch.

The LC optical switch has a very reliable and stable switching mechanism. The LC optical switch does not have mechanically moving parts; therefore it is not subject to mechanical wear and tear. Multilevel digital voltage is applied to the pixel elements of SLM. Each pixel block in a SLM contains thousands of pixel elements; failure of several pixel elements has almost no impact on the switching performance of the LC optical switch. The switching of the LC optical switch also has excellent repeatability (less than 0.05dB). All these features make the LC optical switch have much better stability, reliability and repeatability than MEMS switch.

In comparison, 3D MEMS switch uses mechanical switching. Complex servo system has to be employed to maintain the precision and repeatability of switching. Failure of single switching element may result in the replacement of the entire switch.

Moreover, the LC optical switch's LC on silicon technology is able to benefit from the LCD technology, which has been mature and mass produced in the last decade. This makes assembly of the LC optical switch's components, automatic optoelectronic packaging, performance testing and quality control less challenging tasks. All these problems remain to be solved for MEMS switch.

The LC optical switch is also very cost effective. The cost of The LC optical switch will be much lower than MEMS switch.

4. CONCLUSIONS

In this paper we introduced the design of a LC optical switch. The polarization-independent LC SLM technology makes the design and manufacture of a large-scale reconfigurable optical switch a reality. The performance of the LC optical switch was analysed. Compared with other technologies, especially MEMS, LC optical switch has higher power loss, which needs to be improved. However, LC optical switch has better performance on stability, port repeatability, reliability, power consumption, costs and manufacturability. Moreover, hologram-based nature of the LC optical switch makes it very flexible. Some functions like multicasting, automatic fiber alignment can be realized. With further developments in fabrication technology and LC material, it is expected that LC optical switch will become a strong candidate in the optical switch market.

5. REFERENCES

[1] S. Kuroyanagi, T. Nishi, K. Miyazaki, T. Chujo, and A. Hakata, "Optical Path Cross-Connect System for Large-Capacity and Highly Reliable Optical Transport Networks", *Infrastructure Summit (Networks and Systems),* 2000.

[2] Patrick Berthele, Bruno Fracasso, and Jean-Louis de Bougrenet de la Tocnaye, "Deign and characterization of a liquid-crystal spatial light modulator for a polarization-insensitive optical space switch", *Applied Optics*, Vol. 37, No. 23, pp. 5461-5468, 1998.

[3] Gordon D. Love, "Liquid-crystal phase modulator for unpolarized light", *Applied Optics*, Vol. 32, No. 13, pp. 1986-1989, April 1993.

[4] Thu-Lan and Gordon D. Love, "White-light performance of a polarization-independent liquid-crystal phase modulator", *Applied Optics*, Vol. 38, No. 10, 1999.

[5] Yoriko Morita, Jay E. Stockley, Kristina M. Johnson, Eckhard Hanelt and Frank Sandmeyer, "Active Liquid Crystal Devices Incorporating Liquid Crystal Polymer Thin Film Waveplates", *Japanese journal of applied physics*, Vol. 38, pp. 95-100, 1999.

[6] Agere Systems 5200 Series 64X64 MEMS Optical Switch Module, Preminary Data Sheet, Revision 2, July 2001.

[7] OMM 32X32 Photonic Switch, Preminary Product Brief, November, 2001.

Optical channel waveguides in a chalcogenide (Ga:La:S) glass fabricated through short wavelength (λ=244nm) illumination

Arshad K. Mairaj and Daniel W. Hewak

Optoelectronics Research Centre, University of Southampton, Southampton, SO17 1BJ, United Kingdom
Tel: +44 (0)23 8059 4530, fax: +44 (0)23 8059 3149, e-mail: mkam@orc.soton.ac.uk

Abstract: Chalcogenide optical glass is an interesting material, for both fibre and planar technologies, as it offers possibilities for a wide array of devices suitable for use in both non-linear applications and as IR sources. Channel waveguide structures were directly written into bulk gallium lanthanum sulphide glass via above-bandgap (λ=244nm) illumination. Focused fluence of 1.5 – 150 J/cm^2 from a UV-laser was applied, inducing giant-photocompaction and photochemical changes. A typical channel waveguide had attenuation < 0.3 dB cm^{-1} and was spatially single-mode at 1 μm. The damage threshold for fabricating such devices in this glass system is also discussed.

Key Words: Amorphous semiconductor, Optical writing, Channel waveguide

I. INTRODUCTION

Optical components are essential if the boom in telecommunications is to experience further growth. The biggest breakthrough in fibre optic technology came with the successful development of a long-haul fibre optic amplifier. This significant invention was the Er^{3+}-doped fibre amplifier (EDFA), capable of inherently amplifying optical signals at the coincident wavelength of 1.55 μm [1]. If not for the optical transparency of these 'strands of sand', generation and influencing the behaviour of light in crystals, glass and semiconductors would not have been possible. It might possibly be that silica (SiO_2) is the ultimate inert and stable optical material that will always be relied upon for its ability to transmit light over large distances. However, the inability of optic fibre to assimilate several components and functions onto a single compact platform has had researchers exploring a concept first proposed in 1969 [2].

First conceived by S.E Miller, integrated optics (IO) proposed the incorporation of multiple devices such as transmitters, receivers, modulators, signal amplifiers and lasers onto a single photonic chip. This was generally understood to be the ultimate in miniaturization and functionality. Silica based materials were among the initial demonstrations of passive and active IO devices [3-5]. However, active planar devices based on this material suffer from concentration quenching effects caused by high rare-earth (RE) concentration, which is necessary to compensate for the short interaction length. Despite the dramatic impact silica glass has had on telecommunications, it does have limitations in active planar operation. It can function only weakly as an optic switch and is unable to transmit radiation much beyond 2 μm. The mindset in telecommunications has always been to exceed current state of art and to provide advanced solutions for present and future applications. Research into diverse fabrication techniques for a plethora of host materials that offer capabilities surpassing silica glass led to the development of new non-oxide vitreous materials. Fluoride and chalcogenide glasses (ChGs) have an optical cut-off at longer wavelengths and are manufactured from low-phonon energy materials possessing heavier ions with weak bond strengths [6-7]. The unique optical properties found in ChGs make them very interesting for IO devices. Multiple radiative transitions for RE ions are possible due to low phonon energies (W_p = 325 - 425 cm^{-1}) which are quenched in other amorphous materials. In addition, the high IR transparency, inherent photosensitivity and high non-linear refractive index are only some advantages of ChGs [7]. Notable previous works for planar and channel devices in ChGs can be found in several publications [7-11]. A good overview of planar channel devices in fluoride glasses can be found elsewhere [6]. In this paper, we discuss a method for fabricating channel waveguides in bulk gallium lanthanum sulphide (Ga:La:S) glass. The channel geometry can be beneficial for achieving low laser thresholds, circular spatial outputs and compatibility with optical fibre. For this demonstration, we have directly written channel waveguide structures with above-bandgap illumination provided by a focused UV-laser beam (λ = 244 nm). Section II details channel waveguide fabrication by UV illumination and photoinduced effects resulting in material modification. Some of these effects are present in the form of giant-photocompaction and photochemical modification. Section III discusses the damage threshold for waveguide fabrication and optical characterisation. Finally in section IV, our concluding remarks are given.

II. FABRICATION AND PHOTOINDUCED EFFECTS

The gallium lanthanum sulphide (Ga:La:S) glass system is a vitreous chalcogenide material first discovered in 1976 [12]. Since then, interest has been maintained over the years primarily due to its exceptional and unusual optical properties. These glasses have a wide transmission window between 0.6-7 μm and are classed as an amorphous semiconductor with bandgap energy 2.6 eV (475 nm). The non-toxicity, high transition temperature, excellent REsolubility, low-phonon energy, high non-linearity, highly photosensitive and well documented spectroscopic properties make it an interesting candidate for research into planar waveguide devices [13].

Fabrication of Ga:La:S glass, with a typical molar ratio 65 Ga_2S_3 : 32 La_2S_3 : 3 La_2O_3, was carried out from prepared batches of powders. These glass precursors were loaded into a vitreous carbon crucible while in a controlled atmosphere. The precursors are non-volatile at the glass melting temperature (1150°C for up to 24h) and were heated in an open (argon purged) atmosphere, after which the melt was quenched and annealed. The glass was cut into slabs (20 x 17 x 2 mm) and polished on the top and bottom faces. An optically flat polish is required on the writing surface to minimize scattering of laser light during writing. After laser writing, the sample was polished at both end faces to allow optical coupling.

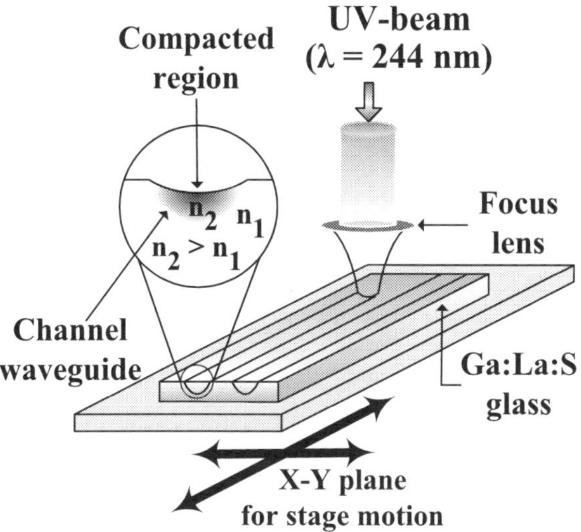

Fig. 1. Schematic diagram of the set-up used to directly write channel waveguides into Ga:La:S glass.

The direct-UV writing set-up shown in Fig. 1 consists of a frequency-doubled UV-laser (Coherent FRED Sabre 500) with 200 mW of CW output at 244 nm. A UV grade spherical fused silica lens (35 mm focal length) provides a focused writing spot from the spatially filtered laser beam. The UV-beam size ($1/e^2$ radius of intensity) is approximately 3.1 mm and the measured spot size of the focused waist is 3.3 μm. The sample, held in place by a vacuum chuck, was attached to a computer controlled translation stage. The stage, which provided the 2D movement of the sample (X and Y dimensions), has a maximum scan velocity of up to 5 cm/s and relative position resolution of 0.1 μm. The quality and dimension of a channel waveguide was determined by both the intensity (I_{UV}) of the focused laser beam and the scan velocity (V_{SCAN}). The average laser fluence (F) applied to the glass surface may then be expressed as [4]:

$$F = \frac{I_{UV}.a}{V_{SCAN}} \tag{1}$$

where a is the spot size. Fig. 1 also indicates photostructural change in the form of giant-photocompaction with material densification. The ability of an irradiated region (channel waveguide) to confine a laser beam suggests a positive change in refractive index ($n_2 > n_1$). For our experiments, the UV spot size was varied between 25 – 50 μm, by adjusting the sample to focus distance, providing laser intensity of I_{UV} = 1.5 – 10.2 kW/cm^2 over the applied area. Scan velocity was varied between 0.2 – 5 cm/s. Hence, a fluence in the range of 1.5 – 130 J/cm^2 was applied to the glass surface. Micrographs of the channel waveguides were obtained using an analytical scanning electron microscope (SEM, JEOL 6400) to which an energy dispersive X-ray microscope (EDAX), allowing compositional analysis, was attached. The channels indicate varying degrees of photocompaction. A channel directly written with UV-laser fluence of 10.8 J/cm^2 had undergone photoinduced changes in the form of giant-photocompaction (1.2 μm). Physical dimensions of the channel were 13 μm by 6 μm. Measurements within the photomodified region revealed variations in elemental ratios, with an increase in lanthanum content and decrease in gallium and sulphur. This photochemical modification contributes to densification by creating a region of raised refractive index, which is the waveguide core. The physical size as well as the chemical change within the photoinduced region varied largely with changes in scan velocity and applied laser intensity, as would be expected. This physical and chemical change in turn varied the attenuation as well as modal size and shape of the waveguide output. The maximum measured refractive index change in these channels was $\Delta n \approx +10^{-3}$.

Amorphous semiconductors (ChGs) can experience structural changes, with modification of physical and chemical states, when exposed to light. Chalcogenide glasses are widely regarded as a "soft semiconductor" because of their inherently quasi-stable atomic structure. The flexible and viscous structure is due to the chalcogen atoms having a two-fold coordination. Electronic mobility as seen in a semiconductor produces asymmetric motion of atoms and this can be paralleled to explain some of the photoinduced effects in ChGs. However, it is certainly plausible that there are contributions from both electronic and thermal phenomena to these photoinduced effects [14]. In the case of As_2S_3, exposure to near-bandgap illumination, macroscopic photoexpansion ($+\Delta V/V \sim 0.5\%$) was observed and could be thermally recovered [15]. This photoelectronic excitation with near-bandgap light with $\hbar\omega \approx E_g$, where $\hbar\omega$ is the photon energy and E_g is the optical bandgap energy of the ChG semiconductor, induces photodarkening and produces structural transformations. Volume changes were even more remarkable with below-bandgap ($\hbar\omega < E_g$) exposure causing giant-photoexpansion ($+\Delta V/V \sim 5\%$), leading to development of As_2S_3 microlenses [16]· Above-bandgap illumination ($\hbar\omega > E_g$) with photon energies lying within the Urbach-tail region can have far more prominent volumetric effects, due to considerably greater material absorption, than that of $\hbar\omega \leq E_g$ illumination. Permanent optical and geometrical changes caused by above-bandgap illumination can potentially be beneficial for fabrication of IO components [14].

III. DAMAGE THRESHOLD AND OPTICAL CHARACTERISATION

It was generally observed that with slower scan velocities (< 0.8 cm/s), for a constant I_{UV}, structures were clearly visible even with the naked eye due to physical damage of the glass. Material damage can lead to the formation of microcracks, which includes generation of bulk or surface cracks, common for fluencies above the damage threshold. For very slow scan velocities (< 0.2 cm/s) ablation was the dominant effect. Below the ablation and damage thresholds such damage of the glass could be avoided, with several kinds of material change being noted. Volume expansion or contraction, material densification and associated refractive index changes can however lead to residual tensile stresses within irradiated glasses [17]. An indication of the range of surface changes that can be induced by direct-UV writing into Ga:La:S glass is shown in Fig. 2.

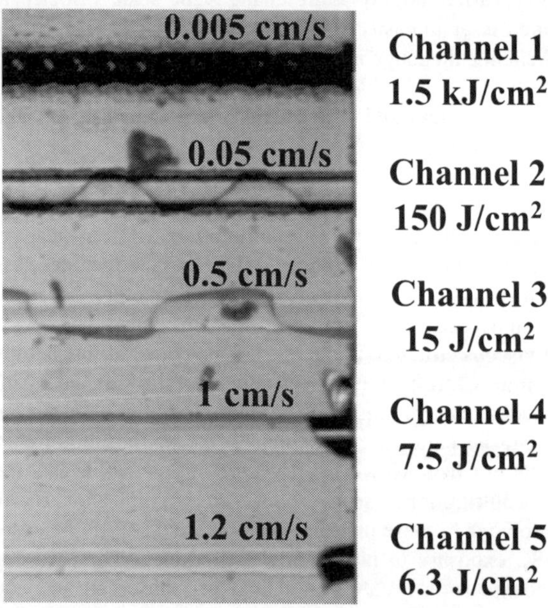

Fig. 2. Optical micrograph of directly written channels with associated scan velocity and applied fluence for constant I_{UV} = 1.5 kW/cm^2.

A general observation is physical dimensions for the resulting structures decrease with increasing scan velocity, for a constant I_{UV}. Furthermore, structures can be damaged either due to ablation (Channel 1) or microcracking (channels 2,3). However if applied fluence is below the damage threshold, structures formed (channels 4,5) will be free from severe physical defects. It should be noted that damage to a channel waveguide could also occur during the polishing process. Channels 4 and 5 show chipping at the end of the waveguide, a defect introduced during the polishing process.

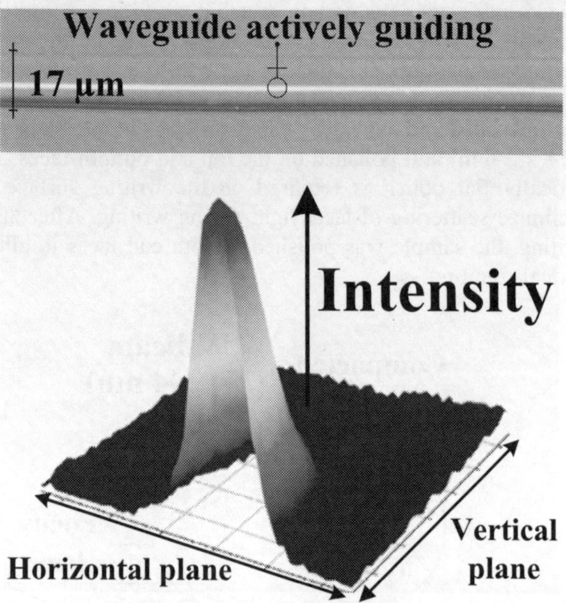

Fig. 3. Optical micrograph of a channel waveguide, actively guiding light, with compaction width 17 μm and surface plot of the near-field TM mode (λ=815 nm).

In Fig. 3 a channel waveguide directly written with fluence of 10.2 J/cm^2 and compaction width 17 μm is shown guiding a laser beam. The channel has a well formed structure and is without any apparent laser induced damage. Great care was undertaken during the polishing process, where the optical waveguide chip was wax bonded on each side to two similar sized Ga:La:S substrates for added polishing stability. It was also observed that the guided mode (λ=815 nm) was in the fundamental spatial-mode and measured modal dimensions were W_x = 7.2 μm by W_y = 8.2 μm for TM polarisation where W_x and W_y refer to the 1/e^2 intensity radii for the guided pump mode profile in the horizontal and vertical planes respectively.

The isolation of a single numerical figure as the lower limit for the damage threshold has proven difficult, as no clear pattern applicable to all write parameters could be identified. The appreciation for this complexity was readily understood once the parameters involved were studied. The key parameters in the direct-UV write process are laser intensity, scan velocity, UV-beam spot size and to a lesser extent surface preparation and homogeneity of glass composition. Investigation into damage threshold was conducted with variation in applied laser intensity (I_{UV} = 1.5 – 3.6 kW/cm^2) and scan velocity (V_{SCAN} = 0.5 – 5 cm/s) while maintaining a constant UV-beam spot size (a = 50 μm). Even at the lowest fluence (1.5 J/cm^2) the onset of surface compaction with depth of 0.2 μm, was observed as shown in Fig. 4. The resulting channel structure was well formed and able to guide light. When V_{SCAN} was in excess of 1 cm/s, channel structures were well formed for I_{UV} = 1.5 – 3.6 kW/cm^2. However, as V_{SCAN} approached low velocities, damage was visible in the form of severe microcracking. This could be due to residual tensile stresses causing fractures during the material relaxation cycle as well as strong absorption of laser energy. The damage threshold seemed to be specific for each applied laser intensity and was very much dependant on scanning velocity.

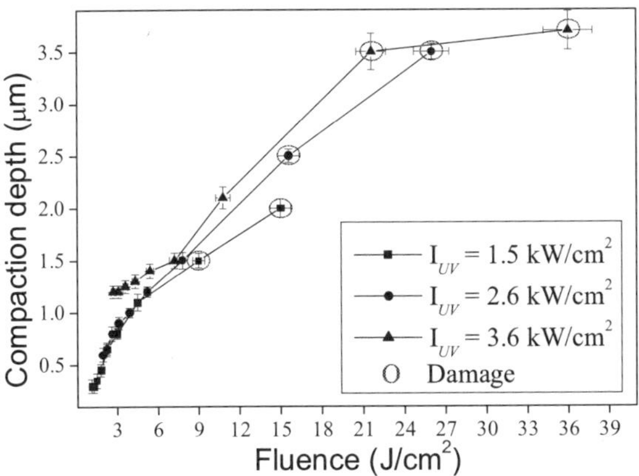

Fig. 4. Compaction depth is plotted against laser fluence for applied laser intensity (I_{UV} = 1.5 – 3.6 kW/cm^2). The graph shows six points of laser-induced damage due to slow scan velocities resulting in either ablation or microcracking.

For I_{UV} = 3.6 kW/cm^2, the channel formed with laser fluence of 10.6 J/cm^2 did not exhibit surface damage and was capable of guiding light. In contrast for I_{UV} = 1.5 kW/cm^2, the channel that was formed with a lower laser fluence of 9.1 J/cm^2 was seen to have undergone channel damage via severe microcracking. This suggests that the damage phenomenon is dependent on scanning velocity more so than it is on applied laser fluence. For a constant I_{UV} = 1.5 kW/cm^2, with applied fluence of 1.5 – 6 J/cm^2, uniform channels without any significant damage were observed. This was also true for I_{UV} = 2.6 kW/cm^2 and 3.6 kW/cm^2 with respective applied fluence of 2.1 – 8.5 J/cm^2 and 3.1 – 10.6 J/cm^2. Waveguide attenuation measurements were TM polarised at 1 μm and a tungsten lamp was used as a light source. For a channel exposed to 3.5 J/cm^2, waveguide attenuation was < 0.5 dB cm^{-1} prior to thermal annealing. The same waveguide had an attenuation of < 0.3 dB cm^{-1} following 30 minutes of thermal annealing at 400 °C.

IV. SUMMARY

Photoinduced changes can be introduced by directly writing waveguides into Ga:La:S glass through exposure to short wavelength light (λ = 244 nm). Photomodifying the physical and chemical properties allowed fabrication of a spatially single-mode channel waveguide with attenuation < 0.3 dB cm^{-1} at 1 μm. Surface compaction was observed for the lowest applied fluence with ablation and microcracking observed for very high fluencies. Photodensification was observed in the guided regions and was verified through EDAX. Maximum measured refractive index change in these channels was $\Delta n \approx +10^{-3}$. Further optimisation of glass quality and laser write parameters is expected, leading towards efficient low-loss devices for use with IO. We have already demonstrated laser operation in a neodymium-doped device [18] through this fabrication technique. The potential for developing highly nonlinear optical waveguides and erbium-doped waveguides with emission in the near-mid-IR seems promising.

REFERENCES

[1] R.J. Mears, L. Reekie, I.M. Jauncey and D.N. Payne, Electronic Letters, vol. 23, page 1026-1028, 1987.
[2] S.E. Miller, Bell System Technical Journal, vol. 48, page 2059-2069, 1969.
[3] K. Shuto, K.Hattori, T. Kitagawa, Y. Ohmori and M. Horiguchi M, Electronic Letters, vol. 29, page 139-141, 1993.
[4] M. Svalgaard, C.V. Poulsen, A. Bjarklev and O. Poulsen, Electronic Letters, vol. 30, page 1401-1403, 1994.
[5] T. Kitagawa, K. Hattori, Y. Hibino and Y. Ohmori, Journal of Lightwave Technology, vol. 12, page 265-270, 1994.
[6] J.L. Adam, Journal of Fluorine Chemistry, vol. 107, page 405-, 20011997.
[7] L.B. Shaw, B. Cole, P.A. Thielen, J.S. Sanghera and I.D. Aggarwal, IEEE Journal of Quantum Electronics, vol. 37, 1127-1137, 2001.
[8] C. Gmachl, H.Y. Hwang, R. Paiella, D.L. Sivco, J.N. Baillargeon, F. Capasso and A.Y. Cho, IEEE Photonic Technology Letters, vol. 13, page 182-184, 2001.
[9] A. Saliminia, A. Villeneuve, T.V. Galstyan, S. LaRochelle and K. Richardson, Journal of Lightwave Technology, vol. 17, page 837-842, 1999.
[10] J.F. Viens, C. Meneghini, A. Villeneuve, T.V. Galstian, E.J. Knystautas, M.A. Duguay, K.A. Richardson and T. Cardinal, Journal of Lightwave Technology, vol. 17, page 1184-1191, 1999.
[11] A.K. Mairaj, P. Hua, H.N. Rutt and D.W. Hewak, Journal of Lightwave Technology, vol. 20, 2002.
[12] A.M. Loireau Lozac'h, M. Guittard and J. Flahaut, Materials Research Bulletin, vol. 11, page 1489-1496, 1976.
[13] Y.D West, T. Schweizer, D.J. Brady and D.W. Hewak, Fibre Integrated Optics, vol. 19, page 229-250, 2000.
[14] K. Tanaka, Current Opinion in Solid State & Materials Science, vol. 1, page 567-571, 1996.
[15] H. Hamanaka, K. Tanaka, A. Matsuda and S. Iizima, Solid State Communications, vol. 19, page 499-501, 1976.
[16] H. Hisakuni and K. Tanaka, Optics Letters, vol. 20, page 958-960, 1995.
[17] N.F. Borrelli, C. Smith, D.C. Allan, T.P. Seward, Journal of Optical Physics B, vol. 14, page 1606-1615, 1997.
[18] A. K. Mairaj, C. Riziotis, A. M. Chardon, P. G.R. Smith, D. P. Shepherd and D. W. Hewak, Applied Physics Letters, submitted July 2002.

Photodiode Module Structure for Easy Alignment Between Fiber and Photodiode Chip

Joonyoung,Jung and Hyungsuk,Ryu
Optical Communication Component Research Center
Electronics and Telecommunications Research Institute
1 Oryong-dong, Buk-gu, Gwangju, 500-712, Korea
Tel : +82-62-970-6596, Fax : +82-62-970-6989, E-mail : jyjung21@etri.re.kr

Abstract: Optical photodiode module structure has been studied for high coupling efficiency and easy packaging. This paper proposes the optical photodiode module structure with focusing grating outcoupler to accomplish high coupling efficiency and easy packaging. A focusing grating outcoupler makes a small spot size beam at free space, so the light that passes through the optical fiber is gone to a optical photodiode chip with a high coupling efficiency by focusing grating outcoupler. Because the optical photodiode module structure proposed this paper has less sensitivity about the fiber position change that maybe happened in the packaging process, we can have a easy packaging process.

Key Words: Photodiode module with focusing grating outcoupler

1. INTRODUCTION

Even though optical photodiode chip have a high quality, Optical photodiode module cannot have a high quality, if the beam coupling efficiency from fiber to optical photodiode chip is not high.

The optical photodiode module structure which is proposed for high coupling efficiency and alignment tolerance uses lens or reflection between fiber and optical photodiode chip. The optical photodiode module structure which uses lens needs a exact alignment between fiber and lens as well as between lens and optical photodiode chip. The optical photodiode module structure which uses reflection is difficult to make because fiber, reflection and optical photodiode chip should be adhere closely to accomplish high quality.

This paper proposes optical photodiode module structure for high coupling efficiency and easy packaging. The optical photodiode module proposed this paper has a high coupling efficiency, because focusing grating outcoupler transfers light from fiber to optical photodiode chip with a high efficiency. This module has a high alignment tolerance, because focusing grating outcoupler focuses light to small spot. So, The optical photodiode module proposed this paper can be packaged easily.

2. OPTICAL PHOTODIODE MODULE STRUCTURE WITH FOCUSING GRATING OUTCOUPLER

Focusing grating outcoupler is not used in optical photodiode structure, but was used optical laserdiode structure a few years ago [1]. This paper proposes optical photodiode with focusing grating outcoupler to get high coupling efficiency and easy packaging.

I will use figure 1 to explain the optical photodiode module structure with focusing grating outcoupler. The optical photodiode module that has high coupling efficiency couples most of light from fiber to photodiode chip. The light passing through fiber core meets focusing grating outcoupler. The roll of focusing grating outcoupler couples the light from fiber core to photodiode chip with a high coupling efficiency and alignment tolerance. When the light passing through fiber core meets focusing grating outcoupler, light goes to the free space and is focused on a small spot size [1].

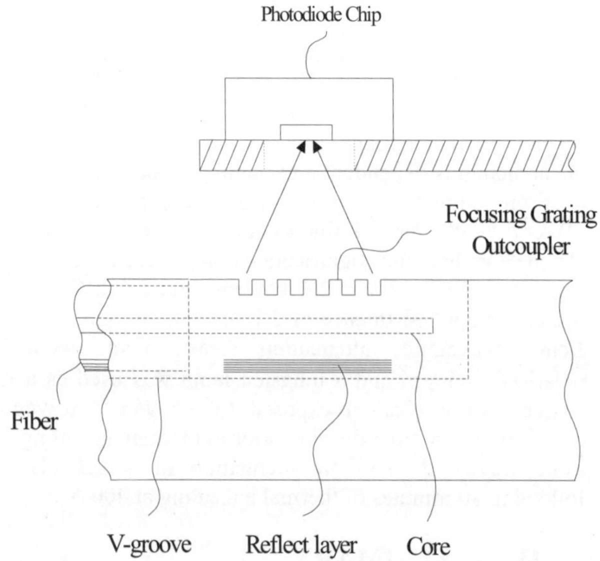

Figure 1. Optical photodiode module structure with focusing grating outcoupler

The active area of optical photodiode chip should be located the place that light is focused to get most of light from focusing grating outcoupler. The light passing through fiber core diffracts towards the substrate. In general, an outcoupling grating diffracts light into the superstrate as well as the substrate. For high efficiency we need to increase the directionality of the focusing grating outcoupler. We use multilayer reflector for high coupling efficiency. By incorporating a multiplayer reflector below the fiber, we can redirect the substrate radiation that add in phase with the surface radiation. By using this technology, we have achieved more than 90% of overall unidirectional coupling efficiency [2]. A fiber is put on the V-groove to put fiber exactly.

3. ALIGNMENT TOLERANCE OF OPTICAL PHOTODIODE MODULE WITH FOCUSING GRATING OUTCOUPLER

For fast transmission, Nowadays active area's diameter of optical photodiode chip becomes smaller than before. We can expect that beam spot size should be smaller than active area of photodiode chip in order to get high coupling efficiency. When we use focusing grating outcoupler, spot size of mode field diameter (MDF) can be made about 8μm ~ 12μm and active area diameter of optical photodiode chip for 10Gbps is about 30μm. So, alignment tolerance between fiber and active area of photodiode chip is high for fast transmission.

If we assume that the beam going from fiber to active area of photodiode chip is circular Gaussian beam, we can calculate the alignment tolerance by integral of the beam profile [3]. We simulate the alignment tolerance of optical photodiode module. We assume that active area diameter of photodiode chip is 30μm and spot size from focusing grating outcoupler is 8μm to 24μm and coupling efficiency target is more than 95%. Alignment tolerance is divided defocus alignment tolerance and decenter alignment tolerance.

First of all, Figure 2 shows defocus alignment tolerance. For example, if beam spot size is 8μm, defocus tolerance is about 11.34μm. If beam spot size is 24μm, defocus tolerance is about 1.79μm. The 8μm beam spot size has better alignment tolerance about 9.55μm than the 24μm beam spot size. The smaller beam spot size, the better defocus alignment tolerance.

Figure 3 shows decenter alignment tolerance. For example, if beam spot size is 8μm, decenter tolerance is about 11.56μm. If beam spot size is 24μm, defocus tolerance is about 1.96μm. The 8μm beam spot size has better alignment tolerance about 9.6μm than the 24μm beam spot size. The smaller beam spot size, the better decenter alignment tolerance. When we use focusing grating outcoupler, spot size of mode field diameter (MDF) can be made about 8μm ~ 12μm. So, we can get a high alignment tolerance for 10Gbps photodiode module.

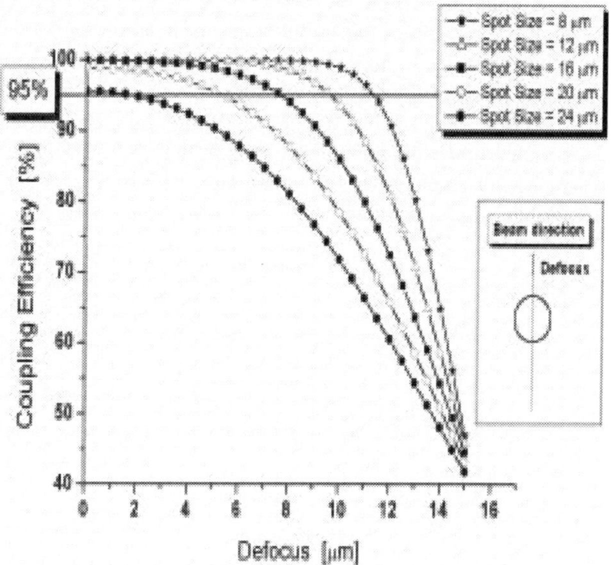

Figure 2. Defocus alignment tolerance

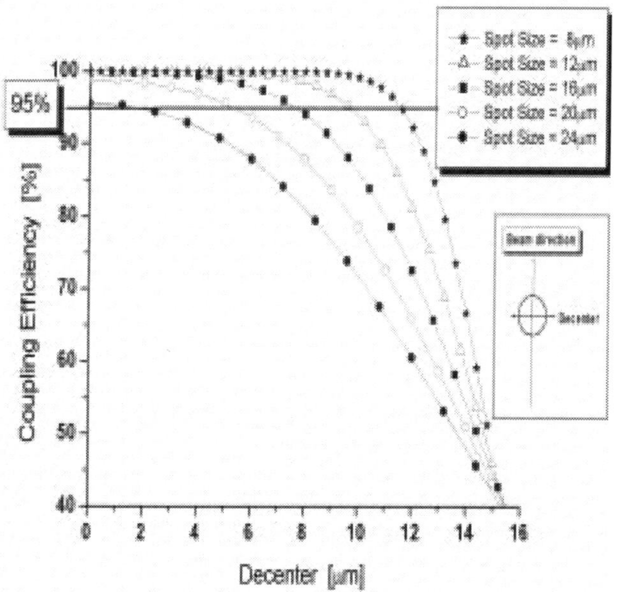

Figure 3. Decenter alignment tolerance

When we package optical photodiode module, we should consider simultaneously defocus and decenter alignment tolerance. We simulate alignment tolerance of optical photodiode module with considering defocus and decenter alignment tolerance simultaneously. We assume that active area diameter of photodiode chip is 30μm and the beam spot size made by focusing grating outcoupler is 12μm and coupling efficiency target is 85% to 100%. Figure 4 shows the result of this simulation. For example, if beam spot from focusing grating outcoupler deviate ±4.84μm from the focus place in the defocus direction, deviation in the decenter direction should be within ±8.38μm for 95% coupling efficiency. If this simulation assume that the beam spot size made by focusing grating outcoupler is 8μm, we can get higher alignment tolerance for 95% coupling efficiency

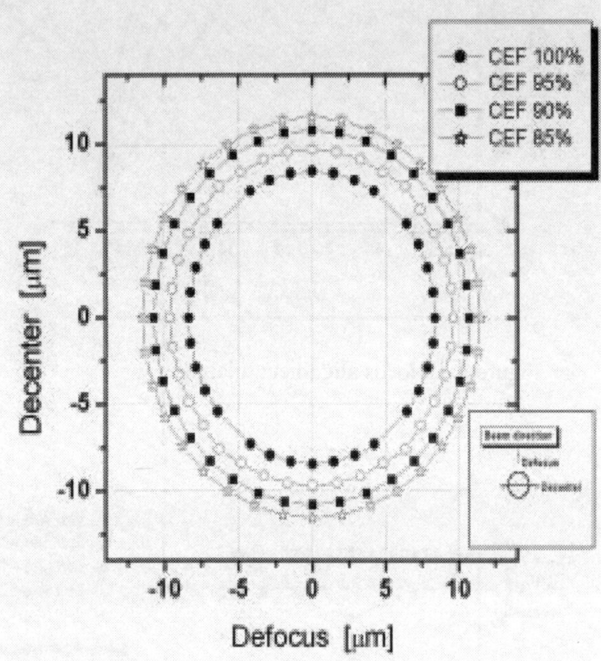

Figure 4. Alignment tolerance in the defocus and decenter direction at coupling efficiency (CEF) of 12μm spot size beam

If we use focusing grating outcoupler to make an optical photodiode module, we can get small spot size and high alignment tolerance between fiber and optical photodiode chip. From this result, we can get easy alignment between fiber and optical photodiode chip and easy packaging process of optical photodiode module. Because the optical photodiode module structure proposed this paper has less sensitivity about the fiber position change that maybe happened in the packaging process.

4. SUMMARY

This paper proposes the optical photodiode module with focusing grating outcoupler. This structure doesn't need exact alignment between fiber and optical photodiode chip. Because focusing grating outcoupler focuses beam on active area of optical photodiode chip with high alignment tolerance. As the result of this, this structure has an easy packaging processing. In addition, this optical photodiode structure is economic photodiode module structure because this optical photodiode structure has less sensitivity about misalignment happened in the packaging process

REFERENCES

[1] Anders Larsson, Niklas Eriksson, Peter Modh, "Grating-Coupled Surface-Emitting Laser with a Hyperbolic Unstable Resonator Producing a Stable Focused Output Beam", *IEEE Photonics Technology Letters., vol 11, no 11, 1999, pp. 1366-1368*

[2] A. Larsson, N. Eriksson, Sigurgeir Kristjansson, Peter Modh, Masahiro Uemukai, et al, "Grating-Coupled Surface Emitters: Integrated lasers, amplifiers, and beam shaping outcouplers", *SPIE., vol 3626, 1999, pp. 190-201*

[3] N.R.Barbeau, "Power deposited by a Gaussian beam on a decentered circular aperture," *Appl. Opt, vol 34, 1995, pp. 6443-6445*

Novel method for pedestal suppression and chirp elimination of high-order compressed solitons by using semiconductor optical amplifier and tunable filter

Zhihong Li (1), Caiyun Lou (2), Yizhi Gao (2), Chao Lu(1), Cheng Tee Hiang (1)
1 : Institute of Communication Research, BLK S2-B3c-23, Singapore 639798, ezhli@ntu.edu.sg
2 : Dept.of Electronic Engineering, Tsinghua University, Beijing, China loucy@ee.tsinghua.edu.cn

Abstract: Numerical simulation about pedestal reduction and chirp elimination of high-order compressed solitons by using SOA and tunable filter was demonstrated. Under certain conditions, the pedestals and chirp can be successfully eliminated without broadenning the pulse.

Key word: Pedestal suppression, chirp elimination, high order soliton, SOA.

1. Introduction

Pico-second pulses with high repetition rate, may play an important role in future ultra-high-speed communication systems. While not all optical pulse sources can meet this requirement. A promising method to achieve picosecond pulse is to employ high-order soliton compression in dispersion-shifted fiber (DSF)[1]. However, there exist the problem that the compression in the DSF would cause a broad pedestal to the compressed pulse, which reduce the energy utilization and also lead to interference noise to OTDM systems after multiplexing and transmission. Further more, evident chirp usually occurs at the center of the compressed pulse, which will reduce the spectrum efficiency of the system and even degradate the performance of transmission and signal processing.

Some techniques, such as intensity discrimination with nonlinear induced birefringence in fiber[2] and various kinds of NOLM[3] were applied to suppress the undesired pedestal. However, these techniques are somewhat complex, and especially for pedestal suppression of picosecond pulses, long fibers (usually several kilometers) should be used to achieve necessary nonlinearity, which would result in more instability. Moreover, to the best of our knowledge, few work concerned about the chirp elimination of the compressed pulse.

Semiconductor Optical Amplifier (SOA) has the advantages of compact, stable, easy to integrate, and has strong nonlinearity as well. In this paper, a novel method was suggested based on spectra filtering by using a SOA and a tunable filter, which is based on the fact that the degree of spectral broadening due to SPM increases with pulse intensity. This method was used to achieve 2-R-all-optical regenerator[4] . In the following numerical simulation we demonstrate that, by properly choose the parameters of SOA and filter, pedestal and chirp of the compressed soliton can also be successfully eliminated by this method.

2. Theory

The schematic diagram of pedestal suppression is shown in Fig1. Pulse generated from the pulse source has Gaussian profile and the pulsewidth is 12ps. Pulse peak power achieved to 119mW after amplified by EDFA, corresponding

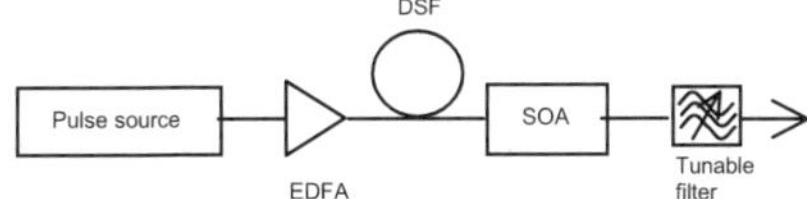

Fig1, Schemetic diagram of pedestal reduction by using SOA and tunable filter.

to the power of the third order soliton to be formed in DSF with dispersion of 2ps/nm/km. In numerical simulation, a Nonlinear Schordinger Equation was used to model the propagation characteristic[5]., and the waveform is plotted in Fig2 after the pulse propagated through 6.3km DSF. The pulse width is compressed to 2.2ps, and the broad pedestal is also evident.

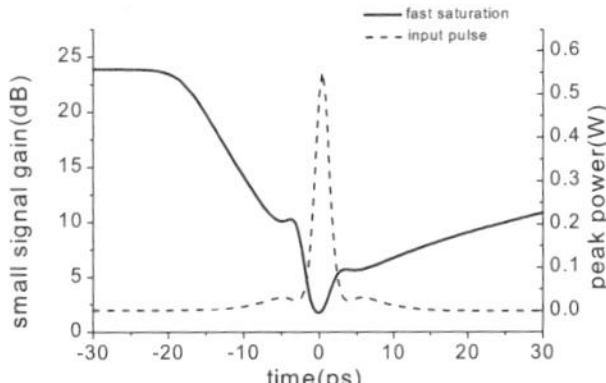

Fig2, SOA gain saturation by the input pulse. solid line:SOA gain dynamics; dash line:pulse waveform after high-order compression.

By taking into account carrier heating, spectral hole burning, two-photon absorption, and ultrafast nonlinear refraction, a model has been obtained [6] which is successful in describing the propagation of strong pico-second optical pulses through an SOA. Relative parameters used in the calculation are listed in table1. Where L is the length of SOA chip; τ_c is the carrier life time; P_{sat} is the saturation power of SOA; α is linewidth enhance factor; ε is nonlinear gain compression factor and G_0 is small signal gain of SOA, respectively.

Table1, parameters of SOA used in simulation

L	τ_c	P_{sat}	α	ε	G_0
500μm	200ps	30mW	5.0	1.05	23.8dB

With the compressed pulse as input, the simulated dynamic gain of SOA is also plotted in Fig2. It can be found that the gain of SOA was saturated by the pedestal at leading edge and decreased rapidly, it reached to the lowest point near the peak of the pulse and then, after a brief recovery, slowly increased.

Fig3 shows the asymmetric amplified pulse with a high pedestal at leading edge that contents more energy, since the leading edge experiences larger gain than the trailing edge. The dash line in the same figure shows the frequency chirp across the amplified pulse, which display a significant frequency downshift (the red shift) on the leading part of the output pulse and a smaller blue shift at the trailing part. These spectral changes are due to SPM effect, which was caused by carrier density variance by the gain dynamics. Due to the high nonlinearity of SOA, the spectrum of the pulse was further broadened by SPM effect. A filter with its central wavelength offset from the peak wavelength of the incoming pulse stream passes only nonlinear generated spectral components lying within the filter pass band. The pedestals, with much low power, generate little SPM and are mostly rejected by the filter. Therefore, central short pulse and pedestals can be discriminated by filtering at either side of the spectrum where the maximum frequency shift occurs.

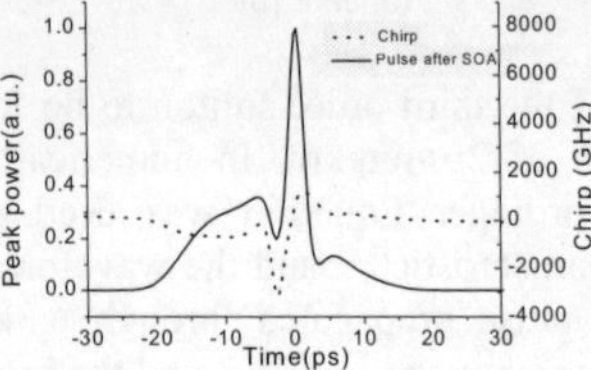

Fig3. Amplified pulse waveform after SOA (solid line) and corresponding chirp profile (dash line).

3. Calculation and analysis

Fig4 shows the simulated pulse waveform after filtering of spectrum from the output port of SOA. In numerical simulation, the FWHM bandwidth of filter is 2.5nm and the offset between the central wavelength of filter and signal is 450GHz at long wavelength side. We can find that the pedestal is almost extinguished and the waveform shows no sign of pedestal down to approximately –30dB below the pulse peak.

To make it more clearly, we compare the pulse waveform with the chirp before filtering by overlapping the curves. It's easy to observe that the filtered part is just located coincide with the part which has the largest frequency offset, while other part of the pulse, however, experience less frequency broadening and be almost filtered out. Just because of this, there's also an offset occur to the central position of the resulting pulse. In this case, it was calculated to be –2.3ps. This would not affect the application of these pulses, since each pulse will experience the same timing offset.

We can also analysis the chirp of the output pulse from the dash line in Fig4. What we mostly concern is just the chirp at the centre of the pulse. Obviously, the chirp is not zero, corresponding to wavelength shifting due to the frequency offset of the filter. Moreover, the chirp change linearly with small slope and the chirp parameter calculated at the center of the pulse is 0.06, compare with the value of 0.13 after high-order compression.

Thus we can say that, after propagation through SOA and filter, the pedestal of high-order compressed pulse can be reduced and chirp characteristic can also be improved.

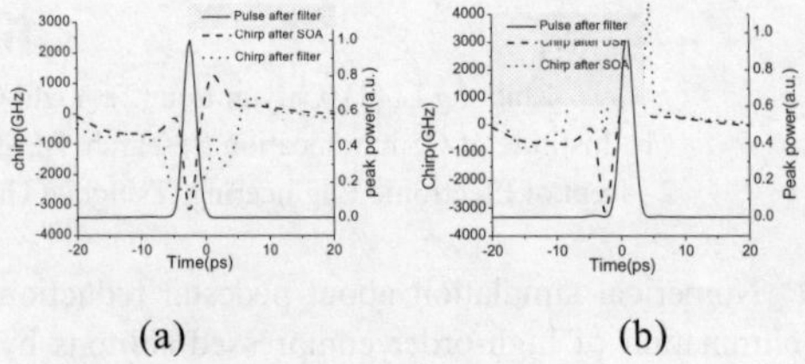

Fig4. Waveform and chirp profile after filtering at (a) long wavelength side; (b) short wavelength side of optical spectrum. Dash line: chirp after SOA; dot line: chirp after filtering.

By using the same filter with central frequency offset of 400GHz at the short wavelength side of the pulse spectrum, resulting pulse waveform and chirp was plotted in Fig5., from which we can find that the pedestal can also be reduced with chirp parameter of 0.03 at the pulse centre.

Please note that the pulsewidth after filtering is only 2.3ps, nearly the same with that after high-order compression. The pulsewidth after filtering is found to be mainly dependent on the filter bandwidth. Thus, by properly choose the parameters of components, the quality of the pulse can be improved without pulse broadening. Moreover, the effect of filter bandwidth, central wavelength and small signal gain of SOA were also investigated numerically in detail and the results will be presented at the conference.

4. Conclusions

Without pulse broadening, the pedestal of the pulse after high-order compression is extinguished by employing a simply configuration composed of a SOA and a tunable filter. Numerical analysis was given in detail. It was found that, besides the pulse shaping improvement, the chirp value in the pulse center could also be reduced.

References

[1] H.F.Liu, Ogawa, Y.; Oshiba, S.; Nonaka, T "Picosecond pulse generation from a 1.3 mu m distributed feedback laser diode using soliton-effect compression " IEEE J. Quantum Electron., Vol 27, pp.1655-1660, 1991

[2] K.A.Ahmed, IOOC'95, FD2-2, pp.99-100, 1995

[3] Laurent Chusseau , "250-fs optical pulse generation by simultaneous soliton compression and shaping in a nonlinear optical loop mirror including a weak attenuation" Optics Letters., pp.734- 736,1994

[4] P.Petropoulos, Monro, T.M.; Belardi, W.; Furusawa, K.; Lee, J.H.; Richardson, D.J., "A highly nonlinear holey fiber and its application in a regenerative optical switch" OFC'2001, TuC3 -T1-3, vol.2

[5] G.P.Agrawal, Nonlinear Fiber Optics Academic Press, third edition. 2001

[6] J. M. Tang, Shore, K.A. "Strong picosecond optical pulse propagation in semiconductor optical amplifiers at transparency" IEEE J-Quantum.Electron. vol. 34, pp. 1263–1269, July 1998.

Session 14E

Optical Networking IV

A Very High-Speed Traffic Aggregator Over GMPLS

Jit Biswas, Qiu Qiang

Laboratories for Information Technology
21 Heng Mui Keng Terrace, Singapore 119613
{biswas, qiu}@lit.org.sg

Abstract: This paper describes the architecture and operation of a Very hIgh-Speed Traffic Aggregator (VISTA) for an Optical Network. The unique feature of the traffic aggregator is that it works in conjunction with the GMPLS control plane of the integrated electro-optical network. The emphasis of our work is on the access to the optical network, at the level of the edge router.

Key Words: Network Processor, GMPLS, Traffic Aggregator, Optical Networking

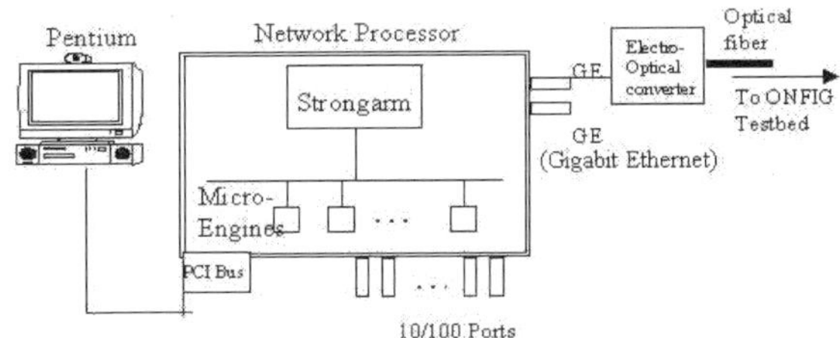

Fig. 1. Electronic Processing of IP Packets at the Edge of the ONFIG Optical Testbed

1. INTRODUCTION

The promise of low-cost network bandwidth emerging out of DWDM technology has given rise to unrealistic expectations of IP traffic bandwidth. While it is true that high capacity optical bandwidth is available through multiplexed wavelengths of light in DWDM networks, to tap this bandwidth for quality IP traffic is still a problem. Quality guarantees must be made to individual streams of packetised IP traffic, which in themselves represent only a fraction of the overall bandwidth available in a single wavelength of light. The solution lies in a variety of approaches ranging from all optical networks [1] to improved means of utilizing traditional TDM (SONET) networks, with newly defined framing protocols such as [2].

We believe that the electronic edge of the network where the transition is made from E to O domain (and vice versa) is a key point for monitoring and control of IP traffic. In this paper we describe a high speed traffic aggregator based on a programmable edge router which is using Network Processor technology, operating at a high speed, and aggregating traffic from multiple high speed LAN segments, multiplexing it onto an optical "uplink" to a core optical network (*Fig. 1*). Various QoS mechanisms such as RSVP [3], Differentiated Services [4], Bandwidth Broker [5] and GMPLS [6]), may be incorporated in programmable edge routers. However for our initial work, we have concentrated only on the GMPLS labeling scheme in order to set up Labeled Switch Paths (LSPs) which are guaranteed quality of service by means of transmission over light paths, all the way to the destination edge router. In related work in the same project [7], we are developing a GMPLS control plane that integrates between a core optical network and router based edge networks which are running the GMPLS protocol.

1.1. GMPLS and TESTBED projects

The ONFIG (Optical Networking Focus Interest Group) project is a collaboration between several research institutions and university departments in Singapore in the area of emerging optical networking. It encompasses both component and systems level work, with application to both electronic and optical parts of the network. The primary focus of ONFIG is on the access part of the optical network, which is recognized as yet unchartered territory. As a vehicle to carry out this R&D, it was decided that a national optical network testbed would be set up. The TESTBED project of ONFIG has a mission to provide a state-of-the-art optical network testbed for participating organisations. The testbed is not biased toward any particular topology and is capable of supporting tree (PON), mesh or ring optical networks through appropriately connected and configured OADMs and OXCs. It is important that the management and control software for the testbed are extensible, to allow incorporation of other types of networking components which may conceivably be introduced in future at the access optical network level.

To provide for the control and management planes for the

This work was partially supported by A-Star under ONFIG GMPLS project grant number 0121080116 testbed, the GMPLS (Generalised Multi-Protocol Label Switching) project was set up. Another objective of the GMPLS project is the development and integration of protocols for the hybrid switching network control and management. Both the electronic and the optical parts of the network are within the scope of this project. The project fits into and integrates with the ONFIG TESTBED project. Thus the testbed is managed by network management software produced by the GMPLS project. Furthermore, the control and signaling aspects of the optical portion and the electronic (data network) portion of the testbed are carried out by the GMPLS signaling software.

1.2. Why traffic aggregation is important

The GMPLS project also provides novel algorithms for traffic engineering and protection / restoration schemes for the integrated metro and access optical network. One of the ways in which the project addresses the particular requirements of the Access Optical Networks, is through traffic aggregation and bandwidth distribution mechanisms.

Traffic aggregation is one of the means of improving resource utilization. The benefits of traffic aggregation are many. It may be used as a means of admission control to handle bursty traffic. It may be used to ensure that VPN (Virtual Private Networks) are all assigned the same control mechanism (such as GMPLS Forwarding Equivalence Class (FEC)), and thereby ensure that all traffic on the VPN gets priority treatment, by mapping onto the same wavelength. It may be used to differentiate between groups of FECs destined for different quality treatment in terms of delivery deadline or handling of failure properties. For example one FEC may be given preferential treatment over others at the time of failure and protection switching.

1.3. Labeling scheme

Traffic aggregation may be achieved via different means. RSVP aggregates traffic at the level of micro-flows. Differentiated services deals with aggregation at the level of Per Hop Behaviours or PHBs, at the core routers. In GMPLS (and its predecessor, MPLS), aggregation is primarily carried out by and hierarchical labeling operations. In this paper we discuss the last approach. We present the architecture and design of a Very hIgh-Speed Traffic Aggregator (VISTA) for GMPLS based optical network. The focus of this paper is on labeling operations that are local to a router. The operation of the distributed label management protocols and algorithms are described in a separate paper [7].

The remainder of this paper is organized as follows. Section 2 introduces the VISTA architecture. The GMPLS extension to VISTA is presented in Section 3. Section 4 concludes this paper.

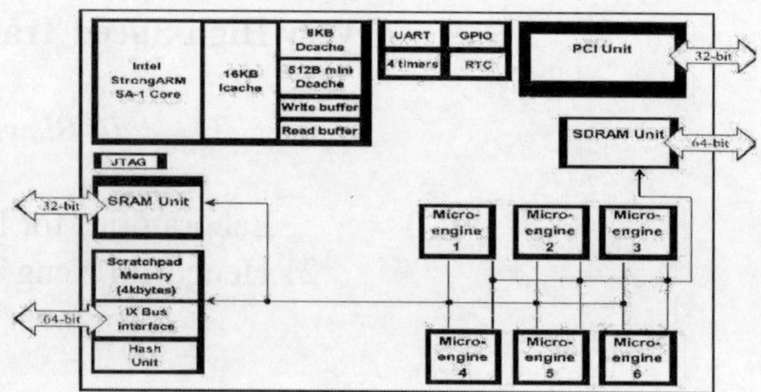

Fig. 2. IXP1200 Evaluation System

2. VISTA ARCHITECTURE

2.1. Hardware Architecture

VISTA runs on a PC using a 1.7GHz Pentium IV processor with the IXP1200 evaluation system (*Fig. 2*) plugged into one PCI slot. The board consists of a IXP1200 network processor chip, 32MB of DRAM, 2MB of SRAM, 4KB of on-chip Scratch memory, a 64-bit 66MHz IX bus, and a set of media access controller (MAC) chips implementing ten Ethernet ports ($8\times 100Mbps+2\times 1Gbps$). The IXP1200 chip itself contains a general-purpose *StrongARM* processor and six special-purpose *MicroEngine* processors all running at 200MHz . Each of the six MicroEngines supports four hardware contexts for a total of 24 contexts.

2.2. Software Architecture

Currently the VERA architecture [8] is used as the framework of VISTA. *Fig. 3* shows the software architecture adopted by VISTA. A *classifier* (C) first reads packets from an input port, and based on certain fields in the packet header, selects a *forwarder* (F) to process the packet. Each forwarder then gets packets from its input queue, applies some functions to the packet, and sends the modified packet to its output queue. All transformations of packets in the router occur in forwarders. Finally, an *output scheduler* (S) selects one of its non-empty output queues, and transmits the associated packet to the output port. The scheduler performs no processing on the packet.

This architecture has two main attributes. Firstly, it provides explicit support for adding new services to the router. Secondly, the architecture does not specify where in the processor hierarchy each component of an algorithm should be implemented: some run on the MicroEngines, some on the StrongARM, and some on the Pentium.

3. VISTA GMPLS EXTENSION

Since VISTA is used as a GMPLS edge router, both the *Control Plane* and *Data Plane* of the GMPLS node should be mapped

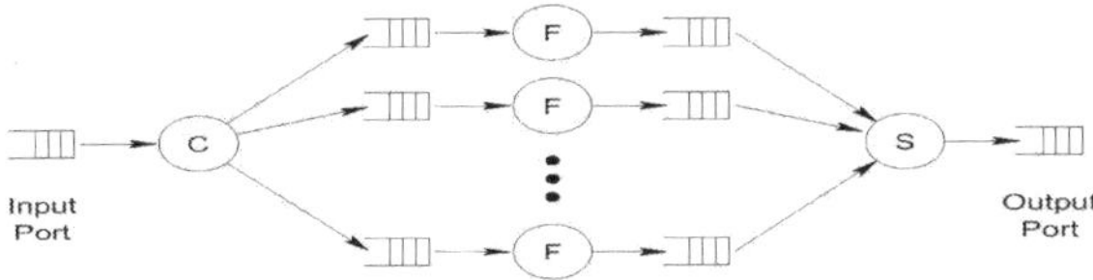

Fig. 3. Basic VERA Software Architecture

to the three processor layers: Pentium, StrongARM and MicroEngines. If such processor hierarchy is considered as an integrated whole, three possible data paths exist. Path A includes the MicroEngines and the *IXP memory*. Path B includes the MicroEngines, the IXP memory, and the StrongARM. Path C includes the MicroEngines, the IXP memory, the StrongARM, the *Pentium memory*, and the *Pentium*. The experiments in [9] show that path A can forward packets at a maximum rate of 1.78Gbps, path B at 269.31Mbps, and path C at 273.41Mbps.

The key feature of GMPLS is to avoid the hop-by-hop decision making by setting up a Layer 2 fast path using labels to move packets quickly along a pre-established path, without deep analysis (routing). Since the packet needs to be examined closely exactly once at entry to the GMPLS network, naturally, the control plane of a GMPLS node will be mapped to Pentium layer, which is on the low line speed path C. After that, the packet is somewhere along the path, and forwarding is based on the simple tagging scheme. So high line speed path A will be used for the forwarding function, that is, the data plane will be mapped to MicroEngine layer. The StrongARM functions as a bridge between the Pentium and the MicroEngine.

3.1. VISTA GMPLS Node Layers

3.1.1. Pentium

The Pentium layer of VISTA runs the Linux OS. The control plane of GMPLS node is loaded into the Pentium layer. The Pentium layer is responsible for populating and maintaining the label forwarding information base (LFIB). An IP Link-state routing protocol, e.g. OSPF, runs in the Pentium to exchange IP routing information with all other GMPLS nodes in the network. The resulting IP routing table provides information on destination network and subnet prefixes used for label binding. The label-binding information, which is maintained in LFIB, then can be distributed using the Label Distribution Protocol (LDP) running in the Pentium.

3.1.2. Pentium Memory

In the Pentium Memory, the complete routing table and the Label Forwarding Information Base (LFIB) generated at the control plane is maintained by the Pentium and can be accessed by the StrongARM at the request from the MicroEngine. The LFIB maintained by a GMPLS node consists of a sequence of Next Hop Label Forwarding Entries (NHLFE). Each entry contains the packet labelling and forwarding information. The 20-bit GMPLS label, which could be decided based on the routing information and perhaps some other criteria, could be used as an index into this array to access the list of NHLFEs corresponding to the label, although in our implementation we have not done this.

3.1.3. StrongARM

The *StrongARM* waits for packets arriving from the MicroEngine and assumes that they are there whenever a MicroEngine cache miss occurs. A packet may be one of the first few packets of a new traffic stream. The StrongARM does a route table lookup to locate the route table entry in the Pentium Memory. The corresponding NHLFE is retrieved based on certain index field in the Routing table entry. The packet will be labeled if the packet is at the entry of a GMPLS network, and forwarded using the next hop information found in the LFIB. The StrongARM updates the MicroEngine cache with the information in the new NHLFE.

3.1.4. IXP Memory

The necessary Labelling and forwarding information will be cached in the IXP Memory by the StrongARM. It makes the fast table lookup possible for the MicroEngine to only access the IXP Memory. Packet forwarding could achieve the line speed of Path A since only the IXP Memory and MicroEngine are involved.

3.1.5. MicroEngine

For each incoming packet, the MicroEngine performs a cache table lookup at the IXP Memory to locate the corresponding labelling and forwarding information. For the packets at the entry to the GMPLS network, which have no label yet, a shim label header is inserted between the link layer and the network layer based on the retrieved information.

If no entry is found in the cache table in IXP1200 Memory , a MicroEngine cache miss occurs. This may occur only for the first few packets of a new traffic stream. Such packets will be passed to the StrongARM to handle and the Cache Table will also be updated by the StrongARM.

3.2. Packet Classification Algorithms

For every incoming packet, a search must be performed in the VISTA's information base to determine what label this packet should be assigned and which next hop the packet is destined for. At the non-edge router of the GMPLS network, the 20-bits label could be used as index into LFIB Table to access the NHLFE. Since only a single memory read is performed, this

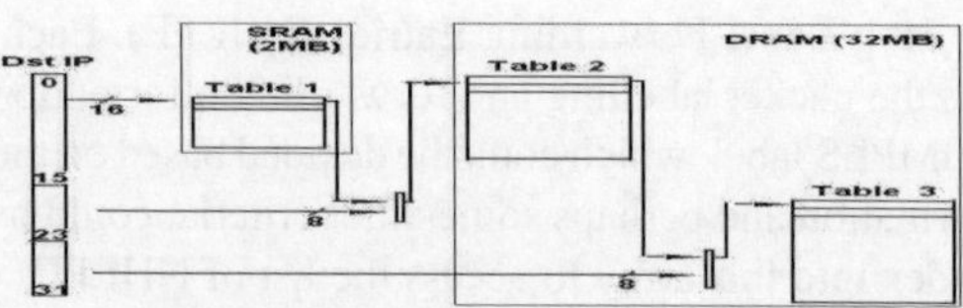

Fig. 4. MicroEngine Cache Table Lookup

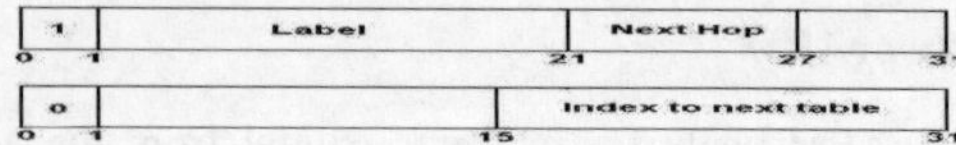

Fig. 5. MicroEngine Cache Table Entry Format

direct mapping is simple and fast. However, since no label has been tagged with the packets at VISTA, which is an edge router of a GMPLS network, the label and next hop information have to be searched, based on the FEC classification information contained in the packet's header. The complexity of such searching will depend on what information was used for packet FEC classification.

3.2.1. Traditional Packet FEC Classification

An FEC is a group of IP packets which are forwarded in the same manner, over the same path, and with the same forwarding treatment. The FEC to which a packet is assigned is encoded as a label. GMPLS can be consider as a packet forwarding technique which uses labels to make data forwarding decisions; therefore, traditionally the FEC for a packet has been determined by performing a conventional longest-match lookup on the destination IP address in the IP header. With traditional packet FEC classification, the label tagged with the packet can be used for forwarding at subsequent hops to avoid conventional routing, which will involve address lookup and heavily degrade the forwarding performance.

In the Pentium Memory, the complete routing and LFIB information is maintained. A Multibit Trie structure is used for the Routing table and the classical routing table lookup method is adopted for the table lookup. A very good description on Multibit Tries may be found in [10]. The table in Pentium Memory is searched only for the first few packets of a new traffic flow and all the necessary information used for labeling and forwarding the packet of the same flow is cached by the StrongARM in the IXP Memory. The MicroEngine uses the cached information for the remaining packets of such flow.

The Multibit Tries [10] hardware scheme could also fit in the IXP Memory perfectly by maintaining the cached information as the three tables shown in *Fig. 4*. The first table is stored in SRAM and contains all possible route prefixes that are up to 16-bits long. This table has 2^{16} entries, addressed from 0.0 to 255.255. Each entry in the table has the format shown in *Fig.5*.

The second and third table was stored in DRAM. The second table stores all route prefixes are between 16-bits and 24-bits and the third one for 24-bits to 32-bits.

For example, a prefix, P, in the routing table in Pentium Memory is cached by StrongARM into the cache table. If P is less than or equal to 16 bits long, it need only be stored in the first table and the first bit of the entry set to '1' to indicate that the remaining 15 bits designate the label and next-hop information; if the prefix P is longer than 16 bits, then the first bit of the entry in the first table is set to '0' to indicate that the remaining field contains a pointer to a set of entries in the second table.

For an incoming packet, the destination address is retrieved from the IP header and the first 16-bits of the address is used as an index into the first table. If the first bit equals '1', then the remaining bits will be used for labelling and forwarding; otherwise, the remaining bits will be multiplied by 256, the product added to the next 8 bits of the original destination address and this value will be used as a direct index into the second table in DRAM. If necessary, similar steps are taken to access a third table.

By such a design, the incoming packet is able to be labelled and forwarded at a very high speed at MicroEngine layer, since in the ideal case, a single SRAM read is required to retrieve the correct labelling and forwarding entry. In most cases, no more than one SRAM plus one DRAM read are required.

3.2.2. Advanced Packet FEC Classification

An FEC might correspond to a destination IP subnet, but it also might correspond to any traffic class that VISTA considers significant. For example isolating traffic for different service may require that each flow should be able to be classified. In such cases, a destination IP address is not enough for FEC classification, and multiple fields of information may be used, e.g., (source_IP, source_port, destination_IP, destination_port). When multiple fields of information are used, it will take a very huge space to cache a complete labelling and forwarding information structure in IXP memory. Therefore, when the length of information used for FEC classification exceeds a certain bit length, the StrongARM simply updates the cache table by one cycle IXP hardware hashing using the multiple fields of information as the key. To avoid severe memory collision, a perfect hashing mechanism could be involved.

3.3. VISTA Performance Evaluation

The control plane in VISTA is implemented in the Pentium layer. Almost any current Linux implementation of signaling protocol, e.g. LDP for MPLS system, could be directly adopted for the VISTA control plane. However, it must be modified with the extensions, such as CR-LDP, to incorporate the GMPLS protocol. In our related work [7], we are carrying out these extensions.

As described in *Sec 3*, the data plane in VISTA, including packets classification and forwarding, is performed at the MicroEngine layer except for the rare MicroEngine cache miss case, in which case, packets are handled by the strongARM. Three data plane processes were implemented for the MicroEngine and assigned to different contexts: receiving process, aggregating process and transmitting process. Both receiving and aggregating process are involved in the data input stage; the transmitting process is involved in the data output stage.

Based on the analysis of instruction latency, the processing of each 64-byte packet, which is the data unit transferred among IXP1200 components, requires 733 MicroEngine cycles at VISTA. This means that a given packet experiences 3665ns of delay as it is forwarded by one context running at 200MHz. This corresponds to a forwarding rate of 0.273Mpps for a single context.

Due to port contention, which occurs when two or more incoming packets destined for the same output port, it is impractical for a single context to forward a packet. Instead, packets are placed into queues, and these queues are serviced asynchronously by a second context. Suppose 24 MicroEngine contexts are evenly divided into two disjoint sets for input and output stages, 12 packets could be processed in parallel. This corresponds to a forwarding rate 0.273 x 12 = 3.276Mpps. Based on the measurements, there is roughly twice as much work to do on input as on output stage. Thus, in theory 4 MicroEngines (16 contexts) for the input stage and 2 MicroEngines (8 contexts) for output stage should give the ability to process 16 packets in parallel, corresponding to a throughput of 4.368Mpps. However since this assumes a perfect parallelism and ignores factors such as bus contention etc., it seems more acceptable to adopt the previous figure of 3.276Mpps as the performance upper bound of our VISTA implementation. Such a maximum packet forwarding rate gives efficient support for 1.677 Gbps of the aggregate link bandwidth. The packet latency caused by MicroEngine cache miss is hard to calculate by simply counting the instruction cycles since it includes the cost of the MicroEngine signalling the StrongARM. Based on the measurement results, one cache miss introduces around 2 ms extra latency on the average.

4. CONCLUSIONS

In this paper we have described the design and architecture of a Very High-Speed Traffic Aggregator which integrates with the GMPLS control plane of an optical network. The labeling engine performs at a reasonable speed, being capable of supporting 1.677Gbps of the aggregate link bandwidth. The main contribution of this paper is that it presents a design and architecture of a packet processing engine (VISTA) operating at the data plane. The signaling protocol of the GMPLS plane operates at a coarser time granularity on the control plane, which operates in the Pentium processor. The manner in which the GMPLS protocol is mapped onto networking edge resources, is an illustrative guide for supporting other IETF protocols and algorithms on top of network processors positioned at the network edge.

5. REFERENCES

[1] All Optical Networks home page - http://www.ll.mit.edu/aon/

[2] "Generic Framing Procedure and Data over SONET/SDH and OTN", *IEEE Communications Magazine*, May 2002, Vol. 40 No. 5

[3] L. Zhang, S. Berson, S. Herzog, S. Jamin, "Resource Reservation Protocol", *IETF RFC 2205*

[4] S. Blake, D. Black, M. Carlson, E. Davies, Z. Wang, W. Weiss, "An Architecture for Differentiated Services", *IETF RFC 2475*, December 1998

[5] "QBone Bandwidth Broker Architecture", http://qbone.internet2.edu/bb/bboutline2.html

[6] Ayan Banerjee, Kireeti Kompella, Yakov Rekhter, et al. "Generalized Multiple Protocol Label Switching: An overview of Routing and management Enhancements", *IEEE Communications Magazine*, January 2001, p 144-150.

[7] Victor Foo, Prashant Agrawal, Luying Zhou and Jit Biswas, "Design and Implementation of a Reliable and Integrated GMPLS-based Control Plane", *Proceedings of ICOCN 2002*, Singapore, November 2002.

[8] S. Karlin and L. Peterson. "VERA: An Extensible Router Architecture", *Proceedings of the 4th International Conference on Open Architectures and Network Programming (OPENARCH)*, pages 3-14, April 2001.

[9] T. Spalink, S. Karlin and L. Peterson. "Evaluating Network Processors in IP Forwarding", *Princeton University Technical Report*, TR-626-00, November 2000.

[10] M. A. Ruiz-Sanchez, E. W. Biersack, and W. Dabbous, "Survey and Taxonomy of IP Address Lookup Algorithms", *IEEE Network*, pp.8-23, Match/April 2001

MAC Protocols Supporting IP Datagrams in a Slotted Packet-Switched WDM Metro Ring Network

K. Bengi, A. Schiffermüller
Institute of Communication Networks, Vienna University of Technology
Favoritenstrasse 9/388, A-1040 Vienna, Austria
email: kemal.bengi@tuwien.ac.at

***Abstract*— Novel distributed medium access protocols for a time-slotted WDM metro ring employing optical packet switching and supporting variable-length data packets such as IP datagrams directly in the optical layer are proposed in this paper. A scalable control-channel based multi-hop WDM ring is considered in which each node is equipped with three transmitters and two £xed-tuned receivers. The new access control scheme relies on a frame-based slot reservation strategy including reservation of successive slots for data packets longer than the given slot size, and immediate access for packets shorter than the slot length.**

I. Introduction

In the LAN/MAN environment, all-optical packet switching can be already achieved with state-of-the-art optical devices. Particularly, optical packet-switched WDM ring network prototypes are being built and experimentally studied in several research laboratories. Generally, in order to avoid collisions on the individual WDM channels of such networks, MAC (Medium Access Control) protocols are necessary to arbitrate the bandwidth access [1]. Several access protocols for all-optical slotted WDM rings have been proposed in the literature, e.g. [2], [3]. One of the major issue, besides the architectural considerations, is the direct support of variable-length IP packets on top of the WDM layer yielding ef£cient IP-over-WDM metro networks. In this context, it is highly desirable to transmit the IP packets without fragmentation and reassembly in order to reduce the processing complexity at the senders and receivers.

Particularly, time-slotted WDM rings are not inherently supporting the transmission and reception of unsegmented variable-length IP packets. Therefore, in this paper, we propose novel access protocols for a packet-switched slotted WDM metro ring supporting variable-length IP packets directly over the WDM layer. The new protocol ef£ciently combines reservation-based and immediate access, thereby also enabling the accommodation of different traf£c types, while only marginally increasing the signaling overhead.

II. Network Architecture

The considered network is based on a ring topology interconnecting M nodes via a single unidirectional optical £ber, as shown in Fig. 1. The bandwidth of the £ber is divided into multiple *data channels* $\{\lambda_1, \lambda_2, \ldots, \lambda_C\}$ and a separate *control channel* λ_0, totally yielding $C+1$ wavelength channels. The data and control channels are in turn segmented into *data slots* and *control slots* of £xed duration carrying data and control packets, respectively. Furthermore, the ring size is assumed to cover a metropolitan area.

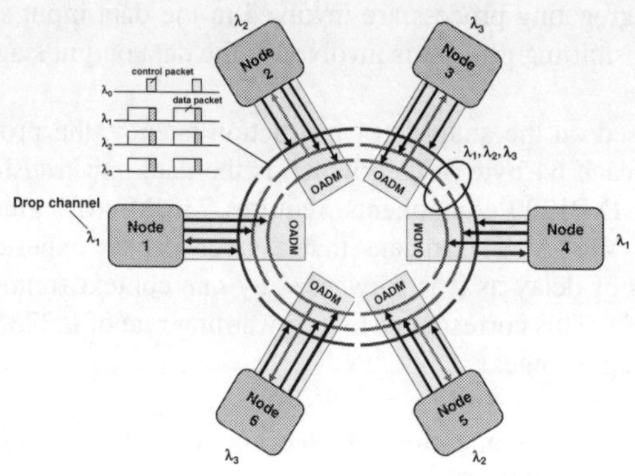

Fig. 1. Time-slotted WDM metro ring ($M = 6$, $C = 3$, only the data channels are indicated within the network schematic).

In our architecture, each node is equipped with one tunable transmitter (TT) and one £xed-tuned receiver (FR) dedicated to the data channels, one £xed transmitter (FT) and receiver (FR) used for the control channel (CC) and an additional £xed-tuned transmitter (FT) tuned to the node's own drop channel (the need for this is explained later). Thus, according to the common notation, we consider a CC–TT/FT^2–FR^2 system. Since the transmission and reception of data packets is performed optically, the nodes may be viewed as optical add/drop multiplexers (OADMs). Consequently, a node may send packets on any wavelength available in the network, while it is allowed to receive packets only on a preassigned wavelength, i.e., on its *drop channel.*

It is further assumed that there are more network nodes than wavelengths in the system, i.e. $M > C$, thus, the channel number is decoupled from the number of nodes as opposed to systems where $M = C$. For symmetry considerations, M is assumed to be an integer multiple of the number of data channels C. We further suppose that the C available data wavelengths are assigned in a cyclic fashion to the M destination nodes, as demonstrated in Fig. 1. Thus, several nodes have to use the same data channel for the reception of data packets. When D denotes the ratio of the number of nodes and channels, i.e. $D = M/C$, exactly D nodes have to share one drop channel. Since we additionally assume *destination release* and *slot reuse* of the transmitted packets, one slot of a channel may be £lled and released several (in maximum D) times within one ring latency, yielding an improved degree of statistical multiplexing and bandwidth ef£ciency within the ring. Moreover, in order to alleviate the head-of-line (HOL) blocking problem observed in the source node's transmission queue, one separate queue is assigned to each available data channel in the network, that is, virtual output queueing (VOQ) is applied. Consequently, there are as many transmit queues as data channels in the system.

III. Access Control

A. General Access Strategy

According to the generic slotted ring access mechanism, slots are continuously circulating around the ring on each channel and can be empty or full with the slot availability being indicated by the control channel. Thus, the corresponding control packet comprises C ¤ags (being "0" or "1") capturing the status of each data channel. When an arriving slot is empty, the node can use this slot to transmit its data packet, otherwise it has to wait until an empty slot arrives. This way, no channel and receiver collisions as well as no packet losses occur on the ring. Furthermore, *destination release* is assumed, thereby enabling spatial slot reuse along the ring. In principle, the so-called *a-posteriori* packet selection strategy is considered in this work. This means, that *£rst* the availability status of the wavelength channels at a slot time is detected, and *then* the appropriate data packet is selected from the corresponding transmission queue at the source node for transmission in the current slot. The packet selection is necessary since only one data packet may be transmitted per slot time due to the single tunable transmitter constraint. Generally, each empty slot on any channel can be used for packet transmission by a source node. However, when more than one channel carries an empty slot in the current slot period, one data packet (or equivalently, one transmission buffer) corresponding to one of the empty channels has to be chosen according to a certain selection rule. Different buffer selection strategies may be applied at the sending nodes for the case that more than one slot is free at a slot instant. In this paper, we rely on the RND scheme [4], i.e., the queue from which a data packet is to be acquired for transmission is selected randomly (following a uniform distribution) in case of buffer contention. Other selection schemes are discussed and analyzed in detail in [1].

B. Direct support of unfragmented IP packets

For accommodating IP packets without fragmentation (at the transmitter) and reassembly (at the receiver), a strong control on reserved slots on the ring is required here. Basically, two basic traf£c types are distinguished in this study, namely *long* IP packets, i.e., longer than the data slot size ($L > L_s$), and *short* IP packets, i.e., shorter than or equal to the slot size ($L \leq L_s$). It is assumed in the following that long IP datagrams are stored in C separate *long-packet buffers*, while short IP packets are handled by C *short-packet buffers* at each node, with all queues treated on a FIFO basis.

For supporting long IP datagrams, i.e., packets which do not £t into one data slot but are to be accommodated by an integer multiple of slots, a reservation-based protocol comprising the reservation and release of successive data slots on the ring is proposed in this paper. The number of reserved consecutive slots naturally depends upon the actual IP packet lengths. According to our protocol, the ring is subdivided into multiple *reservation frames* with the frame size equal to the largest possible IP packet length accommodated in the underlying network. The maximum packet length $L_{\max}$ can be typically predetermined according to an initial traf£c measurement re¤ecting the actual packet length distributions. Moreover, the data slots on the ring are globally numbered and identi£ed by $\sigma \in \{0, 1, \ldots, n_s - 1\}$, where n_s denotes the total number of data slots on the ring (per data channel). Concerning the reservation procedure, each source node is associated with a different reservation frame on each channel, i.e., each node is uniquely assigned a different *initial slot* σ^* which marks the starting time for the full transmission of an IP datagram without segmentation on any channel after successful reservation. Consequently, at least $n_{s,\min} = M \cdot L_{\max}$ data slots are needed on the ring for a proper protocol operation leading to the condition that the ring latency should ideally equal an integer multiple of $n_{s,\min}$. In this paper, we consider two distinct initial slot allocation schemes in terms of the respective nodes. First, the *synchronous allocation*, where a node is assigned the same initial slot σ^* on each data channel according to $\sigma^*_{syn} = (s - 1) \cdot L_{\max}$ with $s \in \{1, 2, \ldots, M\}$ being the source nodes, and second, the *cyclic allocation* in which the initial-slot/node assignment is performed cyclically on the data channels according to $\sigma^*_{cyc} = \{[(s - 1) + (c - 1)] \bmod M\} \cdot L_{\max}$. These different variants are exemplarily shown in Figs. 2 and 3, respectively. As a result, it can be differentiated between a

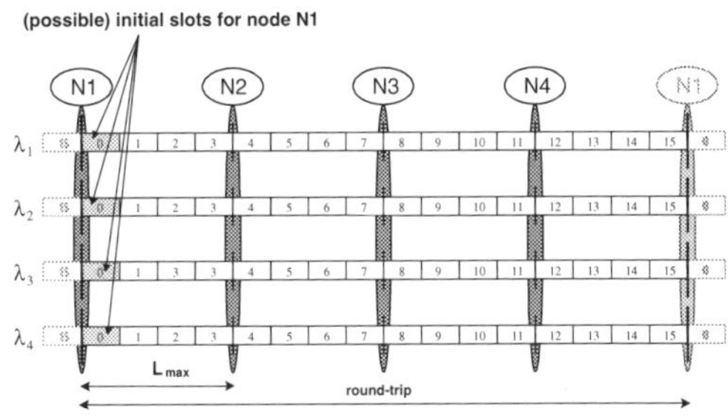

Fig. 2. Synchronous initial slot/node allocation scheme (illustrated for $M = C = L_{\max} = 4$ and related to node N1).

single reservation at a source node per round-trip cycle in the synchronous case (implying a certain selection method among

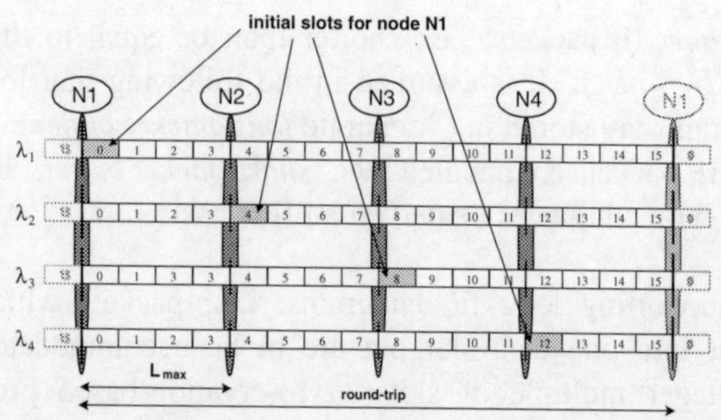

Fig. 3. Cyclic initial slot/node allocation scheme (illustrated for $M = C = L_{\max} = 4$ and related to node N1).

the data packets awaiting transmission in the C long-packet queues), and *multiple* reservations per cycle in the cyclic case. In addition, since the successive access to the reserved slots is occasionally interrupted by in-transit data packets transmitted on the drop channel of the node (as in-transit packets are given higher priority over local traf£c), also one £xed transmitter permanently tuned to the node's own drop channel is to be deployed.

The reservation procedure for long IP packets works as follows. In order to initiate a reservation, a source node waits for a unique *reservation control packet*, whose structure is depicted in Fig. 4. Note that there is just one reservation control

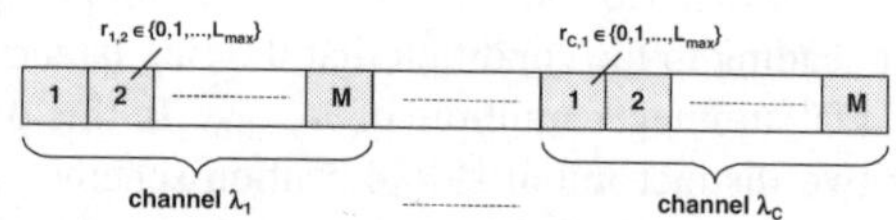

Fig. 4. General structure of the reservation control packet.

packet circulating on the control channel along the ring and that a reservation for an IP packet transmission may solely be performed when a reservation control packet is detected at a node. Furthermore, each node maintains a *slot status table* for keeping track of the current availability of all the data slots on the ring, hence, requiring global synchronization among the nodes. Upon the receipt of this reservation control packet, £rst, the status table of the source node is updated based on the reservation information $r_{c,s} \in \{0, 1, \ldots, L_{\max}\}$ for each channel $c \in \{1, 2, \ldots, C\}$ related to the source nodes $s \in \{1, 2, \ldots, M\}$. From the new status table each node may detect which data slots σ are free for immediate access on any data channel. This information is highly relevant for the support of short packets discussed below. Concerning the signaling overhead, in total $C \cdot M \cdot \log_2(L_{\max})$ bits are needed for building the reservation control packet. Then, the current source node inserts its reservation request(s) according to the associated IP packet length(s) into the corresponding reservation control packet and forwards it towards the downstream nodes. After a complete round-trip delay, this node can be sure that all the other nodes are aware of this reservation, therefore, it resets its dedicated reservation £eld within the reservation control packet to zero, and then solely waits for its assigned initial slot σ^* before starting to transmit the corresponding IP packet on the respective data channel. Note that with this reservation scheme fairness may be inherently guaranteed among the nodes willing to transmit long IP packets. Additionally, in the cyclic initial-slot/node allocation case, the next IP packet waiting for transmission in the corresponding transmit queue can be considered for reservation immediately after the reservation of the previous packet (of course, associated with one round-trip cycle), thus, improving the performance in terms of long packet transmissions. In other words, a certain kind of *look-ahead* scheme (with a degree of 2) and *pipelining* is adopted here. As a result, we obtain three different access schemes for the support of unfragmented IP datagrams, namely the synchronous initial-slot/node allocation scheme (SYNC), the cyclic initial-slot/node allocation scheme (CYCLIC), and the (extended) cyclic initial-slot/node allocation scheme using look-ahead (CYCLIC-E). In conclusion, the long IP packets subject to reservations are obviously better treated in the CYCLIC scheme compared to the SYNC protocol and best treated in the CYCLIC-E scheme. This naturally has a signi£cant impact on the performance related to the transmissions of short IP packets.

Besides the support of long IP packets via the reservation strategy described above, short packets £tting into one data slot are accommodated under the proposed protocol. This is primarily achieved by adopting *immediate access*, i.e., using a slot for packet transmission only when it is unreserved and empty. Thus, the overall bandwidth utilization of the WDM ring is signi£cantly increased, however, at the expense of slightly preferring the long-packets based IP traf£c due to reservations. However, since it is obviously much more dif£cult to transmit a long IP packet without fragmentation, i.e., within contiguous data slots, than to transmit a datagram which is completely £tting into one slot, this approach is justi£ed.

IV. Performance Study

A. Assumptions

The performance analysis in this paper is performed through computer simulations. Concerning the traf£c model used in the simulations, Poisson traf£c with a mean arrival rate λ per node is considered. The traf£c is evenly distributed among the destination nodes, thus, we assume a uniform traf£c scenario. According to some practical traf£c measurements [5], the IP traf£c is made up of four main packet sizes, namely $L = 40$ bytes (ICMP, TCP/IP acknowledgment messages), 522 and 576 bytes (TCP messages), and 1500 bytes (MTU in Ethernet). Therefore, we set $L_{\max} = 1500$ bytes throughout this study. Furthermore, since a data slot payload size of $L_s = 40$ bytes results in an ef£cient protocol operation, we obtain $L_{\max} = 38$ slots. We further assume that the packet sizes given above are uniformly distributed. Concerning the ring length, we assumed minimum ring sizes according to the constraint $n_{s,\min} = M \cdot L_{\max}$. Finally, we use the *random* buffer selection strategy for the SYNC scheme in order to choose only one data channel (and consequently, one initial slot) for transmission reservation at a data slot time. We further consider three different performance metrics throughout this study, namely the network throughput, the throughput and the mean access delay related to long packets. The *network throughput* captures the total number of packet transmissions within the network, while the *long-packet throughput* captures only the throughput caused by the long IP packets. The *long-packet access delay* re¤ects the head-of-line packet delay of long IP datagrams. Additionally, we assume that each transmission queue has an almost in£nite capacity in order to avoid packet losses at the source nodes due to buffer over¤ows.

B. Numerical Results

First, it has been found out that a slot (payload) size of 40 bytes corresponding to the smallest IP packets results in the most bandwidth-ef£cient system performance (not shown here) among different slot sizes L_s. Hence, throughout the following study, we rely on a data slot size of $L_s = 40$ bytes. Figs. 5 and 6 illustrate the node scalability of the considered protocols. In Fig. 5, the network and long-packet throughput is plotted against the number of nodes for 4 channels and overload uniform Poisson traf£c. It can be observed that the network throughput gradually increases until the saturation is reached around 40 nodes. The performance degradations associated with CYCLIC and CYCLIC-E is obvious which is due to the lower probability of obtaining immediate access to unreserved data slots in these schemes. The situation becomes different when considering the behavior of the long-packet throughput in dependence of the number of nodes. The preferential treatment of the long IP packets leads to highly superior results (about 100% improvements) in terms of the CYCLIC and especially the CYCLIC-E scheme relative to the other schemes. Fig. 6 demonstrates

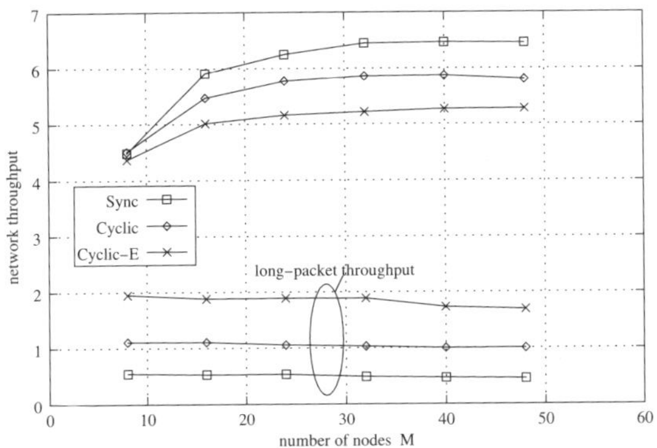

Fig. 5. Network and long-packet throughput versus the number of nodes M ($C = 4$, $L_s = 40$ bytes, uniform Poisson overload traf£c).

the mean access delay associated with long packets given in data slots versus the node count for 4 channels and uniform Poisson traf£c under overload conditions. Note that more

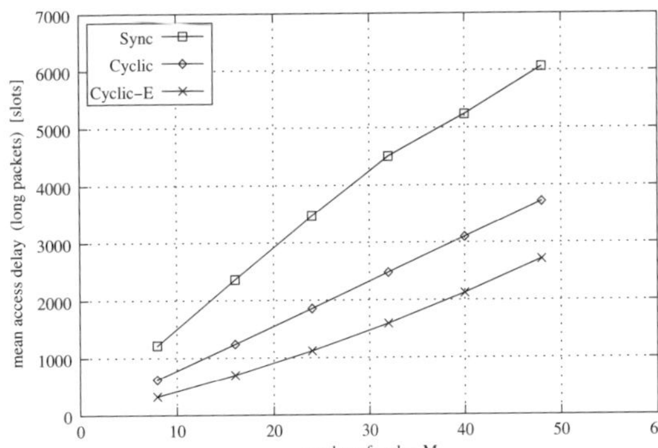

Fig. 6. Mean access delay related to long packets versus the number of nodes M ($C = 4$, $L_s = 40$ bytes, uniform Poisson overload traf£c).

nodes M correspond to more data slots on the ring (since $n_{s,\min} = M \cdot L_{\max}$) which in turn means longer rings and higher round-trip delays. This effect is clearly evidenced in the access delay curves for different node counts. Again the more sophisticated reservation-based protocols, CYCLIC and CYCLIC-E, provide a substantial performance gain over their SYNC counterpart (up to 70% and 140% gains, respectively).

Finally, Fig. 7 depicts the network throughput and long-packet throughput plotted versus the channel number for 24 nodes and overload uniform Poisson traf£c. Most remarkably, the network throughput does not scale linearly with the number of available wavelength channels. In other words, the throughput per channel drops with the channel upgrade of the system, i.e., the respective channels become underutilized. This is mainly due to the single tunable transmitter constraint preventing the simultaneous utilization of several channels at the same time. In terms of the long-packet throughput, clearly, it can be seen that the channel number increase does not improve the throughput performance in terms of the SYNC case since always only one channel is to be selected for reservation and transmission of long IP packets here regardless of how many wavelengths are available.

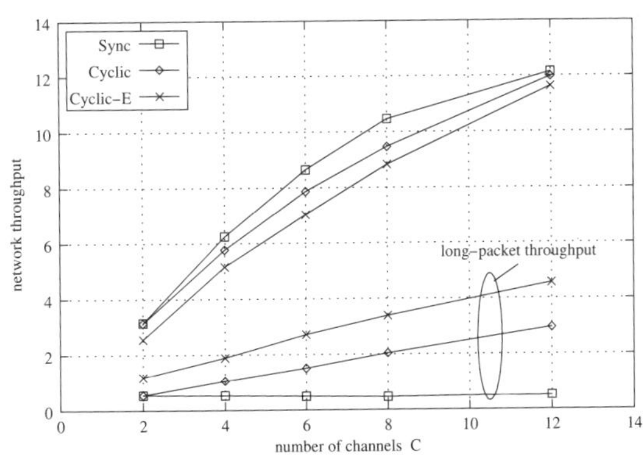

Fig. 7. Network throughput in dependence of the number of channels C ($M = 24$, $L_s = 40$ bytes, uniform Poisson overload traf£c).

V. Conclusions

We proposed novel distributed medium access protocols efficiently supporting unsegmented IP packets in a scalable slotted WDM metro ring, where each node is equipped with three optical transmitters and two fixed-tuned optical receivers. Through the transmission of IP datagrams in their entirety, the receiver hardware is largely simplified. The proposed IP-over-WDM support mechanism combines reservation-oriented transmissions as well as immediate access. The reservation control scheme is further differentiated into three distinct variants. The obtained performance results show that the inherently fair IP packet reservation scheme leads to very low mean packet delays while still achieving high total network throughputs and remarkably good performance for short packets controlled by immediate access.

References

[1] K. Bengi, *Optical Packet Access Protocols for WDM Networks*. Kluwer Academic Publishers, 2002.

[2] M. A. Marsan, A. Bianco, E. Leonardi, M. Meo, and F. Neri, "MAC protocols and fairness control in WDM multirings with tunable transmitters and fixed receivers," *IEEE/OSA Journal of Lightwave Technology*, vol. 14, pp. 1230–1244, June 1996.

[3] A. Fumagalli, J. Cai, and I. Chlamtac, "A token based protocol for integrated packet and circuit switching in WDM rings," *Proceedings of GLOBECOM'98, Sydney, Australia*, pp. 2339–2344, Nov. 1998.

[4] K. Bengi and H. R. van As, "Efficient QoS support in a slotted multihop WDM metro ring," *IEEE Journal on Selected Areas in Communications (special issue on "WDM-Based Network Architectures")*, vol. 20, pp. 216–227, Jan. 2002.

[5] National Laboratory for Applied Network Research (NLANR), *http://pma.nlanr.net/PMA*.

Performance of Distributed-Controlled Dynamic Wavelength-Conversion GMPLS Networks

Eiji Oki†, Daisaku Shimazaki†, Kohei Shiomoto†, Nobuaki Matsuura‡, Wataru Imajuku†and Naoaki Yamanaka†
†NTT Network Innovation Laboratories, ‡NTT Network Services Systems Laboratories, NTT Corporation
3-9-11 Midori-cho, Musashino-shi, Tokyo, 180-8585 Japan
Email: oki.eiji@lab.ntt.co.jp, fax: +81-422-59-6387

Abstract—This paper presents the performance of distributed-controlled dynamic wavelength-conversion (DDWC) optical networks based on our simple route and wavelength assignment (RWA) policy. A source node first chooses an optical-path route by using aggregated optical link-state information, then sends a signaling message on the selected route, after which wavelength assignment is performed. If wavelength conversion is needed at a transit node, the wavelength is adaptively converted by using trunk-type wavelength converters. We show via simulations that the performance of the DDWC network with our simple RWA policy emulates that of the WC network with small number of trunk-type wavelength converters.

Keywords—WDM, optical networks, wavelength conversion, routing protocol, signaling protocol

I. Introduction

Wavelength-division multiplexing (WDM) optical networks are mainly categorized into three types in terms of wavelength-conversion capability. The first type covers wavelength-convertible (WC) networks. The second type covers non-wavelength-convertible (NWC) networks. The third covers limited-wavelength-convertible (LWC) networks.

In a WC network, any wavelength can be converted into any other wavelength at optical cross-connects (OXCs) on an optical path. When an optical path is set up, any wavelength can be used as long as there is at least one available wavelength in a fiber. Therefore, wavelength resources are efficiently utilized. In addition, the routing problem, selecting the best optical path, is simplified. Unfortunately, wavelength converters are expensive. An example of an OXC that has full wavelength conversion capability is shown in Figure 1 (a). Each output port of the switch fabric is equipped with a wavelength converter.

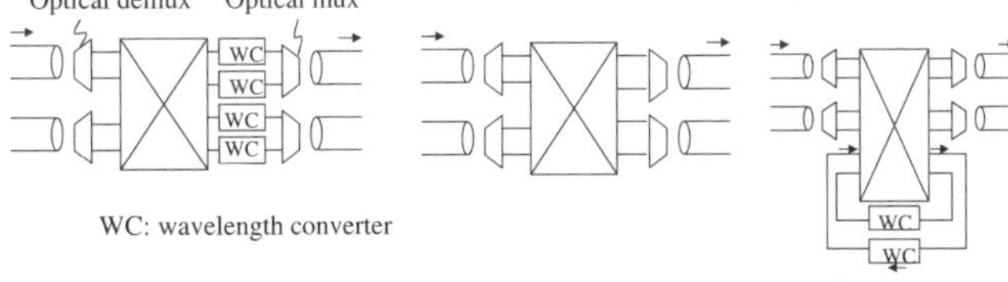

(a) OXC without full-wavelength conversion capability (b) OXC without wavelength conversion capability (c) OXC with trunk-type conversion capability

Fig. 1. Optical cross-connect (OXC) structures with different wavelength conversion capabilities

In an NWC network, wavelength conversion is not performed on any optical path. Since an NWC network does not use any wavelength converters at the OXCs, no wavelength converter are needed.[1] However, an NWC network utilizes wavelength resources less efficiently than a WC network. When an optical path is set up, the routing and wavelength assignment (RWA) problem has to be solved. In general, the blocking probability, the probability that an optical set-up request will be rejected, is much higher in an NWC network than that in a WC network. Figure 1 (b) shows a non-wavelength convertible OXC.

An LWC network is an intermediate solution between WC and NWC networks. An LWC network offers the advantage of WC networks (better wavelength resource utilization) as well as that of NWC networks (lower wavelength-conversion cost). Figure 1 (c) shows an OXC with trunk-type wavelength converters. Such converters are used only when needed. Optical paths in an LWC network experience few or no conversions in wavelength.

Several studies have addressed the effectiveness of an LWC network with trunk-type wavelength converters [3], [9]. They showed that trunk-type wavelength conversion is sufficient for achieving wavelength utilization levels close to those of networks with full wavelength conversion [11]. However, these studies assumed that complete knowledge about the optical-link states in the network was held by a centralized-controlled operation system that collected the optical-link states of each OXC. This centralized approach is rather impractical when the network size is large.

Distributed-controlled traffic engineering (TE) is considered to be a promising way of realizing cost-effective and flexible optical WDM networks. In generalized multi-protocol label switching (GMPLS) networks [2], [7], a source node finds an appropriate route and wavelength based on the optical link-state information collected with a routing protocol such as Open Shortest Path Finding (OSPF) [6], and sets up an optical path by using a signaling protocol such as RSVP-TE [1].

In [10], [12], extensions of signaling protocol of RSVP-TE were introduced to better utilize network resources in NWC and LWC networks based on distributed-controlled traffic engineering. In the proposed extensions, after the route of an optical path is chosen by using the link-state information advertised by OSPF, the wavelength is adaptively set by using extended RSVP-TE. However, in [10], [12], performance evaluations were left as remaining issues. Although several papers [4], [5] addressed the performance of an NWC network with distributed control, no study has investigated the performance of an LWC network with trunk-type wavelength converters when distributed control is adopted.

Several questions are obvious: When distributed-controlled traffic engineering is adopted in an LWC network, how much is the blocking probability improved? How many wavelength conversion resources are needed in an LWC network in order to emulate a WC network? How does the hop length of optical paths impact the blocking probability? This paper provides answers to these questions.

This paper presents a performance analysis of distributed-controlled dynamic wavelength-conversion (DDWC) optical networks that use the extended signaling protocol of RSVP-TE [10], [12] to better utilize the network resources of an LWC network. We introduce a simple routing and wavelength assign-

[1] Although 3R functions are required in NWC, we do not consider them throughout this paper to simplify the discussion on the wavelength conversion cost.

ment (RWA) policy. In our RWA policy, a source node first chooses an optical-path route by using aggregated optical link-state information and then sends an RSVP-TE path message on the selected route; wavelength assignment is then performed. If wavelength conversion is needed at a transit node, the wavelength is adaptively converted by using trunk-type wavelength converters. The advertised information overhead imposed by OSPF and the processing capability needed to solve the complicated RWA problem are saved. We show via simulations that the performance of the DDWC network with our simple RWA policy emulates that of an equivalent WC network but which many fewer trunk-type wavelength converters.

II. Distributed-controlled Dynamic Wavelength-Conversion (DDWC) network

A. DDWC network with simple RWA policy

Figure 2 shows the DDWC network. It consists of OXCs with trunk-type wavelength converters. Each node (OXC) advertises its own optical link states to other nodes and collects optical link-states from other nodes by using the extended OSPF described in [6], [10].

To set up an optical path, a source node has to determine which route and wavelength should be chosen. The problem is the so-called route and wavelength assignment (RWA) problem [13]. If the availability of each wavelength is advertised of all links in the network, it is difficult to scale the network in terms of the number of nodes and the number of wavelengths. This is because the amount of advertised information will rapidly become excessive. In addition, even if a source node knows all link information, the processing power required by the source node to solve the complicated RWA problem can become excessive.

To reduce the amount of advertised information and the required processing power, we introduce a simple RWA policy. A source node first chooses an optical-path route by using aggregated optical link-state information, and then sends an extended RSVP-TE signaling path message on the selected route to the destination node. The path message collects available wavelengths and wavelength converters, if needed, on the route. Finally, the destination node determines which wavelengths should be adopted.

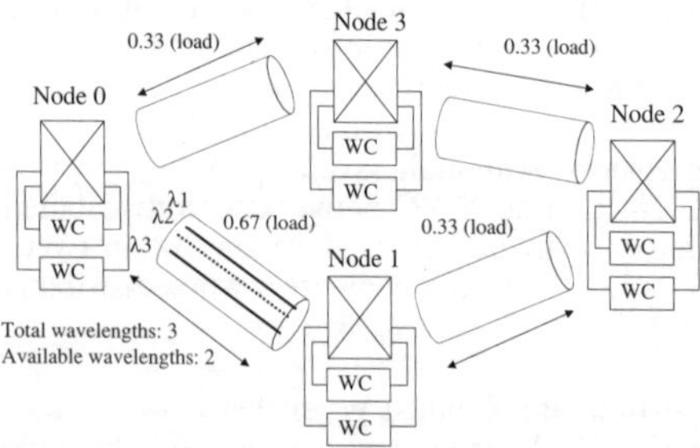

Fig. 2. Distributed-controlled dynamic wavelength-conversion network

B. Optical route selection

Each optical route is selected based on a least-loaded shortest routing policy and the use of aggregated optical link-state information. The load l_i of fiber link i is defined as the ratio of the number of reserved wavelengths (NRW_i) in the fiber link i between neighbor nodes to the number of accommodated wavelengths (NW_i) in the fiber link i (i.e., $l_i = \frac{NRW_i}{NW_i}$).

In the least-loaded routing policy, the link load is taken as link cost. Considering link cost, the shortest route is determined so that $\sum_{i \in \text{route}} l_i$ is minimized.

In OSPF extensions, wavelengths in a fiber link between neighbor nodes are bundled into a traffic-engineering (TE) link. The optical-link information in a TE-link between two neighbor nodes is advertised to other nodes in the network. Therefore, multiple ports of node, each of which corresponds to a wavelength, may be combined into a TE-link. In OSPF extensions, opaque LSA sub-TLVs include maximum reservable bandwidth and unreserved bandwidth types, which sub-TLV types are 7 and 8, respectively [8]. NW_i and NRW_i are expressed using maximum reservable bandwidth and unreserved bandwidth for a bundled TE-link between neighbor nodes as described below.

$$\begin{aligned} NW_i &= \text{maximum reservable bandwidth for a bundled TE-link} \\ &= \text{sum of maximum reservable bandwidth of all component links} \quad (1) \\ NRW_i &= NW_i - \text{unreservable bandwidth for a bundled TE-link} \\ &= NW_i - \text{sum of unreservable bandwidth of all component links} \quad (2) \end{aligned}$$

Note that, each component link corresponds to a different wavelength. The units of the maximum reservable bandwidth and unreservable bandwidth are defined as bytes per second in [6]. Since the values of NRW_i and NW_i are independent of bytes per second, modification of the units or a sub-TLV definition is needed. This modifications was proposed in [10].

C. Extended signaling protocol of RSVP-TE

C.1 AND scheme

After an optical route is determined, the source node sends an extended RSVP-TE path message over the selected route to the destination node. When the path message attempts to set up an optical path at each source/transit node, it carries a set of available labels, i.e. labels that are unreserved on all TE-links from the source node to the transit node. The label set is called the AND label set. The source node sends the AND label set, which is a set of the available labels of it own egress link to the next-hop node. The transit node creates a new AND set by using the transmitted AND set that was sent by the previous node and the available labels of the egress link. If there is no available label in the created AND set, the transit node must perform wavelength conversion. When the wavelength is converted at the transit node, the AND set is reset to be transmitted to the next-hop node on the selected route. If such conversion is not possible, the request for optical path set-up is rejected.

Figure 3 shows two examples of AND usage. In case 1, the available-label set in link A is $\{\lambda1, \lambda3, \lambda4\}$. The available-label set in links A and B is $\{\lambda3, \lambda4\}$, since $\lambda1$ is not available in link B. The available-label set that the path message carries from the source node to the destination node is thus $\{\lambda3\}$, since $\lambda4$ in link C is not available. In this case, no wavelength conversion is required for optical path set-up. The destination node chooses $\lambda3$ to set up the optical path and sends back the reservation message to the source node over the selected route.

In case 2, the available-label set in links A and B is $\{\lambda3, \lambda4\}$. However, since only $\lambda1$ is available in link C, there is no available label in the AND label set that is created by node 2. Therefore, node 2 decides to convert either $\lambda3$ or $\lambda4$ to $\lambda1$ by using a trunk-type wavelength converter. Note that, according to the first fit policy [13], λ3 is chosen.

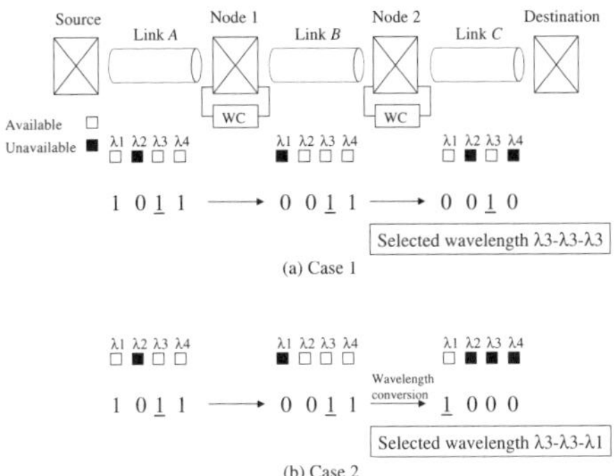

Fig. 3. Examples of AND usage

C.2 ALL scheme

In the AND scheme, wavelength conversion is performed at the transit node only when the label set is empty. On the other hand, in the ALL scheme, the destination node decides which transit node the wavelength conversion is performed at after considering the available labels and wavelength converters at each transit node. For this purpose, the path message carries all information on availabilities of labels and wavelength converters on the selected route instead of the aggregated information as used in the AND scheme.

When a path message attempts to set up an optical path between each source/transit node pair, it carries a set of all available labels for all TE-links on the optical-path route. The label set is called an ALL label set. A destination node receives the ALL label set and decides which labels should be used on the optical path. In the ALL scheme, the destination node has several options when deciding which node should perform wavelength-conversion.

Figure 4 shows an example of the ALL scheme in use. The path message carries all information on the wavelengths available on the links and wavelength converters on the selected route. The example shows that a wavelength converter is available at node 1, but not at node 2. In this example, the AND scheme is not able to set up an optical path. This is because node 2 cannot perform wavelength conversion although it is needed. The ALL scheme, on the other hand, can set up an optical path due to the available wavelength converter at node 1. Therefore, the ALL scheme offers improved blocking performance over the AND scheme at the cost of an increase in information transfer overhead and the cost of performing the sophisticated wavelength-selection algorithm at the destination node.

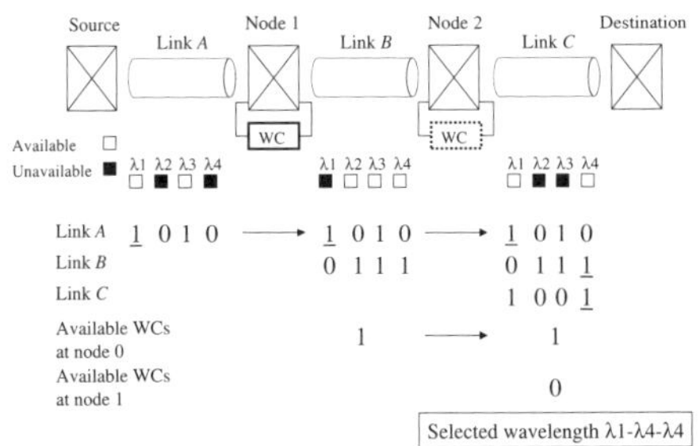

Fig. 4. Example of ALL scheme

III. Numerical results

We evaluated the performance of the DDWC network by using computer event-driven simulations. We investigated the impact on the blocking probability of the DDWC network by several important parameters such as the number of trunk-type converters at each node, x, the optical path hops length, h, and the number of wavelengths in a fiber, w. For this purpose, we assumed the simple network model shown in Figure 5, which has 11 nodes (N = 11). Bi-directional fibers are set between neighboring nodes. We employed the routing and wavelength assignment policy described in Section II. Although we introduced the AND and ALL schemes, we focused on the AND scheme in our evaluation to simply the discussion of DDWC performance. This is because the performance of the ALL scheme may be superior to that of AND scheme so the AND scheme provides conservative numerical results. The simulations assume that traffic demands between all source and destination nodes are the same. Requests for optical-path setup follow a Poisson distribution. The holding time of each source and destination node pair is considered to follow an exponential distribution.

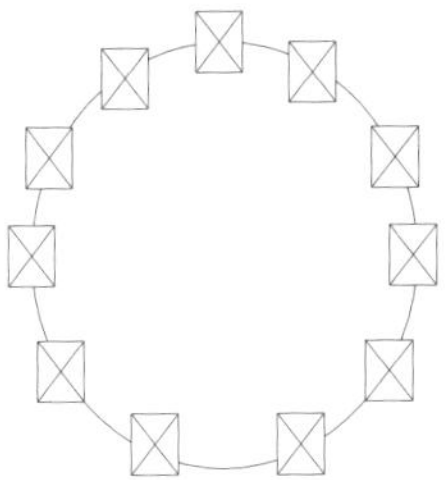

Fig. 5. Network model ($N = 11$)

Figure 6 shows that as the number of trunk-type wavelength converters, x, of the DDWC network increases, the blocking probability approaches that of the WC network. We set $w = 32$. The traffic demand (horizontal axis) is defined as that between each source and destination node pair. In Figure 6, we focused the blocking probability of optical-path requests whose h was 5. The results show that the DDWC network combined with our RWA policy dramatically reduces the wavelength-conversion cost compared to the WC network.

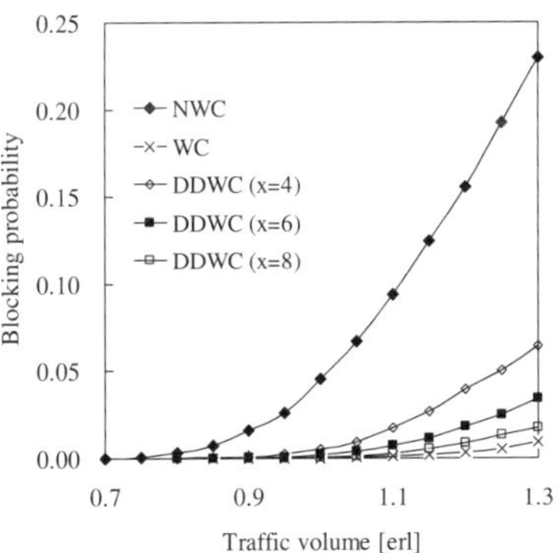

Fig. 6. Blocking probability vs. traffic demand ($h = 5$, $w = 32$)

Figure 7 shows the blocking probability of optical-path requests with $h = 3$. The difference in the blocking probability between the NWC and WC networks is much smaller than that with $h = 5$. Figure 8 shows the blocking-probability versus the optical path hop length for the given traffic volume of 1.3 [erl]. As the number of hops increases, the blocking probabilities become large. Therefore, the impact of x becomes significant. This means that when we design the cost-effective x

to approximately emulate the WC network, we should consider blocking probabilities with large enough h values.

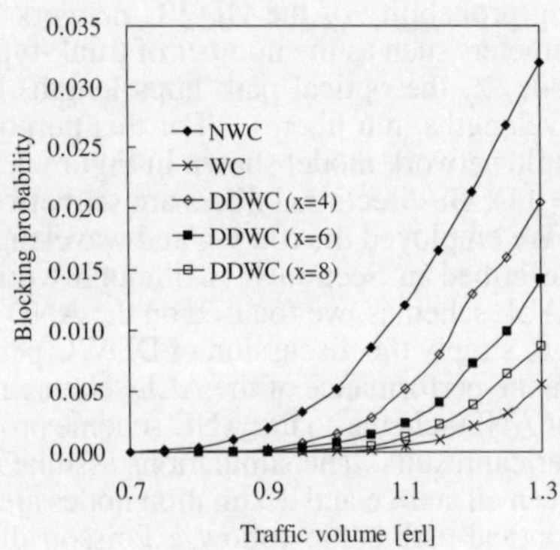

Fig. 7. Blocking probability vs. traffic demand ($h = 3$, $w = 32$)

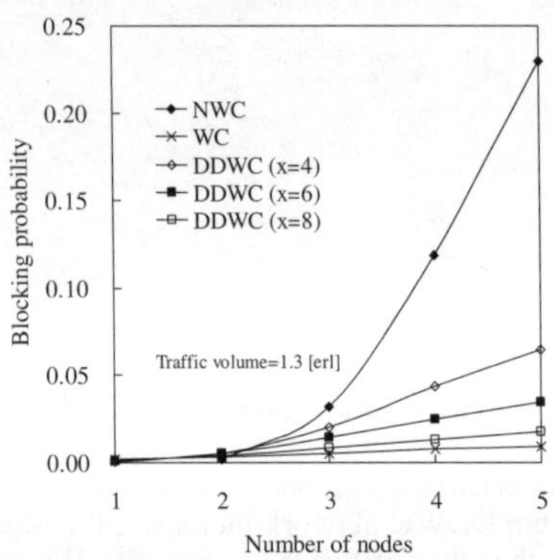

Fig. 8. Blocking probability vs. hop length ($w = 32$)

As the number of trunk-type converters increases, the blocking probability decreases as shown in Figure 9. We define the minimum x, at which the difference in blocking probabilities between the DDWC and WC networks is less than 30%, as the 30%-emulation point e. Arrows in Figure 9 indicates 30%-emulation points for the conditions considered. For example, when the traffic volume is 1.3 [erl], where the blocking probability of the WC network is about 0.01, the 30%-emulation point is $e = 10$. This means that the DDWC network achieves $\frac{pw-e}{pw} \times 100 = 69\%$ cost reduction, in terms of wavelength converters, compared to the WC network while achieving 30% emulation. p is the number of ingress fiber ports in the node. In the network model as shown in Figure 5, p is set to 1. The WC network needs w wavelength converters for each fiber, while the DDWC network needs e trunk-type wavelength converters.

IV. Conclusions

This paper presented the performance of distributed-controlled dynamic wavelength-conversion (DDWC) networks that use our simple RWA policy. Our RWA policy reduces the amount of information that must be advertised by OSPF and the processing capability required to solve the complicated RWA problem. Numerical results showed that the blocking probability of the DDWC network with our simple RWA policy emulates that of the WC network but with fewer trunk-type wavelength converters. In our case study, simulation suggested that the DDWC network achieves 69% cost reduction, in terms of wavelength converters, compared to the WC network under the 30% blocking-probability emulation condition with $w = 32$.

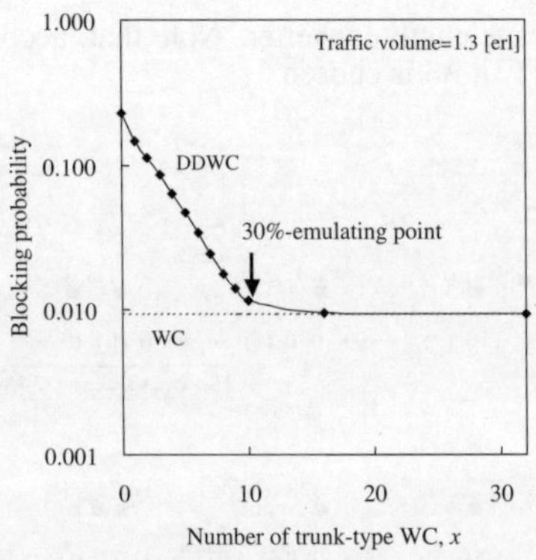

Fig. 9. Blocking probability vs. number of trunk-type wavelength converters ($w = 32, h = 5$)

References

[1] P. Ashwood-Smith et al., "Generalized MPLS Signaling - RSVP-TE Extensions," IETF draft, draft-ietf-mpls-generalized-rsvp-te-07.txt (work in progress)

[2] A. Banerjee, J. Drake, J. P. Lang, B. Turner, K. Kompella, and Y. Rekhter, "Generalized Multiprotocol Label Swtching: An Overview of Routing and Management Enhacements," IEEE Commun. Mag., pp. 144-150, Jan. 2001.

[3] O. Gerstel, R. Ramaswami, and S. Foster, "Merits of Hybrid Optical Networking," Optical Fiber Commun. Conf. (OFC) 2002, Technical Digest, pp. 33-34, 2002.

[4] P.-H. Ho and H.T. Mouftah, "Network Planning Algorithms for the Optical Internet based on the Generalized MPLS Architecture," Proc. IEEE Globecom 2001.

[5] P.-H. Ho and H.T. Mouftah, "A Novel Routing Protocol for WDM Mesh Networks," Optical Fiber Commun. Conf. (OFC) 2002, Technical Digest, pp. 38-39, 2002.

[6] K. Kompella et al., "OSPF Extensions in Support of Generalized MPLS," IETF draft, draft-ietf-ccamp-ospf-gmpls-extensions-04.txt (work in progress)

[7] E. Mannie et al., "Generalized Multi-Protocol Label Switching (GMPLS) Architecture," IETF draft, draft-ietf-ccamp-gmpls-architecture-02.txt (work in progress)

[8] D. Katz, D. Yeung, and K. Kompella, "Traffic Engineering Extensions to OSPF," IETF draft, draft-katz-yeung-ospf-traffic-06.txt (work in progress)

[9] S. Meyer, "Quantification of wavelength contention in photonic networks with reach variation," Optical Fiber Commun. Conf. (OFC) 2002, Technical Digest, pp. 36-37, 2002.

[10] E. Oki, N. Matsuura, W. Imajuku, K. Shiomoto, and N. Yamanaka, "Requirements of optical link-state information for traffic engineering," IETF draft, draft-oki-ipo-optlink-req-00.txt (work in prgress)

[11] R. Ramaswami and K. Siarajan, *Optical Netowrks: A Practical Perspective*, Morgan Kaufmann Publishers, San Francisco, 1998.

[12] K. Sato, N. Yamanaka, Y. Takigawa, M. Koga, S. Okamoto, K. Shiomoto, E. Oki, and W. Imajuku, "GMPLS-Based Photonic Multilayer Router (Hikari Router) Architectuer: An Overview of Traffic Engineering and Signaling Technology," IEEE Commun. Mag., pp. 96-101, March 2002.

[13] H. Zang, J. P. Jue, and B. Mukherjee, "A Review of Routing and Wavelength Assignment Approaches for Wavelength-Routed Optical WDM Networks," SPIE Optical Networks Magazine, vol. 1, no. 1, Jan. 2000.

A Novel Parallel Signaling Protocol For Lightpath Creation in ASON

Tianlong Qi, Xiaoping Zheng, Feng zhang, Hanyi Zhang
Department of Electronic Engineering, Tsinghua University, 100084, Beijing, P.R. China
Phone: 86-010-62772635 Fax: 86-010-62770317 Email: qtl98@mails.tsinghua.edu.cn

Abstract: Fast lightpath creation is very important for ASON (automatically switched optical network), but the traditional protocols such as RSVP-TE or CR-LDP work in a serial style and can't support that service well. This paper proposals a new simple parallel signaling protocol for fast lightpath creation in ASON. The most time-consuming step in lightpath creation is processed in a parallel style and the other operations still work serially as the traditional protocols.The signaling procedure is discussed in detail and the formal description of the protocol is given. An experiment is also setup to test the protocol and the result is compared with hop by hop signaling protocols.

Key Words: parallel signaling, lightpath creation

1. INTRODUCTION

As the fast development of the internet, IP based data traffic will be the dominating services for the next generation communication networks. But the traditional communication network optimized for voice services is not efficient to support IP traffic. On the other hand, Dense Wavelength Division Multiplexing (DWDM) supports multi-gigabit transmission rates per wavelength and is becoming the core technology for the next generation networks. How to control the optical network to achieve an intelligent optical network which can support IP trffice efficiently is still a problem. Many standardization bodies and forums, such as OIF and ITU-T, are working on it. They hope to develop a new control plane which can configure the optical network quickly and efficiently. So the "dumb" transmission network can setup or delete a lightpath rapidly under the control of the carrier Operation systems (OS). To cope with this challenge, the network operators have to adopt some new technologies to provide services faster and more efficiently.

Automatically Switched Optical Network (ASON)[1] is now the most hopeful way to the intelligent optical network. There are three planes in ASON: control plane, management plane and transport plane, which are shown in Fig.1. The new control plane is designed to make network operators control the network more easily. It can provide faster and more reliable lightpath creation and deletion by signaling.

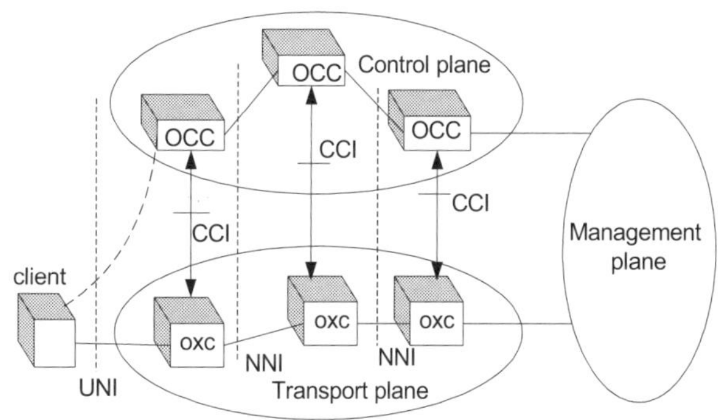

Fig.1 ASON architecture

A lightpath creation can be divided into four steps: The first step is route determination. When the first optical node receives the request of lightpath creation, it will setup a route table according to some route determination algorithm; second, all the nodes in the route table will reserve the resources needed for that lightpath; third, all the nodes do the resource allocation, usually it is a switch operation; fourth, a confirmation message is send to all the nodes so that the data can be transmitted in that lightpath. Obviously, route determination is a necessary process and independent to other three steps. So it is ignored during the following disscussion for the time needed of a lightpath establishment.

There are two signaling protocols for the the lightpath creation[2][3]: CR-LDP or RSVP-TE. But in both protocols, the strategies of doing resource allocation such as optical cross-connects operation for creating a new lightpath work in a serial style. That is to say, each node in the lightpath will do its local resource allocation one after another. One node won't do optical cross-connect operations unless it receives the indication message from its upstream neighbor node. Also it must send out an indication message to inform its downstream neighbour node to begin action after finishing its resourse allocation. So we see that the serial operation will bring a big delay to the process of establishing a new lightpath, especially when the network size becomes considerably big and the number of nodes in an average lightpath becomes large. For example, if we assume the delay caused by a single optical resource allocation is T milliseconds, and if there are totally N nodes at average along a lightpath in a given network, then the time for lightpath creation is N*T milliseconds in sum. If the N is considerably big, this delay will be unacceptable when we need a fast lightpath provision. This problem is originally from the serial resource allocation operations carried with RSVP-TE or CR-LDP. To solve this problem, [5]proposed a new strategy to setup the lightpath in the way of doing parallel resource allocation operation along the nodes in the path. The intermediate nodes will do the resource allocation after it relays the label request. So the time-consuming resource allocation operation is done in a parallel style and the time needed to set up the lightpath is much smaller. However, because the node doesn't verify if the other node succeeds in its resource allocation during the lightpath creation, this strategy will greatly increase the bloking rate when the request is often rejected, that is to say, it only has the good perfomence in a rich resource optical network.

To my knowledge, most lightpath creation failure occurs in the first two steps: route determination and resource reservation. If the resource reservation succeeds, the resource allocation will succeed except that one optical switch disrupt or signaling failure. That is to say, once the resource reservation succeeds and the signaling procedure is successful, the resource allocation will succeed too. So [5] is not a resonable solution, which ignores the biggest failure probability for the problem.

Thus a novel protocol is proposed: Try to verify if the other nodes succeed in resources reservation first. If the reservation succeeds, the last node will send an action request message to all other node along the lightpath so that each node can operate the switch simultaneously. Then the most time-consuming operation can be processed in a parallel style. That is to say, lightpath creation request message is relayed in a serial style but the resources are allocated in a parallel style. During this process, a new type of message must be introduced to indicated if the resource allocation succeed or not.

In section II, the formal description of our proposal is given, including signaling message type, content and state transition diagram. In section III, an experiment setup is explained in detail. Results of the experiment are discussed and compared with CR-LDP. The conclusions are given in the last section.

2. A NOVEL PARALLEL SIGNALING PROTOCOL

We assume that the communication between the optical nodes uses standard primitives, namely request, indication, response, and confirm. The primitives are defined as:

- Lightpath.creation.request (source address, source port number, destination address, destination port number, direction, wavelength, QoS)
- Lightpath.Creation.Response (source address, source port number, destination address, destination port number, direction, wavelength, QoS, positive/negative)
- Resource.Allocation.Response (source address, source port number, destination address, destination port number, direction, wavelength, QoS, positive/negative)

- Lightpath.Creation.Confirm (source address, source port number, destination address, destination port number, direction, wavelength, QoS, positive/negative)
- Lightpath.Release.Request (source address, source port number, destination address, destination port number, direction, wavelength, QoS)
- Lightpath.Release.Response (source address, source port number, destination address, destination port number, direction, wavelength, QoS, positive/negative)
- Lightpath.Release.Confirm (source address, source port number, destination address, destination port number, direction, wavelength, QoS, positive/negative)

Here the term direction is defined to indicte whether the lightpath is unidirectional or bi-directional. QoS defines the quality of service of the lightpath, i.e., the protection level and the priority. Positive /negative is a flag indicating if the request is accepted or rejected.

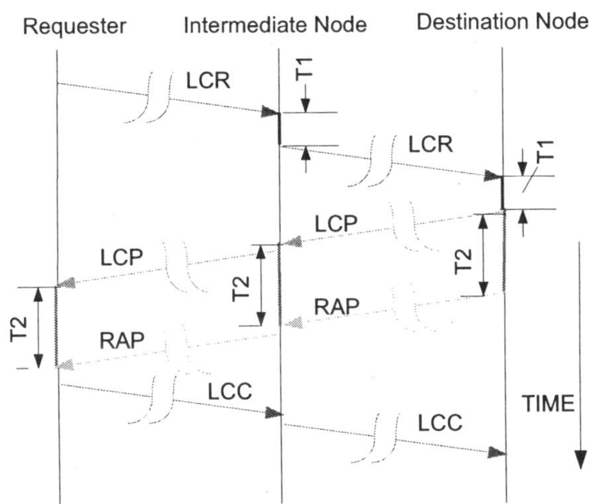

Fig.2 the message sequence of lightpath creation

The normal procedure for lightpath creation is shown in Fig.2. The protocol data units(PDUs) are identified as: Lightpath creation request(LCR), Lightpath creation positive response (LCP), Resource allocation positive confirm(RAP), Lightpath creation positive confirm(LCC), Lightpath release request(LRR), Lightpath release positive response (LRP), Lightpath release positive confirm (LRC). Their relationship with the primitives is easily understood. Strict rules should govern the order in which the primitives may be used. For example, lightpath.creation.response can't be send without receiving lightpath.creation.request. As shown in Fig.2, each node does the resource allocation after it receives positive lightpath.creation.response message. When the destination node finishes its resource allocation operation, it will send out resource.allocation.response message which is introduced as a new type of messsage to clearly indicate if the resouce allocation success or not. So each optical node can process the request and operate the optical switch simultaneously in this procedure. Thus the signaling time is much shorter and the sevice provision is almost real-time. Fig.3 shows the state transition diagram of our proposal protocol, which is based one the CFSM theory[4]. Here a minus sign identifies the transimission of a message and a plus sign means its reception. This transition diagram contains the state transitions of both transmitter(left side) and receiver entity(right side). Here state 0 is defined as the initial state. The term TO means "time out" and RSU means resource state update. From the state transistion diagram, validation of the protocol is easy be done by state exploration.

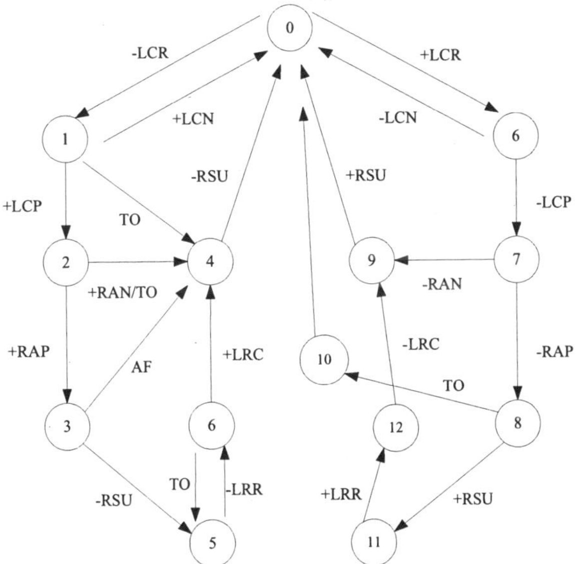

Fig.3 state transistion diagram of new proposal

3. EXPERIMENTAL SETUP AND RESULT

Under the support of both High Technology Project and Natural Science Fund of China, we develop three OXCs and one optical add/drop multiplexer (OADM). With them we can construct a meshed configurable DWDM network as shown in Fig.4, The four optical nodes are located in one room and the fibers between two optical nodes are all 80km long. Each optical node may have one or two IP routers as the client devices. For simplicity, IP routers are simulated by PCs with optical Ethernet adapters, instead of real ones. The OXCs are connected by 10M Ethernet with each other too. So the signaling network is the out-of-band signaling network. We will replace it with OSC according to ITU-T recommendation in the next step work.

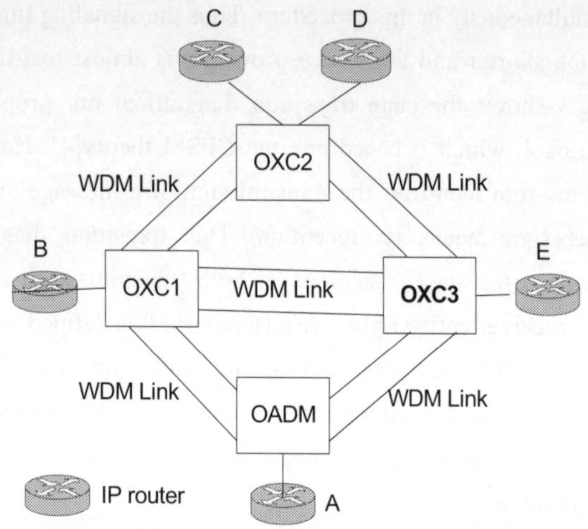

Fig.4 Topology of the experimental network

The system is working according to this procedure: first, a user device sends a request message to the OXC(requester) to setup a lightpath so that it can have a larger bandwidth to the destionation user device. Second, the OXC(requester) will determine the routing according to OSPF protocol and then setup a routing table. Third, the lightpath will be created according to the routing table. Then we can see the lightpath is dynamically eatablished in optical network according to user devices' rquest. Traditional serial style signaling protocol(CR-LDP) and our new proposal is tested respectively and the time for lightpath(include two nodes and three nodes) creation is shown in Fig 5. From Fig 5, we can see that, the signaling time of our proposal is much shorter than traditional serial signaling protocol.

4. CONCLUSION

This paper describes a new simple parallel signaling protocol for fast lightpath creation in ASON. The most time-consuming step in lightpath creation is processed in parallel style and the other operations still work serially as the traditional protocols.The standard primitevs is given and the signaling procedure is discussed in detail. The formal specification of the protocol is shown as the state transition diagram. An experiment is also setup to test the protocol and the result is compared with hop by hop signaling protocols. It is much faster than the traditional policy. Our proposals can dynamically establish a lightpath in ASON more efficiently.

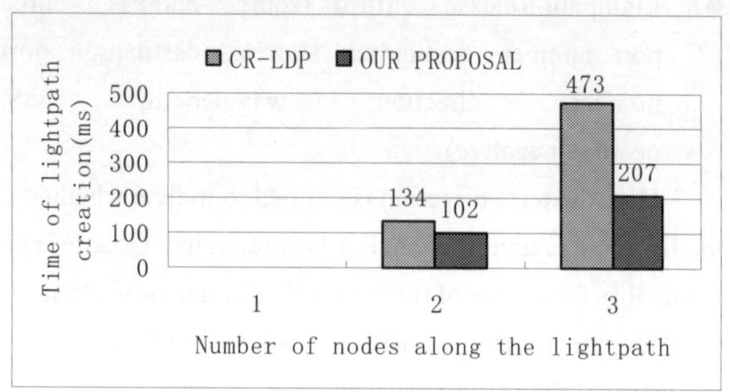

Fig. 5 Time of lightpath creation by two protocols

ACKNOWLEDGMENT

This work is supported by the National Natural Science Foundation of China (6999054, 90104003 and 60132020), the High Technology Project 863 Program (2001AA111051) and Tsinghua-Bell Joined Lab of Optical Network Systems.

REFERENCES

[1] ITU-T G.8080 V1.0 "Architecture for the Automatic Switched Optical Network (ASON)" 2001 OCT.

[2] OIF : "User Network Interface (UNI) 1.0 Signaling Specification", *www.oiforum.com*

[3] L. Andersson, P. Doolan, N. Feldman, A. Fredette, B. Thomas: "LDP Specification", Internet proposed standard, RFC 3036, January 2001.

[4] D. Brand, et al., On Communicating Finite State Machines, Journal on Assoc. Comp. Mach., vol. 30, no. 2, April 1983, pp. 323-342

[5] Hui. Huang, draft-hwang-gmpls-signaling-parallel-00.txt Feb. 2002

System Architecture Comparison of IP Router, LSR, and GMPLS-based OXC from Implementation Perspective

Sun Jianwei, Poo Gee-Swee and Cheng Tee-Hiang
School of EEE, Nanyang Technological University, Nanyang Avenue, Singapore 639798
Fax: 65-67926894, Email: ejwsun@ntu.edu.sg

Abstract: In this article, an overview of IP network development is given. Architectures of Ethernet switch, software-based router, switch router, LSR, and GMPLS-based OXC are presented and compared. The discussion is made from the implementation perspective of network equipment, especially the GMPLS-based OXC system rather than the detail of algorithms or protocols. Arising from the discussion, guidelines are presented for the implementation of GMPLS based OXC system. In particular, it is recommended that GMPLS-based system should support Label switching and IP Layer forwarding for its data traffic over control channels.

Key words: Multi-protocol label switching (MPLS), Label switch router (LSR), Generalized Multi-protocol label switching (GMPLS), Optical cross connect (OXC).

1. INTRODUCTION

In the early years of Internet when Cisco issued its first router in 1986, routers are basically processor centric with limited performance. The rapid development of Internet with the large increase of IP traffic since 1995, demands the production of new breed of packet switching equipment with greater performance. Thus, modern packet switching equipment adopts a new architecture, providing a separation of control plane and data plane. The control plane is based on the usual processor platform, and generates packet forwarding information through routing and signaling protocols to guide the data transmission. The data plane is based on high speed switching platform, to forward and process data traffic with high performance.

Optical network, including WDM or SDH system is introduced to Internet to transmit high volume data traffic. The current approach is to configure and provision the transport channel independently without interoperation with IP layer. IP centric control approach, presented recently by the industry and academia, under the GMPLS framework, is to enhance network flexibility and increase optical network scale to meet the fast development of IP traffic. The ideal network architecture for the core network is believed to be IP/GMPLS/WDM. Under such architecture, OXC, which is the basic building block of optical network, transfers data traffic without the insight of packet headers, but at the optical layer under direction of GMPLS control plane.

Up to now, in the IP network, data switch and transmission equipment used in local area network (LAN), metropolitan area network (MAN) and wide area network (WAN) are basically Ethernet switches, IP routers, Label switch routers. The GMPLS-based optical switches are currently emerging. In the following sections, we shall describe briefly the basic mechanism of network equipment, and compare the simplified system architectures from the implementation perspective. There are basic similarities among all these IP centric devices, from both the operation and implementation point of view. As a consequence of the comparison, we shall present the implementation guidelines for the development of GMPLS-based systems.

2. MECHANISMS AND ARCHITECTURE OF ETHERNET SWITCH

Previously, there are several competing networking technologies for LAN, including Ethernet, Token Ring, FDDI, ATM, etc. Nonetheless, Ethernet is now widely used and becomes the dominant technology, not only in LAN, but also in MAN and WAN.

Ethernet Switch is a LAN interconnection device which operates at the data link layer. By examining the MAC addresses of each received frame, and recording the port on which it was received, the bridge may learn the address of each host station connected via each port. The switch continuously collects MAC addresses for each port. Some of the addresses may be aged out. When an IP packet arrives, the destination MAC is indexed, and the destination port is found, which is the egress port to transfer the frame.

In the topology complex network, the switch uses spanning tree algorithm to construct a connected, loop free forwarding tree. Most managed switches support VLAN functionality. Basically, ports can be configured to support different VLANs. Inside a VLAN, packets are forwarded at Layer 2 based on a transparent bridge protocol. Between VLANs packets are processed at layer 3, the IP layer. So in the VLAN configurable switches, Layer 2 and 3 switches are usually integrated together.

The early implementation of switches is processor centric, that is, the system management, the MAC address learning, aging, and packet forwarding process are based on embedded processor. In recent implementations, all the processes are undertaken by switch fabric, and embedded processor is used for management and running protocols as control plane for the switch fabric. Owing to the architecture of modern switches, there are vendors that provide software suites and switch fabric for Ethernet switches respectively and leave OEM vendors to integrate the whole system [1].

Figure 1 shows the system architecture of Ethernet switch which is VLAN configurable and layer 3 switch capable.

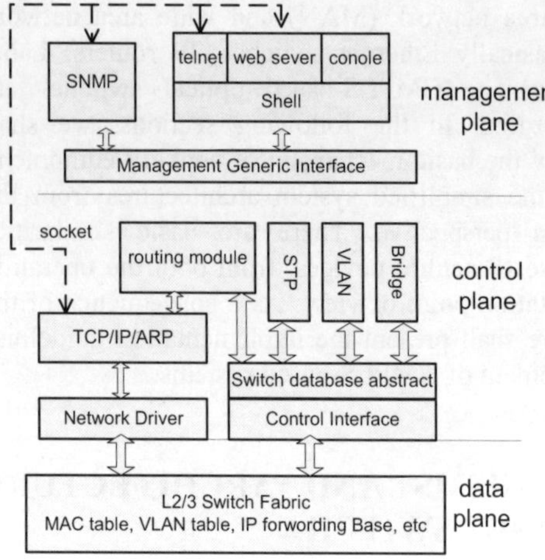

Figure 1 Ethernet switch architecture

3. MECHANISMS AND ARCHITECTURE OF IP ROUTER

Routers are devices to interconnect sub-networks. It collects topology information and calculates routes for certain destination IP address and establishes the routing table accordingly. IP traffic is forwarded to the right interface according the routing table index via a hop-by-hop approach.

Traditional routers are also processor centric, that is, network management, routing protocols, and IP forwarding are all based on the processing power of the processor. Such approach limits the performance of routers and cannot follow the huge increase of IP traffic. Likewise, modern IP routers adopt an approach of control plane and data plane separation. In this approach, the routing protocols and management applications reside on the control plane that is running on a general-purpose processor. The data plane is implemented using some high performance architecture, such as ASIC based switch fabric or programmable network processors, which offer the flexibility to add new services without the changes in hardware platform.

Figure 2.1 and Figure 2.2 show the architectural differences between software routers and modern switch routers. The software router with multiple network interfaces calculates the route using the routing module, while forwarding data traffic by the IP stack. In such a case, the performance of router is limited to the capability of the processor and the efficiency of the IP stack. As to the switch router, data plane and control plane are separated. Routing module and ARP module in the control plane cooperatively generate switch database and write to switch fabric (data plane). Then, switch fabric forwards data traffic according to the current status of the switch database.

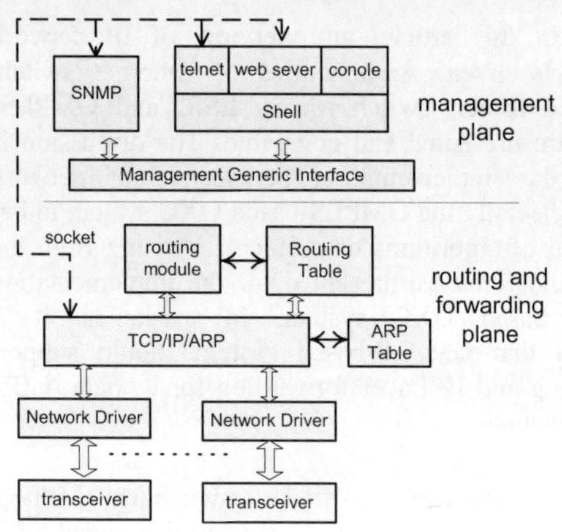

Figure 2.1 software router architecture

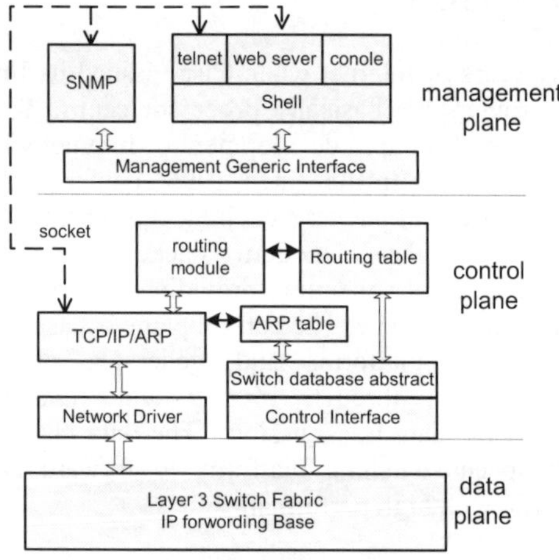

Figure 2.2 Switch router architecture

4. MECHANISMS AND ARCHITECTURE OF LSR

Multiprotocol label switching (MPLS) is a control framework currently being developed to introduce connection-oriented technology to traditional IP networks. MPLS control mechanisms adopt IP routing protocols to calculate routing path for each forwarding equivalent class (FEC), and introduce signaling protocols to establish the LSP along the routing path for each FEC by assigning a series of Label matrix at each nod along the routing path. Each LSR switch the Label encrypted packets from ingress port to egress port with exchange of Labels. The primary signaling mechanism for establishing an LSP in MPLS is the label distribution protocol (LDP). CR-LDP and RSVP-TE are introduced and routing scheme is modified to fulfill Traffic Engineering and network protection purposes. Software based LSR can be implemented by patching the IP stack with MPLS layer below IP layer and above the data-link layer. Concerning the architecture, again, LSR adopts the similar approach to separate data plane and control plane to achieve high performance.

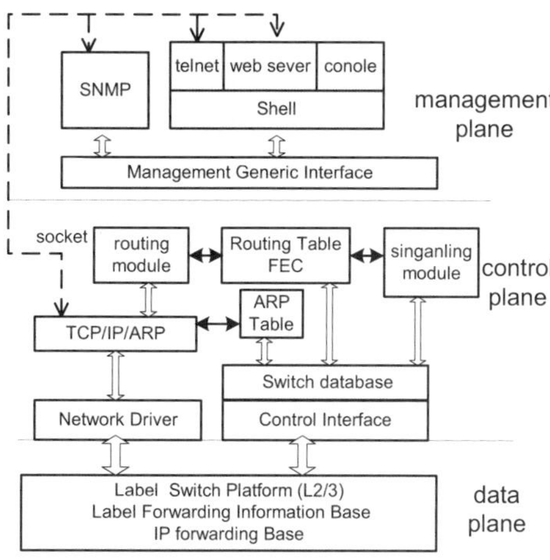

Figure 3 Label Switch Router architecture

Figure 3 shows the simplified architecture of LSR. There are some differences with the switch router. Routing protocols generate routing table and then Forwarding Equivalent Class (FEC) table. At the source or destination node, signaling protocol is initiated to establish a Label Switch Path (LSP) over the MPLS domain for each FEC entry, with a label switch entry added to the Label Forwarding Information Base (LFIB). As to a commercial implementation in the industry, the switch platform should be layer 3 switch (IP forwarding) capable so that those data traffic where no Label Switch Path (LSP) is available, can also be forwarded by the switch platform for higher performance. That is to say, the LSR should also fulfill the functionality of switch router.

5. MECHANISMS AND ARCHITECTURE OF GMPLS BASED OXC

The concept of MPLS can be extended to wavelength-routed optical networks as multiprotocol lambda switching (MPλS) by simply treat the wavelength and fiber as a kind of implicit label with concrete meaning. The premise of Generalized MPLS is that the idea of a label can be generalized to be anything that is sufficient to identify a traffic flow. Currently, much work has been done by the IETF to modify existing routing and signaling protocols in order to support GMPLS, particularly, focusing on enhancements to the Open Shortest Path First (OSPF) routing protocol, which is suitable to generate explicit route, and the Constraint-Based Routing Label-Distribution Protocol (CR-LDP) and Resource Reservation Protocol (RSVP) signaling protocols for the provisioning of light path, and Link Management Protocol for optical data-link management. The proposed enhancements can be found in [2-5].

GMPLS-based system should also support IP forwarding and label switching for the data traffic over control channel. First of all, similar to MPLS network, there are packets, mainly for management and signaling, need to be routed in the GMPLS domain so that communication is viable between any machines inside or outside the GMPLS domain, such as IP router domain and MPLS domain. For example, Network management station can reach any machine from anywhere of the Internet. Also it is necessary for the administrator to access the network nodes of GMPLS domain remotely from anywhere of the Internet. Further more, similar to MPLS domain which support IP forwarding across MPLS domain, GMPLS-based system should also support IP forwarding and label switching so that data traffic can transfer across GMPLS domain without strict request of Light Path.

Figure 4 describes the simplified architecture of GMPLS-based OXC system, considering the functionalities described above. The data plane of Figure 4 supports Label switching and IP forwarding. This switch fabric is actually for control channel of GMPLS system. Although control channel is not necessarily based on switch fabric, it is highly recommended.

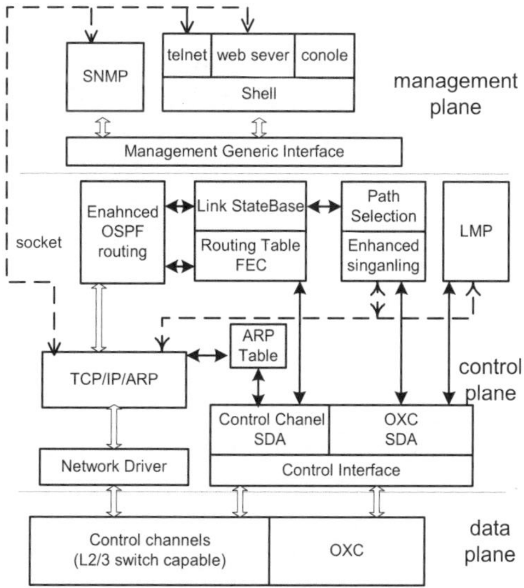

Figure 4 GMPLS based system architecture

The deployment of MPLS and GMPLS in IP network meets the similar problems. Edge devices need to be introduced for both cases. For deployment of MPLS domain in IP network, Label edge routers are introduced, which provide some interfaces to IP domain and some interfaces to MPLS domain. LER will aggregate data from IP domain interface and encapsulates with labels according to certain FEC, then forwards to interfaces of MPLS domain. The definition of FEC can be policy oriented concerning QoS, Diffserv and VPN, etc. On the other hand, at the egress node, IP traffic from MPLS domain will be decapsulated of label overhead, with further check of IP header and forwarded to certain IP domain interfaces following indexes of routing table.

For the deployment of GMPLS, GMPLS-aware edge router should be introduced, which provides some interface to IP or MPLS domain, and some interface to GMPLS domain. Similar with the processes adopted in LER, traffic from IP domain to GMPLS domain is nested to certain wavelength. On the other hand, traffic from GMPLS domain is checked at layer 2 or layer3 and forwarded to certain interfaces of IP domain. Unlike the LER, each GMPLS domain interface is composed of data channel and control channel. Traffic to GMPLS domain, should first be decided which interface to forward, and then split whether to data channel or control channel.

6. IMPLEMENTATION GUIDELINES OF GMPLS-BASED SYSTEM

We introduce the guidelines of the implementation of GMPLS based OXC, after the detailed comparison above. Through the comparison of mechanisms and system architecture of various IP centric network equipments, we have identified that there are great similarities. Owing to the similarity of GMPLS based OXC and LSR and their close relationship, software-based LSR can be a prototype of GMPLS system. Routing and signaling scheme can be further developed based on the architecture of LSR implemented to meet the GMPLS requirements, such as wavelength continuity constraints, protection and restoration mechanisms, etc.

By simply simulating OXC behavior with constraint label set, simulations can be conducted even when there are no OXC deployed in the data plane. This will be useful as a temporary measure when the OXC is not ready or the simple OXC is not capable to demonstrate certain network features.

The deployment of GMPLS-based OXC in IP domain meets similar issues with the deployment of MPLS in IP domain. GMPLS-aware edge devices should be introduced, with the functionalities similar to LER.

For better interoperation and integration with IP domain or MPLS domain, similar with LSR, GMPLS based system should also support L2 (MPLS Layer) and L3 (IP Layer) forwarding for its data traffic over control channels.

REFERENCES

[1] Windriver, “Tornado for Managed Switches 2.0,” Technical White Paper, Version 2.0, Oct. 3, 2001

[2] Adrian Farrel, “MPLS in optical network”, Data connection white paper, Oct. 2001

[3] A. Banerjee, "Generalized Multiprotocol Label Switching: An Overview of Routing and Management Enhancements," IEEE Commun. Mag., Jan. 2001

[4] A. Banerjee, "Generalized Multiprotocol Label Switching: An Overview of Signaling Enhancements and Recovery Techniques," IEEE Commun. Mag., Jul. 2001,

[5] Zhensheng Zhang, “Lightpath Routing for Intelligent Optical Networks,” IEEE Network Mag., July/August 2001.

Session 14F

MEMS Applications

A large-scale optical switching fabric using MEMS 2-axis tilt mirror arrays and free-space optics

Tsuyoshi Yamamoto, Joji Yamaguchi, Nobuyuki Takeuchi, Akira Shimizu,
Eiji Higurashi, Renshi Sawada, and Yuji Uenishi
NTT Microsystem Integration Laboratories, Nippon Telegraph and Telephone Corp.
3-1 Morinosato Wakamiya, Atsugi-shi, Kanagawa 243-0198
JAPAN
Phone:+81 46 240 2097; Fax:+81 46 270 2314; E-mail:t-ymmt@aecl.ntt.co.jp

Abstract: We describe our recent activities for developing a large-scale 3-D MEMS optical switching fabric. The switches, consisting of MEMS 2-axis tilt mirror arrays and free-space optical interconnections, have great potential in the construction of large-scale switching fabrics with more than 1000 ports.

This paper mainly describes the newly proposed MEMS mirror arrays, which consist of single-crystal silicon mirrors supported by high-aspect ratio torsion springs and low-cost, highly accurate two-dimensional optical collimator arrays. A prototype switch having 64ch optical fiber I/O provides a low coupling loss of 3.2 dB and switching time of 14 msec.

Key Words: 3-D MEMS optical switch, OXC, Free-space optical switch

1. INTRODUCTION

The explosive growth of the Internet traffic continuously changes the ground design of telecommunication networks. The future IP backbone network will need a performance level of more than tens of Tb/s within the next decade. Dense-wavelength division multiplexing (DWDM) technology is a solid solution, which promises to provide high-speed, large-scale data transmission systems having a throughput of over several Tb/s. In order to control of the traffic, optical cross-connects and photonic GMPLS routers are going to be introduced in the IP backbone networks. Micro-electro mechanical systems (MEMS) optical switches are increasingly being used in the construction of large-scale, all-optical switching fabrics in the optical cross-connects. Several switching demonstrations, especially those reported by Lucent Bell Laboratories[1-2], have been received with keen interest.

Figure 1 shows the basic configuration of a 3-D MEMS optical switch using free-space optical interconnection. The optical beams passing through the optical fiber are collimated by the microlens array, and reflected twice by a 2-axis tilt mirror array fabricated with MEMS technology. When mirrors are tilted two-dimensionally, the optical beam is deflected two-dimensionally as well. Therefore, any connection between input and output fibers can be achieved by controlling the tilt angle of each mirror. As a result, the switch can handle several channels of optical signals directly without costly optical-electrical or electrical-optical conversion. In general, 3-D MEMS optical switches are attractive as the number of optical input-output ports increases because of their high-density three-dimensional connectivity. Additionally, MEMS technology is suitable for mass production because it is based on silicon LSI fabrication processes. However, the precise alignment and assembly of the optical fiber collimators is time consuming and expensive.

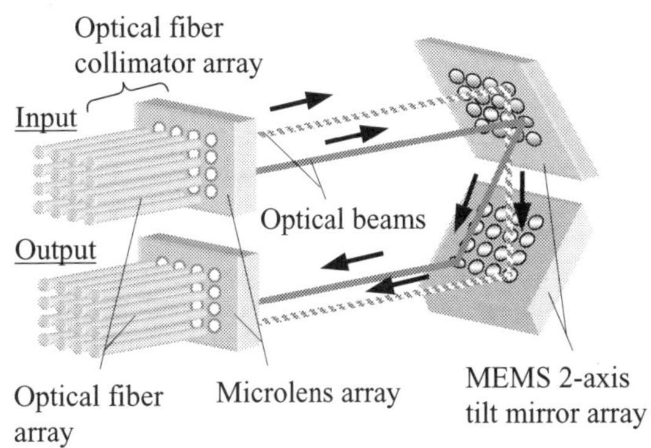

Fig. 1 Basic structure of the 3-D MEMS switch

In this paper, we demonstrate a free-space optical switching fabric comprised of MEMS 2-axis tilt mirror arrays and compact, highly-accurate optical fiber collimator arrays. The MEMS tilt mirrors that constitute the array are made of single-crystal silicon supported by a high-aspect-ratio torsion spring, and are controlled by terraced driving electrodes. The optical collimator arrays are comprised of two-dimensional optical fiber and polymer microlens arrays; the optical fiber arrays consist of optical fibers with micro-ceramic ferrules (as conventionally used in optical connectors) aligned in a polymer substrate having an array of precisely-aligned and positioned holes to fix the ferrules. The polymer microlens array, which is fabricated by plastic-injection molding technique, has a transparency at the wavelength of 1.5-1.6 μm. The MEMS tilt mirror and optical fiber collimator arrays have an 8×8 arrangement with an interval of 1.8mm. We also present in this paper the basic characteristics of our prototype 3-D MEMS optical switch.

2. MEMS 2-AXIS TILT MIRROR ARRAY[3]

Figures 2(a) and (b) show the packaged MEMS 2-axis tilt mirror arrays of both 64 and 1024 pixels, respectively. The 64-pixel array is bonded onto a conventional PGA ceramic package and the 1024-pixel array is also bonded onto a specially-designed ceramic package which consists of a large 70-mm-square mirror chip with 4100 electrode terminals featuring high-density wire-bonds and flexible-print cable direct connections. The mirror arrays were fabricated using SOI (silicon on insulator) bulk micromachining technology[4] and this provided a highly durable mechanical structure.

Figure 3 shows a cross-sectional schematic diagram of the MEMS 2-axis tilt mirror. The MEMS mirror consists of two single-crystal silicon substrates; a mirror and driving electrode substrates. These substrates are processed independently and are bonded to each other. Each mirror is supported by folded torsion springs in two orthogonal axes and is tilted two-dimensionally by controlling the applied voltage between the mirror and electrode substrates. In this mirror, the driving electrode has a 3D terrace structure and this effectively reduces control voltage.

Figure 4 (a) shows the fabricated MEMS mirror. The mirror, supported by the two-axes torsion spring, is 500 μm in diameter and 10-μm thick. The torsion spring, shown in Fig. 4 (b), has an aspect ratio of more than six. This high-aspect ratio spring features strong bending relative to torsion, and provides strong support for the mirror.

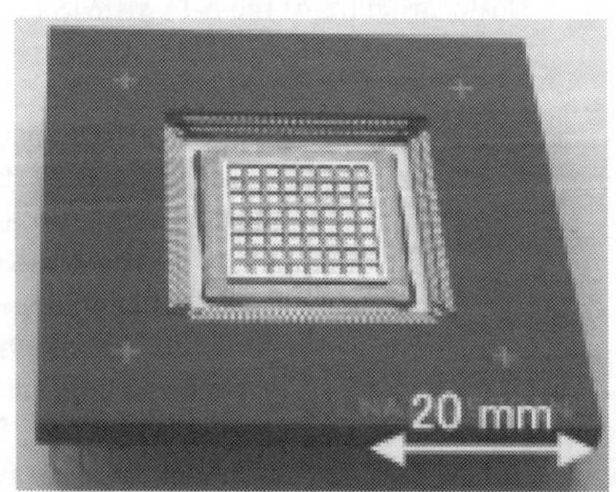

(a) MEMS mirror array having 64 ports

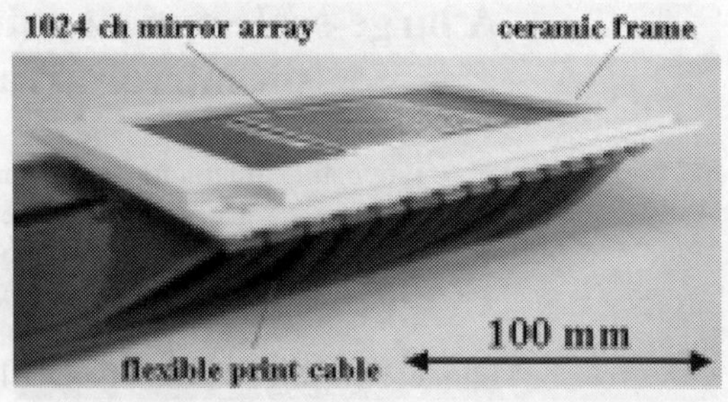

(b) MEMS mirror array having 1024 ports

Fig. 2 Packaged MEMS 2-axis tilt mirror array

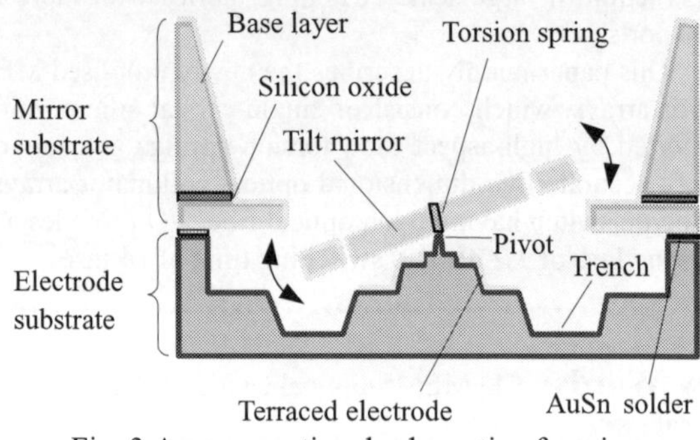

Fig. 3 A cross-sectional schematic of a mirror

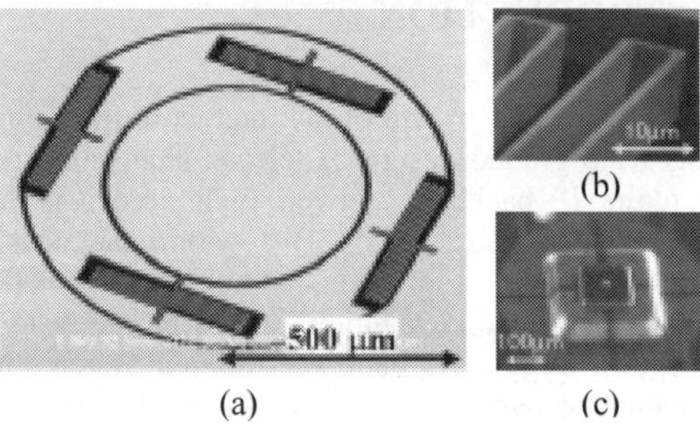

Fig. 4 (a) Single-crystal silicon MEMS mirror and magnified views, (b) high aspect torsion spring, and (c) 3D terrace electrode with center pivot

3. 2-D OPTICAL FIBER COLLIMATOR ARRAYS

Two-dimensional optical fiber arrays are key components in free-space optical switches. However, their assembly has proven to be costly when high-precision positioning is required. Several assembly techniques have been reported. In one such technique, fibers are aligned two dimensionally by using stacked V-grooved substrates[5]. In another, the optical fibers are inserted into holes made in a silicon substrate by wet etching[6]. Such conventional assembly processes may be costly because they require highly accurate, time-consuming fiber handling.

To overcome these problems, we use optical fibers with micro-ceramic ferrules and a polymer substrate having an array of precisely aligned holes. Figures 5 and 6 show the configuration of newly fabricated two-dimensional optical fiber array with 8×8 arrangement and 1.8-mm spacing and its assembly steps. The optical fibers with the ceramic-micro ferrules, which are commercially used in LC/MU-type optical connectors, are inserted into holes in the polymer substrate. Each ferrule end is polished and coated with anti-reflection material by vapor-deposition in advance. The polymer substrate with holes, made of a thermosetting molding compound, was molded by a transfer-molding method based on a common, mass-producible fabrication technique of the ferrule in MT-type optical connectors[7]. The arrays are assembled in a series of simple steps without using any special parts or assembly machinery. First, the polymer substrate is placed on the optical flat [Fig. 6(a)] and the optical fibers with the micro-ceramic ferrules are inserted into holes in the polymer substrate [Fig. 6(b)]. Next, the ferrules are fixed by UV-curable epoxy [Fig. 6(c)], and the optical flat is finally removed.

The microlens arrays with 8×8 arrangement and 1.8-mm spacing are made of transparent polymer with a wavelength of 1550-1600 nm. The microlens array was fabricated by a precise injection-molding method. Each lens in the array has a back-focal length of 3.97mm. Both surfaces of the microlens array also have an anti-reflection coating to prevent multiple reflections in the optics.

Figure 7 shows the fabricated two-dimensional optical collimator array. The optical collimator array is approximately 24×24×10 mm, excluding the optical fibers. The microlens array was aligned to the optical fiber array by monitoring optical beam size and beam position, as measured with an optical beam profiler. When alignment was completed, the two arrays were cemented to each other with a metal spacer by using the UV-curable epoxy.

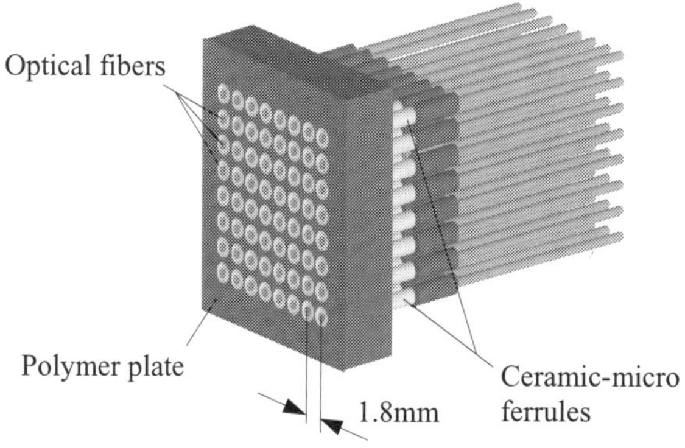

Fig. 5 Structure of the two-dimensional optical fiber array.

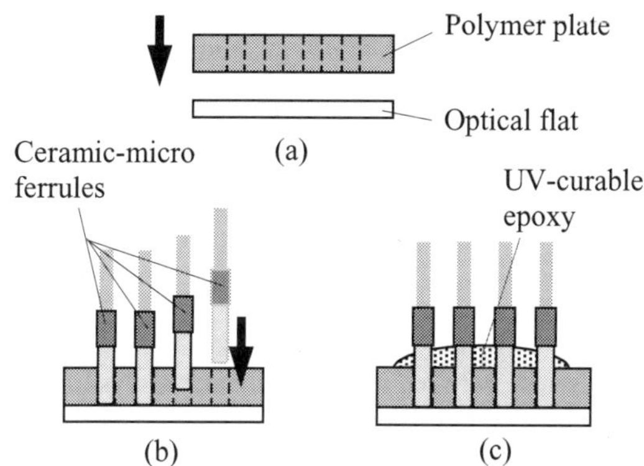

Fig. 6 Assembly steps of the optical fiber collimator array

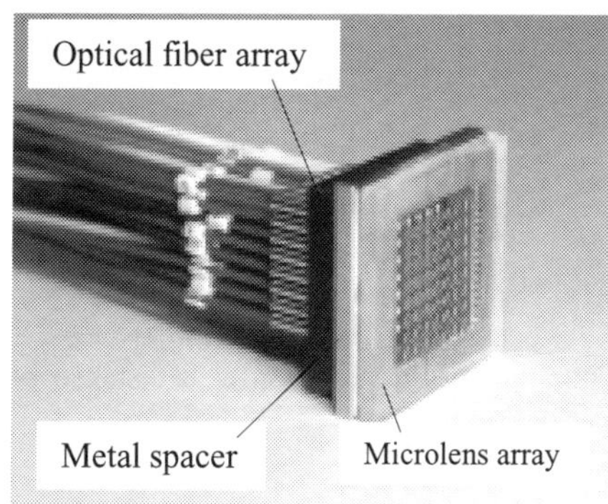

Fig. 7 Photograph of the two-dimensional optical collimator array

4. 3D-MEMS OPTICAL SWITCHING FABRIC

(a) Evaluation of the optical fiber collimator arrays

We first experimentally evaluated the coupling loss between the optical fiber collimator arrays since increasing the coupling loss and crosstalk is the most important issue in conventional systems. Figure 8 indicates coupling loss at different distances between the collimators. It suggests that the minimum coupling loss of 0.54 dB can be achieved at the distance of 300 mm between the collimators. Fig. 9 shows the coupling loss distribution in the collimator array at a distance between the collimators of 300 mm. The three-hatched lines in the figure correspond to the coupling loss when optical fibers and microlenses have misalignments of 0, 2, and 3 μm, respectively. This result indicates that misalignment between the optical fiber and microlens in all channels is within 2.5 μm. Coupling loss among channels is distributed between 0.54-6.91[dB]. This coupling loss deviation can be compensated by controlling the tilt angles of the MEMS tilt mirrors in the switch.

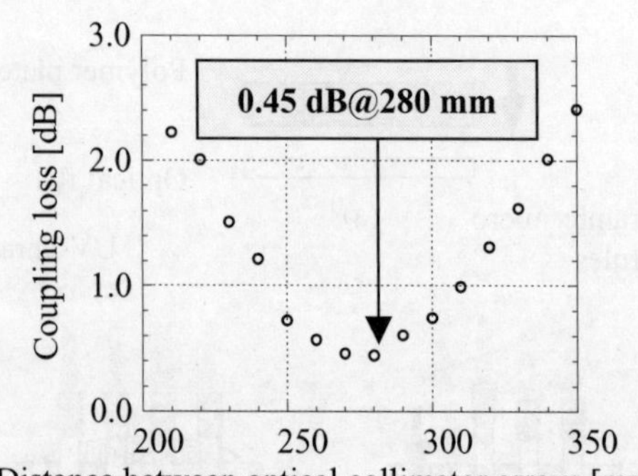

Fig. 8 Coupling loss of faced fiber collimator arrays

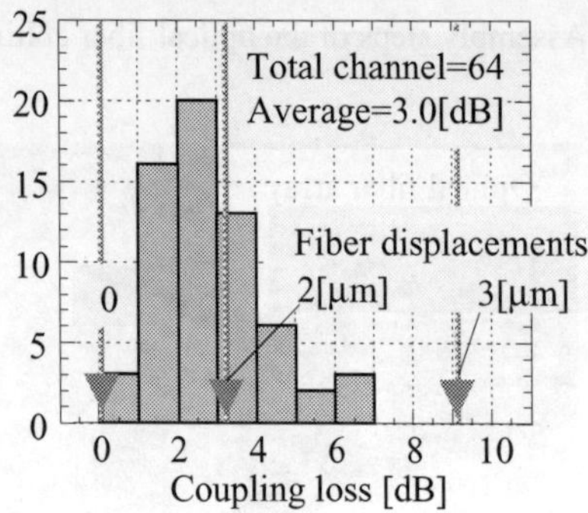

Fig. 9 Coupling loss distribution at a distance between the collimators of 300 mm

(b) Prototype switch

Figure 10 and 11 shows a photograph of our first prototype switch and its CAD drawing. The switch has a 64-input, 64-output structure and consists of two MEMS 2-axis tilt mirrors and optical fiber collimator arrays. The optical system in the switch is designed to be accommodated in a compact housing by using reflection and refraction prisms. The components are aligned and stacked on the metal frames by using UV-curable epoxy. The switch has a hardware volume of approximately 900cc. The measured insertion loss of the module is 3.2 dB, and the switching time between the two ports is 14 milliseconds from the open-loop MEMS mirror control.

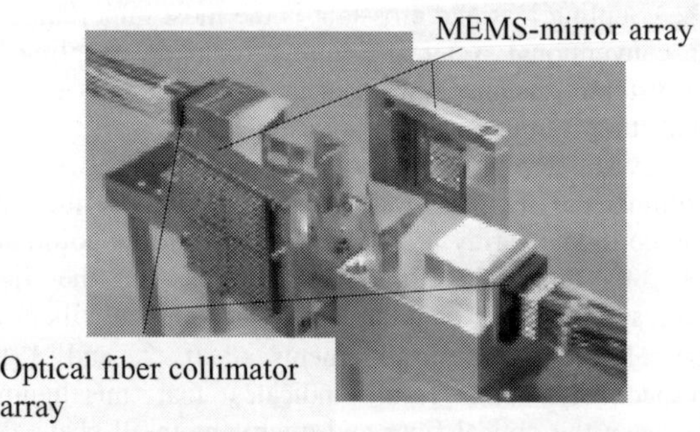

Fig. 10 Prototype 3D-MEMS switch with the MEMS 2-axis tilt mirror arrays and the two-dimensional optical fiber collimators

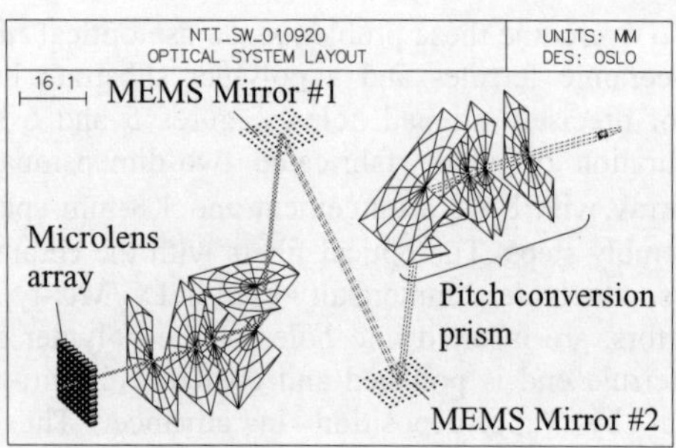

Fig. 11 CAD drawing of the prototype optical switch

5. CONCLUSION

We have presented our recently developed large-scale 3-D MEMS optical switching fabric. The MEMS 2-axis tilt mirror array consists of single-crystal silicon mirrors supported by high-aspect ratio torsion springs that provide for an optically flat mirror and highly durable mechanical structure, and two-dimensional optical collimator arrays (consisting of polymer-based optical fiber arrays and microlens arrays) suitable for mass production. Furthermore, We experimentally confirmed that the prototype switch having 64ch optical fiber I/O provides a low coupling loss of 3.2 dB and switching time of 14 msec.

REFERENCES

[1] D. J. Bishop, C. R. Giles, G. P. Austin, The Lucent LambdaRouter: MEMS technology of the future here today, IEEE Communications Magazine, Vol. 40, No.3, pp. 75-89, 2002.

[2] R. Ryf, et. al., "1296-port MEMS transparent optical crossconnect with 2.07 petabit/s switch capacity", Optical Fiber Communication Conference(OFC2001), paper-PD28-1-3, 2001.

[3] R. Sawada, E. Higurashi, A. Shimizu, and Tohru Maruno, "Single crystalline mirror actuated electrostatically by terraced electrodes with high-aspect ratio torsion spring", 2001 IEEE/LEOS International Conference on Optical MEMS(Optical MEMS 2001), paper-C-1, pp. 23-24, 2001.

[4] J. A. Walker, The future of MEMS in tele-communications networks, Journal of Micro-mechanics and Microengineering, Vol. 10, No. 3, R1-7, 2000.

[5] http://www.ngk.co.jp/ELE/product/01/sheet02.pdf

[6] G. Proudley, C. Stace, H. White, Fabrication of two-dimensional fiber optic arrays for an optical crossbar switch, Optical Engineering, Vol. 33, No. 2, pp. 627-645, 1994.

[7] T. Satake, N. Kashima, M. Oki, Very small single-mode ten-fiber connector, Journal of Lightwave Technology, Vol. 6, No. 2, pp. 269-272, 1988.

Micromirror Switching Device Fabrication for Optical Communication Network Applications

Janak Singh and Ajay Agarwal
Institute of Microelectronics, 11 Science Park Road,
Singapore Science Park II – 117685, Singapore
Fax: (65) 67780136; e-mail: janak@ime.a-star.edu.sg

Abstract: A new process has been presented to fabricate single crystal silicon (silicon) micro mirror switching devices. Silicon torsion and flexural springs have been integrated with a mirror plate for angular displacement amplification. Mirror switching is achieved with a bimorph thermal actuator, which is insensitive to environmental temperature changes and fabricated using surface micro machining layers. The release process combines silicon etching in KOH with sacrificial etching of oxide from backside. The design offers protection for the structure layers from both front and backside. Device demonstrates an angular amplification concept. An angular displacement of 15 degree was achieved.

Keywords: Thermal actuator, silicon mirror, silicon springs, angular amplification, and optical switch

1. INTRODUCTION

In the past few years, fiber capacity for optical communication networks has surpassed every established norm. New technologies such as WDM, DWDM, have fuelled this explosive growth. The demand for transparent optical switching has steadily increased because opaque optical switching using electronics is saturating to its maximum limits [1-2]. Even readjustment of electronic speed limits to higher values has not helped much; it has only created confusions, which has delayed the implementation of new emerging technologies. The increase in data traffic and thus the number of port counts in routers have made the electronics switching very complex and cumbersome. Optical switching where there is no need to convert signal to electronic domain before it is redirected to its desired destination has become very attractive. The routers with optical switching can be much simpler in architecture and easier to handle [1-2].

In the optical communication area, MEMS is a new technology and has to compete with established technologies [3] such as planar lightwave circuit (PLC) and fiber optics for the same segment of optical switching devices. PLC circuits are normally passive and that has made them more reliable [4]. On the other-hand, smaller size of MEMS offers advantages in the number of port counts, dynamic switching, and thus flexibility in selection of channels [5-6]. However, moving parts in MEMS switches make it less durable than PLC devices. Packaging of MEMS switches is another very critical aspect [7]. These issues have made researchers to opt for more durable MEMS components [7-8]. Surface flatness is another important aspect in micromirror devices, which improve the device from alignment point of view [9]. A better mirror surface flatness simplifies fiber alignment and packaging issues and thus improves the power output. Single crystal silicon plate because of high crystalline purity, can be a preferred choice for fabricating a mirror with a very good surface flatness [10-11]. Thermal actuation is a widely used technique for many MEMS applications [12-13]. This is especially useful when a high force or large actuation is required. It is also preferred in applications where large voltage can not be used. A large deflection can be achieved with a voltage as small as 2 volt or less. However power consumption is more in comparison to electrostatic actuation.

This paper presents a process to fabricate an optical switching micromirror device. Silicon springs have been used to obtain angular amplification and thus higher angular deflection for switching. Process design and development have been carried out to integrate silicon springs and mirror plate with thin film bimorph thermal actuator. The thermal actuator is formed of thin composite layers of aluminum and oxide. Symmetrical composition of aluminum thin films in the actuator makes it insensitive to the environmental temperature changes. Silicon springs and micromirror plate are defined by trench etching in silicon. Trenches are filled with sacrificial oxide layer. A combined backside KOH etching and SiO_2 sacrificial etching in BHF (Buffered HF) form the release process. During BHF sacrificial etching, actuator is protected by polysilicon and photosensitive thin films. Process can be utilized to obtain the best of both surface as well as bulk silicon micromachining.

2. DESIGN AND FABRICATION

Device and process designs are carried out for silicon mirror optical switching devices. Concept design of the device and angular amplification is given in Figure 1 (a) and (b) respectively. The actuator is formed of four bimorph beams, which are attached together at the free ends [14]. Torsion and flexural springs are fabricated together with silicon mirror plate as one structure. Solid contact between actuator and silicon parts is provided at the end of flexural spring.

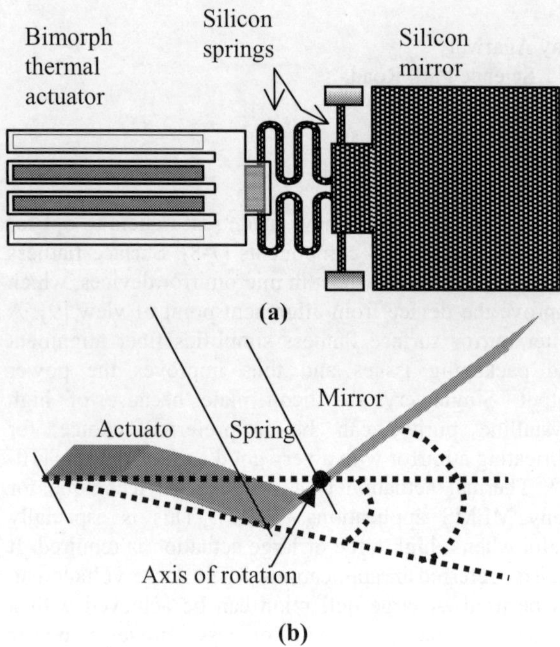

Fig.1. (a) Concept design of thermally actuated micromirror device with springs. (b) Angular amplification concept

Silicon dioxide (SiO_2) is a compressive while aluminum is a highly tensile thin film. As a result of these intrinsic stresses in the two films, actuator will bend towards aluminum side. To keep the actuator straight in surface plane, aluminum film needs to be deposited on the both side of the oxide symmetrically. Four beams in the actuator show the symmetrical composition of aluminum – oxide actuator. Two beams in the center have aluminum on topside while the outer two beams have aluminum on the bottom side of the oxide beams. Deposition conditions and geometrical parameters of the aluminum on the top and bottom should be identical to counter balance the stresses.

Springs are very critical for the successful implementation of angular amplification concept. Geometrical parameters of torsion and flexural springs need to be optimized within yield strength limits to obtain maximum angular displacement. Polysilicon and silicon springs have been designed and simulated to implement in this development. In this paper, silicon springs have been integrated in micromirror device. Thickness of silicon membrane used for silicon mirror as well as for springs is 6 μm. Actuator free end is attached with flexural spring using IC contact process.

Process design: The process, shown in Figure 2, has been designed to integrate bi-morph thermal actuator with micromirror silicon plate. People have used SOI wafers or anisotropic etching of silicon in aqueous alkaline solutions to integrate surface thin films with silicon microstructures [15-16]. One needs to have a large margin for dimensions and also for errors for releasing silicon microstructures using anisotropic etching of silicon in alkaline solutions. So for fine structures such as springs or where high density of devices is required, only anisotropic etching of silicon can not be sufficient. SOI wafers can also be used for integrating silicon with surface films. The process presented here can easily be used with SOI wafers also.

In this article, trench scheme is used to cut through n-diffused silicon membrane. Trenches are filled with silicon dioxide sacrificial layer. Silicon window is opened from backside and stopped short of etching through. Silicon window is etched from backside using KOH aqueous anisotropic etching solution. The etching is stopped at the p-n junction by using three electrodes ECE (ElectroChemical Etch-stop) technique. Actuator films are protected by polysilicon / nitride / photo-resist layers.

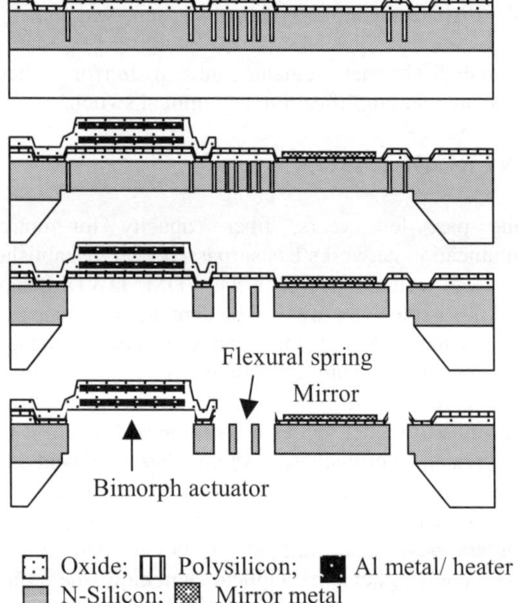

Fig2: Fabrication steps,(1) trench etch and fill, (2) sacrificial oxide deposition and etch (3) polysilicon deposition, (4) bimorph actuator fabrication (Al and SiO_2 deposition and etch steps), (5) poly etch from mirror area (6) ECE metal deposition (global) (7) Si etching in aqueous KOH and ECE steps, (8) BHF SiO_2 etch release, (9) front and backside protection layers removal

Fabrication: The process starts with forming n-p junction by implantation and diffusing phosphorus impurity to a depth of 8 μm in p-silicon substrate wafer. In the next step trenches are etched to define silicon mirror, torsion and flexural springs, which form a part in release scheme also. Sacrificial oxide is filled in the trenches and the surface is planarised. Another silicon dioxide layer is deposited on top of trenches and patterned. This is the sacrificial layer used to free the actuator and springs from silicon substrate. Polysilicon film is deposited on top of the sacrificial silicon dioxide. This layer separates sacrificial mechanisms from rest of the top layers used to form the actuator. Thermal bi-morph actuator is fabricated in the following steps. First aluminum film is deposited and patterned. This forms the bottom metal of the composite. Same metal can also be used to fabricate heaters on the other two beams where metal will be on the topside of the composite actuator. Next, silicon dioxide structure layer is deposited and patterned. This layer is patterned for the four beams of the composite bi-morph actuator. The actuator is attached to the flexural silicon spring by contact fabrication mechanism. Polysilicon is selectively etched from the mirror and ECE contact areas. Cr-Au is deposited for ECE contacts and the film is patterned for 3-electrode ECE process. This step completes the processing on the front side.

On the backside, windows are opened in LPCVD nitride film for anisotropic etching of silicon, in the aqueous solution of KOH. It takes about 15 hrs to complete silicon etching to reach at the p-n junction, where etching is stopped by the 3-electrode ECE technique. After ECE, silicon is further etched using dry RIE process to expose oxide filled trenches. In the next step, front side of the wafer is protected by photoresist (PR) coating and sacrificial oxide is removed in BOE (buffered HF oxide etch) solution. In this step, silicon from under the actuator and surrounding the springs and mirror is also sacrificed. Polysilicon protection layers from backside, and PR and metal layers from front side are etched-off in the last step to completely free the mirror device.

3. RESULTS AND DISCUSSION

Release scheme and fabrication of silicon microstructures are the main elements of the process. Deposition conditions for aluminum thin films in the composite actuator for top and bottom layers need to be identical. There is an oxide deposition in between the two aluminum depositions, and because of this bottom aluminum has to go through an extra temperature cycle of oxide deposition. This can alter the balance of stress in the top and bottom aluminum layers. We have observed a slight bending in the actuators. The change in stress conditions because of the thermal cycle can experimentally be measured and thickness of the aluminum layers can be adjusted to correct it. In the process presented here, we have used 0.5 μm thickness for both top and bottom aluminum and 1.0 μm thickness for the oxide structure layer.

First fabrication samples have been obtained successfully, shown in Figure 3 and Figure 4. The samples shown in Figure 3 represent the angular amplification scheme discussed in this paper. A 10 μm deflection of the actuator end, resulted in about 60 μm deflection of the free end of the micromirror plate. The size of the micromirror is 200 μm x 200 μm as mentioned in Figure 4. The bending values of the micromirror were used to estimate an angular displacement of more than 16°. Theoretical calculations show that a switching speed of 10 msec is achievable with these thermal actuators. Reliability data has not been measured yet, however, use of silicon springs promises a good reliability and durability for the device. In strength, silicon is as good a material as steel on micro-scale [17]. This is an eliminatory device. It can be a very good candidate for 2D and 3D micromirror matrices for high port count optical switches. Figure 4 shows a device without springs. This device can be very reliable because springs have not been used in it. This is more like a scanner device where mirror will change its position to a large extent. To make use of its better reliability in optical switching, some innovative packaging schemes are required.

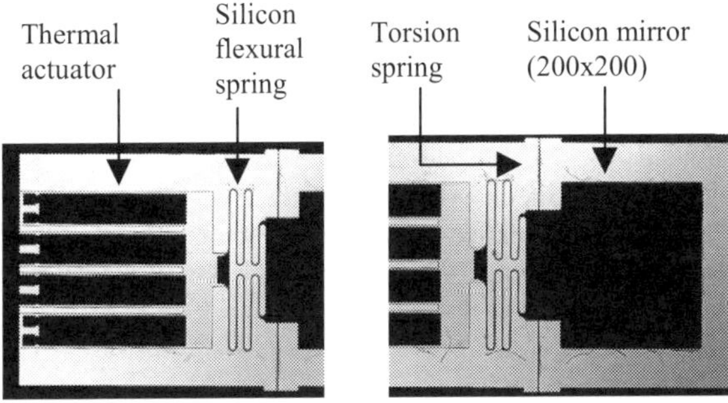

Fig.3: Optical microscope images in transmission mode from backside of a micromirror device.

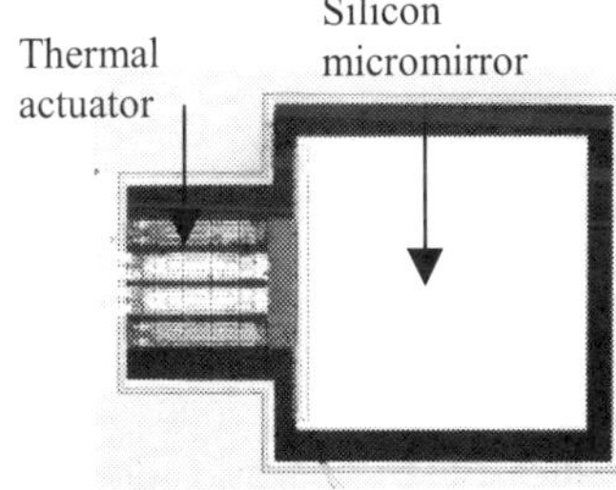

Fig.4: Optical microscope image a micromirror device from front side. This device is without springs

These have been the very first devices obtained after a successful process run. Process needs minor adjustments in release scheme. Some faint lines can be seen in Figures, this is the thin film from the protection layer, which is not etched very cleanly in the release process. A thick photo-resist coating has been included in the release scheme to avoid this release problem. Then in the last step photo-resist can be dissolved to release the structure.

4. CONCLUSIONS

A new mechanism has been presented to integrate silicon with surface-micromachined thin film microstructures. A single-crystal-silicon micromirror device has been developed with angular amplification spring mechanism. Angular amplification concept has been discussed and demonstrated using micromirror device. A thermal bimorph actuator, which is insensitive to environmental temperature variations, has successfully been integrated with silicon micromirror plate and springs. Release process which uses oxide filled trenches together with backside silicon etching in KOH aqueous solution and electrochemical etch stop, is unique. The surface and bulk micromachining integration scheme is quite general from implementation point of view and thus, it can be used for many more MEMS microstructures and devices. In the same process, micromirror scanner type of devices has also been obtained. There are no springs in these devices. Alignment and packaging innovative schemes can be helpful to make use of these more reliable devices in optical communication networks.

ACKNOWLEDGEMENTS

Authors would like to acknowledge the assistance provided by MEMS group at various stages of the development. Special thanks goes to Zou Quanbo, Zeng Wenjiang, Ranganathan Nagarajan, Victor Samper, Pavel Neuzil, Zhang Qing Xin, Mohanraj, Teoh Kum Weng, and Li Hong Bin.

REFERENCES

[1] B. Mukherjee, WDM Optical Communication Networks: Progress and Challenges, IEEE Journal on Selected Areas in Communications, vol. 18, page 1810-1824, 2000.

[2] O. Gerstel, On the Future of Wavelength Routing Networks, IEEE Networks, vol. Nov-Dec, page 14-20, 1996.

[3] W. J. Tomlinson, Requirement, Architectures, and Technologies for Optical Cross Connects, IEEE, 0-7803-5947-X, page 163-164, 2000.

[4] T. Miya, Silica-Based Planar Lightwave Circuits: Passive and Thermally Active Devices, IEEE Journal of Selected Topics in Quantum Electronics, vol. 6, page 38-45, 2000.

[5] S. S. Lee, L. S. Huang, C. J. Kim, M. C. Wu, Free-Space Fiber-Optic Switches Based on MEMS Vertical Torsion Mirrors, J. Lightwave Tech., vol. 17, page 7-13, 1999.

[6] L.Y. Lin, E. L. Goldstein, R W. Tkach, On the expandability of Free Space Micromachined Optical Cross Connects, J. Lightwave Technology, vol. 18, page 482-489, 2000.

[7] L. S. Huang, S. S. Lee, E. Motamedi, M. C. Wu, C. J. Kim, "MEMS Packaging for Micro Mirror Switches", Proc.1998 Electronic Components and Technology Conference and Technology Conference, pp. 592-597, 1998.

[8] M. R. Douglass, "Lifetime Estimates and Unique Failure Mechanisms of Digital Micromirror Device (DMD)", Proc. 36th Annual International Reliability Physics Symposium, Reno, Nevada, pp. 9-16, 1998.

[9] J. T. Nee, R. A. Conant, M. R. hart, R. S. Muller, K. Y. Lau, Stretched-Film Micromirrors for Improved Optical Flatness, IEEE, 0-7803-5273-4/00, page 704-709, 2000.

[10] C. Henri, L. Franck, Fabrication, Simulation, and Experiment of a Rotating Electrostatic Silicon Mirror with Large Angular Deflection, IEEE, 0-78-5273-4/00, page 645-650, 2000.

[11] C. S. B. Lee, R. Y. Webb, J. M. Chong, N. C. MacDonald, Single Crystal Silicon (SCS) MicroMirror Arrays Using Deep Silicon Etching and IR Alignment, IEEE, 0-7803-5273-4/00, page 441-448, 2000.

[12] T. Moulton, G. K. Ananthasuresh, Micromechanical Devices with Embedded Electro-Thermal-Compliant Actuation, Sensors and Actuators A, vol. 90, page 38-48, 2001.

[13] J. Funk, J. G. Korvink, M. Bachtold, J. Buhler, H. Baltes, Coupled 3D Thermo-electro-mechanical Simulations of Microactuators, IEEE, 0-7803-2985-6/96, page 133-138, 1996.

[14] Q. Zou, U. Sridhar, J Singh, R. Nagarajan, Q. Zhang, W. Zeng, X. Qu, H. feng, "Ambient stable thermal bimorph actuator", Proceedings of iMEMS, Singapore, pp. 178-186, 2001.

[15] Y. H. Jin, K. S. Seo, Y. H. Cho, S. S. Lee, K. C. Song, J. U. Bu, An SOI Optical Microswitch Integrated with Silicon Waveguides and Touch-down Micromirror Actuators, IEEE, 0-7803-6257-8/00, page 43-44, 2000.

[16] A. Tuantranont, L. A. Liew, V. M. Bright, J. Jhang, W. Zhang, Y.C. Lee, Bulk-Etched Surface Micromachined and Flip-chip Integrated Micromirror Array for Integrated Applications, IEEE, 0-7803-6257-8/00, page 71-72, 2000.

[17] K. E. Petersen, Silicon as a Mechanical Material, Proc. IEEE, vol. 70, page 420-457, 1982.

A Multiple-Channel Micromachined Variable Optical Attenuator

X. M. Zhang[a], A. Q. Liu[a], C. Lu[a], F. Wang[b], C. Lu[b] and Z. Liu[b]

[a]School of Electrical & Electronic Engineering
Nanyang Technological University
1 Nanyang Avenue, Singapore 639798
[b]Institute of High Performance Computing,
Singapore Science Park II, Singapore 117685
Email: eaqliu@ntu.edu.sg, Tel: (65) 6790-4336; Fax: (65) 6792-0415

Abstract: A low-voltage variable optical attenuator has been developed for applications in dense wavelength division multiplexing systems. The attenuator is fabricated by silicon surface micromachining technology and is then manually assembled and integrated with two single mode optical fibers that act as optical input and output. A 40 × 40 μm^2 mirror coated with gold is driven by a proprietary drawbridge structure to cut partially into the light path between two fibers, enabling the attenuation. The attenuator has a dimension of 0.6 × 1 mm^2 excluding the fibers.

The attenuator has achieved 1.5 dB insertion loss, 45 dB dynamic range and 37 ms response time, and requires only 8 V driving voltage. The wavelength dependence loss is 0.9 dB over C-band (1528 – 1561 nm). The low driving voltage makes it possible to integrate IC control circuits onto the same chip. An array of the VOA can be fabricated to equalize multiple wavelength channels of a DWDM system.

....

Key Words: Variable Optical Attenuator (VOA), Optical MEMS, DWDM, Fiber optical Communication.

1. INTRODUCTION

The development of fiber-optical communication technology has placed new challenges for optical attenuators. Fiber-optical systems operate over a large span of optical power level, from very weak signals after transmitting long distances to very strong signals emitted from high power laser sources, requiring variable optical attenuators (VOAs) that are able to operate over large attenuation range. In the dense wavelength division multiplexing (DWDM) systems, multiple wavelength channels arriving at a node may pass through different paths and experience different losses, their powers need to be equalized before entering the optical amplifier to get flat gain since the gain of each channel depends on the power levels of the other channels. In addition, the powers of wavelength channels vary dynamically due to the add/drop of channels, requiring VOAs that have high-speed response. Conventional opto-mechanical VOAs provide large attenuation range (> 50 dB), but their low speed (~ 1s) limits the applications. Waveguide-based VOAs have high-speed response (~ 1ms), but it is difficult to achieve large attenuation range with low insertion loss. Microelectromechanical Systems (MEMS) technology has been implemented to VOAs to achieve high performance and low cost [1-3]. A mechanical antireflection switch (MARS) attenuator using a deformable quarter wavelength of silicon nitride antireflection layer was developed by Ford *et al.* obtaining 3 μs response and 31 dB attenuation with 35.2 V driving voltage [1]. A deep etched VOA was reported achieving 5 ms response time and 57 dB attenuation range [2]. Surface micromachined VOA with 64 μs response time using 40 V voltage and its application to the wavelength channels in lightwave subsystems was also demonstrated [3].

In this paper, a micromachined free-space VOA using proprietary drawbridge structure is described. The VOA has 1.5 dB insertion loss, 45 dB continuously tunable attenuation range, 37 ms response time from minimum to maximum attenuation and 36.2 ms to return to minimum attenuation. Only 8 V driving voltage is required. Fibers are also packaged.

2. DEVICE DESCRIPTION AND RESULTS

2.1 Schematic arrangement of VOA

The schematic arrangement of VOA is shown in Fig. 1. Two single mode fibers (SMFs) are used as the input and output. A mirror is placed to cut partially into the light path between the SMFs in order to block a certain portion of the light. As a result, the energy coupled to the output fiber is reduced, enabling the attenuating function. When the mirror moves to different positions, the output power changes, resulting in a variable attenuation.

Relation between the attenuation and the mirror position can be estimated by use of free-space diffraction based on

Rayleigh-Sommerfeld diffraction formula and then by mode-overlap integral between the field distribution at the output fiber accepting facet and the fundamental mode of the output fiber [3]. Figure 2 illustrates the calculated diffracted pattern of the Gaussian beams at the mirror's plane (10 μm from the input fiber) and the diffracted beam at the end facet of the output fiber when the mirror edge is positioned at the centre of light path (0 μm). The wavelength is 1550 μm, the waist radius of the fundamental Gaussian mode in the fibers is 5.2 μm, and the distances between the fiber end facets and the mirror are kept at 10 μm. It is shown in Fig. 2 that the beam pattern in the accepting facet of output fiber is distorted, and its energy centre is displaced from the fiber centre. Therefore, only a part of the light energy reaching the output fiber can be coupled into the output fiber.

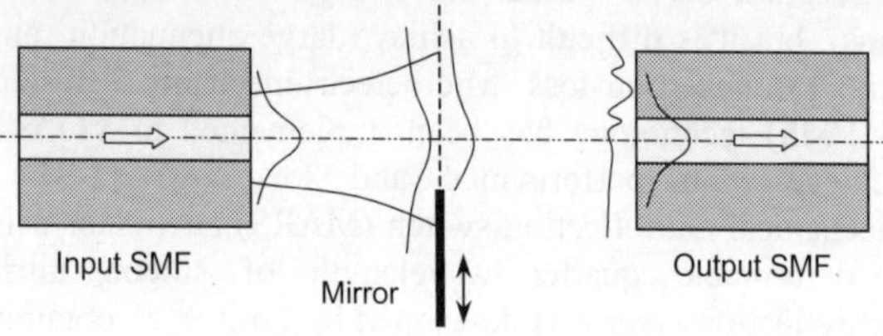

Fig. 1 Schematic diagram of the VOA

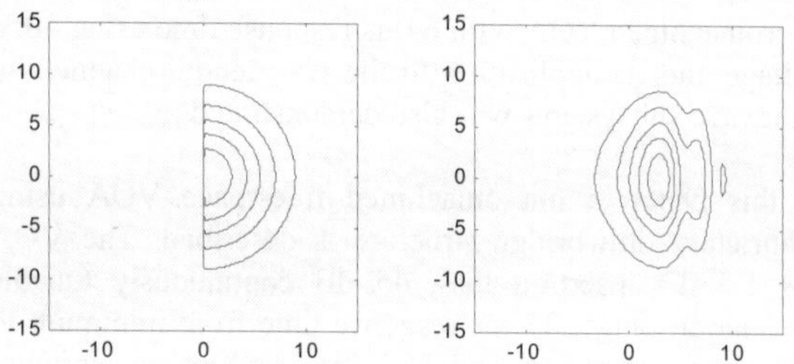

Fig. 2 Diffracted beam patterns calculated at the mirror plane and at the end facet of output fiber.

2.2 Surface micromachined VOA

The MEMS VOA is fabricated using silicon surface micromachining technology and is then assembled, as shown in Fig. 3. The vertical plate is mounted vertically on the mounting plate by the holder. At the end part of the vertical plate, a 40 μm × 40 μm flat mirror is coated with 0.5 μm-thick gold layer. A special designed drawbridge structure composed of a drawing beam and a holding plate that was developed in the previous work [4] is employed to lift up the mirror over the substrate. After fabrication and release, the drawing beam lies on the holding plate, and both are pinned to the substrate by some micro-hinges. In the assembly process, the T-shaped head of drawing beam is latched into the T-shaped hole in the holding plate, forming a triangular support to firmly tilt the mounting plate in a certain angle. Three bending beams connecting the holding plate to the mounting plate are used to concentrate the deformation and to control the stiffness, which in turn determines the response speed and the maximum driving voltage. The actuator is composed of the mounting plate and the electrode on the substrate. When the driving voltage is applied, the electrostatic force attracts down the mounting plate, and the mirror moves as well. The mirror position can be controlled accurately by controlling the voltage.

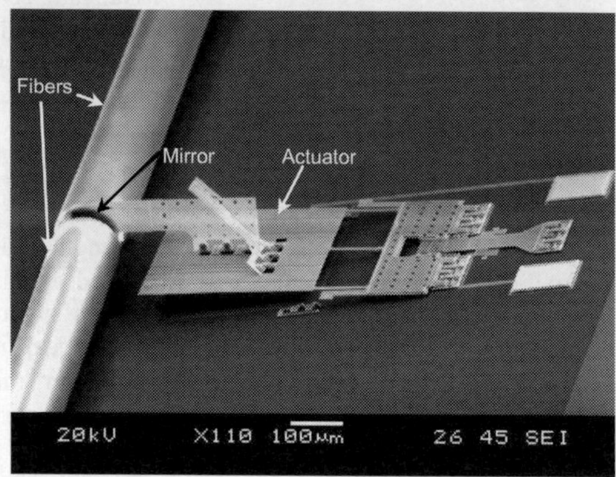

Fig. 3 SEM micrograph of the variable optical attenuator.

In the final step of assembly, two fibers are aligned using translation stages and then fixed to the substrate by use of the adhesive. The distances between the input fiber to the gold mirror and the gold mirror to the output fiber are both 10 μm by aligning the end facets of the two fibers to the alignment lines that are fabricated on the substrate.

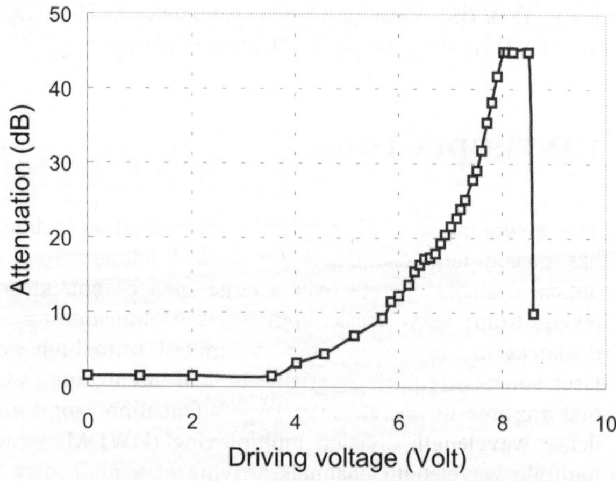

Fig. 4 Static relationship of driving voltage and attenuation of MEMS VOA

2.3 Static relationship of driving voltage and attenuation

Initially, the gold mirror is kept away from the light path, and the light beam from the input fiber passes to the output fiber with a small attenuation, corresponding to the insertion loss of VOA. When a certain driving voltage is applied between the electrode and the mounting plate, the electrostatic force moves down the mirror to induce certain attenuation in the output fiber.

The static characteristic of MEMS VOA measured at 1.55 μm wavelength is shown in Fig. 4. The insertion loss is 1.5 dB. The attenuation does not change obviously until the driving voltage becomes larger than about 3.5 V, probably due to that there are certain initial gap between the mirror and the light path and 3.5 V driving voltage moves the mirror to start to cut into the light path. As a result, the attenuation increases continuously to about 45 dB when the driving voltage increases to 8 V. Further increase of driving voltage does not produce higher attenuation since the signal power becomes too weak to be detected under the background noise (0.018 μW).

As long as the driving voltage reaches the critical value of 8.6 V, the mirror rapidly moves down till the mounting plate touches the substrate. This is due to the snap-down effect that was elaborated in Ref. [4]. The critical voltage is hence named as the snap-down voltage. In the MEMS VOA, the snap-down effect makes the mirror quickly move away from the light path, and the attenuation decreases to a small value, as shown in Fig. 4.

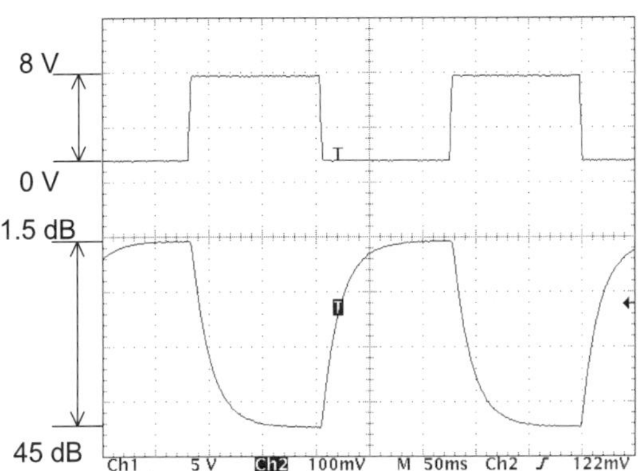

Fig. 5 Dynamic response of MEMS VOA.

2.4 Dynamic response

Dynamic response of the MEMS VOA is shown in Fig. 5. A square signal of 4 Hz applies 0 and 8 V voltages alternatively to the mounting plate and the electrode. The power coupled to the output fiber ranges from 71% to 0.003%, corresponding to the attenuations of 1.5 dB and 45 dB respectively. The 0% - 90% rise time (0 $\rightarrow$ 8 V) is measured to be 37 ms, and the 100% - 10% fall time is 36.4 ms. The rise and fall times are nearly identical.

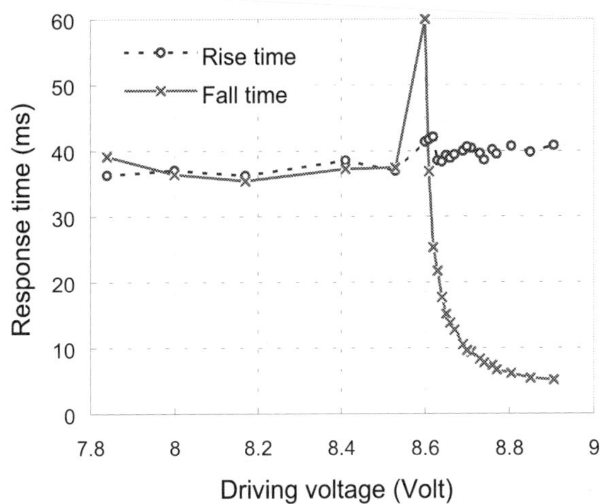

Fig. 6. Variation of response time with the driving voltage.

It is noted that the voltage of 8 V does not induce the snap-down but it is large enough to sweep the full range of attenuation. Further increase of driving voltage does not significantly affect the rise time, as shown in Fig. 6. In contrast, the fall time changes obviously. It is not sensitive to the voltage change when the voltage is lower than the snap-down voltage of 8.6 V, and increases to a high value when the snap-down occurs. After that, the fall time rapidly decreases. In Fig. 6, the fall time is 60.0 ms in the snap-down state, and decreases to 6.1 ms with a voltage increment of only 0.2 V. After that, the fall time becomes insensitive again to the driving voltage. It implies that a fast speed switching is possible if this VOA is used as an ON/OFF optical switch with a driving voltage slightly larger than the snap-down voltage.

2.5 Wavelength dependence of MEMS VOA

Figure 7 shows several attenuation spectra of the MEMS VOA at low attenuation levels by using a tunable laser source (ANDO AQ4321D) to scan over a wavelength range of 1520 – 1620 nm. The attenuation variation (i.e. wavelength dependent loss) is 2.4 dB over the whole range, and less than 0.9 dB over the C-band (1528 – 1561 nm). Although MEMS components are expected to have small wavelength dependence, it is only true when the light beam is operated by the reflection of MEMS mirror. In the MEMS VOA, the near-field diffraction plays a role. It determines the light that propagates to the output fiber, especially when the beam is partially blocked by the mirror. As a result, the attenuation becomes slightly wavelength dependent.

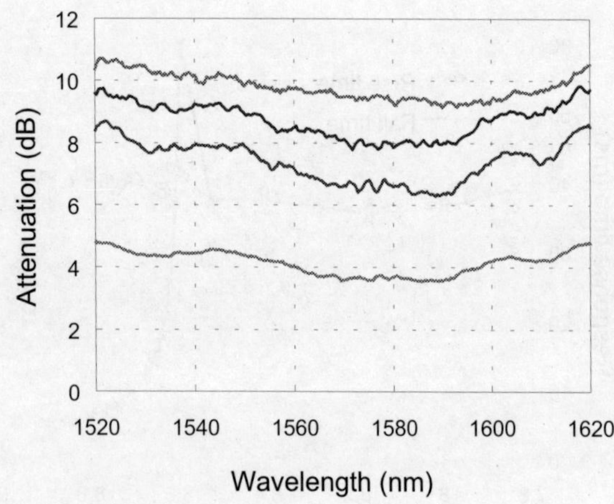

Fig. 7 Wavelength dependence of the VOA.

3. CONCLUSIONS

An optical fiber connectorized, electrostatically actuated MEMS free-space variable optical attenuator having 1.5 dB insertion loss, 45 dB dynamic attenuation range and 37 ms response time has been demonstrated. Only 8 V driving voltage is required to achieve the maximum attenuation. The attenuator is fabricated using surface micromachining technology. The structure can be also used as an optical switch. The fall time can be reduced to several milliseconds by applying a driving voltage slightly beyond the snap-down voltage, implying a fast optical switch. The more information can be found out from

1. http://www.ntu.edu.sg/eee/research/highlights/voa.htm.
2. http://fibers.org/articles/news/4/7/6/1.

REFERENCES

[1] J. E. Ford, J. A. Walker, D. S. Greywall, and K. W. Goossen, Micromachined fiber-optic attenuator with 3 μs response, Journal of Lightwave Technology, vol. 16, no.9, page 1663-1670, 1998.

[2] C. Marxer, P. Griss, and N. F. de Rooij, A variable optical attenuator based on silicon micromechanics, IEEE Photonics Technology Letters, vol. 1, no. 2, page 233-235, 1999.

[3] C. R. Giles, V. Aksyuk, B. Barber, R. Ruel, L. Stulz, and D. Bishop, A silicon Microelectromechanical Systems (MEMS) optical switch attenuator and its use in lightwave subsystems, IEEE Journal of Selected Topics in Quantum Electronics, vol. 5, no. 1, page 18-25, 1999.

[4] A. Q. Liu, X. M. Zhang, V. M. Murukeshan, Q. X. Zhang, Q. B. Zou, and S. Uppili, An optical crossconnect (OXC) using drawbridge micromirrors, Sensors and Actuators A, vol. 91, no 1-2, page 1324-1327, 2002.

Micromachined MEMS Tunable Laser for WDM System Applications

A. Q. Liu, X. M. Zhang, D. Y. Tang, and C. Lu

School of Electrical & Electronic Engineering
Nanyang Technological University
1 Nanyang Avenue, Singapore
Email: eaqliu@ntu.edu.sg, Tel: (65) 6790-4336; Fax: (65) 6792-0415

Abstract: Tunable lasers have wide applications in DWDM systems to save inventory cost and to improve the optical network functionalities. The Microelectromechanical Systems (MEMS) technology has shown strong promise to miniaturize the conventional mechanical tunable lasers with adding merits of high compactness, high speed, batch production and so on. In this paper, external cavity tunable diode lasers using MEMS movable mirrors and rotary gratings as the external reflectors are presented. One tunable laser of 2 mm × 1.5 mm is formed by integration of a surface-micromachined 3D mirror with a diode laser and an optical fiber. A wavelength tuning range of 16 nm is obtained.

Key Words: Tunable laser, external-cavity tunable laser, blazed grating, DWDM, optical MEMS.

1. INTRODUCTION

Tunable lasers have widely applications in Dense Wavelength Division Multiplexing (DWDM) systems to save the inventory volume and to improve the functionality of optical network. Although the wavelength of diode laser can be tuned by changing temperature and injection current, external cavity tunable laser has attracted significant interests since it can provide large tuning range, continuous wave (CW) tuning, high output power and excellent wavelength accuracy while maintaining single longitudinal mode. The conventional opto-mechanical external cavity tunable diode lasers have bulk sizes and slow speed, limiting their applications. The micromachined tunable lasers have been developed using the micromachined mirrors to form the external cavities [1-3]. The precision and stable movement of the microactuators enables fine-tuning of the wavelength. The small size of the micromachined mirrors yields high tuning speed and also makes it feasible to form the extremely-short-external-cavity tunable lasers. In addition, the micromachined tunable lasers yield high compactness, low fabrication cost, low power consumption and easy integration with IC control circuits.

A microfabricated tunable laser diode fabricated by nickel plating was demonstrated by Uenishi et al and achieved a wavelength tuning range of 20 nm [1]. Tunable lasers using micromachined Fabry-Perot (FP) etalons and micromirrors were also reported [2, 3]. An integrated tunable diode laser using a surface-micromachined 3D mirror was also shown in our previous work [4].

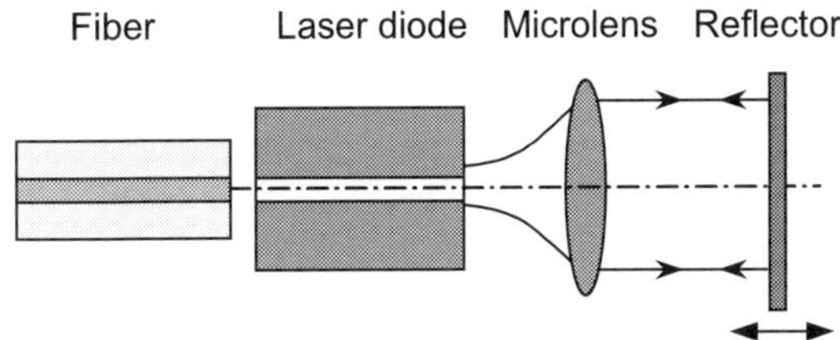

Fig. 1. Schematic configurations of the tunable laser using a translating micromirror as the external reflector.

2. DESIGN OF TUNABLE LASER

The schematic arrangements of MEMS tunable lasers are illustrated in Fig. 1. Either a translating micromirror or a rotary blazed grating can act as the external reflector. When using the micromirror as the reflector, the light beam emitted from one of the end facets of the laser diode is first collimated by a microlens and reflected by the micromirror, and is then coupled back into the lasing cavity of the laser diode. With the translation of micromirror, the phase of coupled light varies. As a result, the output wavelength changes. When the external cavity becomes very short, the microlens is not necessary, simplifying the structure of tunable laser. An alternative way is using a concave spherical mirrors rather than the flat micromirror. The concave mirror can reflect and focus the laser light without needing for any microlens. However, concave mirror is not easy to fabricate using MEMS technology. Circular mirror has to be used. In this laser configuration, the wavelength is mainly determined by the Fabry-Perot cavity formed by the two facets of laser diode and the external mirror. High compactness and large tuning range can be achieved, however, the laser is generally operated in multiple longitudinal modes, and the mode-hopping is difficult to overcome.

One improved method is using a rotary blazed grating as the external reflector. The laser light diffracted by the grating is also focused by the microlens and enters the lasing cavity of diode laser. The grating can be arranged in Littrow mounting [7]. It acts not only as an external reflector, but also as a wavelength selective component to maintain the diode in single mode. Moreover, mode-hopping can be avoided by carefully selecting the length of the cantilever beam. In this laser diode, the output wavelength is determined by the superimposition of the gain bandwidth of the diode, the grating dispersion and the external cavity mode structure. When the external cavity length changes, the possible wavelengths of the cavity modes change accordingly. To avoid mode-hopping, the passband of the grating should follow the change simultaneously. This requirement can be approximately met by selecting a proper length of cantilever beam. In both types of tunable laser, the laser light emitted from the other facet is butt-coupled by a single mode fiber to form the output.

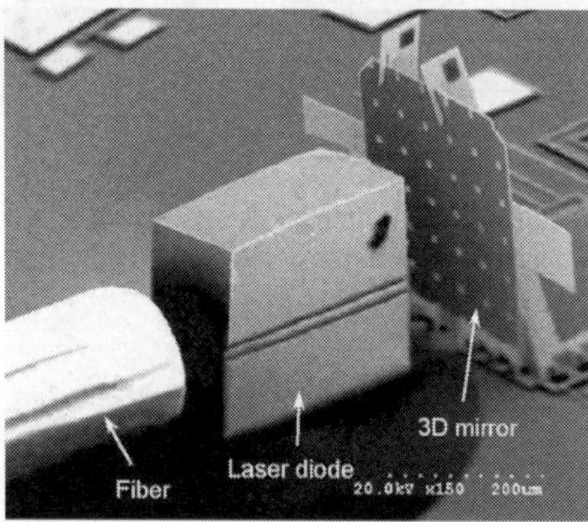

Fig. 2. SEM of the tunable laser using surface micromachined mirror

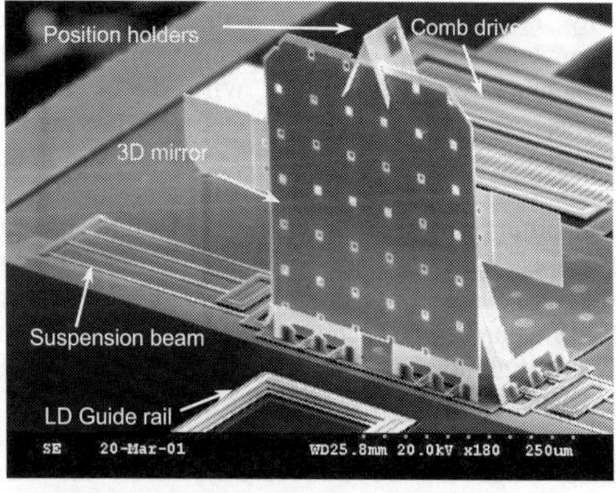

Fig. 3. Surface micromachined 3D mirror.

3. MEMS TUNABLE LASER

A SEM micrograph of the tunable laser is illustrated in Fig. 2. The micromirror as well as the related structures are fabricated by employing the surface micromachining processes. After fabrication and release, the micromirror is lifted up to the vertical position and then held by the position holders. Fig. 3 illustrates an assembled 3D micromirror similar to that shown in Fig. 2 This manual assembly is carried out by use of a probe station. After integration, the laser diode and the fiber are adhered to the substrate. It is noted that the end facet of the fiber and the laser diode facet that faces the 3D mirror are coated with anti-reflection layers, and the 3D mirror is coated with 0.5 μm-thick gold. The separation between the micromirror and the laser diode is initially 10 μm, and the fiber end facet has a 15 μm distance from the laser diode. The tunable laser (including the 3D micromirror and the diode laser) has a size of 2 mm × 1.5 mm × 1 mm.

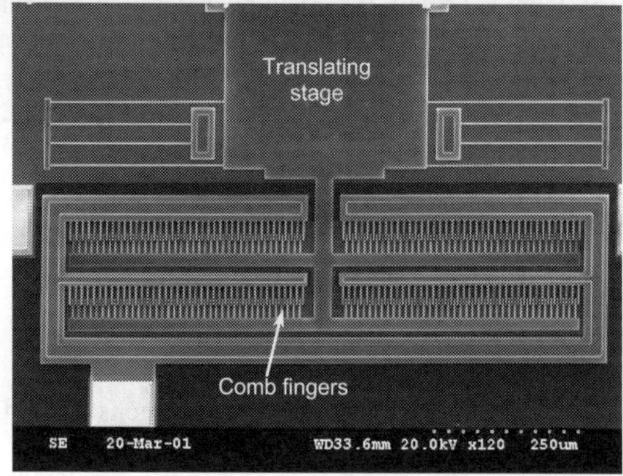

Fig. 4. The comb drive actuator.

The comb drive has two rows of comb fingers, as shown in Fig. 4. There are totally 156 moving fingers. Each finger is 30 μm long, 2 μm wide and 2 μm thick. The finger gap is 2 μm. The moving and the fixed fingers initially overlap by 10 μm to make the comb drive work in its linear range. The static relationship between the displacement of comb drive □ and the driving voltage *V* is governed by [8]

$$\delta = \frac{\varepsilon_0 N h}{g k} V^2$$

where *N* stands for the finger number of the moving comb, *h* and *g* indicate the finger thickness and the finger gap respectively, *k* represents the stiffness of suspension beams.

The relations of comb drive translation and the driving voltage is measured as shown in Fig. 5. The diamond points represent the comb drive displacements in the loading process (i.e. driving voltage increases from 0 to 30 V), and the circular points in unloading process. While the driving voltage is increased from 0, the comb drive does not have obvious displacement until the driving voltage reaches about 9 V. This is due to the stiction between the micromirror structures and the substrate, which can happen if the wet-etching process is used to release the structures. In Fig.5, the data follows the theoretical curve once the stiction is overcome. In the unloading process, there is a hysteresis of about 0.25 μm. In the following loading process, the comb drive does not move until the displacement exceeds the hysteresis region.

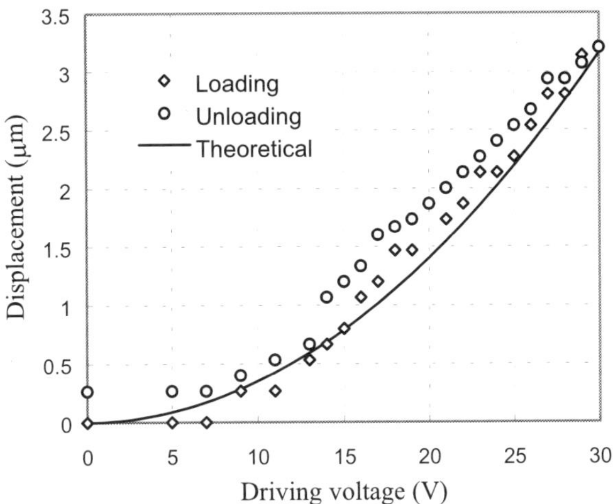

Fig. 5. The actuation relationship of comb drive.

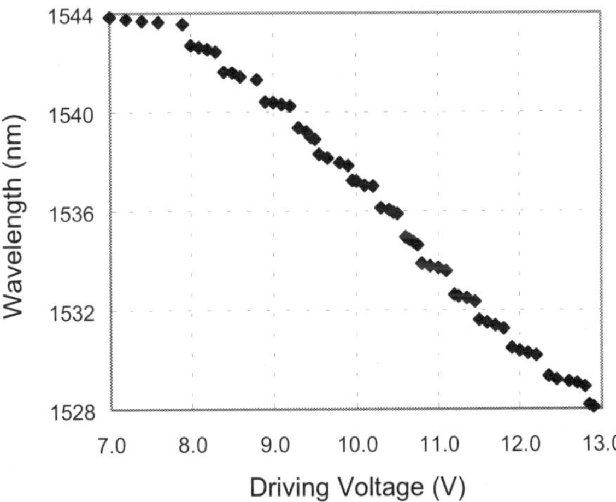

Fig. 6. Variation of wavelength with the driving voltage.

4. CONCLUSIONS

MEMS tunable lasers using moving mirrors and rotary gratings are investigated. A tunable laser of 2 mm × 1.5 mm size has been demonstrated using a surface micromachined 3D mirror as the external reflector. A tunable range of 16 nm has been achieved. Tunable laser using deep-etched circular mirror and rotary blazed grating as the external reflectors have also been developed. The circular mirror one works in multi-longitudinal modes and the mode hopping phenomenon is observed, whereas the rotary grating one works in nearly single longitudinal mode and the mode hopping can be overcome by selecting proper parameters of the grating.

REFERENCES

[1] Y. Uenishi, K. Honma, and S. Nagaoka, Tunable laser diode using a nickel micromachined external mirror, Electronics Letters, vol. 32, no. 13, page 1207-1208, 1996.

[2] Y. Sidorin, M. Blomberg, and P. Karioja, Demonstration of a tunable hybrid laser diode using an electrostatically tunable silicon micromachined Fabry-Perot interferometer device, IEEE Photonics Technology Letters, vol. 11, no. 1, page 18-20, 1999.

[3] M-H Kiang, O. Solgaard, R.S. Muller, and K.Y. Lau, Silicon-micromachined micromirrors with integrated high-precision actuators for external-cavity semiconductor lasers, IEEE Photonics Technology Letters, vol. 8, no. 1, page 95-97, 1996.

[4] A. Q. Liu, X. M. Zhang, V. M. Murukeshan and Y. L. Lam, A novel integrated micromachined tunable laser using polysilicon 3D mirror," IEEE Photonics Technology Letters, vol. 13, no. 5, page 427-429, 2001.

[5] J. D. Berger, Y. Zhang, J. D. Grade, H. Lee, S. Hrinya, H. Jerman, Al Fennema, A. Tselikov and D. Anthon, "Widely tunable external cavity diode laser using a MEMS electrostatic rotary actuator," Proc. LEOS Summer Topic Meet., Copper Mountain, CO, USA, pp. 41-42, 2001.

[6] D. Anthon , J. D. Berger, J. Drake, J. D. Grade, S. Hrinya, F. Ilkov, H. Jerman, D. King, H. Lee, A. Tselikov, and K. Yasumura, "External cavity diode lasers tuned with silicon MEMS," Proc. OFC2002, Anaheim, California, USA, Tu07, pp. 97-98, 2002.

[7] M. C. Hutley, *Diffraction gratings*, London:Academic Press, 1982.

[8] T. Hirano, T. Furuhata, K. J. Gabriel, and H. Fujita, Design, fabrication, and operation of submicron gap comb-drive microactuators, IEEE Journal Microelectromechanical Systems, vol. 1, no. 1, page 52-59, 1992.

MEMS SWITCH FOR RECONFIGURABLE OADM APPLICATION

Q.X. Zhang[b], J. Li[a], S.C. Tjin[a], W.D. Zhong[a] and A.Q. Liu [a]
[a] School of Electrical & Electronic Engineering
Nanyang Technological University, Nanyang Avenue, Singapore 639798
[b] Institute of Microelectronics, 10 Science Park Road
Singapore Science Park II, Singapore 117684

ABSTRACT

We demonstrate a tunable optical add/drop multiplexer (OADM) that uses a micromachined 2 × 2 optical switch and a tunable fiber Bragg grating (FBG). The new OADM can be tuned to add/drop one of the multi input channels dynamically. The insertion loss of the dropped channel and added channel are both 6.04 dB. The transmission loss is 2.08 dB. These losses can be reduced significantly if the loss arising from the circulators and MEMS switch is reduced. The crosstalk between channels is less than -20 dB, and can be further reduced by improving the reflectivity of the FBG used. The tuning speed is in the order of milliseconds. The tuning range is 3.0 nm. The fabricated system is experimentally demonstrated by selectively adding/dropping one of the four adjacent channels with the spacing of 100 GHz. Systems with multi channels being dropped and/or added can be achieved by cascading the proposed structure.

Keywords: Tunable OADM, MEMS optical switch, tunable FBG, WDM, Deep RIE

1. INTRODUCTION

Optical add/drop multiplexer (OADM) is a device that adds and/or drops a subset of wavelength(s) from the transmission system without full opto-electronic regeneration of all wavelengths expressing through the node flexibly. OADMs that can dynamically select desired wavelengths are considered to be the key component in the implementation of flexible wavelength division multiplexing (WDM) ring networks, especially with the increasingly dense channel spacing and transmission speed [1]. The OADMs make it possible to manipulate the traffic on the wavelength basis in the optical layer. Having OADMs at a network node can save cost by reducing the number of required transponders and facilitate network management. It is expected that OADMs will play significant roles in both optical core networks and regional access networks [2].

There are several types of OADM based on different optical devices [3-5]. One type of OADMs is based on back-to-back wavelength demultiplexing and multiplexing [3, 4]. A certain channel is dropped and/or added at each OADM unit according to a specific add-drop plan [4]. Optical switches are necessary to select wavelengths and rearrange them in the spatial domain. This approach does not seem to be very practical because of its high cost. Another type of OADMs are based on the combination of fiber Bragg gratings (FBGs) and optical mechanical/ thermal switches [5]. However, to selectively drop various channels, many switches and FBGs have to be cascaded, and each set only drops a single wavelength. The switching speed of the conventional mechanical/thermal switch is slow, power consumption is high, and the size is large. But the OADMs based on optical circulators (OCs) and FBGs have the advantages of low crosstalk, and temperature and polarization insensitivity. There is no band-width-narrowing when WDM signals pass through many OADM nodes. In view of the maturity of silicon microelectromechanical systems (MEMS) technology, optical MEMS switches can be fabricated with high optical performance and potentially at low cost and high reliability because MEMS technology is highly compatible with the advanced silicon integrated circuit technologies.

In this paper, a MEMS optical switch combined with a tunable FBG is fabricated to increase the switching speed, reduce the cost and improve the reliability. In section 2, the proposed configuration of the OADM is described. The fabrication of the MEMS switch and the tunable FBG is illustrated in section 3. The measurement results and discussions are given in section 4.

2. OADM WITH MEMS SWITCH

Figure 1 shows the proposed wavelength-selective OADM. It consists of an MEMS optical switch, a metal coated tunable fiber Bragg grating (FBG), and two three-port optical circulators (OC). The MEMS optical switch is a 2 × 2 switch, which is located between the two circulators, the tunable FBG is connected to ports 2 and 3 of the switch. The MEMS optical switch determines if there is a wavelength channel to be dropped or not. Which wavelength channel is to be dropped is controlled by changing the Bragg wavelength of the tunable FBG [6]. If there is a channel to be dropped, the switch is set at cross-state. In this case, the WDM signals are fed to the tunable FBG as shown in Fig. 1. A channel whose wavelength is coincident with the Bragg wavelength λ_d of the tunable FBG is reflected back by the FBG and dropped at port b of OC1. All other wavelength channels pass through the tunable FBG and continue their propagation to the second circulator and multiplexed with the added signal which has the same wavelength of the dropped signal λ_d. If no wavelength needs to be dropped, the switch is set at bar-state so that all the input WDM signals pass through the switch without any change except a little loss. Since the FBG is tunable, a channel at different wavelengths can be selectively dropped and added. Combining with the MEMS 2 × 2 switch, it is possible to pass through all of the signals. The switching speed is determined by both the MEMS switch and the tunable FBG.

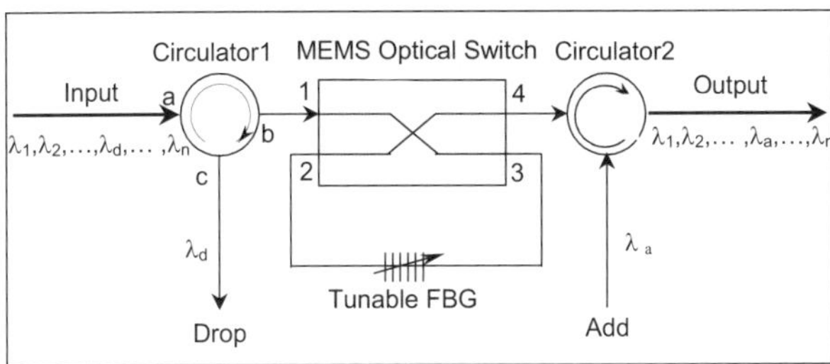

Fig. 1 Schematics of the tunable OADM

The key parameters of a reconfigurable OADM are the tunable range, channel isolation, adding and dropping loss, power consumption and operating speed. They depend on the components' characteristics and the method of combining them. By employing bulk micromachining technology, the optical switch is obtained and has very fast switching speed. The tunable FBG makes use of the thermal effect of the grating instead of piezoelectric transducers (PZTs) tuning. It is easy to control the period of the grating by adjusting the injecting current, to shift the center wavelength. Unlike pieoelectrically tuned FBGs, this method is also relatively insensitive to the excitation frequency [8]. In this work, the tunable range achieved is as wide as 3.0 nm.

3. FABRICATION OF THE MEMS OPTICAL SWITCH AND TUNABLE FBG

The MEMS optical switch and the tunable FBG, which are two key components of the tunable OADM, are fabricated by employing the integrated circuit (IC) technology.

The MEMS optical switch is made by using a deep dry etching technology. The SEM micrograph is shown in Fig. 2. Its size is 1.6 × 1.2 mm^2. The key element of the switch is the micromirror, which is connected to the comb drive actuator through the backbone. The switch is usually at bar-state. The micromirror is outside the beam path. The incident beam in fiber *1* connecting to port b of the first OC transmits directly to port *4*. Because the loss between any two ports of the OCs is 1.0 dB, the passed signals are attenuated by 2.0 dB. When an external voltage of 30 V is applied to the insulated pads of the fixed and the movable combs, the micromirror is driven into the beam path by the electrostatic force. The incident beam from fiber *1* is reflected by the micromirror to port *2* and the signal from port *3* will be redirected to port *4* to multiplex with the other transmission signals at the second OC. The micromirror will launch back to its normal balance position when the external voltage is released as a result of the elastic force induced by the folded suspension beam.

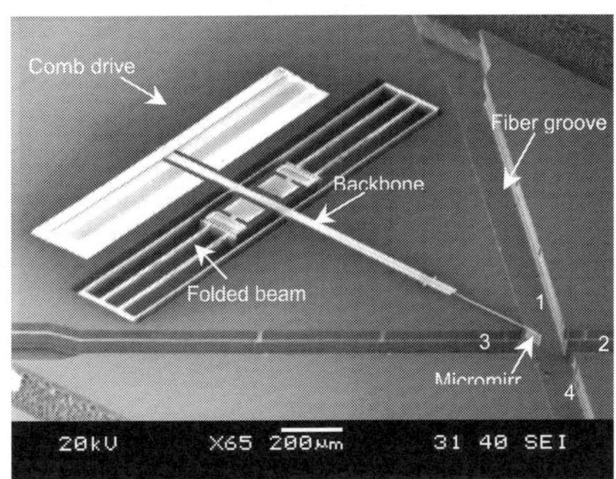

Fig. 2 SEM micrograph of the MEMS optical switch

The anchors support the movable structures and function as the electrodes for the external applied voltage. Because of this, they must be electrically insulated with each other and sufficiently robust. Dielectrical silicon oxide is refilled beneath the silicon layer to satisfy this requirement [6]. The space to be filled is obtained by isotropic etching. The movable structures are released through dry isotropic etch which can prevent the liquid stickness problem during releasing in the surface micromachining and SOI based process. Cost of the normal wafer is significantly lower than that of the SOI wafer. By employing deep RIE technique, high aspect ratio about 20 is achieved. The fiber grooves are obtained at the same time and make it easy to assemble the fibers. The driving voltage is less than 30 volts and switching speed is less than 0.1 ms.

The loss of the switch depends on the longitudinal, transversal and angular offset of the input and output fibers. These factors are fixed during design and assemble. However, when the switch is at cross-state, the micromirror will induce extra loss due to the surface roughness and profile of the micromirror.

By improving the deep RIE process, the verticality of the micromirror is 90°±0.1°. The surface roughness of the mirror is measured with atomic force microscope (AFM). The measurement result is shown in Fig. 3 and the root-mean-square roughness is 19 nm. Therefore, the insertion loss is decreased significantly.

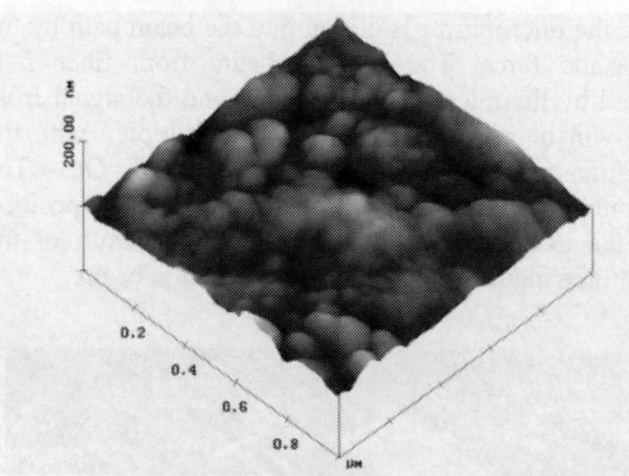

Fig. 3 surface roughness of the micro mirror

The center wavelength of an FBG is dependent on the period Λ and effective refractive index n_{eff} as expressed in formula:

$$\lambda = 2n_{eff}\Lambda \qquad (1)$$

Employing thermal or strain effect, the center wavelength can be changeable. Here, the thermal effect is used. By injecting various current along a resistant material, heat will be generated. Based on this, metal coating FBG employing modified IC technology is used. By modifying the metal deposition process, a very uniform layer is obtained on the FBG. To obtain sufficient heat efficiency, high resistivity metal, titanium is chosen. The thickness of the coating is 1000Å. The length of the tunable FBG is 2.0 cm and the resistance is 5.27 KΩ. By injecting different current to the metal layer, the tunable electrical power is converted to heat power. The induced temperature change will cause the variation in the period of the grating so that the center wavelength of the FBG is tuned. A tuning range as wide as 3.0 nm is obtained and the tuning speed is about 0.65 ms. These properties will be discussed later.

4. EXPERIMENT RESULTS AND DISCUSSION

To investigate the performance of the proposed tunable OADM, the experiment is carried out and the experimental set-up is shown in Fig. 4. The BBS is a broadband light source to generate a broad band light. OSA2 and OSA1 are two optical spectrum analyzers to inspect the dropped and final multiplexed output signals respectively. VS is the voltage source to control the state of the OADM. CS is a current source to tune the center wavelength of the metal-coated FBG. TLS is a tunable laser source to get the added signal.

Figure 5 shows the dropped signals when the current injected to the FBG is zero and 7.0 mA respectively. The center wavelength increases from 1550.92 nm of zero mA to 1551.80 nm of 7.0 mA. A shift as large as 0.88 nm is achieved. And 3 dB bandwidth is not widened. It is still 0.3 nm. There is a deformation at the right edge of the waveform, which is negligible. The crosstalk is less than -20 dB because of the low reflectivity of the FBG, which is only 99%. The crosstalk can be reduced further by improving the quality of the FBG. At the drop port, the selected channel is dropped dynamically. At the same time, the tunable laser source will provide a same wavelength signal and added to the system at the added port as shown in Fig. 1.

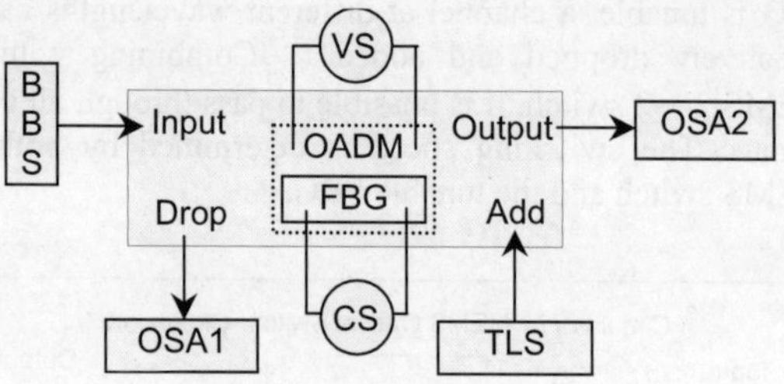

Fig. 4 Experimental set-up

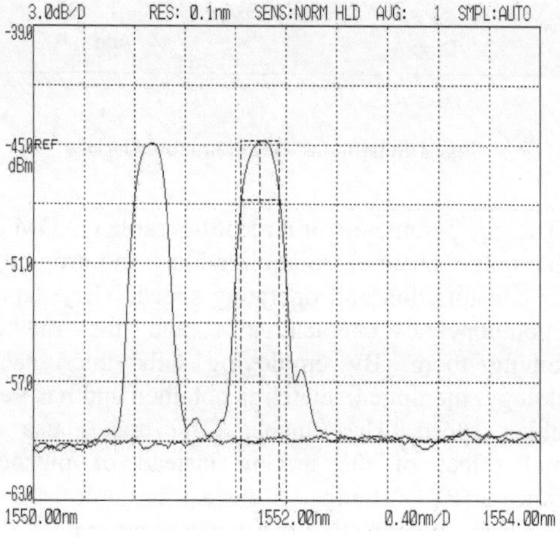

Fig. 5 dropped signals at two different currents

Further shifting could be achieved by increasing the current. Figure 6 shows the relation between the variation of the center wavelength and the square of the injected current, which the temperature change is proportional to. The theoretical curve is also drawn in the same figure. The dots represent the testing results. They match with each other very well. A tuning range as wide as 3.0 nm is realized. The highest current is 12 mA. The power consumption is very small.

When the current is injected to the metal-coated FBG, electrical power is generated, which in turn determines the temperature and the center wavelength of the FBG. Three kinds of thermal consumption occur. They are thermal conduction, convection and radiation. According to the thermal balance equations, it implies that, the heat generated is equal to the heat absorption. The tuning speed is calculated. It is 0.65 ms. The testing result is 0.6 ms. The experiment result agrees with the theoretical value quite well.

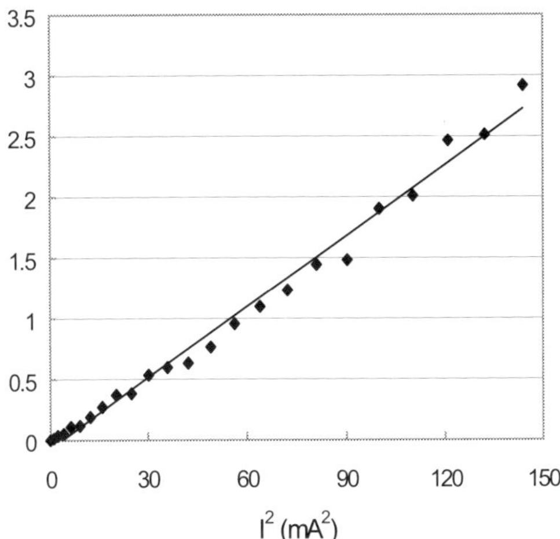

Fig. 6 Center wavelength shift vs. square current

In determining the center wavelength of the tunable FBG, the strain effect is negligible. The center wavelength shift is proportional to the temperature change of the FBG. The relation is expressed as

$$\Delta\lambda = k\Delta T \tag{2}$$

where k is a constant. According to the testing results as shown in Fig. 6, the constant is obtained as 0.01 *nm/K*, which matches with the experimental value.

The losses of the reconfigurable OADM depend on structure and the components in the system, which include the optical circulators, the fiber Bragg grating, and the MEMS optical switch. In this setup, the insertion loss between any two ports of the circulators is 1.0 dB. The isolation is 48 dB. The reflectivity of the FBG is 99%. The insertion loss of the switch is 0.8 dB and 2.0 dB at bar-state and cross-state respectively. Therefore, the insertion loss of the dropped signal is 6.04 dB including 1.0 dB loss from port a to port b and 1 dB for port b to port c of the first OC, 2.0 dB loss from the twice reflection of MEMS optical switch and 0.04 dB loss from the FBG. The insertion loss of the other passing through channels is also about 6.10 dB resulting from the two circulators and twice reflections of the micromirror. When the switch is at bar-state, no wavelength will be dropped. In this case, the loss is 2.08 dB, because the switch introduces 0.8 dB loss and the two OCs each cause 1.0 dB loss.

5. CONCLUSION

A reconfigurable OADM is fabricated, using a MEMS switch and a tunable FBG. By employing bulk micromachining process, the high-speed optical switch is obtained. Tunable FBG is realized by using modified metal deposition. The system has a crosstalk less than –20 dB. The dropping and passing loss are both 6.04 dB. The transmission loss is 2.08 dB. The tuning range is 3.0 nm, which corresponds to four adjacent channels of a DWDM system with 100GHz spacing. Power consumption is very low. It needs a current source which is as small as 13 mA and a voltage as low as 30 volts. Employing MEMS technique, the cost, and the size of the system are reduced substantially compared to the conventional mechanical/thermal optical switch that combines with fixed FBGs or PZT controlled FBGs.

ACKNOWLEDGEMENTS

The author would like to thank Dr Liu Yunqi and Dr. N. Balasubramanian for their useful discussion and express sincere gratitude to all help from MEMS group, the Alpha Cleanroom and Photonic Lab.

REFERENCES

1. Special Issue on Multiwavelength Optical Technology and Networks, IEEE J. Lightwave Technol. 14 1996.
2. C. Pu, L. Y. Lin, E. L. Goldstein, and R. W. Tkach, "Client-configurable eight-channel optical adddrop multiplexer using micromachining technology", Photonics Technology Lett., Vol. 12, No. 12, Dec. 2000, pp. 1665-1667.
3. W. D. Zhong, S. Dods, J. P. R. Lacey and R. S. Tucker, "Reconfigurable multichannel add-drop multiplexer with improved performance", Electronics Lett. Vol. 32, No. 16, Aug. 1996, pp. 1477-1478.
4. K. Okamoto, M. Okuno, A. Himeno, and Y. Ohmori, "16-channel optical add/drop multipleser consisting of arrayed-waveguide gratings and double-gate switches", Elect. Lett., Vol. 32, 1996, pp. 1471-1472.
5. S. K. Liaw, K. P. Ho, and S. Chi, "Multichannel add/drop and cross-connect using fiber Bragg gratings and optical switches", Electronics Lett., Vol. 34, No. 16, Aug. 1998, pp. 1601-1603.
6. A. locco, H. G. Limberger, and R. P. Salathe, "Bragg grating fast tunable filter", Elect. Lett., Vol. 33, no. 25, Dec. 1997, pp.2147-2148.
7. J. Li, Q. X. Zhang, N. Balasubramanian, J. M. Huang, and A. Q. Liu, "Integrated add/drop multiplexer (ADM) using micromachined mirror", Proceedings of SPIE, Asia-Pacific Optical Communications, Beijing China, 2001.
8. H. G. Limberger, N. H. Ky, D. M. Costantini, R. P. Salathe, C. A. P. Muller, and G. R. Fox, "Efficient active Bragg grating tunable filters," in proc. Bragg Gratings and Photosensitivity Conf., 1998, pp. 265-267.

Session 14G

Optical Transmission II

Recent Advanced Development of 40Gbps DWDM Ultra-long Reach Optical Transmission Systems

Le Nguyen Binh

Department of Electrical and Computer Systems Engineering, Monash University

Clayton Victoria 3168 Australia e-mail le.binh@eng.monash.edu.au

Abstract

Recent advances of densed wavelength multiplexed photonic transmission systems are investigated for ultra-long reach (ULR) transport networks, including (i) Optical modulation formats including return-to-zero (RZ), carrier suppressed RZ (CS-RZ), vestigial side-band RZ (VSB-RZ)[3] and their impacts on design and implementation of integrated optic transmitters and associated coding and modulation techniques for dispersion tolerance; thence extending the system span and optical node interconnecting distance (ii) dispersion management techniques chromatic and polarization mode dispersion and compensation techniques (iii) Optical amplification sub-systems including EDFA and Raman distributed and lumped amplifiers and gain equalization and (iv) photonic multiplexing and de-multiplexing sub-systems for the photonic transport networks.

Introduction

Photonic networking technology has progressed significantly over the last decadewith the transmission bit rate reaching at 40 Gb/s, 80 Gbps and 160 Gb/s[1,2]. Simultaneously several optical wavelength-carrier channels are multiplexed over the entire S-, C- and L-bands of the low attenuation 1550 nm window of silica fibers resulting to an aggregate information capacity of tens of Tb/s[1,3,6]. With the frequency spacing between optical carriers of 50 GHz and less the number of information channels, appropriate efficient modulation formats, could reach several hundreds over these bands. However the aggregation of several optical carriers lead to high average optical power and thus the they suffer nonlinear effects, particularly the self phase modulation (SPM), four wave mixing (FWM) and cross phase modulation (XPM). Optical amplifiers must also be used to compensate for the transmission losses by lumped and distributed modes. These would lead to modulation instability. Thus efficient modulation schemes recently attract much interest. Furthermore additional coding such as the Reed-Solomon code is inserted following the transmitters for improvement of the bit-error-rate (BER).

Dispersion effects, chromatic and polarisation mode and nonlinear phase distortion types are critical for optical transmission systems operating at 40Gb/s. Photonic systems operating under linear regime then the nonlinear phase dispersion effect mostly due to the excess optical power of the carriers is to be controlled. Further the dispersion slope, second order dispersion as a function of wavelength and higher dispersion also becomes significant. On the other hand if the photonic transport networks employed soliton transmission then the balance between linear dispersion and nonlinear phase dispersion effects are critical.

This paper presents advances on both regimes of operation of the photonic communications systems. Dispersion management schemes are designed for transmission of pulse trains over several thousand kms. This paper gives only recent advances in muti-Tb/s for photonic systems operating in the linear regime of the transmission medium.

The equalization of optical gain over the ultra-wideband spectrum covering all wavelength carriers are implemented by lumped Raman and or Erbium-doped fiber amplifiers and distributed Raman optical amplifying fiber medium. Practical systems of composite standard single mode (SMF) and non-zero dispersion shifted fibers (NZDSF) are used in fiber cables for transmission. 10Gbps communication systems are installed with provision of upgrading to 40Gbps. The polarisation and chromatic dispersion effects are investigated. A brief background of multiplexing and demultiplexing techniques for 40Gb/s DWDM transport networks are given.

Modulation formats

For multi-Tb/s the modulation serves three principal objectives: firstly to minimize the average aggregate optical power as perceived by the refractive index of the optical transmission medium and secondly to reduce the effective signal bandwidth for efficient spectral density and hence minimising chromatic dispersion effects. Such an optical transmitter in an ultra-long reach systems is shown in Fig. 1. The generation of different modulation formats in various way. However in this paper the modulation formats have been achieved by using a pair of Mach-Zehnder interferometric modulators[3,4] as shown in Fig. 2. The modulation characteristics of these modulation formats for 40Gb/s photonic communications systems are listed in Table 1[9].

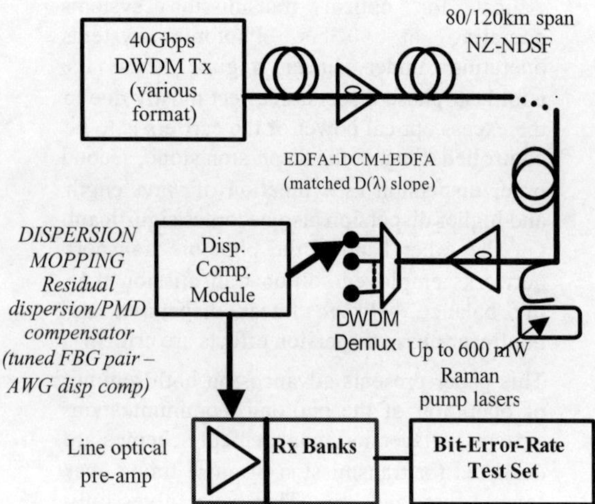

Fig. 1 Typical photonic communications systems for 40Gb/s ultra-long reach

Modulation format	Signal Bandwidth	Remarks
Return-to-Zero (RZ)	80 GHz + carrier	Double side band no carrier suppressed
Carrier Suppressed RZ[2,7,10]	40 GHz	Carrier suppressed – most efficient in reduction of dispersion
Single side band RZ[9]	40 GHz + carrier	No carrier suppressedElectrical phase matching of ultra-broadband signals for modulation.
Vestigial side band RZ[3]	40 GHz	Carrier suppressed – most efficient in dispersion impairment – additional optical filters required.
Chirped RZ[9]	80 GHz + chirped carrier	No carrier suppressed and chirp frequency carriers

Table 1: Modulation format and bandwidth and dispersion impairment/efficiency

Duo-external modulation devices[4] for these transmitters can be implemented by cascading interferometric optical modulators using $LiNbO_3$ or polymeric integrated optic structures [11,12]. The CS-RZ modulation format should be implemented by biasing the MZI for pulse generation at the minimum transmission point and the signals superimposed between the quadrature region of its applied biased voltage - optical intensity transfer characteristics. For the SSB modulation it is much more complicated due the requirement of phase matching of the broadband signals of the modulators with $\pi/2$ phase shifting of the two travelling wave electrodes.

An optical filter with sharp optical rejection band and inserted after the CS-RZ optical modulators would complete a vestigial SSB optical transmitters.

For pre-chirped NRZ or RZ format transmitter an optical phase modulator is need to generate optical phase modulation of the carrier and hence optical chirp effects. The pre-chirped can be proven to be effective in pre-compensation of chromatic dispersion of the transmission medium. The advantages and disadvantages of these transmitters in 40 Gb/s systems would be illustrated in the presentation.

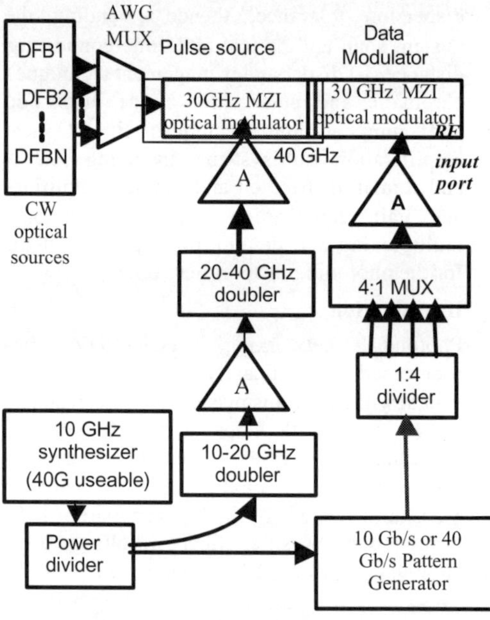

Figure 2 Optical transmitters using different modulation formats – example Return-to-zero carrier suppressed modulator

Dispersion and Gain compensation

Advanced optical fibres, e.g. NZDSF, dispersion compensated fibres with W- profile have been used in ULR DWDM transmission system and networks. Dispersion compensation modules (DCM) incorporated with in-line erbium doped fibre amplifiers (EDFA) are normally inserted between spans to compensate for loss and dispersion[7,10]. The entire band of the EDFA gain spectrum has been used up, i.e. the C- and L-bands. Furthermore the level of multiplexed optical channels must be optimised so that they would not suffer nonlinear distortion effects and accumulated noises from cascaded amplifiers. Thus Raman amplifiers as distributed amplifying media that are the transmission fibres. High-level GeO_2 doped fibres can also be used as lumped amplifiers with different wavelength pumping to flatten the gain coefficients and thus equalisation over the extended spectrum[7,12,14]. The Raman distributed amplifiers can also be used to reduced adjacent pulse interaction in soliton transmission systems [13].

Polarisation mode dispersion is critical in UBB ULH DWDM optical fibre communication systems. PMD compensators using a integrated optic polarisation rotator and a high birefringence fibre delay line with receiver signal filtering feedback to rotate the state of polarisation can be used to compensate for first order PMD and partially compensating higher order PMD[12-15]. These compensators are placed in the post- transmission receiver systems as indicated in Fig. 1. Furthermore compensation of residual post-transmission dispersion by fibre Bragg gratings (in association with optical circulators) must be conducted after the demultiplexing of optical channel. Polarisation dependent losses and dispersion of photonic components in photonic transport networks are also very critical. The control and minimisation of these effects must be carefully considered during the fabrication process.

Optical loss can be compensated using EDFA and/or Raman distributed or lumped optical amplifiers[14]. Figure 3 shows a typical optical system Q for single span and three spans with and without Raman distributed amplification as an additional compensation for 40Gb/s photonic transmission systems with four optical carriers modulated with RZ-CS format.

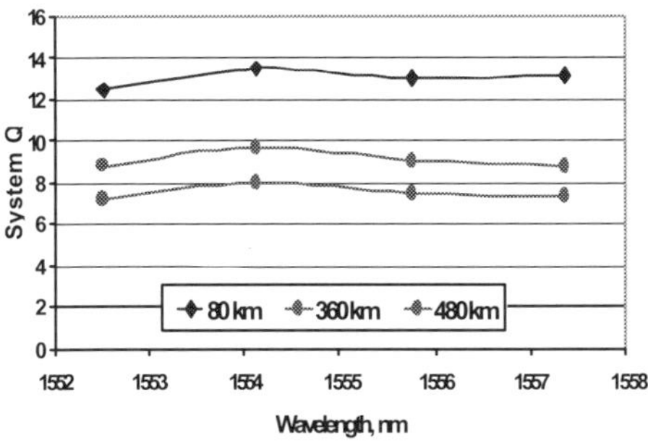

Figure 3 Transmission compensation of optical loss via distributed Raman optical amplification – single span 80 km and 360 km and 480 km with three and four span loss with and without distributed amplification.

Channel multiplexing and demux for 40 Gb/s DWDM transport networks

For ultra-broadband and ultra-wide photonic transport, networks multiplexing, demultiplexing and possibly full or limited wavelength conversion of DWDM optical channels are essential prior to receiving, routing and distributing to other photonic networks. Further all-optical regeneration (2R or 3R) of different modulation formats are critical in these networks. The use of nonlinear optical loop mirror can offer 2R.

With massive parallel transmission and interconnection of UBB and ULH photonic networks, network topologies are to be deployed with extremely high fault tolerance. Photonic ring networks can be structured with high number of photonic cross-connect systems or photonic mesh networks should be used. Wavelength re-use and conversion techniques are important for network flexibility in control and management of wavelength channels. Agile photonic networks should be considered rather than "dump" optic in point-to-point transmission systems currently employed. The 3R all-optical regenerators using photonic signal processing are critical for photonic networks as they would provide reconstructed signals and wavelength channels for further transmission after routing.

Concluding remarks

Multi-tera-b/s photonic communications and networking techniques can be very important for intercontinental and terrestrial information networks. 40Gb/s will play a vital role in these massive photonic transport layers. Advanced modulation techniques would provide simple and spectral effective transmission. Distributed and lumped Raman amplifiers incorporating EDF amplifiers provide flexibility in span design. ULR and ultra-broadband DWDM channels at 40Gb/s would form a super-backbone for all-optical networks operating at lower level, for example 10Gb/s over several thousand km networks that are currently developed for national and local Ethernet networks. The design, modelling and deployment of a future super-photonic transport network backbone remain a challenging problem.

REFERENCES

[1] See for example D. Cotter et al. "Ultra-high bit rate networking: from the transcontinental backbone to the desk top", IEEE Comm. Mag., vol. 35, Apr. 1997, pp. 90-95; L.N. Binh "An overview of 40 Gbps optical transmission and its impact on optical networking", Proc. Workshop on Optical

Networking: Technologies, Architectures and Management", Melbourne Australia, Nov. 2001.

[2] See for example Y. Miyamoto et al. "320 Gb/s (8x40 Gbps) WDM transmission over 360 km with 120 km repeater spacing using carrier-suppressed return-to-zero format", Electronics Lett., vol. 35, No. 23, Nov. 1999; W.S. Lee et al. "2.56 Tb/s (32 x 80 Gb/s) polarization-bit-interleaved transmission over 120 km NDSF with 0.8 b/s/Hz spectral efficiency", Elect. Lett., vol. 37, No.15, July 2001.

[3] M. Suzuki "Multi-terabit long haul WDM transmission with VSB-RZ format", IEEE LEOS Lett., Oct. 2001, pp. 26-28.

[4] See for example E.L. Wooten "A review of lithium niobate modulators for fiber-optic communications systems", IEEE J. Selected Topics Quant. Elect., vol. 6, No. 1, Jan/Feb 2000, pp. 69-82; L.N. Binh and T.W. Chua "Ultra-broadband integrated interferometric optical modulators", Workshop on Appl. Radio Sc., Australia, Feb. 2002; L.N. Binh and P. Keen "Fiber Mode-Locked Lasers at 10G- and 40G- Repetition Rate", Technical Report, Nortel Networks, Advanced Tech. Centre, Harlow U.K., June 2000.

[5] See for example K. Yonenaga and S. Kuwano "Dispersion-tolerant optical transmission system using duo-binary transmitter and binary receiver", IEEE J. Lightw. Tech., vol. 15, August 1997, pp. 1530-1537; E. Forestieri and G. Prati "Novel optical line codes tolerant to fiber chromatic dispersion", IEEE J. Lightw. Tech., vol. 19, No. 11, Nov. 2001, 1675-1684.

[6] L. du Mouza et al., 1.28 Tb/s (32x40 Gb/s) WDM transmission over 2400 km of TeraLight/Reverse TeraLight fibers using distributed Raman amplification", Elect. Lett. Vol. 37, No. 21, October 2001.

[7] W.S. Lee et al., "4 x 40 Gb/s RZ EDTM transmission over 520 km of NDSF at 120 and 160 km span lengths with Raman pre-amplification", vol. 36, No. 8, April 2000.

[8] A. Hodzic et al., "Alternative modulation formats in Nx40Gb/s WDM standard fiber RZ transmission systems", IEEE J. Lightw. Tech., to be published, April 2002. Note: This reference is listed after the acceptance of this paper.

[9] L.N. Binh and T.W. Chua, "Ultra-broadband integrated interferometric optical modulators" Proc. Worshop on Applications of Radio Science, Feb., Australia 2002.

[10] M-C Oh et al., "Recent advances in electro-optic polymer modulators incorporating highly nonlinear chromophore", IEEE J. Sel. Topics Quant. Elect., Sept. 2001, pp. 826-835.

[13] H.N. Ereijei et al. "40 Gb/s long-haul transmission in dispersion-managed soliton system using Raman amplification", Elect. Lett., vol. 37, No.15, Dec. 2001.

[14] D. Dahan and G. Eisentein "Numerical comparison between distributed and discrete amplification in a point-to-point 40 WDM-based transmission with three different modulation formats", IEEE J. Lightw. Tech., April 2002.

[15] F. Heisemann "Integrated optic polarisation transformer for reset-free endless polarisation control" IEEE J. Quant. Elect., vol. 25, 1989, pp. 1898-1906.

[16] L.N. Binh 'Signalling techniques and systems for equalisation of chromatic and polarisation mode dispersion", Technical report, Nortel Networks – Advanced Technology Centre, Harlow, England, October 1999.

FRAMER AND DEFRAMER FOR OC – 192

R. Nakkeeran, C. T. Gururaj, G. Eswar, and S.Swaminathan
Department of Electronics and Communication Engineering
Pondicherry Engineering College
Pondicherry – 605 014.
India
rnakkeeran@hotmail.com

ABSTRACT

Synchronous Optical NETwork, or SONET as it is popularly called, is one of the most advanced telecommunications technology available on the eve of the 21st century. The SONET standard provides a layer of compatibility, which allows for the exchange of bit streams with one common format. SONET equipment can accept country specific bit streams, convert them to the SONET standard, and then pass them across the SONET network with other bit streams. The paper centers on the design of the framer and deframer for OC –192 frames. They operate at data rate of 9.953 Gbps. The core parameters are the A1 and the A2 bytes, which are added during the framing process and then removed during the deframing process. . Since the channel is noise prone, errors can peek in at any place. This added error should not delay or spoil the deframing process. Furthermore, since a 64-bit bus is considered, the framing bits may start from any bit position. So a robust design that should take care of the bit errors and skews is needed. . The architecture put forth is analyzed using various test conditions and is eventually coded using Verilog – HDL.

KEY WORDS:

Framer, Deframer ,OC-192,architecture,Verilog.

1. INTRODUCTION

The optical point-to-point link comprises of two important components, viz. the transmitter and the receiver. The transmitter includes the framer and the other modules needed for the management process. Similarly the receiver has the deframing module and other modules, which decide on the management functions. The current project deals with only the framer module in the transmitter and the deframer module in the receiver. The job of the framer is to dispatch the 192 A1, 192 A2 that would indicate the start of the frame for proper synchronization. The main criterion for the design of the deframer architecture is that it should be robust enough not only to determine the exact sequence even in the presence of errors but also be in a position to determine the start of the frame even if the framing bits exceed the standard sequence i.e. the module should be able to work efficiently if a condition of framing bytes occurring more than 192 is met. The module discussion is done further in the following pages.

2. FRAMER MODULE

2.1 Module Description

By SONET standards, each SONET frame begins with the A1 and A2 bytes defined in the Section overhead portion of the Transport overhead beyond which other bytes exist (including the remaining overhead bytes). A1 has a hexadecimal value of F6 and A2 has a hexadecimal value of 28, again as per SONET standards. In OC – 192, there are 192 A1 bytes followed by 192 A2 bytes that occur in a contiguous fashion.

The framer module generates this continuous sequence of 192 A1s and 192 A2s that are accommodated in their respective byte positions in the OC – 192 frame. The frame is then transmitted once it is complete in all aspects (i.e. completion of all the other overhead information bytes). The module can be fully described in terms of its inputs and output as shown in the table.

For OC – 192, the clock rate required is 9.953 GHz, which is not presently available in the industry in case one bit transfer is assumed. A 64-bit transfer coupled with a clock of 155.53 MHz is used to attain the required bit rate of 9.953 Gbps. As per the above-mentioned method, for each clock, 64 bits of information are transferred. The bit rate eventually becomes 64 X 155.53 Mbps = 9.953 Gbps. A clock of 155.53 MHz is commonly available.

2.2 Inputs and Outputs Description

Table 1 elucidates the various input and output signals that make up the framer module. There are two types of signals that interface the module with other modules; they are the input and the output signals. The clock signal described is a global signal as it is common to all modules.

Table 1. Inputs and Outputs of the framer module

Signal Name	Type (I/O)	Description
Reset	I	Active high asynchronous signal used to reset the framer
Clock	I	Input clock at 155.53 MHz used for synchronization.
Frame _Start _n	I	Active low signal to indicate the starting of the framing process to the framer.
Data out	O	64 bit data (to carry the sequence).
Frame pulse	O	Signal to indicate the completion of the sequence transfer.

This presents an in depth representation of the framer module. It explicitly defines the framer's function and gives its operation under various circumstances as per the frame format defined in the SONET standards. A state diagram eases out the difficulties in synthesizing a Verilog – HDL code for the module.

The framer is a module in the transmitter of an OC – 192 transceiver. A module that recognizes this sequence is known as the deframer. The defamer occupies a promontory position in the receiver of the OC – 192 transceiver. The analysis of this module is presented in consequent paragraphs.

3. DEFRAMER MODULE

3.1 Module Description

The framing bytes, the A1 and A2 bytes that are transmitted as a part of the frame during the transmission process must be recognized during reception. This facilitates the identification of the frame's beginning. As per SONET standards, for OC – 192, the frame's beginning is marked by a continuous sequence of 192 A1 and 192 A2 bytes. Since a 64-bit bus is considered, the frame sequence can start from any bit position. The module must be in a position to recognize the starting position and must then perform the necessary sequence of operations to recognize the frame start.

The module converts the incoming unaligned data (one in which the start of the frame is unknown) into an aligned data by recognizing the frame beginning (from the A1 and A2 bytes). This aligned data is then transferred to subsequent modules through the 64-bit output data bus.

As per the standards, a time period of 250microseconds is given to achieve synchronization (i.e. to achieve alignment). Once again the module can be entirely characterized by the input and output description.

3.2 Inputs and Outputs Description

Table 2. Inputs and outputs of the deframer module

Signal Name	Type (I/O)	Description
Reset_n	I	Active low asynchronous signal used to reset the deframer
Data_in	I	Unaligned 64 bit data
Clock	I	Input clock at 155.53 MHz used for synchronization.
Synch_ out	O	Active high signal to indicate whether the module is in a synchronous state or not.
Data_out	O	Aligned 64 bit data.
Frame_ pulse	O	Signal to indicate the start of the aligned frame. Occurs every 125us

3.3 Internal Description:

The components that make up the module are:

- A 128-bit register is used to store the incoming unaligned information, 64 bits per clock. In the next clock the top 64 bits are pushed down the 128 bit buffer to accommodate the incoming 64 bits.
- 64 comparators each capable of comparing 64 bits. The comparators are arranged in the following manner,
- Comparator0: 0 to 63 bits,
  Comparator1: 1 to 64 bits, and so on until
  Comparator63: 63 to 126 bits

4. ARCHITECTURE

The architecture, which was designed from the state diagram, is the actual implementation of the module using the known digital devices. The architecture is helpful in coding the device using the standard HDLs. Figure 1 gives the architecture of the deframer module.

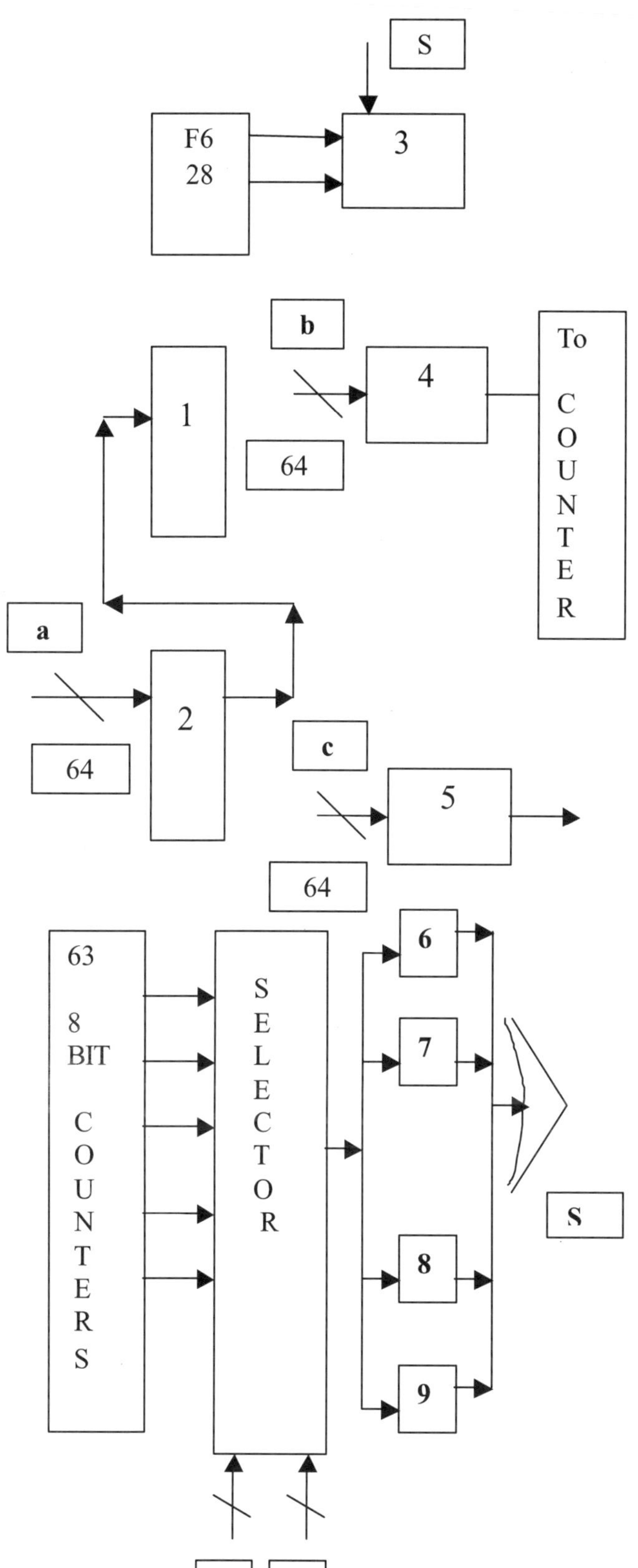

Figure 1. **The Architecture of the Deframer Module.**

LEGEND:

a – data in 64 bit bus, **b-** O to 63 BITS Q[0] to Q[63], **c** – 63 to 127 BITS Q[63]-Q[127]

To counters– To 63, 8 BIT Counters, **c0** – COUNT 0,**c1**- COUNT 1, **c2**- COUNT 2, **c62** – COUNT 62, **c63** – COUNT 63

1 - 64 bit buffer Q [0] – Q [63]
2 - 64 bit buffer Q [63] – Q [127]
3 - Multiplexer
4 - 64-bit comparator (0).
5 - 64 bit comparator (63)
6 - 64 bit comparators for 28_8
7 - 64 bit comparators for F6 28_7
8 - 64 bit comparator for $F6_6 28_2$
9 - 64 bit comparator for $F6_7 28$

5. SIMULATION AND RESULTS

There are close to twenty timing diagrams that are obtained by coding the above architecture in Verilog – HDL along with its various test conditions. Only the ideal cases are shown.

5.1 Timing Diagram For The Framer Module

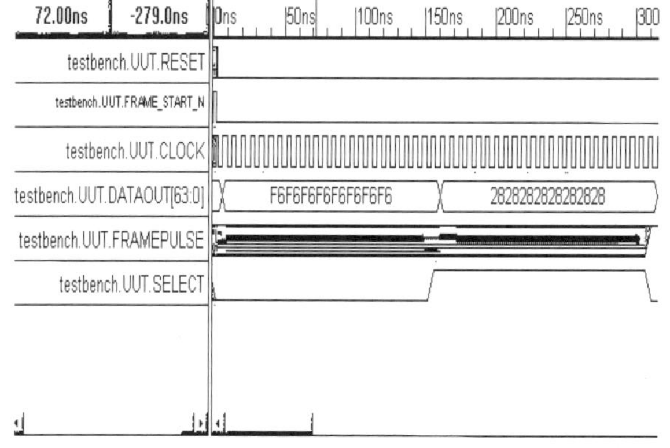

Figure 2. Timing Diagram for Framer Module

5.2 Timing Diagram For Deframer Module

Perfect Case

The timing diagram is for the situation when the framing sequence begins from the 0th bit position.

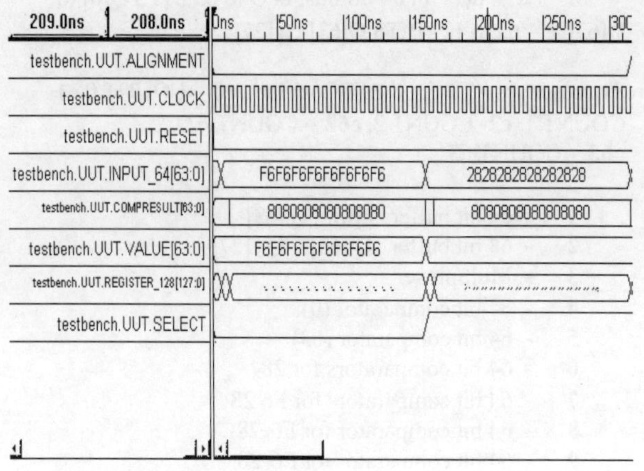

Figure 3. Timing Diagram for Deframer – Perfect Case

6. CONCLUSION

The deframer module forms the most important part of the OC-192 transceiver chip as it is the most difficult to be implemented at such a high bit rate of 9.954 Gbps. The deframer module is considered in detail and the architecture in detail. The complexity of the deframer increases all the more when the operation of the module is considered under error prone environments. The results shown are only sample results of the perfect conditions. The results for the error conditions are not discussed in the present paper.

REFERENCES

Books:

[1] Ming chwan chow," *understanding SONET /SDH – standards and application",* andan, New Jersey, 1996.

[2] Stamatios V.kartalopoulos, *"understanding SONET/SDH and ATM –communication Networks for the next millennium"*, prentice-hall of India, 2000

[3] Andrew S. Tanenbaum," *Computer Networks*", Prentice Hall of India 1996

[4] Eliezer Sternhein, rajvir Singh, Yatin Trivedi,"*Hardware Modeling with Verilog- HDL* ", Automata Publishing Compan.CA1990

[5] Michael John Sabstein Smith," *Application- Specific Integrated Circuits*", Person Education Asia, 2001

[6] Daniel Hyde,"*CSCI 320 Computer Architecture Handbook on Verilog HDL* ", PA, 1995

[7] SONET clocks Specification, ANSI TI.105.09

[8] SONET payload mapping, ANSI TI.105.02

[9] SONET rates and formats ANSI TI.105

WEBSITES:

[10] http://www.cypress.com cypress Semiconductors Private Limited Website

[11] http://www.tektonix.com Tektronix website

[12] http://www.iec.org International Engineering Consortium website

Dispersion Managed Soliton: Perspective & Challenge

Jian CHEN, and Chao LU
Optical Network Division, Institute for Communications Research
18 Nanyang Drive, Unit 230, Innovation Centre Blk 2, Singapore 637723
Tel: +65-6316-0244 Fax: +65-6795-9842 Email: chenjian@icr.a-star.edu.sg

Abstract: *We review the progress and problems in the development of soliton transmission systems. Dispersion managed soliton could be a future candidate as the information carrier. Future research prospect is presented.*

Key Words: soliton; dispersion management; transmission

The new millennium begins with an explosive increase of Internet traffic, which makes the traditional voice traffic no longer dominant as time goes on. The expanding demand for information to stimulate the new global economy and to facilitate everyday life is the main motivation. A most important and ultimate bandwidth demand originating from services such as video on demand and live online games will multiply Internet traffic exponentially. The superior technique for meeting the demands is fiber optic communication systems due to its enormous available bandwidth of more than 30 THz. The worldwide fiber telecommunication system is already in place acting as "information highway", whose bandwidth has just been used only in a little portion of its available spectrum. Thus, the ever-increasing demand for bandwidth has continuously forced telecom operators to upgrade their systems and install new hardware, which results in remarkable progress in the capacity of optical communication systems both in the total bit-rate and in the distance of propagation.

Technically, there are two ways of upgrading the existing optical communication systems to ultrahigh capacity: one is to implement conventional dense wavelength division multiplexing (DWDM) technique, and the other is to increase individual channel bit-rate. Moreover, the information should be sent over a far-reach span with less optical amplifiers or longer amplification spacing. The signal power is in turn required to be higher to keep the necessary signal-to-noise ratio (SNR), but the nonlinear behavior of optical fiber also becomes the main hurdle to the high capacity fibre transmission system without considerable pulse distortion or timing uncertainty.

The primary motivation for solitons [1] was to bring in fiber nonlinearity to compensate for the fiber (anomalous) dispersion, which used to be the main obstacle for high speed over long distance transmission between the era of all-optical amplifier and of dispersion management. By choosing an appropriate pulse power and shape, the effects of group-velocity dispersion (GVD) and self-phase modulation (SPM) can be combated, which results in a stable and robust optical wave-packet entitled as optical soliton transmitted in optical fibres. As higher power is used or needed in soliton systems, the SNR can be sustained over longer transmission distances, compared to linear systems. Moreover, in an ideal lossless optical fibre with a constant dispersion, two solitons that start from well-separated position suffer no reminiscent effect when they collide other than a slight position and phase shifts. This fact is a very important merit of the use of solitons as information carrier in fibers in addition to the fact that the pulse itself is nonlinearly stable and stationary.

For almost 30 years of research, soliton concept has been enthusiastically received in mathematical, physical and telecom research circles. However, it seems that the practical application could not be quickly realized. And even ironically, the records of "superior" soliton transmission experiments have never overtaken the traditional counterparts based on linear transmission concept. There are many problems that limit the practical use of solitons in optical communication: First, since ideal lossless fibre does not exist, ideal soliton also does not exist. The soliton-like pulse that may be used as information carrier is a mixture of soliton and linear dispersive wave; Second, the dispersion amount which could be balanced by SPM is very low. Thus, dispersion-shifted fibre (DSF) is preferred for soliton transmission if dispersion alternating is not considered, but it's difficult for DSF to be used for DWDM systems and for telecom operators to change the embedded optical fiber cable consisting of single-mode fiber (SMF); Third, the transmission system parameters (e.g., optical power and the amplifier spacing, etc.) must be suited to the dispersion, which is not always convenient in installed links. In most case the amplifier spacing should be very short, especially for high bitrate system; Fourth, timing jitter caused by various effects (e.g., amplification noise, neighboring-pulses interaction, inter-channel collision, etc.) can lead to the loss of information even though the wave shape is intact throughout the transmission. This is regarded as Gordon-Haus limit; Fifth, in order to avoid the collision between two adjacent solitons, they should be separated enough (larger than six times pulse width in time). This requirement is a severe drawback for soliton-based communications because it requires bandwidth at least three times wider than linear systems where the pulse separation can be twice or less than the pulse width; Sixth, in DWDM systems, it's difficult to realize average dispersion sufficiently low over all the range of multiplexing, and to use an equal optical amplification with either EDFA (erbium doped fiber amplifier) or Raman amplification for all this range; etc.

As new kinds of dispersion-compensating fibers (DCF) were produced to cancel the dispersion in SMFs and people got the concept of dispersion management (DM), soliton systems were given less attention. In DM systems, it is possible to have both high local and low average dispersion. The high local dispersion disrupts the phase-matching of the different optical frequencies making up a signal, thus reducing the interactions between them, which tends to reduce an effect known as four-wave mixing (FWM) which tends to distort signals and produce intersymbol interference [2]. The low average dispersion reduces its net cumulative effects over long spans of optical fiber. Moreover, the combination of high local and low average dispersion reduces cross-phase modulation (XPM) between channels [3]. The DM system experiments and tests have achieved a great success, and many DM systems were deployed into practical use. But it does not mean the end of soliton as a potential candidate of information carrier in the future. On the contrary, it has been found in recent years that solitons can also be used in DM systems and the term dispersion-managed (DM) solitons has become widespread [4]. It can also estimated that traditional DM systems will possibly migrate to soliton-based DM systems, although some people don't agree with this argument because they have got used to explain what happened in the system conservatively in linear concept. The reasons for this migration are: First, traditional DM systems do not have an effective way to reduce SPM effect, because they only use a strictly chosen initial chirp and dispersion distribution map to balance SPM and residual GVD effect. At present, since the cumulative fiber dispersion can be made zero, the argument to use solitons is reversed. That is to introduce fiber dispersion to compensate for the nonlinearity, because soliton is useful and is the only stationary and stable optical pulse shape in the presence of fiber nonlinearity. So now solitons are differentiated as conventional soliton and DM soliton; Second, both linear DM systems and DM soliton systems need an elaborate choice of proper system parameters; Third, the timing jitter limitation is inherently removed as the average dispersion tends to zero. Moreover, even in conventional soliton systems it has been overcome by sliding frequency filter and synchronous modulation control; Fourth, it has recently become increasingly clear that, for a transoceanic transmission of 20 Gb/s per channel or beyond in optical fibers, return-to-zero (RZ) pulse is the only choice. Coherent or Non-return-to-zero (NRZ) format of transmission has been discarded because of its intolerance to nonlinear effects of fibers and a prerequisite for high-speed optical time-division multiplexing (OTDM) systems. The spectral efficiency of linear transmission systems (used to use NRZ) versus soliton transmission systems (using RZ) will no longer be different. Fifth, Solitons are more robust to poralization mode dispersion (PMD) than linear pulses, and that DM solitons can be even more robust than conventional solitons [5]; etc. We thus conclude, as soon as these specific issues are resolved, a new generation of transmission based on DM soliton waves could become practical.

The future intensive research will surely base on a close view on the worldwide race regarding the most appropriate transmission scheme between linear transmission groups and soliton transmission groups. The linear groups has focused on eliminating nonlinear effects by carefully designing the dispersion map and selecting the initial chirp, and improve system performance by forward error correction (FEC). Meanwhile, the soltion groups are facing a problem to optimize their systems by choosing a trade-off power enhancement parameter to retain the soliton robustness, to reduce soliton interaction and inter-channel collision effects, and to keep its PMD-robustness not deteriorated with the increase of bit-rate simultaneously. Moreover, they are not sure what's the big difference between their systems and the linear ones. The main problem for both of them is, there exists no general mathematical solution for state-of-the-art system models. Unless there will be a breakthrough in the area of solving nonlinear differential equations with fast varying coefficients, they don't know what really happens in the DM transmission system and the general optimization is almost impossible. Whatever the outcome of the above-mentioned race is, intensive research on DM soliton will surely be beneficial to future optical communication systems.

REFERENCES

[1] A. Hasegawa, et al, Solitons in Optical Communications, Oxford, 1995

[2] A. Liang, H. Toda, and A. Hasegawa, Dense periodic fibers with ultralow four-wave mixing over a broad wavelength range, Optics Letters, vol. 24, page 1094-1096, 1999

[3] Hui Rongqing, K.R. Demarest, and C.T. Allen, Cross-phase modulation in multispan WDM optical fiber systems, Journal of Lightwave Technology, vol. 17, page 1018 -1026, 1999

[4] J. Martensson, and A. Berntson, Dispersion-managed solitons for 160 Gbit/s transmission, OFC technical digest, paper MF7, Mar. 2001, California

[5] H. Sunnerud, Li Jie, P.A. Andrekson, and Xie Chongjin, Experimental quantification of soliton robustness to polarization-mode dispersion in dispersion-managed systems, IEEE Photonics Technology Letters, vol. 13, page 118 –120, 2001

Improved Analysis of Stimulated Raman Crosstalk in Wideband Optical WDM System

Md. Golam Mostafa*, Md. Moshiur Rahman, G. M. Haider Ali, Md. Jahangir Alam,
Md. Ataur Rahman Sarker, Mohammad Zulfiker Alam and Mohammed Nazrul Islam**
Department of Electrical and Electronic Engineering, Bangladesh University of Engineering and Technology, Dhaka
Tel: 880-2-8611594, Fax: 880-2-8618377, E-mail: *mgmostafa@ieee.org, **nazrul@eee.buet.ac.bd
Bangladesh

Abstract: An improved model is proposed to approximate the actual Raman gain profile of silica, which in turn is used to calculate the power depletion of the shortest-wavelength channel due to all other channels in the system. In this improved model, the stimulated Raman scattering (SRS) effects from channels at the tail of Raman gain profile are taken into account. A statistical analysis is done for the SRS crosstalk in a wavelength division multiplexed system including modulation statistics. The bit error rate and power penalty are estimated, where the improved model for the Raman gain spectrum is used.

Key Words: Bit error rate, power depletion, power penalty, Raman gain, stimulated Raman scattering, wavelength division multiplexing.

1. INTRODUCTION

Optical Wavelength Division Multiplexed (WDM) system effectively utilizes the enormous bandwidth of an optical fiber and hence meets the demands for large transmission capacity of multimedia communication [1]. However, fiber nonlinear effects limit the performance of a high power optical WDM system. Among the nonlinear effects, the Stimulated Raman Scattering (SRS) is the ultimate power-limiting phenomenon [2]. The SRS causes power transfer from shorter-wavelength WDM channels to longer-wavelength WDM channels and thus induces interchannel modulation between each WDM channel [3]. The optical power limits are calculated by the worst-case assumption that all WDM channels are synchronized and transmitted at ONE level simultaneously [4]. In all the calculations of Raman crosstalk, the experimentally found Raman gain profile of silica reported in [5] is approximated by a triangular function, where the peak Raman gain occurs at 15 THz spacing between channels and no Raman gain is assumed at larger spacing [6]. The results so far reported in literature based on this approximation are sufficiently accurate as long as the bandwidth of the fiber amplifier available currently is concerned [1]. However, for future wideband WDM systems, this approximate model of Raman gain requires to be modified. The probability density function (PDF) of SRS induced crosstalk has been developed in Ref. [7]. But the complete performance evaluation based on the statistical analysis of the Raman crosstalk is yet to be reported.

The objective of this paper is to develop an improved model for the Raman gain profile, and, accordingly, to calculate more accurately the power depletion of the shortest-wavelength channel due to the SRS effects from all possible channels in a wide-band WDM system. A statistical analysis is presented to estimate the bit error rate (BER) of a WDM system based on the improved model for the Raman gain profile. Finally the sensitivity penalty suffered by the system due to SRS crosstalk is evaluated.

2. ANALYSIS OF RAMAN CROSSTALK

In Ref. [6] the actual Raman gain profile of silica is approximated by a triangular function, i.e. the Raman gain, γ, increases linearly with channel spacing and the peak gain coefficient, γ_p, occurs at 15 THz spacing between channels and that there is no Raman interaction beyond this point. However, the actual Raman gain profile of silica shows that Raman gain coefficient is not insignificant beyond 15 THz spacing between channels. Therefore, we introduce an improved model, where the actual Raman gain profile is divided into following three regions of channel spacing.

In Region I (channel spacing $\Delta\nu$ up to 15 THz), the Raman gain profile of silica is assumed to be a linearly increasing function ranging from zero to peak Raman gain coefficient γ_p, which depends on the wavelength.

$$\gamma_i = \frac{(i\Delta\nu)\gamma_p}{1.5\times10^{13}} \quad , \qquad i\Delta\nu < 1.5\,\text{THz} \tag{1}$$

In Region II (channel spacing between 15 THz and 16.5 THz), the Raman gain profile of silica is approximated to be a linearly decreasing function.

$$\gamma_i = \{(-6.087\times10^{-13}\times i\Delta\nu) + 10.06382\}\gamma_p, \; 1.5\,\text{THz} \le i\Delta\nu \le 1.65\,\text{THz} \tag{2}$$

In Region III (channel spacing beyond 16.5 THz), the Raman gain profile of silica is approximated to be an exponentially decreasing function.

$$\gamma_i = 0.22675\gamma_p \exp\left\{\frac{-(i\Delta\nu - 1.65\times10^{13})}{8.64655\times10^{12}}\right\} \quad (3)$$

With this model, we can generalize the analysis of power depletion avoiding any assumption made earlier. In this analysis, we first determine the number of channels that fall into the three regions of Raman gain profile considering the channels to be equally spaced. Power depletions for all three regions D_1, D_2 and D_3, defined as the fractional power lost by the shortest wavelength channel, are then determined as follows:

$$D_1 = \frac{PL_e\gamma_p}{3\times10^{13}A_e}\sum_{i=1}^{N_1-1} i\Delta\nu \quad (4)$$

$$D_2 = \frac{PL_e\gamma_p}{2A_e}\sum_{i=N_1}^{N_2-1}(-6.087\times10^{-13}i\Delta\nu + 10.06382) \quad (5)$$

$$D_3 = \frac{0.22675PL_e\gamma_p}{2A_e}\times \exp\left(\frac{1.65\times10^{13}}{8.64655\times10^{12}}\right)\times \sum_{i=N_2}^{N_i-1}\exp\left(\frac{-i\Delta\nu}{8.64655\times10^{12}}\right) \quad (6)$$

where, P is the input power which is considered to be the same for each channel, L_e is the effective length of the fiber taking the fiber loss into account and A_e is the effective core area of the fiber. Finally the overall depletion is obtained by adding the depletions of all three regions.

$$D = D_1 + D_2 + D_3 \quad (7)$$

The total power depletion can also be expressed in dB incorporating the modulation statistics as

$$D[dB] = K\left[\sum_{i=2}^{N_1}\frac{(i-1)m_i}{1.5\times10^{13}} + \sum_{i=N_1+1}^{N_2}\left(\frac{8.8}{\Delta\nu} - 5.199\times10^{-13}i\right)m_i + \sum_{i=N_2+1}^{N_i} a\exp\left(\frac{x_0 - i\Delta\nu}{b}\right)m_i\right] \quad (8)$$

where $K = \frac{PL_e\gamma_p\Delta\nu}{2A_e}10\log_{10}e$, m_i is the modulation in the ith channel and can assume the value ZERO and ONE with the same probability 1/2.

The depletion D is a random variable due to the randomness of m_i's. The central limit theorem [8] shows that D converges to a Gaussian random variable with mean η given as the sum of the means η_i of the contributions from each channel.

$$\eta = \sum_{i=1}^{N-1}\eta_i \quad (9)$$

Similarly, the variance σ^2 is given by the sum of variances σ_i^2's of the contributions of each channel

$$\sigma^2 = \sum_{i=1}^{N-1}\sigma_i^2 \quad (10)$$

The Probability Density Function (PDF) of D for Gaussian distribution is given by [7]

$$\text{PDF of } D = \frac{1}{\sigma\sqrt{2\pi}}\exp\left[-\frac{(D-\eta)^2}{2\sigma^2}\right] \quad (11)$$

Now the probability of error, or Bit Error Rate (BER), can be found by conditioning on the value of D [9] and minimizing the total probability of error to find the best decision threshold. Thus the BER is given by [10]

$$P_e = \min_{I_{th}}\left(\frac{1}{2}P_{e|0}(I_{th}) + \frac{1}{2}\sum_D P_{e|1,D}(I_{th})P_D\right) \quad (12)$$

where I_{th} is the threshold level of the decision circuit at the receiver, $P_{e|0}$ is the probability of error when a ZERO has been transmitted, $P_{e|1,D}$ is the probability of error when a ONE has been transmitted and the bits transmitted on the other channels are such that the channel under consideration is depleted by D and P_D is the probability of having a depletion D. Finally the penalty for a given number of channels can be calculated from the BER characteristics.

3. DISCUSSIONS ON RESULTS

Figure 1 shows the experimentally found Raman gain profile in Ref. [5] by the dotted line. The triangular approximation of [6] is shown by the dashed line, which signifies that only $M = 1.5 \times 10^{13} / \Delta\nu$ number of channels contribute to the depletion of the shortest-wavelength channel. Our proposed model is shown by the solid line, which cares for the Raman gain beyond 15 THz spacing between channels. Now as the whole range of channel spacing is included in the model for the Raman gain, so there is no need to make any assumption in calculating the power depletion. Rather the power depletion due to channels from the three regions of the Raman gain can be determined individually and then the total power depletion can be found by adding these three contributions. Consequently, the analytical results are expected to be more accurate and generalized.

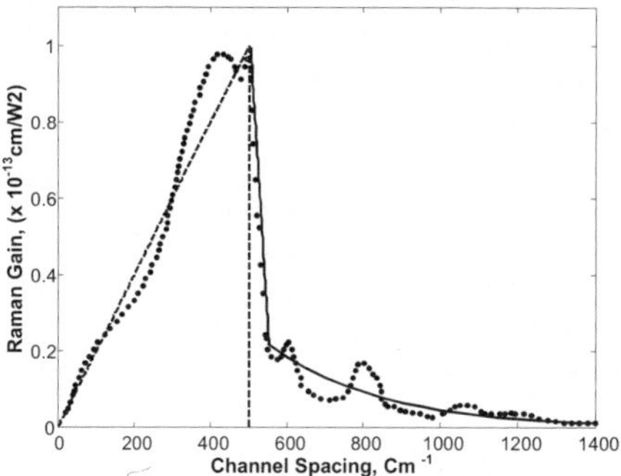

Fig. 1: Experimentally found Raman gain profile (dotted curve), approximation used in [6] (dashed curve) and approximation used in this work (solid curve).

Figure 2 shows the estimated power depletion as a function of the number of channels. The system parameters are chosen to be as follows: the fiber loss coefficient, $\alpha = 0.25$ dB/km, $A_e = 5x10^{-7}$ cm^2, $\gamma_p = 6x10^{-12}$ cm/W^2, launched power = 1 mW, channel separation, $\Delta\lambda = 10$ nm (i.e. $\Delta\nu = 1.3x10^{12}$ Hz at 1.5 μm). The dashed line shows the depletion found in [6] and the solid line shows the depletion found in our analysis. It can be observed that the power depletion increases with the number of channels within the linear regime of the Raman gain profile. The value of depletion then saturates. However, a significant increase in depletion is contributed by the regime beyond 15 THz spacing between channels due to using the improved model of the Raman gain profile. The correction in the power depletion due to the region 2 and region 3 of the Raman gain profile are also shown in this figure.

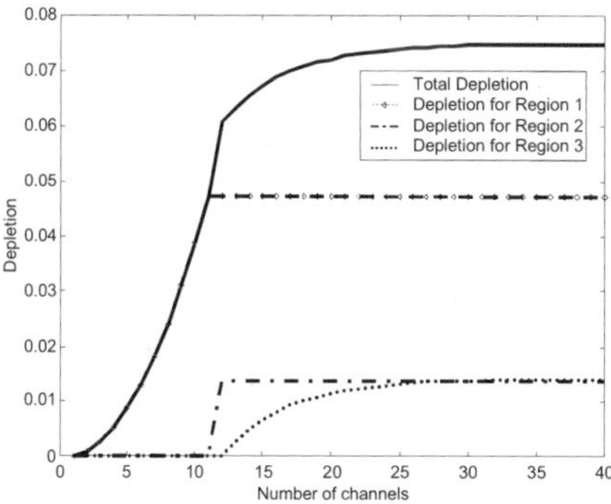

Fig. 2: Overall and individual depletion profile as a function of number of channels for typical system parameters, α= 0.25 dB/km, A_e=5 x 10^{-7} cm^2, $\gamma_p \sim 6 \times 10^{-12}$ cm/ W ^{2}, launched power=1 mW, channel separation, $\Delta\lambda \sim 10$ nm ($\Delta\nu$ ~1.3 x 10^{12} Hz at 1.5 μm).

Figure 3 shows the PDF of D for 50 channels. System parameters used for calculating the power depletion are kept unaltered. The PDF of maximum depletion is then plotted with Gaussian approximation. The mean is obtained near 0.15 dB, which is close to the half value of maximum depletion.

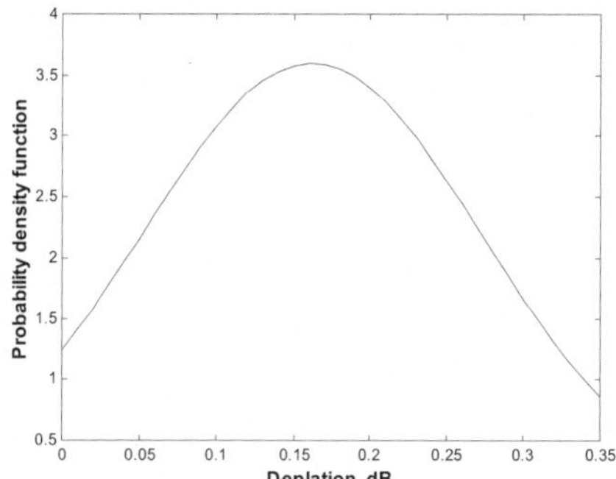

Fig. 3: Gaussian pdf of power depletion for 50 channels.

The BER performance have been evaluated and plotted against the launched power in the channel of interest, holding other channel's launched power constant to keep the fractional crosstalk constant. Figure 4 shows BER curves computed for 10, 15 and 50 channels. The dashed line is the BER curve for the system without SRS crosstalk and the solid lines are with SRS crosstalk. However, with the increase of number of channels the curves are found to shift leftwards.

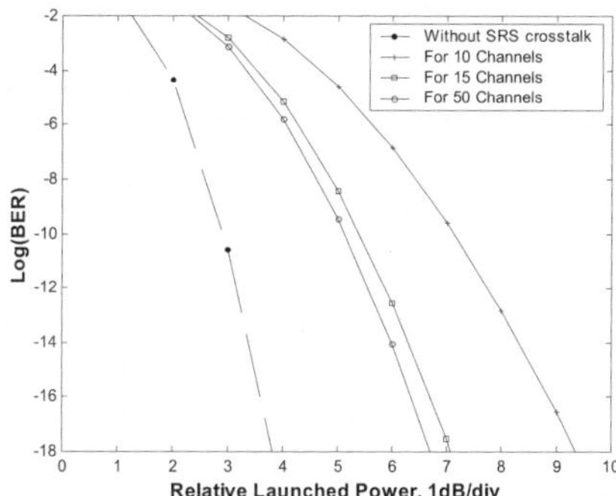

Fig. 4: Bit-error-rate curves versus launched power in the channel of interest for 10, 20 and 50 channels. The dashed line is the BER curve for the system without SRS crosstalk.

Next the BER performance is evaluated for variable bit rate. Figure 5 shows the BER curves for a 50-channel system. As the bit rate is increased, the curves shift rightwards, which implies that the penalty paid by the system is more for higher bit rate.

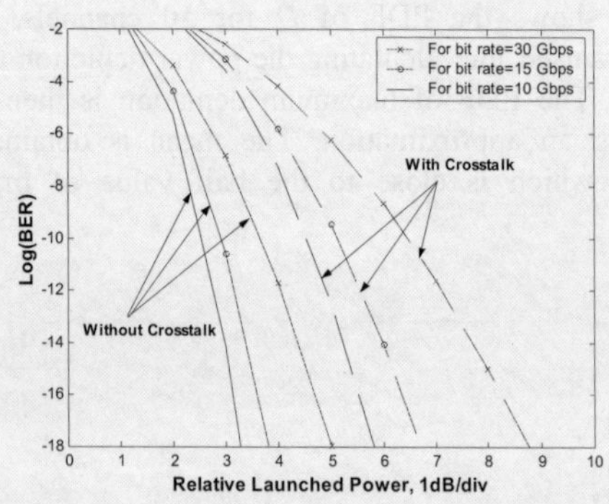

Fig. 5: Bit-error-rate curves versus launched power of a 50-channel system for different bit rate. Dashed lines are the BER curves for the system without SRS crosstalk.

The power penalty for a given number of channels is determined from the BER curves by determining the difference in launched power with and without SRS crosstalk at a given BER. The penalty decreases rapidly with increasing *N*. However, the penalty increases with increased bit rate.

4. CONCLUSIONS

An improved solution to the evaluation problem of the stimulated Raman crosstalk in a wideband WDM fiber optic communication system is presented. The proposed model for the Raman gain spectrum provides a better and more accurate estimation of the power depletion compared to the previous results. The SRS crosstalk is also analyzed statistically. The mean value of power depletion is obtained near the half value of maximum depletion. Therefore the actual penalty is always less than the worst-case penalty. The BER performance and hence the power penalty are evaluated for varying number of channels and bit rates. The analysis presented in this paper can thus quantify the effect of SRS crosstalk in an optical WDM system completely.

REFERENCES

[1] S. Norimatsu and T. Yamamoto, Journal of Lightwave Technology, vol. 19, no. 2, page 159-171, 2001.

[2] S. Tariq and J. C. Palais, Journal of Lightwave Technology, vol. 11, no. 12, page 1914-1924, 1993.

[3] K. P. Ho, Journal of Lightwave Technology, vol. 18, no. 7, page 915-921, 2000.

[4] A. R. Chraplyvy and P. S. Henry, Electronics Letters, vol. 19, no. 16, page 641-643, 1983.

[5] R. Stolen, Proc. IEEE, vol. 68, page 1232-1236, 1980.

[6] A. R. Chraplyvy, Electronics Letters, vol. 20, page 58-59, 1984.

[7] F. Forghieri, R.W. Tkach, and A. R. Chraplyvy, IEEE Photonics Technology Letters, vol. 7, no. 1, page 101-103, 1995.

[8] A. Papoulis, *Probability, Random Variables, and Stochastic Processes,* New York: McGraw-Hill, 1991.

[9] D. Marcuse, Journal of Lightwave Technology, vol. 8, page 1816-1823, 1990.

[10] M. G. Mostafa, M. M. Rahman, G. M. H. Ali, M. J. Alam and M. N. Islam, Proc. International Conf. on Computer and Information Technology, Dhaka, Bangladesh, pp. 234-238, 2001.

A Simulation Platform for Multi-Channel Transmission Systems in Frequency Domain Using Volterra Series Transfer Function

Ken Kai-fu Chang, Le Nguyen Binh
Electrical and Computer System Engineering
Monash University, Clayton, Victoria, Australia 3168

Abstract: We present a novel simulation model based on the proposed segmented Volterra series transfer function (S-VSTF) for advanced optical communication systems. The S-VSTF model is used to obtain analytic solution of the nonlinear Schröndinger (NLS) wave equation. Simulated results are compared with those using the well-known split-step Fourier method and it is confirmed that the model is efficient and accurate for optical pulse transmission systems.

Key Words: Volterra Series, Simulation, Optical fibre, Optic fibre communications systems.

1. MOTIVATION

Fiber-optic communication systems offer transmission capacity reaching tera-b/s over long haul distance have been continuously deployed across the world in inter-city and metropolitan systems. With the availability of ever-increasing optical devices and techniques for building faster and more reliable systems, it has become a daunting task for engineers to conduct proper analysis and designs for their systems. Hence the need for appropriate modelling is of paramount importance since proper analysis can significantly reduce the cost of deploying optical transmission systems on field.

Conventionally, numerical methods such as split-step Fourier (SSF) method are uitilised for simulation of optical transmission systems. It is widely used because of their simplicity and efficiency. However, it is difficult to obtain intuitive knowledge of the characteristics of a system using numerical methods. Design engineers often have to perform trial-and-error procedures to obtain optimal design parameters.

In this paper, a new simulation platform based on the Volterra series transfer function (VSTF) is proposed. The application of Volterra series transfer functions (VSTF) to the analysis and design of optical transmission systems was firstly proposed in 1997 [4] in order to provide a better alternative of simulation platform besides the split-step Fourier (SSF) method. However, the VSTF model has not been widely used due to its inherent limitations, namely, the slow convergence and limited applicability for nonlinear systems. These issues are solved in this work, and improved segmented-VSTF (S-VSTF) models are proposed in order to overcome the limitations of conventional VSTF model. The proposed S-VSTF provides a more accurate solution, and is more applicable for strongly nonlinear systems, which are previously not applicable with the conventional VSTF model. These strongly nonlinear systems include ultra-longhaul optical transmission systems and solitonic systems, etc.

2. Fiber Nonlinearities – Nonlinear Schrödinger (NLS) Equation

The generalized nonlinear Schrödinger wave equation that represents the propagation of the envelop of an optical pulses under the influence of both linear dispersion effect and nonlinear phase evolution can be given by

$$\begin{aligned}
&\frac{\partial A_i}{\partial z}+\frac{\alpha_{0i}}{2}A_i+\beta_{1i}\frac{\partial A_i}{\partial t}+j\frac{\beta_{2i}}{2}\frac{\partial^2 A_i}{\partial t^2}+\frac{\beta_{3i}}{6}\frac{\partial^3 A_i}{\partial t^3}\\
&=j\gamma_i\left[|A_i|^2+2\sum_{j\neq k}|A_j|\right]A_i-a_{1i}\frac{\partial\left[\left(|A_i|^2+2\sum_{j\neq k}|A_j|\right)A_i\right]}{\partial t}\\
&+a_{2i}\frac{\partial A_i}{\partial t}\left[|A_i|^2|A_i|^2+2\sum_{j\neq k}|A_j|\right]+a_{3i}\frac{\partial\left[|A_i|^2|A_i|^2+2\sum_{j\neq k}|A_j|\right]}{\partial t}A_i\\
&+jQ_R A\int_{-\infty}^{\infty}s_r(t-t_1)\left[|A(t_1,z)|^2+2\sum_{j\neq k}|A_j|\right]dt_1
\end{aligned} \tag{1}$$

where $A_i = A_i(t,z)$ represents the slowly-varying complex envelope of the optical pulses of the ith channel; α_i is the linear attenuation factor; β_{1i} is the group velocity or purely the propagation delay; β_{2i} is the linear dispersion factor, β_{3i} is the second order dispersion factor (i.e. the fibre dispersion slope with respect to the optical channel wavelength), γ_i is the SPM factor; a_{1i}, a_{2i} and a_{3i} is the higher order nonlinear coefficients (related to higher order nonlinear effects such as self-steepening, shock formation, and Raman scattering); and finally Q_R is the contribution factor due to the Raman gain effect of the fibre.

The NLS equation cannot be solved analytically due to its nonlinear nature, hence it does not provide intuitive insight of the cross channel inter-modulation effects. These problems would be overcome by the S-VSTF model, which is described in the next section.

3. The VSTF and S-VSTF Models

The VSTF model is a summation of a number of series which can approximate the wave equation in the frequency domain. The linear dispersion effect is represented via the first order kernel, and the nonlinear contribution due SPM and cross phase modulation or Raman scattering effects via the third order and higher order kernels. The transfer function of an optical pulse envelope between the input and output of a transmission fibre in the frequency domain, i.e. $X(\omega)$ and $Y(\omega)$ respectively, can be given by

$$Y(\omega)=\sum_{n=1}^{\infty}\int_{-\infty}^{\infty}\cdots\int_{-\infty}^{\infty}H_n(\omega_1,\ldots,\omega_{n-1},\omega-\omega_1-\ldots-\omega_{n-1},z)$$
$$\times X(\omega_1)\ldots X(\omega_{n-1})X(\omega-\omega_1-\ldots-\omega_{n-1})d\omega_1\ldots d\omega_{n-1} \tag{2}$$

where $H_n(\omega_1,\ldots,\omega_n)$ is the nth-order frequency domain Volterra kernel [1]. The first order and 3rd order kernels can be written as [2]

$$H_1(\omega,z)=e^{G_1(\omega)z}=e^{\left(-\frac{\alpha_0}{2}+j[\beta_1(\omega)-\beta_0]\omega\right)z} \tag{3}$$

$$H_3(\omega_1,\omega_2,\omega_3,z)=G_3(\omega_1,\omega_2,\omega_3)$$
$$\times\frac{e^{(G_1(\omega_1)+G_1^*(\omega_2)+G_1(\omega_3))z}-e^{G_1(\omega_1-\omega_2+\omega_3)z}}{G_1(\omega_1)+G_1^*(\omega_2)+G_1(\omega_3)-G_1(\omega_1-\omega_2+\omega_3)} \tag{4}$$

Other higher order terms can be included if more accuracy is required. As described the Volterra series transfer function of an optical transmission system takes the form of a power series, and naturally its convergence is very critical. This is investigated in the next section and is the central development of this work. The subscripts 1,2 and 3 of the ω terms are dummy variables that should be scanned across the frequency spectra covering all the wavelength channels. In DWDM systems this spectrum can be extended from the S-band to the L-bands, i.e. 1485 nm to 1625 nm.

However, the application of the conventional VSTF model is limited by the slow convergence of higher order kernels. Based on analysis of the convergence characteristics, the S-VSTF model is proposed to overcome the convergence issues. With the S-VSTF approach, the entire transmission link is divided into several segments, each modeled by a single S-VSTF model. In doing so, the transmission distance of each S-VSTF segment is reduced, and hence gives better convergence compared to the conventional VSTF. This can be illustrated in figure 1.

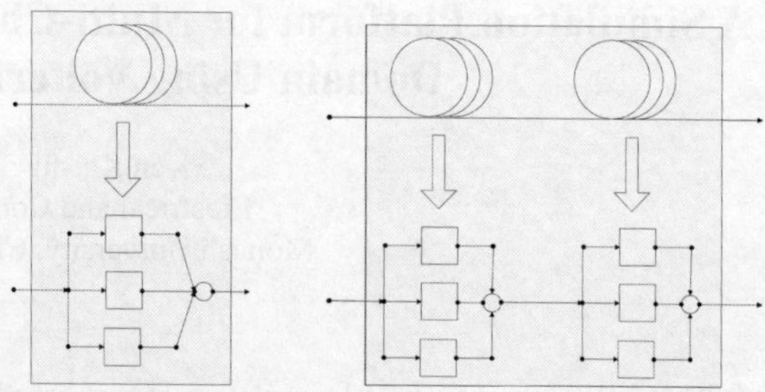

Fig.1 The VSTF model (left) and S-VSTF model (right) of the optic fibre.

Mathematically, the S-VSTF model can be expressed in operator form as,

$$\begin{aligned}[\mathbf{H}]&=[\mathbf{H}_1+\mathbf{H}_3+\mathbf{H}_5]\cdot[\mathbf{H}_1+\mathbf{H}_3+\mathbf{H}_5]\\&=\mathbf{H}_1\cdot\mathbf{H}_1+2\mathbf{H}_1\cdot\mathbf{H}_3+2\mathbf{H}_1\cdot\mathbf{H}_5+H.O.T.\end{aligned} \tag{5}$$

where $\mathbf{H}_1$ is the linear operator; $\mathbf{H}_3$ and $\mathbf{H}_5$ are the third order and fifth order nonlinear operators; $H.O.T.$ stands for higher order nonlinear terms, which are negligible under weakly nonlinear condition. In the frequency domain, Eq. (5) can be written as the follows,

$$\begin{aligned}y(\omega)&=H_1(\omega)H_1(\omega)X(\omega)\\&+2F.C.\left[H_1(\omega_1+\omega_2+\omega_3)H_3(\omega_1,\omega_2,\omega_3)X(\omega_1)X(\omega_2)X(\omega_3)\right]\\&+2F.C.\left[H_1(\omega_1+\omega_2+\omega_3+\omega_4+\omega_5)H_5(\omega_1,\omega_2,\omega_3,\omega_4,\omega_5)\right.\\&\qquad\left.\times X(\omega_1)X(\omega_2)X(\omega_3)X(\omega_4)X(\omega_5)\right]\\&+H.O.T\\&=H_1(\omega)H_1(\omega)X(\omega)\\&+2\int_{-\infty}^{\infty}\int_{-\infty}^{\infty}H_1(\omega)H_3(\omega_1,\omega_2,\omega-\omega_1-\omega_2)\\&\qquad\times X(\omega_1)X(\omega_2)X(\omega-\omega_1-\omega_2)\\&+2\int_{-\infty}^{\infty}\int_{-\infty}^{\infty}\int_{-\infty}^{\infty}\int_{-\infty}^{\infty}H_1(\omega)H_5(\omega_1,\omega_2,\omega_3,\omega_4,\omega-\omega_1-\omega_2-\omega_3-\omega_4)\\&\qquad\times X(\omega_1)X(\omega_2)X(\omega_3)X(\omega_4)X(\omega-\omega_1-\omega_2-\omega_3-\omega_4)\\&+H.O.T.\end{aligned} \tag{6}$$

4. Multi-Channel Pulse Propagation Simulation by VSTF Model and SSF Method

Both the SSF method and the proposed S-VSTF model are used for simulation of pulse propagation in the 400-km of transmission link consists of several NZDSF and DCF fiber segments of 40 km. The differences between these two methods are compared using the normalized deviation criterion:

$$Deviation\ in\ \%=\frac{\int_{-\infty}^{\infty}\left|Y_{VSTF}(\omega)-Y_{SSF}(\omega)\right|^2d\omega}{\int_{-\infty}^{\infty}\left|Y_{SSF}(\omega)\right|^2d\omega} \tag{7}$$

where $Y_{VSTF}(\omega)$ and $Y_{SSF}(\omega)$ are the pulse spectrum at the output obtained from the VSTF method and the SSF method respectively. Fig. 2 shows the evolution of the pulses after propagating over distances of 400 km for the 3-channel simulation setup.

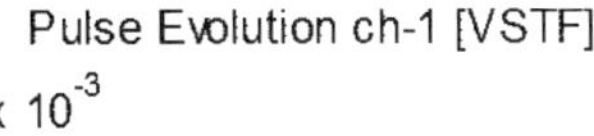

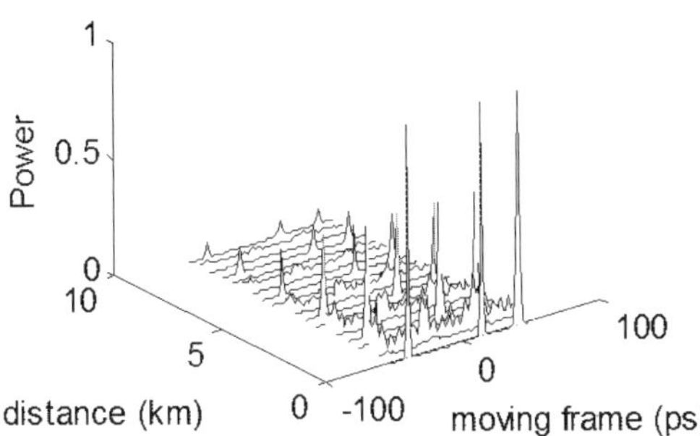

(a) Simulation of Channel-1 with S-VSTF model

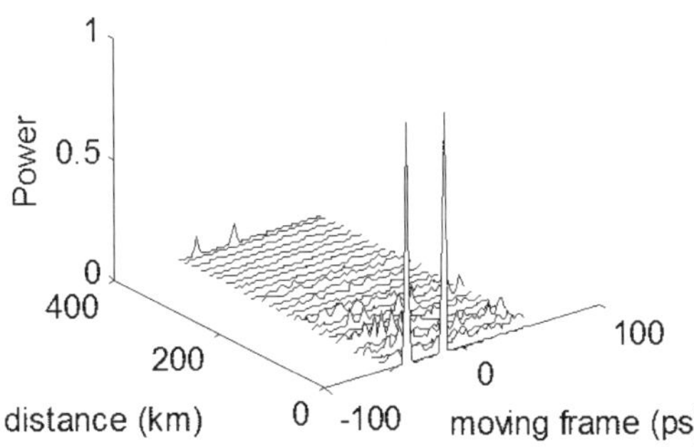

(b) Simulation of Channel-2 with SSF method

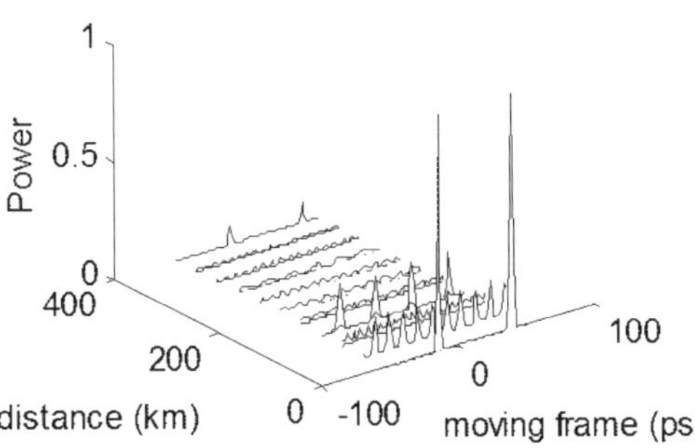

(c) Simulation of Channel-3 with VSTF model

Figure 2. Simulation result obtained from SSF and S-VSTF methods respectively for 400 km of transmission distance and 1 mW of input peak power.

Figure 3 shows the deviation between the simulation results between the SSF method and the VSTF model for all channels.

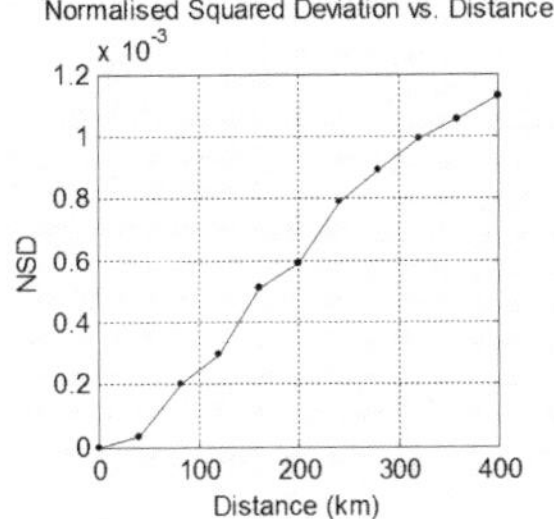

(a) NSD plot for channel-1

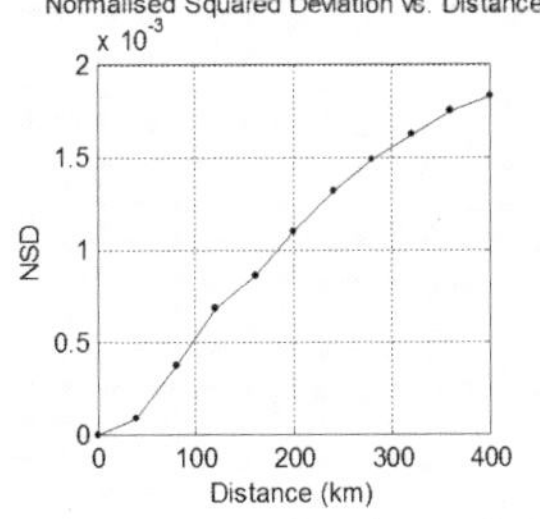

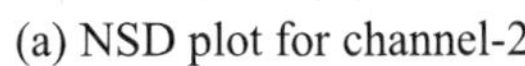
(a) NSD plot for channel-2

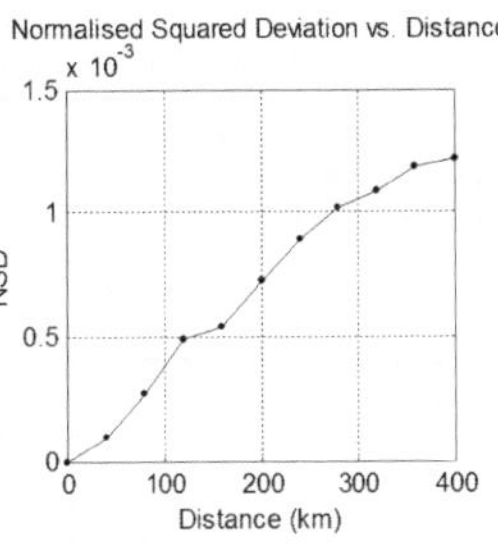

(a) NSD plot for channel-3

Figure 3. The NSD of Channel-1 in the 400-km 3-channel simulation result obtained from the SSF method and the S-VSTF model.

The NSD plots demonstrates the S-VSTF model are capable of providing good results with all the benefits the analytic approach can offer.

3. CONCLUSION

In order to facilitate appropriate analysis, this study has been focused on the application of the proposed S-VSTF models in optical transmission systems. The goal of the study is to provide an alternative simulation platform for the analysis and design of modern advanced optical communication networks. The simulation based on the S-VSTF model has been demonstrated as a simulation platform for study the optical pulse propagation in single-mode fibre optical transmission systems. Additionally, the S-VSTF is a frequency domain-based approach and hence allows us to synthesize the transfer functions of other optical sub-systems such as dispersion compensators, equalizers, filters etc.

REFERENCES

[1] M. Schetzen, *The Volterra and Wiener Theories of Nonlinear Systems*, New York: Wiley, 1989

[2] Kumar V. Peddanarappagari and Matte Brandt-Pearce, “Volterra series transfer function of single-mode fibers”, *Journal of Lightwave Technology*, Vol. 15, No. 12. December 1997.

[3] K. V. Peddanarappagari and M. Brandt-Pearce, “Study of fiber nonlinearities in communication systems using a Volterra series transfer function approach,” *Proc. 13th Annu. Conf. Inform. Sci-Syst.*, Mar, 1997, pp. 752-757.

[4] K. V. Peddanarappagari and M. Brandt-Pearce, "Volterra Series Approach for Optimizing Fiber-Optic Communications System Designs,"*Journal of Lightwave Technolog.*, Vol 16, No. 11, Nov, 1998, pp. 2046-2055..

[5] Martin B. Brilliant, "Theory of the analysis of nonlinear systems", *MIT Research Laboratory of Electronics, Technical Report 345*, March 3, 1958.

[6] A. Samelis, D.R. Pehlke and D. Pavlidis, "Volterra series based nonlinear simulation of HBTs using analytically extracted models", Electronics Letters, 23rd June 1994, Vol. 30, No. 13, p1098-1100.

[7] G. P. Agrawal, *Optical Communication Systems*. New York: Academic. 1997.

[8] Y. Miyamoto, A. Hirano, K. Yonenaga, A. Sano, H. Toba, K. Murata, O. Mitomi, "320 Gbit/s (8×40 Gbit/s) WDM transmission over 367 km with 120 km repeater spacing using carrier-suppressed return-to-zero format", Electronics Lettes, Nov 1999, Vol.35, No.23.

Session 14H

Lasers and Amplifiers II

Erbium doped glass waveguide amplifiers

Edwin Y.B. Pun
Department of Electronic Engineering
City University of Hong Kong
83 Tat Chee Avenue, Kowloon
Hong Kong, P.R. China
Tel : (852) 2788 8609, Fax : (852) 2788 7791, E-mail : eeeybpun@cityu.edu.hk

Abstract: Progress in planar rare earth doped glass devices has been significant over the past few years, and commercial products are now available. In this paper, erbium doped waveguide amplifiers for telecommunication applications are reviewed.

Key Words: Optical amplifiers, Erbium doped waveguide amplifiers

1. INTRODUCTION

Glass materials, such as laser glasses, magneto-optic glasses, and nonlinear glasses, offer interesting possibilities for applications to optically active devices. The motivation for glass lasers was high power applications. The first rare earth doped glass waveguide amplifier was demonstrated in 1972 using sputtering technique [1]. Since then, it takes two decades before significant progress has been made. The bulk of work has focussed on neodymium (Nd^{3+}) or erbium (Er^{3+}) doped glass waveguide amplifiers and lasers. Early work employing Nd^{3+} (λ~1.06μm) demonstrates high efficiency devices, although the application is limited. Recent work concentrates on Er^{3+} and Er^{3+}/Yb^{3+}, because the emission wavelength covers the lowest loss window of optical communication system and will find numerous applications.

2. WAVEGUIDE FABRICATION

Many of the well established deposition techniques commonly used in integrated optics have been used to fabricate rare earth doped glass waveguides and devices. Methods used include sputtering [1,2], ion implantation [3], flame hydrolysis deposition (FHD) [4,5], plasma enhanced chemical vapour deposition (PECVD) [6], vapour phase deposition [7], ion exchange [8-10], and sol-gel [11]. Composite structure has also been used [12]. Most of these deposition techniques are equipment intensive, and are both costly and time consuming. In comparison, ion exchange is a relatively simple and cheap fabrication technology, and has been used to produce commercial planar waveguide devices. However, some drawback remain, for example, the integration with laser on the same substrate is difficult because the heat conductivity of the glass is low.

3. ERBIUM DOPED GLASS WAVEGUIDE DEVICES

Er^{3+} doped and Er^{3+}/Yb^{3+} codoped waveguide devices are very attractive, due to their applications in 1.5μm optical fiber communication systems. Nd^{3+} doped glass waveguide lasers were demonstrated first. This is because the absorption and emission cross-sections in Er^{3+} are substantially lower than those in Nd^{3+}, and Er^{3+} system has lower efficiency because it is 3-level rather than 4-level. Also the Er^{3+} ions tend to cluster that shorten the fluorescence lifetime. Co-doping with Yb^{3+} enhances the Er^{3+} effective pump-absorption cross-sections, and also permits higher Er^{3+} concentrations without clustering.

Despite the successful demonstration of Er^{3+} planar waveguide lasers in 1991 [13], there are still no commercial applications for these devices. For waveguide amplifiers one major advantage is that they can be integrated with other planar glass waveguide devices, and compact, lossless optical integrated circuits can be obtained [14, 15].

Erbium doped waveguide amplifiers (EDWAs) are attractive for telecommunication applications; high gain (30dB), high power (14dBm), and low noise figure (3.1dB) have been demonstrated [16-18]. These devices have been used in 10Gb/s fiber transmission systems. When used as a preamplifier a sensitivity of -35.4dBm was reported. When used as an in-line amplifier, stable long term bit-error rate measurement was reported showing negligible performance degradation. When used as a power booster, negligible degradation was observed at the receiver down to a BER of 10^{-12}.

The gain spectrum of EDWA depends on the glass host used: soda-lime glass has a bandwidth of 17nm, Ge-Al codoped silicate glass has a bandwidth of 40nm, and aluminate-silicate glass has a bandwidth of 50nm. This demonstrates that by careful selection of a proper glass host broad gain spectrum can be achieved with EDWA, and TABLE I summarizes the gain performances of Er^{3+} doped waveguide amplifiers produced by various technologies.

Material	Technique	Gain (dB/cm)	Ref.
P_2O_5:SiO_2	PECVD	0.67	6
P_2O_5:SiO_2	FHD	0.68	5
Al_2O_3	Ion-implant.	0.75	18
Y_2O_3	Sputtering	1.3	19
Silicate	Ion-exchange	2.3	15
Phosphate	Ion-exchange	2.6	20
Phosphate	Sputtering	4.1	21
Soda-lime	Sputtering	4.2	16

TABLE I Gain performances of Er^{3+} doped glass waveguide amplifiers

Tellurite and alkali bismuth glasses were demonstrated recently as suitable hosts for Er^{3+} [22,23]. Intense green (around 525-550nm) and red (around 660nm) emission bands were observed in Er doped alkali bismuth gallate glass under 800nm excitation. At a pump intensity of 15.6W/cm^2, frequency upconversion efficiencies of $2.1x10^{-2}$ and $4.8x10^{-3}$ were obtained for the green and red emissions, respectively. These values are the highest among oxide glasses, and are comparable to those reported for Er^{3+}/Yb^{3+} codoped fluoride glasses. Er^{3+} doped $Na_2O.Ca_3Al_2Ge_3O_{12}$ glasses were also fabricated with excellent properties [24].

Planar glass waveguide amplifiers were fabricated in Er^{3+}/Yb^{3+} codoped phosphate glasses (Fig. 1 and 2). A net gain of ~3.3dB/cm was measured with an input signal wavelength of 1.535μm at a power of –40dBm, for 120mW of 980nm pump power [25].

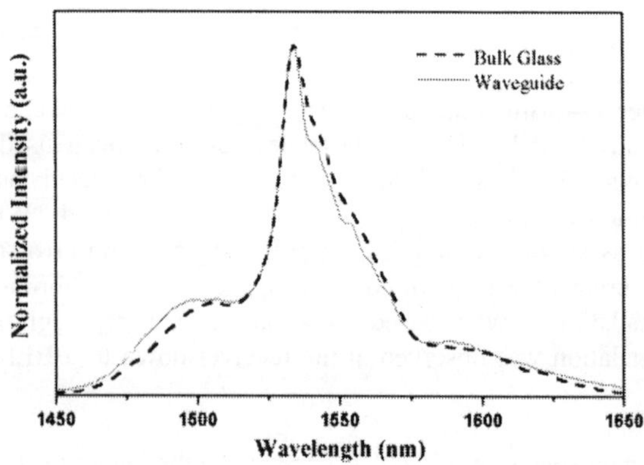

Fig. 1 Fluorescence spectrum of Er^{3+}/Yb^{3+} codoped phosphate glass

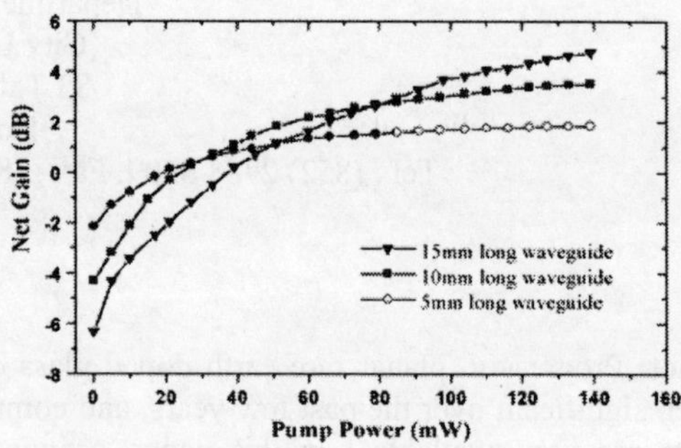

Fig. 2 Net gain versus pump power for a 6μm channel Er^{3+} doped waveguide amplifier

4. CONCLUSIONS

Er^{3+} doped waveguide glass amplifiers exhibit sufficient gain for many applications. Although they are not in the same league as the Er^{3+} doped fiber amplifiers when price-performance is considered, their compactness and cost are attractive properties. Further developments are required, such as background noise reduction, improvement in fiber-waveguide coupling, and so on.

ACKNOWLEDGEMENTS

This work is supported by a Competitive Earmarked Research Grant, Research Grant Council, Hong Kong.

REFERENCE

[1] H.Yajima, Y. Sekimoto, and S. Kawase, Amplification at 1.06μm using a Nd-glass thin film waveguide, Applied Physics Letters, vol. 21, page 407-409, 1972.

[2] J. Shmulovich, A. Wong, Y.H. Wong, P.C.Becker, A.J.Bruce, R. Adar, Er^{3+} glass waveguide amplifier at 1.5μm on silicon, Electronics Letters, vol. 28, page 1181-1182, 1992.

[3] A. Poleman, Erbium implanted thin film photonics materials, J. Applied Physics, vol.82, page 1-39, 1997.

[4] R. Tumminelli, F. Hakimi, and J. Haavisto, Integrated-optic Nd:glass laser fabricated by flame hydrolysis deposition using chelates, Optics Letters, vol. 16, page 1098-1100, 1991.

[5] T. Kitagawa, K. Hattori, K. Shuto, M. Yasu, M. Kobayashi, and M. Horiguchi, Amplification in erbium-doped silica-based planar lightwave circuits, Electronics Letters, vol. 28, page 1818-1819, 1992.

[6] K. Shuto, K. Hattori, T. Kitagawa, Y. Ohmori, M. Horiguchi, Erbium-doped phosphosilicate glass waveguide amplifier fabricated by PECVD, Electronics Letters, vol. 29, page 139-141, 1993.

[7] C. Jacoboni, O. Perrot, and B. Boulard, Vapor phase deposition of rare earth doped PZG glasses, J. Non-Crystalline Solids, vol.184, page 184-189, 1995.

[8] S.I. Najafi, W.J.Wang, J.F. Currie, R. Leonelli, and J.L. Brebner, Fabrication and characterization of neodymium-doped glass waveguides, IEEE Photonics Technology Letters, vol. 1, page 109-110, 1989.

[9] S. Honkanen, S.I. Najafi, P. Poyhonen, G. Orcel, W.J. Wang, J. Chrostowski, Silver-film ion-exchanged singlemode waveguides in Er doped phosphate glass, Electronics Letters, vol. 27, page 2167-2168, 1991.

[10] T. Feuchter, E.K. Mwarania, J. Wang, L. Reekie, J.S. Wilkinson, Erbium-doped ion-exchanged waveguide lasers in BK-7 glass, IEEE Photonics Technology Letters, vol. 4, page 542-544, 1992.

[11] M. Benatsou, B. Capoen, M. Bouazaoui, W. Tchana and J. P. Vilcot, Preparation and characterization of sol-gel derived Er^{3+}: Al_2O_3–SiO_2 planar waveguides, Applied Physics Letter, vol. 71, page 428-430, 1997.

[12] S. Honkanen, S.I. Najafi, and W.J. Wang, Composite rare-earth-doped glass waveguides, Electronics Letters, vol. 28, page 746-747, 1992.

[13] T. Kitagawa, K. Hattori, M. Shimizu, Y. Ohmori, and M. Kobayashi, Guided-wave laser based on erbium-doped silica planar lightwave circuit, Electronics Letters, vol. 27, page 334-335, 1991.

[14] T. Kitagawa, Rare earth doped planar waveguide amplifiers, Proc. OAA'93, page 136-139, 1993.

[15] P. Camy, J.E. Roman, F.W. Willems, M. Hempstead, J.C. Van Der Plaats, C. Prel, A. Beguin, A.M.J. Koonen, J.S. Wilkinson, and C. Lerminiaux, Ion-exchanged planar lossless splitter at 1.5μm, Electronics Letters, vol. 32, page 321-322, 1996.

[16] D. Barbier, M. Rattay, F. Saint Andre, G. Clauss, M. Trouillon, A. Kevorkian, J.-M. P. Delavaux, and E. Murphy, "Amplifying four-wavelength combiner, based on erbium/ytterbium-doped waveguide amplifiers and integrated splitters", IEEE Photonics Technology Letters, vol. 9, page 315-317, 1997.

[17] J.M.P. Delavaux, S. Granlund, O. Mizuhara, L.D. Tzeng, D. Barbier, M. Rattay, F. St. Andre, A. Kevorkian, Integrated optics erbium-ytterbium amplifier system in 10-Gb/s fiber transmission experiment, IEEE Photonics Technology Letters, vol. 9, page 247-249, 1997.

[18] G. N. van den Hoven, R. J. I. M. Koper, A. Polman, C. van Dam, J. W. M. van Uffelen, and M. K. Smit, Net optical gain at 1.53 μm in Er-doped Al_2O_3 waveguides on silicon, Applied Physics Letter, vol. 68, page 1886-1889, 1996.

[19] H.J. van Weerden, H.J.W.M. Hoekstra, P.V. Lambeck, and T.J.A. Popma, Low-treshold amplification at 1.5μm in Er:Y2O3 IO-amplifiers, Proc. ECIO'97, page 169-172, 1997.

[20] D.Barbier, J.M. Delavaux, A. Kevorkian, P. Gastaldo, and M Jouanno, Yb/Er interated optics amplifiers on phosphate glass in single and double pass configurations, in Proc. OFC'95, PD-3, San Diego, CA, 1995.

[21] Y.C. Yan, A.J. Faber, H. de Waal, P.G. Kik, and A. Polman, Erbium-doped phosphate glass waveguide on silicon with 4.1 dB/cm gain at 1.535μm, Applied Physics Letter, vol. 71, page 2922-2924, 1997.

[22] S.Q. Man, E.Y.B. Pun, and P. S. Chung, Upconversion luminescence of Er^{3+} in alkali bismuth gallate glasses, Applied Physics Letters, vol. 77, page 483-485, 2000.

[23] S.Q. Man, S. F. Wong, and E. Y. B. Pun, Erbium-doped potassium bismuth gallate glasses, Journal of the Optical Society of America B: Optical Physics B, vol. 19, page 1839-1843, 2002.

[24] H.Lin, E.Y.B. Pun, S. Q. Man, and X.R. Liu, Optical transitions and frequency upconversion of Er^{3+} ions in $Na_2O{\cdot}Ca_3Al_2Ge_3O_{12}$ glasses, Journal of the Optical Society of America B: Optical Physics B, vol. 18, page 602-609, 2001.

[25] S.F. Wong, E.Y.B. Pun, and P.S. Chung, Er^{3+}/Yb^{3+} codoped phosphate glass waveguide amplifier using Ag^+-Li^+ ion exchange, IEEE Photonics Technology Letters, vol. 14, page 80-83, 2002.

40GHz PULSE GENERATION AND PERFORMANCE ANALYSIS OF ACTIVELY MODE-LOCKED FIBER RING LASER

Baoxi XU, Gaoqiang YUAN, Chong Wei CHUAH, Honggang WANG and Tow Chong CHONG
Data Storage Institute, DSI Building, 5 Engineering Drive 1, Singapore 117608
Tel: 65-8748512
Fax: 65-7778517
E-mail: dsixub@dsi.nus.edu.sg

Abstract *Optical pulse with the repetition rate of 40GHz and pulsewidth of 7ps have been obtained by using of actively mode-locked fiber ring laser with 20GHz bandwidth electro-optical modulator. The performance of the output has been analyzed.*

Introduction

High repetition rate and short pulse laser sources are essential for realizing the higher speed time division multiplexing communication system, soliton transmission system and electro-optical sampling. There are a few method to get short pulse laser output but the active mode-locking of fiber ring laser is one of the promising method to realise the stable and high repetition rate. In the fiber ring laser, usually the cavity length is very long, from tens meter to thundreds meter which correspond to the cavity mode space of the order of megahertz. It means that harmonic mode-locking technology should be used to get gigahertz repetition rate pulse output. One problem in the harmonic mode-locking should be considered is supermode suppression to get stable output. Some researchers have studied on this issue under below 20GHz repetition rate cases[1]~[4]. Few of paper has published the results in 40GHz repetition rate and the performance analysis[5]. In this paper, optical pulse generation with repetition rate of 40GHz and pulsewidth of 7ps using actively mode-locked fiber ring laser are reported and its output performance are analysed.

Experiment setup

The experiment setup is shown in Fig.1. It consists of Erbium Doped Fiber (EDF) with length of 24m pumped by 980nm laser doide, WDM coupler, Isolators, delay line to stabilize the output, Polarisation Controller (PC), $LiNbO_3$ modulator with bandwidth of 20GHz and $V\pi$ of 4.8 voltage and a 10% output coupler. The normal single mode fiber is used in the ring. The output are measured by using of high speed photo-detector (New Focus: 1014, bandwidth 45GHz) and oscilloscope (Agilent: 83484A, bandwidth 50GHz), RF spectrum analyzer (Advantest: R3182, bandwidth 40GHz), optical spectral analyzer (Anritru: MS9710C) and auto-correlator (Femtochrome: FR-103XL/IR). In order to get 40GHz modulation, the bias voltage is controlled to stand on the minimum transmission point.

Experiment results and analysis

The cavity resonance frequency was measured first

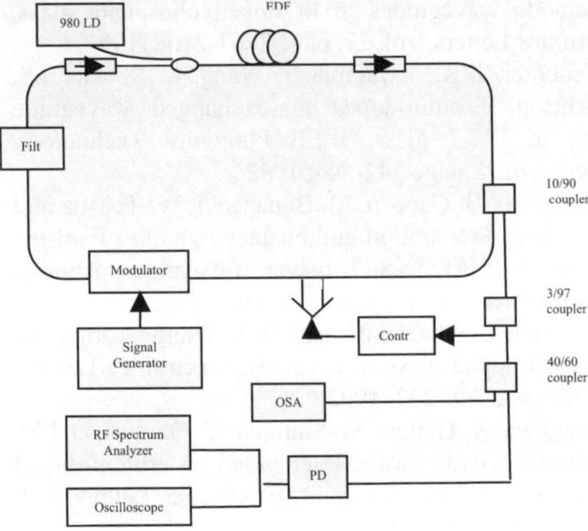

Fig.1 Experiment setup

under 20GHz modulation. The result is 4.48MHz as shown in Fig.2, which corresponds the cavity length of 46.28m if supporting the EDF has the same refractive index. For 40GHz repetition rate, around 8930th harmonic mode-locking should be used.

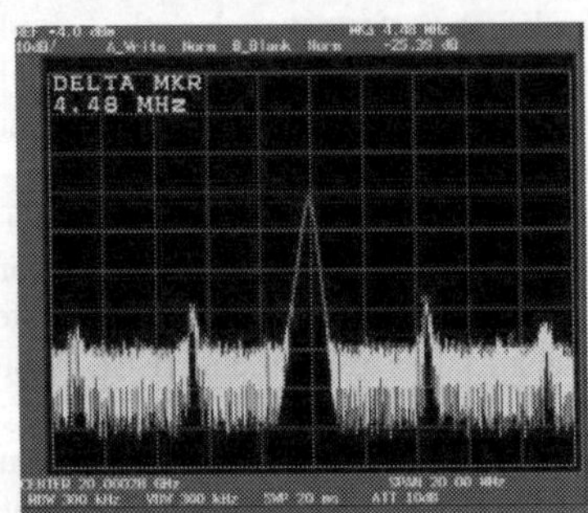

Fig.2 Cavity resonant mode oscillation

The output pulse shape detected by photo-detector is shown in Fig. 3. The pulse repetition rate is 40 GHz and the amplitude is very uniform. The pulsewidth shown in the picture could not be reflected because of the limit bandwidth of photo-detector and oscilloscope. The auto-correlation curve to reflect to

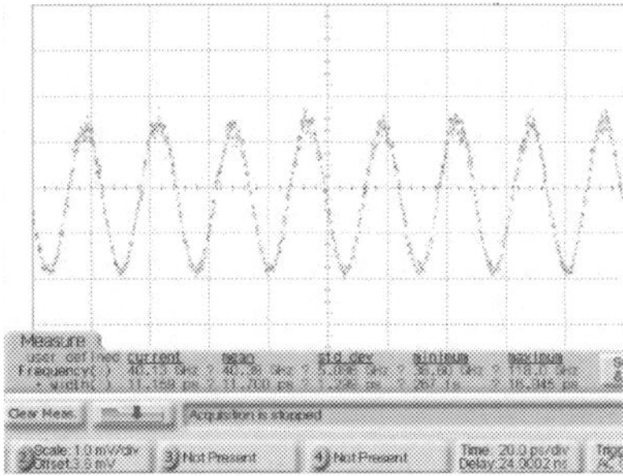

Fig.3 Output optical pulses

pulsewidth and the optical spectrum are shown in Fig.4. The pulsewidth (FWHM) of the auto-correction curve is 10.8ps. Considering the pulse with hyperbolic secant shape, the real pulsewidth is 7.0ps. The spectrum width is 0.37nm (46GHz). The time bandwidth product is 0.322, which is close to the chirp-free time-bandwidth product of hyperbolic secant pulse: 0.315. This result also meets the exploration result of simulation[6]. More than one peak in the autocorrelation curve is caused by the cross-correlation because the pulse repetition rate is very high.

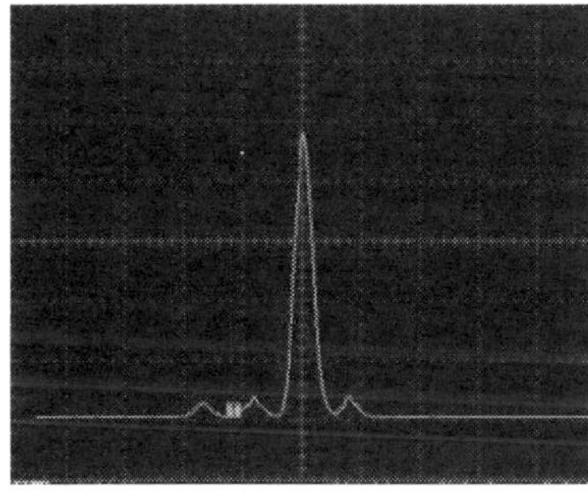

(a)

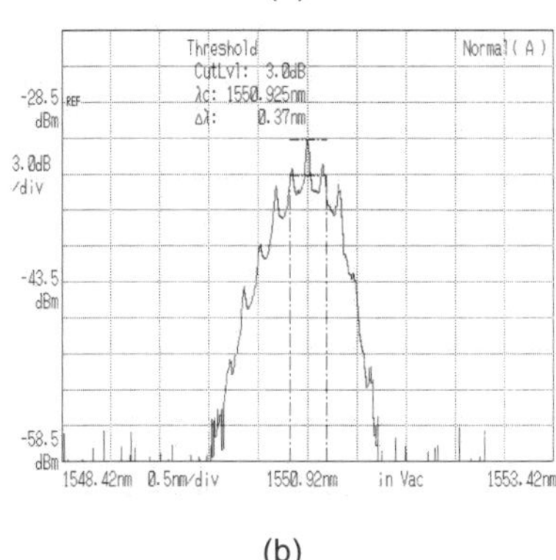

(b)

Fig.4 Auto-correlation curve (a) and optical spectrum (b) of the output optical pulses

The supermode noise is main reason to induce the amplitude noise. In our experiment, there is no supermode oscillation observed (see Fig.5) in our measurement condition. There is also no 20GHz component observed which could be caused by 20GHz modulation of the modulator.

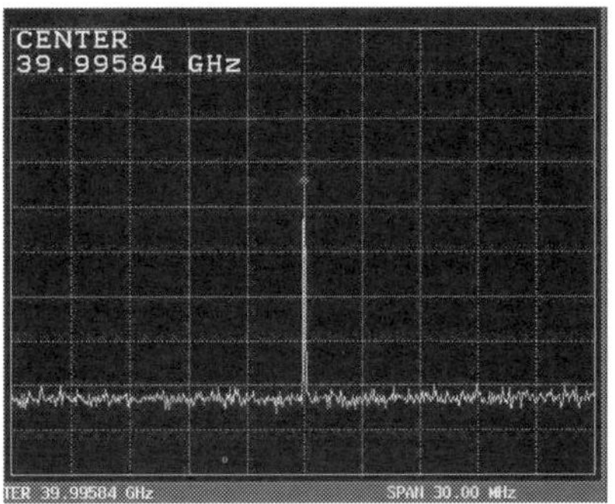

Fig. 5 RF spectrum of the output optical pulses

In order to get the time jitter information, the narrow span range RF spectrum are also measured as shown in Fig. 6. The time jitter is estimated as 110fs.

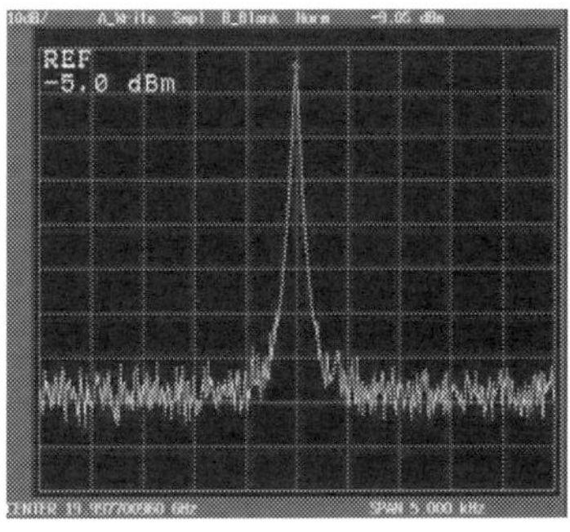

(a)

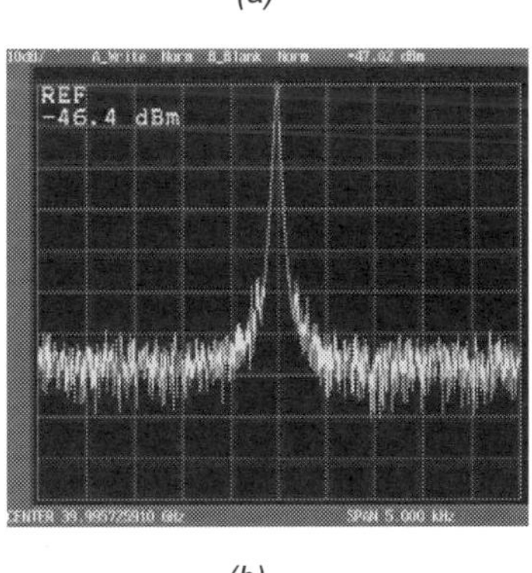

(b)

Fig. 6 Narrow span range of RF spectrum. (a) Output of RF signal generator at 20GHz. (b) Output optical pulse at 40GHz.

The output pulsewidths in different pumping power are measured in certain range and the result is shown in Fig.7. As the pumping power increase, the pulsewidth has a little reduction. The pulsewidth reduction is caused by the Kerr effect in the fiber in higher power, but in our experiment, the change is so little because of the saturation of the EDF amplification.

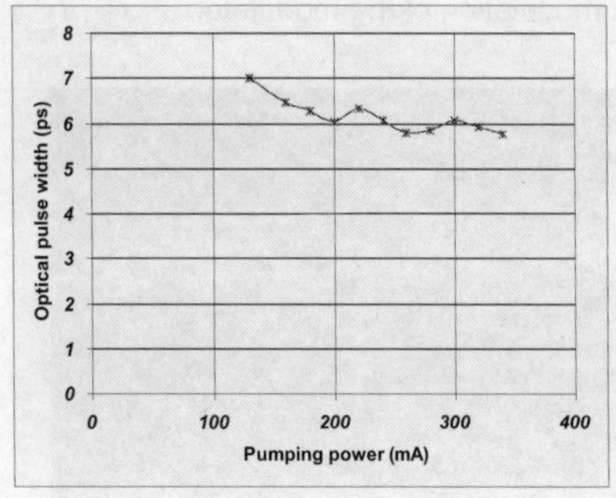

Fig. 7 Dependence of pulsewidth on pumping power

Conclusions

40GHz repetition rate pulses with pulsewidth of 7ps and the time-bandwidth product of 0.322 are obtained from actively mode-locked fiber ring laser under 20GHz modulation. The time jitter of the optical pulse is approximately 110fs. The pulsewidth change with pumping power is also presented.

References

1 M. Nakazawa et al, Electron. Lett, 32(1996),461
2 E.Yoshida et al, Electron. Lett., 34(1998),1753
3 E.Yoshida et al, IEEE Photon. Techn. Lett., 11(1999), 548
4 Jie Li et al, IEEE Photon. Techn. Lett., 12(2000), 1150
5 Th. Pferffer et al. Electron. Lett., 29(1993), 1849
6 Xiupu Zhang et al, IEEE Photon. Techn. Lett., 10(1998), 1103

Self-frequency shift effect in an actively mode-locked fiber laser

Q. Wen, P. Shum, Y. D. Gong, N. Q. Ngo, X. C. Yuan
Network Technology Research Center, School of EEE,
Nanyang Technological University, Singapore 639798
epshum@ntu.edu.sg

Abstract: The self-frequency shift effect is examined in an actively mode-locked fiber laser. We observed, with the insertion of a filter, this effect can be greatly suppressed and the range of gain for stable operation is wider.

Keywords: fiber laser, soliton, mode-locked

1. INTRODUCTION

In recent years, there have been extensive developments of lasers and sources of ultra-short optical pulses due to the widespread increase in their applications [1]-[4]. Among them, an actively mode-locked fiber ring laser is a promising light source since it can generate ultra-short high-quality pulses at high repetition rate. In our work, an actively mode-locked fiber ring laser has been numerically studied to track the evolution of the ultra-short pulses under the self-frequency shift effect. We found that with a filter inside the cavity, the self-frequency shift effect can be greatly suppressed.

2. ANALYSIS

The laser model under study has a ring configuration. It consists of a single-mode, polarization-preserving, erbium-doped fiber pumped by a laser diode, an amplitude modulator, necessary for the realization of mode locking, a tunable filter and an optical isolator to provide unidirectional operation.

The equation used to describe the amplification of pulses is one perturbed Schrödinger equation [5]:

$$j\frac{\partial u}{\partial z} = -\frac{1}{2}\frac{\partial^2 u}{\partial \tau^2} + |u|^2 u + j\frac{1}{2}Gu + j\frac{1}{2}\tau_a G\frac{\partial u}{\partial \tau} - \tau_R u\frac{\partial |u|^2}{\partial \tau} \tag{1}$$

In the equation, $u(z,\tau)$ is the normalized complex amplitude of the pulse, z represents the normalized distance along the direction of propagation, and τ represents the normalized time. The first term in the right hand side of the equation describes the pulse- dispersion spreading effect, and the second term gives rise to the Kerr nonlinear effect – the self-phase modulation. The next two terms represent the active Er^{3+} ions contribution, which are the gain terms. The last term describes the Raman self-frequency shift effect.

After the amplification, the pulse is sent through an amplitude modulator. The transmission function of the modulator is given by

$$\kappa_a = t_a \exp[-d_a \sin(\Omega_a \tau + \phi_a)] \tag{2}$$

where d_a, Ω_a, ϕ_a are the depth, the frequency, and the phase of the amplitude modulation, respectively. In our case, $d_a = 0.3$, $\Omega_a = 0.1$ and $\phi_a = 0$. The coefficient $t_a = 0.4$ is account for the losses, including those associated with the output.

The tunable bandpass filter function is described by

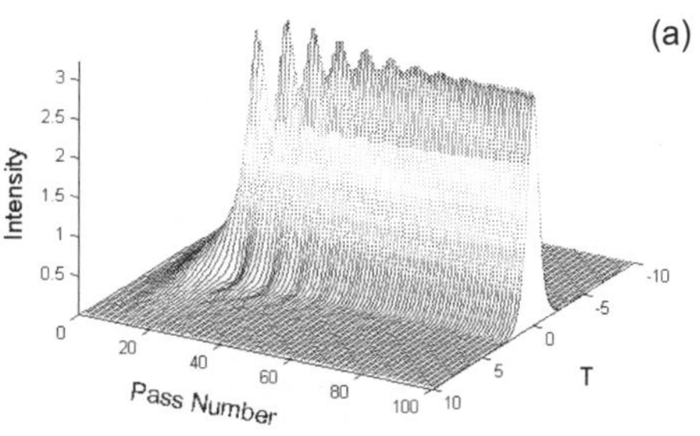

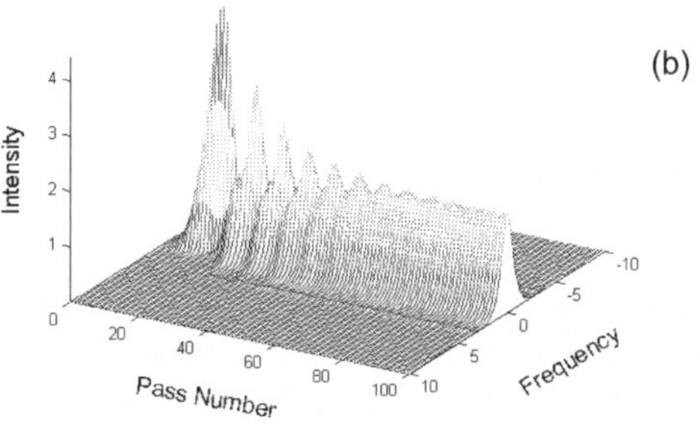

Figure 1. (a) Pulse evolution in the time domain and (b) in the frequency domain with $\tau_R = 0$

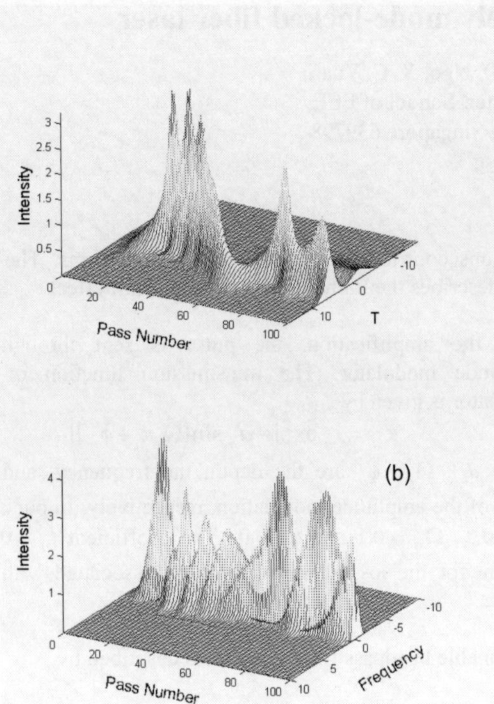

Figure 2. (a) Pulse evolution in the time domain and (b) in the frequency domain with $\tau_R = 0.01$

$$H(f) = \alpha \cdot \exp\left[\left((-1/2)\cdot(f/B_0)^2\right)\right] \quad (3)$$

This is a Gaussian profile, where $\alpha = 0.9$ for insertion loss and B_0 is half of $1/e$ bandwidth of the filter.

These equations have been numerically integrated using the Fourier series analysis method to examine the evolution of pulses inside the cavity for successive trips [6]. The cavity length is set to be 20m, and the normalized pump power is 5 for most cases. We studied the effect of the self-frequency shift on the formation of ultra-short pulses. Since its discovery, considerable attention has been paid to the Raman self-frequency shift effect, which is a higher-order nonlinear effect. This frequency shift is induced due to the delayed nonlinear response of the optical fiber [7].

To investigate the influence of the effect of the self-frequency shift on the pulse formation in our laser model, we first consider a cavity without the tunable filter. When the effect of the self-frequency shift is not taken into account, our simulation results show that steady-state output can be easily generated with desirable characteristics. The pulse evolution in the absence of self-frequency shift term in time domain and frequency domain are illustrated in Figure 1, where the amplification parameter is set to be 2dB. When the self-frequency shift effect is under consideration with a coefficient of 0.01, no stable output can be generated as shown in Figure 2.

Without the suppression of self-frequency shift effect, it is generally difficult to achieve a final steady-state pulse output. Since the self-frequency shift effect tends to shift the spectral peak of the pulse away from the center of the gain profile, it is difficult for the pulse to gain sufficient energy. That is, when the spectral peak of the pulse is shifted away from the center of AM modulator's gain profile, the pulse will experience more loss. If the pulse is shifted so far away that the gain and loss cannot balance each other to reach a steady state, no output will be generated.

However, when a tunable filter with $B_0 = 2nm$ is added to the laser model, no noticeable shift is generated and it is easier for the pulse to reach the steady state. The amplification parameter is 4dB, which is larger than that without the filter, since the filter introduces losses. In Figure 3, the steady-state outputs generated under $\tau_R = 0$ and $\tau_R = 0.01$ are compared. The red shift appeared in Figure 2 has

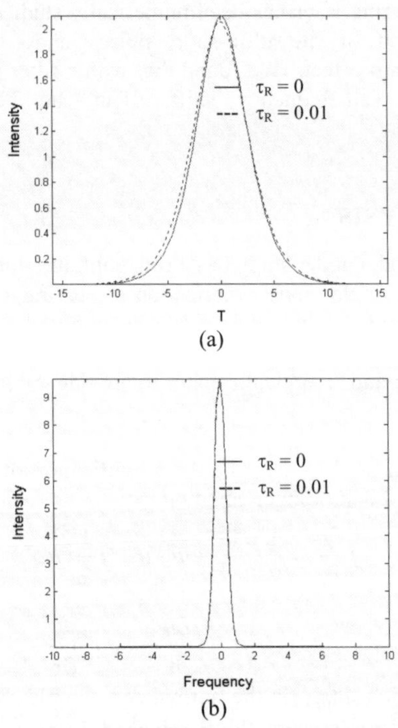

Figure 3. (a) Steady-state output in the time domain and (b) in the frequency domain with filter added

been suppressed. This can be understood through the Raman effect as a kind of intra-pulse stimulated Raman scattering.

Without the filter, the generated pulse's spectral width is large enough that the Raman gain can amplify the red spectral components of the pulse at the expense of the blue components of the same pulse to act as a pump. While with the filter inside the cavity, the pulse spectrum is limited, so that the red shift is greatly suppressed. Another phenomenon worth pointing out is that the range of the laser gain for stable operation is increased by several dB after the insertion of the filter while maintaining the other settings.

3. CONCLUSION

Our results show that even though the self-frequency shift coefficient is small, it can destroy the formation of ultra-short pulse inside an actively mode-locked fiber laser. With the inclusion of a tunable filter inside the ring cavity, the self-frequency shift is greatly suppressed and the range of laser gain for stable operation is broadened compared with those obtained without the inclusion of the filter.

REFERENCES

Journal Papers:

[1]. M. Horowitz, C.R. Menyuk, T.F. Carruthers, and I.N. Duling III, Theoretical and experimental study of harmonically modelocked fiber lasers for optical communication systems, Journal of Lightwave Technology, vol. 18, No. 11, 2000

[2]. J. S. Wey, J. Goldhar, D. W. Rush, et al, Performance Characteristization of a Harmonically mode locked Erbium fiber ring laser, IEEE Photonics Technology Letters, vol. 7, No. 2, page 152-154, 1995

[3]. Y. Li, C. Lou, J. Wu, B. Wu, Y. Gao, Novel Method to simultaneously compress pulses and suppress supermode noise in actively mode locked fiber ring laser, IEEE Photonics Technology Letters, vol.10, No.9, page 1250-1252, 1998

[4]. K. S. Abedin, M. Hyodo, N. Onodera, Active stabilization of a higer-order mode-locked fiber laser operating at a pulse-repetition rate of 154GHz, Optics Letters, vol. 26, No.3, page 151-153, 2001

[5]. E. Marti-Panameno, J.J. Sanchez-Mondragon, and V.A. Vysloukh, "Theory of soliton pulse forming in an actively modelocked fiber laser" IEEE Journal of Quantum Electronics, vol. 30, No. 3, page 822-826, 1994

[6]. H. Ghafouri-Shiraz, P. Shum, M. Nagata, A novel method for analysis of soliton propagation in optical fibers, IEEE Journal of Quantum Electronics, vol. 31, No. 1, 1995

Book:

[7]. G.P. Agrawal, *Nonlinear Fiber Optics*, second edition, Academic Press, 1995

Generation of Pico-second Laser Pulses from a Mode-Locked Fiber Laser

P. Shum, H.L Shao, J.Z. Li
Network Technology Research Centre,
Nanyang Technological University, Singapore 639798
E-mail: shum@ieee.org

ABSTRACT

An actively harmonic mode-locked fiber ring laser was constructed and its stability was analyzed. A chaotic behavior in the actively harmonic mode locked fiber ring laser was observed.

Keywords: fiber laser, mode locking

1. INTRODUCTION

To meet the requirement of rapid expanding for high-speed communication systems, an ultrahigh-speed optical light source is required not only for the development of ultra-high speed optical time domain multiplexing system but also the future ultra-high speed dense wavelength division multiplexed system. Actively mode-locked fiber ring lasers have shown to be an attractive light source to generate ultrashort pulses [1][2][3]. Actively harmonic mode-locked erbium-doped fiber ring lasers are capable of producing narrow Gaussian pulses with multi-GHz bit rates and picosecond pulsewidth in the C-band. They are especially suitable for high speed, long distance WDM transmission [4][5].
For an actively harmonic mode-locked erbium-doped fiber ring lasers, the stability is a key problem because of the environmental influence to the cavity. We analyzed the modulation frequency effect for the laser and adopted a new method to control the cavity length and realized the stable operation of the laser successfully. At the same time, chaotic encryption is a promising and secure method for military communication, so far the speed of chaos fiber laser cannot go beyond 1GHz [6][7], because of the limitation of its relaxation oscillation frequency. We found that if the modulation frequency deviates from the center harmonic frequency with a value of tens of kHz, which is just the system relaxation oscillation frequency of the fiber laser, the system will be in a chaotic state, and therefore it can be applied in secure communications.

2. EXPERIMENTAL CONFIGURATION

The configuration of the actively mode locked fiber ring laser is shown in Fig.1. The loop length of the laser cavity is around 50m, which is comprised of a 12m long erbium doped fiber. A pigtailed InGaAsP laser diode operating at 980nm is used to pump the laser, and the maximum pump output power is 110mW. Unidirectional operation of the laser is enforced by the inclusion of a polarization-dependent isolator (P-I) in the cavity. A Fabry-Perot filter with 3dB bandwidth of 1nm is inserted in the cavity to control the wavelength (FSR range from 1530 to 1560nm). A $LiNbO_3$ Mach-Zender amplitude modulator with a half-wave voltage of 4.6 V was used to realize the harmonic mode locking. A polarization controller is installed before modulator in order to achieve the maximum modulation depth. A 980/1550 wavelength-division-multiplexing (WDM) coupler is used to couple the pump light into the ring cavity. The output was taken from a 10:90 ratio output coupler.
Mode-locking can be achieved in the any wavelength from 1530nm to 1560nm. Fig.2 shows the RF spectrum when fundamental repetition rate mode locking is achieved. It shows that the fundamental repetition rate is 4.24MHz, which corresponds to the cavity length of approximately 50m.
By carefully adjusting the modulation frequency and polarization controller, a 2.503GHz (around 590th harmonic) repetition rate soliton-like pulse was observed, as shown in Fig.3(a) (measured by Tektronix CSA8000 Communications Signal Analyzer with bandwidth 50GHz and 45GHz bandwidth detector and amplifier). The pulse width is 39.0 ps. Fig.3(b) is its corresponding RF spectrum, it has strong signal frequency at 2.5GHz and low background noise.

3. MODULATION FREQUENCY

From the center stable modulation frequency, if we deviate the modulation frequency several kHz there is still clear pulses trace but the output is not so strong. If we vary the modulation frequency upwards and downwards for several kHz, the output pulses will enter a chaotic range. The output pulses trace is in chaos and the pulses have different amplitudes.

If we continue to deviate the modulation frequency upwards and downwards for another tens of kHz, the harmonic mode locking will be lost and no soliton-like pulse will be formed at output. The output trace with 100kHz frequency deviation will lost clear soliton-like profile. If the frequency deviates more, the optical pulse will disappear totally.

We summarized the all output pulse regime with the deviated frequency in Fig.3, the center is the stable mode-locking regime with 5khz width. The harmonic center frequency is 2.503GHz. 24kHz to 26kHz range is the chaotic regime. If the modulation frequency deviated more than the above regime, we will not obtain any optical pulse.

A ideal actively mode-locked soliton fiber ring laser should be working in the stable mode-locking regime. But a common problem exists in such actively mode-locked soliton fiber ring laser system is that its output is inherently unstable because of the fluctuations of polarization state and cavity length caused by mechanical vibration and temperature variation. So some methods need to be applied to stabilize the laser source operation [8-13]. From the above RF spectra, we can find that there is very low background noise in stable mode-locking regime. We can take the noise amplitude of RF spectrum as the feedback signal. The diameter of PZT cylinder is 30mm, and 10m DSF fiber was wired on it. The maximum driving voltage for PZT is 140V, and this leads to the maximum cavity length compensation value of 100 μm. Once the computer feedback system turned on, then the laser system can always maintain the stable output pulses.

When we deviate the modulation frequency tens of kHz (this value is different for different laser cavity), then the laser enter the chaotic state. From the RF spectrum we found that the relaxation oscillation was activated, which is shown in Fig.4(a), its extended from DC to 200kHz with 19kHz spacing. Reference[8] used this to feedback the cavity length, but we found this is not an effective way. The relaxation frequency is difficult to be activated. The range of the signal frequency under the same pump and modulation signal is shown in Fig.4(b), it has the same frequency spacing, and it is the sum frequency of both the harmonic frequency and relaxation oscillation frequency. In the mean time, period bifurcation and period doubling phenomena were also observed. So we can conclude that the chaotic pulses can be produced at any frequency which is just the sum of the harmonic frequency and the relaxation oscillation frequency. If two such lasers have the same parameters and the modulation signal can synchronize, then the chaos synchronization can be formed and high speed chaos communication can be realized. However, the higher the frequency, the more difficult we can match the frequency.

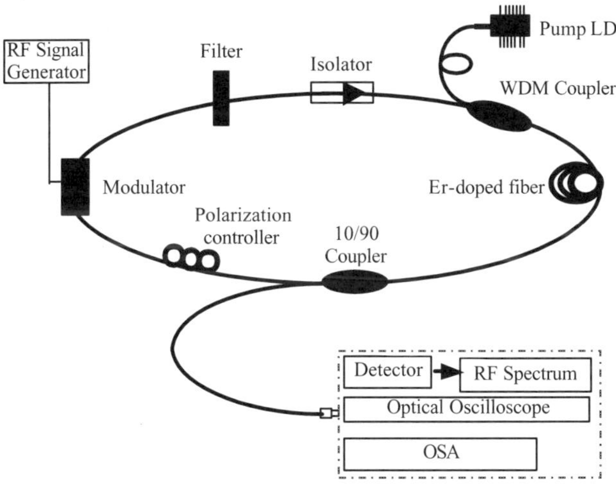

Fig.1 Configuration of actively mode- locked fiber ring laser

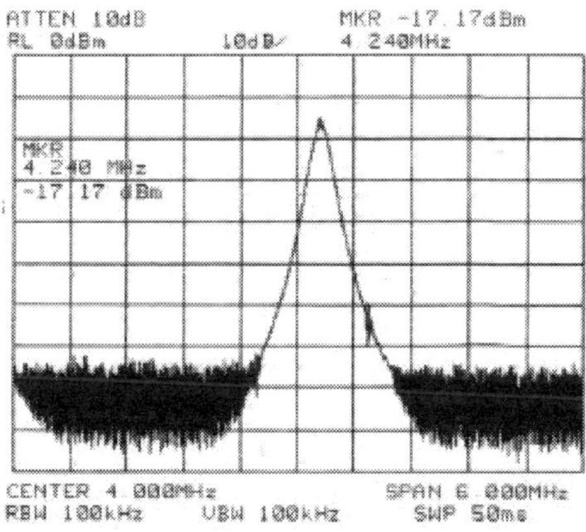

Fig. 2 RF output spectrum obtained at mode-locking condition.

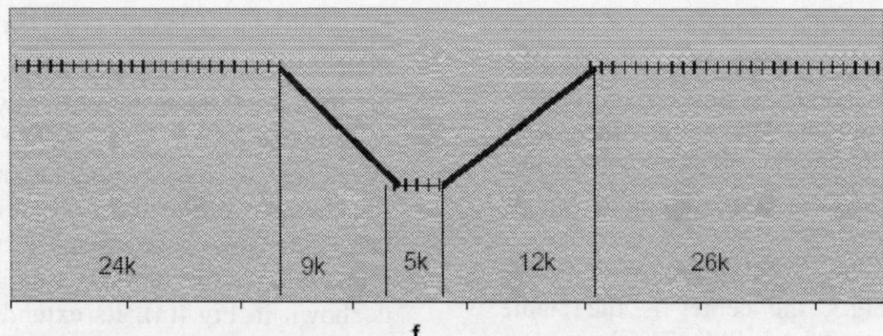

Fig.3 Modulation frequency range and pulse output at center frequency 2.503GHz

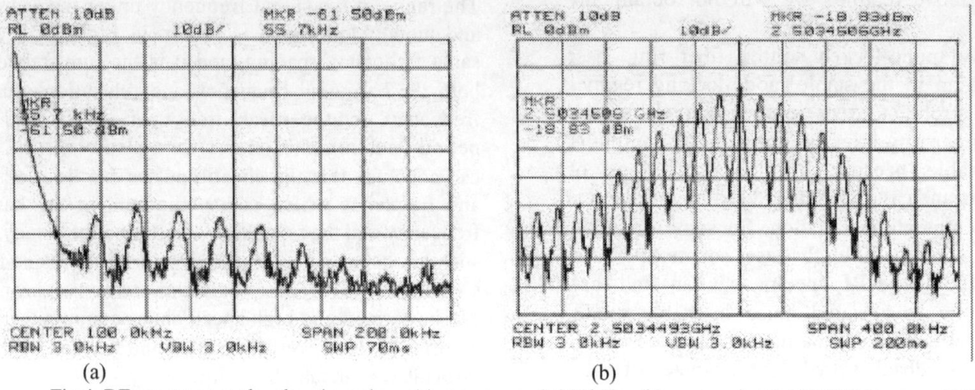

(a) (b)

Fig.4 RF spectrum at the chaotic regime (a) scan range:0-200kHz; (b) centered at 2.50345GHz, span 400kHz

4. CONCLUSION

An actively mode-locked 2.5GHz fiber laser was established. The relationship between modulation frequency and output pulses is analyzed, and stable mode locking and pulses output were realized by the cavity length feedback system. If the modulation frequency is at the frequency which is the sum of harmonic frequency and the relaxation oscillation frequency, the chaotic pulses can be produced from the actively harmonic mode-locked fiber laser. Two same laser source can be established to realize the chaotic communication.

REFERENCES

[1] J.S.Wey, J.Goldhar, D.W.Rush, et al. Performance Characteristization of a Harmonically mode locked Erbium fiber ring laser. IEEE Photonics technology letters. 7(2), 152~154, 1995

[2] Li Yunhua, Lou Caiyun, Wu Jian, Wu Boyu, Gao Yizhi. Novel Method to simultaneously compress pulses and suppress supermode noise in actively mode locked fiber ring laser. IEEE Photonics technology letters. 10(9), 1250~1252, 1998

[3] Kazi Sarwar Abedin, Masaharu Hyodom, Noriaki Onodera. Active stabilization of a higer-order mode-locked fiber laser operating at a pulse-repetition rate of 154GHz. Optics Letters, 26(3), 151~153, 2001

[4] D.J.Jones H.A.Huas, E.P.Ippen. Subpicosecond solitons in an actively mode locked fiber laser. Optics Letters , 21(22), 1818~1820, 1996

[5] Jian Yao, Jianping Yao, Yong Wang *et al.* Active mode locking of tunable multi-wavelength fiber ring laser. Optics Communications, 191, 341~345, 2001

[6] Liguo Luo, P.L. Chu, T. Whitbread, R.F. Peng. Experimental observation of synchronization of chaos in erbium-doped fiber lasers. Optics Communications 176, 213–217, 2000

[7] Sungchul Kim, Byoungho Lee, and Dong Hwan Kim. Experiments on Chaos Synchronization in Two Separate Erbium-Doped Fiber Lasers. IEEE PHOTONICS TECHNOLOGY LETTERS, 13(4), 290~292, 2001

[8] H.Takara, S.Kawanishi and M.Saruwatari. Stabilisation of a mode locked Er-doped fiber laser by suppressing the relazation oscillation frequency component. Electronics Letters, 31(4), 292~293, 1995

[9] R.Hiyan, O.Deparis, O.Pottiez, P.Megret, M.Blondel. Stabilization of actively mode-locked Er-doped fiber lasers in the rational-harmonic frequency-doubling mode-locking regime. Optics Letters, 24(15), 1029~1031, 1999

[10] X.Shan, D.Cleland and A.Ellis. Stabilising Er fiber soliton laser with pulse phase locking. Electronics Letters, 28(2), 182~183, 1992

[11] M.Nakazawa, E.Youshida and Y.Kimura. Ultrastable harmonically and regeneratively mode locked polarization-maintaining erbium fiber ring laser. Electronics Letters, 30(19), 1603~1605, 1994

[12] X.Shan, T.Widdowson, A.D.Ellis and A.S.Siddiqui. Very simple method to stabilize mode locked erbium fiber lasers. Electronics Letters, 32(11), 1015~1016, 1996

[13] R.Kiyan, O.Deparis, O.Pottiez, P.Megret and M.Blondel. Stabilisation of actively mode locked Er-doped fiber laser by minimizing interpulse noise power. Electronics Letters, 34(25), 2410~2411, 1998

Simulation of Multi-wavelength backward-pumped Broadband Fiber Raman Amplifiers

Fei Lin*, Yandong Gong, Ping Shum
Network Technology Research Center, Nanyang Technological University
50 Nanyang Ave., S2-B3-C23 Sch. of EEE, Singapore 639798
Email: afeilin@msn.com, G02048743@ntu.edu.sg

Abstract: The highly nonlinear interactions in multiple-wavelength pumped fiber Raman amplifiers (MW-FRAs) make the performance simulation extremely time-exhaustive. So the efficient models are necessary. Here we carried out simulation with Average Power Method and discussed the numerical results.

Keywords: Optical amplifier, simulation, Raman

1. INTRODUCTION

Broadband fiber Raman amplifiers (FRAs) are particularly attractive for DWDM system due to their advantages over erbium-doped fiber amplifiers (EDFAs). For example, Raman amplifiers can operate through the low loss window (1.1μm~1.7μm) under appropriate pumps compared with EDFAs (which are limited in 1.55μm region with relative narrow bandwidth), offer the extra-broad bandwidth (>80nm) and distributed amplification. More important, FRAs provide distinctive flexibility in bandwidth designs by multiple pumps arrangement [1], [2], which are more efficient than gain-equalizing by filters for EDFAs. But there are still some difficulties in design of broadband FRAs.

To multiple-wavelength pumped FRAs (MW-FRAs) for broadband WDM system, some nonlinear interactions experienced by signals include stimulated Raman scattering (pump-to-pump interactions, signal-to-signal interactions), amplified spontaneous emission (ASE) noise, Rayleigh back scattering and so on. These interactions severely complicated the performance simulation of MW-FRAS, especially the backward-pumped MW-FRAs. So the efficient models are necessary.

In this work, we carried out the performance simulations with Average Power Method (APM) and discussed the numerical results.

2. AMPLIFIER MODEL AND AVERAGE POWER METHOD (APM)

The propagation equations governing forward and backward power evolutions in MW-FRAs that included mentioned interactions (such as stimulated Raman scattering, Rayleigh back scattering and temperature dependent spontaneous Raman emission) can be written as [1]

$$\frac{dP^{\pm}(z,v)}{dz} = \pm\left[B(z,v)P^{\pm}(z,v) + C^{\pm}(z,v)\right] \qquad (1)$$

Where $P^{\pm}(z,v)$ denotes the forward and backward powers with frequency v at distance z. and where

$$B(z,v) = -\alpha(v) + \sum_{\varsigma} g(\varsigma,v)\left[P^{+}(z,\varsigma) + P^{-}(z,\varsigma)\right] + \sum_{\varsigma} g(\varsigma,v) K_{eff} S(\varsigma,v) \qquad (2)$$

$$C^{\pm}(z,v) = \sum_{\varsigma} \frac{g(\varsigma,v)}{2}\left[P^{+}(z,\varsigma) + P^{-}(z,\varsigma)\right] K_{eff} S(v,\varsigma) + \gamma(v) P^{\mp}(z,v) \qquad (3)$$

In above equations:

$$g(\varsigma,v) = \begin{cases} g_r(\varsigma,v)/(K_{eff}A_{eff}), & v < \varsigma \\ -g_r(\varsigma,v)(v/\varsigma)/(K_{eff}A_{eff}), & v > \varsigma \end{cases} \qquad (4)$$

$$S(\varsigma,v)=\begin{cases}\left(1+\dfrac{1}{e^{h(v-\varsigma)/kT}-1}\right)2h\varsigma\Delta\varsigma & v>\varsigma, Stokes\\ -\dfrac{1}{e^{h(\varsigma-v)/kT}-1}2hv\Delta v & v<\varsigma, Anti-Stokes\end{cases} \quad (5)$$

$\alpha(v), \gamma(v), g_r(\varsigma, v), k, T, A_{eff}$ and K_{eff} denote fiber loss, Rayleigh back scattering coefficient, Raman gain coefficient, Bolzman's constant, temperature of the fiber, effective area of the fiber and relative polarization factor between signals and pumps (K_{eff}: 1~2), respectively. Here we considered the anti-stokes effect because this effect can't be neglected for DWDM system with small-frequency separations (~100 GHz).

According to the APM applied in EDFA [3], [4], replaced $B(z,v)$ and $C^{\pm}(z,v)$ with their corresponding length averaged values ($\langle B(z,v)\rangle$ and $\langle C^{\pm}(z,v)\rangle$), then the analytical solution of power can be written as

$$P^{\pm}(z,v)=\left[P^{\pm}(0,v)\pm\frac{\langle C^{\pm}(v)\rangle}{\langle B(v)\rangle}\right]\exp^{\pm\langle B(v)\rangle z}\mp\frac{\langle C^{\pm}(v)\rangle}{\langle B(v)\rangle} \quad (6)$$

So we got the length average values of powers in the amplifier with length of L as

$$\langle P^{\pm}(\varsigma)\rangle=\frac{P^{\pm}(L,\varsigma)-P^{\pm}(0,\varsigma)}{\ln\left(P^{\pm}(L,\varsigma)\pm\dfrac{\langle C^{\pm}(\varsigma)\rangle}{\langle B(\varsigma)\rangle}\right)-\ln\left(P^{\pm}(0,\varsigma)\pm\dfrac{\langle C^{\pm}(\varsigma)\rangle}{\langle B(\varsigma)\rangle}\right)}\mp\frac{\langle C^{\pm}(\varsigma)\rangle}{\langle B(\varsigma)\rangle} \quad (7)$$

with boundary values of $P^{+}(0,\varsigma)$ and $P^{-}(L,\varsigma)$. The equation (6) and (7) should be iteratively solved until results are self-consistent.

For better accuracy, the length of entire Raman fiber amplifier can be divided into a series of concatenated elemental amplifier sections. And the above processes can be proceeded in each section.

3. NUMERICAL RESULTS & DISCUSSION

We simulated a 50km Raman silica amplifier with 100 signal channels from 1520nm to 1600nm, and with power of –3dB/Channel. Six backward pumps with evenly space between 1420nm to 1500nm had the same power of 100mW. A_{eff} and K_{eff} for fiber were 50μm^2 and 2 respectively. The Rayleigh back scattering coefficient and Raman gain Coeffient were taken from Ref. [1] and [5], correspondingly. The loss profile was also taken from Ref. [5].

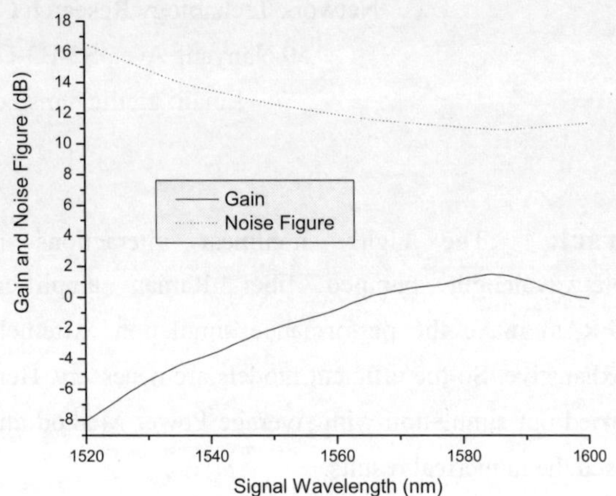

Fig. 1. Gain and noise figure of 6 evenly-space-pumped FRA.

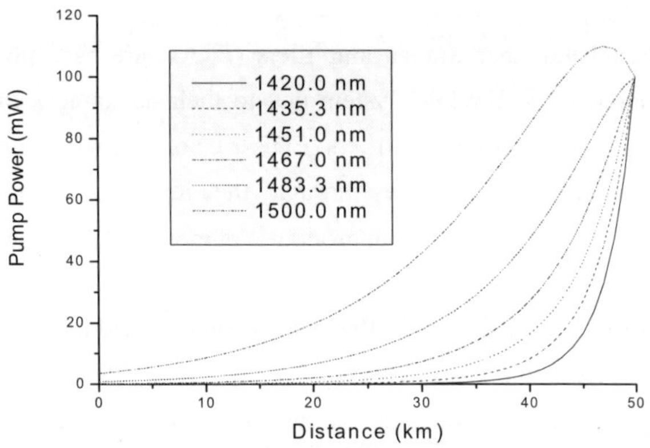

Fig. 2. Power evolution of pumps.

This Raman amplifier has gain variation of 9.6 dB (see Fig. 1) due to pump-to-pump power transfer (see Fig. 2) and signal-to-signal power transfer. From Fig. 2 we can find that the 1500nm pump was amplified in the initial stage, which was the result of pump-to-pump power transfer. And as we can see from Fig.3, after a long distance of propagation, the signal power transferred from short wavelengths to long wavelengths (including signal-to-signal and pump-to-pump power transfer), which caused relative higher gain in long signal wavelengths. Because of backward pump scheme, the signal experiences loss at the beginning of the fiber, which degraded signal and increased bit-error rate. Recently, second-order pumping [6] has been investigated to improve it.

From Fig. 1 and Fig. 3 it can be seen that the interactions between signals and pumps were mainly focused on the last 10km of fiber. So the short signal wavelengths experienced more loss than long wavelength due to signal-to-signal power transfer in the initial stage of transmission. And in the final stage of transmission, the pumps brought more noise to short signal wavelengths than long signal wavelengths. It caused the higher noise figure for short signal wavelengths (in Fig. 2).

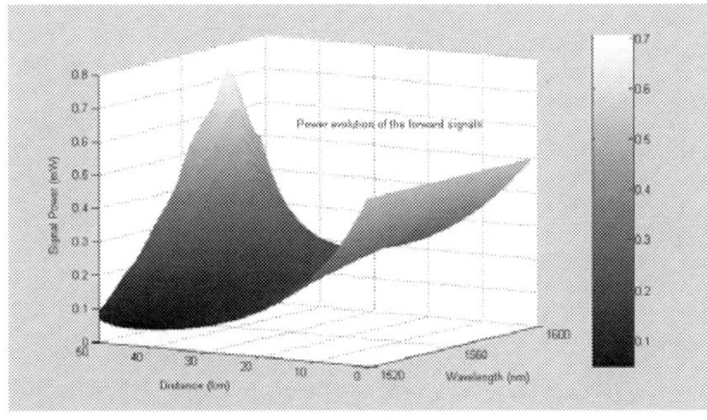

Fig. 3. Power evolution of signals.

Obviously, the performance of this amplifier is not good because of large gain variation. Our recent work in gain-flattening of Raman amplifier enabled us to balance the power transfer among signals and pumps by adopting appropriate powers and wavelengths for pumps. An optimal pump scheme was obtained for 6 pumps. The pump wavelengths and powers were shown in the inset of Fig. 4 and the gain and noise figure were shown in Fig. 4. The amplifier had output gain of 5 dB and a small peak-to-peak gain ripple of 0.39 dB. Fig. 5 showed the power evolution of forward signals.

In conclusion, we carried simulation of MW-FRAs with efficient models and discussed the Raman process.

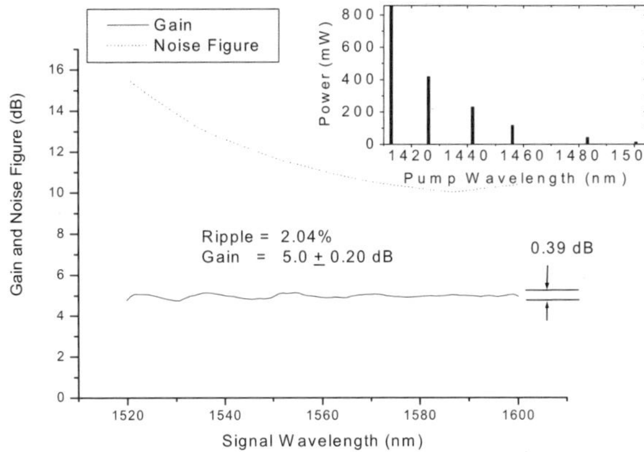

Fig. 4. Gain and Noise Figure of optimal designed amplifier.

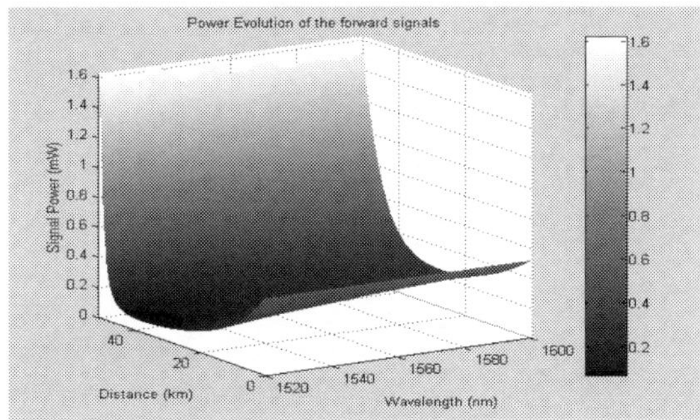

Fig. 5. Power evolution of signals.

REFERENCES

Journal Papers:

[1] H. Kidorf, K. Rottwitt, M. Nissov, M. Ma, and E. Rabarijaona, Pump interactions in a 100-nm bandwidth Raman amplifiers, IEEE Photon. Technol. Lett., vol. 11, page 530-532, 1999.

[2] X. Zhou, C. Lu, P. Shum, and T. H. Cheng, A simplified model and optimal design of a multiwavelength backward-pumped fiber Raman amplifier, IEEE Photon. Technol. Lett., vol. 13, page 945-947, 2001.

[3] T. G. Hodgkinson, Average power analysis technique for erbium-doped fiber amplifiers, IEEE Photon. Technol. Lett., vol. 3, page 1082-1084, 1991.

[4] T. G. Hodgkinson, Improved average power analysis techniqure for erbium-doped fiber amplifiers, IEEE Photon. Technol. Lett., vol. 4, page 1273-1275,1992.

Books:

[5] G. P. Agrawal, *Nonlinear Fiber Optics,* Academic Press, San Diego, 2001.

Proceeding Papers:

[6] K. Rottwitt, A. Stentz, T. Nielson, P. Hanson, K. Feder and K. Walker, "Transparent 80 km bi-directionally pumped distributed Raman amplifier with second order pumping", in ECOC 1999,Technical Digest, **2**-144,1999.

AUTHOR INDEX